ENDPAPER 2

Predators, such as birds, tend to overlook insects that resemble their background or inedible objects. A mantis from Costa Rica *(a, upper left)* matches the leaves on which it stalks other insects as food. In Malaya, the nymph of the flower mantis *(b, upper right)* waits for prey among pink orchids or, in this photo, may create its own attractive blossom on a leaf. The bark of a tree in Mali is the hunting territory for an African mantis *(c, lower left).* The crouched position and marginal fringes of hair eliminate shadows. The insect's outline is further obscured by two disruptive dark bars. While the preceding mantises rely on stillness to escape notice, this dead-leaf mantis *(d, lower right)* from Costa Rica quickly darts away from its enemies on the forest floor. (a, b, c *photos, courtesy of Edward S. Ross;* d *photo by Howell V. Daly.)*

INTRODUCTION TO
INSECT BIOLOGY AND DIVERSITY

INTRODUCTION TO INSECT BIOLOGY AND DIVERSITY

Howell V. Daly
Professor of Entomology
University of California, Berkeley

John T. Doyen
Associate Professor of Entomology
University of California, Berkeley

Paul R. Ehrlich
Bing Professor of Population Studies
Stanford University

Illustrated by Barbara Boole Daly

McGraw-Hill Book Company
New York St. Louis San Francisco Auckland Bogotá Düsseldorf Johannesburg London Madrid Mexico Montreal New Delhi Panama Paris São Paulo Singapore Sydney Tokyo Toronto

INTRODUCTION TO INSECT BIOLOGY AND DIVERSITY

1 2 3 4 5 6 7 8 9 0 V H V H 7 8 3 2 1 0 9 8

This book was set in Baskerville by Typothetae Book Composition. The editors were James E. Vastyan and Michael Gardner; the designer was Elliot Epstein; the production supervisor was Charles Hess.
Von Hoffmann Press, Inc., was printer and binder.

Library of Congress Cataloging in Publication Data

Daly, Howell V.
Introduction to insect biology and diversity.

Bibliography: p.
Includes index.
1. Entomology. I. Doyen, John T., joint author.
II. Ehrlich, Paul R., joint author. III. Daly, Barbara Boole. IV. Title.
QL463.D34 595.7 77-25852
ISBN 0-07-015208-X

To the memory of our friend
Robert L. Usinger

CONTENTS

Part Four INSECT DIVERSITY

PREFACE

In his autobiography, Robert L. Usinger (1972) declared he had "the best of all possible lives" as an entomologist. His enthusiasm was so infectious that the senior author of this book readily agreed to his suggestion that a comprehensive textbook on insects as living creatures be written—not realizing that entomology was on the eve of an explosive growth of information. Fortunately John Doyen and Paul Ehrlich were persuaded to join in the task, and with their contributions the book was completed. Usinger's interests ranged throughout entomology, and he did not hesitate when a line of research led him into the biology of bats in South America, fossil amber deposits in Alaska, or a full-scale scientific expedition to the Galápagos. In writing this text, we have tried to preserve his attitude of free inquiry and have departed from the traditional boundaries of entomology whenever important relationships have needed additional explanation.

This text provides an introduction to the study of insects. It is intended for students who have completed at least a basic course in biology. Those without this background should prepare themselves by reading about animal physiology, genetics, and the Phylum Arthropoda in a textbook on biology or zoology, such as Storer et al. (1972).

Our text is designed for modern courses that emphasize the major features of insects as living systems. The lectures in such courses are commonly devoted to insect structure and function, behavior, ecology, and the applied aspects, and the laboratory work is focused on the recognition of different kinds of insects. Parts One to Three of this text are thus appropriate as readings for the lectures, and Part Four supplies information on the insect groups and keys for their identification.

In the usual sense, this is not a text on applied entomology [see, for example, Metcalf, Flint, and Metcalf (1962)]. Yet the management practices now being developed for beneficial and harmful insects are increasingly sophisticated. They will require detailed knowledge of insect biology to be understood, implemented, and improved by future entomologists. In this special sense, we hope our text also provides a foundation for professional training.

The treatment of insect biology begins with the structure and function of individual insects, progresses through the relationships among members of a species, and then describes in turn the different relationships that insects have to their environment. We have tried to limit the terminology and discussions to those topics required for a broad knowledge, but not at the expense of oversimplifying complicated subjects. Nor have we avoided controversial topics. The study of the world's most varied animals *is complex.* With the application of new physical, chemical, and mathematical techniques for the analysis of insect biology, this complexity is increasing.

Part Two on population biology is based in part on *The Process of Evolution,* second edition (1974), by P. R. Ehrlich, R. W. Holm, and D. R. Parnell. Certain passages are reproduced here by permission of the publisher, McGraw-Hill Book Company. We include Part Two because our students have wanted to understand the evolutionary mechanisms underlying the great variability of insects. We trust this early introduction will encourage students in the biological and agricultural sciences to seek additional courses dealing with the properties of populations. Furthermore, it is becoming evident that the long-term success of any pest management strategy depends on how well the management planners understand population dynamics, genetics, and coevolution.

Part Four surveys the orders and families of insects. To a large extent, the principles of modern ecology and evolution are based on studies of insects. Mimicry, competition, and coevolution are only three out of many important concepts that have been developed largely through an understanding of the biology of various groups of insects. We have attempted to integrate this information into Part Four, so that the various taxa may be viewed in a broad biological context. Throughout, we have cross-referenced discussions of biological phenomena to more detailed treatments in other parts of the book.

The keys include nearly all the families of insects known to occur in North America. However, they are not designed to accommodate exceptional species that are normally encountered only by specialists. Likewise, extremely difficult taxa that require special preparation or collection techniques are mentioned, but not keyed. Discussions of the large endopterygote orders are organized by suborder and superfamily. These are important taxonomic categories whose members usually share many biological features.

We wish to acknowledge the generous assistance of many colleagues in preparing this text. Excellent photographs were offered for our use by Thomas W. Davies, Howard E. Hinton, Kenneth Lorenzen, Robert W. Mitchell, Joseph H. Peck, Jr., George O. Poinar, David Rentz, Carl Rettenmeyer, Edward S. Ross, Frank E. Skinner, and Gerard M. Thomas. The following persons gave us valuable information, key references, or specimens: Paul H. Arnaud, Herbert G. Baker, Byron Chaniotis, Frank R. Cole, David Durbin, Wyatt Durham, Mark Eberle, Deane P. Furman, Steve Hendrix, Marjorie A. Hoy, Carlton S. Koehler, E. Gorton Linsley, Richard Merritt, P. S. Messenger, Robert O. Ornduff, William Peters, George O. Poinar, Helmut Riedl, Edward Rogers, Ray F. Smith, William E. Waters, and Ward B. Watt. We especially thank E. I. Schlinger for access to his personal library. Persons who read parts of the text and gave us the benefit of their special expertise are John R. Anderson, John G. Baust, George W. Byers, Robert Cruden, George C. Eickwort, Rosser Garrison, Norman E. Gary, Kenneth S. Hagen, Bernd Heinrich, John T. Hjelle, James Johnson, Richard A. Kawin, Jarmila Kukalova-Peck, John F. Lawrence, Werner J. Loher, James J. McGivern, W. W. Middlekauff, D. R. Minnick, Bernard Nelson, Rudolph L. Pipa, Jerry A. Powell, David Rentz, Vincent H. Resh, Edward S. Ross, Edward L. Smith, Sigurd L. Szerlip, Yoshinori Tanada, Catherine A. Tauber, S. Salman Wasti, and David L. Wood. Paul A. Rude and Edward Rogers assisted with preparation of the manuscript and illustrations.

Howell V. Daly
John T. Doyen
Paul R. Ehrlich

INTRODUCTION TO

INSECT BIOLOGY AND DIVERSITY

PART ONE

INSECTS AS ORGANISMS

INTRODUCTION 1

In this first chapter we will briefly view the biology of insects from a variety of aspects: the relative numbers of species and the limits to their geographic distribution; the kinds of places in which they live and the food they eat; elementary facts about their anatomy; major features of their evolution and classification, and factors in their success as terrestrial organisms; and we will conclude with a discussion of insects in relation to human welfare. This will provide some sense of the scope of insect life and introduce you to the first of the many terms, facts, and concepts that comprise a working knowledge of entomology.

Insects are the earth's most varied organisms. Almost exactly half (50.8 percent) of the species of living things and 72 percent of all animals are insects (Fig. 1.1). Of all the animal phyla, only the arthropods and the chordates have succeeded extensively in adapting to life in dry air. Insects now inhabit virtually all land surfaces of the globe except the extreme polar regions and the highest mountain peaks. Wherever they occur they tend to dominate the small fauna, being rivaled only by another group of arthropods, the mites, in some habitats.

The limits to the geographic distribution of insects are still incompletely known. In Antarctica, for example, insects are represented by two species of flies, a species of bird flea, and several species of lice that parasitize birds and seals. In the Arctic, at least 300 species of insects, mostly flies, live in the Canadian islands north of the 75th parallel. At 6000-m altitude (19,685 ft) in the Himalaya surprising numbers of insects have been found as permanent residents. Insects have also been found living in deep caves, hot springs, salt lakes, and pools of petroleum. About 3 percent of all species of insects live in freshwater, and perhaps 0.1 percent are found in the marine intertidal zone. One genus of water striders, *Halobates* (Family Gerridae), lives permanently on the surface of the open ocean, and many other gerrids occur on the surfaces of tropical seas closer to land. Only a few kinds of insects are found in the offshore waters of deep lakes, and they are missing entirely in the deeper waters of the ocean.

The vast majority of insects, therefore, are terrestrial. They occur in an enormous variety of habitats, including arid deserts. Of course, they must have access to water for drinking or in food to restore that lost by evaporation from their body surfaces. As in other small terrestrial organisms, the surface area of their body is relatively large in proportion to their volume, so the risk of desiccation is always great. This loss is reduced in insects by a thin wax coating on the body surface and by certain water-conserving mechanisms associated with excretion and respiration. Insects also avoid unfavorably dry environments by their behavior, and many species satisfy their water requirements by metabolizing carbohydrates.

Some small invertebrates and algae survive drought by suspending metabolic activity and becoming dehydrated. Insects usually die if their water content falls below about 20 percent. One exceptional insect, however, is able to tolerate almost total dehydration. The larvae of immature stages of the African midge *Polypedilum vanderplankei* (Family Chironomidae) live in pools

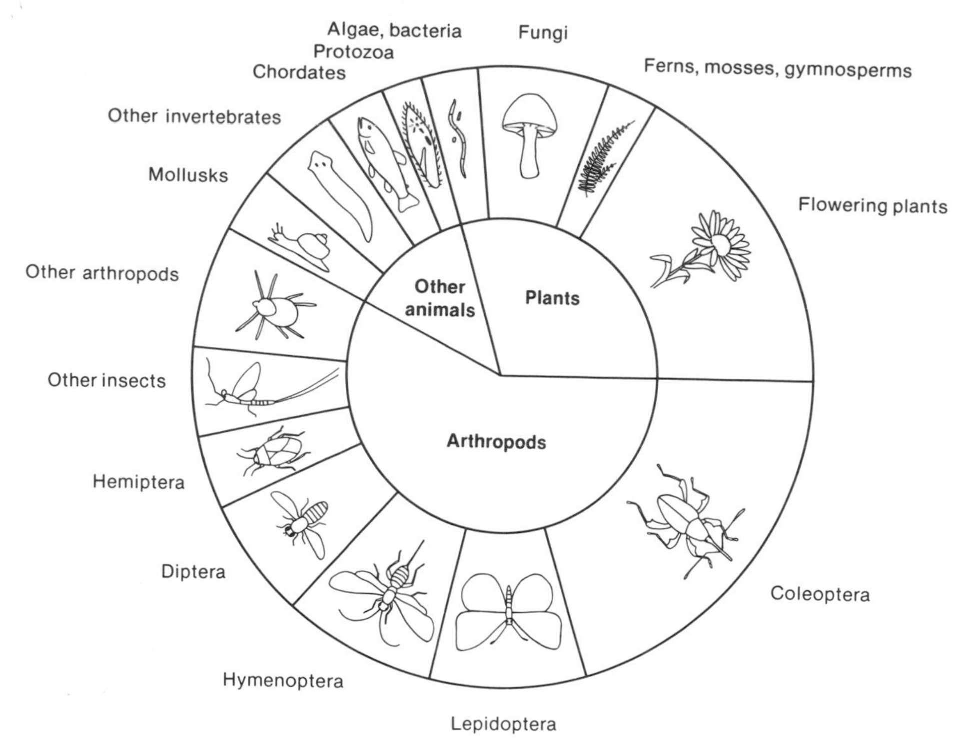

Figure 1.1 Relative numbers of species of arthropods, other animals, and plants. Each 1° equals 4200 species.

that periodically become dry. The larvae can survive drought for several years by becoming dehydrated. In this state they are resistant to extremes in the physical environment. For example, Hinton (1977) has immersed them in liquid gases at −270°C and heated them to 102°C for a minute without injury.

The critical upper limit in temperature for many insects is in the narrow range of 40 to 45°C, but some survive even above 50°C. The lower limits for activity and survival are much broader. Some tropical species become inactive and do not live for long at 5°C, yet some temperate species are normally active on snow at temperatures barely above 0°C. Insects of the temperate and colder regions survive winter by finding suitable shelter and by physiological adjustments that prevent desiccation and freezing. Some species can tolerate the formation of ice crystals in their tissues and survive −35 to −40°C. These are the most complex animals to tolerate freezing.

HABITATS AND FOOD HABITS. Insects can be found in forests, grasslands, deserts, cultivated lands, urban areas, bodies of fresh and salt water, and flying in the air. In other words, they occupy all major habitats within the limits of geography and physical environment just discussed. Specifically they are found in or on soil, decaying plant or animal matter, living plants, animals, or other insects, or water. Many insects are able to occupy more than one habitat and feed on different foods during different stages of their lives.

In soil, insects are especially rich in variety in the litter of leaves and dead plant matter. Some penetrate the upper layers of soil by following natural crevices or entering mammal burrows, while others are adapted to digging burrows themselves. Rocks and other "cover objects" provide shelter for ground-dwelling forms. Decaying logs are occupied by many types of insects, especially beetles and termites. The "subcortical" habitat beneath the bark of dead and, less frequently, living trees is favored by beetles. Animal waste in the form of dung or carcasses on the ground is highly attractive to certain *scavenging* beetles and flies, and to the insects that prey on the scavengers.

Insects that feed on green plants are termed *phytophagous.* All parts of green plants are attacked: roots, trunks, stems, twigs, leaves, flowers, seeds, fruits, and sap in the vascular system. Insects either feed externally by chewing tissues or by sucking sap or cell contents, or they feed internally by boring into the plant's tissues. The sucking insects, especially Hemiptera, are the only animals that are able to extract sap in quantity from the vascular systems of plants. Likewise, except for some tropical bats and some birds such as hummingbirds, insects are the main consumers of pollen and nectar in flowers. Insects also live among and feed on primitive plants such as fungi, algae, lichens, mosses, and ferns.

Insects feed on most other kinds of terrestrial animals. Insects that kill other insects are termed *entomophagous.* Of these, the *predators* kill their prey more or less immediately, while *parasitoids* feed externally or internally in their host for some period before finally killing it. Insect predators and parasitoids also devour other small invertebrates such as snails, millipedes, spiders, and earthworms. Mammals, birds, reptiles, and amphibians are *parasitized* by bloodsucking insects and usually not killed. Some parasites live on the host (biting and sucking lice, adult fleas), or burrow in its flesh, or inhabit the alimentary or respiratory tract (fly maggots). Other parasites, e.g., mosquitoes, visit the host only to suck blood.

In the aquatic environment are scavenging, phytophagous, and predatory insects, but few parasitoids and true parasites. Some species are able to skate about supported by the surface film, while others float in the film or swim for brief periods below the surface and return for air. Certain groups are adapted to remain beneath the surface indefinitely and obtain their oxygen from water by means of gills. Of these, some are free-swimming, some crawl about on vegetation and the bottom, and some burrow into the bottom gravels, sand, or silt. Some insects are the only truly aquatic animals capable of prolonged flight—a distinct advantage for inhabitants of temporary bodies of water.

Some insects construct shelters about their bodies of rolled leaves or plant debris. Many that have a resting pupal stage in the life history construct a special cocoon of silk or a chamber in soil or decayed wood. Shelters such as these provide protection against desiccation and reduce detection by enemies. Social ants, termites, and bees may construct elaborate nests that permit some degree of control over temperature, humidity, and light, as well as provide a defensible fortress against enemies.

By their feeding activities and being fed upon, insects have evolved complex trophic relationships among themselves and with other animals, plants, and microbes. The origin and diversification of flowering plants, terrestrial vertebrates, and disease agents of plants and animals are due in part to the interactions of these organisms with insects. The blossoms of flowering plants that we enjoy were evolved to attract pollinators, primarily insects. Spices and flavorings originated in plants as chemical defenses against plant-feeding insects. The abundance of insects as food on land probably lured the ancestors of the reptiles to forage far from water, thus leading to the adaptations that made the reptiles the first fully terrestrial chordates. The first birds, mammals, and primates were also primarily insectivores. Lastly, it was not until insects began feeding on the blood of animals or the sap of plants that certain disease agents, especially viruses, could be transferred directly from the vascular system of one host to another.

ELEMENTS OF INSECT ANATOMY. A typical adult insect has a segmented body with an external skeleton, or *exoskeleton,* composed of chitin (a nitrogenous polysaccharide) and protein. The exoskeleton provides not only strong support and protection for the body but also a large internal area for muscle attachments. It is subdivided into plates, or *sclerites,* that are separated by joints or lines of flexibility and are movable by muscles. In proportion to weight the tubular construction of the body segments and appendages gives relatively great resistance to bending because of the mechanical strength of a hollow cylinder.

The adult body is divided into three main regions: head, thorax, and abdomen (Fig. 1.2). The *head* bears a pair of large *compound eyes* and as many as three simple eyes, or *ocelli,* a pair of sensory *antennae,* and the feeding appendages or mouthparts surrounding the mouth. The *thorax* is composed of three segments, each bearing a pair of *legs.* The last two segments may also bear a pair of *wings.* The *abdomen* is composed of no more than 10 or 11 visible segments and lacks appendages except for a pair of *cerci* and the reproductive *external genitalia* that may be present near the tip.

Internally the body cavity, called the *hemocoel,* is filled with muscles, viscera, and the blood or *hemolymph.* The tubular *alimentary canal,* or *gut,* extends from the mouth to the anus at the extreme tip of the abdomen. The gut is divided into three regions: *foregut, midgut,* and *hindgut.* At the junction of the last two regions are the filamentous *Malpighian tubules,* which are excretory in function. The paired *salivary glands* open beneath the mouth in a single *salivary duct.* The *central nervous system* consists of a large *brain* in the head in front of the mouth, and a *ventral nerve cord* of segmental *ganglia* extending through the body with nerves to each segment. The hemolymph is circulated by a *dorsal vessel* that is differentiated into a chambered *heart* in the

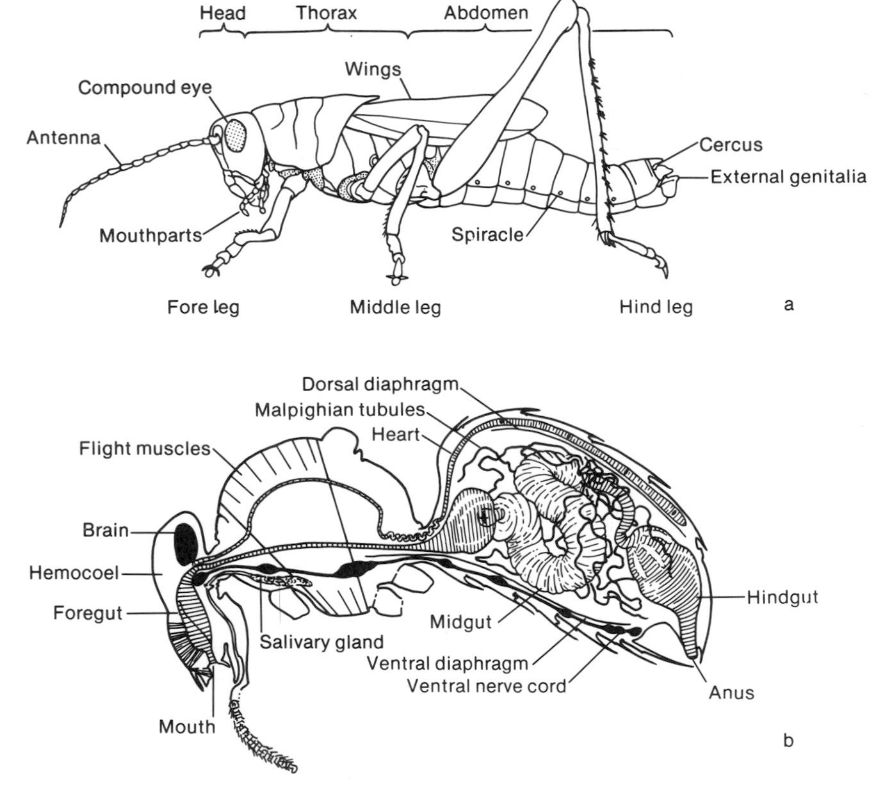

Figure 1.2 Elements of insect anatomy: *a,* external anatomy of the grasshopper *Romalea microptera* (Acrididae); *b,* internal organs of the honeybee *Apis mellifera* (Apidae). (a *redrawn from original by M. Lynn Morris, by artist's permission;* b *redrawn from Dadant and Sons,* 1975, *by permission of Dadant and Sons, Inc., Hamilton, Ill.*)

abdomen and a tubular *aorta* in the thorax and head. Transverse membranes, the *dorsal* and *ventral* diaphragms, subdivide the abdominal hemocoel. Large clusters of fat cells commonly create a *fat body,* in which food reserves are stored. Paired openings, or *spiracles,* are situated on each side of the last two thoracic segments and the first eight or fewer abdominal segments. These openings lead to a system of tubular *tracheae* that convey oxygen in air to the internal tissues.

The male and female reproductive organs are located in the abdomen. In each sex the paired gonads usually open to the exterior by a single, tubular duct in association with the external genitalia. The duct of the male opens beneath on the ninth abdominal segment and is surrounded by an intromittent organ, or *aedeagus,* that transfers the sperm directly to the female. The duct of the female usually opens beneath on the eighth abdominal segment. The female usually has specialized appendages on the eighth and ninth abdominal segments that form an *ovipositor* for placing the eggs in soil, plant tissues, or other media.

ELEMENTS OF INSECT CLASSIFICATION AND EVOLUTION. The hardest task in beginning a study of entomology is learning the scientific names of the different kinds of insects. Recall that the hierarchic system for classifying the animals includes at least six categories: Phylum, Class, Order, Family, Genus, and Species. You will be mainly concerned with learning the "higher classification" of insects, i.e., the names of the classes, orders, and common families. The identification of genera and species is often difficult and requires special training. To represent accurately the evolutionary lineages of insects, authorities have added intermediate categories, such as superclass or subclass, to the higher classification. Unfortunately, authorities are not in universal agreement on the choice of names for these categories or on the best arrangement. In this text we treat 32 orders and generally follow the classification of Britton et al. (1970). Our classification is outlined in Table 1.1. It will be helpful to remember that the names of the various categories used here have the following endings:

Subclass (-ota)
 Infraclass (-ptera)
  Division (-ota)
   Superorder (-pteroidea)
    Order (-ptera, -ura, or -odea, except Collembola, Archeognatha, Odonata)
     Superfamily (-oidea)
      Family (-idae)
       Subfamily (-inae)
        Tribe (-ini)

The names of genera and species are italicized, and, in this text, the family is indicated in parentheses.

To assist you in obtaining an overall view of insect diversity, let us briefly trace the key steps in their evolution and match these steps with the higher classification, as shown in Table 1.1. In the broad sense an insect is a terrestrial arthropod that has three pairs of walking legs. This distinguishes insects from the arachnids (spiders, ticks, mites, and scorpions), which have four pairs of walking legs, and from the myriapods (millipedes and centipedes), which have more than four pairs. Insects presumably evolved from arthropods such as the millipedes and their allies by a reduction in the number of legs. The six-legged condition, however, probably evolved more than once from the same ancestral line. For this reason, the six-legged arthropods are placed in the Superclass Hexapoda (meaning "six-legged organisms"), and two classes are recognized: Class Entognatha for small, six-legged soil-dwelling arthropods that have their mouthparts retracted in a cavity in the head (hence the name Entognatha); and Class Insecta for the insects in the strict sense. Insects do not have their mouthparts retracted in the head. The word "insect" is derived from the Latin *insecare,* meaning "to cut into," referring to the bodies of some insects that are almost cut in half by constriction at the neck or waist. The first insects were wingless, and some wingless representatives survive today from an early period in insect evolu-

TABLE 1.1. Approximate Size and Diversity of Classes and Orders of Hexapodous Arthropods (common names in parentheses)

Category	Taxon	No. of families		No. of species	
		World	NA	NA	World
Class	Entognatha				
Order	Protura	3	3	18	170
	Collembola (springtails)	5	5	315	1,700
	Diplura	4	3	25	660
Class	Insecta				
Subclass	Apterygota				
Order	Thysanura (silverfish)	5	4	25	330
	Archeognatha (bristletails)	2	2	25	250
Subclass	Pterygota				
Infraclass	Paleoptera				
Order	Ephemeroptera (mayflies)	17	15	600	2,500
	Odonata (dragonflies, damselflies)	24	10	425	4,900
Infraclass	Neoptera				
Division	Exopterygota				
Superorder	Orthopteroidea				
Order	Blattodea (roaches)	5	5	35	3,500
	Mantodea (mantids)	8	1	20	1,800
	Isoptera (termites)	6	4	41	2,000
	Plecoptera (stone flies)	8	6	300	1,500
	Dermaptera (earwigs)	7	4	18	1,200
	Embioptera (web spinners)	8	3	9	2,000
	Phasmatodea (walkingsticks)	2	1	12	2,500
	Grylloblattodea	1	1	5	12
	Orthoptera (grasshoppers, katydids)	13	8	1,000	20,000
	Zoraptera	1	1	2	22
Superorder	Hemipteroidea				
Order	Psocoptera (book lice, bark lice)	17	11	150	1,700
	Phthiraptera (biting lice, sucking lice)	16	9	380	3,000
	Thysanoptera (thrips)	5	5	500	4,000
	Hemiptera (true bugs)	104	69	11,000	55,000
Division	Endopterygota				
Superorder	Neuropteroidea				
Order	Neuroptera (lacewings, ant lions)	17	11	270	4,300
	Megaloptera (alderflies, dobsonflies)	2	2	50	300
	Raphidioptera (snake flies)	2	2	19	80
	Coleoptera (beetles)	135	110	29,000	280,000
	Strepsiptera	5	4	60	300
Superorder	Mecopteroidea				
Order	Mecoptera (scorpion flies)	7	4	85	400
	Diptera (flies)	117	98	16,000	85,000
	Siphonaptera (fleas)	17	7	200	1,400
	Trichoptera (caddis flies)	24	19	1,000	5,000
	Lepidoptera (moths, butterflies)	102	70	10,000	113,000
Superorder	Hymenopteroidea				
Order	Hymenoptera (ants, bees, wasps)	73	67	17,000	108,000
	Totals:	762	564	87,700	704,700

tion. These are placed in a separate subclass, the Apterygota (meaning "without wings"). The silverfish is a familiar example (Fig. 4.5*a* through *d*).

The winged insects are placed in the Subclass Pterygota (meaning "with wings"). The first winged insects were unable to flex their wings flat over their backs. Some survivors from this stage in evolution are also alive today. These are the dragonflies (Fig. 21.1) and mayflies (Fig. 20.1*a*), and they are grouped together in the Infraclass Paleoptera (meaning "with an ancient type of wing"). Insects that can fold their wings flat over their backs (see, for example, Fig. 22.1*a*) are classified as the Infraclass Neoptera (meaning "with a modern type of wing"). Within this infraclass, two groups can be distinguished on the basis of the manner in which the wings develop. In the Division Exopterygota, the wings develop externally and are visible on the young insect as small wing pads—hence the prefix *exo-* (Fig. 4.5*e* through *i*). Insects in the Division Endopterygota have wing rudiments that develop internally during the early life of the insect. The life history is divided into stages that are strikingly different in form and habits: larva, resting pupa, and winged adult. This type of life history is called complete or *holometabolous metamorphosis* (Fig. 4.6).

The orders of the Exopterygota may be divided into two groups or superorders: Orthopteroidea (meaning "orders resembling the Order Orthoptera") and Hemipteroidea ("orders resembling the Hemiptera"). The Orthoptera are grasshoppers, crickets, and katydids, and rank as the sixth largest order in number of species. The Hemiptera are the true bugs, such as stinkbugs and aphids, and rank as the fifth largest order.

The orders of the Endopterygota are divided into three superorders: Neuropteroidea, Mecopteroidea, and Hymenopteroidea. The largest order of the Neuropteroidea is not Neuroptera (lacewings, ant lions, and their allies) but the Coleoptera, or beetles. The superorder is so named because at one time the Neuroptera were considered ancestral to the neuropteroid group of orders. The Coleoptera are the largest of all insect orders; included in it is the largest family of insects, the Curculionidae, or weevils, with over 60,000 species. The Mecopteroidea are similarly named after the Mecoptera (scorpion flies) because the order probably resembles the ancestral stock that gave origin to the Lepidoptera (butterflies and moths) and Diptera (true flies). These orders are the second and fourth largest orders, respectively. The last superorder, Hymenopteroidea, includes only the Order Hymenoptera (ants, bees, wasps, and their allies). The Hymenoptera are the third largest order.

FACTORS IN THE SUCCESS OF INSECTS. The factors that are probably the most significant in the success of insects are the following: (1) a highly adaptable exoskeleton; (2) colonization of the terrestrial environment before the chordates; (3) small body size; (4) high birthrate and short generation time; (5) highly efficient power of flight; and (6) life history with complete metamorphosis. To a large extent these factors are interdependent and must be considered in combination.

The arthropods were the first animal phylum to overcome the problems of locomotion, respiration, and water conservation in a terrestrial environment. Their success can be traced to their adaptable exoskeleton. The jointed, paired walking legs were suited for locomotion on land as well as in water. Insects and some arachnids solved the problem of respiration without losing water by evaporation: a wax covering was added to the exoskeleton, and the respiratory surfaces were restricted to tubular invaginations of the exoskeleton, the tracheae. The tracheae open to the exterior by spiracles that admit air but reduce the loss of moisture. Thus equipped, insects had ample opportunity to occupy various habitats on land long before any other animal phylum offered serious competition.

In comparison to the calcareous endoskeleton of chordates, the chitin-protein exoskeleton is a distinct asset to small terrestrial animals such as insects. The tubular design and chemical composition combine strength with light weight. Small size itself has several physiological and ecological advantages. Muscles can be used more effectively because their power increases as the square of their cross-sectional area, whereas body volume (and weight) increases as the cube of body dimensions. Because of their size insects are able to occupy an enormous variety of small places that are not accessible to larger animals. A single piece of shelf fungus on a dead tree, for example, is sufficient to provide food and shelter for several generations and hundreds of individuals of fungus beetles.

Small size also permits short generation time, because less time is required to grow to maturity. When combined with a high birthrate per generation, an increased number of generations per year increases the potential for genetic changes in populations. Such populations, if reproductively isolated from one another, may differentiate into separate species more frequently than animals with longer generation times and lower reproductive rates.

The appendages of the exoskeleton are greatly varied in structure and function in different groups of insects. The mouthparts may be adapted for taking solid or liquid food; the legs may be adapted for walking, jumping, clinging, grasping, swimming, or digging; and the ovipositor may be modified according to the media into which the eggs are deposited. Wings are also parts of the exoskeleton. Insects are the only invertebrates and historically the first of any animal group to possess wings. The power of flight has allowed insects to escape unfavorable habitats and to colonize new habitats, sometimes at great distances. Flying insects are at an advantage in escaping enemies, and in finding mates, food, and places to lay eggs. Indeed, except for the Coleoptera, all the large endopterygote orders are strongly dependent on flight for most of the requisites of life.

The majority of the insect species are in the Division Endopterygota (Fig. 1.1, Table 1.1). In addition to all the features discussed above, the endopterygotes have a complete metamorphosis. The larva is specialized for feeding and in many instances can reach food that is inaccessible to the adult. For example, larvae of certain species are able to bore into stems or leaves or into the bodies of other animals that the winged adult is unable to enter. This specialization also reduces the competition between the larva and the adult for the same resources.

INSECTS AND HUMAN WELFARE. The relationships of early humans to insects were similar to those between insects and other primates and mammals. Insect parasites fed on humans, annoyed them, and transmitted diseases among them. In common with other mammals, humans have acquired host-specific parasites, e.g., the head and body louse, *Pediculus humanus,* and the crab louse, *Pthirus pubis* (both in the Family Pediculidae). Insects also destroyed stored food and shelters, and articles made of wood, plant fibers, and animal hides. In turn, insects were eaten by primitive peoples, sometimes as a regular part of the diet. Honey was widely sought in both the Old and New World. Honeybees are native to the Old World, and stingless bees produce honey in the tropics of the New World.

With development of agriculture and cities, humans came into cooperation and conflict with insects. Despite the devastating effects of some insects as destroyers of crops and wooden structures, and as carriers of diseases, it is generally agreed that the majority of insects are directly or indirectly beneficial to human society. In terms of actual dollars, insects contribute more of value to the agricultural economy of the United States than they subtract in damage. Certain species are of direct benefit because they provide useful products, pollinate crops, or can be used as natural enemies to reduce the populations of insect pests or weedy plants. Insects serve as food for many game birds, mammals, and fish, especially trout, on which our recreational hunting and fishing industry depends. The indirect benefits of the work of insects in recycling nutrients in streams and soils, in pollinating wild flowers, and in regulating the populations of plants and animals in our wildlands are impossible to evaluate.

BENEFICIAL INSECTS. McGregor (1976) lists 54 crops in the United States that are dependent on or are benefited by insect pollination. The total value of these crops in 1972 was estimated to be over $11.5 billion. Honeybees are the main pollinators, but some other kinds of bees and other insects are effective pollinators on certain crops. In 1975 in California, beekeepers themselves received over $25 million from the sales of products, services, and bees to other beekeepers (honey, $10.2 million; beeswax, $459,000; pollination services, $8.5 million; package bees and queens, $5.9 million).

Other valuable products of insects are silk and shellac. Silk is secreted from the salivary glands of the larva of the silkworm moth, *Bombyx mori* (Bombycidae) when it spins its cocoon. Beginning in China, the silkworm has been in culture for over 6000 years and no longer exists as a wild insect in the natural state outside human care. Metcalf estimated the annual value of raw silk in the world to be $200 million to $500 million in 1962. Shellac is used for finishing wood and as a component of many industrial products. It is a secretion of the lac insect, *Laccifer lacca* (Coccidae), which feeds on trees in Burma and India. The resinous material is collected, refined, and dissolved in denatured alcohol to make the familiar clear paint. Metcalf et al. (1962) estimate the annual sales of shellac in the United States to be $10 million to $20 million.

The control of plant and animal pests by natural enemies is termed *biological control.* Populations of organisms are kept in check by a combination of physical and biotic factors, a phenomenon sometimes called the "balance of nature." In the natural terrestrial environment, insects play a significant role in this regard by feeding on other insects, mammals, and vascular plants, and by transmitting diseases. If insect and weed pests of agriculture were not checked naturally by entomophagous and phytophagous insects, it would be difficult to grow crops at all. In biological control, such natural enemies are manipulated to maximize their regulatory effects and reduce pest populations.

Most insect and weed pests of agriculture in the United States originated in foreign countries. To find the natural enemies, it is necessary to locate the native home of the pest and to unravel the details of its biology and enemies. Colonies of enemies are brought to special laboratories in the United States under quarantine to test for harmful attributes and to eliminate any diseases or enemies of the desired species. The insects are then bred in large quantities for release. In some instances the release is *inoculative,* i.e., the intent is to establish self-sustaining populations that will continue to provide control indefinitely. In other cases, the release is *inundative* and intended to bring about immediate control without necessarily establishing resident populations.

The cost/benefit of biological control is difficult to estimate because some of the benefits are not readily assessible in dollars. Biological control agents normally cause no environmental pollution or harm to other organisms. They are usually chosen to attack a specific pest. Once the natural enemies are established, the control is essentially permanent and at no further cost. Only rarely, if ever, does the pest evolve appreciable resistance to the natural enemies, thus requiring a renewed effort. In terms of savings over previous losses (total losses in products and control costs), Huffaker and Messenger (1976) cite savings totaling $274,942,000 in the biological control of seven agricultural pests in California for the years 1928 to 1973. It is estimated that about $30 has been saved in California for each dollar spent in biological control research by the University of California.

Theoretically, biological control could be applied to any agricultural pest, but this management strategy remains underdeveloped because of the relatively slower rate of control and the lower market value of products with minor defects. The rising cost for the development of chemical pesticides (currently estimated at $15 million to $20 million each), the cost of repeated applications (machinery, labor, fuel), the evolution of resistance by pests, and the undesirable effects of pesticides on natural enemies and other organisms, however, are making biological control more attractive.

INJURIOUS INSECTS. About 3000 species of insects are directly destructive or harmful to human welfare. These include insects that (1) damage crops directly or by transmitting diseases, thus lowering their yield and the value of the product; (2) destroy products in storage; (3) damage wooden structures; (4) attack, parasitize, annoy, and transmit diseases to domestic animals and reduce their value; and (5) attack, parasitize, annoy, and transmit diseases to humans. Here we are limited to discussion of but a few examples.

The most destructive agricultural pests are those that feed on stored grains. Damage to grains in storage is especially costly because energy and other resources have already been invested in producing, harvesting, and storing the crop. Two beetles, the rice weevil, *Sitophilus oryzae,* and the granary weevil, *S. granarius* (Curculionidae), are the most harmful of the grain insects. They attack rice, wheat, corn, oats, barley, sorghum, and other grains. With over two-thirds of the world's population dependent on rice as a staple food, these weevils must be considered the most destructive insects in the world.

Various species of the moth genus *Heliothis* (Noctuidae) are of worldwide importance as pests of crops in the field. In the United States, *H. zea* is known by several common names according to the crop attacked by the larvae: corn earworm, tomato fruitworm, or cotton bollworm. This species also attacks tobacco, beans, alfalfa, and various garden plants. In 1975 over 66.9 million dollars in agricultural losses were attributed to *H. zea* in California, making it the single most destructive species in the state. *Heliothis virescens,* the tobacco budworm, is similar in habits and appetite.

Grasshoppers and locusts (Acrididae) are of great importance in the more arid regions of the world. They devour nearly all kinds of plants, both cultivated and wild. The migratory locusts, such as *Schistocerca gregaria* and *Locusta migratoria* of Europe, Asia, and Africa, can cause total devastation of edible plants when they land in swarms and feed. In the United States four species of *Melanoplus* and *Camnula pellucida* do 90 percent of the grasshopper damage of rangelands and cultivated plants. During the late nineteenth century, *M. sanguinipes* developed migratory forms and moved in vast swarms from areas east of the Rocky Mountains to the Mississippi Valley and Texas. Since that time environmental conditions have not favored the formation of the migratory phase, but the species continues to be a pest over wide areas of the semiarid West.

In timberlands insects attack trees at every stage of growth, from seeds to mature trees, as well as the freshly cut logs and lumber in storage. Graham and Knight (1965) state that "more wood has been destroyed by insects, fungi, and fire than has ever been cut and used." They estimate the total yearly losses due to insects for the United States and coastal Alaska to be 8.6 billion board feet (a board foot equals the volume of a board 12 × 12 × 1 in).

The most important forest insect pest in the Eastern United States is the gypsy moth, *Porthetria dispar* (Lymantriidae), that was introduced from Europe during the last century (Endpaper 3*b*). Although the moth was well known as a forest pest, a naturalist who was experimenting with silkworms brought eggs of the moth from Europe to his home in Medford, Massachusetts. Apparently eggs or some young larvae escaped his care in 1868 or 1869 and started an infestation near his house. The destructive caterpillars went largely unnoticed, except by his neighbors, until 1889, when the first severe outbreak occurred. Fruit and shade trees in Medford were completely defoliated and the citizens subjected to hordes of migrating caterpillars. Concerted efforts to control the moth began in 1890 with an appropriation of $50,000 by the Massachusetts legislature. Early research revealed that the caterpillars would experimentally eat over 400 different species of plants, many of which were of economic value. The pest has continued to spread and now occurs over most of New England and in various places in eastern New York, Maryland, northern New Jersey, eastern Pennsylvania, southeastern Quebec, and near Lansing, Michigan. Larvae have been captured as far away as California. The young larvae are carried considerable distances on wind. The egg masses of 100 to 1000 eggs, normally laid on trees, may be laid in vehicles, thus increasing the chance for dispersal over long distances. The favored hosts of the caterpillars are oaks, gray birch, and poplar. Conifers are also attacked when growing in mixed stands with hardwoods. To date, over $100 million has been spent by state and federal government agencies to combat the gypsy moth by means of research, control, and preventive measures.

The most important pests of living coniferous trees are bark beetles (Scolytidae). The beetles attack in large numbers and overcome the resistance of mature trees. The loss of mature trees is especially detrimental to forestry management because the optimal schedule for long-term harvesting is disrupted.

In the urban environment insects attack not only living ornamental and garden plants but also the framework of houses and their contents. Virtually any article of plant or animal origin can be damaged or destroyed by insects, including wooden structures and furniture, clothing and furs, food, leather goods, books and paper products, and drugs and tobacco. The most common home-infesting insects are termites, cockroaches, ants, fleas, wasps, flies, silverfish, powderpost beetles, and various pantry pests. Furthermore, people are usually annoyed by the sight of live insects in their homes, even when the insects are harmless intruders from outdoors. It has been estimated that the operators of urban pest control agencies in the United States do an annual business of $1.3 billion to control and prevent infestations of insects and rodents.

Insects that transmit disease agents are called *vectors.* The most important vectors of human diseases are the mosquitoes. They transmit the microbes that cause malaria, yellow fever, dengue, encephalitis, and filariasis. These diseases not only kill but also cause sicknesses that result in lost working days and reduced working efficiency. Malaria is considered the most important human disease. Even though it has declined in the United States, malaria continues to cause increasing mortality and debilitation elsewhere in the world. Some other important vectors and the diseases they carry are the human body louse (epidemic relapsing fever, typhus); several species of fleas (plague, murine typhus); tsetse flies, or *Glossina* spp. (Muscidae), of Africa (trypanosomiasis or "sleeping sickness"); and kissing bugs, or *Triatoma* spp. (Reduviidae), of the New World (trypanosomiasis or "Chagas' disease").

PEST CONTROL. Extensive plantings of single crops, i.e., monocultures, favor the rapid population growth of phytophagous insects and mites and of plant diseases. Prior to World War II, agricultural pest control was a combination of several methods because no single approach was usually satisfactory. Cultural methods such as early or late planting avoided the seasonal appearance of certain pests. Destruction of crop residues and plowing exposed overwintering pests to adverse physical factors. Varieties of plants that were known to be resistant to pest attack were often selected for cultivation in certain areas. The role of natural enemies was widely acknowledged. In 1888 the purposeful introduction of the Australian lady beetle, *Rodolia cardinalis* (Coccinellidae), to control the cottony-cushion scale, *Icerya purchasi* (Coccidae), in California, proved that such enemies could be manipulated successfully. Chemical pesticides were limited in variety and effectiveness, and some were toxic to humans and domestic animals. As early as 1897, pests were known to develop resistance to insecticides. Yet for annual crops, where prompt relief from pests meant increased production and profits, pesticides were generally expected to be the most economical approach in the future.

In 1942, the chlorinated hydrocarbon DDT (*d*ichloro*d*iphenyl*t*richloroethane) was demonstrated to have major medical importance in the military effort of World War II. It was effective against body lice, mosquitoes, and flies, and had a low acute toxicity for humans and mammals. Furthermore, the residue had a persistent action against insects.

After the war, DDT and other newly developed synthetic organic insecticides were shown to be highly effective on a wide range of agricultural pests. For a period of about 10 years DDT was produced and used at a rate of some 100,000 tons per year. Pesticides were frequently applied according to a fixed schedule of calendar dates during the growing season and without evaluation of the actual damage taking place. Nonchemical methods of pest control were largely neglected.

The success of insecticides, however, began to be offset by several factors. Pests evolved resistance to specific insecticides, thus requiring stronger and more frequent doses or new synthetic compounds. Species that had previously not been pests were becoming economically important because their natural enemies were eliminated by pesticides. Lastly, evidence began to accumulate that DDT and its breakdown products were widely distributed in soil and water, and had entered the food chains of other animals, including humans. Food chains may act as "biological amplifiers," concentrating a toxic substance in higher amounts at each trophic level. The chlorinated hydrocarbon insecticides were especially likely to be concentrated because of their high solubility in fatty substances and their low solubility in water. In 1962, Rachel Carson argued the case against reliance on pesticides in her book *Silent Spring*. Ten years later, DDT was restricted in the United States to uses requiring special authorization by the Environmental Protection Agency. (For details on the uses and misuses of pesticides, see Ehrlich et al., 1977.)

The reported instances of resistance to pesticides continue to mount. Whereas in 1948, 14 species of insects and mites exhibited resistance in some populations, the total in 1975 was 305 species, plus 59 species in which resistance had not been fully confirmed by laboratory tests. Included are nearly all the major agricultural and medically important pests. Resistance is known to all types of insecticides, especially to DDT and other chlorinated hydrocarbons such as the cyclodienes (e.g., chlordane), the organophosphates (e.g. diazinon), and the carbamates (e.g., carbaryl). Resistance to one insecticide may confer resistance to chemically related compounds, and resistance may be evolved in the same population to several different classes of chemicals. In a few species, no insecticide now available is sufficient to give economical control.

Ray F. Smith and others (see, for example, Apple and Smith, 1976) have urged a multidisciplinary approach that is called *integrated pest management.* Based on ecological principles, it integrates all kinds of available control methods against all types of pests (insects, mites, nematodes, plant diseases, weeds) into a complete program. Thus entomologists, nematologists, plant pathologists, and weed scientists as well as agricultural economists are being encouraged to collaborate on pest problems. The intent is to provide an effective and economical approach that optimizes farm profits while protecting both public health and the quality of the environment. It is hoped that such a diversified but coordinated approach will give long-term success in pest control.

Chemical pesticides will continue to be needed in situations where other control measures fail or can be enhanced by pesticide treatment. It is clear that pesticides are extremely valuable tools. The effective life of an insecticide can be prolonged by proper techniques of application at a frequency that reduces the evolution of resistance by pests. Furthermore, the application can be planned according to estimates of economic damage and at times that avoid mortality of pollinating insects and natural enemies of pests. Efforts should be made to develop compounds that are toxic to specific groups of pests.

Recent attention has focused on various aspects of insect biology that offer new tactics for control. Four such tactics deserve mention here. (1) Certain microbial agents of disease in insects are now regularly used in control. (2) Striking success has resulted from the release of males of the screwworm fly, *Cochliomyia hominivorax* (Calliphoridae) that have been sterilized by x-rays or gamma rays. The sterile males mate with wild females, resulting in infertile eggs. (3) Synthetic mimics of the normal growth hormones that regulate insect development may prove to be "third-generation pesticides." The hormone mimics disrupt insect growth while presumably not affecting vertebrates. (4) The sex attractants that are released by insects to attract the opposite sex can be used to lure pests to traps or toxic baits. The attractants are especially valuable in monitoring the first appearance of a pest and the population buildup.

Most, if not all, such new approaches to pest management are subject to the same limitation that was encountered during the pesticide era: insects are capable of rapid evolution of resistance when placed under strong selection pressure. Cases of resistance to hormone mimics and, at least in the laboratory, to microbial insecticides are now on record. The sterile-male technique has encountered problems because the laboratory flies evolved characteristics that were less attractive to wild females. Some insects reproduce without mating, i.e., parthenogenetically. Parthenogenetic strains of a pest would be resistant to the use of sterile males or sex attractants for control. Thus, natural evolutionary processes tend to undermine pest control efforts. Only by a thorough understanding of these processes can we expect to develop long-term programs in pest management.

SELECTED REFERENCES

General texts that also provide keys to the identification of insect families are Borror et al. (1976), Britton et al. (1970), Comstock (1940), Essig (1942, 1958), Richards and Davies (1957), and H. H. Ross (1965). Brues et al. (1954) is entirely devoted to classification. The illustrated handbook by Borror and White (1970) is especially attractive but has no keys. Texts that emphasize biology are Fox and Fox (1964), Horn (1976), Romoser (1973), and Wigglesworth (1964). Natural history is the theme of Brues (1946), Frost (1942), Hutchins (1966), and Imms (1971). Insect biology is illustrated in color by Farb (1962) and Linsenmaier (1972). Applied aspects of entomology are treated by Apple and Smith (1976), Ebeling (1975), Little (1963), and Metcalf et al. (1962). The use of insects to control pests is discussed by DeBach (1964, 1974), Huffaker (1971), Huffaker and Messenger (1976), and Swan (1964). History of entomology is reviewed by Essig (1930), Mallis (1971), and R. F. Smith et al. (1973). Insect ecology has been recently treated by Price (1975), and insect structure and function are dealt with by Bursell (1971), Chapman (1969), Snodgrass (1935, 1952), and Wigglesworth (1972). Collected readings in entomology are provided by Barbosa and Peters (1972). Study techniques and useful tips are given by Oldroyd (1970) and Peterson (1964).

2 THE INSECT BODY

This chapter provides an introduction to the general body plan and external anatomy of insects. The description of each part of the anatomy begins with the features seen on common adult insects, such as grasshoppers, and concludes with a brief discussion of some distinctive modifications of other insects. The functional relationships of the parts and the anatomical characteristics of each order of insects are reserved for later chapters.

The application of anatomical terms to different insects requires the determination of *homology*. Recall that this is the structural likeness of anatomical parts due to a common evolutionary origin, i.e., a structure possessed by two or more organisms is homologous because the genes responsible were inherited from a common ancestor. Homologies are established in insects by matching the structures with regard to segmental position, location on the segment, time of embryonic development, size, shape, and the association of other structures whose homologies have been determined. The structures named in this chapter are usually recognizable by simple inspection, yet some insects possess uniquely modified parts whose homology is unknown.

Another kind of homology, called *serial homology,* is applied to the same structures on different segments of an individual. For example, each body segment of an arthropod may have a pair of serially homologous, jointed appendages. In insects these appendages are modified into mouthparts, legs, certain types of gills, and external organs for reproduction. The serial homology of structures on different segments is sometimes not immediately apparent; careful study of embryology is then required. Considerable argument, for example, surrounds the nature of the antennae. Are they true appendages, like legs, or special sensory structures peculiar to the head? To date this question has not been resolved.

Before proceeding further, you may wish to review the following terms that denote anatomical position:

anterior, cephalad, or cephalic: Toward or pertaining to the front end or head.

posterior, caudad, or caudal: Toward or pertaining to the hind end or tail.

dorsal or dorsad: Toward or pertaining to the upper surface or back.

ventral or ventrad: Toward or pertaining to the lower surface or belly.

lateral or laterad: Toward or pertaining to the side or outer part.

median, medial, mesad, or mesal: Toward or pertaining to the middle or inner part.

proximal, basad, basal, or base: Toward or pertaining to the part nearest the body or base of a part.

distal, distad, apical, or apex: Toward or pertaining to the part farthest from the body or part.

longitudinal: Oriented parallel to the length or anteroposterior axis of the body, or in the case of an appendage, the direction parallel to a line extending from base to tip.

transverse: Oriented perpendicular to the longitudinal axis of the body or appendage.

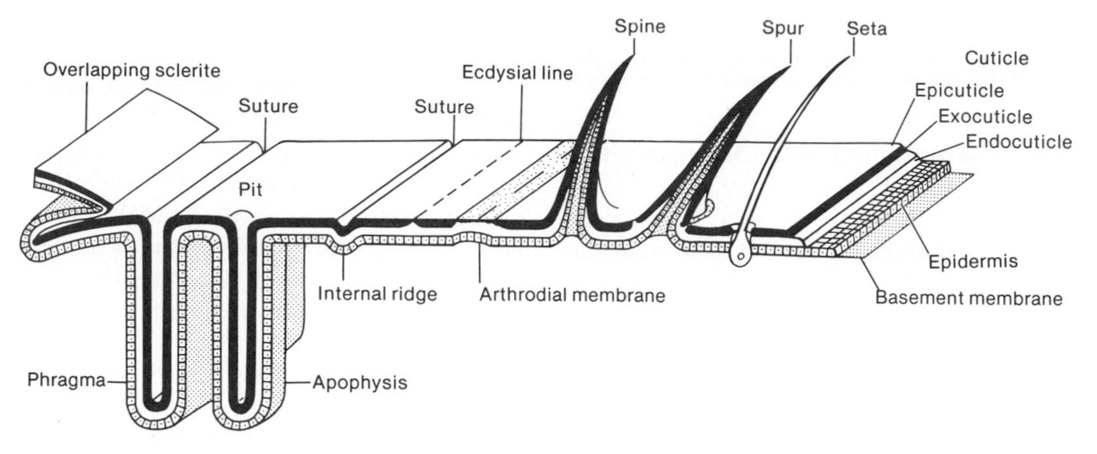

Figure 2.1 Diagram of integumental features. (*Modified from Metcalf et al., 1962, by permission of McGraw-Hill Book Company.*)

PROPERTIES OF THE EXOSKELETON. The body wall of an insect is called the *integument.* The outer layer is a noncellular *cuticle,* chemically composed of chitin and proteins that are secreted by the inner layer of *epidermal cells* (Fig. 2.1). In some areas the cuticle is chemically tanned, or sclerotized, by the epidermis to form rigid, platelike *sclerites.* Areas of the cuticle between sclerites remain tough and flexible, creating membranous joints or articulations. The sclerites and membranes together provide an external support, or *exoskeleton,* for the insect body.

Single cells of the epidermis may be modified into hairs, or *setae.* The seta-forming cell is the *trichogen,* and it is accompanied by a socket-forming cell, or *tormogen.* Setae are greatly varied in both form and function. The common types are hairlike or *simple* setae, featherlike or *plumose* setae, and platelike *scales. Poison setae* are hollow and filled with toxic fluids for defense. Setae of diverse shapes are innervated by the peripheral nervous system and function as sensory organs called *sensilla* (Figs. 3.1*a,* 6.5*a* through *d*).

Many cells are involved in local outgrowths or ingrowths of the integument. Rigid outgrowths are commonly called *spines.* Some outgrowths, especially on the legs, are movable and called *spurs.* Rigid fingerlike ingrowths of the integument associated with muscle attachments are commonly called *apodemes* or, if larger and armlike, *apophyses.* The latter term is sometimes used for external protuberances, but in this text we restrict it to invaginations. Large apophyses in the head and thorax buttress the exoskeleton and provide areas for muscle attachment. These apophyses function as the *endoskeleton.* Platelike invaginations associated with the dorsal flight muscles are *phragmata.*

The integument may also be infolded linearly to produce internal ridges for strengthening the body wall or for muscle attachment. The internal ridge is usually traced externally by a line or groove. Virtually any line in an insect is loosely called a *suture,* regardless of whether it is a line of flexibility (e.g., an articulation) or a line marking an internal ridge, or phragma. Sutures are used as convenient boundaries to delimit areas of the exoskeleton. In the strict sense, a suture is a seam along which two plates have fused. The lines between plates of the vertebrate skull are properly called sutures. The cuticle of the insect integument, however, is a continuous layer; adjacent sclerites usually become unified in the course of evolution without leaving a trace. A "suture" transversing a sclerite, therefore, should not be interpreted to mean that the adjacent areas were once separate. Snodgrass (1963) has argued that all groovelike sutures should be called *sulci.*

EVOLUTION OF THE BODY REGIONS. The organization of an insect's body is more readily understood if we recall that insects and other arthropods evolved from the segmented worms, or Annelida. Snodgrass (1935) proposed a sequence of steps leading to the insects that is instructive in this context. In the earliest stage (Fig. 2.2*a*), the wormlike ancestor has a cylindrical body with the tubular intestine running nearly the full length. Between the mouth and anus the body is divided transversely into segments. In front of the mouth is an unsegmented part called the *prostomium.* The last part of the body encircles the anus and is the *periproct.* This is the endpiece and also is not a true segment. In a second hypothetical stage (Fig. 2.2*b*), the prostomium acquires a pair of light-sensitive receptors, or *eyes,* and a pair of sensory antennae. Each true body segment develops a pair of movable, ventrolateral appendages. In the third stage (Fig. 2.2*c*) the ancestor is an arthropod with jointed legs and enlarged sensory apparatus. Evidently it has become terrestrial, because most of the segments have a pair of openings, called *spiracles,* leading to the internal air tubes, or *tracheae,* of the respiratory system. The appendages of the last true segment are called *cerci.*

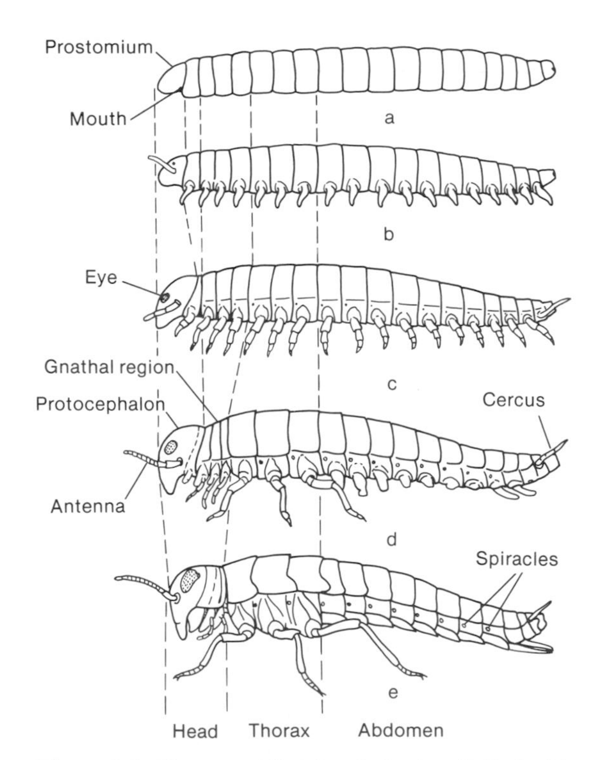

Figure 2.2 Diagrams of the steps in the hypothetical origin of the insect body from a wormlike ancestor. See text for explanation. (*Redrawn from Snodgrass, 1935, by permission of McGraw-Hill Book Company.*)

The features of the insects are foreshadowed in the fourth stage (Fig. 2.2*d*) as regions of the body become specialized. The modification of segments into functional units is called *tagmosis.* At this stage, four such units, or *tagmata,* can be recognized: *protocephalon* (prostomium plus one to three segments), *gnathal region* (three segments), *thorax* (three segments), and *abdomen* (eleven segments plus *periproct*).

The factual basis for the hypothetical protocephalon is found in the complex structure of the insect brain. The exact number of segments that have become fused to form the insect head is a matter of much debate. Snodgrass has defended the simple theory that the prostomium housed the primitive brain. This was a mass of nervous tissue in front of the mouth. As the sense organs on the prostomium enlarged, that part of the brain innervating the eyes became the *protocerebrum* and the part innervating the antennae became the *deutocerebrum.* The segments behind the mouth each had a pair of primitive *ganglia,* or masses of nerve cells. Each ganglion innervated, respectively, the appendage on its side of the body. The ganglia communicated transversely within each segment by nervous *commissures,* and longitudinally from segment to segment by nervous *connectives* (Fig. 2.3*a*).

Snodgrass speculates that the appendages of the first postoral segment became sensory in function and shifted forward. The ganglia also moved forward to either side of the mouth and became more closely associated with the brain (Fig. 2.3*b*). In the insects, the appendages atrophied and were lost in the course of evolution, yet the ganglia of the first segment still persist in

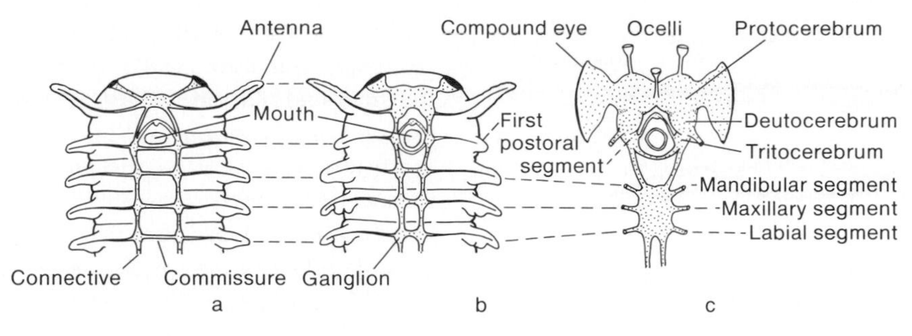

Figure 2.3 Diagrams of the steps in the hypothetical origin of the insect brain and head segmentation. See text for explanation. (*Redrawn from Snodgrass,* 1935, *by permission of McGraw-Hill Book Company.*)

insects as the third part of the brain, or *tritocerebrum.* The tritocerebral commissure is evidence of its former segmental position behind the mouth (Fig. 2.3*c*). In Crustacea the appendages of this segment still persist as the second pair of antennae.

The appendages of the next three segments are specialized for feeding. This is the gnathal region. The three segments following are modified for walking and became the thorax. The last of the four tagmata, the abdomen, contains no more than 11 true segments plus the periproct. Most of the visceral organs are located here. With the walking appendages concentrated in the thorax, the appendages of the abdomen became modified for other functions or atrophied.

In the final stage (Fig. 2.2*e*), the body regions have been reduced to three by the tagmosis of the protocephalon and gnathal region to form the *head* tagma. The head combines the sense organs and brain with the gnathal appendages to form a center for sensing the environment and ingesting food. The thorax is the center for locomotion and houses the large muscles that attach to the leg bases, as well as others that act on the thoracic skeleton to move the wings. The abdomen is the center for the reproductive organs and intestines. By pumping movements, the abdomen also aids airflow in the tracheal system.

SECONDARY SEGMENTATION. Among arthropods a distinction is made between *primary segments,* which correspond to true embryonic segments, and *secondary segments,* which are functional subdivisions. The trunk segments of annelid worms and soft-bodied insect larvae are clearly defined by transverse constrictions on which the principal longitudinal muscles insert. These are primary segments, and the transverse folds to which the muscles attach are true intersegmental boundaries. By the contraction of the muscles, the soft segmental walls are deformed and wriggling movements produced (Fig. 2.4*a*).

With the evolution of a rigid exoskeleton, movement between segments is restricted to transverse lines of flexibility just in advance of the primary intersegmental lines. The functional subdivisions of the body no longer coincide with the primary segments, and are accordingly called *secondary segments* (Fig. 2.4*b*).

The dorsal plate of a secondary segment is the *tergum,* and the ventral plate is the *sternum.* The transverse, muscle-bearing ridge that marks the primary intersegmental fold is the *antecosta.* Externally, the ridge is traced by the *antecostal suture.* The margin of the tergum anterior to the antecostal suture is the *acrotergite,* and the corresponding sternal margin is the *acrosternite.*

THE HEAD

The feeding appendages, the most important sense organs, and the brain are contained in the head. As the first, or most anterior, body division, the head leads the way when the insect moves forward. It is in a position to detect the changing physical and chemical properties of the environment. Sensations of images in color, moisture,

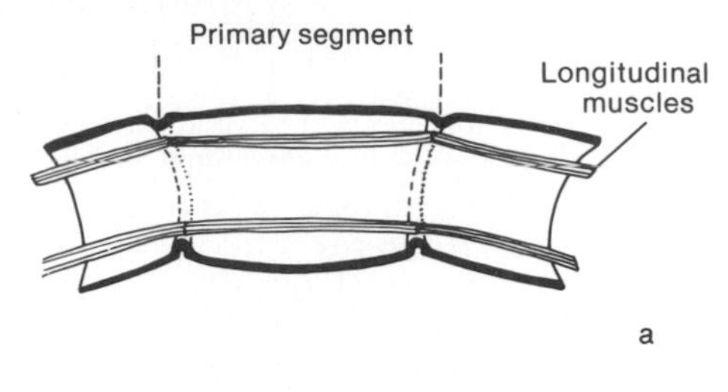

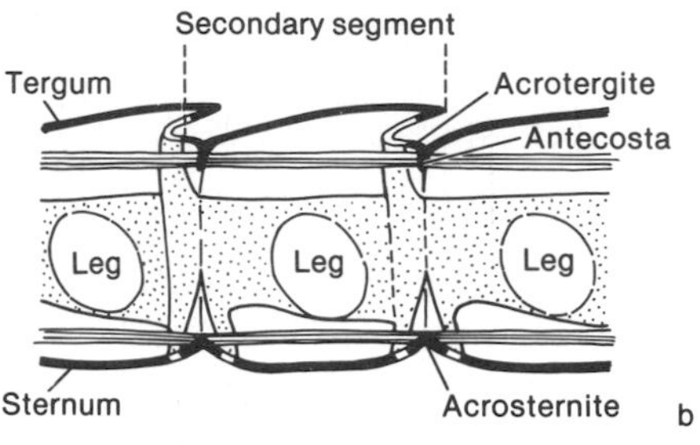

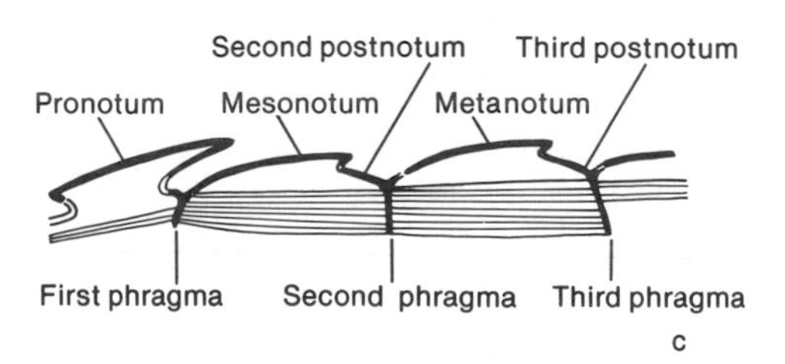

Figure 2.4 Types of body segmentation: *a*, primary segmentation; *b*, secondary segmentation; *c*, secondary segmentation and formation of dorsal phragmata in pterothorax. (*Redrawn from Snodgrass,* 1935, *by permission of McGraw-Hill Book Company.*)

touch, sounds, odors, and flavors travel but a brief distance to the brain, where the insect's behavioral reactions are controlled. The feeding appendages by which the body is fueled are appropriately located in the midst of this battery of sensing devices.

The head articulates with the body at the membranous neck, or *cervix.* Small *cervical plates* are sometimes present at each side in the cervical membrane. These function in the extension and retraction of the head. The rear of the head opens into the cervix and the body beyond by the *foramen magnum.* Through this opening pass the ventral nerve cord, salivary duct, foregut, aorta, and various neck muscles.

Among the various insects, three positions of the head can be distinguished. The original position relative to the body is with the mouthparts directed downwards, immediately ventral to the head capsule. This is the *hypognathous* position of grasshoppers (Fig. 2.5*d*), roaches, and many other insects. In the *prognathous* position, seen in some beetles (Fig. 13.2*b*), the mouthparts are directed forward and project anterior to the eyes. The last is the *opisthognathous* position possessed by Hemiptera (Fig. 34.1*b*). Here the sucking beak is directed toward the rear, beneath the thorax.

CRANIUM. The sclerotized head capsule, minus the appendages, is the *cranium.* In the course of evolution the distinction between the head segments has been lost almost entirely. Most of the lines seen now on the cranium represent secondary modifications unrelated to segmentation. Some lines trace internal ridges that strengthen the exoskeleton, while others are actually lines of weakness that allow the cuticle to split during ecdysis. Only the postoccipital suture, described below, is believed to be a segmental boundary. Here a strong internal ridge has been retained between the maxillary and labial segments, serving as a place for the attachment of neck muscles that move the head.

The *compound eyes* are easily recognized as the large paired organs located dorsolaterally on the cranium (Fig. 2.5*a*). The surface of the eye is covered with minute facets, each of which represents the lens of an individual eye unit, or *ommatidium.* Situated between the compound eyes on the face are as many as three simple eyes, or *ocelli.* Each has a single lens. One is usually present on the midline, and the others are placed above and at each side.

On some insects a pale line, shaped like an inverted Y, can be seen medially on the top of the head. This is the *ecdysial line,* along which the old cuticle first splits when it is shed by immature insects. One or two ventral ecdysial lines occur on the underside of the head of many endopterygote larvae. The lines serve no function in the adult.

The cranium has an internal, sclerotized structure called the *tentorium* (Fig. 2.5*b*). This endoskeleton is created by paired invaginations

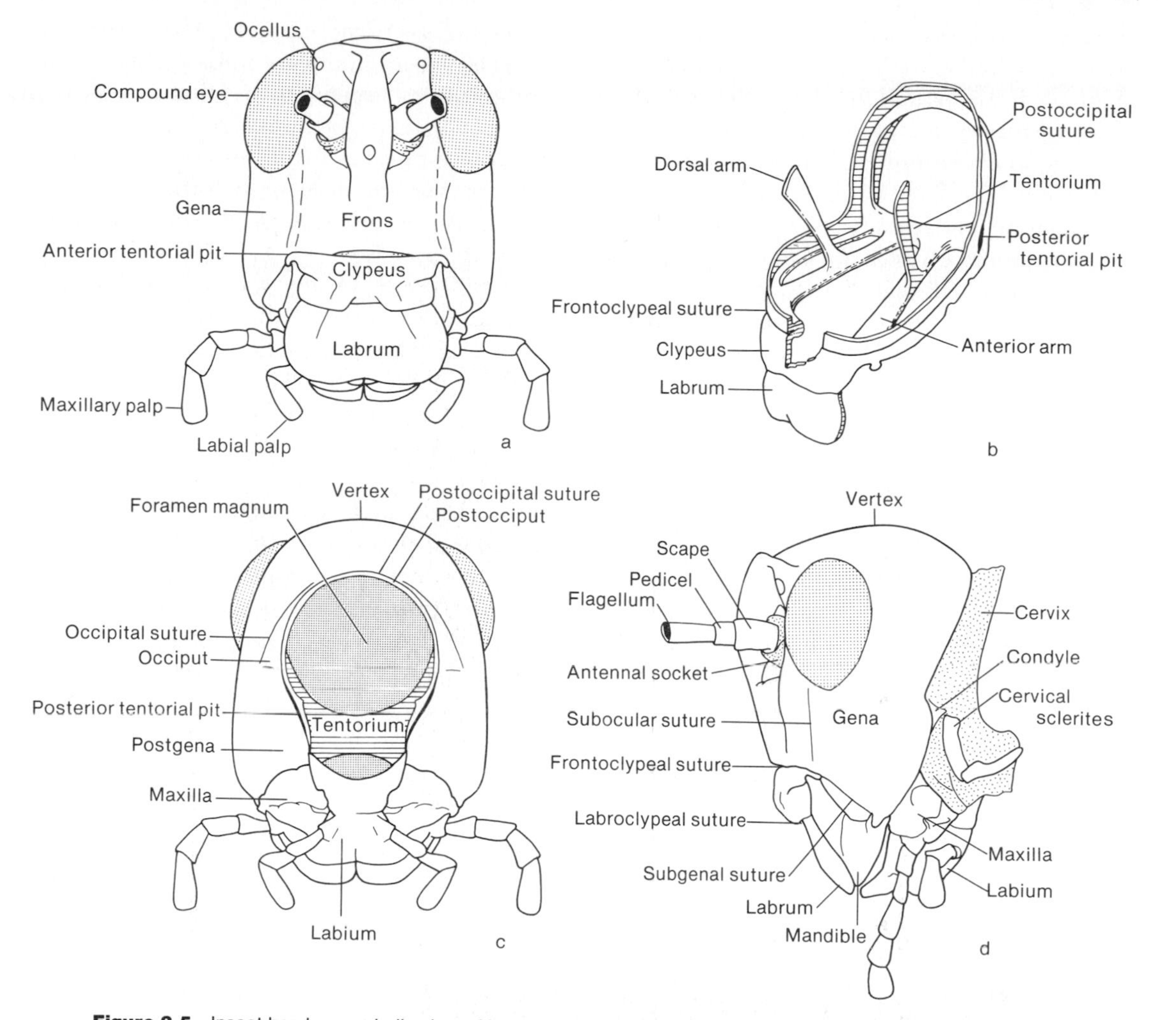

Figure 2.5 Insect head: *a*, cephalic view of head of *Romalea microptera* (Acrididae); *b*, diagram of tentorium of a grasshopper with most of the cranium removed; *c*, caudal view of head of *Romalea; d*, lateral view of same. (b *redrawn from Snodgrass*, 1935, *by permission of McGraw-Hill Book Company.*)

of the integument, at both the front and rear of the head, that fuse together inside the head. Externally the invaginations are indicated by pits or narrow slits. The *anterior tentorial pits* mark the entry of anterior invaginations that become the internal *anterior tentorial arms*. Similarly, the *posterior tentorial pits* mark the invaginations of the *posterior tentorial arms*. The arms meet medially to form the *tentorial bridge*. In some insects, the anterior arms each have a *dorsal arm*. The tentorium is variously shaped in different insects. It strengthens the cranium and provides attachment for muscles to the antennae and the gnathal appendages.

Names have been given to different areas of the cranium. In some instances sutures serve as convenient boundaries between areas, but elsewhere the limits are undefined. This should not cause confusion because the names refer merely to topographic areas, not to discrete sclerites.

Beginning at the top of the head and descending in front, the names of the facial areas are as follows: the summit of the head, between and behind the compound eyes, is the *vertex*. The next area, the *frons,* is between the antennae and the compound eyes. No definite boundary exists between the frons and the vertex. Below the frons is the *clypeus.* The clypeus may be defined as the area of the cranium to which the labrum is attached. The *labroclypeal suture* is the line of articulation between the labrum and clypeus. The frons and clypeus are sometimes separated by a suture extending between the anterior tentorial pits. This is the epistomal or *frontoclypeal suture.*

Returning to the top of the head and descending on one side (Fig. 2.5*d*), the *gena* is the area extending from below the vertex and behind the compound eyes, to the ventral edge of the cranium. The gena is not delimited above from the vertex, but along other boundaries sutures may exist. It may be separated anteriorly from the frons by a *frontogenal suture* or a subocular suture in *Romalea,* and from the clypeus by a *clypeogenal suture.* Ventrally, a *subgenal suture* may extend between the anterior and posterior tentorial pits, separating off a marginal area, the *subgena.* Posteriorly, the gena may be delimited by the *occipital suture.*

Turning now to the back of the head, the areas can be seen best if the head is removed at the neck, or cervix (Fig. 2.5*c*). Care should be taken in cutting the cervical membrane to avoid damage to the posterior rim of the cranium. At each side near the base of the labium are the slitlike posterior tentorial pits. Extending from the posterior pits and running dorsally around the foramen magnum is the *postoccipital suture.* The area anterior to the suture is the *occiput.* The labium is suspended at each side from the lower edges of the postocciput, indicating that this is the tergum of the labial segment. The narrow sclerotized rim of the cranium, posterior to the suture, is the *postocciput.* The postocciput also bears the occipital condyles on which the cervical plates articulate. Returning to the occiput, an occipital suture, or at least an incomplete ridge, may separate the occiput dorsally from the vertex and laterally from the genae. The lower part of the occiput, near the posterior tentorial pits, is called the *postgena.*

MODIFICATIONS OF THE CRANIUM. The sutures and areas of the cranium vary considerably in extent and position, largely as a result of modifications of the mouthparts and associated musculature. It is entirely possible for distinctive, new areas of the cranium to develop in certain groups of insects. A special terminology is often applied to these areas.

The faces of Hemiptera (Suborder Homoptera) and Psocoptera (Fig. 32.2*a*) are thus altered by the development of a sucking pump. The prominent bulge between the eyes that houses the muscles for this pump is called the *postclypeus,* and the smaller area below, the *anteclypeus.* In reality this is a secondary modification of the entire frontoclypeal region where the original frontoclypeal boundary has been obscured.

Insects such as Orthoptera lack sclerotization of the integument beneath the foramen magnum and posterior to the labium. In other words, the throat behind the mouthparts is membranous. In hypognathus insects, at least two modifications of this area can be seen: the formation of *hypostomal* or *postgenal bridges.* Recall that the subgenal suture runs near the edge of the cranium and terminates in the posterior tentorial pit. That part of the suture posterior to the mandible is also known as the *hypostomal suture,* and the subgenal area thus delimited is the *hypostoma.* Recall also that the adjacent area of the occiput is the postgena. In caterpillars of Lepidoptera, sclerotic lobes of the hypostoma extend mesally onto the throat, but do not meet. In adult Diptera and some Hymenoptera the hypostomal lobes unite to form a hypostomal bridge beneath the foramen magnum (Fig. 2.6*b*). In other adult Hymenoptera and Hemiptera (Suborder Heteroptera), the postgenae also unite, forming a postgenal bridge (Fig. 2.6*a*). In the formation of both types of bridges, the posterior tentorial pits remain unaltered in position.

In prognathus insects such as some Coleoptera and Neuroptera, the underside of the head is substantially modified by the forward position of the mouthparts. The postgenae are usually enlarged, and the posterior tentorial pits are shifted forward. The posterior pits, however, still retain

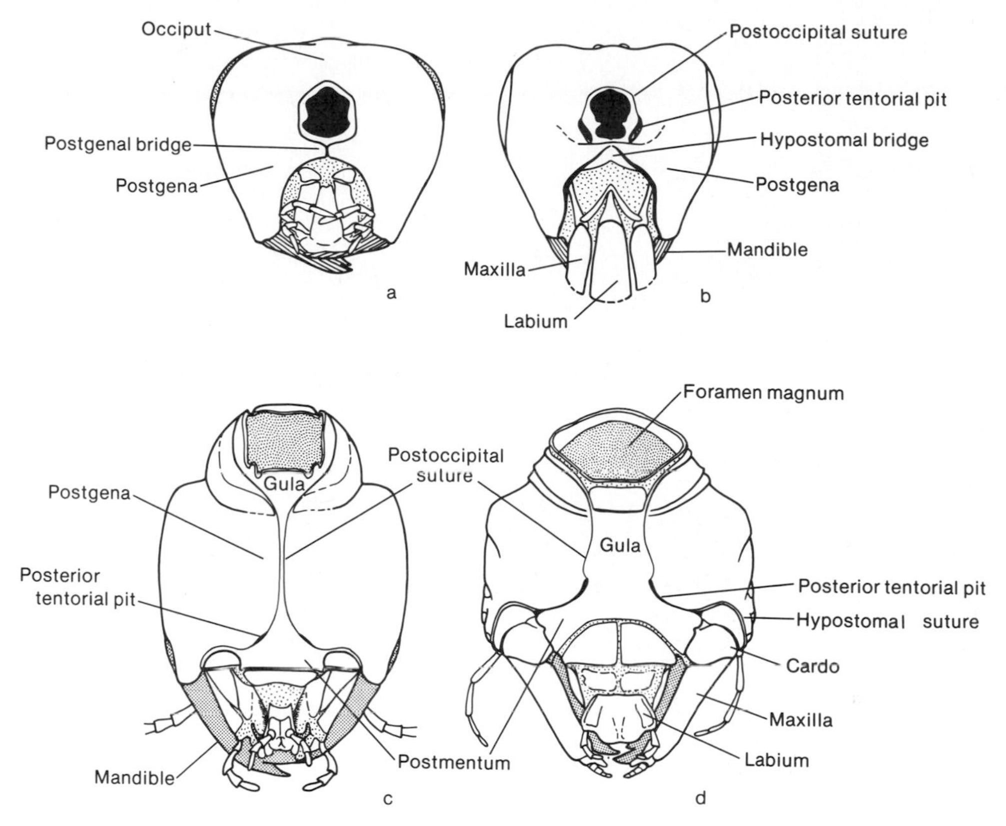

Figure 2.6 Modifications of the posteroventral region of the head; *a,* postgenal bridge of a wasp, *Vespa* spp. (Vespidae); *b,* hypostomal bridge of the honeybee, *Apis mellifera* (Apidae); *c,* narrow gula of a beetle, *Staphylinus* spp. (Staphylinidae); *d,* broad gula of the dobsonfly larva, *Corydalus* spp. (Corydalidae). (*Redrawn from Snodgrass,* 1935, *by permission of McGraw-Hill Book Company.*)

their connections at each side with the postoccipital suture. The midventral area of the head between the pits corresponds to the membranous throat of Orthoptera. In Coleoptera and Neuroptera, this area is secondarily sclerotized to form the *gula* (Fig. 2.6*c* and *d*). The gula is continuous with the postocciput and sometimes also the labial base. The parts of the postoccipital suture that delimit the gula laterally are correspondingly renamed the *gular sutures.*

ANTENNAE. The antennae are paired, segmented appendages that articulate with the cranium between or below the compound eyes. Three parts can usually be recognized: the basal, or first, segment is the *scape;* the second segment is the *pedicel;* and the remaining segments constitute the *flagellum* (Fig. 2.7). The whole antenna is moved by muscles from the head that insert on the scape. In the *antennal socket* the antenna may pivot on an articular process, or *antennifer.* The flagellum is moved by muscles from the scape that insert on the pedicel but not on the segments beyond. In accordance with its sensory function, the surface of the flagellum is supplied with many sensory receptors that are innervated by the deutocerebrum of the brain. In addition, within the pedicel is a mass of sense cells that detects movements of the flagellum. This is called *Johnston's organ* (Fig. 6.5*f*). Exceptions to the foregoing account are found in the Collembola and Diplura, where the flagellar segments have intrinsic muscles and Johnston's organ is absent. Protura lack antennae entirely.

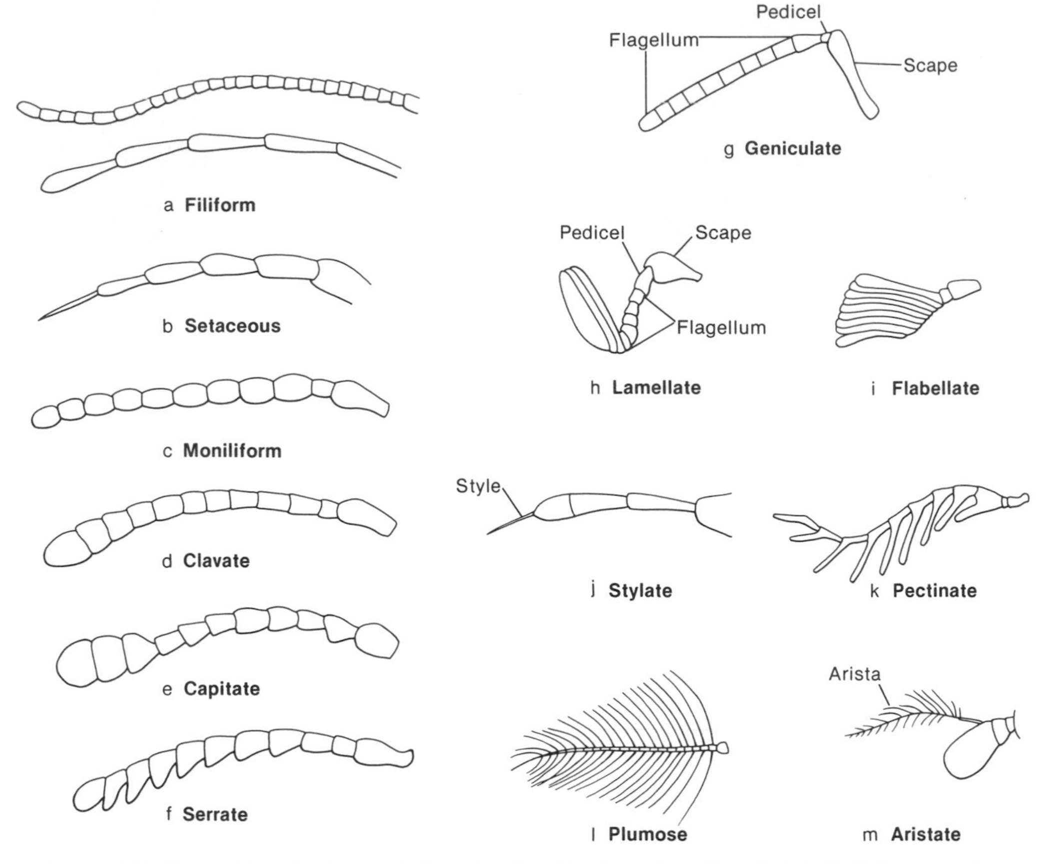

Figure 2.7 Types of insect antennae. (a through g, j, and l *redrawn from Metcalf et al.,* 1962, *by permission of McGraw-Hill Book Company.*)

TYPES OF ANTENNAE. Special terms for different shapes of antennae are indicated in Fig. 2.7.

MOUTHPARTS. The mandibulate, or chewing, type of mouthparts is the basic type from which the specialized types have been derived. The mandibulate type consists of an upper lip (or labrum), paired mandibles, paired maxillae, a lower lip (or labium), and a median tonguelike hypopharynx (Fig. 2.8). The mouthparts enclose the preoral cavity, within which are the true mouth and the opening of the salivary glands.

The *labrum* is a flap that closes the preoral cavity in front (Fig. 2.5*a*, 2.8*b*). It articulates with the cranium by the membranous *labroclypeal suture.* The inner surface forms the roof of the preoral cavity and may have a median lobe called the *epipharynx* (Fig. 2.8*a*). Muscles from the head capsule insert near the base of the appendage and provide a variety of motions. The labrum is innervated by the tritocerebrum of the brain.

The *mandibles* are heavily sclerotized jaws that represent limbs that have lost their distal segments (Fig. 2.8*c*). The inner edge is modified for biting. Toward the tip the edge has cutting teeth, or incisors; at the base is a grinding surface, or *molar.* Each mandible has two basal points, or *condyles,* where it articulates with the cranium. The posterior condyle is larger and fits against

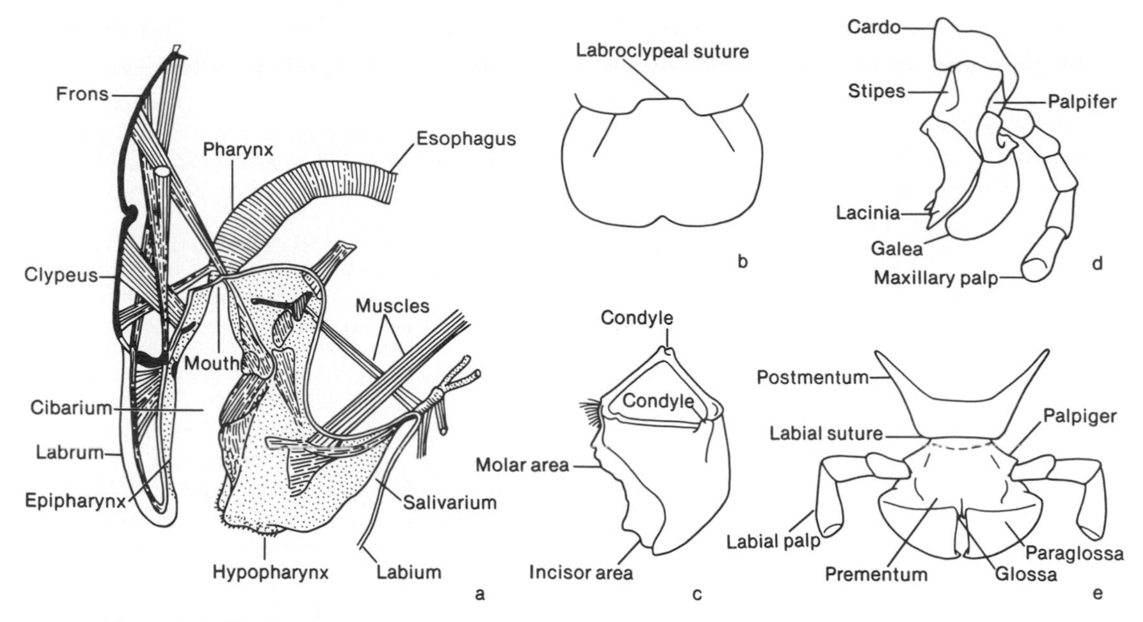

Figure 2.8 Mouthparts and preoral cavity: *a,* diagram of preoral cavity and lateral view of hypopharynx; *b,* labrum of *Romalea microptera* (Acrididae); *c,* mandible of same; *d,* maxilla of same; *e,* labium of same. (a *redrawn from Snodgrass,* 1935, *by permission of McGraw-Hill Book Company.*)

the gena or postgena. This was the original articulation point of the limb base. The anterior condyle is a secondary modification and fits against the base of the clypeus, thus forming a hinge line that permits the mandibles to swing laterally. Powerful adductor muscles from the cranium insert at the base of the mandibles on the inner side of the hinge line and by their action bring the opposing mandibles together in a pinching motion. Smaller abductor muscles, attached to the outer margin of the mandibles outside the hinge line, open the mandibles. The mandibles are innervated by the subesophageal ganglion.

The *maxillae,* a second pair of jaws, are much less massive and retain the segmentation of an appendage (Fig. 2.8*d*). Each maxilla has a limb case that articulates with the cranium by a single condyle. The base is divided into a proximal sclerite, or *cardo,* that bears the condyle, and a distal sclerite, the *stipes.* At the apex of the stipes are two lobes: an inner *lacinia,* which may be toothed, and an outer *galea.* Lateral to the galea is the maxillary *palp,* which represents the distal part of the limb and is sensory in function at the apex. The area on the stipes where the palp articulates may be differentiated into a segmentlike structure called the *palpifer.* Muscles from the cranium insert on the cardo and stipes, producing various motions of the maxilla about the single articulation. The lacinia, galea, palp, and palpal segments are also individually movable by muscles. The innervation of the maxilla is from the subesophageal ganglion.

The *labium* closes the preoral cavity to the rear. The structure the labium resembles that of the maxillae except that the labial appendages are fused medially with each other and basally with the segmental sternum (Fig. 2.8*e*). The lobed apex of the labium, called the *ligula,* consists of an inner pair of lobes, the *glossae,* and a lateral pair, the *paraglossae.* The glossae and paraglossae correspond, respectively, to the maxillary lacineae and galeae. Similarly, the labial *palp* corresponds to the maxillary palp and is likewise sensory in function. The labial palp may be borne on a segmentlike *palpiger.* The sclerite to which the ligula and palps attach is the *prementum.* The prementum articulates basally with the *postmentum* by the *labial suture* and contains all the muscles of the ligula. The postmentum in turn may be subdivided into a distal sclerite, or *mentum,* and a proximal sclerite, the *submentum.* The labium is complexly musculated such that the ligular lobes, the palps, and their segments, and the prementum are individually movable. Innervation of the labium is provided by the subesophageal ganglion.

The *hypopharynx* is a large median lobe situated between the mouth and the labium (Fig. 2.8*a*). The *cibarium* is a special chamber in front of the mouth that is formed by the anterior surface of the hypopharynx. On the posterior side, the *salivary duct* opens in a cavity, or *salivarium,* between the hypopharynx and the labium. The hypopharynx is movable by muscles from the tentorium that insert on sclerites in the wall of the hypopharynx. Nerves are supplied by the subesophageal ganglia.

TYPES OF MOUTHPARTS. The structure and function of insect mouthparts vary according to the nature of the food eaten. As a consequence, they not only provide important characters to be used in taxonomic classification, but also indicate the feeding relationships of the insect in its ecological community. The latter is of special importance to human welfare. The successful transmission of a disease, for example, may be determined by the minute structure of an insect's beak. Or an insecticide residue on a leaf surface may or may not be ingested by a pest, depending on the exact manner of feeding.

Some insects do not feed in the adult stage and the mouthparts are atrophied and functionless. Examples are Ephemeroptera and at least some members of the following orders: Plecoptera, Trichoptera, Megaloptera, Lepidoptera, Coleoptera, Diptera, and Hemiptera (male scales).

Various schemes for classifying functional mouthparts have been proposed. Here we have adopted distinctive categories and listed examples and exceptions, with brief descriptions of the mechanisms. Additional information can be found elsewhere in this text under the treatments of the different taxonomic groups.

A. *Entognathous* mouthparts. The mandibles and maxillae are recessed and largely hidden from view by lateral folds of the head. Examples: Protura (Fig. 18.1*c,d*), Diplura (Fig. 18.2*b*), and Collembola.

B. *Ectognathous* mouthparts. The mandibles and maxillae are visible or secondarily recessed, but lateral folds of the head are absent. Examples: all insects.

   1. *Mandibulate* mouthparts. The feeding appendages are more or less complete (Fig. 2.8), freely movable, and not united in a beak. The mandibles are usually conspicuous and suited for seizing objects or chewing solid food. Those of predators are fitted with sharp apical teeth for catching prey, while the mandibles of plant feeders have more obtusely angled cutting teeth together with basal grinding molars. Mandibles are also used in defense, courtship, and the construction of nests or shelters. Most adult and immature insects have mandibulate mouthparts or some modification of this type. The exceptions are listed under the suctorial and degenerate types below. Unusual modifications of the mandibulate type deserve mention:

      a. Odonata. The labium of the nymph is greatly enlarged (Fig. 21.4), hinged at the labial suture, and held retracted beneath the head. By means of quick forward thrusts, prey are seized in the pincherlike labial palps and brought to the mandibles near the mouth.

      b. Hymenoptera. The adult mouthparts are basically mandibulate, but the maxillae and labium are united in a mobile, extendible structure (Fig. 46.2*b*). The glossae are fused medially, forming a lapping organ, so that liquids can be taken readily in addition to solid food. An elongation of the maxillolabial structure forms the sucking tongue of certain families of bees [see below under 2, *b*, (2)].

      c. Hymenoptera, Lepidoptera, Trichoptera. In the larvae the maxillae, labium, and hypopharynx are united into a lower lip with the opening of the salivary duct at its tip. In these orders the salivary glands produce the silk used in making cocoons or larval shelters. The palps and other distal parts are reduced or wanting in some groups, but the mandibles are usually retained.

d. Isoptera (Fig. 24.2*b*), Megaloptera (Corydalidae, Fig. 36.1*b*), Coleoptera (Lucanidae), and Hymenoptera (Formicidae). Adults of certain insects in these taxa have greatly enlarged mandibles that are not used for feeding. The jaws of male Lucanidae actually bite less severely than the small jaws of the female. Their function is unknown. The other taxa listed use the mandibles primarily for defense.

2. *Suctorial* mouthparts. In this type, one or more feeding appendages are modified into a tubular organ for taking liquid food. Insects in different, unrelated orders have independently evolved suctorial, or haustellate, mouthparts, so the exact mechanisms vary. Parts are commonly reduced or missing. The cibarium and pharynx, either alone or in combination, provide a sucking pump in most insects of this type. Note that suctorial mouthparts are present in both the immature and adult stages of exopterygote orders listed below, whereas among the endopterygotes only one stage of the life history has developed suctorial mechanisms.

a. *Piercing-sucking* mouthparts include one or more appendages that are sharp at the apex and suited for piercing the surface of plant or animal bodies. Saliva is usually injected while feeding. Insects that attack vertebrates often have anticoagulants in the saliva to aid the flow of blood. Examples of insects with piercing-sucking mouthparts are listed below:

(1) Thysanoptera. The mouthparts are sometimes called *rasping-sucking,* because sharp stylets are used to rupture leaf cells and the exuding sap is sucked up by a cibarial pump (Fig. 35.1*d*). Thrips also pierce and suck. The mouthparts form a short conical beak and are uniquely asymmetrical. Inside the beak the right mandible and inner processes of the two maxillae are slender and sharp, forming piercing stylets. Thrips feed mostly on leaves, seeds, fungi, or pollen, but some are predatory (Chap 35).

(2) Hemiptera. An elongate proboscis is formed by the mandibular and maxillary stylets and the labium (Fig. 34.2*e,g*). The labium ensheaths the stylets but does not penetrate the host. The maxillary stylets interlock and create two canals by matching their opposing grooves. The anterior canal leads to the cibarial pump and mouth, and is appropriately named the "food canal." The posterior, or salivary, canal carries saliva under pressure from a syringelike organ within the hypopharynx. At each side of the united maxillae are the separate mandibles. By alternate thrusts, the mandibular blades pierce the host's tissues and provide an entry for the maxillae. The vascular systems of both plants and vertebrates are penetrated by the mouthparts of Hemiptera.

(3) Phthiraptera (Suborder Anoplura). The mouthparts of sucking lice are concealed in a long pouch beneath the foregut in the head (Fig. 33.2*a* through *c*). Three stylets, possibly representing the united maxillae, the hypopharynx, and the labium, make up the piercing organ. Suction is applied to the feeding wound by a muscular region of the foregut just behind the mouth and including the pharynx. Sucking lice feed only on mammals.

(4) Siphonaptera. The adult mouthparts of fleas lack mandibles. The piercing organ is formed by the bladelike maxillary laciniae and stylet-shaped epipharynx (Fig. 43.1*b*). Both the maxillary and the labial palps are present. The latter ensheath the laciniae but do not penetrate the wound. Once the skin is penetrated, the blood is sucked up by a cibarial and pharyngeal pump. Fleas bite mammals and birds.

(5) Neuroptera and Coleoptera (Dytiscidae). The predatory larvae of these in-

sects have each mandible elongate, sickle-shaped, and grooved on the inner surface. Each maxilla of the larval Neuroptera is similarly elongated and fits against the mandibular groove to form a closed canal (Fig. 38.1*c*). The body of the insect victim is pierced by the opposing mandibles, and fluids are extracted. The cibarium of Dytiscidae and the pharynx of Neuroptera provide the sucking pumps.

(6) Diptera (Culicidae, Tabanidae). The piercing organ of adult female "biting flies" consists of six stylets: labrum-epipharynx (enclosing the food canal), hypopharynx (containing the salivary duct), paired mandibles, and paired maxillae (Fig. 42.3*b*). The labium ensheaths the piercing stylets but does not penetrate the feeding puncture. Suction is provided by a cibarial pump. Mosquitoes (Culicidae) force the slender piercing organ into the skin, leaving the blunt labium at the surface. Blood is taken directly into the food canal. Horseflies (Tabanidae) cut the skin surface with their broader, bladelike stylets. The apex of the labium is greatly enlarged in two lobes called the *labellum.* The undersurfaces of the lobes are traversed by fine grooves, called *pseudotracheae.* These lead mesally to a cleft within which the labral food canal opens. The pseudotracheae direct the flow of blood to the tip of the labrum, and the blood is sucked into the canal.

(7) Diptera (*Stomoxys, Haematobia, Glossina*). The piercing organ of the stable flies, horn flies, and tsetse flies is a secondary modification of the sponging type described below for adult Diptera. Both sexes of these flies bite mammals. The labium is slender and stiff, with the labellum reduced in size and provided at the tip with rasping denticles. The labrum and hypopharynx are partly enclosed by the labium, so that the combined appendages form the piercing organ.

(8) Lepidoptera (certain Noctuidae, Pyralidae, Geometridae). The piercing organ of the fruit-piercing moths of Asia is a secondary modification of the nonpiercing proboscis of adult Lepidoptera described below. The proboscis is formed by the elongation of the maxillary galeae (as in Fig. 44.2*c*). These are interlocked to form a tubular food canal. The apex of the piercing proboscis is sharp and with hundreds of erectile barbs that aid in penetration. By alternate movements of the galeae, the tip is forced into fruit. Some species also visit the eyes of mammals for eye discharges and take blood from wounds. The unique *Calpe eustrigata* (Noctuidae) pierces the skin of mammals and sucks blood.

b. *Nonpiercing-sucking* mouthparts. The following examples are insects that take water or nutrients dissolved in water, such as the nectar of flowers (Fig. 13.1) or the honeydew (plant sap) excreted by aphids and leafhoppers.

(1) Lepidoptera. Adult butterflies and moths take nectar or honeydew with a long proboscis that is coiled beneath the head when not in use. The proboscis represents the galeae, greatly elongated and tightly interlocked to form a tube (Fig. 44.2*c*). Suction is created by the pumping action of muscles inserted in front of and behind the mouth. The other mouthparts are reduced or wanting, except the labial palps and the smaller maxillary palps.

(2) Hymenoptera (Megachilidae, Apidae, Anthophoridae). The bees in these families have the maxillolabial structure elongated and are appropriately named the "long-tongued bees" (Fig. 46.2*b*). The long galeae and labial palps cover the glossal tongue to form a tubular proboscis. The pharynx is modified into a sucking pump. By this means nectar can be withdrawn from deep inside flowers.

(3) Diptera. Most adult flies that do not suck blood have a flexible proboscis consisting of the labrum (enclosing the food canal) and slender hypopharynx (containing the salivary duct) (Fig. 42.3*c*). Together these fit in an anterior groove on the labium. The labium is enlarged apically into the bilobed labellum which functions as a sponging organ. The structure of the labellum is described above for the Tabanidae. Liquid food taken up by the labellum is directed into the food canal and sucked up by a cibarial pump. Small particles of food, such as pollen, can also be eaten by some flies. Maxillary palps are present, but mandibles and labial palps are missing. Flies of the genus *Melanderia* (Dolichopodidae), however, have unique horny processes on the labellum that function like mandibles to grasp prey.

3. *Mouth hooks of maggots.* Larvae of Diptera (Division Cyclorrapha) have the head highly reduced and invaginated into the thorax, leaving only a neck fold to form a conical snout or functional head (Fig. 42.4*c*). Within the snout is a preoral cavity, or atrium, from which projects a pair of mouth hooks. The hooks move vertically and arise from the lips of the atrium. No trace of the normal feeding appendages remains. The hooks are secondary modifications of the atrium and are unique to fly maggots. Within the prothorax are sclerotized pharyngeal plates, that together with the hooks, make up the cephalopharyngeal skeleton. Maggots feed on a variety of soft or semiliquid foods, including decaying matter as well as living tissues of plants and animals.

THE THORAX

The second body division, or thorax, is modified for locomotion. The three segments and their ventrolateral appendages are designated from front to rear as follows: *prothorax,* bearing the *forelegs; mesothorax,* bearing the *middle legs;* and *metathorax,* bearing the *hind legs.* Internally each segment has a segmental ganglion of the ventral nerve cord. A pair of lateral spiracles is present at the anterior edge of each of the last two segments. In some insects the mesothoracic pair actually opens on the prothorax, but this is a secondary condition.

In the adult pterygote, a pair of wings may be located dorsolaterally on the mesothorax and metathorax. The two wing-bearing segments are called the *pterothorax.* Wings are absent in the prothorax of insects living today, but certain extinct insects had winglike appendages on the prothorax.

THORACIC NOTA. The terga of thoracic segments are called *nota* to distinguish them from abdominal terga. Accordingly the nota of the respective segments are designated the *pronotum, mesonotum,* and *metanotum.*

The pronotum is usually a simple plate, but in some insects it is greatly enlarged to cover the head or other parts of the body (Figs. 22.1, 34.9*a*). In the grasshopper the pronotum extends back over the pterothorax and has several transverse sutures (Fig. 2.9*a*). The sutures are related to the musculature (which is peculiar to the prothorax) and do not correspond with sutures on the pterothoracic nota.

The pterothoracic nota are each divided into a wing-bearing sclerite, the *alinotum,* and a phragma-bearing sclerite, the *postnotum* (Fig. 2.10*a*). The *phragmata* are internal transverse plates that provide attachment for the large longitudinal flight muscles. Each alinotum is usually traversed by one or more sutures. These trace the position of internal supporting ridges or lines of flexibility. The extent and position of the sutures vary greatly according to the nature of the flight mechanism. Commonly the alinotum is divided anteriorly by a transverse *prescutal suture* that delimits an anterior *prescutum.* Posteriorly a V-shaped *scutoscutellar suture* separates an anterior *scutum* and a smaller, posterior *scutellum.* The lateral edges of the alinotum are modified for the articulation of the wings. At each wing base are

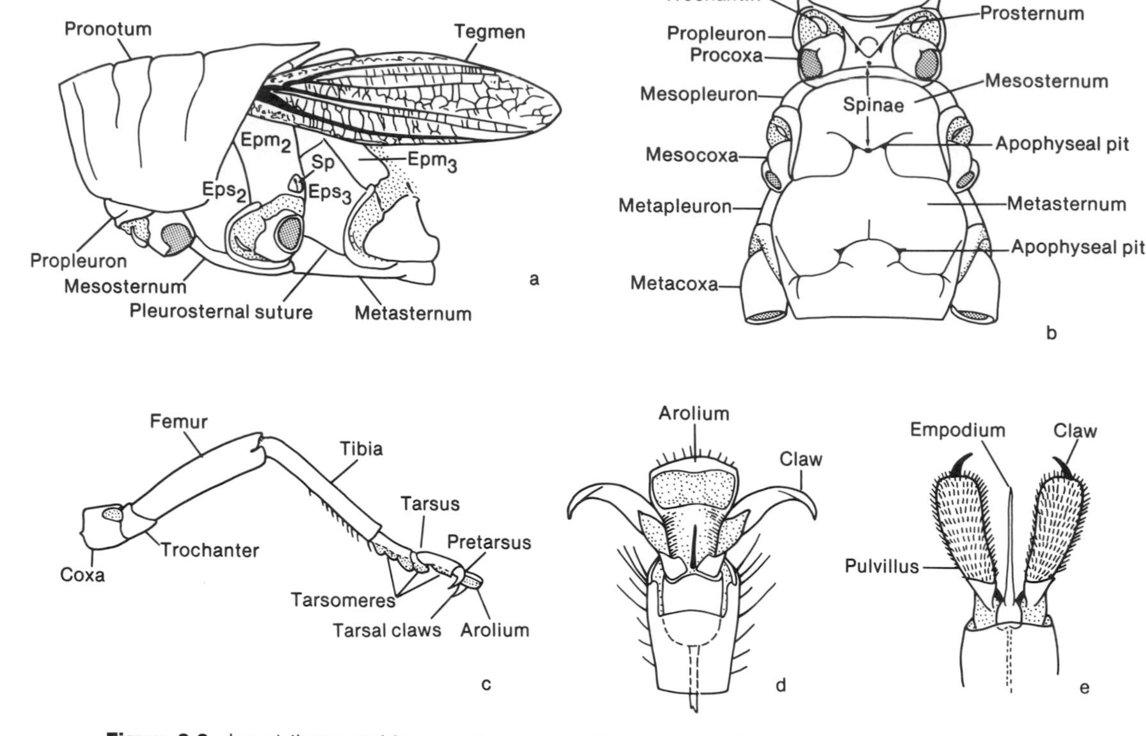

Figure 2.9 Insect thorax and legs: *a*, lateral view of thorax of *Romalea microptera* (Acrididae); *b*, ventral view of same; *c*, foreleg of same; *d*, pretarsus of a roach, *Periplaneta* spp. (Blattidae); *e*, pretarsus of an asilid fly, ventral view. *Epm*, epimeron; *Eps*, episternum; *Sp*, spiracle. (d, e. *redrawn from Snodgrass*, 1935, *by permission of McGraw-Hill Book Company.*)

anterior and *posterior notal wing processes* (Figs. 2.10*a*, 2.11*a*).

To understand the origin of the postnota and phragmata it is necessary to recall the terminology of secondary segmentation (Fig. 2.4*c*). The postnotum corresponds to the acrotergite, and the phragma corresponds to the antecosta. The acrotergite and antecosta anterior to the pronotum were apparently lost during the evolution of the cervix, leaving the pronotum as a simple plate. The narrow acrotergite and the *first phragma* anterior to the mesonotum are continuous with the alinotum to the rear, just as expected in a secondary segment. In the metathorax, however, the *second postnotum* and its *second phragma* are detached from the metathoracic alinotum and associated anteriorly with the mesothorax. Likewise the *third postnotum* and *third phragma* are detached from the first abdominal tergum and associated anteriorly with the metathorax. As a result of this reassociation, the first abdominal tergum is left as a simple plate.

THORACIC STERNA. The ventral plate of a thoracic segment is the *eusternum*. In the intersegmental regions between the eusterna of the prothorax and mesothorax, and the mesothorax and the metathorax, may be small, separate sclerites. These are named *spinasterna* because they carry a median, internal apodeme called the *spina*. One or both of the spinasterna may fuse anteriorly with the eusterna of the preceding segments (Fig. 2.9*b*). The eusterna plus spinasterna is designated segmentally as the *prosternum, mesosternum,* and *metasternum*. Each eusternum may have a pair of large invaginations, *sternal apophyses*. The apophyses may be well separated and their positions externally marked by a pair of furcal or

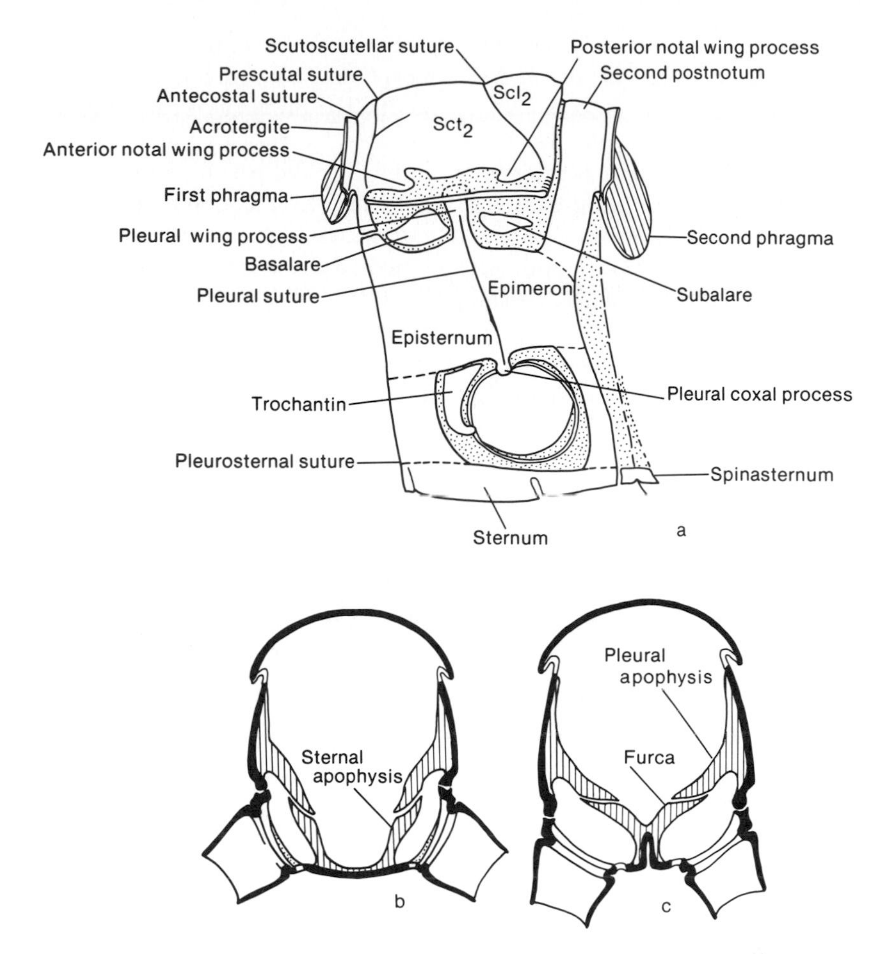

Figure 2.10 Diagrams of the pterothorax: *a*, lateral view of a mesothoracic segment; *b*, cross section of a pterothoracic segment, showing separate sternal apophyses; *c*, same, showing apophyses fused medially to form furca. (*Redrawn from Snodgrass,* 1935, *by permission of McGraw-Hill Book Company.*)

apophyseal pits (Fig. 2.9*b*, 2.10*b*) or fused medially to form a *sternal furca* (Fig. 2.10*c*). The longitudinal, midventral line that traces the invagination of the furca is the *discrimen.* The lateral boundaries of the eusternum are usually fused partly or entirely with the pleuron. The exact limits are often indeterminate, but sometimes a *pleurosternal suture* can be identified.

THORACIC PLEURA. The lateral wall of a thoracic segment between the notum and sternum is the *pleura* (Fig. 2.10*a*). Sclerites of the pleura are *pleurites.* When the pleurites are unified in a solid plate to provide support and points of articulation for the legs and wings, the plate is called the *pleuron.* The segmental designations are *propleuron, mesopleuron,* and *metapleuron.*

The pleuron of a pterothoracic segment has a dorsal *pleural wing process,* on which the wing articulates, and a ventral *pleural coxal process,* which provides an articulation for the coxa, or leg base. Extending between the two pleural processes is a strong internal *pleural ridge.* This provides mechanical support for the pleuron. The ridge may be invaginated at one point to form an internal arm, or *pleural apophysis.* The arm is connected by muscles or elastic cuticle to the sternal apophysis, and together the apophyses form a thoracic endoskeleton. Externally the position of the ridge is traced by the *pleural suture.*

The pleural suture divides the pleuron into two areas, an anterior *episternum* and posterior *epimeron.* Small sclerites beneath the wing base are the *epipleurites.* The epipleurite above the episternum is the *basalare,* and that above the epimeron is the *subalare.* The small crescent-shaped sclerite anterior to the coxa is the *trochantin.*

The propleuron is smaller and simpler in structure than a pterothoracic pleuron. The wing process and epipleurites, of course, are lacking, but the pleural suture may be present.

MODIFICATIONS OF THE THORAX. The pterothoracic segments are similar in structure and nearly equal in size in those insects that provide muscular power about equally to both pairs of wings. Odonata and Orthoptera, for example, fly with the wings unattached to each other. The pterothoracic segments of insects that fly with the wings coupled together tend to exhibit an enlargement of the mesothorax and a reduction in the size of the metathorax. This is connected with the dominant role assumed by the forewings in flight. In the Diptera, where the hind wings no longer function in flight, the metathorax is virtually absent (Fig. 42.2). An opposite trend is seen among insects that fly mainly with hind wings. The metathorax is thus enlarged in Dermaptera, Phasmatodea, Strepsiptera, and Coleoptera (Fig. 39.1*b*).

LEGS. The true paired appendages of the thoracic segments are the legs. Each pair has the same general structure, but they can be modified for different functions. Beginning at the base, the segments of the leg are named as follows: *coxa, trochanter, femur, tibia, tarsus,* and *pretarsus* (Fig. 2.9*c*). Each segment is individually movable by muscles. The tarsus is divided into no more than five subsegments, or *tarsomeres.* The tarsomeres are commonly called "segments," but with the exception of the muscles attaching to the basal tarsomere, they do not have individual muscles. Hence the entire tarsus is considered the morphological equivalent of the other leg segments. The pretarsus is the smallest segment. It is movable by muscles that originate in the tibia and femur and pass to the pretarsus by a slender apodeme. In the grasshopper the pretarsus is represented by a pair of *ungues,* or *tarsal claws,* and a median *arolium* (as in Fig. 2.9*d*).

MODIFICATIONS OF THE LEGS. The various named modifications of legs for locomotion include the following: *ambulatory,* or walking, legs (Fig. 39.1*b*); *cursorial,* or running, legs (Fig. 22.1); *fossorial,* or digging, legs (Fig. 12.1); *saltatorial,* or jumping, legs (hind legs, Fig. 29.3); and *swimming* legs (Figs. 34.8*c*, 39.4*d*). Legs modified for grasping are *prehensile,* and for seizing prey, *raptorial* (forelegs, Fig. 23.1*a*). The legs of bees are used to collect and transport pollen from flowers. Worker honeybees and bumblebees have a pollen basket, or *corbiculum,* of long setae on the hind legs (Fig. 46.17*c*). The females of many other species of bees have a brush, or *scopa,* of plumose setae on the hind legs (Fig. 46.17*d*). Larvae of advanced Diptera (Fig. 42.5*c*,*d*) and Hymenoptera (Fig. 46.3*b*) lack thoracic legs and are called *apodous.*

The coxa normally articulates dorsally with the pleural coxal process. In many endopterygotes the coxa also has a ventral articulation with the sternum. In Neuroptera, Mecoptera, Trichoptera, Lepidoptera, and Diptera, the coxa has a conspicuous posterior subdivision called the *meron.* In an extreme development the meron is actually incorporated into the pleural wall of Diptera (Fig. 42.2*a*).

The trochanter is usually a single segment, but in Odonata it is divided into two subsegments. In parasitic Hymenoptera, the base of the femur is divided into a short subsegment that resembles a second trochanter (Fig. 46.9*b*). Wasps with the arrangement are commonly said to have two trochanters. Trochanters are missing entirely on one or more pairs of legs in some insects. In the saltatorial hind leg of grasshoppers, for example, the trochanter is fused with the femur. The tibia of many insects is armed with large movable *tibial spurs* near the apex (Fig. 29.3*c*).

The number of tarsomeres varies from none to five. The basal tarsomere, often longer than the others, is named the *basitarsus.* The undersurface of the tarsus may be modified to grip smooth surfaces. The tarsi of Orthoptera and some Coleoptera have pads, or *tarsal pulvilli.* Male beetles of the Family Gyrinidae even have suckers on the foretarsi for clinging to the smooth bodies of their mates. The entire tarsus and pretarsus of larval Lepidoptera and Coleoptera (Polyphaga) is represented by a single claw.

The pretarsus of Diptera has a pair of lateral lobes, or pulvilli, under the claws. The arolium is usually absent, but a spinelike median structure, the *empodium,* is present (Fig. 2.9*e*). The tarsal claws may be reduced to a single claw, for example in Anoplura or nymphal Ephemeroptera. In the mammal-infesting lice, the single-clawed legs are modified for gripping the hairs of their host (Fig. 33.3*c,d*). Both claws are reduced in Thysanoptera, but the arolium is much enlarged into a sucker.

WINGS. The evolutionary origin of wings is not definitely known. Theories concerning their origin are discussed in Chap. 17. During ontogenetic development, the wings grow as saclike expansions of the lateral body walls. The upper and lower layers of the sac partly fuse, leaving a system of narrow, blood-filled channels that are

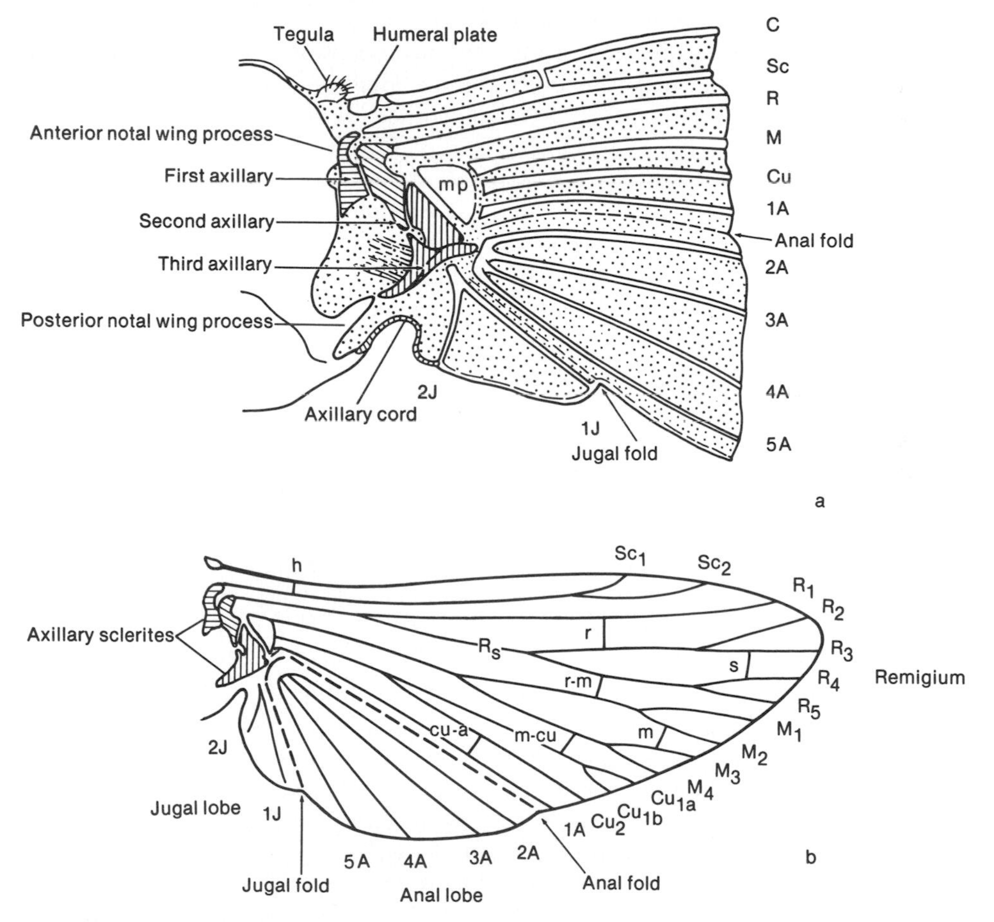

Figure 2.11 Diagrams of the base and regions of the wing and generalized venation: *a,* wing base, showing articulation and bases of major veins; *b,* diagram of venation. See text for explanation of abbreviations. (a *modified from Snodgrass,* 1935, *by permission of McGraw-Hill Book Company.*)

continuous with the body cavity. Tracheae and nerves grow into the channels. At maturity, most of channels are transformed into *wing veins* with rigid sclerotized walls that support the fused or membranous portions.

At the base of the wing are small articulatory sclerites (Fig. 2.11*a*). These include the three *axillary sclerites (Ax),* the *humeral plate (HP),* and the *median plate (mp).* The main, or *longitudinal,* wing veins are named as follows, beginning at the anterior or leading edge of the wing: *costa (C), subcosta (Sc), radius (R), media (M), cubitus (Cu), anal (A),* and *jugal (J).*

The scheme shown in Fig. 2.11*b* was developed by J. H. Comstock and J. G. Needham, and is called the Comstock system (Comstock, 1918). Although there is not universal agreement among entomologists on the homologies of the veins from order to order, nor even agreement on the basic number of veins (see Hamilton, 1972; Kukalova-Peck, 1977), the Comstock system remains the most widely used scheme. We have adopted it for general use in this text.

The longitudinal veins commonly fork, and the branches are indicated by subscripts. The radius often has a main anterior branch, R_1 and a main posterior branch, the *radial sector* or R_S, that is further subdivided into R_2, R_3, R_4, and R_5. When two or more veins are thought to have fused together to form a single vein, the names of the veins are joined by a plus sign (see, for example, $Sc + R_1 + R_S$ in Fig. 44.12*b*). Some longitudinal veins are joined by *cross-veins* that are usually designated according to the longitudinal veins involved (Fig. 2.11*b*): *humeral* (h), *radial* (r), *sectorial* (s), *radiomedial* (r-m), *medial* (m), *mediocubital* (m-cu), and *cubitoanal* (cu-a).

The membranous areas enclosed by veins are called *cells* and are formally named according to the longitudinal vein just anterior to the cell. For example the veins M_2 and M_3 define a cell, cell M_2. This may be divided by a median cross-vein to form a proximal cell, first M_2, and a distal cell, second M_2. For taxonomic purposes the cells may be informally referred to as *basal, marginal,* or *submarginal,* etc. (Fig. 46.6*d*), or be given special names such as *discal* (Fig. 44.10*c*), depending on the taxonomic group. A *closed* cell is completely surrounded by veins, whereas an *open* cell has one side bordering the wing margin.

The area of the wing anterior to the anal vein and its branches is called the *remigium* (Fig. 2.11*b*). This is the area, stiffened by veins, that delivers the power of the wing stroke against the air. The anterior edge close to the base of the wing is the *humeral* region. In the area of the anal veins, the wing is more flexible and may be expanded into a fanlike vannal or *anal lobe.* An *anal fold* may separate the anal lobe from the remigium. A smaller lobe at the posterior base of the wing is sometimes present; it is called the *jugal lobe* and is separated from the anal lobe by the jugal fold.

MODIFICATIONS OF THE WINGS. Wings that are transparent are called *membranous.* Wings may be modified to serve protective functions. Thickened, leathery wings such as the forewings of grasshoppers are called *tegmina* (Fig. 29.1). The hard forewings of beetles are the *elytra* (Fig. 39.1*a*). Hemiptera (Suborder Heteroptera) have only the bases of the forewings thickened. These are called *hemelytra* (Figs. 34.1*a,* 34.4*a*).

Wings may be reduced partly or entirely in size or discarded after use. Wings that are functional but are shed after flight are said to be *deciduous* (compare alates and dealates, Fig. 7.1). Insects with short wings are *brachypterous* (Figs. 1.2*a,* 22.2); those that never develop wings are *apterous* (Figs. 25.1, 33.3). The small structures representing the reduced hind wings of flies (Fig. 42.2*a,c*) and male coccids are *halteres* (Fig. 34.12*b*). The same term is sometimes used for the reduced forewings of Strepsiptera (Fig. 40.1*a*).

An evolutionary tendency exists for the two pairs of wings to function as a single unit. The simplest coupling is made by the overlap of the adjacent margins. For example, the jugal lobe of the forewing overlies the humeral region of the hind wing in Megaloptera (Corydalidae) and many Neuroptera. In *jugate* coupling, the jugal lobe is fingerlike in certain moths and clasps the hind wing (Fig. 44.3*b*). In *frenate* coupling of other moths, the humeral region of the hind wing has one to several large bristles, called the *frenula* (Fig. 44.3*a*). These are engaged by a hooklike flap or a cluster of special bristles called the *retinaculum,* located near the anterior base of the forewing. In the *amplexiform* coupling of butterflies, a linking structure is absent and the broad overlap of wing margins provides the coupling (Fig. 44.10*a*,*b*). The hind wings of Hymenoptera (Fig. 46.2*a*) and some Trichoptera have a row of hook-shaped setae, or *hamuli,* along the leading edge of the hind wing. The hamuli engage the folded posterior edge of the forewing. This is called *hamate* coupling.

The number of veins and crossveins in wings tends to become reduced in the more advanced orders. The homologies of the veins with the basic pattern are sometimes controversial. A standardized scheme has usually been established within an order, but differences in terminology may exist between orders.

Minute flying insects literally swim through the air. Their wings are usually reduced in size and fringed with long setae. Examples of such wings are found on thrips (Fig. 35.1*a*), certain parasitic wasps of the Superfamily Chalcidoidea, and beetles of the Family Ptiliidae.

THE ABDOMEN

The last body division, or abdomen, contains the viscera (Fig. 2.12*b*). Included are most of the alimentary canal and dorsal circulatory vessel,

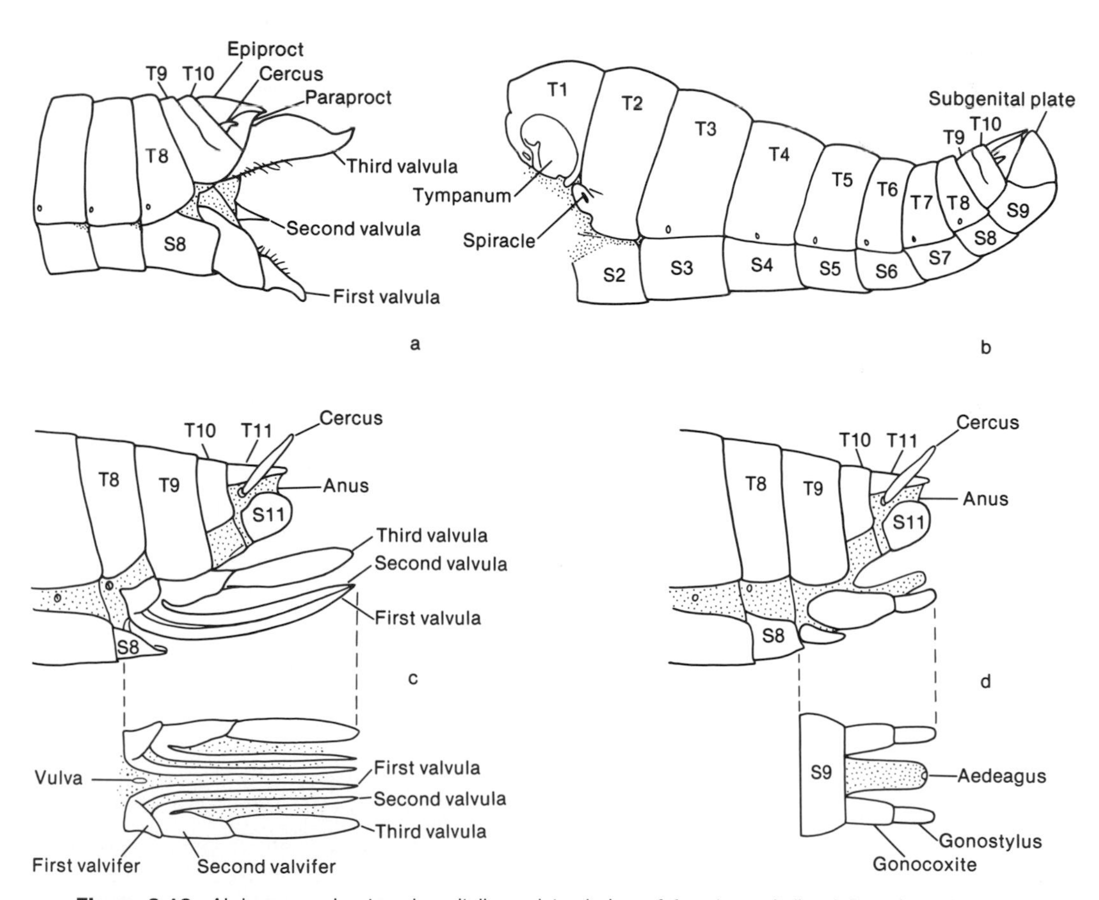

Figure 2.12 Abdomen and external genitalia: *a,* lateral view of female genitalia of *Romalea microptera* (Acrididae); *b,* same of male abdomen and genitalia; *c,* diagram showing general structure of female genitalia; *d,* same of male genitalia. (c *redrawn from Snodgrass,* 1935, *by permission of McGraw-Hill Book Company.*)

as well as the Malpighian tubules, fat body, and reproductive organs. No more than 11 segments can be counted in the abdomen of adult Insecta, and frequently only 10 or 9 are clearly evident on close examination. The first 8 segments may each have a pair of lateral spiracles and internally a ganglion of the ventral nerve cord. Ganglia of the ninth and remaining segments are indistinguishable; they have fused with the ganglion of the eighth segment (Fig. 6.3).

Paired appendages, homologous to legs, are absent on the first seven, or *pregenital,* segments of adult pterygotes. On the eighth and ninth segments of the female and the ninth of the male, the appendages are modified as external organs of reproduction, or *genitalia.* These segments are known as the *genital* segments. The tenth and remaining segments are the *postgenital* segments. The eleventh may have a pair of lateral appendages called *cerci.*

Except where appendages are present, each abdominal segment typically has the *tergum* and *sternum* as single plates. The plates overlap posteriorly and often exhibit the features of secondary segments. Externally near the anterior margin is the antecostal suture that delimits the anterior acrotergite, and traces the internal ridge or antecosta. Muscles attach to the antecosta and to lateral apodemal lobes of the sterna. On the first abdominal segment of the grasshopper is a pair of lateral acoustical organs, or *tympana* (Fig. 29.2*b*).

OVIPOSITOR. The paired appendages of the eighth and ninth segments of females may be modified into an appendicular *ovipositor* for depositing eggs (Fig. 2.12*c*). On segment 8 the sternum may be present, but it is absent in segment 9. The ovipositor consists of basal *valvifers,* on which muscles insert, and elongate *valvulae.* The eighth segment bears the *first valvifers* and *first valvulae,* and the ninth segment bears the *second valvifers* and the *second* and *third valvulae.* The valvifers correspond to the paired coxae on each segment. The first and second valvulae correspond to median lobes of the respective coxae. The third valvulae correspond to apical lobes of the coxae on the ninth segment. The first and second valvulae form a shaft that can penetrate soil or plant or insect tissues. Eggs issue from the genital opening on the eighth segment, pass down a channel formed by the valvulae, and are deposited in the ovipositional medium. The third valvulae usually form a sheath that does not penetrate the medium. In grasshoppers and their allies the ovipositor is differently constructed (Fig. 2.12*a*). Here the organ is formed by the first and third valvulae. The first valvulae are ventral, the third valvulae are dorsal, and the second valvulae are small, median in position, and concealed by the other valvulae.

MALE GENITALIA. The appendages of the ninth segment of males are modified to form a copulatory organ (Fig. 2.12*d*). The basic structure of many insects consists of a median intromittent penis, or *aedeagus,* and lateral clasping *parameres.* The ninth sternum may be present as the *subgenital plate.* The parameres are commonly two-segmented and are thought by most morphologists to be modified legs. The proximal segment is the *gonocoxite,* and the distal segment, if any, is the *gonostylus.* These sclerotized structures engage the genital opening of the female and maintain the connection during intromission.

The aedeagus is formed by a median fusion of median coxal lobes and is often partly sclerotized. The ejaculatory duct opens apically. During copulation the membranous portion of the aedeagus everts into the vagina of the female and discharges the sperm and seminal secretions from the ejaculatory duct. The male genitalia of grasshoppers, seen externally, consist of a ventral, cup-shaped subgenital plate and a dorsal, membranous *pallium* (Fig. 2.12*b*).

POSTGENITAL SEGMENTS. The segmentation beyond the genitalia may be difficult to determine. The eleventh segment, when present, surrounds the anus and may bear paired appendages or cerci. Sclerotized plates, if any, on the eleventh segment are divided into a dorsal *epiproct* and lateral *paraprocts* below the cerci. In the grasshopper both sexes have the ninth and tenth terga fused and the small cerci situated between the epiproct and paraproct (Fig. 2.12*a*,*b*).

MODIFICATIONS OF THE ABDOMEN. The number of visible segments is often reduced in the endopterygotes by the invagination of the genital and postgenital segments into the apex of the abdomen. In some cases the first segment is reduced or fused with the second, or in advanced Hymenoptera, the first segment is incorporated into the posterior wall of the thorax and known as the *propodeum* (Fig. 46.1). Also in the Hymenoptera, the second and sometimes the third segments are constricted into a *petiole,* leaving the remaining segments as the *gaster* (Fig. 46.9*c*).

Paired appendages are present on the pregenital segments of Entognatha and the immature stages of certain pterygotes. The abdominal appendages of most Entognatha are clearly homologous with legs. Those of Archaeognatha, for example, have paired, flattened coxae, each within a stylus that represents the distal part of the limb (Fig. 19.1*a*). The abdominal appendages of pterygotes are much more highly modified and less readily identified as limbs. Examples include the abdominal tracheal gills of nymphal Ephemeroptera (Fig. 20.4*b* through *d*) and larval Megaloptera (Fig. 36.2) and the fleshy abdominal *prolegs* of larval Lepidoptera (Fig. 44.5*a*,*b*) and Hymenoptera (Fig. 46.3*a*).

The appendicular ovipositor is well developed in certain Odonata, Orthoptera, Hemiptera (Suborder Homoptera), Thysanoptera, and Hymenoptera. The appendicular ovipositor is absent in many other insects, e.g., Coleoptera, Diptera, or Lepidoptera, but is replaced in these orders by special modifications of the terminal abdominal segments.

The size and shape of the ovipositor are closely correlated with the nature of the medium into which the eggs are inserted. Insects that deposit their eggs in living plant tissues usually have stout valvulae with sawlike ridges. Parasitic wasps may use long, slender valvulae to insert eggs into insect hosts that are in burrows deep in solid wood (Fig. 46.9*a*).

The ovipositors of predatory wasps, ants, and bees are modified into an organ for defense, the *sting* (Fig. 46.2*d*). The second valvulae are fused to form a stiff piercing organ, with the first valvulae locked beneath in sliding grooves. Glands associated with the ovipositor release toxic secretions into the wound created by the valvulae. In the stinging Hymenoptera, the ovipositor no longer functions to deposit eggs; they are merely released from the vaginal opening on the eighth segment.

The male genitalia of insects are so greatly varied that they are used by insect taxonomists to classify species. Special terminologies have been applied in some orders where the structures are complex and the homologies uncertain. The genitalia are especially complicated in Diptera, where in some groups the terminal abdominal segments and genitalia actually undergo a clockwise rotation of 90°, 180°, or even 360° during pupal development. Male Odonata have the sterna of segments 2 and 3 modified as a secondary genitalia for the transmission of semen to the female (Fig. 21.2*b*,*c*). Male Ephemeroptera have paired intromittent organs on the seventh segment. These correspond to the paired openings of the lateral oviducts on the seventh abdominal segment of the female.

The cerci are appendages of the eleventh abdominal segment. They may be long, slender, and subsegmented as in Thysanura, Archaeognatha, and Ephemeroptera (Fig. 20.1*a*); shorter and with fewer subsegments as in Blattodea (Fig. 22.3); or one-segmented as in Orthoptera (Fig. 29.3). The cerci of Dermaptera (Fig. 26.1*a*,*c*) and Diplura (Japygidae) (Fig. 18.2*e*) are stout and forcepslike. The epiproct is greatly elongated into a cercuslike, *median caudal filament* in Thysanura (Fig. 19.2), Archaeognatha (Fig. 19.1), and Ephemeroptera (Fig. 20.1*a*). In nymphal Odonata (Suborder Zygoptera) the epiproct and paraprocts are greatly enlarged into tracheal gill plates (Fig. 21.5*b*).

SELECTED REFERENCES

The text by Snodgrass (1935) continues to be the best general reference for insect morphology from an evolutionary viewpoint. Other articles of general interest are Snodgrass (1952, 1958, 1963). An entertaining account of this distinguished entomologist is provided by Thurman (1959). An excellent condensed introduction to insect anatomy is given by Richards and Davies (1957) in their revision of Imm's text. Recently, Chapman (1969) has provided a comprehensive treatment of insect structure from a functional viewpoint. Torre-Bueno (1962) supplies a general glossary, and Tuxen (1970*a*) treats terminology of the genitalia. Reviews of specific body regions are given by Matsuda for the head (1965) and thorax (1970), by Hamilton for wings (1971, 1972), and by Scudder (1971) for genitalia.

THE INTEGUMENT

3

The organ most suitable to begin with in a microscopic study of the insect body is the integument. It is physically the largest of all organ systems and exhibits the greatest diversity in structure and in function. In each of its manifestations, three components can usually be identified: an outer, noncellular *cuticle;* a single layer of cells, or *epidermis;* and an inner sheet of connective tissue, or *basement membrane* (Figs. 2.1, 3.1*a*).

The most easily seen portion of the integument is the hardened and jointed cuticle which covers the outside of the body and forms an external skeleton, or *exoskeleton.* As will be explained later, the exoskeleton is extremely resistant to decomposition. When protected from pests, dried specimens of insects may outlast the pins and cabinets of the museum and usually outlast the collectors and curators. When preserved as fossils, pieces of the exoskeleton may remain intact. The chemical component chitin, for example, has been detected in the fossil of an extinct eurypterid arthropod from the Silurian Period, over 405 million years ago. The special arrangements of the hairs and other features of the exoskeleton provide nearly all the diagnostic characters for the recognition of the species of insects. This means that easily a million or more distinctive designs of exoskeletons exist. Taxonomists are indeed fortunate that the exoskeleton is so simply preserved and that so much information concerning the insect's habits can be gained from it.

In addition to functioning in the support of the internal organs, the exoskeleton is involved in such important activities as feeding, locomotion, and reproduction. The physical properties as well as the intricate shapes of the mouthparts permit the exploitation of a wide variety of foods. Similarly, the walking and flight mechanisms depend in large degree on the mechanical features of the exoskeleton. In reproduction, parts of the skeletal genitalia provide the means by which the male may transmit sperm to the female and the female, in turn, may deposit the eggs in special places.

The exoskeleton determines the form and maximum size of the insect. At intervals during growth, portions of the cuticle are digested and a new, larger cuticle is deposited. This process is termed the *molt,* and the actual shedding of the old cuticle is called the *ecdysis.* At times of molting, much of the old exoskeleton is reclaimed and the chemical components are used again. During starvation some of the cuticle is also reclaimed. For this reason, the exoskeleton is actually a kind of food reserve.

The exoskeleton also serves many of the same functions as the skin of other animals. It is the main barrier to the loss of water, a problem which is particularly acute in small terrestrial organisms with a relatively large evaporative surface area in proportion to their volume. The entrance of disease organisms and the penetration of chemicals such as insecticides are impeded by the exoskeleton. Assaults by predators are not only resisted by the tough outer covering, but also avoided by cryptic or warning coloration. The sense organs of the nervous system are intimately connected with specialized areas of the exoskeleton because all the stimuli from the environment must be transmitted through the exoskeleton to the receptors beneath.

The integument includes not only the conspicuous exoskeleton but also extensive internal portions. The tracheal system, functioning in respiration, is created by invaginations of the integument. The ultimate subdivisions of these tubular ducts number literally in the millions and supply oxygen to virtually every cell of the body. The elimination of carbon dioxide is accomplished through the tracheae and also through the general body surface. In the alimentary canal, the foregut and hind gut are formed by invaginations of the integument during development. The integument also lines the ducts leading from the internal reproductive organs, providing not only passageways and storage vessels for the gametes, but also glandular secretions which function in their protection and transmission. Finally, a rich variety of other glands, including the salivary glands, is derived from cells of the epidermis. Their products function in nearly every phase of the insect's life history by supplying enzymes for digestion, materials for shelter and support, and other chemical compounds for communication, defense, and offense. Among these integumental secretions, human beings have found some of great value, e.g., silk, shellac, and beeswax.

ORIGIN OF THE INTEGUMENT. Of the integument's three major components, only the epidermis is cellular. In the embryo, the epidermis is derived from the ectoderm, as described in Chap. 4. The cuticle is periodically secreted by the epidermis, both daily as additional thin layers and at longer intervals when the cuticle is replaced throughout nearly all the body during molting. Certain parts of the cuticle are thought to be secreted by oenocytes (cells of ectodermal origin in the blood). The basement membrane applied to the inner surface of the epidermis is believed to be secreted by certain hemocytes (blood cells) in the hemolymph and possibly by other cells of the body.

CUTICLE. The discussion below deals with the cuticle as it exists in the exoskeleton. The special features of the cuticles found elsewhere in the body will be treated with the appropriate systems. The cuticle may equal up to half the dry weight of the insect's body, and always exists external to the epidermis. When sections are taken and viewed microscopically, it is seen to be subdivided in layers. These can be shown to have special chemical and physical properties and to be secreted by the various cells of the epidermis. Two major divisions of the cuticle are the thin outer *epicuticle,* no more than about 4 μm thick and usually lacking chitin, and the inner *procuticle* (combined *exocuticle* and *endocuticle*), which may measure a millimeter or more in thickness and contains chitin. Each division may have sublayers, depending on the species of insect and the location of the cuticle on the insect's body.

The *epicuticle* is created by secretions from several different sources which are deposited outside the procuticle. At least three layers may be recognized: the outer *cement layer,* the middle *wax layer,* and the inner *protein epicuticle.* The cement layer, or tectocuticle, is derived from certain modified epidermal cells, called *dermal glands,* which empty their secretions on the outer surface through gland ducts penetrating the procuticle. The cement is thought to protect the wax layer from abrasion and absorption by foreign objects. When present, this layer also determines the surface properties of the cuticle. The copious secretions of the lac insect (*Laccifer lacca*), which yield shellac, are probably quite similar in origin to the cement produced normally and in much smaller quantities on the body surfaces of other insects.

The wax layer is of special importance in preventing the loss of water from the insect. Even thin layers of wax or lipids are sufficient. Wax may permeate the cement layer or appear on the outside surface. In such cases a hydrophobic surface is created which has important consequences for the insect in contact with water, e.g., the ability of an insect to be supported rather than trapped by the surface tension of water.

This layer is also important in the resistance to invasion by pathogens such as fungi, and is probably the main barrier to the penetration of insecticides. The wax is produced by the epidermal cells and probably passes through the pore canals (see further on under "Epidermis") of the procuticle to the epicuticle in most insects. Filaments 60 to 200 A in diameter, called *wax canals,* are found in the pore canals and in epicuticle. These may actually be liquid crystals created by the molecular interaction of the waxes or lipids and water, and may serve as routes for the secretion. The secretions vary in chemical composition and physical properties; the greasy surface of the cockroach and the firm honeycombs of the beehive are both products of the wax-producing cells of the epidermis.

The protein epicuticle is the least understood and possibly the most important layer in the cuticle. A zone beneath the wax layer has often been called the *cuticulin* layer. It is only slightly thicker than a cell membrane but may function as a critical barrier between the new and old cuticles during molting and may determine the maximum expansion of the cuticle which is possible before another molt. Although polyphenols are detected in the protein epicuticle, no distinct layer of this composition can be demonstrated.

The procuticle is commonly divided into two layers, an outer hardened and often darker *exocuticle* and an inner, flexible *endocuticle,* which is usually lighter in color. Both are believed to be of much the same composition, viz., proteins and chitin linked in a cross-grid arrangement to form glycoproteins. Unbound proteins and possibly also unbound chitin are present as well. The procuticle has a high water content. A variety of proteins exists in the procuticle (the water-soluble proteins are collectively called *arthropodin*) and may contribute more than half the dry weight. One of these, a rubberlike protein called *resilin* (from the Latin *resilire,* to jump back), may occur in almost pure form and confers an exceptional elasticity to certain wing articulations and muscle insertions.

The second major component of the procuticle, chitin, is thought to be widely distributed among organisms, being reported in fungi, sponges, hydroids, bryozoans, brachipods, mollusks, annelids, and arthropods. Unfortunately, certain tests for chitin are not specific, and no test is now available which can be used on histological sections. Many of the organisms mentioned in earlier reports will have to be reexamined to determine the presence of chitin. Pure chitin has been described as a colorless, amorphous solid, insoluble in water, dilute acids, dilute alkalis, alcohol, and all organic solvents. It may be dissolved in concentrated mineral acids, but is rapidly degraded. In chemical composition chitin is similar to cellulose, a structural compound widespread in plants.

The glycoprotein chains create *microfibers,* which, in turn, are arranged in a three-dimensional pattern to produce *lamellae.* Some lamellae are deposited at times of molting, and additional lamellae may be added each day as a distinct *growth layer.* The characteristics of the growth layers are influenced by temperature and by the relative lengths of day and night.

The hardness of the exocuticle is largely independent of its color. Albino mutants of insects, for example, may have hard cuticles lacking dark pigments. The process of hardening, called *sclerotization,* involves the tanning or cross-linking of the proteins to form *sclerotin* in the outer zone of the procuticle. Sclerotin alone may give a dark color to the cuticle. This hardened zone becomes the exocuticle and is resistant to digestion during molting. The chitin chains, bound to the tanned proteins, are part of the network but are not responsible for the special hardness of the exocuticle. For this reason a hard piece of the exoskeleton is correctly termed "sclerotized" rather than "chitinous." In view of the difficulty in demonstrating chitin in small samples, the latter term should only cautiously be used.

Although calcium is present in the cuticles of crabs and other crustaceans, it is rarely incorporated in the cuticle of insects. When present in insects, calcification does not modify the hardness of the sclerotized parts. By geological standards, some sclerotized cuticles can scratch the mineral calcite. On a scale of 10 ranked by increasing hardness, these cuticles would be placed third; i.e., in hardness they are between a thumbnail and a copper penny. In southern California, the beetle *Scobicia declivis* (Bostrichidae) easily chews through the lead sheathing of aerial telephone cables, earning it the name "lead-cable borer." Other insects can penetrate thin sheets of tin, copper, aluminum, zinc, or silver. Sclerotized cuticle is, however, not indestructible, and its surface together with the epicuticle may show extensive abrasion and wear, especially in burrowing insects. Renewal of the cuticle at each molt may serve not only for increases in body size, but also to restore the valuable properties of the cuticle.

Articulations between the sclerites, or areas of exocuticle, are created by reduction in the thickness of the sclerotized zone. The flexible endocuticle then predominates at such joints, or *arthrodial membranes* (Fig. 2.1). The exocuticle may also be broken into wedgelike blocks to permit bending. Elastic properties may be conferred to hinges by the presence of resilin.

EPIDERMIS. The many functions of the epidermis are best discussed in detail elsewhere under the appropriate sections. Among these are the digestion of the old cuticle and deposition of the new cuticle during molting; wound repair; determination of growth and form; development of special glands and sensory devices; and the production of color and color changes.

Epidermal cells usually have a distinct nucleus, one to several nucleoli, mitochondria, and desmosomes between adjacent cells. The physical routes by which the epidermal cells pass their secretions to the procuticle and the epicuticle are the *pore canals* (Fig. 3.1). These are minute channels, 1.0 to 0.15 μm in diameter, extending from the surface of the epidermal cell to the interface between the epicuticle and the procuticle. They are not lined with the plasma membrane of the cell, but filamentous structures of unknown function pass from the cell through the pore canal. The density of canals may be quite high, ranging from 15,000 to 1,200,000 per mm^2. The canals are not straight but seem to be coiled about an axis perpendicular to the cuticular surface or curve with the microfibers in each cuticular lamella.

BASEMENT MEMBRANE. A layer of connective tissue ensheaths all the internal organs and the internal surfaces of the integument. Besides a supportive function, this membrane may be an important barrier between the cells of the various organs and the hemolymph. If so, the idea that the cells of the insect body are directly bathed by the blood in an "open circulatory system" is incorrect. When beneath the epidermis, the connective tissues are called the basement membrane and consist of an amorphous granular layer up to 1 μm in thickness. In some cases a fibrous structure is apparent in insect connective tissues. In certain chemical and physical properties, these fibers resemble collagen of the vertebrate connective tissues.

THE PROCESS OF MOLTING

The integument of insects undergoes cyclic activities throughout the life of the insect. The most conspicuous changes are associated with the periodic shedding of the old cuticle. The growing insect is restricted in size by the maximum area to which the cuticle will expand. The new cuticle that is laid down beneath the old is highly wrinkled, thus providing potentially greater area for expansion after the molt. The major growth in the size of the exoskeleton is therefore limited to times of molting and is discontinuous. Changes in the form of the exoskeleton and in patterns of setae and the regeneration of lost parts are likewise limited to the molt. In addition to these changes in morphologic characteristics, the valuable physiological properties of the cuticle that may become weakened by age and wear must be renewed, and the molt serves this purpose as well. Between molts, the epidermis con-

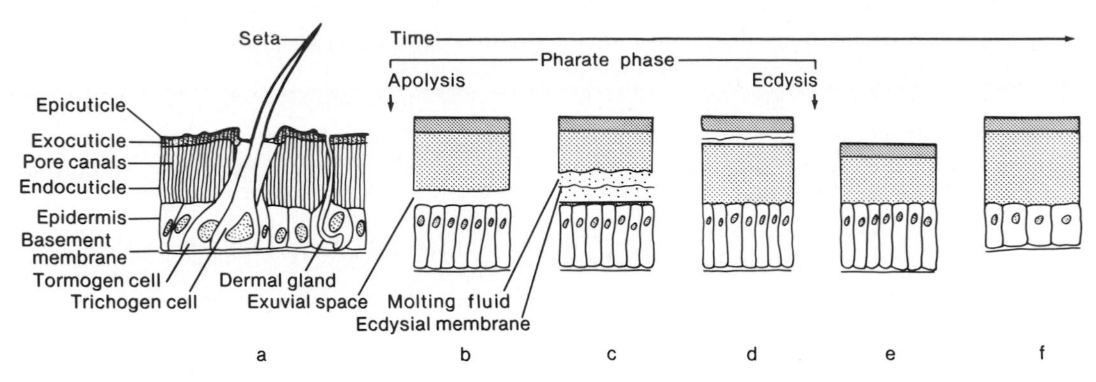

Figure 3.1 Structure of the integument and the molting cycle. See text for explanation.

tinues to be active in secreting additional layers of endocuticle and in wax production.

APOLYSIS. The events which trigger the molting cycle and the hormonal control of molting are treated elsewhere. Here we will concern ourselves with the integument itself, and our attention is directed to the interface between the epidermis and cuticle. This is an area of exceptional physiological activity and can be studied only with great difficulty. The onset of molting is signaled by *apolysis,* or the detachment of the epidermis from the old cuticle. This detachment may possibly be created by a retraction of the cells or by a change in the consistency of the thin subcuticle that presumably binds the cuticle to the epidermis. The space created is termed the *exuvial space.* Just prior to the detachment, several preliminary events occur. The epidermis changes from a thin layer of polygonal cells to a thick columnar layer. Mitotic figures, indicating cell divisions, may also be seen at this time. Lastly, an additional layer of cuticle, called the *interzone cuticle,* is deposited. After the appearance of the exuvial space, a membrane forms between the old cuticle and the epidermis. This *ecdysial membrane* is probably derived from the interzone cuticle previously secreted. Also in this space is deposited the molting fluid from the epidermal cells. The fluid may exist in this space for some time before becoming active (Fig. 3.1*c*).

The secretion of the new cuticle begins with the outer portion, or cuticulin layer, of the protein epicuticle. This first layer is believed to be critical in determining the surface features and area of the new cuticle. Then the remainder of the protein epicuticle is deposited, followed by the procuticle. Once the protein epicuticle is in place, the molting fluid becomes active. The main function of the fluid is the digestion of the old cuticle. Proteases and chitinases dissolve up to 90 percent of the old cuticle, leaving only the resistant epicuticle and exocuticle, together with the ecdysial membrane. The products of digestion are absorbed and later incorporated in the new procuticle by the epidermis. How these products pass through the new epicuticle is not known. As the old cuticle is digested, the new procuticle increases in thickness. The lamellae, characteristic of the procuticle, probably are the result of cyclic secretory activities of the epidermal cells. The pore canals are initiated with the first layer and become progressively longer with increasing thickness of the procuticle. By means of these canals the cells remain in communication with the epicuticle-procuticle interface.

Sclerotizing of the outer zone of the procuticle to form the exocuticle may begin now in certain areas of the insect's body or be delayed in other areas until after ecdysis. The tanning and hardening of the cuticle take place from the outside inward. To accomplish this, polyphenols diffuse outward from the epidermis to the surface through the pore canals where they are oxidized to quinones. These are the sclerotizing agents, and they diffuse inward to create the exocuticle.

Just prior to the shedding of the old cuticle (now called the *exuviae*), a layer of wax is deposited on the surface of the new epicuticle. The molecular mechanisms by which a hydrophobic substance such as wax may pass through the hydrophilic procuticle are not known. The pore canals may provide the exit passageway, but wax is secreted on some cuticles lacking these canals. Finally, a cement layer is deposited over the wax layer by the dermal glands.

ECDYSIS. The insect is now ready to shed the exuviae by the process of *ecdysis.* By taking in air or water in the gut, the insect increases its body volume and this, together with muscular contractions, splits the old cuticle along preformed lines of weakness. Certain areas of the cuticle, even within the boundaries of sclerites, have the exocuticle greatly reduced. Once the endocuticle is dissolved by the molting fluid, only the epicuticle remains along these lines. Such *ecdysial lines* (Fig. 2.1) provide the initial lines for breakage of the old cuticle. A lubricant may be demonstrated on the inner surface of the exuviae and probably aids in the escape of the emerging insect. It is mucilaginous in nature and may be derived from the molting fluid.

Once freed of the exuviae, the cuticle is expanded to the limits predetermined by the epicuticle. Blood pressure is increased by swallowing air or water and by contractions of various parts of the body. Certain regions of the cuticle are plastic and expand under this pressure. By this means a dramatic change in body size and shape takes place over a relatively brief time. This is especially marked at the last molt, when the insect's wings are inflated to full size. After expansion the cuticle rapidly hardens and darkens. These processes are apparently under the control of hormones different from those which initiate the molting cycle. One hormone, named *bursicon,* has been identified in the fly *Sarcophaga* (Calliphoridae) and is secreted in the blood by the median neurosecretory cells of the brain and from the combined thoracic and abdominal ganglia. In the presence of this hormone, the hardening and darkening of the new cuticle begin.

After the molting cycle is over, additional layers are added to the endocuticle by the epidermis, and wax may continue to be produced.

THE PRODUCTION OF COLOR

Beautiful colors, or even the general darkening of the cuticle, are created by processes that may be quite apart from the events leading to hardening. In some species of wild bees, for example, certain areas of the cuticle may be perfectly transparent and reveal the colors of the tissues beneath. Yet they are as rigid as the adjacent black cuticle. Except for special mutant forms or some subterranean species, all insects exhibit colors. The functions of these colors in survival are treated in Chap. 15 under protective coloration. Here color will be discussed from three points of view: (1) the anatomical location of the color-producing structure; (2) the manner in which the color is generated; and (3) changes in color during the life of the insect. With regard to the first topic, most colors are produced by the cuticle. When the cuticle is translucent or transparent, the underlying epidermal cells, hemolymph, or other internal tissues may produce color. Both the cuticle and the deeper tissues may contribute to the production of a single pattern of color.

White light is composed of various wavelengths of light, commonly identified as red, orange, yellow, green, blue, indigo, and violet. When an object or insect absorbs all but the red portion of the spectrum, we perceive a red color, etc. The selective reflection of certain colors may be due to the physical structure of the object, producing *structural* colors; or it may be the result of selective absorption by certain chemical compounds, producing *pigmentary* colors; or a color may be produced by a combination of both structure and chemical compounds together. When the cuticle contains the pigment or physical structure for color, the colors often remain lifelike in museum specimens. On the other hand, when the color resides in the cells or hemolymph beneath the cuticle, one may expect fading and discoloration as the internal organs decompose after death.

STRUCTURAL COLORS. To produce a structural color, white light is altered in one of three ways: by the interference, diffraction, or the scattering of light waves. Interference colors are familiar as the iridescence of a thin film of oil on water or a soap bubble. Both the upper and lower surfaces of the film reflect light. The thickness of the film will determine the distance these reflecting surfaces are separated. Since light reflected from the lower surface must travel a minute distance farther to reach the eye of the observer than the light from the upper surface, the former is said to be retarded. When waves are retarded by an odd number of half-wavelengths they will be out of phase with the waves of the same length reflected from the upper surface. In this case the two reflections cancel each other and only that part of the spectrum with the waves in phase will be seen. The colors vary because of the varying thickness of the film. Changes in the angle of observation will also change the thickness traversed by the retarded waves and likewise produce iridescence.

Interference colors are widespread among the insects. Least conspicuous, but extremely common, is the play of iridescent colors on the transparent wings of insects. The membranes of the wings have the same effect as thin films. More spectacular are the butterflies and moths whose scales are intricately modified to produce striking colors.[1] The blue of the *Morpho* (Nymphalidae) arises from a series of vertical vanes on each scale. Each vane has a series of delicate ribs spaced at the correct distance for canceling or transmitting without reflection all wavelengths but the blue which is reflected. A series of superimposed laminae on each scale of the diurnal moth *Urania* (Uraniidae, Geometroidea) has a similar effect. Lamellate scales or lamellae within the cuticle itself may produce interference phenomena in certain metallic beetles. When underlain by dark pigments, the interference colors are especially sharp, since stray reflections from internal organs do not detract from the purity of the reflected colors.

Only one insect, the beetle *Serica sericea* (Scarabaeidae), produces color by diffraction. A parallel series of minute grooves, or striae, are evenly spaced at about 0.8 μm apart. These produce an iridescence on the elytra when viewed in direct sunlight. This is actually a diffraction grating which operates by reflection. An artificial impression of the close-spaced striae, made of collodion and peeled from the elytron, will exhibit the same iridescence.

The last type of structural color, that produced by scattering, is not iridescent and does not change with the angle of observation. Minute particles, less than 0.6 nm in diameter, reflect more of the short waves of light than long waves, and produce a blue color. The sky is blue, for example, because dust particles and atmospheric gases reflect more blue and violet than other colors. Without the impurities in the atmosphere to reflect these colors the sky would appear black. Such a color is called *Tyndall blue* after an early student of sky color. Among the dragonflies, granules in the epidermal cells produce a blue color. Granules in cells in combination with yellow pigments will give green. Waxy secretions on the surface of the cuticle in other species of dragonflies also give the Tyndall effect. Larger granules reflect all wavelengths of light, producing a structural white. Depending on the arrangement of the reflecting surfaces, a dull, pearly, or silvery white can be produced.

PIGMENTARY COLORS. The second major category, pigmentary colors, includes a much wider range of colors and a diverse assortment of chemical compounds which serve as pigments. These colors are not metallic or iridescent, but otherwise vary enormously. They are often excretory products which are no longer involved in biochemical pathways but which now serve an important function in the ecology of the insect. Pigments may also be absorbed from the food of the insect.

[1]See Endpaper 1*a*.

One of the most common of the pigments among animals is melanin. The colors produced by this pigment vary from black and brown to light brown and yellow. The synthesis of each color pigment is incompletely known. Nearly all colors in this range were formerly attributed to melanin. Some are now known to be the result of other chemical compounds, such as sclerotin. Melanogenesis of the black pigment involves the oxidation of tyrosine by the enzymatic action of tyrosinase, plus several additional steps, one of which requires air by way of the tracheal system. Once formed, melanin is insoluble in ordinary solvents, but it can be bleached if a transparent exoskeleton is desired for study.

Tyrosine and tyrosinase are present in insect blood but somehow kept apart. When an insect is injured, the blood that is exposed to air often blackens. Such melanization in experimental insects is usually fatal and may be prevented by excluding oxygen from the surgery or by blocking the action of tyrosinase. The hereditary lack of tyrosinase produces the familiar albino mutant.

Melanogenesis may be partly influenced by the environment. For example, insects reared in the laboratory at lower temperatures often show increased melanization. In nature, an insect with two or more generations per year often displays distinctive colors associated with the seasons during which pigments were developed. The interaction of the environment and melanogenesis is not easily unraveled. The presence of darker populations of an insect in a geographic region of lower temperatures cannot be simply explained as entirely an environmental effect. Such individuals may be genetically predisposed to greater melanin deposition, and the darker colors may have survival advantages peculiar to the geographic region, irrespective of the lower temperatures. High humidity and darker colors have also been associated in the study of geographic variation of insects, but oddly there is little evidence for a direct physiological influence. Some environmental factors have indirect effects through the nervous or endocrine systems: certain wavelengths of light received by the eyes of an insect in the pupal stage may promote melanin deposition or, as in the migratory-phase locusts, the darker gregarious phase is associated with crowding.

Yellow, orange, and red colors are produced by carotenoid pigments derived from alpha and, more commonly, beta carotene which occurs in plants. Insects are believed to be unable to synthesize these pigments and therefore must depend on plants for their supply. Some predators acquire their pigment via plant-feeding victims. The pigments are absorbed and deposited in the epidermis or fat cells or in secretions such as silk or wax.

Green colors which closely resemble the chlorophyll of leaves are produced by many kinds of insects.[1] Thus far no experiments have shown that the green plant pigment is absorbed and used directly for the color. In several different insects, however, the green is produced by a combination of a yellow carotenoid and a blue or blue-green pigment belonging to a group known as a bilin or bile pigment. The origin of the latter is not clear, but it is probably synthesized by the insect. Curiously, a diet of fresh green leaves seems to be important for the synthesis, but it is thought that some component other than chlorophyll is the needed constituent.

[1]See Endpaper *2a*.

SELECTED REFERENCES

Introductions to the integument are found in Chapman (1969), Richards (1951), and Wigglesworth (1972). The volume edited by Hepburn (1976) contains many articles on the integument, as well as a tribute to A. Glenn Richards, who has made major contributions. Reviews of specific aspects of the integument include structure and formation by Locke (1974), chemistry by Hackman (1974), permeability by Ebeling (1974), epidermal glands by Noirot and Quennedy (1974), and silk structure and secretion by Rudall and Kenchington (1971). Fox and Vevers (1960) provide a general introduction to animal colors. Cromartie (1959) and Fuzeau-Braesch (1972) review insect pigments, and Rowell (1971) discusses color changes.

CONTINUITY OF THE GENERATIONS: DEVELOPMENT AND REPRODUCTION

4

In this chapter the major events in the life history of insects are traced. After an introduction giving the sequence of key events, we begin with the egg as it leaves the female and follow development through the immature stages to sexual maturity. Then we return to the reproduction system, discuss the structure and function of the male and female organs, and complete the life history with a description of different types of reproduction.

The details of insect development and reproduction, in terms of both structure and physiology, have been investigated in depth in only a small fraction of species. Perhaps at one time a general scheme could be considered applicable to most insects, but recent studies with advanced techniques and a wider selection of species, reveal great diversity. Here we will provide a framework to which the interested reader can add further information from other sources.

The majority of insects are diploid animals that reproduce sexually and are *oviparous,* i.e., they lay eggs. Haploid eggs and sperm are produced by meiosis. The eggs of the Class Insecta and the Order Diplura are rich in yolk, i.e., *centrolecithal.*

The early cleavage, or division, of the nuclei in the fertilized egg does not involve division of the large yolk mass. This is known as *meroblastic* or *superficial* cleavage. The eggs of Collembola, like those of the myriapods, have less yolk (*microlecithal*), and the early cleavage is *holoblastic,* i.e., the egg is completely divided. Eggs of Protura have never been found.

Among the Pterygota the mates find each other by sight, sounds, or chemical sex attractants. Following courtship, sperm are transferred by the male to the female directly and internally by a copulatory organ (Fig. 4.1). The Apterygota and Entognatha have an indirect or external transfer of sperm, not involving copulation. Apparently in all hexapods sperm are stored by the female and some are released to fertilize the eggs as they pass out the genital tract. Compared to the sperm wastages of other animals, this means of ensuring fertilization is highly efficient. One sperm of the several that may enter the egg unites with the egg pronucleus to form the diploid zygote. The penetration of more than one sperm, known as *polyspermy,* is unusual among animals. Most other animals prevent entrance of more than one sperm, presumably because more are likely to cause abnormal development.

During *embryonic* development, the zygote repeatedly divides by mitosis, forming layers of cells by *gastrulation* that appear to correspond to the germ layers of other animals. The outer layer, or *ectoderm,* gives rise to the nervous system, oenocytes, and the integument in all its manifestations. The inner layer, or *mesoderm,* develops into the muscles, fat body, connective tissues, dorsal vessel, and gonads. Small, paired, coelomic cavities arise within the segmented mesoderm. Before embryonic development is completed, however, these cavities are lost, becoming incorporated into the heart lumen, the hemocoele, and certain of the internal genital ducts. The sex cells form at an early stage in the embryo

Figure 4.1 Cerambycid beetles of the genus *Moneilema* copulating on cactus. Direct transfer and storage of sperm are important adaptations that permit insects to reproduce in terrestrial environments, including deserts. (*Photo courtesy of and with permission of Kenneth Lorenzen.*)

and later migrate to the gonads. The midgut epithelium, Malpighian tubules, and blood cells are derived from tissues that may correspond to endoderm, but this is controversial.

The eggs of Endopterygota tend to develop more rapidly than those of Exopterygota. At the end of embryonic development in all hexapods except Protura, the body is complete in segments. This is called *epimorphic* development, in contrast to the *anamorphic* development of Protura, in which segments are added after hatching from the eggs. Embryonic development ends and *postembryonic* development begins when the insect hatches, or *ecloses,* from the egg.

Growth in size and differentiation of adult characteristics take place stepwise over a series of molting cycles. The postembryonic development of an insect can be most conveniently divided into time intervals, or *stadia,* by the ecdyses. The insect during any one stadium is called an *instar* and is designated by numbers. For example, the first instar is the insect after eclosion from the egg and before the first ecdysis. Toward the end of an immature instar, the molting cycle begins anew with apolysis. During the interval between apolysis and ecdysis, the insect is actually enclosed in the old cuticle. This is the *pharate,* or "cloaked" phase of the instar, during which the size and form of the next instar are determined (Fig. 3.1).

Arthropods, amphibians, and certain other animals may evolve a distinctive, actively feeding, sexually immature stage that is generally called a *larva.* The changes in body form from the larva to the sexually mature adult, or *imago,* are broadly termed *metamorphosis.* The Exopterygota generally undergo more molts to reach adulthood, and the larval stage is not greatly different from the adult. Endopterygotes require fewer molts, but the larval stage is strikingly different from the adult. When the insect reaches sexual maturity, molting ceases in the Pterygota, but continues in certain of the Apterygota and Entognatha.

EMBRYONIC DEVELOPMENT

EGG STRUCTURE. The eggs of insects are usually elongate and ovoid, with rounded ends (Fig. 4.2*a*), but other shapes also occur. Eggs of exopterygotes tend to be slightly larger than those of endopterygotes, but some of the largest eggs are those of large carpenter bees, *Xylocopa* (Anthophoridae), which measure as much as 16.5 × 3.0 mm. From the outset eggs are bilaterally symmetrical and have a dorsal and a ventral surface.

The outer shell, or *chorion,* is secreted by the follicle cells of the ovary. Passages are provided for the sperm to enter, for respiration, and in some, for water absorption. At the anterior end is the *micropyle* of one or more holes through the chorion that admit sperm. *Aeropyles* are pores of one to a few microns that permit the entry of oxygen and the exit of carbon dioxide, with rather little loss of water. In eggs that are laid in water or moist places such as decomposing organic matter or in areas that are temporarily submerged, the chorion is often modified into a *plastron.* This is a spongy meshwork in the chorion that retains a film of gas and resists wetting. The gas film operates as a physical gill, permitting oxygen dissolved in water to pass into the gas and then into the embryo (Fig. 5.5*a,b*). Special areas called *hydropyles* are sometimes present on terrestrial eggs to absorb water.

Beneath the chorion is the *vitelline membrane,* which is secreted by the *oocyte*, or egg cell, itself in most insects. Most of the substance of the egg is *yolk,* consisting of lipid, protein and carbohydrate. A yolk-free island of cytoplasm surrounds the female nucleus, which is usually located peripherally in an anterodorsal position. A thin yolk-free *periplasm* often surrounds the egg beneath the vitelline membrane.

FERTILIZATION. When the egg is laid by the female, it is usually in the metaphase of the first maturation division of meiosis. During the exit of the egg from the female's reproductive tract, sperm are released near the micropyles, and one to several enter the egg (Fig. 4.2*b*). After penetration of the sperm, the egg nucleus completes meiosis, resulting in two or three polar bodies and the haploid female pronucleus. The latter migrates toward the interior of the egg, and the polar bodies remain grouped at the surface of the oocyte. The female pronucleus unites with one sperm pronucleus to produce the diploid zygote. The other sperm degenerate.

FORMATION OF THE BLASTODERM, GERM BAND, AND EXTRAEMBRYONIC MEMBRANES. The early cleavages, or mitotic divisions, of the zygote result in nuclei, each surrounded by an island of cytoplasm but without a cell membrane (Fig. 4.2*c*). Such nuclear units are called cleavage *energids.* The energids spread through the yolk, emerging into the periplasm and covering the yolk surface (Fig. 4.2*d*).The energids now become cells, with the addition of individual cell membranes, and form the *blastoderm,* one cell thick (Fig. 4.2*e,f*). Some cells remain in the yolk or return to it as *vitellophages* that digest yolk for the embryo.

Along the midventral line, the blastoderm thickens into a single layer of columnar cells, producing the *germ band,* while the remainder of the blastoderm becomes a thin membrane (Fig. 4.2*g*). Also at this time, a cluster of cells migrates

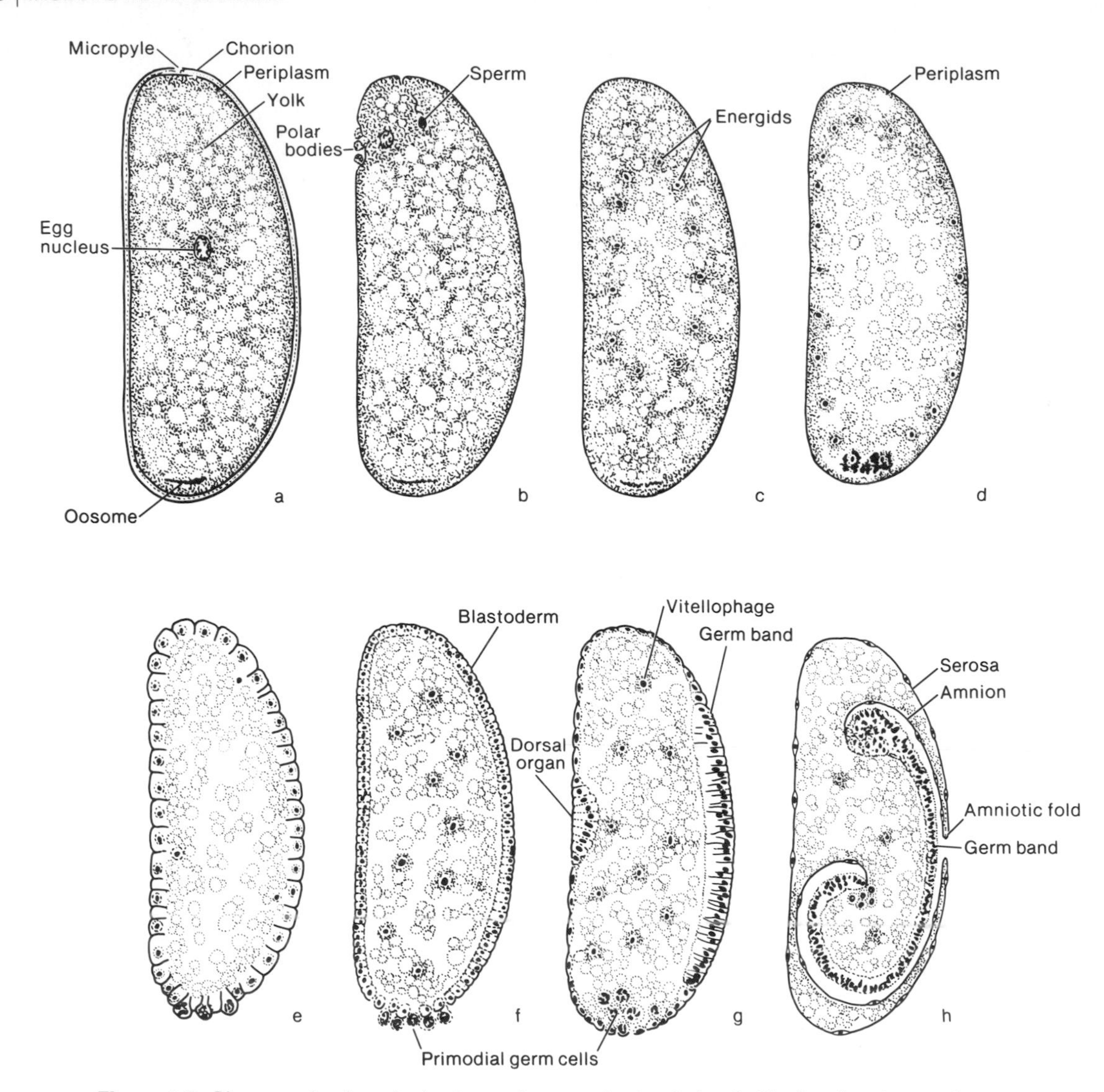

Figure 4.2 Diagrams of embryonic development: *a*, egg structure before fertilization; *b*, entrance of sperm; *c*, cleavage of zygote; *d*, migration of energids; *e*, beginning of blastoderm formation; *f*, blastoderm complete; *g*, formation of germ band; *h*, elongation of germ band. (*Redrawn from Johannsen and Butt,* 1941, *by permission of McGraw-Hill Book Company.*)

through a special region at the posterior pole of the egg, called the *oösome,* and becomes the sex or *primordial germ cells.* These can usually be identified through development and will give rise to eggs and sperm.

The extraembryonic membranes now develop as the germ band elongates, thickens anteriorly, and sinks caudally into the yolk (Fig. 4.2*h*). The margins of the band fold over, bringing the blastoderm into contact midventrally. Fusion takes place such that the blastoderm forms an outer membrane, the *serosa,* and the infolded margins of the germ band form an inner membrane, the *amnion.* The space enclosed by the latter is the *amniotic cavity* that is characteristic of the Class Insecta, although incomplete in Apterygota. The cavity is absent in the Entognatha and Myriapoda. A local thickening in the dorsal serosa of some insects at this time is the *primary dorsal organ.*

FORMATION OF THE GERM LAYERS, NEURAL GROOVE, AND SOMITES. As the amniotic cavity is taking shape, the germ band develops a median furrowlike invagination. This is the *gastral groove* (Fig. 4.3*a*), so named because it corresponds to the first step in gastrulation, seen in other animals. The opposite (lateral) edges of the furrow fuse, restoring the continuity of the outer layer of the germ band and leaving the former trough internally as a tube that later flattens into an inner layer two or three cells thick. The outer layer, still one cell thick, is the ectoderm. The inner layer widens and divides into lateral bands of mesoderm and a *median strand.* The latter may correspond to the endoderm of other animals.

The embryo now forms a second midventral furrow, the *neural groove,* bordered at each side by ridges, or *neural crests* (Fig. 4.3*b*). Anterior and posterior to the neural groove, median pitlike invaginations appear that will become, respectively, the *stomodaeum* or foregut, and the *proctodaeum* or hindgut. Internal to each invagination the median strand forms a mass of cells, called respectively the *anterior* and *posterior midgut rudiments.*

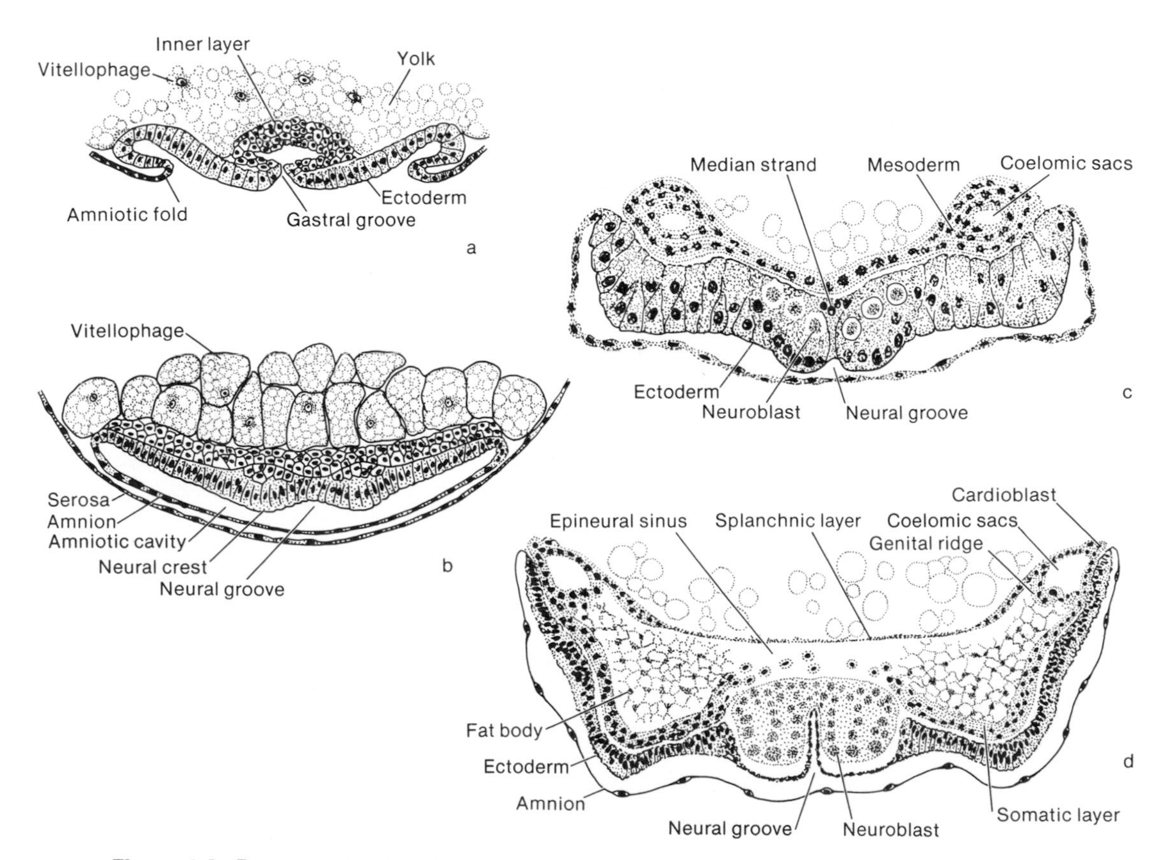

Figure 4.3 Formation of embryonic grooves and mesoderm differentiation, shown by cross sections: *a,* gastral groove (note beginning of amnion and serosa at sides); *b,* neural groove, amnion complete; *c,* formation of coelomic sacs; *d,* formation of epineural sinus. (*Redrawn from Johannsen and Butt,* 1941, *by permission of McGraw-Hill Book Company.*)

Meanwhile, the ectoderm and mesoderm begin to show transverse divisions into segments, or *somites,* especially in the region of the head and thorax. The lateral bands of mesoderm separate into paired, segmental blocks, each lateral block of which develops a cavity or sac (Fig. 4.3*c*). The mesodermal sacs correspond to the main body cavity, or *coelom,* of other animals.

BLASTOKINESIS AND APPENDAGE FORMATION. The caudal sinking movement of the embryo that leads to the formation of the extraembryonic membranes continues after the membranes are complete and segmentation has begun. This movement is *blastokinesis* (Fig. 4.4*a* through *d*). For unknown reasons, the embryo rotates in various ways within the egg. Commonly the rotation is such that the posterior end moves upward and forward, bringing the ventral surface of the embryo beneath the dorsal surface of the egg. The movement is then reversed and the embryo returns to its original position.

In the course of blastokinesis, paired buds appear on each somite that are the appendage rudiments of the labrum, antennae, gnathal segments, legs, and abdominal appendages, if any. The rudiments are formed by outgrowths of ectoderm into which mesoderm and sometimes a part of the coelomic cavities extend. The labrum is in front of the mouth or stomodaeal opening, but the antennae are behind at each side and later migrate forward. Between the antennae and the first gnathal appendages, or mandibles, is a pair of rudiments belonging to the "intercalary" segment. This segment and the appendages later disappear, but the segmental ganglia become the tritocerebrum of the brain. The appendage rudiments of the first abdominal segment are sometimes enlarged in the embryo to form glandular *pleuropodia,* but these also later disappear.

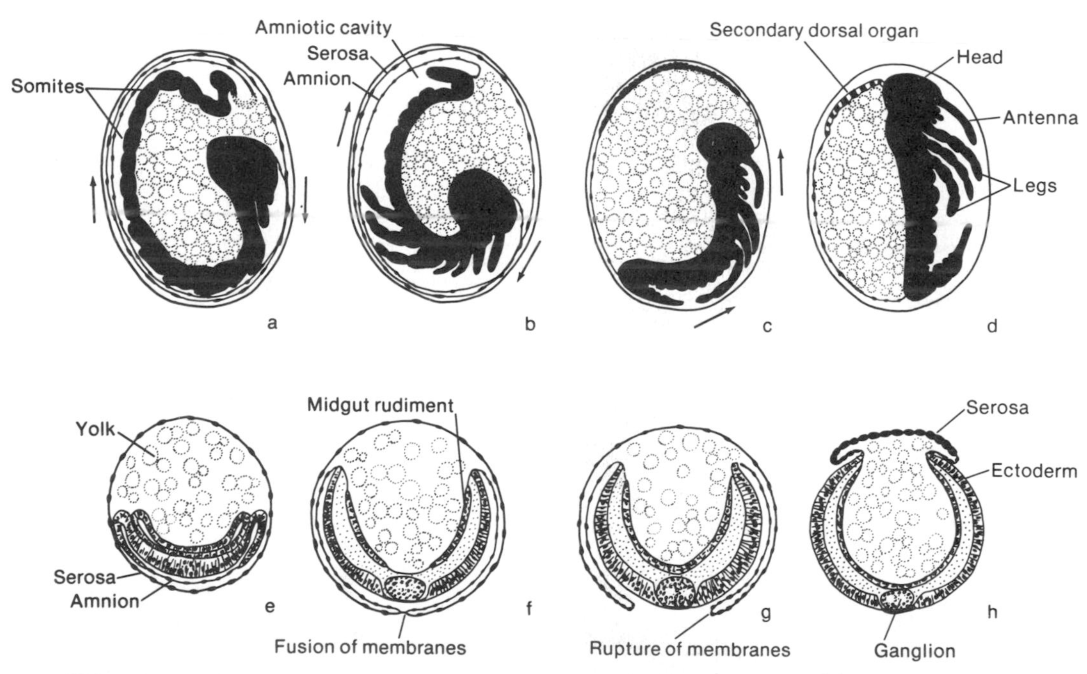

Figure 4.4 Steps in blastokinesis (lateral views) and dorsal closure (cross sections) of the embryo: *a,* beginning of blastokinesis (amnion and serosa fused near head), *b,* embryo rotates caudad; *c,* rotation reverses, embryonic membranes split apart to form single membrane, serosa thickens; *d,* serosa forms secondary dorsal organ; *e,* embryo with membranes intact; *f,* fusion of amnion and serosa, dorsal growth of ectoderm; *g,* rupture of embryonic membranes; *h,* dorsal closure. (*Redrawn from Johannsen and Butt,* 1941, *by permission of McGraw-Hill Book Company.*)

DORSAL CLOSURE OF THE EMBRYO. Up to this point, the embryo's body has been open above to the yolk. When the embryo has reached its maximum length, the amnion and serosa reunite near the head (Fig. 4.4*f*) and split open along the union (Fig. 4.4*g*). The membranes shrink from the embryo in such a way that the serosa sinks dorsally into the yolk, forming a *secondary dorsal organ* (Fig. 4.4*h*), and the amnion, now unfolded, forms a temporary dorsal closure over the yolk. The amnion is then replaced as the ectoderm grows upward and over the yolk to make the final closure. The dorsal organ, now inside the embryo, is absorbed by the yolk.

DIFFERENTIATION OF THE ECTODERM. The ectoderm will develop into the body wall or integument, oenocytes, endocrine and exocrine glands, and the nervous system. The paired neural crests become segmented, and within each somite an inner layer of cells or *neuroblasts* develops at each side of the neural groove. The outer layer of ectoderm remains continuous with the body wall. The neuroblasts give rise to nerve cells that separate from the body wall and form the segmental ganglia, transverse commissures, and longitudinal connectives of the nervous system.

After the neural ridges are segmented, the spiracles appear as paired lateral invaginations on the mesothorax, metathorax, and first eight abdominal segments. The invaginations fork and connect anteriorly and posteriorly to create the tracheal system. Additional pairs of spiracles are seen in some insects on the labial segment, prothorax, and ninth and tenth abdominal segments, but these later disappear. Near the spiracles of the abdominal segments, groups of ectodermal cells become incorporated into the fat body and are the *oenocytes.*

DIFFERENTIATION OF THE MESODERM. After the coelomic sacs are formed within the mesoderm, a central cavity is created, beginning first with spaces at each side medial to the sacs. This cavity is the *epineural sinus* (Fig. 4.3*d*). The mesoderm is now divided into (1) an upper or *splanchnic layer,* separating the yolk from the epineural sinus; (2) a ventral or *somatic layer,* beneath the epineural sinus; (3) a ventrolateral part, medial and adjacent to the coelomic sacs, that becomes the *genital ridge;* and (4) a cluster of cells, the *cardioblasts,* situated lateral to the coelomic sacs, at the junction of the splanchnic and somatic layers.

During subsequent development the mesoderm differentiates as follows: The splanchnic layer will form the muscles of the viscera. The somatic layer further divides into two layers: (1) a ventral layer, next to the ectoderm, which gives rise to the skeletal muscles; and (2) a dorsal spongy layer from which the fat body develops. The primordial germ cells that became visible earlier have now migrated to the region of the ninth or tenth abdominal segment. Here the cluster divides and the halves move to opposite sides, where they penetrate the mesoderm and enter the genital ridges. The gonads develop from the ridges in which the germ cells are lodged. Genital ridges on other segments degenerate. Thc cardioblasts are carried dorsally and medially by the developing ectoderm and mesoderm during dorsal closure. Middorsally the cardioblasts from each side meet to form the tubular heart. The aorta is formed separately by the median walls of the antennal coelomic sacs.

DIFFERENTIATION OF THE MEDIAN STRAND AND MIDGUT RUDIMENTS. When the epineural sinus is complete, the median strand begins to break down except at the extreme ends where the midgut rudiments are located. The free cells then disperse and are thought to have various fates, possibly forming blood cells, or contributing to the formation of the midgut, or disintegrating in the yolk. It is also possible that the blood cells actually are derived from the margins of the coelomic sacs.

The midgut rudiments, situated at the stomodaeal and proctodaeal invaginations, begin to grow as ribbons of cells toward each other below the yolk and above the splanchnic layer of mesoderm. In the middle of the embryo the ribbons fuse, then expand into a sheet which ultimately surrounds the yolk as a tube at about the time of dorsal closure (Fig. 4.4*f* through *h*). Thus the midgut is formed between the stomodaeum and proctodaeum. The blind end of the proctodaeum develops elongate evaginations that become the Malpighian tubules. Near the time of eclosion the blind ends of the stomodaeum and proctodaeum break down to provide a continuous alimentary canal.

POSTEMBRYONIC DEVELOPMENT

TYPES OF POSTEMBRYONIC DEVELOPMENT. All hexapods except Protura exhibit *epimorphosis,* i.e., the first instar has a complete number of body segments upon eclosion. Protura differ in this respect and grow by *anamorphosis,* or the addition of segments after eclosion. The first instar has 9 abdominal segments, and segments are added in subsequent instars until the adult number of 12 is reached.

As explained in Chapter 3, the integument undergoes repeated molting cycles during the life of an insect. The number of molts may be as few as 4 or 5 or may exceed 30. Only at times of molting can the exoskeleton be increased in size or be altered structurally. Other organ systems are less clearly synchronized with molting, but changes in their size and organization are under the same hormonal controls.

Among the insects three general types of metamorphic development can be recognized, based on the degree to which the larva deviates from the imaginal form. In the most extreme metamorphosis, the tissues of virtually the entire larva are destroyed and replaced by imaginal tissues.

Ametabolous development literally means "without change" (Fig. 4.5*a*). The living hexapods with this type of growth are all primitively apterous: Collembola, Diplura, Thysanura, and

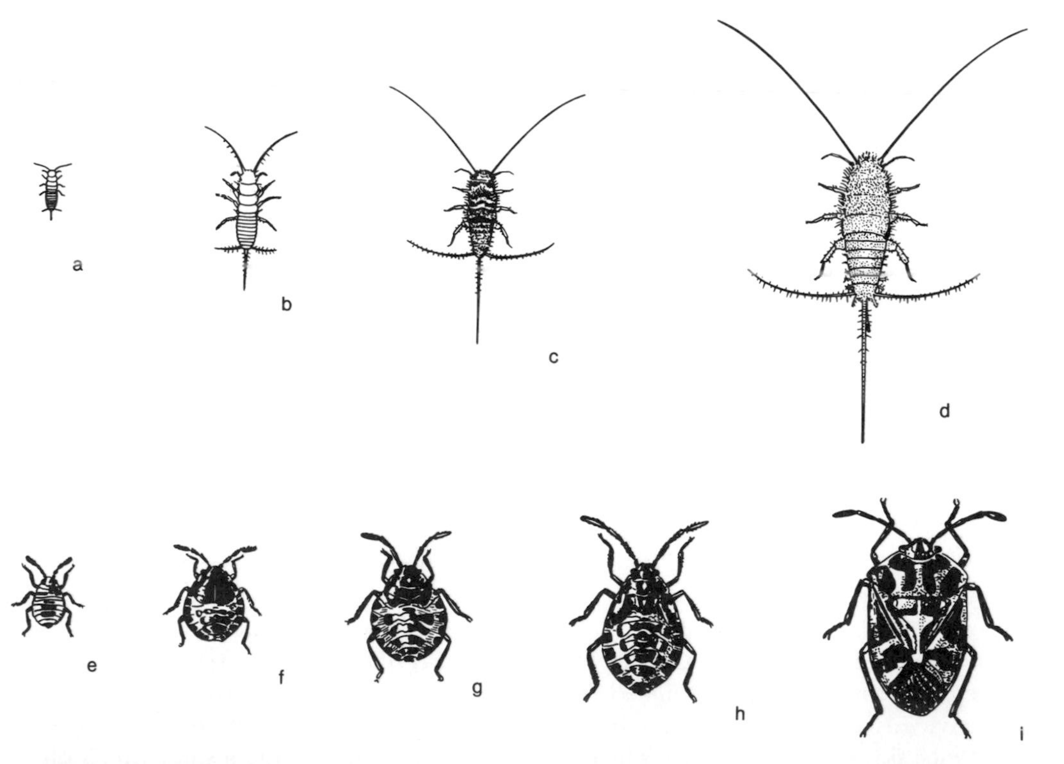

Figure 4.5 Types of life histories: *a–d,* ametabolous development, illustrated by the firebrat, *Lepismodes inquilinus* (Lepismatidae) (*a–c,* nymphs; *d,* imago); *e–i,* hemimetabolous development, illustrated by the harlequin bug, *Murgantia histrionica* (Pentatomidae) (*e–h,* nymphs; *i,* imago). (*Redrawn from Metcalf,* 1962, *by permission of McGraw-Hill Book Company.*)

Archeognatha. The larvae change little except in size and proportions as they grow from instar to instar and mature sexually. Once adulthood is reached in these orders they continue to molt an indeterminate number of times until death.

Hemimetabolous, or incomplete, metamorphosis involves a partial change from larva to imago. The most conspicuous alteration is the appearance in the imago of functional wings and external genitalia. The larvae usually possess well-developed exoskeletons with legs, mouthparts, antennae, compound eyes, and, except in Hemipteroidea, ocelli and cerci. These structures are carried over without much change into the imago. The transition in form is gradual up to the last molt; each successive larval instar increasingly resembles the form of the imago. Wings develop as external pads on the pterothoracic nota and then expand enormously at the last molt (Fig. 4.5*e* through *i*).

Hemimetabolous metamorphosis is characteristic of Paleoptera and nearly all Exopterygota. Most Exopterygota have larvae and adults that live in the same places and eat the same food. Except for the wings and genitalia, the alterations at maturity are usually slight. The larvae of hemimetabolous exopterygotes are commonly called *nymphs.* The larvae of Odonata (Fig. 21.5), Ephemeroptera (Fig. 20.4), and Plecoptera (Fig. 27.1*b*), however, are aquatic, and the adults are aerial. Consequently the change is marked, involving the loss of gills and modification of the mouthparts as well as the acquisition of wings and new body proportions. The larvae of these three orders are called *naiads.*

All living pterygotes differ from the living ametabolous orders in that once sexual maturity is reached, molting ceases. Furthermore, functional wings are restricted to the imago of all living pterygotes except the Ephemeroptera. In this order the first winged instar, or *subimago,* is usually sexually immature. After a brief aerial existence, the subimago molts and the second winged instar is the sexually active imago.

In *holometabolous,* or complete, metamorphosis the immatures are fittingly called "larvae" because they undergo histological reorganization before maturing sexually (Fig. 4.6). All Endop-

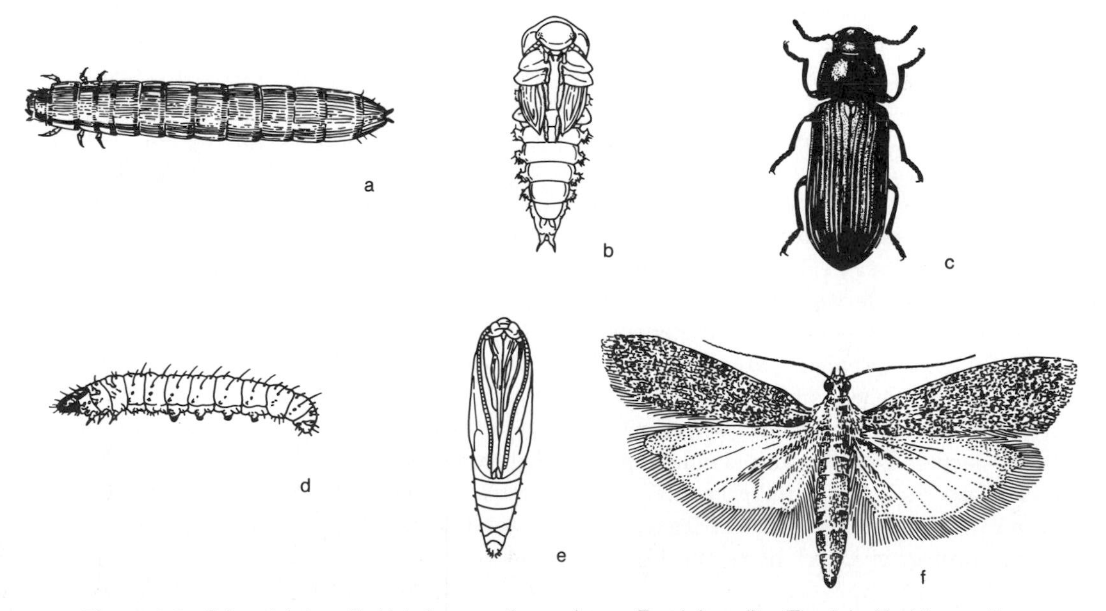

Figure 4.6 Holometabolous life histories: *a–c,* the mealworm, *Tenebrio molitor* (Tenebrionidae) (*a,* eruciform larva; *b,* adecticous exarate pupa; *c,* imago); *d–*f, the Mediterranean flour moth, *Anagasta kuhniella* (Pyralidae) (*d,* eruciform larva; *e,* obtect pupa; *f,* imago). (*Redrawn from Metcalf et al.,* 1962, *by permission of McGraw-Hill Book Company.*)

terygota and certain exceptional Exopterygota (Thysanoptera, Fig. 35.2; Aleyrodidae, Fig. 34.10*c–e*; and male Coccidae, Fig. 34.12*b*) have holometabolous development. The larvae molt into a nonfeeding, inactive instar, the *pupa,* before the final molt into the imago. Near the end of the last larval instar, the insect may cease feeding; it is then known as a *prepupa* or *pharate pupa.* The molt into the pupal instar is called *pupation.* During pupation and in the pupa, the larval tissues are partly or entirely destroyed and replaced by imaginal tissues.

In contrast to the hemimetabolous exopterygotes, the holometabolous exopterygotes have the development of wings delayed until the last one to three larval instars (Fig. 35.2). At this time the wings develop externally and the body undergoes various degrees of histological reorganization.

In addition to having concealed wing buds, endopterygote larvae differ in other respects both from hemimetabolous larvae and from their own imaginal instars. Megaloptera (Fig. 36.2) and Raphidioptera (Fig. 37.1*b*) exhibit the least difference between larvae and imagoes, while in Diptera the headless, legless maggots are totally unlike adult flies (Fig. 42.5). In general, larvae differ from imagoes in being soft-bodied and in having simplified sensory, feeding, and locomotory structures. The larval antennae are reduced to but a few segments in all endopterygotes except beetles of the Family Helodidae. Compound eyes are present in Mecoptera (Fig. 41.1*c*) and certain Lepidoptera (Suborder Zeugloptera), but in other larvae the light receptors consist of lateral clusters of single-faceted stemmata (Fig. 17.4*b*). Ocelli are absent except in certain Mecoptera. The mouthparts and legs are often reduced in structural complexity. Legs are absent in larval Diptera and Hymenoptera (Suborder Apocrita). Pregenital appendages are present on the abdomen as gills in Megaloptera or as fleshy prolegs in Lepidoptera and Hymenoptera (Suborder Symphyta).

During the last larval instar, the insect usually seeks a pupation site and often secretes a silken cocoon or constructs a special cell of earth or wood particles. The functions of the cocoon or cell are probably to reduce loss of moisture, provide protection from ice crystals in winter, and to avoid predators. Certain specialized endopterygotes do not construct pupal shelters.

Following the larval-pupal apolysis the insect closely resembles the imago. The larval features of the exoskeleton have been replaced by incipient imaginal structures such as long antennae, compound eyes, ocelli, mouthparts, legs, and miniature wings. Pregenital abdominal appendages of the larva, if any, are absent, and rudimentary genitalia are present.

After the pupal-imaginal apolysis the wings grow enormously in size within the wing sacs of pupal cuticle. Internally the alimentary canal and nervous system are reorganized, the growth of the endoskeleton apophyses is completed, and the musculature assumes its final distribution and attachment. At the pupal-imaginal ecdysis, the body usually contracts in volume as the wings are inflated by blood pressure. Following ecdysis a fluid substance, called the *meconium,* is sometimes excreted.

TYPES OF ENDOPTERYGOTE LARVAE. Three general types can be distinguished as follows: *Campodeiform* larvae have prognathous mouthparts, elongate flattened bodies, long legs, and usually some kind of caudal appendages. Larvae with these characteristics resemble the Campodeidae (Fig. 18.2*a*) of Diplura and are so named. They are usually active predators. Examples are found in Neuroptera (Fig. 38.1*b*), Megaloptera (Fig. 36.2), Rhapidioptera (Fig. 37.1*b*), Trichoptera (Rhyacophilidae, Fig. 45.2*a*), and many Coleoptera (Dytiscidae, Fig. 39.4*b*; Carabidae; Staphylinidae). *Eruciform* larvae have hypognathous mouthparts, cylindrical bodies, short thoracic legs and abdominal prolegs, and reduced or no caudal appendages (Fig. 4.6*a,d*). Examples are catepillars of Lepidoptera (Fig. 44.5*a,b*) and larvae of Mecoptera (Fig. 41.1*c*) and Hymenoptera (Suborder Symphyta, Fig. 46.3*a*). Larvae of this type are rather inactive and feed primarily on plant materials. *Vermiform* larvae are wormlike and cylindrical, and lack locomotory appendages. Most larvae or maggots of Diptera (Fig. 42.5) are of this type, as are the larvae of Siphonaptera (Fig. 43.1*d*) and certain wood-boring beetles (Fig. 39.9*c*) and aculeate Hymenoptera (Fig. 46.3*b*). Other types of larvae are described under the various orders in Part IV.

HYPERMETAMORPHOSIS. In some insects the larvae pass through two or more instars that differ markedly in appearance. The first instar is often campodeiform and actively seeks food or a host. The subsequent instars may be relatively inactive and grublike. Examples are commonly parasitic and are found in Neuroptera (Mantispidae), Coleoptera (certain species of Carabidae and Staphylinidae; Meloidae and Rhipiphoridae), Strepsiptera, Diptera (Bombyliidae, Acroceridae, Nemestrinidae, and some Tachinidae and Calliphoridae), Hymenoptera (certain Chalcidoidea), and certain ectoparasitic Lepidoptera (Cyclotornidae and Epipyropidae). When the first instar has legs, as in Coleoptera, Mantispidae, or Strepsiptera, it is called a *triungulin* (Fig. 40.1*b,c*). The legless *planidium* larvae of Diptera and Hymenoptera use long "walking" setae or jumping movements for locomotion (Fig. 46.4).

TYPES OF ENDOPTERYGOTE PUPAE. Pupae with movable mandibles are called *decticous.* Such pupae are characteristic of Megaloptera, Neuroptera, Mecoptera, Raphidioptera, Trichoptera, and certain Lepidoptera (Suborders Zeugloptera and Dachnonypha). The pupal mandibles are used by the pharate adult to cut through the cocoon or gnaw out of the pupal cell. The other appendages are free from the body in a condition called *exarate* (Fig. 4.6*b*). In some of the above taxa the legs are used by the pharate adult to walk or swim before the pupal cuticle is shed.

Pupae without movable mandibles are *adecticous.* The pupae of this type may be exarate, or have the appendages cemented to the body in a condition known as *obtect* (Fig. 4.6*e*). Exarate adecticous pupae are found in Hymenoptera and Coleoptera. To escape the cocoon or cell the insect first sheds the pupal cuticle and then uses the adult mandibles and legs to exit. In Strepsiptera and certain Diptera (Cyclorrapha) the pupa is exarate, but it is enclosed in a *puparium,* or sclerotized cuticle of the last larval instar. Obtect pupae are found in certain Lepidoptera (Suborders Ditrysia and Monotrysia, Fig. 44.5*e*) and Diptera (Nematocera, Orthorrapha). Backwardly directed spines on the cuticle of some obtect pupae help to force the emerging insect out of the cocoon or cell.

HORMONAL CONTROL OF GROWTH, DIFFERENTIATION, AND REPRODUCTION. The postembryonic development and sexual maturation of insects are regulated by *hormones.* These are substances secreted by one part of the body that have a profound effect on another part of the body. The organs that secrete the developmental hormones are associated with and under the control of the nervous system. Such organs, called *endocrine,* include the following (Fig. 6.2):

1. The *neurosecretory cells* are located in the pars intercerebralis region of the brain and secrete the "brain hormone," or *ecdysiotropin.* The chemical identity of ecdysiotropin is unknown, but it is thought to be protein.

2. The paired *corpora cardiaca* are located behind the brain and connected to the brain's neurosecretory cells by nerves. The corpora cardiaca receive ecdysiotropin from the brain and later release it to the hemolymph. The corpora cardiaca also produce hormones that regulate blood-sugar (trehalose) concentration, fatty acid content in blood, heartbeat, and Malpighian tubule movement.

3. The paired *ecdysial glands,* or "prothoracic glands," are located in the thorax or head. Upon stimulation by ecdysiotropin in the hemolymph, the ecdysial glands begin secretory activity. Correlated with this activity are the initiation of molting and the differentiation of imaginal tissues. A group of steroid hormones named the *molting hormones* or *ecdysones* is known to be responsible for the onset of molting and differentiation. Ecdysone is synthesized from cholesterol and exists in several forms. The most active of these in morphogenesis seems to be β-ecdysone. It was once thought that β-ecdysone was secreted directly by glands. Now there is evidence that the glands secrete a precursor, α-

ecdysone, that permits β-ecdysone to be synthesized elsewhere in the body.

4. The paired *corpora allata* are located behind the brain and innervated by nerves from the brain that pass through the corpora cardiaca and also from the subesophageal ganglion. The corpora allata directly secrete three terpenoid hormones named the *juvenile hormones.* The effects of the hormones in larvae are to inhibit metamorphosis and to favor the expression of larval characteristics during the molting cycle. In the adult the hormones stimulate the fat body to produce yolk proteins. The hormones are known chemically and have been synthesized. At least three compounds are responsible for the natural hormonal effect in insects.

The neuroendocrine organs are ultimately under the control of the central nervous system. Depending on the kind of insect involved, certain events serve to stimulate molting. External stimuli such as a pattern of changing temperature or light, or internal cues such as the stretching of the abdomen after feeding, are received by the brain, which then causes ecdysiotropin to be released by the corpus cardiacum. Ecdysiotropin stimulates secretory activity in the ecdysial glands, and by an unknown pathway, molting hormone increases in the body. This in turn stimulates the epidermis to begin a molting cycle. The exact means of control of the corpora allata is not understood. Nervous connections, neurosecretions, or the composition of the hemolymph, singly or in combination, might serve to regulate the secretion of juvenile hormones.

The growth and differentiation of the developing insect are regulated by the amount of juvenile hormone present during a molting cycle. When molting is stimulated in a young larva by the molting hormone, sufficient juvenile hormone is present to cause the cells to produce another larva. The insect then increases in size without metamorphosis.

In hemimetabolous insects, juvenile hormone disappears in the last larval instar, and at the next or final molt, the insect differentiates into the adult. In holometabolous insects, the level of juvenile hormone declines in the last larval instar, and at the next molt the insect transforms into the pupa. The final molt to the adult takes place in the absence of juvenile hormone.

In adult pterygote insects, the ecdysial glands degenerate and molting ceases. The corpora allata, however, renew their activity. The juvenile hormones now stimulate the reproductive organs, i.e., they are *gonadotropic.* In females of some insects the hormone is required for yolk proteins to be manufactured in the fat body and to be deposited in developing eggs. In other insects, secretions from the neurosecretory cells of the brain, in combination with corpora allata hormone, provide this effect on yolk deposition. In males, the production of fluids by the genital accessory glands is dependent on stimulation by corpora allata secretions. In certain male Lepidoptera spermatogenesis begins in the pupa. Here the molting hormone has gonadotropic effects on the testes and controls spermatogenesis.

In the ametabolous Thysanura, sexual maturation occurs gradually. Once adulthood is reached the neurosecretory cells of the brain, ecdysial glands, and corpora allata remain active. Molting cycles, initiated by the brain, occur regularly. If females are mated within a few days after ecdysis, the corpora allata become active and yolk is deposited in the eggs. If the females are not mated, the eggs are resorbed at the onset of the next molting cycle. Thus molting and reproduction alternate in their demands on the body reserves.

COMPOUNDS IN OTHER ORGANISMS WITH HORMONE ACTIVITY IN INSECTS. β-Ecdysone has been identified in crayfish and will induce molting in other crustaceans as well as spiders and ticks. It is probably the normal molting hormone of many arthropods. Surprisingly, both α-ecdysone and β-ecdysone, as well as over 30 related compounds with similar hormonal effects on insects, have been isolated from plants. Ferns and yews are the major sources of the active substances, now called *phytoecdysones.*

Extracts with juvenile hormone activity have been isolated from vertebrates, bacteria, yeast, protozoa, and plants, especially the balsam fir.

The active compound in balsam fir was discovered by a curious accident. Laboratory colonies of the European linden bug, *Pyrrhocoris apterus* (Pyrrhocoridae), failed to mature when brought from Europe to Dr. Carroll Williams's laboratory at Harvard. A careful search revealed that the paper towels in the cages contained a "paper factor" with juvenile hormone activity. This was traced to the balsam fir, which is the principal source of paper pulp in Canada and Northern United States. Towels in Europe apparently lack the paper factor. The factor was isolated, identified, and named *juvabione.* The compound has proved to be exceptionally specific in its action, affecting only bugs of the Family Pyrrhocoridae.

The significance of natural compounds in plants that have hormonelike activity in insects is not clear. It is possible that the compounds disrupt the normal growth of plant-feeding insects, but the evidence thus far is contradictory. For example, ferns are relatively free of insect attack, but mulberry leaves, also rich in ecdysone, are eaten by silkworms without effect.

Chemicals structurally related to the juvenile hormones have also been synthesized, and some of these are more active than the natural hormones. The possibility that these may be used as specific insecticides is being investigated, and the prospects in some cases are quite promising (Williams, 1967).

GENERATION TIME AND LONGEVITY. The numbers of generations per year in insects is called *voltinism. Univoltine* insects have one generation per year, *bivoltine* insects have two, and *multivoltine* insects have more than two per year. Some insects require one or more years to complete their life cycle; but a year or less is probably adequate for most.

The number of generations per unit time tends to be inversely related to body size. In other words, smaller insects have more generations than larger insects during the same period. This is only approximate because other factors such as moisture, nutrition, and temperature greatly influence the rate of growth.

Insects that live in dry environments and eat nutritionally deficient food require longer to mature. Wood-boring beetles (Cerambycidae, Bostrichidae, Buprestidae), wasps (Siricidae), and moths (Cossidae) take 1 to 3 years. Records of 23 years for a cerambycid and 22 years for a bostrichid were obtained when the beetles emerged from wood that had been dried and made into furniture. The familiar 13- and 17-year periodical cicadas (Cicadidae) spend most of their life in soil sucking the juice of roots.

The adult life of insects is devoted to mating and reproduction. Adults may last only a few hours as in male Strepsiptera (Chap. 40), some species of Ephemeroptera, or some species of the *Clunio* midges (Chap. 11). On the other hand, univoltine insects that diapause in the adult stage may have adults that live for almost a year. Among the longest-lived adults are the reproductive castes of social insects. Queen honeybees normally live 2 to 3 years, and up to 5 years or more have been recorded. The maximum age of an ant queen is 18 years and of a termite queen, 12 years. Tenebrionid beetles also are long-lived as adults. Marked individuals have been recovered in the field after 3 years, and it is possible they live for up to 10 years.

ARRESTED DEVELOPMENT: QUIESCENCE AND DIAPAUSE. Development or reproduction of insects may be interrupted by periods of cold or heat, drought, or a scarcity of food. When activity is suspended at the onset of the unfavorable period and resumed immediately afterwards, the temporary arrest is called *quiescence.* During quiescence the hormonal stimulation of growth or reproduction remains unaffected. A few chilly days in spring, for example, may lower metabolism, but growth continues with the return of sunny weather.

Development or reproduction may be arrested on a long-term basis in winter or summer, however, by the failure of the endocrine organs to secrete appropriate hormones. A prolonged dormancy regulated by hormones is called *diapause.* Depending on the species of the insect, diapause may occur in any stage of the life. In eggs, larvae, and pupae the diapause is an interruption of growth and development; in adults the diapause is a period of no reproduction. The stage at which diapause occurs is genetically determined, but whether it will occur or not may be influenced by environmental conditions.

An *obligate* diapause takes place regardless of the environment. Univoltine insects commonly have an obligate diapause in an immature stage, as for example in many Lepidoptera of the Temperate Zone. The onset, or *induction,* of a *facultative* diapause is dependent on certain environmental cues, e.g., a winter diapause can be triggered by the shorter days and longer nights of autumn. Bivoltine and multivoltine insects usually have a facultative diapause. During spring and summer, development and reproduction continue without interruption. The brain senses the photoperiodic cues, possibly directly and not through the eyes, and the neurosecretory cells cease to stimulate the appropriate endocrine organs. As a result, in larvae and pupae ecdysone is not produced and molting and the associated developmental changes do not occur. In adults the corpora allata, also under neurosecretory control, become inactive and reproduction is not stimulated.

The onset of diapause usually begins at a time when environmental conditions are still favorable for continued growth. Diapause is recognized by a variety of symptoms. Active insects become relatively inactive, but not necessarily immobile or unresponsive to prodding. Oxygen consumption falls as the rate of metabolism is lowered and no longer is correlated with temperature. Water loss decreases. Thus in diapause a resistant state is attained before the beginning of unfavorable weather or food shortage.

The *intensity* of diapause and its probable duration may be judged by the extent to which metabolism is reduced. The period during which the insect is in diapause varies according to the seasonal changes in environment, the insect's response to the environment, and internal processes that are curiously called *diapause development* even though development in the usual sense has stopped. This terminology is intended to indicate that a physiological process, the nature of which is unknown, must be completed before diapause is terminated. Once diapause is broken, the brain resumes neurosecretory activity and development continues or, in adults, the reproductive functions are restored. Additional details on the role of the environment in diapause induction and termination are given later (Chap. 11).

Diapause in the egg stage may take place late enough to be controlled by the young larva's neuroendocrine system. In the embryo, however, the nervous system is not developed. Diapause in grasshopper eggs takes place presumably in the absence of neuroendocrine factors. In the silkworm, *Bombyx mori* (Bombycidae), a *diapause hormone* is involved. It is secreted by the female pupa and causes her developing eggs to diapause. Some genetic races of silkworms have an obligate diapause in the eggs of each generation; others have a facultative diapause. The eggs begin development in the female pupa. In the facultative races, the environment experienced by the female influences the type of eggs she will lay. If the female embryo is exposed to high temperatures of about 25°C, or to long days of 14 hours or more, during early larval life, then in the pupa a diapause hormone is secreted by the subesophageal ganglion that causes the developing eggs to diapause. The eggs are dark brown and will not hatch until they have been chilled at 0°C for several months. Lower temperatures and shorter days during female development inhibit the secretion of hormone, and nondiapausing, white eggs are laid that soon hatch if kept warm. Thus females reared under spring conditions lay eggs that will hatch and produce a second generation. The latter, reared under summer conditions, lay eggs that will survive the winter in diapause.

MALE REPRODUCTIVE SYSTEM

MALE ORGANS. The functions of the male organs of insects are to produce sperm and to transfer the sperm to the female of the same species (Fig. 4.7*a* and *b*). The mesodermal gonads are the *testes,*

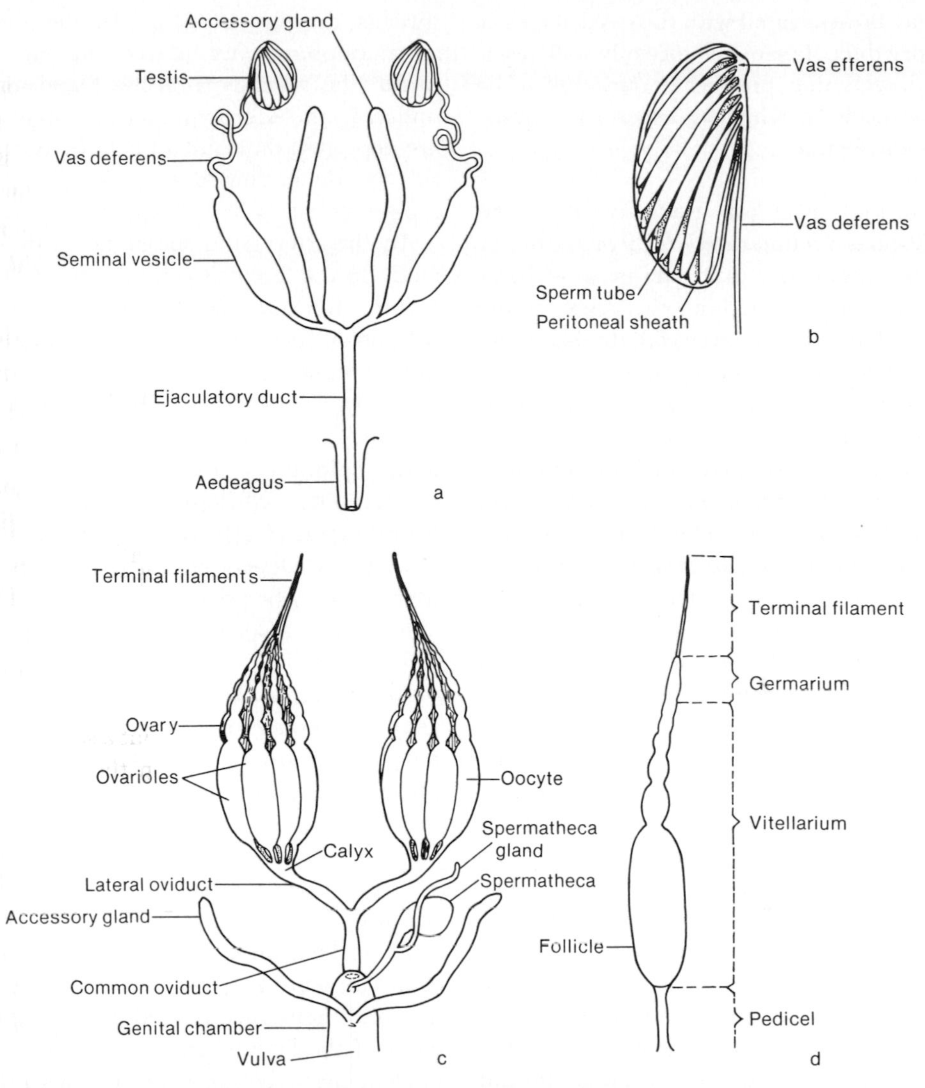

Figure 4.7 Diagrams of general structure of internal reproductive organs: *a* male system; *b*, testis; *c*, female system; *d*, ovariole. (*Redrawn from Snodgrass,* 1935, *by permission of McGraw-Hill Book Company.*)

which are paired but may be secondarily united in a single body. *Spermatogenesis,* or the production of sperm, takes place in the *sperm tubes.* In insects with a short adult life, spermatogenesis may be completed as early as the larval or pupal instars, while in long-lived adults sperm may be produced in the adult. The sperm tubes may be single as in some Apterygota, Coleoptera, and Diptera, or multiple as in other insects. In the latter case, the number is usually less than the number of ovarioles in the females of the same species. The sperm tubes of each testis are usually held together by a *peritoneal sheath.*

Mature sperm pass from the sperm tube through a short *vas efferens* to the *vas deferens.* The vas deferens is usually mesodermal in origin, but in Diptera it is ectodermal. It may be dilated into a *seminal vesicle* where sperm are stored prior to their exit via the ectodermal *ejaculatory duct. Accessory glands* of mesodermal or ectodermal origin may be associated with the vas deferens or ejaculatory duct. These are especially well developed in insects that produce a *spermatophore,* or jellylike capsule in which the sperm are transferred to the female.

SPERM-TUBE STRUCTURE AND SPERMATOGENESIS. The sperm tube has a cellular *epithelial sheath* of one or two layers. Inside the tube, the germ cells, or *spermatogonia,* are located at the apex in the *germarium.* Here also is a large cell, the *apical* cell. The spermatogonia multiply mitotically. The cells from consecutive divisions remain clustered and enclosed in a cellular capsule or *sperm cyst.* Within the cyst the spermatogonia produce *spermatocytes* which undergo meiotic division to form the haploid spermatids. The spermatids then differentiate into flagellated *spermatozoa* or sperm. Insect sperm are quite slender and long. Their narrow diameter is probably correlated with the diameter of micropyles in the egg chorion, an opening that must be kept small to reduce water loss.

The several steps in spermatogenesis can be seen in successive zones arranged lengthwise in the sperm tube. Near the base where the vas efferens is located, the cyst disintegrates and the sperm are released, still grouped in bundles. The sperm pass into the vas deferens and are stored in the seminal vesicles until ejaculation.

TRANSFER OF SPERM. The primitively apterous hexapods transfer sperm to the female indirectly. The sperm are contained in a spermatophore and deposited on the substrate or on threads to be picked up by the female. This method of transfer requires humid conditions to prevent the drying of the spermatophore. In Collembola and Diplura the sperm are contained in a stalked spermatophore that is deposited by the male on the substrate. In some Collembola the female picks up the spermatophore with her vulva in the absence of the male; in other species the spermatophore is picked up after courtship and in the male's presence. Diplura males deposit the spermatophore only in the presence of the female. In Thysanura and Archeognatha, the male spins threads during courtship. In the former the spermatophores are placed on the substrate under the threads and the female is guided under. In the Archeognatha the spermatophores are deposited on a thread held by the male and the female is guided to pick up the spermatophores.

In the Pterygota, sperm are transferred directly to the female's genital tract during copulation. Internal fertilization results in less wastage of sperm and is much less dependent on humid conditions. The spermatophore is retained in many orders: Orthoptera, Blattodea, Mantodea, Dermaptera, Psocoptera, Neuroptera, Lepidoptera, and some Diptera (Suborder Nematocera). Mecoptera, and most Diptera, transfer sperm freely without a spermatophore to the female. Depending on the species, spermatophores may be present or absent in Hemiptera, Trichoptera, Coleoptera, Hymenoptera, and Neuroptera.

Once the spermatophore is deposited in the genital opening, the sperm are freed by the partial rupture or digestion of the spermatophore. The sperm then migrate or are moved to the spermatheca. The empty spermatophore is dissolved in the vagina or eaten by the female.

Traumatic or *hemocoelic insemination* occurs in the Strepsiptera; in the bedbugs, Cimicidae, and some of their allies, the Anthocoridae and Polyctenidae; and in some Nabidae (Hemiptera). The male genitalia are used to pierce the integument of the female, and the sperm are injected into the hemocoel. In some bedbugs a *spermalege,* or specialized organ of the integument, is developed where the puncture normally is made.

FEMALE REPRODUCTIVE SYSTEM

FEMALE ORGANS. The functions of the female organs of insects are to produce eggs and to ensure their fertilization and placement in the environment (Fig. 4.7*c* and *d*). The mesodermal gonads are the paired *ovaries,* each of which may be single or divided into *ovarioles. Oogenesis,* or the production of eggs, takes place in the ovarioles, beginning in some insects in the pupa or last nymphal instar. The number of ovarioles is commonly four to eight on a side, depending on the species, and is correlated with the number of eggs produced. In Lepidoptera the number is usually four on each side. The highly fertile queen ants or termites, for example, may have more than 1000 on each side.

At *ovulation* the eggs pass from the ovariole and follow ducts toward the exterior. The paired *lateral oviducts,* each with a *calyx,* lead to the single *common oviduct,* which empties into the *genital chamber.* Attached to the genital chamber are *accessory glands* and the *spermatheca* with its *spermathecal gland.* Fertilization takes place in the genital chamber before the eggs are oviposited. The external opening of the reproductive tract is the *vulva.* Except for the lateral oviducts, which may be partly mesodermal, the other ducts and glands are ectodermal in origin and are lined with cuticle. The arrangement of the female tracts varies in the Lepidoptera (see Fig. 44.4) in which separate openings for copulation and oviposition may be present.

OVARIOLE STRUCTURE AND OOGENESIS. The ovariole is essentially a tube, attached anteriorly in the body by a *terminal filament.* The combined filaments of the ovarioles of each ovary may form a *suspensory ligament.* The wall of the tube is the *tunica propria,* which is lined internally by a *follicular epithelium.* The ovariole is divided along its length into the *germarium, vitellarium,* and *pedicel.* In the germarium are the *oogonia,* which are derived from the embryonic primordial germ cells. By mitotic divisons of the oogonia, the oocytes are produced, and these pass into the vitellarium. Each oocyte becomes surrounded by follicular cells, forming a cystlike *follicle.* Within the follicle, yolk is deposited in the oocyte during *vitellogenesis* and the chorion is added. As the follicles swell, arranged one behind the other, the vitellarium elongates and becomes beaded in appearance. The first or most advanced follicle is separated from the opening of the pedicel by a *follicular plug.*

Two major structural types of ovarioles are recognized: *panoistic* and *meroistic.* The former is the simple ovariole just described and found in Entognatha, Apterygota, Paleoptera, most Orthopteroidea, Thysanoptera, and in one endopterygote order, Siphonaptera. The meroistic type has additional cells, or *trophocytes,* associated with each oocyte. These are usually derived from the oogonia and function in vitellogenesis.

Meroistic ovarioles are further divided into *polytrophic* and *telotrophic* types. In the former the trophocytes accompany each oocyte into the vitellarium and are included in the follicle. Polytrophic ovarioles are found in the exopterygotes Phthiraptera, Dermaptera, and Psocoptera, and in most endopterygotes except Siphonaptera and Coleoptera (Adephaga). In the telotrophic ovarioles, the trophocytes remain in the germarium and are connected with the oocytes by *nutritive cords.* Ovarioles of this type are found in Hemiptera and Coleoptera (Suborder Polyphaga).

The oocytes rapidly increase in size as protein, carbohydrate, and lipid yolk bodies are formed by vitellogenesis. Recall that this process is under endocrine control, often by the corpora allata. The nutrients that make up the yolk are derived from food eaten by the female or, if the adult does not feed, from food eaten and stored in the fat body during the immature instars. The process of vitellogenesis is complex and not fully understood. Some of the protein components of yolk are synthesized in the fat body and enter the oocyte from the hemolymph by way of the follicle cells. Protein synthesis inside the oocyte may involve additional nucleic acids contributed by the trophocytes or follicle cells. Lipids and the glycogen found in the yolk of many insects are transported to the oocyte by the trophocytes and follicle cells, especially the latter.

When yolk deposition is complete, the chorion is secreted around the oocyte by the follicle cells. The chorion is layered like a cuticle with respect to the follicle cells: next to the follicle cells is protein, and the layer next to the vitelline membrane of the oocyte is lipoprotein. The sculpturing and various passages through the chorion are created by the follicle cells. The vitelline membrane is usually considered to be the cell membrane of the oocyte, but in some Diptera it also is secreted by the follicle cells.

OVULATION AND FERTILIZATION. At ovulation the follicular covering breaks down and the egg pushes past the follicular plug into the pedicel and to the calyx beyond. The degenerate follicular covering forms a reddish or yellowish mass which persists for some time in the ovariole. The presence of the colored mass can be used to discriminate between *parous* females that have ovulated and *nulliparous* females that have not. In those insects which lay their eggs in groups, such as many Orthopteroidea, the eggs may be accumulated in the pedicels or oviducts before fertilization and oviposition.

As the eggs pass into the genital chamber from the common oviduct, they are usually oriented in a manner to receive sperm directly from the opening of the spermathecal duct. The mechanism of sperm release involves the special muscles that are associated with the spermatheca. In most insects the release is probably stimulated by ovulation or the passage of eggs in the oviduct. In Hymenoptera, however, the female can control the release so that some eggs are fertilized and others are not. The latter become males.

OVIPOSITION. The segmental position of the genital opening varies among the orders. All insects have a common oviduct except the Ephemeroptera, in which the lateral oviducts open separately just behind the seventh sternum. In Dermaptera the short common oviduct opens behind the seventh sternum. Those insects with an ovipositor derived from the appendages of the eighth and ninth segments usually have the genital opening on the eighth segment. The great majority of insects, however, have the appendicular ovipositor reduced or replaced by special modifications of the abdominal apex: most Odonata, Plecoptera, Phthiraptera, Thysanoptera (Suborder Tubulifera), Coleoptera, most Neuroptera, Mecoptera, Trichoptera, Lepidoptera, and Diptera. In these insects the opening is in the eighth or ninth segment. Many have the caudal abdominal segments tubular and telescopic, with the genital opening at the apex. This permits the eggs to be placed in crevices or cemented to leaves or other surfaces. Before the eggs are released, they may receive adhesive substances or a protective coating from the accessory glands. Groups of eggs of Mantodea (Fig. 23.1*b,c*) and Blattodea (Fig. 22.1) are enclosed in a protective case, or *ootheca*, derived from the accessory (also called colleterial) glands.

Oviposition behavior may be quite simple, as in Phasmatodea, which merely drop the eggs at random. More complex behavior involves the selection of appropriate sites through the use of various sense organs, including those on the ovipositor or abdominal tip. The sites are usually near the food required by the first instar of the offspring. Depending on the food habits of the species, the favored site may be in or on water, soil, decaying organic matter, certain species of live plants or animals, or other insects. Oviposition may occur only at certain times during the 24 hours. All the eggs may be laid at once, as in some Ephemeroptera, or singly or in small groups over a considerable period. Social relationships, involving the special care of eggs and offspring, are discussed in Chap. 7.

SEX DETERMINATION

Sex in insects, as in other organisms, is commonly determined by *heterogamy*, or the production of gametes of two types. The types differ in genic sex determiners and may also differ in the presence and number of sex chromosomes. The exact nature of the sex-determining mechanism is not yet clear, but the ratio of males to females in insects is usually 1:1. Sex chromosomes have been identified in many orders. Sex chromosomes that occur in pairs in one sex are designated X; those that occur as single chromosomes are Y; and the missing chromosome of a pair is 0. Males are heterozygous XY and XO and females homozygous XX in most Diptera, Neuroptera, Mecoptera, Hemiptera, Odonata, Coleoptera, and Orthoptera. Females are the heterozygous XY or XO and males homozygous XX in Lepidoptera and Trichoptera.

In Hymenoptera, fertilized eggs develop into females and unfertilized eggs into males. The females are therefore diploid and the males haploid. The sex ratio may deviate considerably from 1:1. This mechanism of sex determination, called *haplodiploidy,* is also known in some Thysanoptera and *Micromalthus* (Coleoptera). Male Coccidae (Hemiptera, Suborder Homoptera) are also haploid, but apparently this is a secondary condition because the male zygote is initially diploid.

TYPES OF REPRODUCTION

Thus far we have discussed the most common type of reproduction, oviparity, in which eggs are deposited by the adult female shortly after fertilization. A variety of other reproductive methods, however, exists among the insects; these are listed below.

OVOVIVIPARITY. In this type the eggs are retained in the female's genital tract until embryonic develoment is complete. The female then deposits a larva or nymph instead of an egg. The embryo is nourished solely by the yolk initially deposited in the egg. Examples are found in some Thysanoptera, Blattodea, Coleoptera, Diptera (some Tachinidae and Muscidae), and other taxa.

VIVIPARITY. In contrast to the above, the embryo of viviparous insects is fed by the female after development has begun. Three types have been recognized: (1) *Adenotrophic viviparity* is found in the tsetse fly, *Glossina* (Muscidae), and several other ectoparasitic flies, Hippoboscidae, Nycteribiidae, and Streblidae. One embryo at a time is carried by the female and deposited as a larva, which soon pupates. After the egg yolk is consumed, the larva ecloses from the thin chorion and is fed orally by special glands of the genital chambers. (2) *Pseudoplacental viviparity* involves eggs that have little yolk and the embryo is nourished by the wall of the genital tract, but not orally. A placentalike organ for the transfer of nutrients may be developed by maternal or embryonic tissues or both. Examples are found in Hemiptera (Polyctenidae, Aphidoidea), Dermaptera (Suborders Arixeniina and Hemimerina), and *Diploptera* (Blattodea) and *Archipsocus* (Psocoptera). (3) *Hemocoelous viviparity* occurs in all Strepsiptera and in the paedogenetic larvae of certain Diptera (Cecidomyidae) to be discussed below. In these insects the ovaries disintegrate and the embryos in their egg membranes take their nourishment directly from the maternal tissues. The female is consumed as a result.

PAEDOGENESIS. Reproduction by larval insects occurs in the beetle *Micromalthus* (Micromalthidae) and the flies *Miastor, Mycophila,* and certain other Cecidomyiidae. The ovaries become functional, and the eggs develop parthenogenetically. The life cycles of these species are quite complex, involving a number of paedogenetic generations and occasional imagoes that may not be reproductive.

PARTHENOGENESIS. Development without fertilization occurs in at least some species in all orders that have been investigated except the Odonata and Hemiptera (Suborder Heteroptera). The types of parthenogenesis are classified by the sex produced, by the cytological mechanism, and by the frequency of occurrence in the species. According to the sex produced, three types are recognized: (1) *Arrhenotoky* is the parthenogenetic production of males. This is the sex-determining mechanism in all Hymenoptera, some Thysanoptera, and *Micromalthus* (Coleoptera). (2) *Thelytoky,* the parthenogenetic production of females, is the most common type. (3) *Amphitoky,* the production of both sexes, is known in certain aphids and cynipid wasps.

The cytological mechanisms fall into two main types: (1) In *apomictic* parthenogenesis the egg fails wholly or in part to undergo meiosis, resulting in no reduction in chromosome number and no opportunity for new gene combinations. Except for mutations, the offspring retain the genes of the mother. This is the most common type of mechanism in insects. (2) In *automictic* parthenogenesis, the egg undergoes meiosis and the diploid condition is restored in a variety of ways, i.e., by fusion of two cleavage nuclei, two polar bodies, or a polar body with the egg pronucleus. The last is probably the most frequent method. A given lineage will become increasingly heterozygous or homozygous depending on which nuclei fuse.

Parthenogenesis may be *facultative,* i.e., the eggs if fertilized develop normally and if not fertilized develop parthenogenetically. Thus the eggs of Hymenoptera are facultatively arrhenotokous. *Obligatory* parthenogenesis involves only thelytoky; males are rare or altogether lacking in the population. Such populations are often morphologically closely similar to a normal bisexual population and apparently arose as a parthenogenetic race. Without males or the need for mating, a thelytokous race can rapidly reproduce but at the expense of genetic variability. Examples are found in some Phasmatodea, Lepidoptera (Psychidae), and Coleoptera (Curculionidae). *Cyclical* parthenogenesis or heterogony involves the alternation of parthenogenetic and bisexual generations. Examples are found in the Hemiptera (Aphidoidea), Hymenoptera (Cynipidae), and Diptera (paedogenetic Cecidomyidae). Here the advantages of both methods are combined in an annual cycle, and the life histories may be complex. In aphids usually one or more parthenogenetic generation takes place in spring, followed by a bisexual generation that produces overwintering eggs. Cynipids alternate between a spring parthenogenetic generation and a summer sexual generation. In some species of aphids and cynipids, one type of female produces both males and females by amphitoky, while in other species the females are of two types: male producers and female producers. Adults of the paedogenetic cecidomyids appear when the fungus on which they feed becomes dry or otherwise unsuitable.

Most species of aphids live throughout the year on one species of plant. The typical cycle begins in the spring when the eggs hatch to produce the first generation, or fundatrices. These become alate females that reproduce by parthenogenetic viviparity. Several successive generations are produced by this means until autumn. As colonies become crowded, individuals may fly to other plants of the same species and establish new colonies. In autumn, alate males and apterous oviparous females are produced parthenogenetically. These mate, and the females lay eggs that overwinter. Some species of aphids, however, live on woody plants (primary hosts) in the autumn, winter, and spring, and fly to herbaceous plants (secondary hosts) in the summer. For example, the annual cycle of the apple-grain aphid, *Rhopalosiphum padi,* is shown in Fig. 4.8. The fundatrices (*a*) are found in the spring on various trees, including apple, pear, and plum. The fundatrices and most of the second generation (*b*) are apterous, but by the third generation, alates are produced (*c*). These migrate to various grasses such as the grains, wheat, oats, and corn, or weedy grasses. New colonies of apterous aphids (*d*) are established by parthenogenetic viviparity. When the colonies become crowded, alates (*e*) are again produced, and they fly to new grasses. In autumn, alate males (*g*) and females (*f*) are produced that fly to the primary host. The females produce apterous females (*h*) that mate with the alate males. The apterous females are oviparous and lay eggs (*i*) that overwinter in diapause. The role of light in controlling some aspects of the aphid's life history is discussed in Chap. 11.

POLYEMBRYONY. In certain Strepsiptera and Hymenoptera (Chalcidoidea, Braconidae, Dryinidae), a single egg results in two or more individuals. These insects are endoparasites of other insects. Polyembryony permits a large number of offspring, sometimes several thousand, to emerge from oviposition on one host (Fig. 14.3*b*).

FUNCTIONAL HERMAPHRODITISM. Individual insects are sometimes found or experimentally produced that have both male and female characteristics.

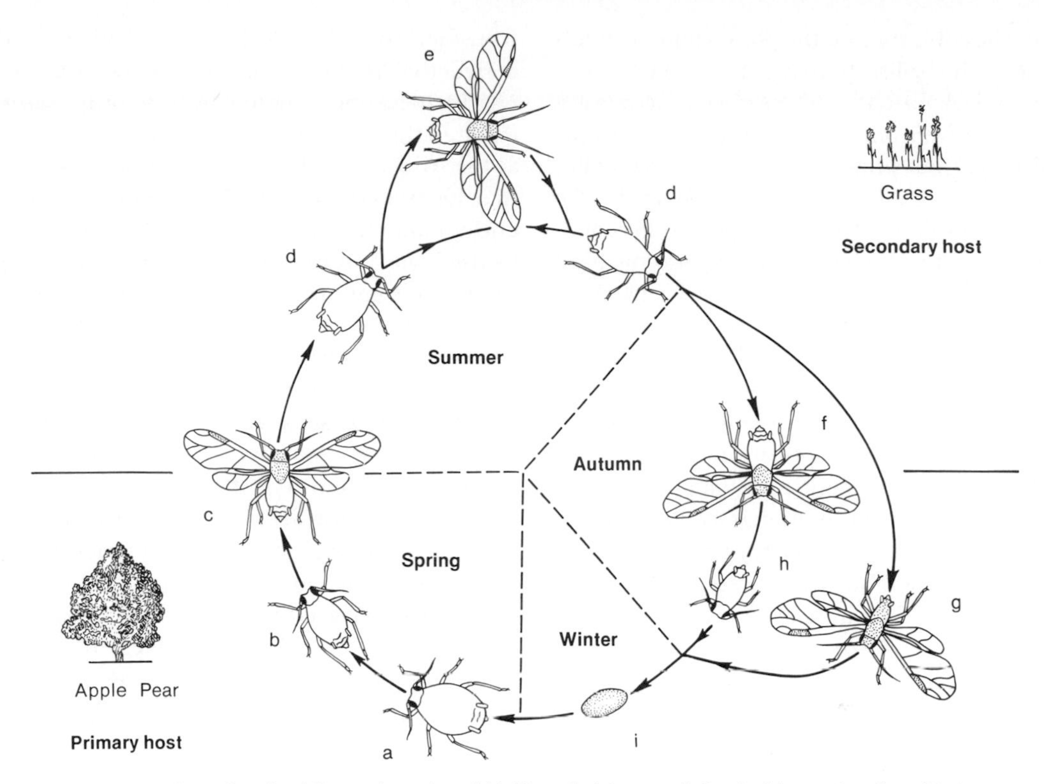

Figure 4.8 Annual cycle of the apple-grain aphid, *Rhopalosiphum padi.* See text for explanation. (*Redrawn from Dixon,* 1973, *by permission of Edward Arnold Publishers, Ltd., London.*)

Gynandromorphs are genetic mosaics of male and female tissues derived from a zygote plus other, genetically different, nuclei. Tissues of *intersexes* have the same, but unstable, genetic composition, resulting in the differentiation of male and female features. Gynandromorphs and intersexes are not normal in any insect species. In the cottony-cushion scale *Icerya purchasi* (Coccoidea) and related species, however, a *functional hermaphroditism* does occur normally. Both male and female gonads develop in the female scale, and the eggs are self-fertilized. Haploid male scales are rarely produced, and no pure females are known.

PHASE POLYMORPHISM

We conclude this chapter with a discussion of phase polymorphism in those species of acridids known as locusts: *Anacridium aegyptium, Dociostaurus maroccanus* (Moroccan locust), *Locusta migratoria* (African migratory locust), *Nomadacris septemfasciata* (red locust), and *Schistocerca gregaria* (desert locust).

Polymorphism in the broad sense means the presence of two or more forms of an organism. The seasonal forms of aphids described above are examples of polymorphism. In locusts, individuals of each species may develop into one of a spectrum of adult forms or *phases,* depending on the circumstance under which the young nymph or hopper grows to maturity. The extreme forms are called the *solitary* phase and the *gregarious* phase. The visible differences are in coloration, proportions of body parts, and behavior, but these are only reflections of underlying physiological differences.

Hoppers of the solitary phase of *Schistocerca,* for example, are uniformly colored green, whereas the gregarious hoppers have a black pattern on a yellow or orange background. In the adults, the coloration of the phases is more or less reversed: the solitary phase has a more pronounced dark pattern and the gregarious phase a less pronounced one. Furthermore, male adults of the gregarious phase continue to change color, but the solitary adults do not. Structurally, the head of the solitary locust is narrow, the pronotum is convex or arched in profile, and the hind femur is relatively long. In the gregarious phase the head is wider, the pronotum is depressed in profile, and the hind femur is relatively shorter.

In terms of behavior, the solitary hoppers spend relatively less time walking and walk at a slower rate, whereas the gregarious hoppers form groups that march together for long periods and at a faster rate. The most dramatic difference is in the adult behavior. The solitary adults remain in the vicinity of their birthplace, but the gregarious adults aggregate in immense swarms and take flight. Swarms of *Schistocerca* may have up to 10,000 million or more individuals. Over a season, a swarm may migrate for distances totaling up to 2000 mi. The direction of flight is with the prevailing winds that carry them to areas of recent rainfall and new plant growth. To sustain this flight, locusts may consume their body weight in green plants in a day. Large swarms may weigh up to 50,000 tons or more. When a large swarm settles on the ground, vegetation and crops are completely defoliated.

Locust plagues are among the most feared of natural disasters. Locusts are figured in Egyptian tombs as early as the sixth dynasty (2345–2181 B.C.), and the Bible contains vivid references to them (e.g., Book of Joel). Famine and disease epidemics with high mortality often followed the devastation of crops. It has been only through international cooperation and research that we are beginning to understand the biology and ecology of locusts.

Until 1921, entomologists considered the migratory locusts to be several species that were distinct from solitary acridids. About this time and almost simultaneously, J. C. Faure in South Africa, V. Plotnikov in Central Asia, and Boris Uvarov in the Caucasus of Russia came to the same independent conclusion: each species of locust may exist in two phases, of which one is the destructive migratory form and the other is a relatively harmless, nonmigratory form. Largely through the research of Uvarov and his associates at the Anti-Locust Centre in London, it is known that locusts are influenced in their development and sexual maturation by the presence of other locusts (see Uvarov, 1966). When hoppers are reared in isolation they develop into the solitary phase, but when as few as two hoppers are reared together the development is shifted partly toward the characteristics of the gregarious phase. Up to a certain density, increases in the number of hoppers reared together increases the shift toward characteristics of the gregarious phase. Beyond a certain density, no further change is noticed. As a result a continuous series of adult forms can be reared experimentally.

Changes also take place in the adult stage when locusts are placed together. If sexually immature gregarious females and males are placed with sexually mature males, their rate of maturation is increased, and the young males also change color from pink to yellow. Loher (1960) found that the epidermal cells of mature males secrete a volatile pheromone that is stimulating to other locusts, even at a distance. The pheromone stimulates the yellowing of males and the maturation of both sexes. Crowding of hoppers and adults, therefore, increases the sensory stimuli (probably including tactile, visual, and auditory stimuli as well as olfactory stimuli) that, acting through the endocrine system, promote the development of the gregarious phase.

Certain geographic areas, known as "outbreak areas," have been identified from which migratory swarms regularly emerge. These are usually mixtures of habitats favorable to oviposition and to food and shelter. At intervals several years apart, however, fluctuations in weather lead to a reduction in the size of the favorable areas. When such a reduction follows a period of population buildup, the locusts are concentrated and the gregarious phase is produced.

Similar phase changes in response to crowding are known in about two dozen other species of acridids, but the gregarious phase is less regularly produced. As noted in Chap. 1, *Melanoplus sanguinipes* (Rocky Mountain locust) developed migratory forms during the late nineteenth century in North America. Now the outbreak areas presumably have been altered by agriculture, and the species is no longer migratory. Physiological changes in response to density are also known in phasmatids and caterpillars.

SELECTED REFERENCES

The classic texts are by Johannsen and Butt (1941) on general insect embryology and by Hagan (1951) on viviparous insects. Recently both embryology and development are reviewed by various authors in the volumes edited by Counce and Waddington (1972, 1973). A phylogenetic perspective is given by Anderson (1973). Special reviews are given certain aspects of eggs and embryology by Hinton (1969*b*, 1970) and Tremblay and Caltagirone (1973); postembryonic development by Agrell and Lundquist (1973), Ashburner (1970), Berridge and Prince (1972), Chen (1971), Edwards (1969), Gilbert and King (1973), Hinton (1963*a*, 1971), Ilan and Ilan (1973), Lawrence (1970), Smallman and Mansingh (1969), Thomson (1975); hormonal action by Chippendale (1977), Menn and Beroza (1972), Novák (1975), Staal (1975), Wigglesworth (1970), and Willis (1974); and aging by Rockstein and Miquel (1973). Genetics of sex determination is thoroughly discussed by White (1973). Environmental and physiological effects on sex are reviewed by Bergerard (1972). Anatomy of the internal reproductive organs is described in detail by Snodgrass (1935) and Chapman (1969). The physiology of reproduction is comprehensively reviewed by Engelmann (1970). Other useful references on reproduction are Baccetti (1972), Davey (1965), Jones (1968), Leopold (1976), Schaller (1971), Telfer (1975), and deWilde and de Loof (1973*a* and *b*).

5  MAINTENANCE AND MOVEMENT

This chapter is devoted to the nutrition of insects and the various organs that are involved in the general flow of nutrients and oxygen into the insect's body, the utilization of energy to produce movements, and the elimination of waste products (Fig. 5.1). After a discussion of the nutrients required by insects, we treat the structure and function of the organs as follows: the alimentary canal, digestion, and absorption; the tracheal system and respiration; the circulatory system; the excretory system; and the muscles and locomotion of insects.

NUTRITION

Nutrition includes both the kinds and amounts of chemical constituents needed for building tissue and energy, as well as the physiological transformations, or metabolism, of these components. Knowledge of insect nutrition is of great practical significance in rearing large numbers of insects, so-called "mass rearing," for release in various pest-management strategies. Here we are concerned with origin, quality, and quantity of the specific nutrients: amino acids, vitamins, fatty acids, sterols, sugars, minerals, water, and certain other substances. Considering the infinite number of possible dietary combinations, the remarkable similarity throughout the animal kingdom in general nutritional requirements is evidence of the ubiquity of the same biosynthetic systems. In the study of insect nutrition, a sufficient number of different kinds of insects (about 40 species) has been surveyed to indicate a common need for essentially the same chemical nutrients. Yet the specific dietary requirements for a given species in terms of origin, molecular structure of nutrients, and relative amounts remain difficult to define. This is not only because of technical problems in experimentation, but also because of the dynamic nature of internal metabolism and the ecological relationships involved in feeding and nutritional symbiosis.

ANALYSIS OF NUTRITIONAL REQUIREMENTS. The analysis of requirements has been approached along several lines of inquiry. A comparison of ingested natural food with the egested feces is a logical first step. Yet the chemical composition of such a diet is poorly known, intestinal microbes may transform the food to other substances which are used by the insect, and that part of the excreta which is released (some may be stored) will be contaminated by microbial waste.

The next steps include devising chemically defined diets and eliminating other organisms that may contribute nutrients to the insect under investigation. *Meridic* diets consist of known chemical constituents with only one or a few substances of unknown structure or purity. *Holidic* diets are compounded from purified ingredients of known chemical structure. Though such a diet may be nutritionally adequate, insects may refuse to eat normally unless a suitable texture or taste is provided; the physical and chemical phagostimulants must be simulated, or some food substance of uncertain composition must be added for growth. Thus plant extracts or thickening agents are commonly added to a chemically defined mixture. Chemically defined diets have been formulated for the roach *Blattella germanica* (Blattellidae); the aphids *Acyrthosiphon pisum, Aphis fabae,* and *Myzus persicae;* the flies *Agria housei* (Sarcophagidae), *Drosophila melanogaster* (Drosophilidae), *Musca domestica* (Muscidae), and *Aëdes aegypti* (Culicidae); the beetles *Tribolium confusum* (Tenebrionidae), *Tenebrio molitor* (Tenebrionidae), and *Oryzaephilus surinamensis* (Cucujidae); and the wasp *Itoplectis conquisitor* (Ichneumonidae).

Microorganisms are frequently present in the intestinal tracts of insects, in their hemolymph, or even in special organs called *mycetomes* (Chap. 16). Many are believed to be symbiotic and are passed to succeeding generations by special devices, especially when the insect normally feeds consistently on nutritionally inadequate diets. For example, *Glossina* (Muscidae), *Rhodnius* (Reduviidae), *Cimex* (Cimicidae), and *Pediculus* (Pediculidae) feed exclusively on blood in all stages of life. Blood is deficient in water-soluble vitamins. Microbes are believed to supply these and other missing nutrients. Such insects, deprived of their normal microbes, fail to grow or may die, but will complete development if the diet is supplemented with vitamins. Female mosquitoes, on the other hand, suck blood only as adults and acquire sufficient nutrients during larval feeding to complete the life cycle. Plant-feeding Hemiptera that suck plant juices exist on a diet deficient in nitrogen and vitamins. Microorganisms probably supply vitamins and may aid in recycling nitrogen from waste products inside the insect. Wood-feeding termites depend on flagellate Protozoa in the hindgut to digest cellulose and release carbon compounds, probably acetic acid. Special experimental cultures of insects devoid of other organisms may be prepared for nutritional studies. Such cultures are termed *axenic.*

Insects may store nutrients to be utilized in a subsequent stage of the life history or even in the next generation. Eggs taken from well-fed parents may contain enough reserves of certain essentials to permit the resulting insects to survive on an experimentally deficient diet. Effects on offspring of the maternal nutrition are known in ants, locusts, and cockroaches. The use of stored food is a major evolutionary strategy in many Lepidoptera, Ephemeroptera, and some Diptera (Oestridae and Gasterophilidae) of which the adults do not feed. Adults of other Lepidoptera, Diptera, and Hymenoptera are also dependent to a large extent on reserves built up by the larvae, but additional nutrients are needed for egg production and longevity of females, as well as carbohydrates for energy.

Some nutrients can be replaced by others, such as many sterols for cholesterol, where the

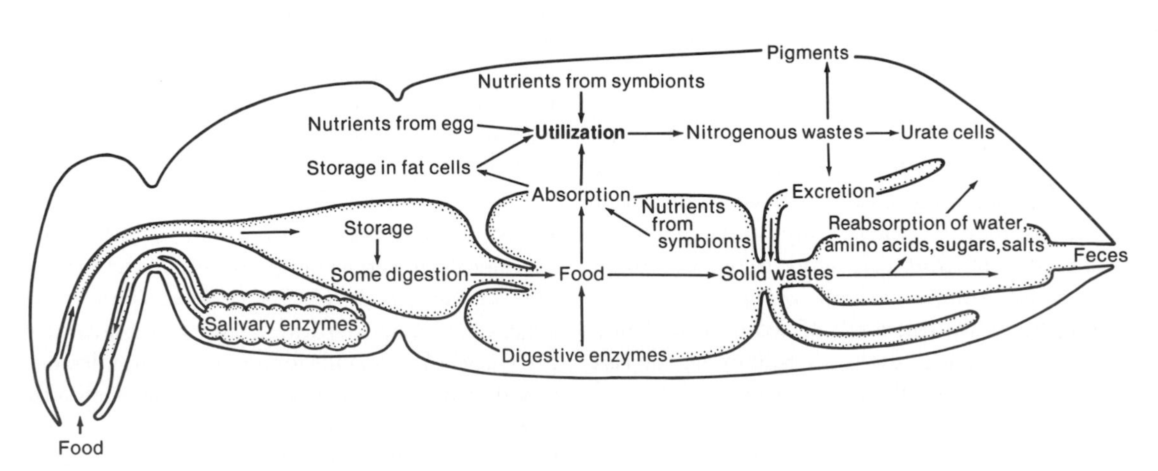

FIGURE 5.1 Diagram of routes of nutrients and wastes in an insect.

requirement is actually for only a part of the molecule. Substitutes or combinations may also replace certain functions of essential nutrients or have a "sparing action" when a structural similarity exists. Utilization of a diet depends not only on the chemical constituents, but also on the quantity of each with respect to others. Excesses of a single constituent may inhibit growth; deficiencies in one can decrease the use of others. The ability to synthesize essential nutrients is comparable to that in mammals with the exception of sterols, certain fatty acids, and possibly fat-soluble vitamins. Mammals can synthesize cholesterol, but insects require a sterol in their diet. A need for vitamins A (or precursors) and E has been recently demonstrated in insects. It is not known if insects can synthesize them. Insects do not appear to be able to synthesize polyunsaturated fatty acids such as linoleic and linolenic acids. Diptera and perhaps some Coleoptera require additional dietary nucleic acids for proper growth.

From an ecological and evolutionary viewpoint, dietary requirements are subject to change. Natural selection will favor genetic strains, for example, of a phytophagous insect that is able to thrive and reproduce on the unique combination of nutrients supplied by a specific host plant. When the success of the insect becomes detrimental to the survival of the plant population, the host may evolve defensive responses such as protective structures, lowered dietary value to the insects, antifeeding compounds, or toxins (Chap. 13).

WATER. Water is obtained directly by drinking or from moisture in food, or it is produced during oxidative metabolism and conserved. For example, pests in dry, stored grains have a higher water content than the food they eat. Some insects are able to absorb water from moist atmosphere, but this mode of acquisition is generally prevented by the waxy epicuticle; the risk of water loss is usually greater than the gain.

ENERGY. Energy is available from a variety of sources, mostly carbohydrates in plant tissue, nectar from flowers, or honeydew (plant sap) excreted by aphids, etc. Glucose, fructose, and other hexoses are utilized, but pentoses and hexose sorbose usually are not. Energy from sugars is often stored in the fat body as glycogen. Larvae of Diptera require little sugar during growth, but adult Diptera and Hymenoptera require sugar to fuel the flight muscles. Insect flight muscle converts more energy per unit weight than any other animal tissue. For this reason flies, wasps, and bees frequent flowers for sucrose-rich nectar. Long-distance fliers (many Lepidoptera, locusts, Coleoptera, and the belostomatid bug, *Lethocerus*) usually consume their stored sugars and then utilize fat. The latter is suited as a reserve energy source because it yields twice as many calories as would the same weight of carbohydrate. Fat is frequently stored in the fat body of larval insects for later use during pupation, energy for nonfeeding adults, or development of eggs in females.

AMINO ACIDS. Amino acids are required for building the tissues of a growing insect. Of the 21 amino acids in plants and animals, insects require the same essential 10 amino acids as the white rat: arginine, histidine, isoleucine, leucine, lysine, methionine, phenylalanine, threonine, tryptophan, and valine. Some growth-promoting response has been obtained in certain insects by including in the diet one or more of the following: alanine, aspartic acid, cystine, glutamic acid, glycerine, hydroxyproline, proline, serine, and tyrosine. All the above amino acids listed are the L isomers, but some insects can substitute with the D isomers of methionine, phenylalanine, and histidine.

LIPIDS. Lipid requirements include fatty acids and sterols. Reserve energy sources in the form of fat, mentioned above, are derived from either sugars or dietary saturated and monosaturated fatty acids. Fatty acids for energy are not deemed essential because alternate sources of energy are available. Polyunsaturated fatty acids, on the other hand, are involved in the formation of phospholipids of all cell membranes and are dietary essentials for most insects because insects are peculiarly unable to synthesize them. Dietary linoleic and linolenic acids usually satisfy this need, which tends to manifest a deficiency late in the insect's life (at pupation in many Lepidoptera; in reduced fecundity of adult boll weevils) or in the next generation because the eggs are inadequately provisioned (cockroaches). Diptera have a unique lipid metabolism not dependent on polyunsaturates. Their major fatty acid is palmitoleic, and their phospholipids are of the ethanolamine phosphoglyceride type.

In contrast to mammals, insects are unable to synthesize sterols. Besides other functions, a sterol is required for the production of the growth hormone ecdysone. Foods vary in the predominant kind of sterol: zoophagous insects normally acquire cholesterol and 7-dehydrocholesterol, whereas omnivores and phytophagous insects fill this requirement with plant β-sitosterol or microbial ergosterol. Regardless of the origin, the presumed biogenetic pathway to ecdysone involves cholesterol to 7-dehydrocholesterol. Cholesterol is ordinarily adequate in an artificial diet, but some insects with highly specific diets have apparently lost the ability to convert cholesterol: the fly *Drosophila pachea* requires the cactus sterol, schottenol, from its host *Lophacereus schotti*.

VITAMINS. Vitamins are substances essential for living processes but are not used for energy or basic structures. They may be usefully classified in two broad categories: water-soluble and fat-soluble. Until relatively recently, insects were thought to need only water-soluble vitamins. Among the fat-soluble vitamins are D (calciferol), involved in calcium metabolism; E (α-tocopherol), an antisterility factor for mammals; A (retinol) or precursors, involved in carotenoid visual pigments; and K (phylloquinone, etc.), essential for clotting mammalian blood. Vitamin D is not required because insects lack calcareous skeletons and are apparently unable to use it even as a sterol. The clotting of insect blood differs from that of mammals and does not involve vitamin K. Vitamin E deficiencies in insects have been difficult to demonstrate consistently, possibly because the adverse effects are manifested late in the insect's life during reproduction. Evidence is now at hand that both egg and sperm production, as well as larval growth, may be affected. A need for vitamin A has long been suspected because the visual system of insects includes carotenoid light-sensitive reactions. Yet until recently, little or no physiological effect could be detected in insects reared in its absence. Now morphological and visual defects have been definitely confirmed in several species.

Water-soluble vitamins that are essential to many insects are seven B-complex vitamins: biotin, folic acid, nicotinic acid, pantothenic acid, pyridoxine (B_6), riboflavin (B_2), and thiamine, (B_1), plus ascorbic acid (C) and choline. Some insects additionally require the following, or respond by increased growth or survival: carnitine (B_T), cyanocobalamin (B_{12}), inositol, and lipoic acid. Another growth-promoting substance, ribonucleic acid (RNA) or an equivalent mixture of components, is essential or increases growth in several species of Diptera and Coleoptera.

MINERALS. The mineral requirements of insects are approximately the same as those of vertebrates, because minerals, of course, cannot be synthesized. Thus the salt elements that are essential to at least some, but not necessarily all, insects that have been tested include calcium, chlorine, copper, iron, magnesium, manganese, phosphorus, potassium, sodium, sulfur, and zinc. In comparison to mammals, insects require less calcium (not needed for the skeleton), less iron (not needed for hemoglobin), and a higher proportion of potassium to sodium.

ALIMENTARY CANAL, DIGESTION, AND ABSORPTION

The alimentary canal of an insect is simply a tube passing through the body that provides the special internal environment for food to be mechanically and chemically disintegrated and brought to the vicinity of absorbent cells. Despite the diversity in physical consistency and composition of food, the gut exhibits certain general anatomical features throughout the insects (Figs. 5.2, 5.3).

GENERAL STRUCTURE AND FUNCTION. The gut is attached at each end to the body wall, but otherwise is positioned in the hemocoel largely by the pressure of adjacent organs and flexible tracheae. The three major divisions may be traced to their embryonic origins: (1) an anterior invagination of ectoderm, forming the *stomodaeum* or *foregut* (Fig. 5.2*a*); (2) a connecting growth, probably derived from endoderm, forming the *mesenteron,* or *midgut;* and (3) a posterior invagination of ectoderm, forming the *proctodaeum,* or *hindgut.* The wall of the alimentary canal is composed of a single layer of cells covered by a basement membrane. The lumina of the ectodermal derivatives, the foregut and hindgut, are lined with an intima (thin cuticle) that is continuous with the integumental cuticle. These portions of the gut are properly classed as organs of the integument, and the cell layer is homologous with the epidermis. The intermediate section, the midgut, is differentiated by its endodermal origin and the lack of an adherent intima protecting the epithelial cells. In place of an intima, however, is secreted periodically in some insects a loose film, the *peritrophic membrane,* that, in addition to other functions, protects the delicate cells.

Muscular connections between the gut and exoskeleton are limited to near the mouth and pharynx, where food is wrested from the outer world. Once within the gut lumen, the food is moved, mixed, and sometimes further crushed or triturated by the action of variously layered muscles surrounding the gut. The proventriculus

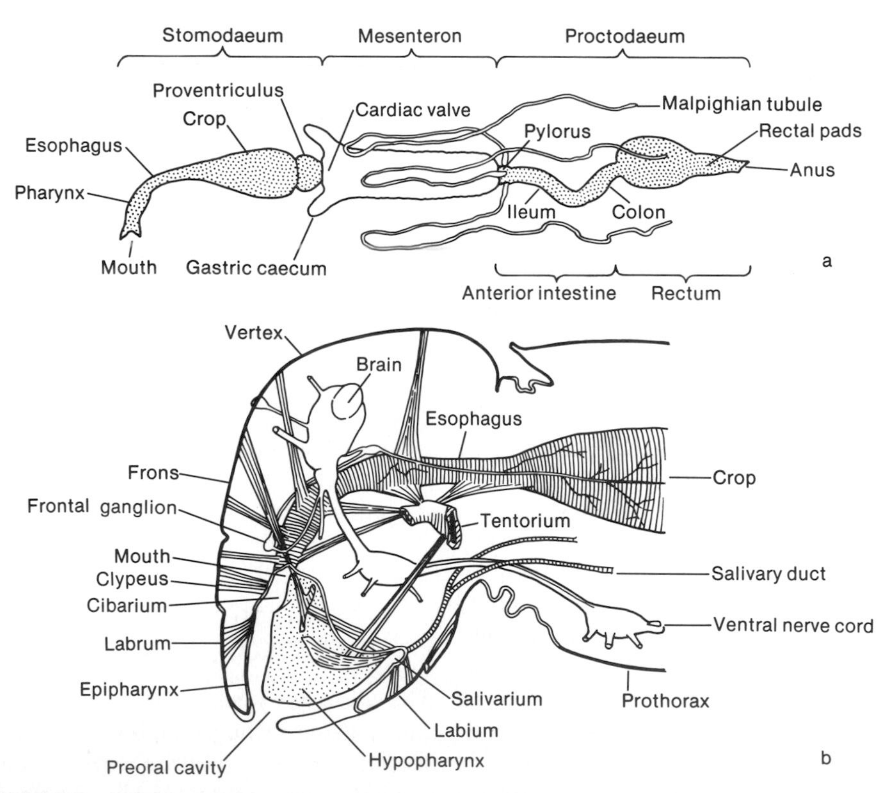

FIGURE 5.2 Alimentary canal: *a*, general structure; *b*, mouth and foregut. (*Redrawn from Snodgrass,* 1935, *by permission of McGraw-Hill Book Company.*)

and cardiac valve between the foregut and midgut, and the pyloric valve between the midgut and the hindgut regulate the flow of food (Fig. 5.3*b,c*). Anteriorly, the peristaltic movements of the gut are controlled by sensory and motor innervations from the stomodaeal nervous system (Chap. 6) to the foregut and, in some insects, the midgut. The hindgut receives nerves from the posterior abdominal ganglia of the ventral nerve cord. The scarcity of sensory nerves probably means that insects rarely suffer stomachaches. Tracheae from visceral trunks supply oxygen to the alimentary canal according to the metabolic needs of the different regions. Inserted at the juncture of the midgut and hindgut, in the same place as the pyloric valve, are the filamentous excretory organs called *Malpighian tubules*. Nitrogenous wastes are secreted from these tubules into the hindgut.

The functional organization of the gut parallels the anatomical subdivisions. Food is ingested by the mouth and pushed caudad by the muscular pharynx into the esophagus and on to the crop. Here the food is stored and often partly digested by enzymes in the saliva or regurgitated from the midgut. Absorption is prevented by the intima. In passing to the midgut, the food particles may be further mechanically degraded by the grinding action of the proventriculus. The midgut is the principal site of enzyme production and action. The digested products filter through the peritrophic membrane and are absorbed by the epithelium. The residue is moved caudad to join the secretions of the Malpighian tubules beyond the pyloric valve. Before this substance is egested as feces, the hindgut completes the selective reabsorption of useful components such as water, sugars, salts, and amino acids.

The alimentary canals of primitive insects such as Collembola as well as many larvae of Endopterygota are simple, direct tubes with but slight modifications even in the regions of the cardiac and pyloric valves. More advanced chewing insects, including Orthoptera, Odonata, Hymenoptera, and many Coleoptera, have an enlarged crop and proventicular specializations to handle particulate meals, plus frequent caecal pouches and enlarged rectal pads. The ingestion and passage of a fluid diet poses other engineering problems. Some fluid-feeders, such as some Diptera and Lepidoptera, possess lateral, bladderlike diverticula protruding from the crop that receive liquids. The midgut is enlarged in Siphonaptera and many Hemiptera (Suborder Heteroptera). The most elaborate arrangement is the filter chamber of many Hemiptera (Suborder Homoptera) which has evolved to shunt excess plant sap from the anterior part of the midgut to the region of the pyloric valve and Malpighian tubules.

ANATOMY OF THE ALIMENTARY CANAL. To discuss the alimentary canal in detail, we begin with the space enclosed by the mouthparts, the *preoral cavity* (Fig. 5.2*b*). Recall that the true mouth is recessed and hidden from external view. The cavity is closed anteriorly by the labrum and clypeus. The inner wall of these sclerites forms the roof of the cavity, or epipharynx. To each side the cavity is limited by the mandibles and maxillae. The floor or posterior wall is formed by the labium. Projecting into the cavity is the hypopharynx.

The mouth opens into the preoral cavity above the hypopharynx and below the epipharynx. The portion of the preoral cavity in front of the mouth is the cibarium. It is here that food is passed to the true mouth. Muscles from the clypeus which attach to the epipharynx exert a dilating action on the cibarium. The preoral cavity also contains another chamber beneath the hypopharynx and above the labium. This is the salivarium, where the saliva is emptied into the cavity from the labial or salivary glands.

The first distinctive region of the foregut is the *pharynx*. This heavily muscled organ is situated behind the mouth. In addition to the aforementioned dilator muscles from the clypeus and frons, the pharynx has an inner layer of longitudinal muscles and an outer layer of circular muscles. Acting together, these muscles push the food backward through the esophagus to the crop. The *esophagus* is merely a tubular portion of

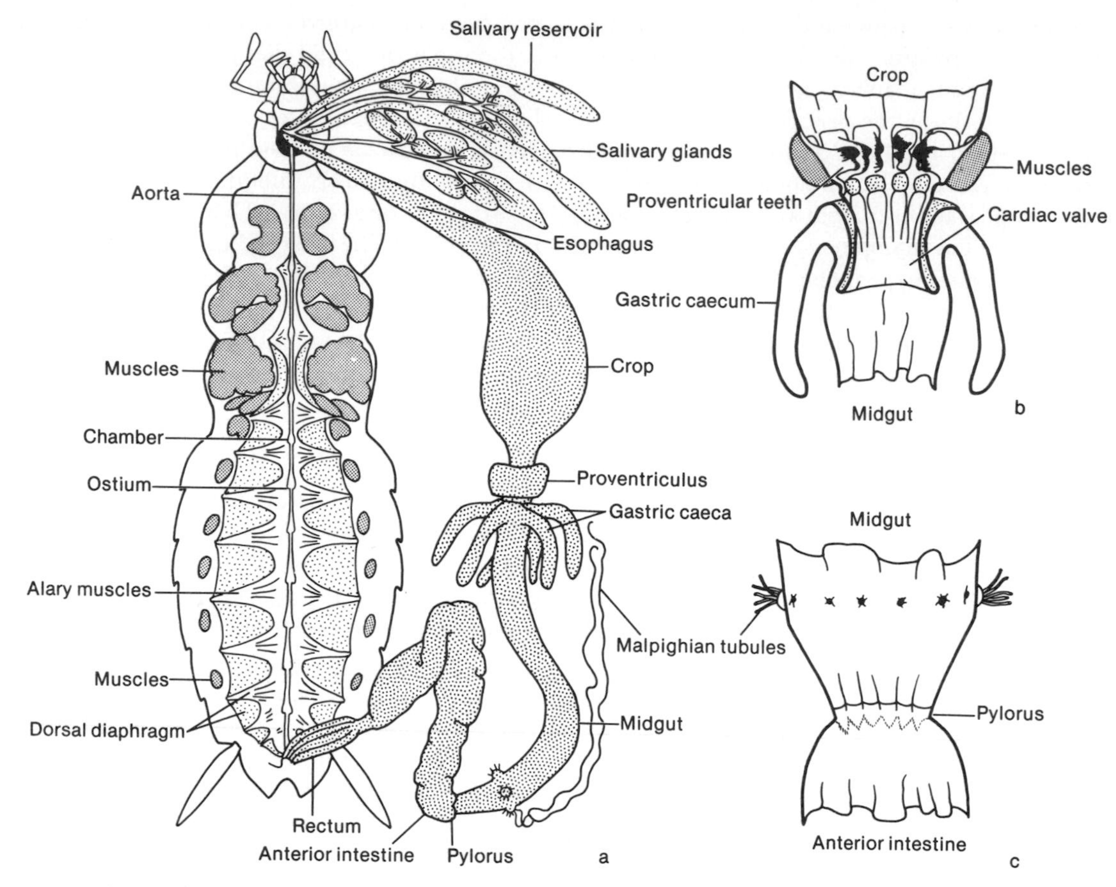

FIGURE 5.3 Alimentary canal and dorsal vessel of the roach *Periplaneta americana* (Blattidae): *a*, ventral view of dissected roach with fat body, reproductive organs, and most Malpighian tubules removed; *b*, interior of proventriculus and cardiac valve; *c*, interior of pylorus.

the foregut leading to the saclike *crop* (Fig. 5.3*a*). No definite boundaries define the limits between the crop and the esophagus; they merge imperceptibly into one another. All parts of the foregut are lined with a flexible, but generally impermeable, cuticular intima. No enzymes are produced by the epidermal cells, but the salivary enzymes as well as enzymes from the midgut can act on the food while it is temporarily stored in the crop. Absorption of the digested products, however, is sharply limited by the intima. For this reason, the foregut is probably no more absorptive than a portion of the outer integument.

The release of food from the crop is controlled by an intricate organ, the *proventriculus*. Like the pharynx, the proventriculus may have a heavy muscular coat. Internally the cuticle is often shaped into heavy teeth, plates, or spines, but in fluid-feeding insects and many larvae, it is a simple sphincter. The teeth act as a filter, regulating the passage of food posteriad and permitting digestive enzymes from the midgut to enter the crop. They are also believed to grind the food further into smaller particles. The last part of the foregut is invaginated in the midgut. This is the *cardiac valve* (Fig. 5.3*b*), but the extent to which it actually functions as a valve is questionable. Special secretory cells between the valve and the midgut wall are partly responsible for the formation of the peritrophic membrane.

Except in some true bugs, the midgut, or ventriculus, is rarely subdivided into distinctive regions which can be recognized externally. The midgut of the bugs is associated with the hindgut in a filter chamber. The details of this arrangement will be discussed below (Fig. 5.7*c*). In insects such as grasshoppers, true bugs, and larval scarab beetles, hollow, fingerlike growths may be present at the anterior or posterior ends of the midgut. These are the gastric caeca and are known to shelter symbiotic bacteria. When compared with their positions in the foregut, the positions of the muscular layers are reversed, the circular layer being innermost and the longitudinal layer outside. Internally, the epithelium of the midgut may be differentiated into distinctive histological regions. The columnar epithelial cells are especially active, since the secretory processes which yield the digestive enzymes often involve the eventual disintegration and rejuvenation of the cellular lining. Next to the lumen, the cells characteristically are covered with short, dense, hairlike filaments which in their sections are visible as a striated border. The cells may be classified histologically as (1) goblet cells, enclosing deep invaginations with a striated lining; (2) secretory cells which produce digestive enzymes, with or without obvious vacuoles, vesicles, or apparent disintegration; and (3) regenerative cells, distributed singly (Diptera, larval Lepidoptera), in groups of smaller cells called *nidi* (Orthoptera, Odonata, Plecoptera), or in papillate crypts which project from the outer surface of the midgut (Coleoptera). Secretory cells lost during enzyme production or by molting are presumably replaced by regenerative cells. Goblet cells may represent declining secretory cells. Substances may be accumulated in epithelial cells and discharged during molts, so that excretory functions are performed. Artifacts introduced during histological preparation have greatly complicated the interpretation of the secretory and molting cycles of epithelial cells.

At the junction of the midgut and hindgut are slender outgrowths, the *Malpighian tubules.* These are closely associated with the anterior end of the hindgut and serve as a convenient external marker to distinguish the posterior end of the midgut. Also in this area is the pyloric valve (Fig. 5.3*c*). The valvelike function is best developed in caterpillars and beetles.

The hindgut may have several recognizable sections: (1) an *anterior intestine,* composed of the *ileum,* which is larger in diameter, and a narrow posterior portion, the *colon;* and (2) the *posterior intestine,* or *rectum,* which is the final section before the anus. The ileum is modified to contain symbiotic microbes in termites and larval scarab beetles. Each section of the hindgut is lined with cuticle, but this lining is permeable to water, salt, sugars, and amino acid molecules. Together with the Malpighian tubules, the ileum and rectum form parts of the excretory system. The epidermis of the *rectal pads* is specially developed into columnar cells that are heavily tracheated to supply oxygen for active metabolism. Here the feces are dehydrated, and useful ions and amino acids are reclaimed. The sparse muscular coat of the hindgut is arranged in the fashion of the midgut, with the circular muscles outermost. In the rectum the waste is dehydrated and compressed into a pellet which bears the shape of the rectal walls.

In the aquatic, immature stages of dragonflies, tracheated gills project into the rectum and serve to acquire oxygen from water circulated in the enlarged rectum. Forceful contractions of the rectum also eject the water and propel the nymphs forward.

SALIVARY GLANDS. The enlarged integumental glands of the head function in a great variety of life processes, but are discussed here because they contribute generally to ingestion, movement of food, and digestion. The largest and most common glands which secrete saliva are the paired *labial glands* (Fig. 5.3*a*). The single, median salivary duct opens in the salivarium between the labium and hypopharynx. The glandular cells extend back into the thorax, or even to the abdomen in some species, like clusters of grapes, and may be differentiated into regions. In addition to saliva, the modified labial glands are responsible for the silk secreted by larvae of

Lepidoptera, Trichoptera, and Hymenoptera. These fibrous proteins, including collagen structures, are used in construction of cocoons, webs, and nets. Other glands of the head found in various insects are (1) *mandibular glands,* which are the main salivary glands of larval Lepidoptera and are also known to secrete the pheromone called *queen substance* in the queen honeybee (Chap. 7); (2) *maxillary glands,* which are small and probably provide lubricants for the mouthparts; and (3) *pharyngeal glands,* which are best developed in worker honeybees for production of the food royal jelly, fed to larvae.

Functionally the saliva of the labial glands moistens and lubricates the mouthparts, dissolves food, transports flavors to gustatory (taste) sensilla, contains enzymes that act on the food both before and after ingestion (Fig. 5.1), and also serves as an excretory organ. The housefly ejects saliva on solid food, and the partly digested liquid is eaten. Enzymes such as amylase (which converts starch to maltose) or invertase (which converts sucrose to glucose and fructose) are commonly present. The honeybee adds salivary invertase to nectar that is carried back to the hive in the crop. The sucrose-rich nectar is converted to honey by dehydration of the resultant glucose and fructose mixture. Aphids, mirids, a leafhopper, and a lygaeid bug are aided in the penetration of their piercing-sucking mouthparts into plants by a pectinase in the saliva. This enzyme hydrolyzes the pectin in the cell walls of plant tissues. Hemiptera (Suborder Homoptera) and the stinkbugs (Pentatomidae) and their allies secrete a salivary sheath around their feeding punctures. The sheath presumably functions to retain the turgor pressure of the plant's vascular system, thus enhancing the insect's sucking mechanism.

Bloodsucking insects, such as the tsetse fly *Glossina* (Muscidae), some mosquitoes, and parasitic Hemiptera, facilitate ingestion by the inclusion of anticoagulins and antiagglutinins in the saliva. Blood clots would otherwise plug the delicate canals of the mouthparts. Digestive enzymes are notably absent in their saliva, perhaps to avoid disturbing the host while they feed.

Paralysis and extraintestinal digestion of prey are accelerated when assassin bugs (Reduviidae) inject hyaluronidase into prey. This spreading agent probably dissociates the cells of the prey and hastens the action of digestive enzymes which are concurrently injected. The bug then ingests the liquefied meal. The venomous saliva can also be defensively ejected at vertebrate enemies.

The injection of saliva by insects into plant or animal hosts provides a natural entry for pathogens. Complex cycles of transmission have been evolved by viruses and microbes which take advantage of the insect vector's feeding behavior (Chap. 16).

PERITROPHIC MEMBRANE. This is a thin, loose film in the midgut and hindgut containing both chitin and protein. Structurally it is a fibrous network, or a laminar or amorphous sheet partially permeable to both enzymes and the products of digestion. The membrane presumably functions to protect the exposed midgut epithelium from abrasion by food particles, and to lubricate and facilitate movement of food; it may also act as a filter controlling the movement of molecules as well as microbes between the gut lumen and epithelial cells. The lubricating role of the membrane seems likely, because insects lack mucus, which serves this purpose in other animals. The protective function is indicated by the general presence of the membrane in insects that ingest particulate food and the general absence in fluid-feeders. Exceptions in both cases exist. Orders which commonly possess the membrane include Collembola, Thysanura, Ephemeroptera, Odonata, Orthoptera, Isoptera, Neuroptera, Coleoptera, Hymenoptera, Diptera, and larvel Lepidoptera. It is absent in Hemiptera, adult Lepidoptera, and carabid beetles.

The membrane is secreted by either or both of two methods. Secretory cells located between the invaginated cardiac valve and the anterior end of the midgut extrude a tubular film around the incoming food bolus. In Diptera this is continuously secreted and results in a single membrane.

Other insects, including Ephemeroptera, Orthoptera, Odonata, Coleoptera, and larval and adult Hymenoptera, exhibit multiple membranes created by periodic delamination of the epithelial surface. Some of these orders and Dermaptera, Isoptera, and larval Lepidoptera produce the peritrophic membrane by a combination of cardiac press and midgut delamination. The ensheathing film continues caudad with the food and is ultimately discarded, still in place around the fecal pellets.

DIGESTION. As already mentioned, chemical breakdown of food may begin before ingestion by means of enzymes ejected in saliva or regurgitated from the midgut. The labial glands and midgut epithelium are the primary sources of digestive juices. Triturition by the mouthparts and proventriculus mechanically reduces the sizes of food particles and exposes larger surface areas of food to digestion in the crop and midgut. The hydrolytic action of the catalytic enzymes disintegrates the complex organic constituents in the food and stepwise simplifies the molecular structure to a degree absorbable by the epithelial cells of the midgut.

Proteases of insects often resemble trypsin of mammals and are active in neutral or alkaline environments. They split proteins at the peptide linkages. Differences in electrophoretic mobility (Chap. 9) and substrate preferences indicate that a number of components with proteolytic activity exist in individual insects. Peptidases, but not pepsin, have been identified in digestive juices and epithelial cells. These enzymes further hydrolyze the products of protein digestion, the peptones and polypeptides, to render dipeptides and free amino acids.

Structural proteins in vertebrate tissues, such as collagen of connective tissues or keratin in wool, hair, and feathers, are highly resistant to degradation by common proteolytic enzymes. This is evident when a dead bird or mammal decays. The feces of blowfly larvae feeding on such corpses contain a collagenase which attacks the collagen and elastin of muscle tissues, resulting in extraintestinal digestion. Bird lice (Mallophaga), the larvae of clothes moths (Tineidae), and dermestid beetles possess a keratinolytic enzyme which attacks keratin. A strong reducing agent is able to break the sulfur bonds of keratin because poor tracheation in the insect's midgut provides a low oxidation-reduction potential. Once the peptide chains are free, the keratinase is effective.

Lipases hydrolyze the triglycerides of plant and animal fats, yielding fatty acids and glycerol. Beeswax is ordinarily resistant to digestion, but the larval wax moth is able to utilize some wax components. At an optimal pH of 9.3 to 9.6, the wax is hydrolyzed by enzymes originating either from the larva's cells or from intestinal bacteria or both.

Carbohydrases are well known among the insects. Commonly identified are α- and β-glucosidase; α- and β-galactosidase, β-h-fructosidase, and amylase. The hydrolysis of a variety of glycosides, polysaccharides, and oligosaccharides seems to be possible by rather few enzymes. The nutritional requirements are usually limited to simple sugars, especially glucose, which is available from many sources. Unusual diets, however, are associated with special enzymes. Woody plant tissue is regularly exploited as a food source by larvae of some wood-boring anobiid and scarabaeoid beetles; termites; wood-feeding roaches; and silverfish. Cellulase, hemicellulase, lignocellulase, and lichenase have been demonstrated in insects, some or all of which have been identified in members of the above list. Microbes in the intestine doubtless play a role in producing these enzymes. Termites and wood roaches are entirely dependent on flagellate protozoa housed in the ileum. Bacteria in the midgut caeca of Scarabaeoidea beetles are thought to digest cellulose. Some of these enzymes, however, are directly secreted by insects, e.g., the larvae of some cerambycid beetles. Among wood-boring insects, the ability to digest cellulose to varying degrees permits partitioning of the food resource: some thrive only on the more easily digested, but scarce, starch and sugar in the wood (Lyctidae, Bostrichidae); others use also the hemicelluloses (Scolytidae); and a third group consumes the whole cell wall, minus the lignin (Anobiidae, Cerambycidae).

Not surprisingly, the array of enzymes produced by an insect or its intestinal biota matches the array of dietary components at the chemical level. Within the gut the insect is able to exert some control over the physical and chemical conditions which lead to digestion. A favorable temperature can be regulated by behavioral or metabolic activity. Hydrogen-ion concentration or pH is more highly correlated with taxonomy than with foodstuffs (midgut juices of Lepidoptera, 8.4 to 9.8; Coleoptera, 8.4 to 9.6; Diptera, 6.8 to 7.8; and Orthoptera, 5.6 to 7.2), but also differs along the length of the gut in an individual insect. Special oxidation-reduction potentials may be created to permit unusual digestive processes. Complex buffering systems involving mixtures of ions from various sources may stabilize the pH against the influence of incoming food. These factors combine to permit the insects to exploit an enormous range of food materials.

ABSORPTION. The midgut epithelium and the hindgut are the principal absorptive regions. Only limited evidence demonstrates absorption by the crop, which is protected by a waxy intima. The rate of absorption involves not only the passage of ions and molecules through cell membranes but also, and possibly more importantly, the release and transport of these substances to the vicinity of the cells. Movements of the gut and the release of food from the crop have been ascribed to neural controls centered in the stomodaeal nervous system or to the effects of osmotic pressures of the meal.

Monosaccharide sugars, commonly glucose, rapidly pass through the permeable epithelial membrane without an active transport mechanism. In insects a steep osmotic gradient is created by converting the simple sugar to the disaccharide trehalose. In the migratory locust this conversion takes place in the fat body which is disposed about the midgut. In the silkworm, oligosaccharides may be absorbed and further hydrolyzed in the epithelium.

Amino acids are similarly absorbed by passive diffusion processes related to the uptake of water into the hemolymph. In *Rhodnius* (Reduviidae) even proteins that are not fully degraded to amino acids may be absorbed and then digested further by the midgut cells.

The absorption of lipids is more difficult to analyze because it is possible that these molecules could pass through the cuticular intima of the crop or between epithelial cells of the midgut, in addition to the usual route via the epithelial cells. Lipoidal material has been seen as droplets in cells of both the midgut and crop. Although cholesterol has been recently demonstrated to be absorbed by the crop of a cockroach, the midgut is known to be the major absorptive region. The fats need not be fully degraded before absorption. No evidence exists that an active transport mechanism is involved.

The rectal pads of the hindgut are specialized for absorption of water and inorganic ions. Of unusual interest is the phenomenon that the movements of these molecules appear to be independent.

TRACHEAL SYSTEM AND RESPIRATION

Oxygen for cellular respiration in insects is physically transported by an internal system of air-filled tubes, the *tracheae,* to within a few cell diameters of each cell (Fig. 5.4*a*). In contrast to oxygen transport in vertebrates, which involves the pigment hemoglobin in red blood cells, neither respiratory pigment nor blood is involved in most insects. The respiratory system of terrestrial vertebrates has clearly evolved from that of aquatic ancestors, but the aerial systems of insects and certain other arthropods are thought to be peculiarly terrestrial adaptations that may have independently evolved in several arthropod stocks. Tracheae are ramifying and interconnecting tubular invaginations of the integument that arise from paired, lateral openings, the *spiracles,* and ultimately terminate in fine fluid-filled tubules, the *tracheoles* (Fig. 5.4*f*).

Small insects, such as Protura and most Collembola that live in damp places, or some aquatic larvae of Chironomidae (Diptera), lack a developed tracheal system and respire through the cuticle. Likewise, small Hymenoptera larvae living as internal parasitoids of other insects may depend on cutaneous respiration. In larger, free-living, terrestrial insects, the demand for oxygen cannot be met by the body surface, partly because the area is inadequate for sufficient transfer, but more importantly, because the thicker cuticle permits only a negligible diffusion of oxygen. The tracheal system creates an enormous internal surface of permeable cuticle, estimated in the fifth-instar larva of the silkworm to include 1.5×10^6 tracheoles.

In trachea diffusion plays the dominant role in oxygen transport and is a millionfold more rapid in the gas phase than in tissue. Calculations of oxygen consumption, diffusion rates, and measurements of tracheae indicate that diffusion alone could largely supply even active flight muscle. The distance that oxygen must travel over a gradient from atmospheric concentration to respiring tissue can be greatly diminished and the gradient steepened by ventilatory movements of drafts of air through the system.

The passage of air into the tracheal system is regulated by the spiracles. In some apterygotes, such as the springtail *Sminthurus* (Sminthuridae), the opening is simply a hole in the body wall. The rate of gas diffusion through this pore, as through the stomata of plant leaves, is a function of the perimeter rather than the cross-sectional area. Elaborate modifications of the spiracle, therefore, are possible without necessarily reducing the intake of air. In most insects the tracheal orifice is recessed in the integument, creating a small chamber, or *atrium,* which communicates with the exterior. The atrium may be situated on a separate sclerite, the *peritreme,* and provided internally with filtering hairs or projecting, valvelike lips, closable by muscles (Fig. 5.4*c*). The latter are characteristic of thoracic spiracles and are not found on the abdomen. The atria of abdominal spiracles remain open, with a closing device located inside at the tracheal orifice. The opening is pinched shut by the action of an occlusor muscle on one or two sclerotic bars. When the muscle relaxes, the elastic bars spring open, or they are pulled apart by a second muscle, the dilator. Spiracular muscles are innervated from their segmental ganglion or the ganglion immediately anterior. Tracheae leading from the spiracles may be connected longitudinally by *lateral trunks* and transversely by *tracheal commissures.*

The histological parts of tracheae are homologous to corresponding parts of the integument: an inner, cuticular intima secreted by a surrounding layer of epidermal cells that are bounded outside by a basement membrane. The intima is thin, permeable to gases, and marvelously flexible, because of fine, spiral thickenings called *taenidia* (Fig. 5.4*e*). The ultrastructure of the intima is still being debated. The epicuticle contains at least a cuticulin layer and probably a hydrofugic wax layer next to the lumen. Corresponding to the external procuticle is a chitin-protein layer which is largely (if not entirely in smaller tracheae) involved in constructing the taenidia. At each molt, portions of the intima are routinely shed along with the exuviae. During metamorphosis of flying insects the entire system is reorganized to accommodate the increased demand for oxygen by the flight muscles. New tracheal branches or fusions are developed, and old tracheal pathways are not replaced, yet the design of major trunks and anastomoses is consistent in taxonomically related forms and may be traced from the immature stages to the adult.

The morphogenesis of taenidia has been ascribed to physical strains on the new cuticulin layer as it is secreted by the epidermal cells. With increases in its surface area within the restraining epidermal tube, the plastic intima is buckled into a spiral configuration. Other investigators emphasize the role of the epidermal cells in creating the pattern.

The tubular network of tracheae may be interrupted at intervals by enlarged dilatations, or *air sacs* (Fig. 5.4*b*). Taenidia may be discernible on the walls of some sacs as if the trachea is merely inflated, but other cavities are

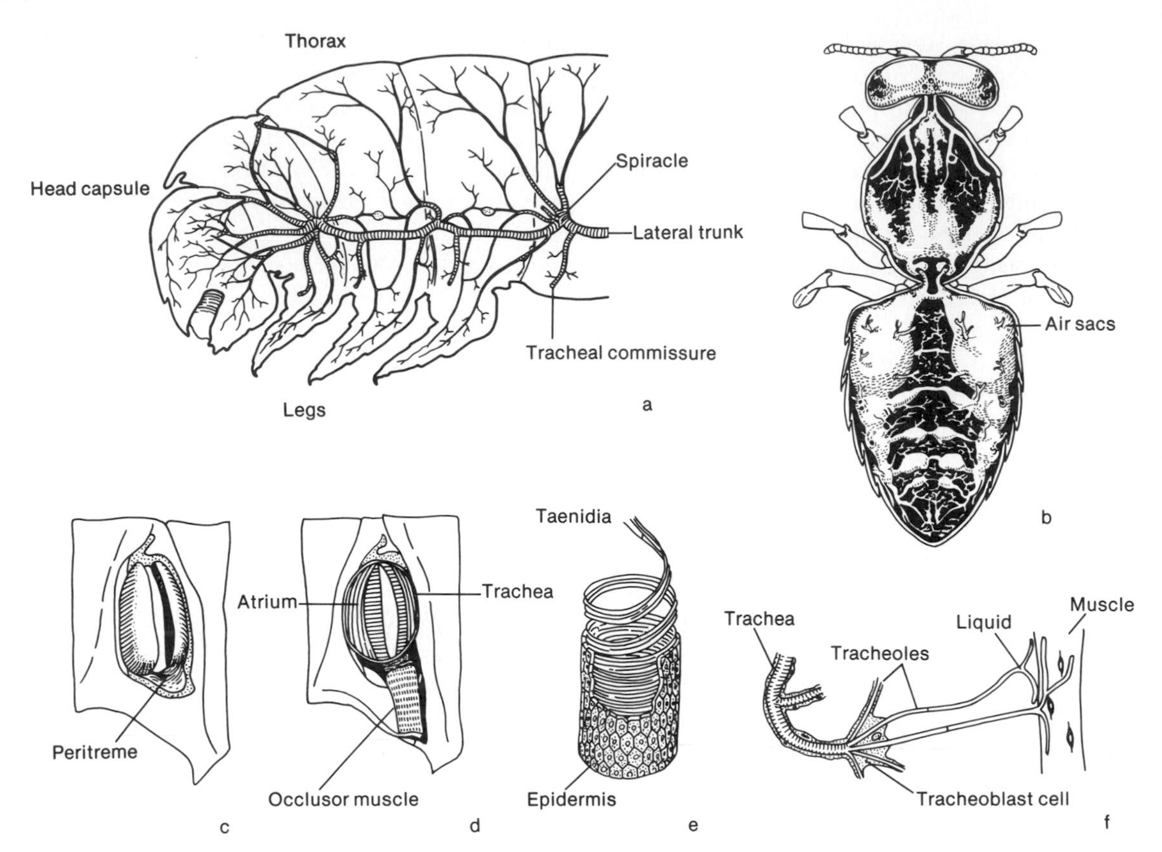

FIGURE 5.4 Tracheal system: *a*, tracheation of the head and thorax of a noctuid caterpillar; *b*, air sacs in a honeybee; *c*, external view of closing mechanism of a thoracic spiracle; *d*, internal view of same; *e*, structure of a trachea; *f*, structure of a tracheole. (*Redrawn from Snodgrass,* 1925, 1935, *by permission of McGraw-Hill Book Company.*)

reinforced by reticulate thickenings. The sacs are collapsible, a feature of importance in ventilation and one that also permits internal organs to expand in the body cavity, as when the ovaries are enlarged by eggs or the gut is engorged with food.

Associated with the tracheae in Diptera, Lepidoptera, and Hymenoptera are *peritracheal glands,* which consist of several cells each and are distributed pairwise in thoracic and abdominal segments. The cells are thought to secret exuvial fluids during molting and possibly hydrofugic (water-repellent) lipids which coat the intima. Of special note are the large, polytene chromosomes in the cells that undergo "puffing" or enlargement at certain times.

TRACHEOLES. Through repeated branching, the tracheae are reduced in diameter and terminate in tiny subdivisions, the *tracheoles* (Fig. 5.4*f*). These are tubules up to 350 μm length and 1 μm in diameter, tapering to 0.2 to 0.1 μm. Whereas the tracheal tube is enclosed in a multicellular coat, the epidermis, the tracheole is intracellular in a *tracheoblast cell.* The intima of the tracheole exhibits taenidia and a cuticulin layer, and may be exempt from renewal at each molt. When tissues are experimentally deprived of tracheae or organs are implanted in a new host, tracheoles migrate to the affected tissues and establish an oxygen supply. The same processes are presumably involved in normal development and injury repair. Although the general surface of the tracheal intima probably is diffusible by gases, the tracheole is believed to be the major site of oxygen transfer. Organs of high metabolic activity are richly tracheated. For example, the flight muscles of fast-flying insects are enveloped in tracheae and air sacs. The profuse tracheoles indent the plasma membranes of muscles to bring oxygen to the mitochondria deep in the fibers and only 3 μm part.

Tracheoles are often filled with a lymphlike liquid. During hypoxia this is withdrawn against a capillary force by the tracheoblast cell, bringing gas to the tracheolar tip. The physical mechanism of imbibition is not clear but is correlated with changes in osmotic pressure of surrounding tissues. A porous structure observed in the tracheolar cuticulin layer may facilitate this process. Oxygen moves the final distance from the tracheoles to respiring cells by diffusing through tissue. The carbon dioxide produced by respiration exits by the reverse course, hastened by a permeability through tissue 36 times faster than that of oxygen but diffusing slightly slower than oxygen in gaseous phase. The greater mobility of carbon dioxide in tissue favors some loss of this gas directly through the cuticle, especially at less sclerotized intersegmental membranes.

VENTILATION. The distance gases must diffuse is shortened by ventilation of the tracheal system. Motions of the body or churning of internal organs incidentally move air through the interconnected major trunks. Air movement through the system is greatly accelerated by pumping movements of the telescoping abdominal segments, creating inspiratory and expiratory flows. When the spiracles remain open, air may enter and exit tidally through the same orifice. The closing valves attached to spiracles in most insects permit coordinated opening and closing of the tracheal orifices, resulting in unidirectional flow. Ordinarily air is taken in through the anterior spiracles and expelled via the posterior spiracles. Air sacs increase the volume of air pumped and may have other mechanical functions in ventilation, since they are partly collapsible. During flight, abdominal pumping is inadequate to supply the flight muscles. Movements of the thoracic segments, in conjunction with open spiracles, ventilate the air sacs and tracheae surrounding the muscles.

Ventilatory movement of air in and out of the insect's body entails some risk of water loss. It is therefore advantageous for the spiracles to remain closed, with the occlusor muscle contracted, until breathing is demanded. Spiracles may be controlled separately or in concert by the central nervous system. Cellular respiration leads to lowered levels of oxygen and higher levels of carbon dioxide. Changes in both gases will reduce impulses from the central nervous system to the occlusor muscle (thus relaxing the muscle and opening the spiracle by its elastic hinge) or where dilator muscles are involved, increase impulses to the dilator to stimulate opening. Occlusor muscles are also influenced directly by carbon dioxide and will relax when the local concentration is high.

HEMOGLOBIN. Hemoglobin functions in the respiration of only three insects studied thus far: the bloodworm, *Chironomus* (Chironomidae), with molecular weight of the pigment, 31,400; the backswimmer, *Anisops pellucens* (Notonectidae); and the horse-botfly larva, *Gasterophilus intestinalis* (Gasterophilidae). The hemoglobin of insects contains two heme groups, half as many as the vertebrate pigment. In each case, hemoglobin provides a temporary respite from oxygen deprivation. The bloodworm lives in stagnant or polluted water deficient in oxygen. Movements of the larva in its burrow refresh the cutaneous supply and charge the hemoglobin with oxygen, which is not released unless the tension in tissues is low. The uptake of oxygen by hemoglobin is more rapid than diffusion in the hemolymph; hence recovery from hypoxia is aided and aerobic respiration is sustained under poorly oxygenated conditions. The horse-botfly maggot is similarly situated in a liquid, oxygen-poor medium: a horse's stomach. Hemoglobin in the hemolymph, and later in enlarged tracheal cells, permits more rapid uptake and storage of oxygen derived from passing gas bubbles. The bug

Anisops utilizes hemoglobin not only to extend its period under water, but also to sustain a favorable specific gravity which allows submerged floating.

TYPES OF RESPIRATORY SYSTEMS. The structure, number, and position of spiracles vary taxonomically: the apterygotes exhibit the most diverse arrangements and provide some insight to the evolution of the system. As already mentioned, most Collembola and Protura altogether lack tracheae. Some springtails, i.e., *Sminthurus,* have a single pair of simple openings on the cervix, between the head and thorax, leading to branched tracheae at each side. The proturan *Eosentomon* has two thoracic pairs, likewise leading to independent tracheal tubes. *Campodea* of the Diplura is similar with three thoracic pairs of spiracles, but the system of *Japyx* has four thoracic (the mesothorax and metathorax each have two pairs) and seven abdominal pairs, making a total of eleven, the maximum for any insect. Moreover, the spiracles are connected at each side by a longitudinal trunk, thus permitting movement of air lengthwise, but not transversely.

The Machilidae are relatively less advanced, with two thoracic and seven abdominal spiracles individually leading to mostly independent tracheae. The Lepismatidae, on the other hand, exhibit features which also characterize many of the winged insects: a pair of spiracles on each of the mesothorax, metathorax, and first eight abdominal segments; plus longitudinal trunks and transverse segmental commissures. Ventilating drafts of air through this interconnected system are forced by movements of the body and internal organs.

Three major types of respiratory systems are recognized: (1) The *holopneustic* respiratory system described for the Lepismatidae is found in the immature stages and adults of many terrestrial Orthopteroidea, Hemipteroidea, and some Hymenoptera. (2) The *hemipneustic* system is derived by the loss of one or more functional spiracles and is characteristic of larval Neuropteroidea. The hemipneustic system is subdivided into the *peripneustic* arrangement prevalent among terrestrial larvae (metathoracic spiracle nonfunctional) and the *oligopneustic* type of many Diptera that is specialized for life in water or liquid media (in the *amphipneustic* type the prothoracic plus posterior abdominal spiracles are functional; in the *metapneustic* type only the last pair of abdominal spiracles is functional). (3) The *apneustic* system of submerged aquatic insects has no functional spiracles but does have a closed tracheal system. Oxygen enters the body by cutaneous diffusion, either over the general body surface or in special, integumental gills. This system is found in immature stages of Ephemeroptera, Odonata, Plecoptera, Trichoptera, Sialidae (Megaloptera), and many larval Diptera.

AQUATIC RESPIRATION. Nearly all aquatic insects (Table 12.1) have a gas-filled tracheal system. The few exceptions are the young fly larvae of *Chironomus* (Chironomidae) and *Simulium* (Simuliidae), and the young caterpillars of *Ancentropus* (Pyralidae). In some groups the system is *open* and air can be taken inside the body, but in other groups the system is *closed* (apneustic) to the entry of air. In either case, many aquatic insects have the body surface well supplied with tracheae and are able to obtain at least part of their oxygen by *cutaneous respiration.*

Closed tracheal systems are found only in the immature stages. Some, such as many larval Diptera, depend mainly on cutaneous respiration, but others have special *tracheal gills* that augment respiration. These are thin-walled outgrowths of the integument that are richly tracheated. Oxygen in water diffuses through the cuticle and into the gas-filled tracheoles. Gills may also be important in the elimination of carbon dioxide. Platelike gills are found in Ephemeroptera (Fig. 12.3) and Odonata (Suborder Zygoptera, Fig. 21.5*b*). Filamentous gills are also found in Ephemeroptera (Fig. 20.4*b*) and are characteristic of Plecoptera (Fig. 27.1*b*), Megaloptera (Fig. 36.2), Neuroptera (Sisyridae, not visible in Fig. 38.2*b*), Trichoptera (Fig. 45.2), and certain Coleoptera (e.g., Gyrinidae, Helodidae, Psephenidae), Diptera (*Phalacrocerca,* Tipulidae), and Lepidoptera (Nymphulinae, Pyralidae). The rectum of Odonata is thin-walled and tracheated as a rectal gill. In general, gills occur most commonly on the sides of the abdomen, less commonly on the thorax, and rarely on the head.

Adult and immature insects with an open tracheal system obtain their oxygen by frequent trips to the surface for air, or by tapping air spaces in submerged plants, or by carrying a bubble or film of gas with them while they are submerged. The oligopneustic system of larval Diptera is suited to obtaining air by returning to the surface. The spiracular opening is usually surrounded by hydrofuge areas or hairs that prevent entry of water into the spiracle.

Air spaces in plants are punctured by the specialized respiratory tubes of various unrelated insects: certain larval beetles (Donaciinae, Chrysomelidae; *Noterus,* Noteridae; *Lissorhoptrus,* Curculionidae) and flies (*Chrysogaster,* Syrphidae; *Notiphila,* Ephydridae; *Mansonia* and *Taeniorrhynchus,* Culicidae).

Bubbles of air are carried beneath the surface by many aquatic bugs and beetles. Such temporary air stores are held in place by special hairs or are carried in cavities beneath the wings or elytra. The spiracles are situated to open into the air bubble. All aquatic Hemiptera in North America that swim beneath the water carry air stores. The manner of obtaining air varies, e.g., through a caudal siphon (Nepidae, Fig. 34.8*b*), caudal flaps (Belostomatidae), abdominal tip (Naucoridae, Notonectidae), or the pronotum (Corixidae). Aquatic beetles of the Suborder Adephaga carry air under the elytra and renew the store by breaking the surface film with the tips of their elytra and abdomen. Aquatic members of the Suborder Polyphaga have elytral air stores and usually also a coating of hydrofuge hairs on the ventral surface that holds a film of gas. Hydrophilid beetles break the surface with an antenna.

The bubble that is taken below serves not only as a store of atmospheric oxygen and as a hydrostatic organ, but also as a *physical gill.* The latter functions as follows: The gas in the bubble initially has the composition of air, viz., about 20 percent oxygen, 79 percent nitrogen, and less than 1 percent carbon dioxide. The carbon dioxide produced by the insect rapidly diffuses into the water and is of negligible importance as a gas in the bubble. As oxygen is used by the insect, the content in the bubble falls below that in the surrounding water. Oxygen will then diffuse into the bubble from the water. The bubble of gas persists for a time because nitrogen diffuses out of the bubble relatively slowly. As a result, the insect is able to obtain considerably more oxygen than was originally contained in the bubble.

A remarkable modification that utilizes the principle of the physical gill is the *plastron.* In this device, a thin film of gas is held on the body surface indefinitely by dense hairs or a fine cuticular meshwork (Fig. 5.5). The European bug *Aphelocheirus* (Naucoridae) has fine, short hairs over much of the body surface. The density of hairs is 200 to 250 million per square centimeter. The hairs are bent at the tips in an arrangement that supports the gas/water interface and resists compression of the gas film. The bug and similarly equipped elmid and dryopid beetles can remain submerged for months.

The aquatic pupae of certain beetles (*Psephenoides,* Psephenidae) and flies (some Tipulidae, Blepharoceridae, Deuterophlebiidae, Simuliidae, Empididae) have long cuticular processes developed from the peritreme and atrial region of one or more spiracles. These processes are called *spiracular gills.* The gill is associated with a plastron that is supported by a cuticular network rather than by dense hairs. Likewise, the chorion of the eggs of some insects has a plastron that allows the embryo to respire when the egg is flooded or covered by a film of moisture.

The hair or cuticular plastron functions equally well when the insect or egg is out of the water, because the spiracles or embryonic respi-

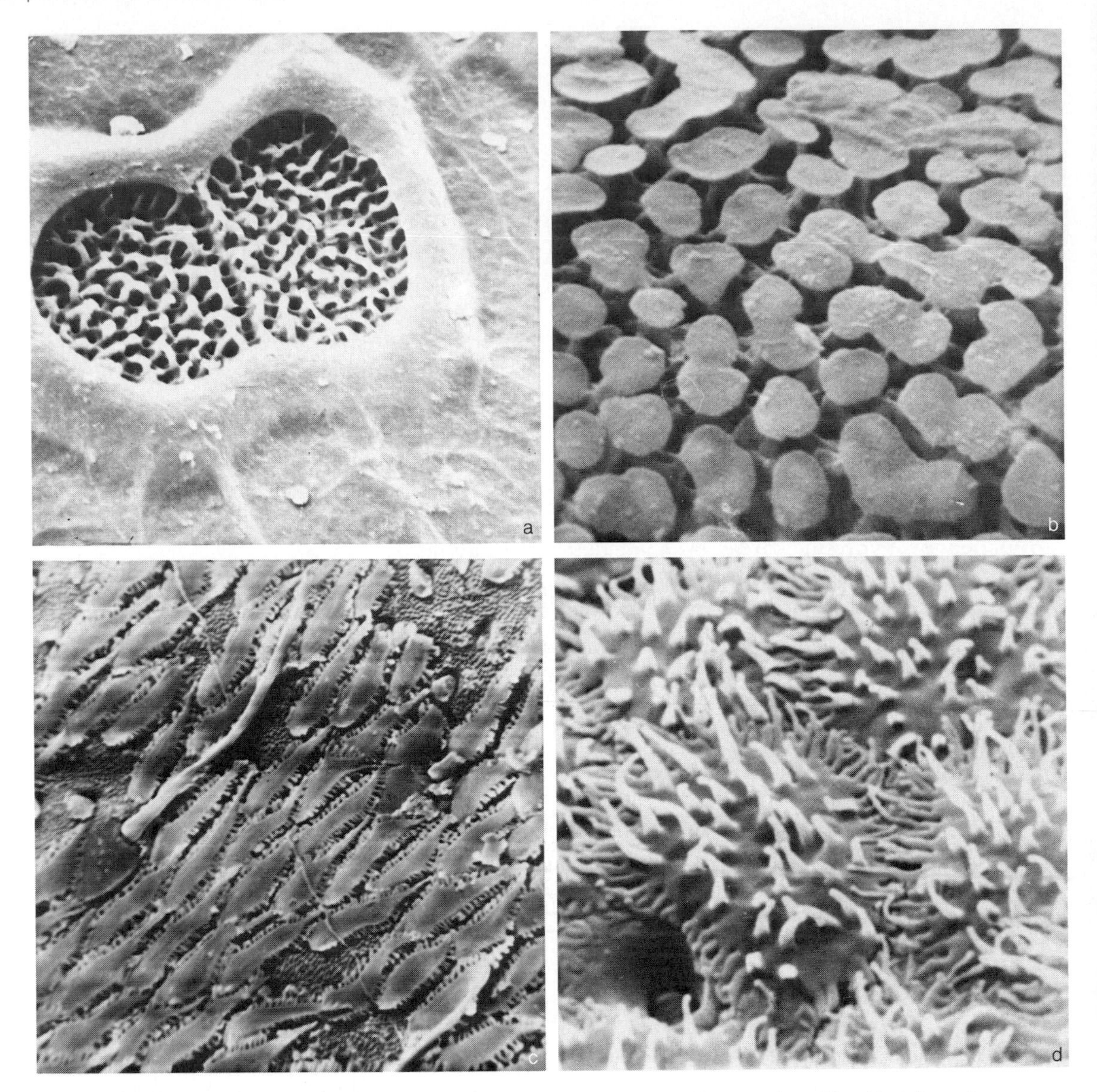

FIGURE 5.5 Examples of plastrons: *a*, plastron crater of eggshell of fly, *Musca sorbens* (Muscidae); *b*, respiratory horn of egg of fruit fly, *Drosophila melanogaster* (Drosophilidae); *c*, metasternum of aquatic beetle, *Tylctelmis mila* (Elmidae); *d*, elytron of aquatic beetle, *Portelmis nevermanni* (Elmidae). (*Scanning electron microscope photos courtesy of and with permission of Howard E. Hinton.*)

ratory organs are then in direct contact with air. In water deficient in oxygen, however, the physical gill fails, because oxygen no longer diffuses into the bubble. For this reason, insects with plastron respiration are typically found in swiftly flowing streams or other well-oxygenated waters.

CIRCULATORY SYSTEM

The fluid, or *hemolymph,* that circulates in the bodies of insects differs in many respects from that of vertebrates. It is a single fluid with cells, not divisible into blood and lymph systems, that moves freely around the internal organs without a pressurized, closed network of blood vessels. The cells, or *hemocytes,* are highly variable in size, shape, numbers, and functions. The plasma is biochemically unique in the regulation of osmotic pressure by amino acids and organic acids instead of inorganic ions. Neither hemocytes nor plasma is significantly concerned with the transport of oxygen and carbon dioxide in most species. Insect hemolymph is usually clear or straw-colored or variously pigmented yellow, blue, green, or rarely, red with hemoglobin.

The blood is limited to the major cavity of the body, the hemocoel. Insect hemolymph does not, however, directly bathe the cells of the body, as is often claimed. The inner surface of the integument and all the organs are ensheathed in a basement membrane of connective tissue. This membrane probably is interposed between the hemolymph and all other tissues. Thus situated, the exchange of materials between the body cells and the hemolymph may be regulated by the properties of the basement membrane.

DORSAL VESSEL. The circulation of the hemolymph is not haphazard but follows a pattern of flow. Motion is imparted to the fluid by movements of the body segments and of the gut, and by the peristaltic contractions of the *dorsal vessel* (Fig. 5.6*a*). This is a tube developed middorsally from the embryonic mesoderm and extending almost the full length of the body. In the abdominal region the vessel is closed at the posterior end but perforated by paired segmental openings, or *ostia.* Insects may have as many as twelve pairs. These have valvelike flaps projecting into the lumen of the vessel. The portion of the dorsal vessel with the ostia is called the heart. It is slightly inflated and especially so between the pairs of ostia. These dilations create the chambers of the heart. The

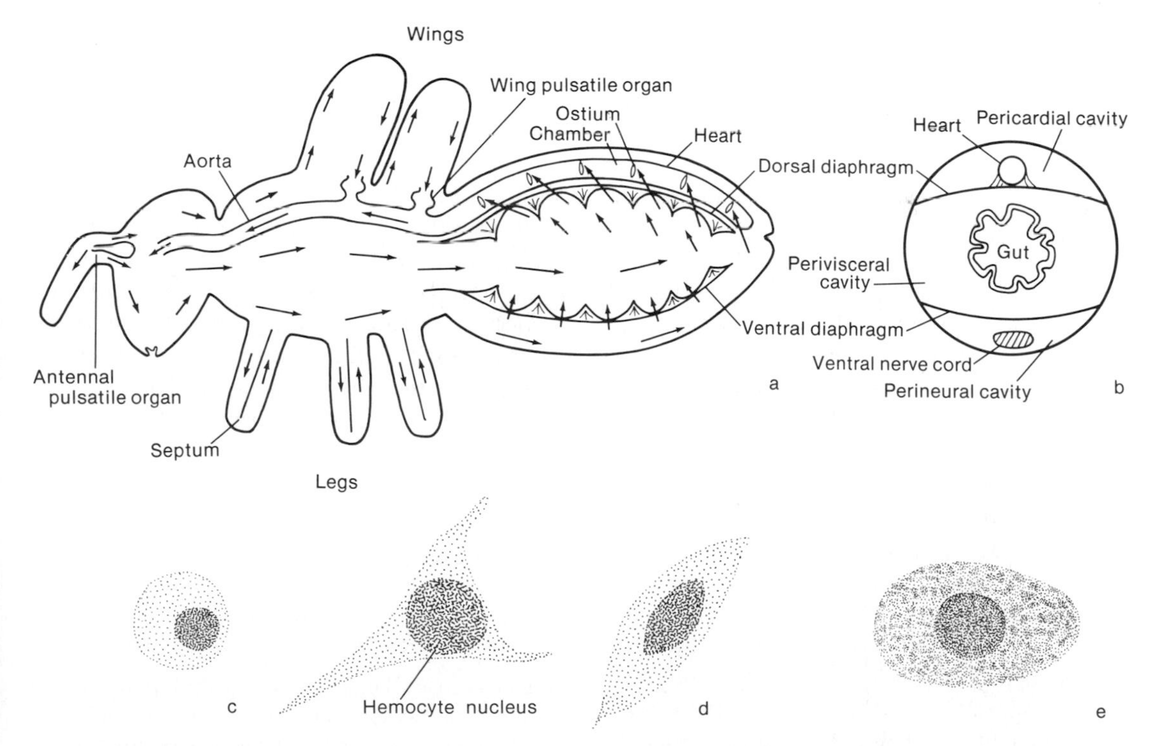

FIGURE 5.6 Circulatory system and hemocytes: *a*, diagramatic longitudinal section of general structure and direction of circulation; *b*, cross section of abdomen; *c*, prohemocyte; *d*, plasmatocytes; *e*, granular hemocyte. (a *modified from Weber,* 1933.)

heart is suspended middorsally in the abdomen by a transverse membrane of connective tissue and thin muscles. This *dorsal diaphragm,* with its *alary muscles* (meaning "winglike" in shape), is not complete but is interrupted segmentally by openings. By this means blood is allowed to pass from the vicinity of the gut or *perivisceral cavity* to the space above the heart, the *pericardial cavity.* Also in the abdomen is a *ventral diaphragm,* formed by an incomplete sheet of connective tissue. This membrane extends transversely, usually above the ventral nerve cord, and partly separates the perivisceral cavity from the ventral *perineural cavity.*

MOVEMENT OF HEMOLYMPH. Muscles embedded in the fibrous tissues of the heart wall provide the sequential contractions which impel the blood forward. As the wall of a chamber begins to contract (the systolic phase), the ostia close and their inner flaps occlude the rear chamber. Thus, blood is moved forward to the next chamber. When the wall relaxes (the diastolic phase), the ostia open and the flaps permit blood to enter from the outside and from the chamber to the rear. The flow of blood is normally toward the head, but reversals can be observed. The rate of heartbeat varies widely with age, species, the state of nervous excitement, and environmental factors such as temperature. The initiation of contraction may come from an independent but diffuse "pacemaker" in the heart or may be partly or entirely controlled by the central nervous system.

The blood is conveyed to the head by a continuation of the dorsal vessel, or *aorta.* This part of the vessel is distinguished by the absence of ostia and the generally smaller diameter. It is not supported by the dorsal diaphragm, but it may be suspended by a vertical sheet, or *dorsal septum,* in the thorax. The aorta opens beneath the brain and above the pharynx. Blood is released here and moves posteroventrad, entering the appendages of the head and the thorax. Flow into and out of these cylindrical structures is made possible by a septum which divides the appendage lengthwise into two channels. Blood also enters the wing veins and flows in an orderly, but variable, pattern throughout the wing and returns. Local movement of the blood is aided by *accessory pulsating membranes.* After passing around the leg muscles and flight muscles in the thorax, the blood returns to the abdomen. The movement here is upward from the perineural and perivisceral cavities through the dorsal diaphragm and back to the heart. Although there is almost no blood pressure, complete mixing time in an insect has been estimated at from 8 to 10 or up to 30 minutes. In humans the time is 2 to 4 minutes.

HEMOCYTES. The cells or hemocytes of insect blood vary greatly in size, cytological details, and the numbers that one can see in a sample of hemolymph (Fig. 5.6). Some of these differences arise from the several methods of preparing the delicate cells for study. More importantly, hemocytes increase in numbers by mitotic division. Several distinctive kinds of cells reproduce in this manner, but some types seem to be derived by transformation from other types. Direct observation of the origin and development of hemocytes has not been possible until recently because the culture of insect tissues in vitro has been largely unsuccessful. Repeated attempts have been made by different scientists to classify the bewildering variety of hemocytes, often with the result that confusion has been compounded by new names. A welcome stability has followed the efforts of Jones (1962) and Arnold (1974). Three types are sufficiently distinct cytologically and functionally to be widely recognized in different insects. These are:

1. *Prohemocytes:* These hemocytes are small, round or ellipsoidal, and have large, granular nuclei and little cytoplasm. The staining reaction is strongly basophilic. Prohemocytes are often observed dividing mitotically and are believed to be the main source of the other hemocytes (Fig. 5.6*c*).
2. *Plasmatocytes:* These hemocytes are highly variable in size and shape, and the volume of cytoplasm is about equal to that in the nucleus, or greater. The staining reaction is moderately basophilic. Plasmatocytes are derived from prohemocytes, and

intermediate forms exist between the classes. Plasmatocytes may also divide mitotically. They are usually amoeboid and phagocytic (Fig. 5.6*d*).

3. *Granular hemocytes:* These are round or disklike, are variable in size, and have a small nucleus surrounded by a large amount of cytoplasm, the conspicuous granules of which react positively to periodic acid-Schiff reagent (PAS). The origin and function of the granular hemocytes are unclear (Fig. 5.6*e*).

Additional categories of cells have been recognized in some, but not all, insects:

4. *Adipohemocytes:* These are round or ovoid, nonmotile cells, with a small nucleus placed to one side in a large volume of lipid-rich, granular cytoplasm. The cytoplasm exhibits lipoid droplets and vacuoles, and PAS-positive granules. Of the basic types listed above, adipohemocytes most resemble granular hemocytes.
5. *Spherule cells:* These are nonmotile cells related to granular hemocytes, but with large rounded inclusions or spherules which fill the cell and may obscure the nucleus. The spherules may be PAS-positive.
6. *Oenocytoids:* Cells also related to granular hemocytes, of variable size and shape, nonmotile, and with a small, round nucleus eccentrically placed in a large amount of basophilic or neutrophilic cytoplasm. The cytoplasm may include granules or, in certain flies, crystals.
7. *Cystocytes:* Also called coagulocytes, these cells respond to contact with foreign surfaces, e.g., after injury, by either or both of the following reactions: long processes are extruded and collectively the cells form a meshwork, or cytoplasmic materials are secreted which are believed to promote coagulation of the surrounding plasma. The coagulocytes sometimes disintegrate during these reactions. In either event, a clot is developed which stops hemorrhage and supports wound repair. The reaction is often so rapid that the cells are observed only after the reaction has taken place.

The kinds and numbers in the hemolymph "picture," or hemogram, of an individual insect also vary greatly, depending on age and physiological condition. Most insects examined possess an average number of 20,000 to 30,000 hemocytes per cubic millimeter. The numbers gradually increase during larval life, decline during pupation, and again increase in the early adult, declining thereafter. The molting cycle is usually associated with an apparent drop in numbers just before ecdysis. This is correlated with an increase in blood volume which aids in the expansion of the new cuticle. Blood volume falls and cell numbers rise after ecdysis. The causes of short-term fluctuations probably include not only losses and gains in actual cell numbers, but also changes in blood volume leading to dilution or concentration, and the temporary adherence of some cells to tissues so that they are not in circulation.

WOUND REPAIR AND CLOTTING. The risk of bleeding is greatly reduced among insects because the exoskeleton protects against minor injuries, and since blood pressure is lacking, hemorrhage is less likely. However, mechanisms exist in the hemolymph of many, but not all, insects for forming clots and repairing wounds. Coagulation of hemolymph can occur so rapidly that the steps involved have not been carefully analyzed. Essentially two processes have been observed: cystocyte agglutination and plasma coagulation. These can occur independently, but more commonly cystocytes are associated with clot formation. The reaction of cystocytes to contact with foreign surfaces has been described above. Unlike blood platelets of mammals, the cells do not invariably disintegrate. Other hemocytes may become entrapped in the meshwork or plasma gel, but they are not believed to participate actively. The biochemical reactions of plasma coagulation have been insufficiently studied, but it is known that substantial amounts of protein become involved. At least seven factors have been tentatively identified in the blood of honeybees which correspond to human coagulation factors. Ironically, the hemolymph of honeybees does not clot.

COMPOSITION OF THE PLASMA. The cell-free fluid, or plasma, of the hemolymph contains hormones, nutrients absorbed by the gut and produced by internal symbiotes, metabolic wastes, and water. The volume of hemolymph may vary widely. At times it is even difficult to obtain a sample. When water is readily available, the volume may increase as a reservoir, contributing up to 94 percent of the body weight. Volume ordinarily increases prior to ecdysis to provide the hydrostatic pressure for shedding the exuviae and expanding the new cuticle. After ecdysis, the volume decreases.

In vertebrate blood the inorganic cations (positively charged ions) of sodium, potassium, calcium, and magnesium, together with the anions (negatively charged) chloride, bicarbonate, phosphate, and sulfate, are regulated in critical proportions. This is because the total concentrations, especially of sodium and chloride, determine osmotic pressure. The balance of cations is essential to the integrity of cell membranes, and certain ions determine pH of the blood plasma. These biochemical constituents are also found in the plasma of insects, but the proportions shift markedly as the evolution of various groups is traced, leading to the replacement of inorganic ions by amino acids and other small organic molecules as osmolar effectors. The high concentration of amino acids is the most distinctive feature of insect blood. Up to 50 times the concentration in human serum has been measured.

The apterygote *Petrobius maritimus* (Machilidae) resembles other arthropods and many animals in utilizing sodium and chloride as the primary ions regulating osmotic pressure. In the paleopterous orders Ephemeroptera and Odonata and the neopterous orders Blattodea and Hemiptera (Suborder Heteroptera), sodium continues to have a major role, accompanied by the other cations, and choride is the major anion, together with small amounts of inorganic ions which are reduced relative to other plasma constituents. In a strikingly different composition, magnesium replaces sodium as the major cation, postassium level is high, and the phosphate concentration greatly increases in the Phasmatodea. The endopterygote orders Megaloptera, Neuroptera, Mecoptera, Diptera, and some Coleoptera retain sodium as the major cation among the others present, but the anions are reduced and replaced by increased numbers of amino acids and other organic compounds. The increased role of small organic molecules is fully developed in the Lepidoptera, Hymenoptera, and many Coleoptera, where amino acids replace inorganic ions as the main osmotic effectors.

The proportions of inorganic ions in the Lepidoptera and phytophagous Hymenoptera are nevertheless of interest because potassium or magnesium (or both) replaces sodium as the major cation. Differences in the concentrations of cations have been attributed to diet: zoophagous insects would acquire high amounts of sodium from their food, and phytophagous insects would acquire high amounts of potassium and magnesium. Exceptions exist to this simple rule because physiological patterns of ion regulation have evolved in various orders which override individual dietary habits.

Nutrients in the blood include amino acids, sugars, and lipids. Large amounts of amino acids are present as mentioned above, both from the diet and from synthesis, and are utilized for growth and metamorphosis. The main carbohydrate is the sugar trehalose, although glucose and some other sugars have been found in quantity in the honeybee and certain flies. Glycogen is present in small amounts, but most of it is stored in the fat body. Glycerol, on the other hand, is accumulated in the plasma of certain insects as an antifreezing agent (Chap. 11). Lipids and proteins are transported via the plasma to the fat body for storage. The protein content is comparable to that in vertebrate blood. Among the proteins is a remarkable variety of enzymes. Aside from their roles in metabolism in the plasma, plasma proteins have provided useful taxonomic information on the evolutionary relationships of different species. Organic acids, especially those serving as substrates in the tricarboxylic acid cycle (citrate, α-ketoglutarate, succinate, fumarate, malate, and oxalacetate), are most prevalent in larval endopterygotes. Acid-soluble organic phosphates are synthesized and found in high concentrations in certain moths. Levels of waste end-products of protein metabolism in plasma—including uric acid, allantoin, allantoic acid, urea, and in some immature aquatic insects, ammonia—are sometimes high. The plasma may be colored by various pigments, including α-carotene, riboflavin, chromoproteins, etc., but probably not chlorophyll. Hemoglobin has been discussed previously as a respiratory pigment in three insects.

EXCRETORY SYSTEM

Excretory processes expel or render harmless those metabolic wastes and other substances that are harmful, and at the same time they maintain water balance (Fig. 5.7*a*). Excretion thereby plays the key role in physiological homeostasis (i.e., the maintenance of a constant internal environment) by regulating body chemistry. In contrast to defecation (i.e., physical elimination of food not digested or absorbed), excretion involves the movement of waste molecules across one or more plasma membranes. The principal wastes are nitrogenous compounds resulting from protein metabolism and the breakdown of nucleic acids. Also included as waste are excessive amounts of other substances, such as water or salts. Most of the excretory products in insects are regulated by the Malpighian tubules and the rectum acting in concert.

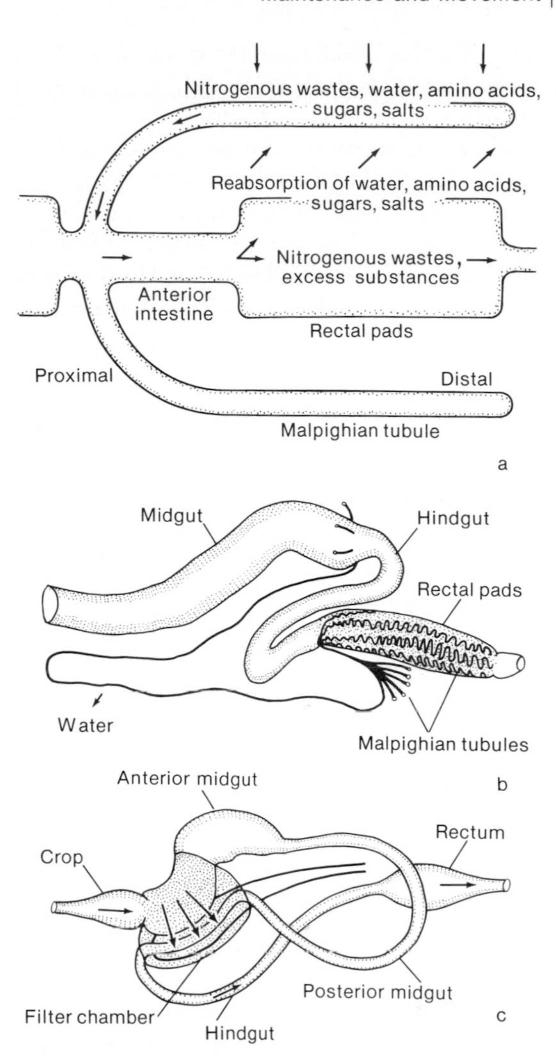

FIGURE 5.7 Excretory systems: *a*, generalized diagram of flow of nitrogenous wastes in the insect excretory system; *b*, cryptonephric system of *Tenebrio molitor* larva (Tenebrionidae) with five of the six tubules partly removed; *c*, diagram of filter chamber of Homoptera. (c *modified from Weber,* 1933.)

NITROGENOUS WASTE. Nitrogenous end products may be discussed in three groups: (1) ammonia, (2) urea, and (3) uric acid, allantoin, and allantoic acid. Ammonia is a direct product of the oxidative deamination of amino acids. Containing 82 percent nitrogen, it is an efficient waste product, but it is also highly toxic in water and highly soluble. Terrestrial animals must conserve water; hence ammonia is usually transformed to the less toxic urea (e.g., in mammals, amphibians, some reptiles) or uric acid (e.g., in birds). Among insects, ammonia is the primary waste of the aquatic immature stages of the predatory dragonfly *Aeshna cyanea* (Aeshnidae), the alderfly *Sialis lutaris* (Sialidae), and the meat-eating fly maggots of *Lucilia sericata* (Calliphoridae). Ammonia is of secondary importance in the larval carpet beetle, *Attagenus piceus* (Dermestidae), whose diet is rich in the protein keratin, and the bloodsucking adult mosquitoes, *Aëdes aegypti, Anopheles quadrimaculatus,* and *Culex pipiens* (Culicidae). Urea, containing 46 percent nitrogen, is the primary waste of *Attagenus,* and a secondary product of the phytophagous bug *Dysdercus fas-*

ciatus (Pyrrhocoridae) and the above mosquitoes. Most insects excrete uric acid, or one of its breakdown products allantoin and allantoic acid, or a combination of these substances. These are the least soluable, as well as the least toxic, end products, and they contain less nitrogen (32 to 35 percent). Some uric acid is derived from the breakdown of nucleic acids, but the bulk of these wastes must be synthesized by energy-consuming pathways which are not as yet clearly understood.

Earlier generalizations that uric acid alone was the common waste of insects must be modified in view of recent analyses. Among the exopterygote insects, uric acid forms virtually all the nitrogenous waste in Orthoptera and Dermaptera, but mixtures have been found dominated by allantoic acid in the roach *Blatta orientalis* (Blattidae) and by allantoin in the walkingstick *Carausius morosus* (Phasmatidae). Waste in Hemiptera (Suborder Heteroptera) is predominately allantoin or mixtures with uric acid or with urea in *Dysdercus.* The bloodsucking bug *Rhodnius prolixus* (Reduviidae) excretes primarily uric acid. Honeydew excreted by plant-sucking aphids and coccids contains large quantities of amino acids, but these are mostly passed unchanged from the plant sap and are not excretory products in the present sense. The proportions of nitrogenous wastes are more varied in the endopterygote insects because of differences in diets and habitats between larvae and adults of the same species. Uric acid is the major waste in many Coleoptera, Hymenoptera, Diptera, and Lepidoptera, but allantoin predominates in certain species of the first three orders, and allantoic acid prevails in certain Lepidoptera. Comparison of adults, larvae, and pupae of Lepidoptera indicates that uric acid is the major pupal waste, but caterpillars and adults of the same species may switch from uric acid to allantoic acid and vice versa. Even during larval life the proportions may vary greatly.

WASTE STORAGE. In certain insects some of the nitrogenous waste is not expelled but is stored in a harmless state or utilized elsewhere in the body. Uric acid is deposited in special *urate cells* of the fat body in the roach *Periplaneta americana* (Blattidae), adult *Culex* (Culicidae) mosquitoes, the silkworm *Bombyx mori* (Bombycidae), and the wasp *Habrobracon juglandis* (Braconidae). The rate of accumulation of uric acid thus stored provides an index of aging. Adult males of *Blattella germanica* (Blattellidae) store uric acid in accessory sex glands in amounts up to 5 percent of the insect's weight. This is discharged as a white covering on the spermatophore during mating. The white color of crystalline uric acid is used as a pigment in the epidermis of some insects or conspicuously in the wing scales of butterflies in the Family Pieridae.

WASTE ELIMINATION. Most of the nitrogenous waste is eliminated from the insect's body as part of regulatory processes involving other substances of small molecular size. The organs responsible are the *Malpighian tubules* (named for the seventeenth-century insect anatomist M. Malpighi) and the hindgut, especially the *rectum.* The filamentous, blind tubules insert and empty at the anterior end of the hindgut.

The walls of the Malpighian tubules and rectum are both one cell in thickness. Thus an internal and an external plasma membrane are situated between the hemolymph and ducts leading to the exterior. The cells of the Malpighian tubules are richly tracheolated and bounded by a basement membrane. The plasma membrane next to the hemolymph is deeply and complexly infolded; the luminal membrane is microvillate and associated with long mitochondria. The tubules are usually free in the hemocoel and may move by means of slender muscles placed lengthwise in a spiral fashion. Malpighian tubules are lacking in Collembola, *Japyx* (Diplura), and most aphids; are rudimentary in certain Protura, Diplura, and Strepsiptera; and are only two in number in Coccoidea and some larval parasitoid Hymenoptera. Other insects usually have multiples of two, commonly having a total of six. Fifty or more tubules are formed in some Ephemeroptera, Plecoptera, Odonata, Orthoptera, and aculeate Hymenoptera.

The composition of the hemolymph is regulated in the following manner (Fig. 5.7*a*). Water, salts, sugars, amino acids, and nitrogenous wastes pass into the lumen of the Malpighian tubules, beginning at the tip or distal end. The filtrate then moves through the tubule to the hindgut and rectum. Along this route useful components are selectively reabsorbed back into the hemolymph while the waste and surplus are voided. The mechanisms for the movement of molecules across the plasma membranes are not fully understood. A recent hypothesis is that the movement of molecules through both plasma membranes of the tubule involves an energy-consuming active transport of potassium ions, possibly stimulated by sodium ions on the hemolymph side and together with active transport of sodium ions on the luminal side. This is called the "potassium pump." The anionic chloride and phosphate ions, water, sugars, and amino acids are thought to be carried along with this flow by one of several passive mechanisms. The fluid is nearly iso-osmotic with the hemolymph when it is initially secreted.

The parasitic *Rhodnius* passes large amounts of urine following a blood meal. The secretion of the Malpighian tubules increases about 1000-fold under the influence of a diuretic hormone. In this insect, active transport of sodium and chloride ions are believed mainly responsible for the flow of fluid into the lumen. As demonstrated with dyes, large organic molecules are also passed and probably involve a separate transport system.

Selective reabsorption may begin in the proximal portion of the tubule or the anterior parts of the hindgut, but the major site of activity is the rectum in most insects. The cuticular intima of the hindgut functions as a molecular sieve, allowing penetration of smaller molecules but preventing passage of larger organic molecules secreted by the Malpighian tubules. In this way the larger molecules are believed to be excluded from reabsorption. The excretory fluid is also acidified, thus promoting precipitation of uric acid. By mechanisms not yet clarified, useful amounts of water, amino acids, salts, and sugars are absorbed by the enlarged, columnar cells of the rectal pads. Potassium ions are rapidly returned to the hemolymph against an osmotic gradient.

In larvae and adults of Coleoptera and larval Lepidoptera, the distal ends of the Malpighian tubules may be held in contact with the rectum. This *cryptonephric* arrangement of the tubules has been carefully studied in the mealworm, *Tenebrio molitor* (Tenebrionidae). The adults and larvae live in dry, stored grains. The rectal complex is specially developed to dehydrate feces before elimination, thus conserving water. Water is even extracted from the humid air around the feces in the rectum. The six tubules are held in contact with the rectum by a multilayered perinephric membrane and an outer membrane (Fig. 5.7*b*). This arrangement allows the tubules to take up much of the water and ions extracted by the rectum. The fluid is transported proximally into the free portion of the tubule, where it is passed to the hemolymph. The rectum is thus assisted in withdrawing water from feces against a steepening osmotic gradient.

Insects that feed directly on the sap in the vascular tissues of plants have the opposite problem: excessive fluid and the need to eliminate water rapidly to avoid dilution of the hemolymph. In the Cicadoidea (Hemiptera, Suborder Homoptera), the posterior midgut, Malpighian tubules, and anterior hindgut are looped forward to contact the thin-walled anterior midgut (Fig. 5.7*c*). The area of contact is enclosed in a membrane. This is called a *filter chamber.* Water passes directly through the walls of the anterior midgut to the anterior hindgut, with little or no liquid flowing through the remaining portion of the midgut.

MUSCLES AND LOCOMOTION

Insects move from place to place in virtually every natural medium on the earth's surface except solid rock, ice, or rarely, salt water. By their feeding and burrowing activities, they move through the flesh of animal hosts, the tissues of plants, including solid wood, and the leaf litter, soil, or subsoil of the earth. On the surface of the ground they wiggle, hop, walk, or run, often with surprising agility. In the aquatic environment some skate on the surface, propelled by their legs or by chemicals which alter the surface tension. Under the surface they swim with the legs or wings, with body movements, or by jets of water expelled from the anus. The best speeds and the longest distances are achieved in the air. Insects are the only invertebrates to fly actively and to control the duration and direction of their flights. The only other invertebrates to fly are the ballooning spiders, which, however, are passive drifters. To judge by the fossil record, insects had wings and were presumably flying in the Upper Carboniferous Period over 150 million years before the first flying reptiles (Pterosauria) in the Jurassic. Much later the birds and mammals (bats) each independently evolved flying mechanisms.

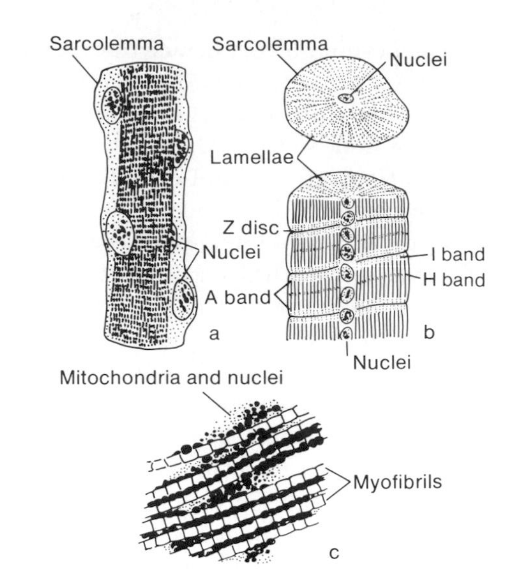

FIGURE 5.8 Types of muscles: *a*, larval; *b*, tubular; *c*, fibrillar. (*Redrawn from Snodgrass,* 1935, *by permission of McGraw-Hill Book Company.*)

The power for the locomotory and other movements is supplied by the muscular system. Internal organs, such as the gut, heart, or Malphighian tubules, also have sets of muscles. All muscles of insects are *striated,* i.e., each fiber has a cross-banded appearance when viewed under the microscope (Fig. 5.8). In contrast, vertebrates and many other invertebrates have both the striated type and a second type, smooth muscle, which is composed of single spindle-shaped cells.

The muscular system of insects differs in several important respects from that of vertebrates. The most conspicuous difference is the relation to the skeleton: the muscles of insects are inside the skeleton, while those of the vertebrates are outside. As a consequence of the tubular construction, which resists bending, and the general absence of calcium salts, the insect skeleton compared to the vertebrate skeleton is relatively lighter and stronger and provides greater space and attachment areas for the muscles. The attachment of muscles to the skeleton is not by connective tissues, as in the vertebrates, but by *tonofibrillae.* The origin of the tonofibrillae is obscure: they may be continuations of the muscle fibers or special secretions of the epidermal cells. The attachments may appear to terminate on the basement membrane or epidermal cells or extend through the cells into the cuticle as far as the epicuticle. At the point of attachment the cuticle may be invaginated into a tendonlike apodeme. When composed of normal cuticle such apodemes lack elasticity, but when the special cuticular protein resilin is present, a highly resilient tendon is formed.

Although insects perform feats of unusual strength, the power of their muscles is closely similar to that of vertebrates. Power is measured by the cross-sectional area of the muscle in relation to the greatest load it can lift. The explanation for their relative strength lies in the effective use of lever systems and the small body size. As the body of an organism decreases in size, the volume or mass decreases as the cube root, whereas the cross-sectional area decreases as the square root. If the absolute power of a muscle per unit of area remains the same, an exactly scaled reduction in the size of an organism will result in relatively lighter loads for the smaller muscles. Miniaturization of organisms is rarely linear, but the general relationship remains valid. Reductions in the sizes of insect muscles involve reductions in the numbers of fibers rather than reductions in the sizes of the fibers. A tiny insect may have only a single fiber functioning in the position occupied by hundreds of fibers in a large insect.

STRUCTURE OF MUSCLE FIBERS. Each fiber is the product of many cells which share a common plasma membrane and an outer sheath, the *sarcolemma.* This outer covering may be invaginated into the fiber in certain places, forming a system of intracytoplasmic *intermediary tubules* (called the IT or T, for transverse, system). Tracheoles may reach the interior of the fiber without penetrating the sarcolemma by entering these tubules. It is hypothesized that a chemical transmitter substance may pass from the nerve endings to the deeper regions of the fiber by means of the T system. Within the fiber are the contractile *myofibrils* traversing the length of the fiber. These are striated and may exist as flat sheets (lamellae) or as cylinders. Under great magnification with an electron microscope each fibril is seen to be composed of protein filaments of two sizes. The thick filaments are myosin, and the thin filaments are actin. In cross section each actin filament of an insect muscle is situated exactly between two myosin filaments. In vertebrate muscle the actin filament is between three myosin filaments. When the muscle contracts, the two sets of filaments are believed to slide past each other. The molecular basis for this movement is not yet clear. The distribution of the two kinds of filaments also provides an explanation for part of the banded, or striated, appearance. The I bands represent only actin filaments, the H band only myosin filaments, and the A band represents both together. The Z line is disklike in shape, but its ultrastructure is uncertain. Dispersed among the myofibrils are numerous nuclei and mitochondria and various amounts of a sarcoplasmic (endoplasmic) reticulum.

Individual fibers are grouped together into a "muscle unit" which has a separate tracheal supply. A single muscle unit or several units may be separately innervated by branches of one, two, or three motor axons (Chap. 6), making a motor unit. Finally, a single motor unit or several motor units may be grouped closely together into a functional "muscle" which is anatomically discrete from other such units, has definite areas of attachment, and exerts a pulling force on the attachments in a definite direction. Unlike vertebrate muscles, insect muscles lack connective tissues among the fibers. Ordinarily a muscle has one of its attachments, or its origin, on an immovable part of the skeleton and its insertion on a movable part such as a segment of an appendage or an articulating sclerite. Some are bifunctional, i.e., under certain circumstances the structures at either end may be moved. For example, some thoracic muscles function during walking by moving the leg bases and also function in flying by moving sclerites at the opposite ends of the muscles. The large indirect flight muscles move large portions of the thoracic skeleton simultaneously. The total number of muscles and the arrangements vary with the different kinds of insects. In a caterpillar the number may be more than three times the 529 present in humans.

HISTOLOGICAL TYPES OF MUSCLES. Although all insect muscle is striated, evolutionary changes in structure and function have led to several distinctive kinds of muscle. Differences become pronounced even in a single individual of a holometabolous insect, in which larval musculature differs histologically from that of the adult and the flight mechanism may include muscles with unique histological and physiological properties. Intermediate conditions in the anatomy of muscles may be discovered as the ontogeny and evolution of a taxonomic group are traced. For the purposes of discussion, five basic types can be recognized: larval, visceral, tubular, microfibrillar, and fibrillar.

1. *Larval* muscles, known from immature bees and flies, are distinguished by minute myofibrils with the nuclei located externally beneath the sarcolemma (Fig. 5.8*a*).
2. *Visceral* muscles are composed of parallel or anastomosing fibers similar to larval muscles in

construction, but they have larger myofibrils and are associated with the internal organs of adult insects.

3. *Tubular* fibers are so named because the myofibrils are arranged in lamellae around a central canal filled with nuclei (Fig. 5.8*b*). These are the muscles in the appendages and between sclerites in many insects. The flight muscles of some Orthoptera are tubular and resemble the tubular muscles elsewhere in the body. The flight muscles of Odonata, however, represent an extreme modification of the tubular design; the lamellae are greatly thickened, and the mitochondria measure up to 7 μm.
4. *Microfibrillar* muscles have myofibrils measuring 1 to 1.5 μm in diameter and are found in the flight musculature of certain Orthopteroidea and in the neuropteroid orders Neuroptera, Raphidioptera, Lepidoptera, and Trichoptera. Microfibrillar and fibrillar muscles possess threadlike myofibrils with the nuclei and mitochondria scattered among the fibrils.
5. *Fibrillar* muscles have myofibrils measuring 1.5 to 5.4 μm in diameter (Fig. 5.8*c*). These coarsely fibered muscles are also pinkish when freshly dissected. They are the main source of power for the flight mechanism of the orders Hymenoptera, Diptera, Coleoptera, and certain Hemipteroidea.

The last three types are involved with movements of the exoskeleton. Visceral, larval, tubular, and microfibrillar muscles respond to a nerve impulse by contracting *synchronously* with the impulses. Fibrillar muscles contract many times for each nerve impulse and are said to be *asynchronous* with the impulses.

WALKING. The bodies of arthropods other than insects usually have many segments, each with a pair of ventrolateral appendages. The anterior appendages are modified for sensory and feeding functions, and several pairs may be devoted to reproduction, but most of them are locomotory. Insects are unique among the arthropods in having only three pairs of locomotory appendages. Each leg must function in two ways: (1) to support the body's weight above the substrate, and (2) to thrust against the substrate in a propulsive direction. The best explanation for the evolutionary choice of six legs lies in the fact that a minimum of three legs can provide support in a stable tripod arrangement while the other three legs are stepping forward (Fig. 5.9). The insect walks by shifting from alternate tripods of leg contact with the substrate. The sequence or rhythm of leg movements changes as the insect increases speed. At times, more than three legs may be on the substrate at once. Wilson (1966) proposed the following model to explain various locomotory gaits observed in most, but not all, insects:

1. Legs step forward in a sequence which begins at the rear. No leg steps forward until the leg behind is in a supporting position.
2. Legs of the same segment alternate in stepping.
3. The time required to extend the leg remains constant.
4. The time required to thrust the leg against the substrate decreases as the frequency of steps increases.
5. The time interval between steps of the hind and middle legs, and between the steps of the fore- and middle legs remains constant, but the interval between steps of the forelegs and hind legs decreases as the frequency of stepping increases.

The rhythmic movements of insect legs are understood to originate in the central nervous system and to be modified by feedback from sensory receptors in the legs. The segmental ganglia control their respective legs but are coordinated by intersegmental nervous coupling.

The forelegs, in front of the center of gravity, exert their force in a pulling direction, while the middle and hind legs, which are near or behind the center of gravity, exert a pushing thrust. The propulsive thrust of the hind legs is greatly developed as a jumping device in Orthoptera (Figs. 29.1, 29.3), many Homoptera (Fig. 34.9*a*, *b*), fleas (Fig. 43.1*a*), and flea beetles (Alticinae, Chrysomelidae). The middle legs of Encyrtidae (Hymenoptera) are modified for jumping. Jumping is also accomplished in other ways: click beetles snap their bodies at the joint between the prothorax and mesothorax, and springtails

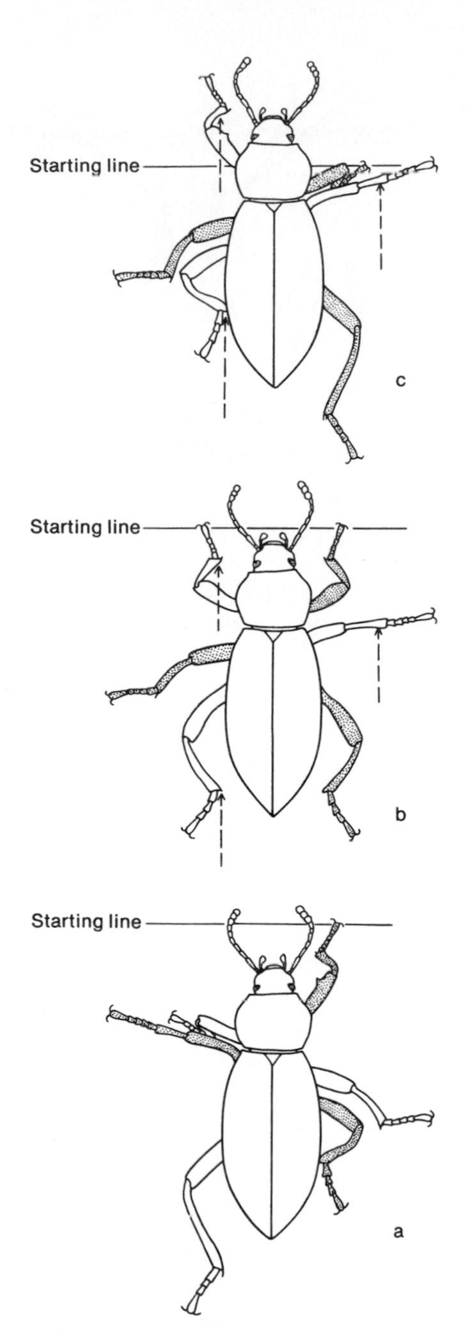

FIGURE 5.9 Diagram of walking movements of the beetle *Eleodes dentipes* (Tenebrionidae). Shaded legs are supporting the beetle as the unshaded legs step forward in the sequence *a* to *c*. At *c* the unshaded legs assume support of the body and the shaded legs begin stepping forward.

utilize an abdominal appendage (Figs. 18.3, 18.4*a*). The skeletomuscular mechanism of jumping in the locust has been investigated by Hoyle (1955 a, *b*). The height and length of the jump are determined by the angle of takeoff and the force of the hind legs against the substrate. The best jump by a 1.5-g female was 30 cm in height and 70 cm in length. The extensor muscle of the femur is attached inside the femur in a herringbone pattern which is visible externally. This increases the effective cross-sectional area and hence increases the force, but reduces the length of contraction. The muscle is inserted by an apodeme to the tibia above a dicondylic joint. This lever system and the long tibia effectively transmit the short, powerful contraction into a thrust estimated to be nearly 17 times the locust's weight. The strain on the apodemes is so great that they break if the legs are prevented from opening.

FLIGHT. The origin and evolution of wings are discussed in Chap. 17. Here we are concerned with the skeletomuscular mechanism. Flight in an insect such as the locust involves two major groups of muscles acting on a thoracic box which is partly rigid and partly elastic. The wing base rests on the pleural wing process in the same manner as a lever on a fulcrum (Fig. 5.10*a*). An upward movement of the thoracic notum lifts the wing base and forces the wing proper downward (Fig. 5.10*c*). Likewise, depression of the notum lowers the wing base, forcing the wing upward (Fig. 5.10*b*). At the anterior edge of the wing base is located a small sclerite, the basalare, and at the posterior base of the wing is the subalare (Fig. 2.10*a*). Muscular action on the basalare depresses the leading edge of the wing in a motion called *pronation* (Fig. 5.10*a*). Similarly, a pulling force on the subalare depresses the trailing edge in *supination*. The flight muscles are divided into "indirect" muscles, attached broadly within the thorax and affecting the wing indirectly by deformation of the thoracic box; and "direct" muscles, inserting directly at or near the wing base. The longitudinal indirect muscles extend between skeletal phragmata and, by their

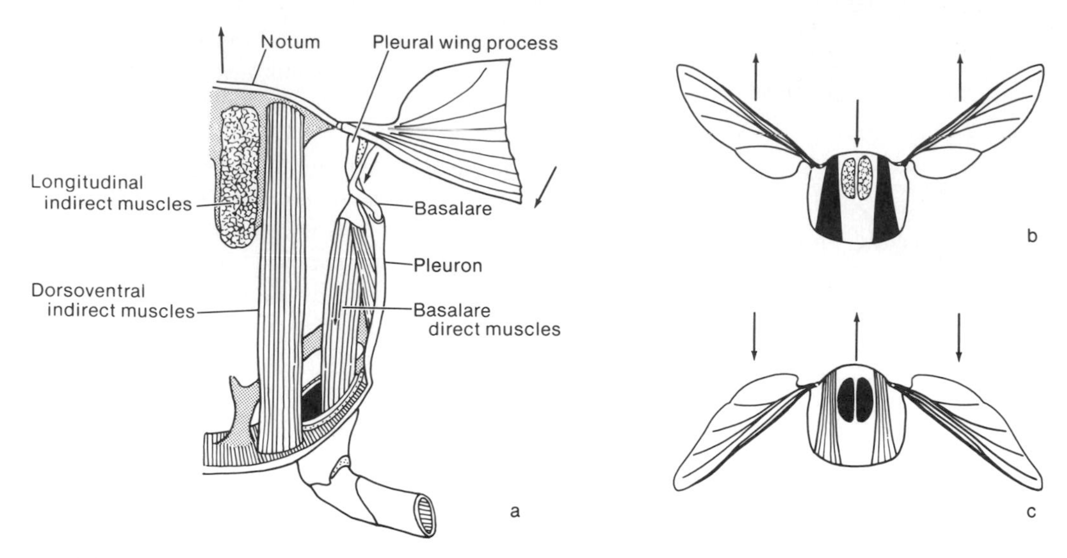

FIGURE 5.10 Flight mechanism: *a*, cross section of a mesothoracic segment during the downstroke and pronation of the wing; *b*, wing upstroke and supination; *c*, wing downstroke and pronation. (*Redrawn from Snodgrass,* 1935, *by permission of McGraw-Hill Book Company.*)

contraction, raise the center of the notum. This provides the force which results in the wing's moving downward. The dorsoventral indirect muscles operate as antagonists and depress the notum by their contraction, raising the wings. A simple up-and-down movement would not, however, generate the forward thrust needed for flight. Propulsive action is created by pronation of the wing on the downstroke and supination of the wing on the upstroke. Direct flight muscles attaching to the basalare and subalare accomplish these twisting motions on the plane of the wing as it is raised and lowered by the indirect muscles. In the locust, the hind wing traverses a slightly greater amplitude and leads the forewing, thus avoiding its turbulence. The elastic properties of the thorax and antagonistic muscles are greatly enhanced by special deposits of pure resilin on the pleural wing process.

The general design of the flight mechanism of the locust is found throughout the Orthopteroidea and some Neuropteroidea. The muscles of roaches are of the tubular type, but other Orthoptera exhibit transitional states between tubular and microfibrillar types. Microfibrillar muscle is found in the neuropteroid orders Neuroptera, Raphidioptera, Lepidoptera, and Trichoptera. The wingbeat frequency and, consequently, the speed of flight in all these orders are limited to the rate at which the nervous system can send impulses to each of the participating muscles in a coordinated fashion, i.e., about 50 per second. Yet the wingbeat frequencies of bees and flies is much greater, on the order of 100 to 300 beats per second. With wings shortened experimentally and at a high temperature, the midge *Forcipomyia* was rated at a frequency of 2218 strokes per second. The high frequencies of Diptera, Hymenoptera, Coleoptera, and some Hemipteroidea are the result of fibrillar flight muscles contracting asynchronously or many times for each nerve impulse. In these insects the main source of power for the wings resides in one thoracic segment and is often limited to one set of longitudinal and one set of dorsoventral muscles operating antagonistically. The direct muscles to the basalare and subalare may be fibrillar in some insects and contribute their usual roles, but in the most advanced fliers, these muscles no longer are responsible for the pronation and supination of the wing. The complex path of the wing is now determined by the configuration of the articulatory sclerites at the base of the wing and the adjoining thoracic processes. You can demonstrate this for yourself in a freshly killed bee or fly. Depression of the mesonotum will automatically elevate and supinate the wings; or raise the mesonotum with a pin and watch the depression and pronation. The direct muscles of these insects are tubular and serve to modify the wing stroke over many beats.

SELECTED REFERENCES

Introductions to the structure and function of the various internal organs are given by Chapman (1969), Snodgrass (1935), and Wigglesworth (1972). Reviews are provided on certain aspects of the alimentary canal and digestion by House (1974*b*), Miles (1972), and Richards and Richards (1977); nutrition by Dadd (1973), Heinrich (1973), House (1974*a*), and Vanderzant (1974); tracheal system and respiration by Hinton (1969*a*, *b*, 1970), Keister and Buck (1974), P. L. Miller (1974*a*, *b*), and Whitten (1972); blood and circulation by Arnold (1974), Crossley (1975), Florkin and Jeuniaux (1974), Grégoire (1974), Jones (1974), McCann (1970), and T. A. Miller (1974); excretion by Maddrell (1971), and Stobbart and Shaw (1974); muscles and locomotion by Hoyle (1974), Hughes and Mill (1974), Johnson (1969), Maruyama (1974), McDonald (1975), Nachtigall (1974*a*, *b*), Pringle (1974), Rainey (1976), Rothschild et al. (1973), Sacktor (1970), D. S. Smith (1965), Usherwood (1975), and Weis-Fogh (1975); biochemistry by Agosin and Perry (1974), Bridges (1972), Candy and Kilby (1975), Gilmour (1961, 1965), Ilan and Ilan (1974), Kaplanis et al. (1975), Linzen (1974), Robbins et al. (1971) and Sacktor (1974); cell structure by Satir and Gilula (1973) and D. S. Smith (1968); and tissue culture by Brooks and Kurtti (1971) and Maramorosch (1976).

6 RECEPTION OF STIMULI AND INTEGRATION OF ACTIVITIES

In this chapter we describe the last of the organ systems, the nervous system, and then shift our emphasis to the varied relationships that insects have to their environment. Beginning with the structure and function of nerve cells, we progress through the major subsystems, including the sense organs by which insects detect changes in their external world. The responses of insects to internal and external stimuli are discussed in the context of the physiological state of the individual, the specific anatomical characteristics, the capacity of the nervous system to exhibit complex behavior, and the adaptive significance of the behavioral acts. After an introduction to the kinds of behavior exhibited by insects, we conclude the chapter with a detailed discussion of communication.

ORGANIZATION OF THE NERVOUS SYSTEM

NEURONS AND NEURONAL CONNECTIONS. Electric impulses are rapidly transmitted from one part of the insect to another by specialized cells called *neurons.* The nucleated cell body of a neuron is the *perikaryon* (Fig. 6.1*a*). Extending from the perikaryon are one or more slender processes, the fibers or *axons,* by which the neuron connects with other neurons and with other tissues in the body.

The usual point of contact, or *synapse,* between connecting axons is at the tips of the axons. The tips are finely divided into *terminal arborizations* and separated by a gap. Nervous impulses are transmitted from the *presynaptic* axon across the gap to the *postsynaptic* axon by a chemical transmitter substance. The postsynaptic or receptive axon is also called the *dendrite.* In insects as in some other animals, the impulse probably crosses the gap by the quick release of the chemical

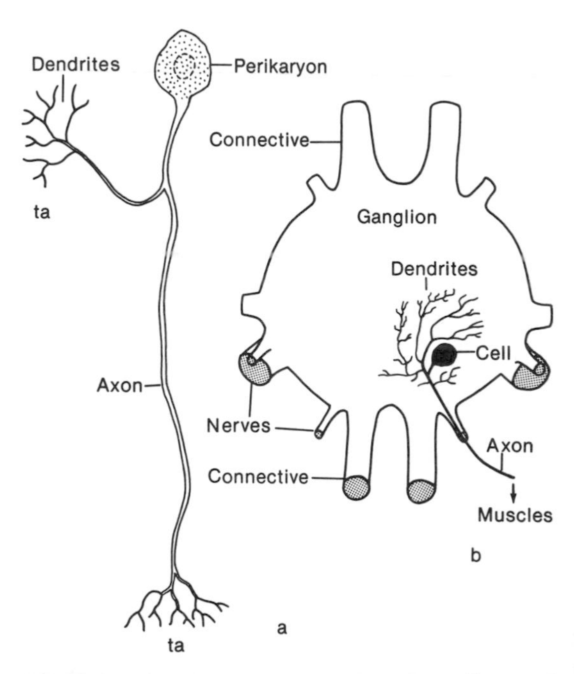

FIGURE 6.1 Structure of neurons and ganglia: *a*, diagram of a monopolar neuron; *b*, diagram of a monopolar motor neuron inside a thoracic ganglion. *ta*, terminal arborizations. (a *modified from Snodgrass,* 1935, *by permission of McGraw-Hill Book Company;* b *simplified and modified from Altman and Tyrer,* 1974.)

acetylcholine. This transmitter substance is then destroyed by the enzyme acetylcholinesterase. Certain insecticides, such as the organic phosphorus compounds, derive their toxic effect by acting as inhibitors of the enzyme. Except at the synaptic junctions, individual neurons are largely surrounded by *glial* cells, which probably function to prevent contact between neurons. Some axons, however, have points of contact other than at the tips, and form axon-axonal synapses.

Neurons that have one axon issuing from the perikaryon are called *monopolar.* The single axon is at least forked so that one branch serves as a dendrite. Some monopolar neurons, however, are highly branched and complexly interconnected with other neurons. Neurons having two separate axons extending from the perikaryon are called *bipolar.* Again, one serves as the dendrite and the other leads to the next neuron. Multipolar neurons with more than two axons are rare in insects (stretch receptors; see further on under "Mechanoreceptors").

Nerve impulses commonly begin in a sense organ, pass to the central nervous system, and then out to an appropriate *effector,* i.e., a muscle or gland that provides a response when stimulated. In the simplest pathway the impulse begins in the dendrite of a bipolar *sensory* neuron closely associated with a sense organ. The sensory perikaryon is usually located just beneath the sense organ, and it communicates with the segmental ganglion by a long sensory axon. The neuron initiates impulses only in response to certain kinds of stimuli, depending on the structure of the sense organ and the special sensitivity of the dendrites. Thus only certain mechanical disturbances, light waves, temperature, or chemicals elicit an impulse.

Within the ganglion the sensory axon transmits impulses either directly by one synapse to a monopolar *motor* neuron or indirectly by two or more synapses involving other monopolar neurons called the *internuncial* neurons or interneurons. The speed of transmission of an impulse is delayed by each synapse. The motor neuron then completes the arc by sending the impulse over a long motor axon to stimulate the effector.

The nervous stimulation of muscle fibers in insects differs from that of mammals. In the latter the axon terminates in a single motor end plate, and the transmitter substance to the muscle is acetylcholine. Motor axons of insects are branched, providing many endings on a muscle fiber. A muscle may receive stimulation from two or three different kinds of motor axons. Stimulation by a *fast* axon produces a "fast" response in the form of a brief powerful contraction. Each stimulation by a *slow* axon results in a small contraction, but when the stimulation is increased in frequency a "slow" contraction of increasing force is produced. Slow axons thus give greater control over movements. The third type of axon, if present, may inhibit the slow contraction of muscle. The transmitter substance in the neuromuscular junctions is not known; apparently it is not acetylcholine.

Most nervous pathways are probably much more complex than the simple reflex arc. The various segmental responses to environmental stimuli are regulated by the brain and by the interactions of other segmental ganglia. Synapses may involve more than two neurons, and a given neuron may have more than two synapses. Communication between ganglia is by internuncial axons. Some of these, called giant fibers, are larger in diameter and span many segments without intervening synapses, thus bringing distant segments of the body into direct communication for rapid responses.

NERVES AND GANGLIA. On a larger scale, the bundles of sensory and/or motor axons that issue from the segmental ganglia are called *nerves.* A *ganglion* is a mass of nervous tissue (Fig. 6.1*b*). The word is used loosely to apply not only to a single large mass but also to smaller masses that may be present inside. Thus each segmental ganglion is formed by the median fusion of a pair of embryonic ganglia. The central portion is the *neuropile,* consisting entirely of axons and processes of their associated glial cells. Included are sensory, motor, and internuncial axons and their various synaptic junctions. The cell bodies of the neurons and glial cells are located only around the outside of the neuropile. The paired longitudinal connectives between ganglia are bundles of internuncial axons plus sensory and motor axons that communicate with other segments. The ganglia, nerves, and connectives are surrounded by a sheath composed of an outer, noncellular *neural lamella* and an inner, cellular *perineurium.* The sheath not only serves a supportive function but also maintains a chemical environment around the axons which is favorable for impulse transmission.

The nervous system is divided into four subsystems: (1) *sense organs,* including the light, chemical, temperature, and mechanical receptors; (2) *peripheral nervous system,* including the sensory and motor neurons that communicate with the integument and muscles (this system is not discussed further); (3) *central nervous system,* including the brain and the segmental ventral ganglionic chain that generally regulate the bodily activity; and (4) *stomatogastric nervous system,* including several small ganglia that regulate the foregut, the midgut, and several endocrine glands.

CENTRAL NERVOUS SYSTEM

The central nervous system of insects includes the *brain,* situated above the mouth, and the *ventral nerve cord,* beginning behind the mouth. The evolution of the central nervous system was discussed earlier (Fig. 2.3). Recall that the basic structure of the system consists of paired embryonic ganglia connected transversely by commissures and longitudinally by connectives.

BRAIN. The brain has three ganglionic masses (Fig. 6.2): (1) The *protocerebrum* is the largest and most complex region. Laterally it receives the nerves from the optic lobes of the compound eyes; dorsally, it receives nerves from the ocelli. Internally, the protocerebrum has a pair of mushroom-shaped neuropiles, the *corpora pedunculata,* which are important centers for the integration of behavior. These bodies are especially large in social Hymenoptera. Medially is the *pars intercerebralis,* a complex cellular mass of neurosecretory cells and ordinary neurons that communicate with the corpora cardiaca of the stomatogastric system. Three other neuropiles are present in the protocerebrum: the median *protocerebral bridge* and *central body,* and the ventral, paired *accessory lobes* that are connected by a commissure. (2) The *deutocerebrum* has motor and sensory axons to the antennae. The paired neuropiles, or *antennal lobes,* are connected by a commissure. Recall that both the protocerebrum and deutocerebrum were primitively in front of the mouth. (3) The *tritocerebrum* has sensory and motor axons to the labrum and to the frontal ganglion of the stomatogastric system. The paired neuropiles, or *tritocerebral centers,* are connected by the *tritocerebral commissure,* which is visible behind the esophagus. The tritocerebrum was primitively the first ventral ganglion behind the mouth and secondarily became incorporated into the brain.

VENTRAL NERVE CORD. The ventral nerve cord is a chain of segmental ganglia connected anteriorly with the tritocerebrum by the paired *circumesophageal connectives.* The ganglia of the three gnathal segments are fused into a compound mass named the *subesophageal ganglion.* Motor and sensory nerves extend to the mandibles, maxillae, labium, salivary glands, and cervical muscles. The three ganglia of the thoracic segments serve the legs and flight mechanism and are usually the largest ganglia of the ventral nerve cord. Sensory and motor nerves extend to the various sense organs and muscles. In the abdomen are no more than eight ganglia. The last is a large, compound ganglion serving the eighth and following segments, including the genitalia and cerci.

An evolutionary tendency exists for segmental ganglia to fuse anteriorly into compound ganglia (Fig. 6.3). In the grasshopper, for example, the first three abdominal ganglia are fused with the metathoracic ganglion. In some advanced He-

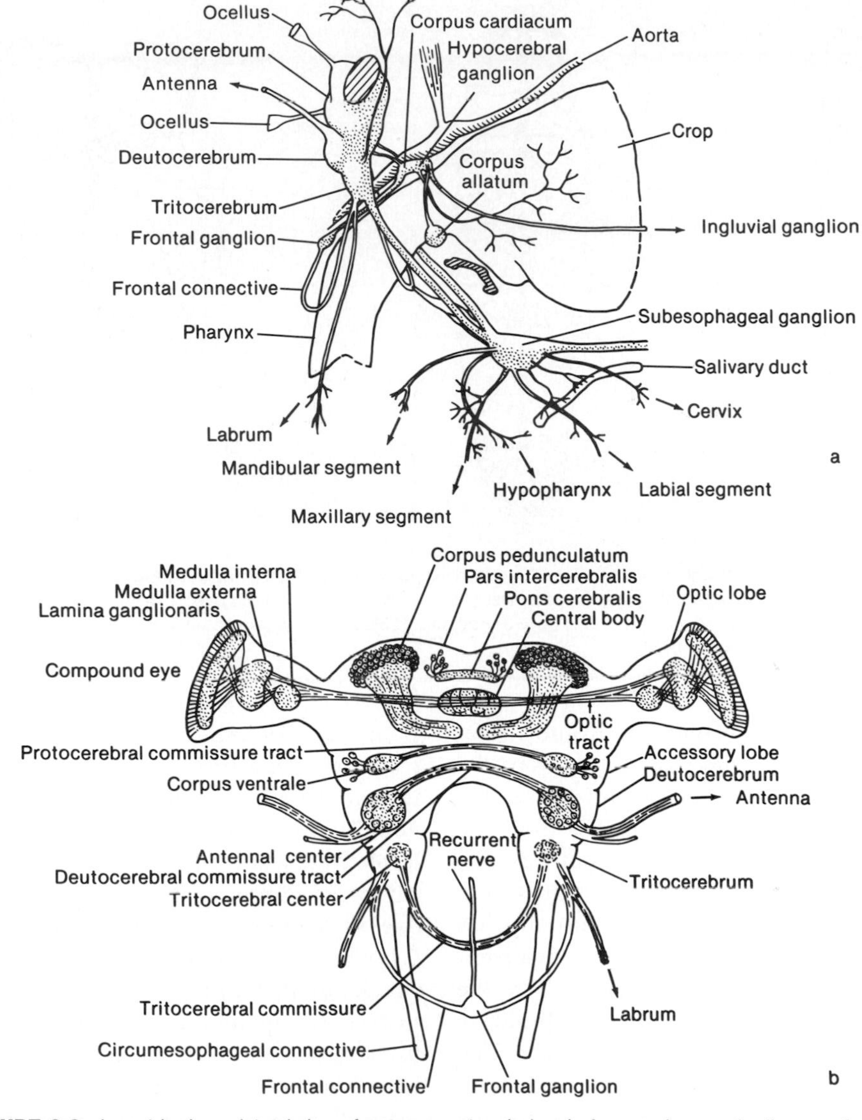

FIGURE 6.2 Insect brain: *a*, lateral view of nervous system in head of a grasshopper; *b*, diagramatic cross section of brain, showing ganglia and commissures. (*Redrawn from Snodgrass,* 1935, *by permission of McGraw-Hill Book Company.*)

miptera, Diptera, and Coleoptera most or all of the ganglia of the thorax and abdomen are fused into a single compound ganglion, with nerves extending to the respective segments. This arrangement presumably permits more rapid communication among the ganglia by reducing the lengths of internuncial axons.

STOMATOGASTRIC NERVOUS SYSTEM

The stomatogastric system includes two median ganglia connected to the brain and a pair of ganglia on the surface of the foregut. (1) The *frontal ganglion* is situated medially in front of the

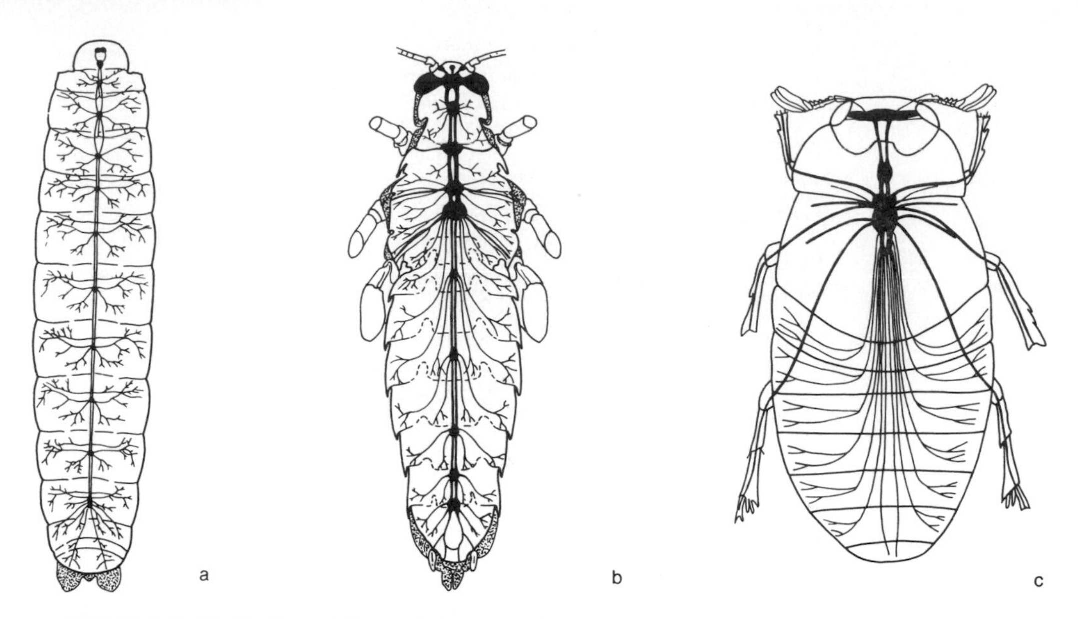

FIGURE 6.3 Modifications of the ventral nerve chain: *a*, system of a caterpillar, *Malacosoma americana* (Lasiocampidae); *b*, system of a grasshopper, *Dissosteira carolina* (Acrididae); *c*, system of a beetle, *Lachnosterna fusca* (Scarabaeidae). (a *and* b *redrawn from Snodgrass,* 1935, *by permission of McGraw-Hill Book Company;* c *redrawn from Packard,* 1898.)

pharynx, usually just behind the frontoclypeal suture. It receives the *frontal connectives* from the tritocerebrum and the *recurrent nerve* from the hypocerebral ganglion. Motor nerves from the frontal ganglion innervate the foregut. (2) The *hypocerebral ganglion* is situated behind the brain, between the aorta and foregut. It usually has three connections: anteriorly by the recurrent nerve to the frontal ganglion; posteriorly to the ingluvial ganglion; and dorsally to the brain by way of the neuroendocrine corpora cardiaca. The hypocerebral ganglion also sends nerves directly to the esophageal wall. (3) The paired *ingluvial ganglia* are located posteriorly on the foregut. They control gut movements through nerves to the gut wall, proventriculus, and midgut.

The neuroendocrine corpora cardiaca and the corpora allata are closely associated with the stomatogastric system (Chap. 4).

SENSE ORGANS

INSECT EYES. Light is detected by specialized light-sensitive cells that are grouped into eyes (Fig. 6.4). The cells are typically clustered in a *retina* beneath a transparent lens-shaped area of the cuticle called the *cornea.* Two general kinds of visual organs are recognized: (1) *Compound eyes* are located dorsolaterally on the head, and each is composed of one to many thousands of closely packed photoreceptive units, the *ommatidia* (Fig. 6.4*b*). Each ommatidium has its own corneal lens, which usually forms a hexagonal facet on the eye surface. Compound eyes are directly connected to the optic lobes of the brain. (2) *Simple eyes* each have a single round corneal lens with several to many light-sensitive cells beneath. Simple eyes are of two types: (*a*) *Dorsal ocelli* (Fig. 6.4*e*) are located dorsomedially and number no more than three. They are innervated by the median part of the protocerebrum and commonly occur in addition to the compound eyes. (*b*) *Stemmata,* or lateral ocelli, are located laterally and number usually less than a dozen at each side. They are innervated by the optic lobes and occur in the place of compound eyes (Fig. 6.4*f*).

Protura (Fig. 18.1*b*) and Diplura have no discrete eyes. Compound eyes are found on at least some adults of all other orders except Siphonaptera, in which the lateral eyes, if present, are simple (Fig. 43.1*a*). Compound eyes are also found on the immature stages but they are missing in the larval Endopterygota (except for the Mecoptera). The lateral eyes of larvae are stemmata (Fig. 17.4*b*). Ocelli occur together with compound eyes, but both may be secondarily lost in various groups of insects, especially in apterous insects such as the parasitic Phthiraptera, female scale insects, cave insects (Fig. 12.2*b*), or worker castes of some termites and ants.

COMPOUND EYES AND VISION. The general structure of an ommatidium (Fig. 6.4*b*) includes two lenses, the *corneal lens* and *crystalline cone,* and a set of six or eight sensory, or *retinula,* cells. Surrounding the crystalline cone are *primary pigment cells,* and around the retinula are *secondary pigment cells.* The biconvex cornea is secreted by two modified epidermal cells, the *corneagenous cells.* The cone is composed jointly of four cells, and by their internal structure, light from the cornea is focused on the retinula. The elongate cells of the retinula are packed together and modified

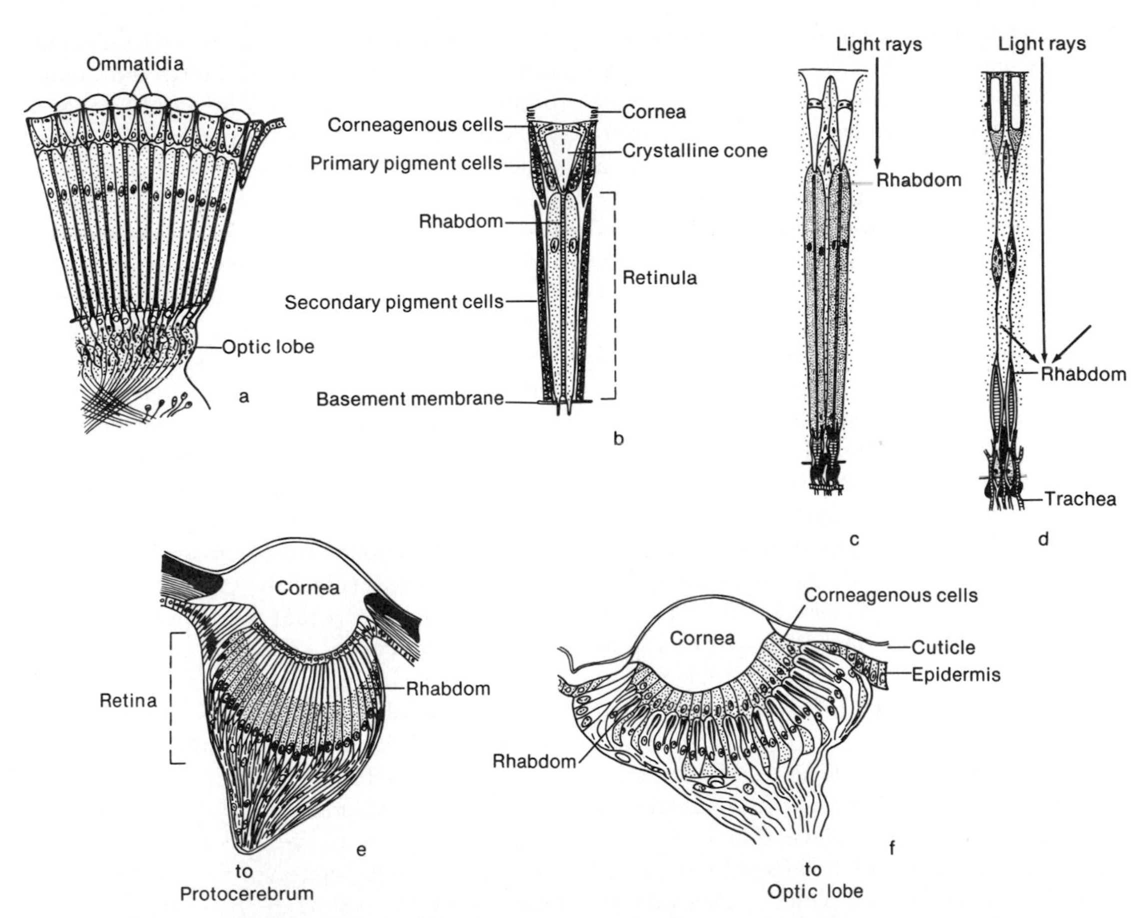

FIGURE 6.4 Structure of insect eyes: *a*, section of compound eye; *b*, ommatidium; *c*, ommatidium of the apposition type; *d*, dark-adapted ommatidium of the superposition type; *e*, dorsal ocellus; *f*, stemma or lateral ocellus. In *c* light rays that pass through the cornea of an ommatidium strike only the rhabdom beneath, but in *d* the light rays pass into adjacent ommatidia and strike other rhabdoms. (*Redrawn from Snodgrass,* 1935, *by permission of McGraw-Hill Book Company.*)

around the central axis into a *rhabdom.* Each cell contributes a *rhabdomere* of tightly packed microtubules oriented at right angles to the central axis. Microtubules within a rhabdomere are parallel to one another and usually also parallel with those on the opposite side of the axis. Various modifications of this arrangement occur. In the honeybee, for example, the eight retinula cells are fused pairwise into four compound units. The opposite units have parallel microtubules such that adjacent units have microtubules at right angles. Each retinula cell has a separate axon to the optic lobe. Within the lobe the number of pathways is reduced to the protocerebrum.

The nerve impulse is presumably initiated in the rhabdomere by the reaction of a photopigment to light. The reaction is probably similar to that in mammals, because a retinal-protein complex has been identified in insect heads. In a single insect, different photopigments may be sensitive to as many as three different portions of the spectrum. Thus a basis for color vision exists, but the nervous system must be able to process the information.

Through behavioral and electrophysiological studies it is known that many insects do discriminate colors. The spectrum visible to insects, however, differs from that of humans. In honeybees, the spectrum is shifted away from red toward the shorter wavelengths. By a combination of ultraviolet, blue, and yellow sensitivities, honeybees behaviorally discriminate among at least four colors: ultraviolet (300 to 400 nm), violet-blue (400 to 480 nm), blue-green (480 to 500 nm), and green-yellow-orange (500 to 650 nm). Red is invisible to bees, but not to some butterflies.

Some insects are also able to discriminate the plane of polarized light coming from the sky. A light wave vibrates in a plane that is at right angles to the line of travel. Polarized light is light vibrating in a single such plane, whereas direct sunlight has planes of vibration in all directions around the line of travel. Depending on atmospheric conditions the proportion of light that is polarized increases in certain regions of the sky up to about 70 percent. How insects detect the plane of polarization in a given patch of sky is not known, but it is possible that the orientation of the microtubules in the rhabdom is important in this regard.

IMAGE FORMATION IN COMPOUND EYES. Insects evidently detect the form and movement of objects in their environment, but the mechanism of image formation is not understood and is controversial. Each ommatidium receives light from a small spot in the total visual field. According to the *mosaic theory* of insect vision, each ommatidium responds only to the intensity of light from its own limited view, not to the details of the tiny image, if any, formed inside the ommatidium by the lens system. The overall image of the visual field is then constructed as a spotted mosaic of the different responses of the ommatidia to the different light intensities. Images of two types have been identified: an *apposition* image is formed when each ommatidium is stimulated only by light passing through its own lens system, while a *superposition,* or overlapping, image is formed when ommatidia also receive light passed by the lenses of adjacent ommatidia.

The compound eyes of some insects form images only by apposition. Structurally the ommatidia of these insects are shielded from stray light inside the eye by a complete jacket of pigment cells, and the rhabdoms are situated just beneath the lenses (Fig. 6.4*c*). Apposition eyes are found among insects that are active during the strong light of day.

The mosaic image formed by apposition gives the highest resolution of detail. In terms of visual acuity, such insects can distinguish two objects that are separated by an angle of about 1°, which corresponds to the smallest angular separation of ommatidia. Human eyes by comparison can distinguish objects separated by as little as one-ninetieth of 1°. Although the visual acuity of insects would seem adequate for seeing objects in some detail, behavioral experiments with honeybees demonstrate that their discrimination of form is poorly developed. The compound eye, however, is well adapted for the detection of the movement of objects because successive ommatidia are stimulated. The distance to an object can be judged where the visual fields of the paired eyes overlap. The position of the object seen by both eyes is thus triangulated. The wide separation of the eyes in predators such as mantids and dragonfly naiads probably improves distance perception.

The compound eyes of other insects may form either apposition or superposition images. Most insects of this type are Lepidoptera, Coleoptera, and Neuroptera that are active at twilight or night. The exact structure of the ommatidia varies from group to group, but in general the rhabdoms are relatively short and separated from the lens system by a gap, or clear zone (Fig. 6.4*d*). The migration of the secondary pigment cells permits the eye to form both types of images over a wide range of light intensities. During day the pigment and to some extent the individual pigment cells move proximally into the clear zone and limit the passage of light from the lenses to single rhabdoms. This is the *light-adapted* eye; the image is formed by apposition. At dusk, the pigments and secondary pigment cells contract distally, permitting light from adjacent ommatidia to stimulate the rhabdoms. The *dark-adapted* eye then forms images by superposition. Some question exists whether an image, if formed, is sensed in certain types of dark-adapted superposition eyes. The detection of overall light intensity, however, would seem to be increased because stray light rays within the eye are not blocked by pigment. In some moths, tracheal tubes are arranged between and parallel to the ommatidia, forming a *tapetum* that may aid in reflecting light onto the rhabdoms.

OCELLI AND STEMMATA. Simple eyes have a single, biconvex, cuticular lens that focuses light on the rhabdoms of sense cells beneath (Fig. 6.4*e*, *f*). In ocelli the image is formed below the level of the sense cells so that no form can be detected. Ocelli are quite sensitive, however, to low intensities of light. They probably function to detect daily changes in light intensity and generally to stimulate the nervous system accordingly. On the other hand, the stemmata of larvae are visual organs. The lens of a stemma forms an image on the rhabdoms. When several stemmata are clustered at each side of the head, the mosaic image is sufficient to detect gross features of the environment. In some larvae, the stemmata detect the plane of polarized light.

MECHANORECEPTORS. Mechanical distortion of the body is detected by a variety of simple sense organs, or *sensilla*. In the usual arrangement, the dendrite of a single bipolar neuron is attached to a movable part of the body, often by a minute cuticular sheath called a *scolopale*. Movements of the body part in response to mechanical stress, touch, wind, or vibrations in air, water, or solid substrates initiate nerve impulses that are transmitted to the central nervous system. Some sensilla respond to disturbances from the environment, while others respond to the position of one part of the body relative to another, thus functioning as *proprioceptors*. These are important in maintaining posture and in orienting to gravity.

Four types of mechanoreceptors are recognized: (1) *Trichoid* sensilla are hairlike setae with a single dendrite attached at the base. Some project well above the cuticle surface and serve as *tactile*, or touch, organs (Fig. 6.6). (2) *Campaniform* sensilla are oval, domelike areas of cuticle that raise or lower as the adjacent exoskeleton is mechanically strained (Fig. 6.5*b*). The dendrites detect movements of the dome. (3) *Chordotonal* organs are completely internal and formed by units or *scolopidia* consisting of three cells: a bipolar neuron, a scolopale cell, and an attachment cell. The dendrite sometimes ends inside the scolopale in a process that resembles a cilium. One or more scolopidia cluster into a chordotonal organ and are variously attached to the epidermis. Some are suspended between two points in membranous areas or between leg joints, thus permitting detection of movements

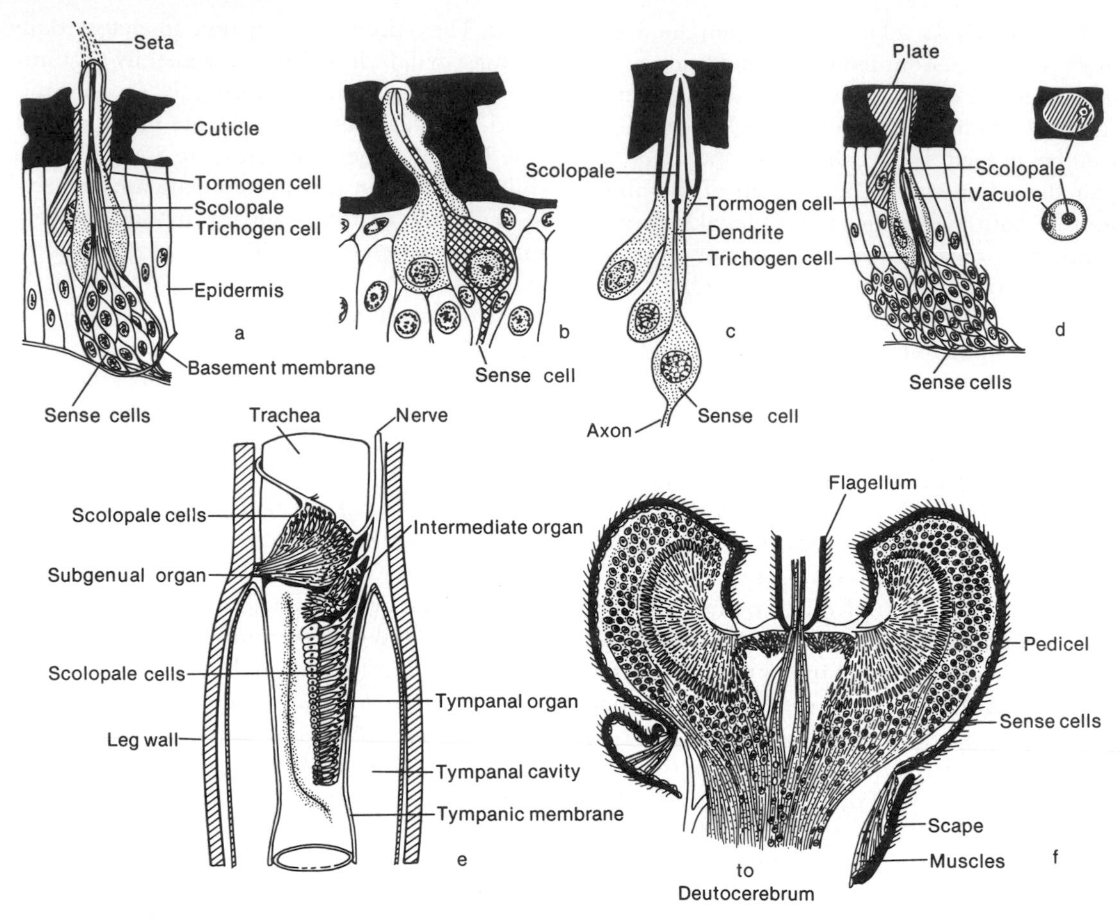

FIGURE 6.5 Sensilla: *a*, trichoid chemoreceptor; *b*, campaniform mechanoreceptor; *c*, coeloconic chemoreceptor; *d*, plate organ; *e*, chordotonal acoustical organ in tibia of right leg of tettigoniid; *f*, Johnston's organ in antenna of the fly *Chaoborus*. (*Redrawn from Snodgrass,* 1935, 1925, *by permission of McGraw-Hill Book Company.*)

between the points of attachment. *Subgenual* organs of the tibia, however, have a single attachment and do not span a joint. They are especially sensitive to vibrations in the substratum. *Johnston's organ* is a chordotonal organ in the antennal pedicel (Fig. 6.5*f*). By its attachment to the base of the flagellum, movements of the flagellum are detected. *Tympanal* organs are specially designed for the reception of sound (Fig. 6.5*e*). A chordotonal organ is attached to a thin cuticular membrane or *tympanum* that freely vibrates when struck by sound waves. (4) *Stretch* receptors are attached to connective tissue or muscles, registering the tension of such soft tissues, especially in the abdomen. These receptors differ from other sensilla in having multipolar rather than bipolar neurons.

All four types of sensilla may function as proprioceptors. Groups of trichoid sensilla located at each side behind the head on the cervical sclerites register the position of the head with respect to the body. Similar groups at leg joints register the positions of the limbs. Campaniform sensilla are widely scattered over the body surface at points of stress and bending. The halteres of Diptera and insect wings in general are well supplied. Strategically placed chordotonal organs and stretch receptors permit detection of movements of the limbs, segments, and viscera. The positions of the antennae are detected by Johnston's organ.

Vibrations are detected mainly by trichoid sensilla and chordotonal organs. Hairs in the cerci of cockroaches, for example, respond to sound waves of low frequency as well as to wind currents. Subgenual organs are found in many insects but are lacking in Archeognatha, Coleoptera, and Diptera. In male Culicidae and Chironomidae (Diptera), the antennae have fine, long hairs that vibrate in response to the flight tone of the females. This stimulates a mating response. The hair vibrations move the flagellum, which in turn stimulates Johnston's organ in the pedicel. The same response can be evoked with a tuning fork that vibrates at 100 to 800 Hz.

The special acoustical receptors of insects are the tympanal organs found in various unrelated taxa and in different places on the body. The 6.7*a–c*), and Pyraloidea and Geometroidea (Lepidoptera Fig 44.6*f*); on the tibia of the forelegs 6.7*a–c*), and Pyraloidea and Geometroidea (Lepidoptera Fig 44.6*f*); on the tibia of the forelegs in Gryllidae and Tettigoniidae (Orthoptera Fig. 29.2*a*); and on the mesothorax of the bugs *Corixa* (Corixidae) and *Plea* (Pleidae), and the metathorax of Noctuoidea (Lepidoptera Fig. 44.6*e*).

The Lepidoptera listed above are night-flying moths. Their tympanal organs are used to detect and avoid approaching bats, the sounds of which can be sensed at about 30 m. In the Noctuoidea the organ has only two scolopidia at each side, and these respond to sounds well over 100 kilocycles per second. The tympanal organs of Orthoptera and Hemiptera are used in mating behavior. Sounds produced usually by the male, or by both sexes, are detected and serve to bring mates together. In *Plea* the organ has but one scolopidium per organ, but in others the organ may be extremely complex, involving up to 1500 scolopidia in Cicadidae. These organs respond to a wide range of sound frequencies, generally less than 20 kilocycles per second. The average upper limit for humans is about 14 kilocycles per second. Insects discriminate poorly, if at all, among the different frequencies, but respond mainly to the intensity and duration of sound and the intervals between sounds. Insects can detect an interval as brief as 0.01 second, whereas the human ear fails to detect intervals less than 0.1 second.

CHEMORECEPTORS. The sensilla that respond to chemicals in air or liquids are modified setae. Typically the dendrites of several to many bipolar neurons pass through one scolopale, divide into fine branches, and terminate at one or more minute pores through the cuticle of the hair. The nerve endings are possibly in direct contact with molecules of air or liquids or separated by a receptive chemical substance. Each neuron of the cluster responds to a different, but possibly overlapping, array of chemical compounds that are important in the insect's life. In addition to specific sensitivities, the neurons may also respond to general irritant substances, such as ammonia fumes.

Setae of four types are commonly associated with chemoreceptor neurons: (1) *Trichoid* sensilla may serve both as mechanoreceptors and chemoreceptors (Figs. 6.5*a*, 6.6). Such hairs on the labella and legs of blowflies (Calliphoridae) have a mechanoreceptor attached to the base and the dendrites of a few chemoreceptor neurons situated at a single pore at the tip. Trichoid sensilla are commonly gustatory (taste) receptors. (2) *Basiconic* sensilla are peglike and thin-walled, have many pores, and project above the cuticle. They are the most common type of olfactory (smell) receptors, being found on the antennae of most insects, as well as on the mouthparts and other places on the body. (3) *Coeloconic* sensilla have the sensory peg sunk in a cuticular pit (Figs. 6.5*c*, 6.6). These are found on the antennae and mouthparts of some insects. (4) *Plate organs* are found on the antennae of aphids and honeybees (Figs. 6.5*d*, 6.6). Each is a round, flat cuticular plate perforated by pores. *Sensory pits* contain many sensilla in a subcuticular pit. Such pits have been identified on the third antennal segment of Diptera (Suborder Cyclorrapha) and labial palpi of Lepidoptera and Neuroptera.

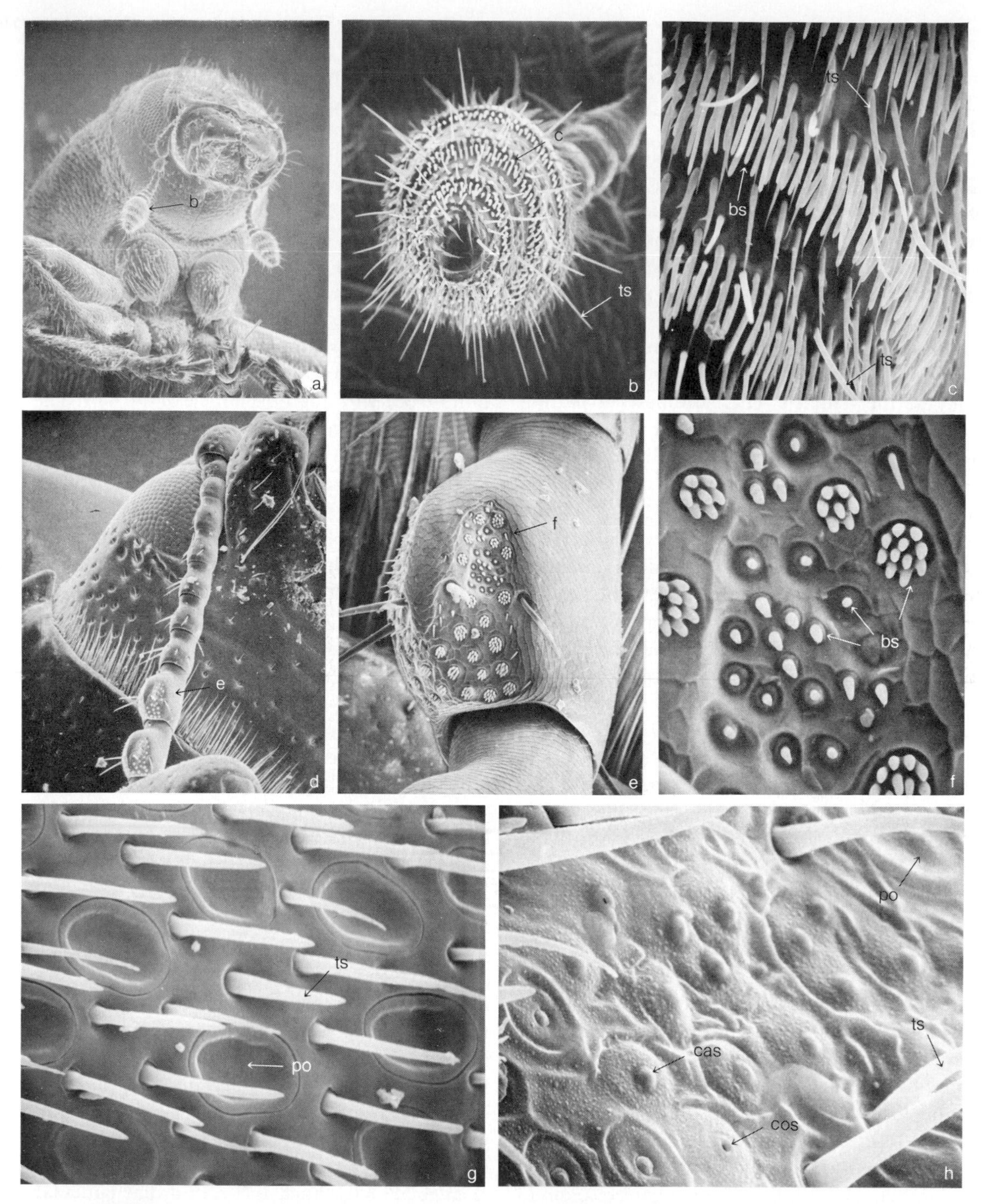

FIGURE 6.6 Antennal sensilla: *a*, head of the bark beetle *Pseudohylesinus* spp. (Scolytidae), showing antennae (arrow marks position of photo *b*; 37X); *b*, antennal club of same showing sensory bands (arrow marks position of photo *c*; 304X); *c*, sensilla on club of same, blunt basiconic sensilla form dense bands (743X); *d*, head of a male beetle *Temnochila virescens* (Ostomidae), showing antenna (arrow marks position of photo *e*; 32X); *e*, ninth antennal segment enlarged (arrow marks position of photo *f*; 169X); *f*, basiconic sensilla of several types whose functions are unknown (878X); *g*, eleventh antennal segment of the worker honeybee, *Apis mellifera* (Apoidea; 1439X); *h*, apical (twelfth) antennal segment of same (2483X). Basiconic sensilla, *bs*; campaniform sensilla, *cas*; coeloconic sensilla, *cos*; plate organ, *po*; and trichoid sensilla, *ts*. (*Photos* a *through* f *courtesy of and with permission of Clyde Willson*; *photos* g, h *courtesy of and with permission of Kim Hoelmer*).

Chemoreceptors are found on the antennae, mouthparts, legs, and ovipositor. They play an important role in many activities. The detection of odors, or *olfaction*, is involved in locating food and sites for oviposition, and in finding mates through chemical sex attractants. Social insects recognize colony members by odor. The ability of insects to detect minimal concentrations of chemicals varies with the physiological state of the insect and, of course, with the specific chemical compound or mixture of compounds. In tests with honeybees, involving many odorous compounds, the general conclusion is that the threshold concentration detected is on the average about equal to that of humans. Bees do sense certain substances important to them at lower concentrations, such as the odors of beeswax and certain glandular secretions.

Certain key compounds that are crucial in the life of an insect may be sensed at unbelievably low concentrations. Calculations for the male silk moth, *Bombyx mori,* indicate that single molecules of the sex attractant bombykol produce impulses in the appropriate antennal chemoreceptors. The moth responds behaviorally to as few as 200 bombykol-induced impulses per second. This is barely above the theoretical minimum needed to transmit information in a system that already has 1600 impulses per second of spontaneous nerve activity or "noise."

A distinction is sometimes drawn between olfaction of odors and the taste or *contact chemoreception* of substances in liquid. Though such a distinction may be useful in terrestrial insects, it is irrelevant in aquatic forms. Furthermore, the sensilla and physiological mechanisms are the same. By behavioral and electrophysiological analyses, insects have been shown to discriminate among sweet, salt, acid, and bitter compounds in solution and to exhibit a sense for water. The minimum taste thresholds are generally lower in insects than in humans. A convenient behavioral response has aided in surveying chemicals in various concentrations. When sugar solutions are applied to the tarsi of flies, butterflies, and honeybees, they respond by extending the proboscis. The proboscis is not extended for water or solutions too weak to detect. As a consequence responses to sugars have been extensively studied.

Honeybees respond positively to only seven, and marginally to two, of 30 substances that taste sweet to humans. The seven are sucrose, glucose, fructose, melezitose, trehalose, maltose, and α-methyl glucoside. Of these, five are naturally found in the bee's diet: nectar contains the first three and honeydew includes the first five. The last two sugars, as well as the 23 tasteless or nearly tasteless sugars, are not in the bee's natural food. Sucrose solutions of 0.06 to 0.12 M can be discriminated from water by starved bees with their mouthparts. Their antennae are much more sensitive. With behavioral training, bees respond to sugar solutions applied to their antennae in concentrations as low as 0.0001 M.

The trichoid sensilla on the tarsi and labella of blowflies can be stimulated individually with solutions. They accept a much greater variety of carbohydrates than honeybees. When it is applied to the tarsi, the acceptance threshold for sucrose in solution is 0.01 M in *Phormia regina,* but as low as 0.0006 M in *Calliphora erythrocephala. Pyrameis* butterflies respond by tarsal stimulation to 0.00008 M sucrose when starved. The threshold for humans at 0.02 M is 250 times poorer by comparison.

BEHAVIOR

What an insect does in response to external and internal stimuli depends not only on the nature of the stimulus, but also on the insect's physiological state. The response to food may depend not only on the taste of food and hunger, but also on the hormonal milieu within which the nervous system functions. At the onset of molting, for example, feeding may cease altogether. Thus an insect's behavior is controlled by two interrelated systems: nervous and endocrine. Both function to sequence behavior and prevent conflicting behavior so that the survival of the individual and the propagation of the species are favored. Both communicate directly to various tissues; the nervous system communicates by physical contact, and the endocrine system by hormones in the hemolymph. The small size of insects is probably an advantage in this internal communication. The systems differ in that the nervous system functions on a moment-to-moment basis, integrating specific sensory and motor impulses, while the endocrine system reacts more slowly and has long-term effects in a variety of tissues, including the central nervous system itself.

The behavior of insects under natural conditions is a complex mixture both of stereotyped responses to stimuli that are controlled and coordinated by the central nervous system, and of learning. Segmental ganglia may be largely independent in controlling the responses to certain stimuli. Stepping movements or reflexes of the legs, for example, are controlled directly by the respective thoracic ganglia but are coordinated by the inhibition and stimulation of the brain and other ganglia. Stereotyped responses are programmed into the nervous system by heredity but the "program" is not rigid. Considerable flexibility may exist in the timing, coordination, and sequencing of acts, depending on sensory input.

TAXES. A complex but stereotyped movement that orients the whole body to an environmental stimulus is called a *taxis*. Commonly this is an orientation to gravity (*geotaxis*), to light (*phototaxis*), to wind currents (*anemotaxis*), or to sound (*phonotaxis*). An orientation may be either *positive* (toward the stimulus) or *negative* (away from the stimulus). A positive phototaxis, for example, involves orienting the body so that both eyes are equally stimulated in front. The orientation to light may also involve positioning the body so that the light is dorsal or ventral. The *dorsal light reaction* is common among aquatic insects, providing a means to orient to the earth's surface under circumstances where the effect of gravity on proprioceptors is reduced by the buoyancy of water. The backswimmer *Notonecta* (Notonectidae), so named because it swims upside down (Fig. 12.3*d*), appropriately has a *ventral light reaction*. Orientation at a fixed angle to sunlight is called the *light compass reaction*. When it is used as a means for navigation to and from a given point, compensation must be made for the sun's movement. Honeybees, some ants, and a few other insects have such a compensatory mechanism, probably involving their sense of time or circadian rhythm (see Chap. 11). Orientation to a celestial body combined with compensation for its movement is an *astrotaxis*.

Although a taxis is a stereotyped behavior, a given taxis is not an inevitable response to a given stimulus. Taxes must be under the control of the nervous system to benefit the insect. Depending on the circumstances, a taxis can be switched on or off; the orientation switched from positive to negative or vice versa; and one stimulus replaced by another. An ant, for example, leaving the underground colony on a foraging trip at the surface is negatively geotactic and positively phototactic, and follows a compass direction by astrotaxis. On her return, the "sign" is switched in the astrotaxis, permitting her to navigate back to the nest, and on arriving becomes positively geotactic and negatively phototactic. Positive geotaxis and negative phototaxis, and each with opposite signs, are widely coupled among insects. In honeybees, the sun and gravity are interchanged as stimuli for orientation (Fig. 6.8).

INSTINCTS. Many of the activities of insects are stereotyped sequences of behavior of greater or lesser complexity. Such sequences are loosely called *instincts*. Examples include feeding, migration, mating, oviposition, nest building, cocoon spinning, and movements during ecdysis. Here the physiological and developmental state of the insect, especially the hormonal milieu, is important in determining whether a stimulus will trigger, or release, a given program of behavior. The program may be complex and flexible, including alternate sequences of behavior depending on the response of the mate, suitability of food, or other circumstances.

LEARNING. A change in behavior as a result of past experience is *learning*. Learning can be shown to occur even in the segmental ganglia of insects.

Headless cockroaches and grasshoppers can be trained to lift a leg to avoid an electric shock. Such elementary forms of learning may prove to be widespread and may be hitherto undetected parts of instinctive patterns of behavior.

The most important centers for learning are the brain's mushroom bodies, or corpora pedunculata. In insects that exhibit complex behavior and learning, the bodies are larger in volume and have more cells and more synapses than in insects with simpler and clearly stereotyped behavior.

As a comparison of learning ability, Schneirla (1946) tested eight white rats and eight *Formica* ants in similar mazes with six blind alleys. The rats mastered the maze in 12 runs, but the ants required 30 or more runs. When the maze was reversed, the rats were able to transfer their previous experience and reduced the time needed to learn the maze. The ants performed in the maze as if it were entirely new.

COMMUNICATION. The behavior interactions of insects among themselves and with other organisms usually involve some sort of communication. *Communication* occurs when an organism provides a visual, tactile, auditory, and/or chemical signal that influences the physiology or behavior of another organism. The distance over which the communication takes place may be such that only one sense will effectively receive the signal. This is *long-range* communication, in contrast to *short-range* communication in which more than one sense may be involved.

Instances of communication may be broadly divided into those that occur between members of the same species and those that occur between members of different species. The former, or *intraspecific,* cases of communication are for the purpose of attracting or recognizing mates, aggression or defending territories, assembling or dispersing aggregations, or, in social species, for various cooperative activities and to control caste development and behavior.

Communication between different species, i.e., *interspecific,* may be divided further into those that benefit the sender, those that benefit the receiver, and those that result in a mutual benefit. (1) In the first relationship the sender is usually under threat of attack by a predator and produces one or a combination of warning signals to discourage attack (see Chap. 15 for additional details of this strategy). Other instances in which the sender benefits involve signals that confuse the communication of a predator or competitor. High-frequency clicks produced by the tymbals of night-flying Arctiidae moths (see below) probably confuse the echolocation system of bats. (2) In the second relationship, the sender may become a victim when some aspect of its activities signals its presence to a predator or parasitoid. Or a potential victim escapes by sensing the enemy's presence. The detection of the sounds of bats by the tympanal organs of moths is an example. In either case the receiver of the signal is the beneficiary. (3) In the last type of relationship, the insect pollination of flowers supplies an example. The flower provides visual, tactile, and chemical signals that attract insects. The insects obtain food, and the flower achieves cross-pollination (see Chap. 13).

As we shall see below, communication may become quite complex: a given signal may be sensed both by members of the same species and by members of different species; and different species of organisms may produce an identical signal.

COMMUNICATION BY SIGHT. Insects that are active by day commonly depend on visual signals. For example, the color patterns, size, and motion of female butterflies are known to attract males of the same species. Visual communication, however, is not restricted to diurnal insects. In Lampyridae (Coleoptera) the winged males and wingless, larviform females communicate by rhythmic flashes of light from abdominal luminescent organs. Large, well-tracheated cells in the organs generate light by oxidation of luciferin in the presence of the enzyme luciferase. The reaction is highly efficient; 98 percent of the energy is released as light. A product of the reaction inhibits luciferase and stops further reaction. Nerve impulses stimulate light production, probably by removing the inhibition of luciferase. The females of *Photuris fairchildi* in Nova Scotia mimic the flashing frequency of species in three other genera and eat the males that are attracted (Buschman, 1974).

COMMUNICATION BY SOUND. Of all the animals, only arthropods and vertebrates communicate by producing vibrations in air, water, or solid substrates. In insects the vibrations may be created as a by-product of some other activity, especially flying. We have already mentioned how the flight tone of female mosquitoes attracts males, thus serving an intraspecific function. Loud buzzing by bees and wasps serves interspecifically as an effective warning signal that is even imitated by other insects. Vibrations in wood or soil are created by striking the body against the substratum. The death-watch beetle *Xestobium* (Anobiidae) bores in furniture and wooden houses. At sexual maturity it strikes its head agains the burrow walls, producing a rapping sound that at night influences the behavior of humans as well. Many species of termites drum their heads in response to disturbances of their colonies.

Special devices for creating sound are prevalent in some taxonomic groups, especially Orthoptera and Hemiptera (Fig. 6.7). Vibrations are generated mostly by one of two mechanisms: movement of roughened surfaces together or vibration of a membrane. In the first type of mechanism, one surface may be modified into a scraper and the other into a file (Fig. 6.7*d–f*). The movement of the scraper on the file is called *stridulation.* Some adult as well as immature insects stridulate to produce arrhythmic, rasping warning sounds. These are heard for only a short distance or sensed by contact. The devices are simple and may be formed wherever two parts of the body normally rub together.

In Orthoptera stridulation by adults can be heard at long range and serves to bring the sexes of the same species together for mating. The devices for sound production and reception are complex, and the sounds of each species have a characteristic frequency, rhythmic spacing, and duration. In Gryllidae and Tettigoniidae, one or both forewings have a resonant area of cuticle that vibrates as the scraper of one wing strikes the teeth of the file on the opposite wing. Tympanal organs of hearing are located in the fore tibiae (Fig. 29.2*a*). Receptive females orient to the calling songs of the males and are attracted to them. Acrididae rub a row of pegs on the inner side of the hind femur against raised veins of the forewings, causing the wings to vibrate. The tympanal organs are on the first abdominal segment (Fig. 29.2*b*). Both males and females produce long-range calling songs and approach each other.

Sound-producing membranes called *tymbals* are situated in the metathorax of Arctiidae (Lepidoptera) and in the abdomen of the following Hemiptera: male Auchenorrhyncha; both sexes of Cercopidae and some Cicadellidae and Cicadidae; and some Pentatomidae. Tymbals are best developed in male Cicadidae (Fig. 6.7*a–c*), in which they produce long-range calling songs. In these insects the tymbal is a thin resilient area of cuticle situated dorsolaterally on the first abdominal segment. Contraction of a fibrillar muscle attached to the center causes the tymbal to click inward. When the muscle relaxes, the elastic cuticle then clicks outward. Large air sacs in the abdomen resonate with the tymbal vibrations and increase the intensity of sound. The tympanal organs are located posterolaterally in the same segment. Orthoptera and Cicadidae are unique among insects in producing rhythmic "songs" and in the synchronous singing of many individuals to form a "chorus."

COMMUNICATION BY CHEMICALS. Chemical messengers are used for communication throughout the animal kingdom and in algae and fungi. The chemicals, either single compounds or mixtures, are usually active in minute quantities and are usually species-specific. Those that function intraspecifically are called *pheromones.* Pheromones may be grouped broadly into *primer* pheromones that have a long-term effect on the physiology and development of the receiver that is mediated by hormones, but may not have an immediate

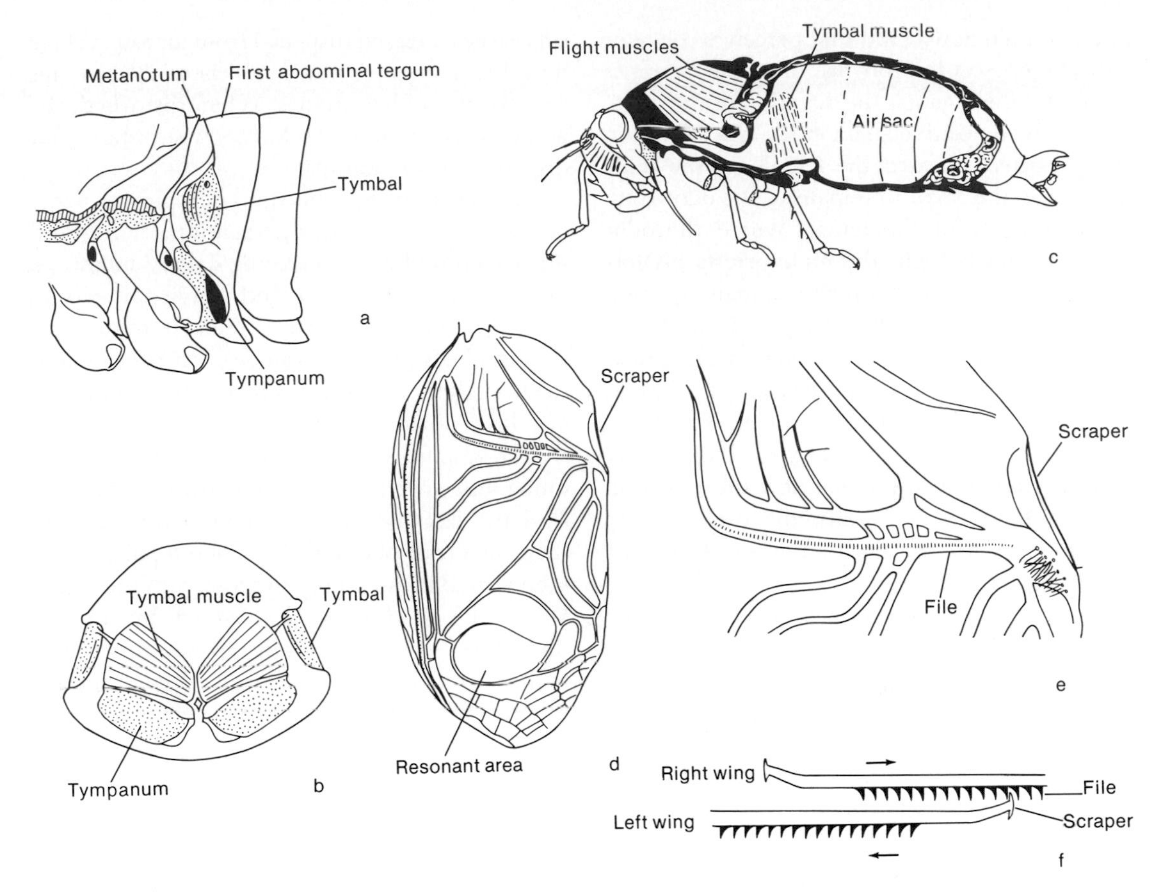

FIGURE 6.7 Sound-producing organs: *a*, lateral view of anterior abdomen of male cicada, *Okanagana vanduzeei*, showing sound-producing organ (tymbal) and sound receptor (tympanum); *b*, cross section of same; *c*, longitudinal section of *Magicicada septendecim* (Cicadidae) showing large resonant air sac; *d*, ventral view of right wing of the cricket *Acheta assimilis* (Gryllidae); *e*, enlargement of stridulatory organ of same; *f* diagram of relationship of right wing and left wing during stridulation. (c *redrawn from Snodgrass,* 1935, *by permission of McGraw-Hill Book Company.*)

effect on behavior; and *releaser* pheromones that have immediate, reversible effects on the receiver's behavior. Among insects, primer pheromones are well known in social species, where they are involved in caste determination and the inhibition of reproduction by workers (see Chap. 7). In other insects, especially those that aggregate, primer pheromones function to bring adults into reproductive synchrony. For example, the presence of mature males of the locust *Schistocerca gregaria* increases the rate of sexual maturation of immature males. Experimental evidence indicates that the substance responsible for promoting maturation is transmitted by air and by contact to the young adults and affects their corpora allata (Chap. 5).

Releaser pheromones are greatly varied in function and widespread taxonomically. The best known are the *sex attractants,* which bring the sexes together (Jacobson, 1972). In the most common situation, odors are released at a certain time of day or night by the female as a long-range signal to advertise her receptivity to mating. The odors are carried away from her by wind currents, forming an odor trail that becomes wider and less concentrated with increasing distance. A male downwind that perceives the odor with chemoreceptors on his antennae becomes active and flies against the direction of the wind, oriented by a positive anemotaxis. The flight is a zigzag pattern toward the female as the male turns back and forth to remain in the odor trail. In the vicinity of the female where the odor concentration is high, the male arrests his forward flight and searches for the female by other senses, especially sight. Courtship and copulation are then stimulated by close-range communication involving visual, tactile, or auditory signals, alone or together with pheromones. The long-range sex attractant released by the female may continue to excite the male, and the male, on close contact, may produce his own releaser pheromone or aphrodisiac that makes the female more receptive.

The foregoing sequence of sex attraction is by no means universal. In some insects males produce long-range sex attractants, in some the females produce aphrodisiacs, and in some, one sex produces pheromones or *aggregating scents* that attract both sexes to a common place (see behavior of bark beetles, below).

The active compounds in the sex attractants of a number of insects have been isolated and identified, and some have been synthesized (Jacobson, 1972). The common names for the active synthetic compounds include grandlure for the boll weevil, *Anthonomus grandis* (Curculionidae); disparlure for the gypsy moth, *Porthetria dispar* (Lymantriidae); riblure for the red-banded leaf-roller moth, *Argyrotaenia velutinana* (Tortricidae); hexalure for pink bollworm moth, *Pectinophora gossypiella* (Gelechiidae); bombykol for the silkworm moth, *Bombyx mori* (Bombycidae); trimedlure for the Mediterranean fruit fly, *Ceratitis capitata* (Tephritidae). Sex attractants are extremely useful in pest management (Beroza, 1976; Wood et. al, 1970). Specific pests can be attracted to poisonous baits or traps, or the mating behavior can be disrupted.

Intraspecific pheromones also have nonsexual functions. In some insects, the aggregations of immatures or of females at oviposition sites are probably assembled and maintained by chemical releasers. Rapid dispersal from an assemblage may likewise be signaled by chemicals, in this case by *alarm* pheromones. When disturbed, the aphid *Myzus persicae* discharges a secretion that disperses nearby aphids.

Social insects, especially ants, employ an array of chemical signals for aggregation, alarm, and other communications among colony members. Odor trails that recruit and guide foragers to food are known in certain ants, termites, and stingless bees (*Trigona,* Apidae). On returning to the nest after discovering a food source, a forager ant lays down a minute trace of a volatile, species-specific chemical, or trail pheromone. Other ants are stimulated to follow the odor trail from the nest to the food, and on their return add their own trail pheromone. A rich food source is soon visited by an increasing number of foragers. When the food becomes covered by ants or diminishes in size or quality, some ants return without contributing to the trail. Unless renewed by successful foragers, the trail declines in attractiveness because the odor is rapidly dissipated to a level below the ant's threshold of perception. Thus the foragers readily shift in appropriate numbers to exploit new food sources.

Chemicals involved in *interspecific* communication are divided into allomones and kairomones (Brown et al., 1970). *Allomones* are substances released by the sender that evoke a response in the receiver which is beneficial to the sender alone or mutually favorable to both. *Kairomones* are substances that, when transmitted, benefit the receiver. Allomones include the venoms injected by the sting of female aculeate Hymenoptera; defensive secretions of nasute termites and various beetles and true bugs; and toxic chemicals in the body, such as cardiac glycosides in the monarch butterfly, that discourage ingestion by vertebrate predators (Chap. 15). Plants also produce allomones, such as the scents of flowers that attract pollinators, as well as the phytoecdysones that protect against plant-feeding insects. Kairomones include odors or tastes that attract and stimulate attack, be it an insect victim or a plant host. The same chemical can function as an allomone or kairomone, depending on the circumstances. For example, secondary plant substances (Chap. 13) discourage attack by some plant-feeding insects but are stimulants to feeding by others.

AGGREGATION BEHAVIOR OF BARK BEETLES. A complex system of chemical signals has been revealed in the mass attack on living trees by the bark beetles *Ips, Dendroctonus,* and *Scolytus* (Scolytidae) (Wood, 1972; Borden, 1974). Coniferous trees resist attack by expelling or drowning boring insects in sap. This resistance is overcome by bark beetles when they attack simultaneously in large numbers and introduce a fungus that interrupts the vascular system of the tree. In the usual sequence of events, beetles of both sexes emerge from the old host or overwintering site and fly for some period before orienting to a new host. Visual and possibly odor cues guide the beetles toward trees. Attacks by "pioneer" beetles (males of *Ips;* females of *Dendroctonus*) are initiated, and feeding and gallery excavation occur in the bark of acceptable hosts, probably as a direct result of chemical feeding stimuli in the bark and/or phloem. During this initial boring activity, an aggregation pheromone or a species-specific mixture of such pheromones is produced in the hindgut and liberated with the feces. The compounds ipsenol, ipsdienol, and *cis*-verbenol, alone or in combination, are pheromones of various species of *Ips; exo*-brevicomin, *trans*-verbenol, and frontalin are found in *Dendroctonus* species; multistriatin has been identified in *Scolytus multistriatus.* These compounds are probably derived from bacteria and mycangial fungi that are closely associated with the beetles as symbionts in the gut or gallery (Chap. 16). The waste, or frass, from the borings is thus made attractive at a long range to other beetles. Thus the pioneers attract more beetles of both sexes. By their boring activity the aggregation pheromone concentration increases. In some cases certain volatile compounds from the host tree itself also contribute to the overall attractiveness of the tree, e.g., myrcene for *D. brevicomis* and α-cubebene for *S. multistriatus.* The total aggregation pheromone now consists of products (probably microbial) from the beetle's gut, from microorganisms in the gallery, and from the tree.

Mates are located and recognized by both sound and odor. *Ips* females are attracted to frass apparently marked with a male pheromone, and they stridulate before being permitted entry into the gallery occupied by a male. Males of *Dendroctonus pseudotsugae* are similarly attracted to frass marked with female-produced pheromones, and they also stridulate at the entrance to her gallery before entering. The factors that terminate a mass attack are not well understood. Possibly the aggregation pheromone declines in production or one or more compounds from either or both sexes increase in concentration and interrupt the chemoreception of the pheromone.

Birch and Wood (1975) discovered that *Ips pini* and *I. paraconfusus* both attack *Pinus jeffreyi* and *P. ponderosa* at the same season in the Sierra Nevada of California, but not the same individual trees. Apparently the volatile compounds produced by male *I. paraconfusus* inhibit the response of *I. pini* to its own pheromone and vice versa. Thus the first males to arrive at a given tree effectively prevent colonization by the other species. The pheromone simultaneously acts as an allomone against a competitor.

The aggregations of bark beetles attract a large number of other insects which feed on the beetles, or on other insects, or on the weakened host tree. Certain of the predators and parasitoids of bark beetles are specifically attracted by one or more of the constituents of the aggregation pheromones. In this situation, the pheromone compounds function simultaneously as kairomones (Borden, 1974).

LEARNING AND COMMUNICATION IN HONEYBEES. Aristotle noticed that a bee forages on one species of flower on each foraging trip and is attended by other bees on her return to the hive. These simple observations describe aspects of what has proved to be a remarkable ability for learning and communication. Much of our knowledge of honeybee behavior has resulted from the investigation of Prof. Emeritus Karl von Frisch of the University of Munich and his associates. For his efforts, he shared a 1973 Nobel Prize in Physiology or Medicine with N. Tinbergen and K. Z. Lorenz (see von Frisch, 1974).

A worker honeybee begins foraging for pollen and nectar in flowers after about 3 weeks spent as an adult in the hive (Chap. 7). Once she has learned the characteristics and location of flowers of a given plant species, she usually forages on that species even when other kinds of flowers are present. Bees from each colony usually visit several different species of plants simultaneously and many species throughout the season. Such constancy to flowers of a single plant species by individual insects is known among certain other flower-visiting insects (Chap. 13), but the variety of senses used and the extent of learning involved are known best for the honeybee.

In an extensive series of experiments which began prior to 1914, von Frisch demonstrated that bees could be trained to seek a sugar solution and to associate specific stimuli with this food source. The following experiment for color discrimination is typical. A table was placed near a hive. A blue square of cardboard was placed flat on the table and surrounded by gray cards of different shades. On each square was placed a glass dish. The dish on the blue card was then filled with sugar solution. After a few hours during which the bees foraged on the food, the cards and dishes were removed and replaced by clean cards and empty dishes arranged in a different pattern. The bees returned and landed on the blue card, demonstrating that they discriminated the blue color from all shades of gray. This indicated that they have true color vision, and that they had learned to associate the color with food. In similar experiments, von Frisch trained bees to one or a combination of odors, flavors, geometric figures, and colors within their range of discrimination (1971). He also trained bees to visit a feeding station at a specific time. Some bees were even trained to seek food at three to five separate times of day, thus demonstrating a time sense in bees.

Evidence of learning by punishment rather than reward was observed by Reinhardt (1952). When inexperienced bees enter alfalfa flowers directly for nectar, they frequently touch a part of the flower that causes the sexual column of the flower to snap upward and to strike the bee under the head. This is known as "tripping" the flower. Furthermore, the bee's proboscis is sometimes caught momentarily by the tripped flower so that she must struggle to free herself. Reinhardt watched the behavior of inexperienced bees visiting alfalfa and concluded that some bees soon learned to avoid tripping the flower by approaching from the side; other bees learned less readily or not at all. The latter bees usually quit foraging on the flowers entirely. In alfalfa fields, most nectar-foraging bees are skilled "side workers" that rarely trip the flowers.

The release of the tripping mechanism is necessary for the pollination of alfalfa flowers, and hence seed production. To increase seed production, beekeepers may purposefully select colonies with inexperienced honeybees to serve as pollinators of alfalfa. The alfalfa leaf-cutter bee, *Megachile rotundata* (Megachilidae), and the alkali bee, *Nomia melanderi* (Halictidae), are not adversely affected by the flower mechanism. Both species are managed as pollinators to increase alfalfa seed production (Bohart, 1972).

Honeybees and probably many other pollinating insects are thus able to learn to recognize the species-specific combination of flower characters of different plant species. They also learn the time when nectar and pollen become available each day and how to operate or avoid the complex and varied floral mechanisms.

We have seen that ants can recruit and guide other individuals to a food source by laying a chemical trail. Certain termites and some social stingless bees of the genus *Trigona* do likewise. Worker bees of *Trigona*, when returning to the nest from a food source in the forest, deposit droplets of odorous secretion from the mandibular glands on vegetation at 2- to 3-m intervals. This "odor trail" guides new foragers to the site. Odor trails are especially efficient in marking surface routes over broken terrain or in the three dimensions of a tropical forest.

Honeybees, however, have evolved a complex signaling system that communicates the distance and direction to food sources up to about 14 km from the hive. On returning to the hive from a rich food source about 100 m or less away, the worker inside the dark hive moves excitedly in small circles on the vertical comb (Fig. 6.8*a*). This kind of "round dance" lasts for several seconds or up to a minute before she moves to a different area on the comb where the dance is repeated. Although it is dark, some bees follow the movements and with their antennae they detect the flower odor on the waxy surface of the forager's body. The forager may also offer some of the nectar she has collected and carries in her crop. A rich concentration of sugar is highly attractive and stimulating to other bees that are as yet inexperienced in foraging. In experiments with

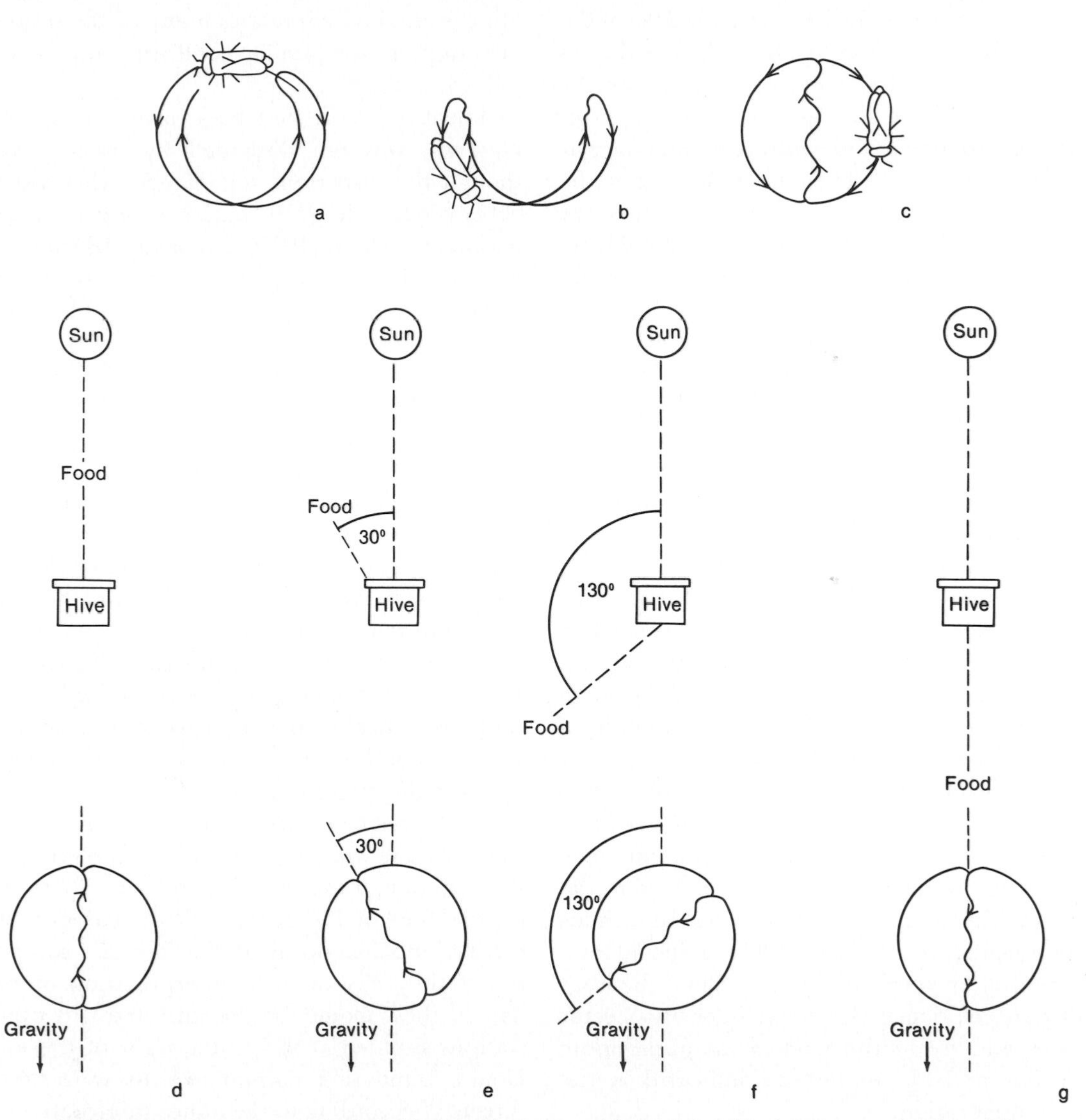

FIGURE 6.8 Dance communication of the honeybee: *a*, round dance; *b*, transition between round and wag-tail dance; *c*, wag-tail dance; *d–g*, diagrams of wag-tail run (on vertical surface of comb in dark hive) in relation to the direction of food from the hive and the compass direction of the sun.

feeding stations located in all directions from the hive, von Frisch demonstrated that bees recruited by a round dance search in all directions for food of the same odor and nectar flavor as that collected by the dancer. After several successful foraging trips, these newly recruited bees begin to dance too. More bees are recruited and the number of foragers rapidly increases. When the food supply diminishes, the frequency of dancing declines.

As the food site is moved toward and beyond a distance of 100 m from the hive, the bee dances change in form (Fig. 6.8*b*). During part of the dance she makes a straight run, vigorously wiggling her abdomen and emitting sound pulses at about 32 per second. Then she circles back to the starting point and repeats a "wag-tail" run of the same duration (Fig. 6.8*c*). The time spent during each run and the total number of pulses are directly proportional to the distance traveled to the food source. At 100 m the time of a single wag-tail run is about 0.25 second, and at 1200 m, 2 seconds. Thus, with increasing distance the "wag-tail" runs take longer, and fewer runs are complete within a given period.

The direction she moves during the run on the vertical comb has a consistent orientation with respect to gravity (Fig. 6.8*d–g*). This is related to the direction from the hive to the food with respect to the compass direction of the sun, when it is more than about 2° from the zenith position. Recall that positive phototaxis and negative geotaxis are commonly linked. For example, if the forager flew toward the compass direction of the sun in route to the food, her wag-tail run is oriented vertically up the comb, and her head points up during the run. If she flew at an angle to the right of the sun, the run is oriented at the same angle to the right of vertical. In various experiments, von Frisch (1967) gathered evidence that inexperienced bees "follow" the wag-tail dancers, detect the flower odor and nectar flavor, and fly to the food in the appropriate direction and for the distance indicated by the dance orientation.

Wenner (1967, 1971) and associates criticized the design of previous experiments and argued that their own experiments indicated odor (specific food odor, Nassanoff gland scent, and overall odor of the food area) was the main stimulus, both for recruitment and for the finding of food at some distance from the hive. They doubted that other bees used the dance information, and relied instead on their memory of the location of favored flowers. For example when bees have been trained to visit a sugar syrup scented with peppermint, the experimental release of peppermint in the hive stimulates many of the trained bees to fly to the familiar food site without first following dancers.

The hypothesis that bees have a language, however, was not discarded by others, even though the important role of odor was widely acknowledged from the outset. In ingenious experiments, Gould (1975) demonstrated that distance and directional information is indeed passed from the dancer to other bees. When bees dance on the comb in a light beam of sufficient brightness, they will orient to the light instead of gravity. The bees that follow the dancer also orient to the light and correctly interpret the direction to the food source. If, however, the ocelli of the dancer are covered by paint she becomes much less sensitive to light and a much brighter light is required to stimulate her to reorient to it. In Gould's experiments, the light was adjusted to not affect the ocelli-covered dancer, but to cause reorientation of the recruits to the light. Suppose the wag-tail run of her dance is vertically upward, correctly indicating a food station directly toward the sun. If the light beam is directed from the left at 90° from the vertical, i.e. horizontally, then the recruits with normal ocelli will orient to the light and interpret the direction of the food station to be 90° to the right of the sun. If they rely only on her odor, the recruits should appear at the dancer's station; but if they rely on their interpretation of her dance, they should be "misdirected" to other stations situated at 90° to the right of the sun. Gould found that normal recruits were "misdirected" according to the dance information of an ocelli-covered dancer. The light could be placed at various angles and the recruits appeared at the predicted stations with an error of about 11.9°, or 31 m at 150 m, and 4.2°, or 29 m at 400 m. The reasons why the more distant stations are indicated with less error are unknown.

Recruits probably use either the dance information and odor or odors alone in finding food. The dance may be important in initially recruiting new foragers to a new, distant food source, after which time they return to the food by their memory of landmarks, distance, various odors, and the compass direction. For the last, they utilize an astrotaxis and compensate for the sun's movement.

Bees may follow a dancer and not be recruited. This is especially true of bees that have never foraged. They may follow a number of dancers before responding. By following the dance, the recruit learns the distance, direction, flower odor, and nectar flavor, but not the color or shape of the flower or the landmarks and obstacles en route. In contrast to the chemical trails of stingless bees in forests, the dance of honeybees seems most effective in fields or open woodlands, where the line of flight is direct and foraging is done at low elevations above the ground.

SELECTED REFERENCES

Introductions to the structure and function of the nervous system are provided by Chapman (1969), Bullock and Horridge (1965), Snodgrass (1935), Treherne and Beament (1965), and Wigglesworth (1972). Major references on the nervous system and/or behavior are by Browne (1974), Carthy (1965), Dethier (1963, 1976), Fraenkel and Gunn (1961), Horridge (1975), Johnson (1969), Roeder (1967), and Wynne-Edwards (1962). Reviews are provided for certain aspects of the sense organs by Goldsmith and Bernard (1974), Goodman (1970), Hodgson (1961, 1974), McIver (1975), Schwartzkopff (1974), and Slifer (1970); central nervous system by Huber (1974), T. A. Miller (1975), Parnas and Dagan (1971), and Pichon (1974); physiology of behavior by Alloway (1972), Barth and Lester (1973), Browne (1975), Camhi (1971), Eisenstein (1972), Gelperin (1971), Howse (1975), Hoyle (1970), Johnson (1974), Kennedy (1975), Kring (1972), Markl (1974), Truman and Riddiford (1974), and D. M. Wilson (1968); sound communication by Alexander (1957, 1962, 1968), Bennet-Clark and Ewing (1970), Bentley and Hoy (1974), Haskell (1974), Michelson and Nocke (1974), Roeder (1965, 1966, 1967), and Wenner (1971); bioluminescent communication by Lloyd (1971), and McElroy et al. (1974); chemical communication by Beroza (1976), M. C. Birch (1974), Blum (1969), Hölldobler (1971), Jacobson (1972, 1974), Roelofs (1975), Roelofs and Carde (1977), Schneider (1974), Shorey (1973), Topoff (1972), and Wood et al. (1970). Host selection by bark beetles is reviewed by Borden (1974) and Wood (1972). Sensory physiology and behavior of honeybees are treated by Esch (1967), Gould (1975), Lindauer (1961), and von Frisch (1967, 1971, 1974). Other references including behavior are cited in Chap. 7 and Part III.

7 SOCIAL RELATIONSHIPS

In this chapter we will be concerned with those cooperative relationships among members of the same species that are called *social* and are beyond relationships directly involved in sexual behavior. Social interactions form the basis for important evolutionary strategies in diverse animals, both invertebrate and vertebrate, including humans. Of all nonhuman animals the most complex societies are those of insects. Familiar examples are termites, ants, hornets, and honeybees. They are all the more remarkable because each represents an independent lineage of social evolution from essentially solitary ancestors. At this point we need to define a series of grades or levels of *presocial* organization, of which one or a combination may have led to true, or *eusocial,* societies.

LEVELS OF SOCIAL RELATIONSHIPS. The great majority of insects are *solitary,* i.e., interactions among adults are largely limited to sexual behavior and competition, and adult-offspring contacts are limited to activities associated with oviposition. For example, the sexes of Phasmatodea meet during copulation and remain joined for long periods, sometimes for more than a day. Later the female, aloft in vegetation, drops her eggs to the soil litter without further care. The more advanced solitary species exhibit some preparation for the safety of offspring, even though they do not directly interact with the immatures. Many species of bees and wasps, for example, are solitary. They construct individual chambers or "cells," which are then provisioned with food for larvae. This is called *mass provisioning,* because the adult does not return to add food once the egg has hatched.

Adults of *subsocial* insects protect and/or feed their own offspring for some period of time after hatching, but the parent leaves or dies before the offspring become adults. This parent-offspring relationship is one possible route to true social behavior. Careful observation has revealed a surprising number of species which exhibit various degrees of maternal and/or paternal care. Among these are some species of the following: cockroaches, crickets, earwigs, mantids, jumping plant lice, web spinners, thrips, 13 families of true bugs, treehoppers, nine families of beetles, and certain bees and wasps, plus some other arthropods such as spiders. Protection of young during critical early stages has obvious advantages to insects as well as to some amphibians, reptiles, and fish, and to all birds and mammals.

Another route from solitary to true social behavior is via interactions of adults of the same generation. This behavior is broadly termed *parasocial.* A simple type of parasocial relationship is the *communal* colony of a group of female bees or wasps, in which fortuitously or regularly the females inhabit the same nest burrow, but in which each constructs and provisions cells for her own eggs. The frequent presence of at least one of the females in the communal burrow provides added defense to the colony against natural enemies. The more advanced or *semisocial* colonies are characterized by a reproductive "division of labor," i.e., some females reproduce and others do not. Some females are mated and lay eggs, thus serving as queens. Others of the same age and size may mate or not, but they fail to lay eggs and they become workerlike in behavior. The latter may be said to be altruistic, because they sacrifice their reproduction in favor of the queens. Some bees are regularly semisocial. Other species of truly social bees and wasps pass through a semisocial stage during colony development.

True social, or *eusocial*, behavior is attained in only two orders of insects: Isoptera and Hymenoptera. These societies are characterized as follows: (1) members cooperate in caring for young, (2) a reproductive division of labor exists, and (3) at least two generations overlap so that the offspring aid the parents in the colony's work. Many, but not all, social insects exchange liquids from the mouth or anus among members of a colony. This is called *trophallaxis* and is important in food distribution.

BIOLOGY OF TERMITES

All termites live in eusocial colonies with castes of both sexes. Almost certainly they evolved from cockroaches, with which they share many anatomical and physiological similarities. The present-day cockroach, *Cryptocercus punctulatus* (Cryptocercidae), may give us some clues to the behavior of the termites' ancestors. These roaches exist as subsocial groups in cavities in decaying wood. They derive nourishment from cellulose by means of flagellate protozoa that live symbiotically in the roaches' intestine. Flagellates are transmitted among roaches and to the next generation by anal trophallaxis. Although the circumstances under which true sociality first arose in termites are not clear, it seems probable that termites evolved from subsocial roaches during the early Mesozoic Era (Table 17.3) and were the first eusocial insects.

A living survivor of early termites is the single species *Mastotermes darwiniensis* in northern Australia. For a relic from the past, these termites retain a surprising vigor that is evidenced by their immense colonies, omnivorous habits, and unrivaled destruction of man-made structures. In common with roaches, *Mastotermes* shares five-segmented tarsi, an anal fan in the hind wing, eggs laid in oothecalike rows, and the same taxonomic families of symbiotic flagellates, including the genus *Trichonympha* (Chap. 16).

Termites are specialist feeders on cellulose, the most abundant organic compound in terrestrial habitats. Because of this a large part of the stored energy in an ecosystem is available to them. Wood of living trees is usually not attacked, but sound or decaying dead wood, twigs, fresh and dead grass, leaves, seeds, humus, and dung are attacked by various species and disintegrated by physical chewing and action of intestinal symbionts. In tropical regions, termites are ecological analogs of earthworms in the Temperate Zone, playing a crucial role in tropical soil development and erosion. Because wood is a building material and source of many useful products for humans, termites are major pests in both temperate and tropical regions. They also damage dried food, fabrics, rubber, hides, wool, linoleum, and insulation materials.

The four Families Mastotermitidae, Kalotermitidae, Hodotermitidae, and Rhinotermitidae are dependent on intestinal flagellates to digest cellulose. On the basis of anatomical and biological characteristics, these families are considered primitive in comparison to the Family Termitidae. The latter does not depend on flagellates but probably utilizes spirochete bacteria in the intestinal flora for the same purpose. The Termitidae are primarily tropical and include more than three-quarters of all species in the order.

Fungi are a normal component of the termite's habitat. One African subfamily of the Termitidae, the Macrotermitinae, cultivates the fungus *Termitomyces* on spongy combs built of the insects' feces (Fig. 24.3). The fungal combs are continually eaten and replaced. Ingestion of feces in this manner is not unusual when it is remembered that other termites regularly practice anal trophallaxis. Plant material in the original feces is apparently further degraded by the fungus, resulting in breakdown of lignin. Cellulose is thus freed and exposed to further bacterial action.

The ingested fungi probably also contribute to the termites' diet. This symbiotic association with *Termitomyces* is paralleled in the New World by fungus-growing ants, Tribe Attini, and their relationships to several fungi, mostly basidiomycetes.

CASTES OF TERMITES. The reproductive castes of termites (Figs. 7.1, 24.1) are:

1. *Primary Reproductives:* The only individuals that are fully winged after the last molt. They correspond to the alate, imaginal instar of other insects. Correlated with flight are a sclerotized and often pigmented body, compound eyes, and frequently a pair of ocelli. After the nuptial flight the wings break at preformed lines of weakness near the base, leaving characteristic wing stumps. After insemination, the female's body undergoes extensive changes leading to sustained egg production. The

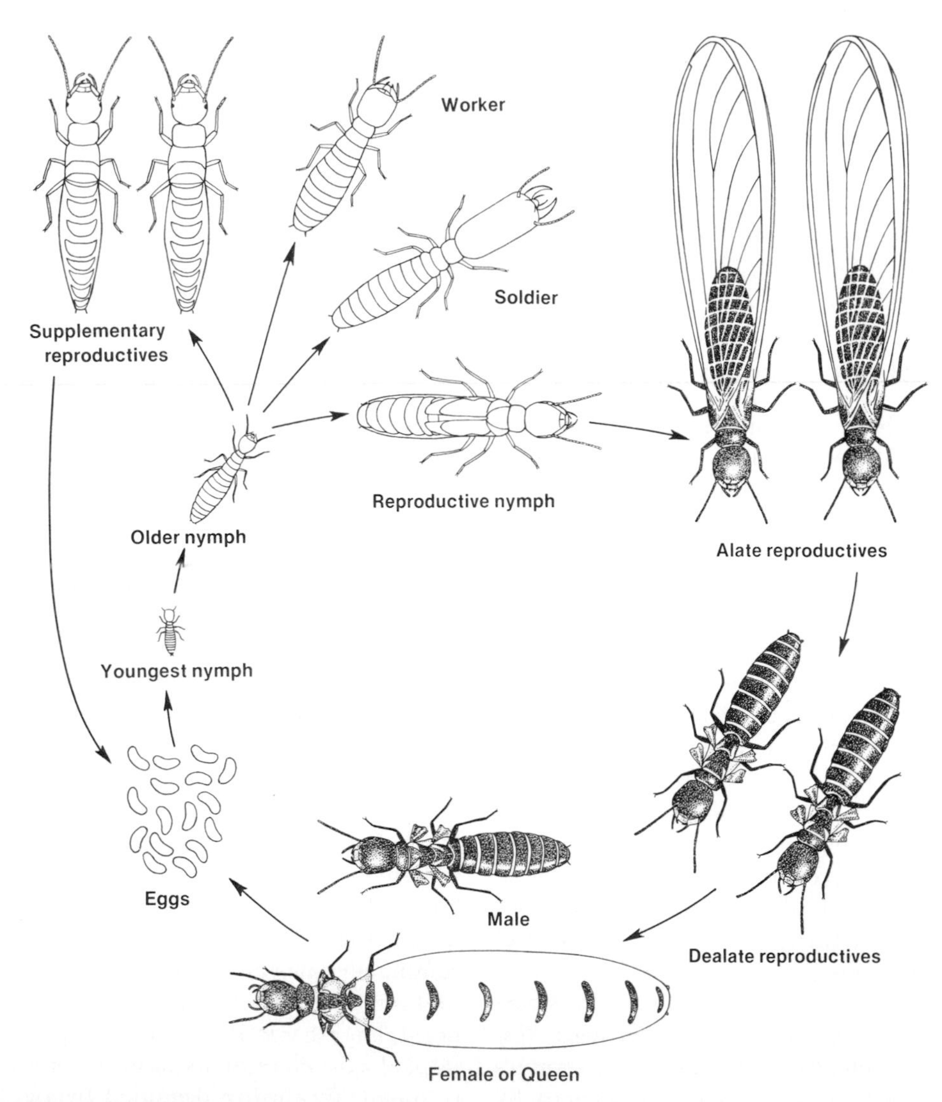

FIGURE 7.1 Diagram of life histories of castes in a termite colony such as that of *Reticulitermes hesperus*. [*Redrawn from Kofoid (ed.), 1934; copyright 1934 by The Regents of the University of California; reprinted by permission of the University of California Press.*)

swollen abdomen of queen Termitidae, for example, may reach 9 cm in length. The male remains small. Usually only one functional primary reproductive of each sex is in a colony, and both are relatively long-lived in comparison to the sterile castes.

2. *Supplementary Reproductives:* Imaginal individuals capable of producing eggs and sperm, but whose development has been partly inhibited. In comparison to primary forms, the pale body is less sclerotized and the compound eyes are reduced. Functional wings are not developed at the last molt, but those individuals with wing pads are termed *secondary reproductives,* and those without wing pads are called *tertiary reproductives.* One or more supplementary reproductives of the appropriate sex replace a dead or declining primary reproductive.

The apterous, sterile castes are (Figs. 7.1, 24.2):

1. *Workers:* Most members of a termite colony can be loosely referred to as workers and are usually of either sex. They perform all the duties of foraging, nest construction, and tending the young. The mandibles are not exceptional but are suitable for gnawing and are strongly muscled. Most workers are pale and sightless, but tropical Hodotermitidae, which forage during the day for grasses, are pigmented and possess compound eyes. Some scientists limit the worker caste to a more or less distinctive group of older instars that have lost the capacity to develop further into soldiers or reproductives. Such worker castes are found almost entirely in the advanced Family Termitidae. The workers are already different after the first molt and may be divided into subcastes, with one sex predominating in each. The "workers" of more primitive families are properly termed *pseudergates* or "false workers." These older nymphs without wing pads may later differentiate into soldiers or reproductives or may remain "arrested" and workerlike throughout the rest of their instars. Pseudergates may also be derived from other partly differentiated castes by regressive molts resulting in loss of wing pads. Soldiers and sexually functioning reproductives are final instars and do not regress to pseudergates.

2. *Soldiers:* Individuals with special development of the head for defense. One type found in some species has enlarged mandibles. A second type, known as the *nasute,* has a nozzlelike frontal projection from which is secreted a defensive fluid. In some species the soldiers have both large mandibles and the nasute head. Soldiers cluster about openings when a nest is damaged and engage intruding ants or other enemies. Soldiers may be of both sexes, but in advanced termites, soldiers may be all females or all males, depending on the taxonomic group.

CASTE DETERMINATION IN TERMITES. How is caste determined? Early investigations sought to distinguish between intrinsic or genetic mechanisms *versus* extrinsic or environmental influences. The research of Lüscher (1961) on a relatively primitive termite, *Kalotermes flavicollis* (Kalotermitidae), points conclusively to the latter explanation: eggs are equipotent, capable of developing into any caste depending on the nest environment. Yet in the Termitidae, early caste differentiation and sex-specific castes suggest an additional genetic basis. Unfortunately species of Termitidae have not been studied as thoroughly as *K. flavicollis.*

Determination of the caste of an individual *K. flavicollis* involves the familiar hormones of insect development plus pheromones from caste-determined members of the colony. Not unexpectedly, ecdysone promotes differentiation from pseudergate to the reproductive caste. On the other hand, low doses of juvenile hormone in a pseudergate result in no further differentiation, but high doses oddly promote differentiation into soldiers. Judging by the volume of the corpora allata just following a molt, the pseudergate at this time has the lowest level of juvenile hormone and is most likely to differentiate toward a reproductive if ecdysone is present. Proportions of the hormones are presumably regulated by the brain, which in turn is responsive to pheromones ingested by anal feeding from other termites. The pheromones have not been chemically identified, but experimental evidence points to (1) a substance produced by the female reproductive that inhibits female pseudergates from becoming functional queens, possibly by suppressing their secretion of ecdysone after molts; (2) two substances produced by the male reproductive, viz., a corresponding male inhibitory substance, and another substance that stimulates female pseudergates to become functional queens; (3) a substance produced by soldiers that inhibits pseudergates from becoming soldiers. Thus, recruitment of pseudergates into each caste is controlled by the presence of individuals already caste-determined. Excess numbers are destroyed by cannibalism.

LIFE HISTORY OF SUBTERRANEAN TERMITES. To illustrate the life cycle of a colony of termites, let us examine the rhinotermitid, *Reticulitermes hesperus,* a common and especially destructive species in the Western United States (Fig. 7.1; Kofoid et al., 1934). Other species of *Reticulitermes* are found in almost every state and are equally important as structural pests. These are known as subterranean termites because colonies are nearly always in contact with damp earth by means of closed tunnels. When decaying logs are rolled, pieces of lumber turned over, or, worst of all, when part of a house collapses, the presence of *Reticulitermes* is revealed by clusters of hundreds of white insects the size of rice grains. New colonies are established by a male and female primary reproductive. After the autumn swarming flight, the reproductives flutter to the ground and by flexing the abdomen break off their wings at the lines of weakness near the wing base. Females raise the abdomen and release a sex attractant. Males running on the ground orient to the odor and locate females. The pair then move off in tandem with the female in the lead. On finding a crevice under or in a piece of wood, the two enlarge the cavity and plaster the walls with feces and wood bits. Copulation follows the first efforts at nest construction and is repeated at intervals for the rest of the reproductives' lives. Initially fewer than 10 eggs are laid and carefully cleaned to prevent mold. The young nymphs are fed from the mouth or anus of the parents. By the second instar they have acquired intestinal protozoa. Growth of the population is slow, and members are long-lived. Workers are estimated to live 3 to 5 years, and reproductives probably live longer. The first clearly differentiated castes are apterous, nonreproductive workers and soldiers. After no less than 3 or 4 years under favorable conditions, the colony produces its first large numbers of new primary reproductives. Fully winged in the seventh instar, they remain in the nest until early autumn rains soak the soil. Then on a bright warm day, they emerge by the thousands. During a weak flight of 200 m or so, they mingle with reproductives from other colonies in the vicinity.

Large colonies may also reproduce by a budding process whereby a part of the constituency becomes isolated and develops its own supplementary reproductives. Outlying food sources, such as houses, may become rapidly infested in this way. Supplementary reproductives of the appropriate sex also replace primary reproductives which die or lose physiological control of the colony.

BIOLOGY OF SOCIAL HYMENOPTERA

Eusociality has emerged repeatedly in Hymenoptera: once or twice to give rise to Formicidae, or ants; once or twice in Vespidae, or hornets and paper wasps; possibly once in Sphecidae, because the neotropical *Microstigmus comes* seems to be eusocial; and at least several times in Apoidea, or bees, of which honeybees, stingless bees, bumblebees, some allodapine or small carpenter bees, and some Halictidae or sweat bees are eusocial. Social behavior in this order is an activity among adult females. The larvae are helpless and rarely contribute to colony welfare. Males are usually winged and function solely in reproduction.

Why are Hymenoptera predisposed to evolve social behavior? The search for common factors among the multiple origins has not yielded a simple explanation, but three features may have uniquely set the stage for recurring sociality. (1) The mandibulate mouthparts of adults are versatile in nest construction and brood care, as well as useful in tasks of obtaining food of diverse kinds. But other orders of insects are mandibulate. (2) We find well-developed nest building and various levels of presocial behavior by long-lived female parents that could lead to eusocial colonies. But other orders exhibit presocial behavior without further social evolution. (3) The haplodiploid reproductive system creates genetic relationships in a family that favor evolution of altruistic behavior among sisters (Hamilton, 1964). Again, a few other nonsocial insects have the same reproductive method, but taken in combination with the circumstances listed above, haplodiploidy may provide the essential mechanism.

Consider the genetic relationships in a hymenopteran family. Unfertilized eggs develop into males; fertilized eggs yield females. Male offspring receive one of each of the paired chromosomes from the mother (resulting, incidentally, in a situation where a son has a grandfather, but no father). The mother therefore shares half of her genes with each son. The daughters have half their chromosomes from the mother and half from the father. Accordingly, the mother shares half of her genes with each daughter. Of special importance is the genetic relationship among sisters sired by a single father. They all share the same paternal chromosomes, so that half their chromosomes are the same. Among the maternal chromosomes, they are likely on the average to share half. This means that an additional one-fourth are likely to be the same. On the average, therefore, sisters share three-fourths of their genes in common. The average fraction of genes shared is called the coefficient of relationship.

Altruism among insects means reduced or no reproduction in favor of another individual and may also involve self-sacrifice in the defense of the colony. In order for genes for altruistic traits to increase, the altruist must actually promote the survival of genes like its own by increasing the numbers of closely related individuals who will reproduce. Altruistic sisters (related at three-fourths) who live with their mother and who increase the survival of their reproductive sisters by caring for them, defending the colony against enemies, and foraging for food for the colony are increasing altruistic genes more effectively than if they cared equally for their own daughters (related at one-half).

Within the Hymenoptera the favorable genetic relationship among sisters may well contribute to the evolution of altruism in social groups where the sisters assist the mother in the production of more sisters, some of whom will reproduce. On the other hand, the genetic relationship is not favorable in semisocial groups, even when sisters are involved, because an altruistic sister would be helping the production of nieces (related at three-eighths) rather than her own offspring (related at one-half). Evidence exists that unrelated females may sometimes establish semisocial groups, thus further lowering the coefficient of relationship. Similarly, if the female parent of a eusocial group is inseminated by several males or if several females contribute female offspring, the average genetic relationship among the group of female offspring is rapidly lowered. It remains to be shown that local inbreeding might raise the general level of genetic relationship to a point where the coefficients are again favorable.

Thus with certain exceptions yet to be resolved, a pervasive tendency exists in the order to evolve sterile sisters if they coexist with their mothers in a eusocial relationship. Another feature of the haplodiploid system, viz., control of the sex ratio by the controlled release of sperm to fertilize eggs, permits the numbers and seasonal appearance of males to be regulated.

BIOLOGY OF ANTS

The Family Formicidae, or ants, are the second oldest group of insects to evolve social behavior. Two workers of a primitive ant, *Sphecomyrma freyi,* were found preserved in amber (fossil tree resin), dating from about 100 million years ago in the Upper Cretaceous Period. The slightly older fossil termite, *Cretatermes carpenteri,* is considered relatively advanced among modern Hodotermitidae. On the basis of this evidence from the fossil record and comparative anatomy, ants seem to be just beginning to differentiate from wasps in the Cretaceous, while termites were already diversifying. The subsequent evolutionary radiation of ants, however, is unequaled among all social insects.

Currently, 7600 species and 250 distinct genera have been described; possibly twice as many species will be known ultimately. The greatest variety exist in the Tropics, but ants are found north to the arctic tree line and south to the southern tips of the continents. Most oceanic islands within these latitudinal extremes are also inhabited by ants. Furthermore, Wilson (1971) estimates that 1 percent of all insects are ants, i.e., 10^{15} individuals.

Most ants are predaceous on other arthropods, taking supplementary food from plant sources—the nectar of flowers, sap, or sap-derived honeydew excreted by certain plant-feeding insects. Some species of ants thrive almost exclusively on honeydew milked from aphids or scale insects that the ants protect in a mutualistic relationship (Way, 1963), while other species gather and store seeds, i.e., are seed predators. Others form a symbiotic relation with certain shrubs or trees, protecting against phytophagous mammals and insects and competing plants in return for shelter and food structures that the plant specially produces for the ants. Like fungus-growing termites of the Old World, the neotropical Tribe Attini cultivate and eat fungi. The ants' fungal gardens, however, are grown on freshly cut leaves and flowers.

The early evolution of this successful family is not well understood. One naturally seeks to identify ancestors among stinging wasps, which are similarly predatory and anatomically akin. Among allied Scolioidea are wingless females of Tiphiidae, which strongly resemble the fossil worker of *Sphecomyrma.* Although this may be an instance of convergent similarity due to reduction of the flight mechanism, the antlike wasp and the wasplike ant are temptingly close. Missing in tiphiids, however, is evidence of clear subsocial behavior. Tiphiid females sting their prey and simply lay an egg on it. Certain other nonsocial wasps in the Sphecidae progressively provision their brood by repeatedly returning to their nests with prey for their larvae. It is this kind of behavior that should lead to the "partially claustral" method of colony foundation seen among the primitive ants. The fertilized female ant, or queen, lays eggs in a "cloister," or chamber, and occasionally forages for prey to feed her brood. Fully claustral colony foundation is an advanced trait. The queen never leaves her chamber; she feeds her brood with saliva or trophic eggs enriched with nutrients from her fat stores and histolyzed flight muscles.

Males are winged and have well-developed sense organs, such as compound eyes and antennae. They function only in reproduction.

CASTES OF ANTS. The castes of ants are convergently similar to those of termites, but among ants only the female sex is involved.

1. *Queens:* The queens, or functional reproductives, have a fully developed flight mechanism when they emerge from the pupal stage. After a mating flight, the wings are shed and the flight muscles histolyzed (Fig. 7.2*a*). Queens may resemble larger workers in the possession of three ocelli and compound eyes, but the queens usually have more ommatidia in the eyes. Colonies of some species regularly have more than one queen.

2. *Workers:* The most numerous individuals in the colony are distinguished from the queen by their smaller size and greatly reduced thorax, which never develops a flight mechanism. (Figs. 7.2*b*, 46.14*b*). In those species with a more or less continuous range of worker sizes, the largest are called "major workers," the intermediate forms are "media workers," and the smallest are "minor workers." The number of ommatidia varies with size or may be lacking in subterranean species. Some species lack media workers, resulting in a distinct gap between majors and minors. The majors are then called "soldiers." Such large workers may have disproportionately larger heads and mandibles, in keeping with their defensive function.

CASTE DETERMINATION IN ANTS. The determination of caste can be viewed in the following way: In the absence of modifying influences, a diploid

FIGURE 7.2 Army ants, *Eciton: a*, queen laying eggs; *b*, five ants cooperate in carrying tail of scorpion. (*Photos courtesy of and with permission of Carl W. Rettenmeyer.*)

female egg will develop into a queen. The anlagen or imaginal disks, of the larva are divided into two sets: (1) a dorsal group, including wing buds, incipient gonads, and ocellar buds; and (2) a ventral group, including leg buds, mouthparts, and central nervous system. In the morphogenesis of a queen, both sets develop with equal vigor. However, an interwoven set of environmental and/or maternal factors may inhibit or fail to promote (presumably via the endocrine system) the complete development of the dorsal set. Such an individual becomes a worker.

What are the environmental factors responsible? Certain experiments indicate that pheromones secreted by a functional queen inhibit development of more queens among her brood. Other observations indicate that nutrition of larvae probably varies both in quantity and quality as the colony grows. A large foraging force may provide optimum food, or the queen's inhibitory pheromones may be diluted, or both. In any event, new queens are produced from large colonies. The critical experiments are lacking to evaluate the possible factors. Other environmental influences are chilling temperatures. Brood produced in the fall at temperate latitudes become dormant and later chilled. Dormancy is broken by the warmth of spring, and surviving larvae are able to grow rapidly, metamorphosing into queens.

Maternal effects include the size of eggs laid by the queen. During periods of high egg production, eggs average smaller and produce mostly workers. Younger queens also tend to lay eggs that result in workers.

The annual cycle of an ant colony usually includes production of winged males and queens. These emerge synchronously from neighboring nests in response to environmental cues. The males form large aerial swarms, often over a conspicuous object or above a tree. Females fly into the swarm and are clasped by the males. The copulating pair may fall to the earth before disengaging. The female sheds her wings and excavates the chamber mentioned above, wherein she rears the first brood. The workers then take over brood care, expand the nest, and forage for food. Queens may live up to 18 years, depending on the species and vicissitudes of the colony. Colonies of more primitive ants usually are hostile to individuals from other colonies. The unique odor of a colony is probably the basis for recognition. Some advanced species permit individuals of different colonies to mingle freely.

LIFE HISTORY OF THE ARGENTINE ANT. A familiar species which deviates from the normal life history is the Argentine ant, *Iridomyrmex humilis.* Commercial traffic from South America distributed the ants to countries in the Northern and Southern Hemispheres within latitudes of 30 to 36°. In North America the first colonies were seen in 1891 in New Orleans, Louisiana. They are now widely distributed in the Southern United States and in California, where they first appeared in 1905. Native ants are regularly displaced by Argentine ants, often by physical combat.

Nectar and honeydew from plant-sucking Homoptera are more than 99 percent of the diet of these ants. This search for sweet fluids has earned the Argentine ant a place among the important household and agricultural pests. Massive numbers may invade well-kept kitchens or restaurants for food, often in late summer and after heavy rains. Beehives are entered and honey is taken from the combs. Mealybugs, scale insects, and other honeydew-producing Homoptera are tended and protected by the ants. The homopterous pests of citrus are protected by ants from attack by predators and parasitoids, thus greatly reducing the effectiveness of biological control programs.

In southern California, the annual cycle has been described by Markin (1970). The seminomadic colonies do not have clearly defined limits or individuality. In a given area nests are excavated in soil at favorable sites, but these are parts of an extended colony. Workers move freely from one nest to another. Nests are abandoned when flooding or drying renders a site unsuitable, and are reestablished. When the general area is unfavorable, occasionally a single large nest is made or the colony moves 100 m or more to a new area. From December through February, the colony weight, or biomass, is 90 percent workers, with few immature stages. Eggs laid by the queens increase in frequency in late February and early March. These are destined to become reproductives. Adult males appear in mid-March, followed by queens in mid-April. Both reach a peak abundance about mid-May, with the queens generally scarce, but the males may make as much as 15 percent of the colony biomass. Mating usually takes place entirely in the nest shortly after the queens emerge. Normal flights of males and queens have been observed, but more commonly only the males disperse from the nest. Their flights in early July take place long after the queens have matured, so that further mating is unlikely. The inseminated queens shed their wings and begin laying eggs. The number of queens is probably not increased after mid-June, at which time they may comprise as much as 3.5 percent of the biomass. The number may be as high as 8 per 1000 workers. The worker population now increases until October, when the colony begins to diminish. In January and February, workers attack queens and dismember them, reducing the number of queens in the colony before the next year's cycle.

BIOLOGY OF THE HONEYBEE

The best-known and most useful social insect is the honeybee, *Apis mellifera.* Possibly it has been studied more than any other insect. The native home of the honeybee is in the Old World, where they have been exploited throughout human history. Colonies in natural cavities, such as a hollow tree, or in hives made by human beings were killed to extract stored honey and wax. Only relatively recently, in the 1500s, did European beekeepers devise a hive with movable wooden frames so that individual combs could be removed without destruction of the entire colony. Apiculture is now the most widely practiced agricultural activity in the world. The role of the honeybee as a pollinator of crops is becoming increasingly important in North America. During the earlier stages in the development of agricultural lands, adjacent wild areas were usually inhabited by native pollinators. Under modern intensive cultivation, these pollinators have been eliminated, and honeybees now are regularly transported from crop to crop for this function (McGregor, 1976).

The common honeybee is one of four distinct species or species complexes in the genus *Apis.* All are properly called honeybees because they are eusocial and construct wax combs for rearing brood and storing honey. All except *A. mellifera* are confined to the eastern Palearctic and Oriental regions. *Apis mellifera* originally inhabited Western Asia, Europe, and Africa south of the Sahara Desert.

RACES OF EUROPEAN HONEYBEES. Several distinctive races of honeybees evolved in Europe during the Pleistocene, when advancing glaciers fragmented their distribution. The bees survived in separate refugia in what are now Spain, Italy, Austria, and the Caucasus. The respective races are *A. m. mellifera,* the German, Spanish, or English honeybee; *A. m. ligustica,* the Italian honeybee; *A. m. carnica,* the Carniolan honeybee; and *A. m. caucasica,* the Caucasian honeybee. Consciously or not, beekeepers in these temperate regions have selected for productivity and ease of management, so that European bees are relatively tractable.

When colonies of European settlers were established abroad, honeybees were taken from the home countries. In the New World, *A. m. mellifera* was the race initially introduced. As communication and transportation increased, beekeepers exchanged queens and colonies around the world in an effort to improve productivity in the new climates. The Italian honeybee became especially favored and now is the predominant commercial honeybee in North America.

AFRICAN HONEYBEES. In Africa, *A. m. adansoni* evolved under quite different conditions. Here it was beset by a variety of natural enemies, including human beings, and existed in a tropical climate with dry seasons. The African honeybee is consequently highly aggressive in the defense of its nests and able to undertake migrations. The high productivity in the Tropics has attracted bee breeders on several occasions to make hybrids with European stocks. These experiments had gone largely unnoticed until a recent episode had calamitous results (Michener, 1975).

European bees in tropical South America are unsatisfactory honey producers. In 1956, African honeybees were taken to southern Brazil for the careful breeding of a strain incorporating higher productivity with the desirable features of European bees. In 1957, protective screens which confined the African queens were mistakenly removed and 26 queens with workers escaped. Hybrids were formed between the African bees and the resident European bees. In the Old World, however, *A. m. adansoni* and the European races had evolved apart for at least 10,000 years, separated by the Sahara Desert. As a result, free cross-mating between the races was reduced by about a third, owing to the apparent preference of reproductives for their own race. Such hybrids as were formed perpetuated mostly African characteristics.

The Africanized bees directly compete for food and nest sites, even invading the hives of European bees and displacing them. Large numbers of wild colonies were also established in small cavities not suitable for European bees. As the Africanized bees migrated annually into new areas of Brazil, it became evident that the European bees were being completely replaced by a much higher population of Africanized colonies. Beekeeping became difficult or impossible. Reports appeared in the world press of deaths and injury to humans and domestic animals. Although some differences in the chemistry of bee venom may exist, the danger is created by the large numbers of bees that sting and pursue a victim. The Africanized bees are now found throughout most of tropical South America east of the Andes. The northward advance reached Venezuela in 1976 and, in time, may reach the Southern United States. Southward in Argentina, the Africanized bees are unable to survive the cold winters of the temperate climate. A similar temperature barrier may prevent northward invasion in the United States beyond the Southern and Southwestern states.

In southern Brazil, large numbers of Italian queens were distributed to beekeepers in a program to reduce the aggressive behavior of Africanized bees by further hybridization and by killing aggressive colonies. This has been apparently successful in producing bees that are manageable and still productive. The beekeeping industry has recovered in this area. The genetic approach offers the best means of controlling the undesirable traits of the bees, but over vast regions of tropical America this is impractical. Possibly the northward movement through Central America can be checked because the route is narrow and more easily saturated with European bees.

DRONES. All the adult honeybees in a colony are winged. The males, or drones, are produced during the spring and summer seasons. They are distinguished by large eyes and their larger body size. Drones are fed throughout their 4- to 5-week lives by the worker bees and are unable to take nectar from flowers. Aerial excursions are taken from the hive in search of queens, with which they mate in the air. The male genitalia are uniquely constructed to evert violently and be torn off during copulation. Successful drones die after mating. At summer's end, unsuccessful drones are expelled from the hive, also to die.

FEMALE CASTES. Females in the colony include two castes (Fig. 7.3):

Queen: The functional reproductive is slightly larger than a worker and has a longer abdomen that extends well beyond the folded wing tips. She never forages for food, but does leave the hive for one or more mating flights early in her life and during periodic swarming. Ordinarily, only one queen is present at a time in a colony. She produces all the eggs that are fertilized and most of the unfertilized eggs. These may total 600,000 over the 2 to 3 years of her normal life-span (Butler, 1954).

FIGURE 7.3 Queen honeybee, *Apis mellifera* (Apidae), on comb surrounded by worker bees. (*Photo from Dadant and Sons, 1975, courtesy of and with permission of Kenneth Lorenzen and Dadant and Sons, Inc., Hamilton, Ill.*)

Workers: Females whose reproductive organs are atrophied to the extent that copulation is prevented but which can occasionally produce eggs. Some of the unfertilized, or drone, eggs in a colony come from this source. (An exceptional race of honeybees restricted to the Cape region of South Africa has laying workers whose eggs develop by thelytokous parthenogenesis into females.) The sterile workers perform all the foraging, defense, wax secretion and comb construction, brood care and feeding, and regulation of the physical environment, and also locate new sites during swarming. Colonies normally contain 60,000 to 100,000 workers. During the active foraging seasons, a worker may live 5 to 6 weeks. Workers reared in the fall and overwintering may live 6 months or more.

CASTE DETERMINATION IN HONEYBEES. The comb is made of wax secreted by glands on the abdominal sterna of workers. The bees shape the scale-like pieces of wax into cells which are hexagonal in cross section. Human engineers have imitated this design for creating strong, light-weight structures. The bees' cells are used as a nursery or for the storage of honey or packed lumps of pollen. Most of the cells are closely similar in size. In the brood area the queen lays a single, fertilized egg in each cell to produce workers. When food is plentiful, larger cells near the margin of the comb receive unfertilized eggs to produce drones. New queens are reared in much larger cells which are round in cross section and are specially constructed to hang from the surface or lower edge of the comb. The brood cells remain open while the larvae are repeatedly fed by the workers. The determination of caste among the females is directly related to differences in food.

For the first 2 days, drone and worker larvae are fed a mixture of glandular secretions and crop fluids from workers functioning as nurse bees. This "bee milk" includes (1) clear hypopharyngeal gland secretions plus clear crop fluids, and (2) a lesser amount of white mandibular gland secretions and some hypopharyngeal secretions. The white secretions diminish and disappear in the food on the third day, and afterwards the larvae are fed "bee bread," consisting of the clear fluids variously mixed with pollen and honey. A female larva destined to become a queen is fed throughout her life on "royal jelly," which is composed of equal parts of the clear and white foods. These secretions include basic nutrients such as sugars, amino acids, proteins, nucleic acids, vitamins, and cholesterol plus a fatty acid, *trans*-10-hydroxy-2-decenoic acid, which is not pheromonal but is chemically related to a pheromone secreted by queen bees. The quantity, quality, and timing of the food, rather than an obvious pheromone, seem to divide worker-destined and queen-destined larvae at an early age. Intermediates are rarely found in nature but can be experimentally produced.

Once the larva has reached full size, it defecates for the first time. A cocoon is spun that separates the larva from its feces. The workers cap the cell with wax. The larva now transforms to a pupa and then to an adult inside the closed cell. After the final molt, the new bee chews its way out of the cell to join the colony. Three weeks are required for growth from egg to adult.

LIFE OF THE WORKER HONEYBEE. A worker bee spends about three more weeks inside the hive before becoming a forager, or field bee. Much time is spent resting or patrolling the comb. Young worker bees first clean the hive and feed larvae with secretions from their hypopharyngeal and mandibular glands, which are well developed at this time. With age these glands normally decline in activity and change secretions. The hypopharyngeal gland later secretes invertase, which converts the sucrose in nectar to the glucose and fructose of honey, and the mandibular gland secretes an alarm pheromone, 2-heptanone, when the worker later serves as a guard or forager. As the nursing function decreases, the sternal wax glands may become active if comb building is prevalent in the hive. She also begins short flights from the hive. After the wax glands decline, activity shifts to the entrance, where the worker may guard the hive before finally becoming a forager. For the remaining 1 to 3 weeks of her life, she collects nectar and pollen from flowers (see "Learning and Communication in Honeybees," Chap. 6). Plant resins are also collected and used to seal openings in the hive. Beekeepers call the resins *propolis* or *bee glue.*

ANNUAL CYCLE OF A COLONY. The annual cycle begins when the spring flowers provide abundant food. The queen begins laying fertilized eggs that greatly augment the worker force. Normally the increased food supply, stored provisions, and high population inside the crowded hive lead to swarming. Special cells are constructed in which new queens are reared. Egg production by the old queen declines, as does foraging by the workers. Before the new queens emerge, the old queen, part of the worker force, and some drones leave the hive. The mass flight, or swarm, is spectacular. The bees may periodically settle in a cluster on a tree limb or other object. Scout worker bees fly from the swarm in search of suitable new quarters. The scouts return and dance on the cluster, indicating the distance and direction to the favored site. Once the swarm has occupied the new hive, wax combs are constructed, food is collected, and the old queen begins to restore the worker population. In the original hive, the first new queens to emerge fight physically for the role as queen. Pupal queens are destroyed by the workers. The survivor flies from the hive several times over several days and may be inseminated by a drone each time. The sperm are stored in her spermatheca for use during the remainder of her life. Her new colony has the benefit of existing combs and stored food. The colony may produce swarms more than once during favorable seasons. In the fall when flowers no longer supply enough food, brood rearing declines and drones are expelled. More propolis is added to reduce cold drafts. The colony clusters together on combs filled with stored honey during cold weather. The bees utilize honey as heat-giving energy, remaining alert through winter.

SELECTED REFERENCES

The recent books by Wilson on social insects (1971) and sociobiology in general (1975*b*) provide excellent general introductions. Evolution of social behavior is discussed by Alexander (1974), Crozier (1977), Dawkins (1976), Eberhard (1975), Hamilton (1964), and Trivers and Hare (1976). Michener (1974) gives a comprehensive review of the varieties of social behavior among bees. Two volumes by Krishna and Weesner (1969, *a* and *b*) are devoted to the biology of termites. Popular books include Skaife (1961) on termites; Goetsch (1957) and Schneirla (1971) on ants; H. E. Evans and Eberhard (1970) on wasps; Butler (1959) and Alford (1975) on bumblebees; and Butler (1954), Ribbands (1964), and Dadant (1975) on honeybees.

PART TWO

POPULATION BIOLOGY OF INSECTS

NUMBERS OF INDIVIDUALS IN A POPULATION

8

An insect population may be defined most simply as any group of insects one wishes to consider. Thus one may refer to the population of insects in a square meter of forest litter. The type of population that most entomologists think of first, however, is a group of similar individuals, a group made up of the same kind, or species, of insect. Thus one might consider all the *Blattella germanica* (Blattellidae) cockroaches in Texas as a population, or just those *B. germanica* inhabiting a single kitchen in a home as a population.

Insect population biology deals with those properties of insect populations which are not properties of individual insects in the population. Some of these properties of single-species populations are:

1. Population size
2. Population density
3. Distribution in space (at a given point in time)
4. Variation in one or more measurements at a particular time
5. Social behavior
6. Population genetics
7. Evolution

These properties of populations are not properties of single individuals in the population. An individual cannot have "population size" or "population density" (the number of individuals per unit area), or be in more than one place at a time.

Populations show the property of variation between individuals at a particular time. For instance, at this moment, the population of checkerspot butterflies, *Euphydryas anicia* (Nymphalidae), living at Cumberland Pass at almost 4000-m elevation in the Colorado Rockies shows variation from individual to individual in a multitude of characteristics, including wing length, the shape and size of various spots on the wings, the structure of the genitalia, and the amino acid composition of various enzymes.

There are a myriad of other examples where such variation has been documented. Populations of alate aphids *Pemphigus populi-transversus* in the Eastern United States vary in such characters as forewing length, head width, thorax length, tibia length, the dimensions of antennal segments, and the number and characteristics of antennal sensilla (Sokal and Rinkel, 1963). Populations of the ambush bug *Phymata americana* (Phymatidae) vary in the dimensions of their thorax and their color pattern (Mason, 1973); and one population of the grasshopper *Moraba scurra* (Acrididae) in Australia contains three kinds of individuals with respect to the sequences of their "CD" chromosomes (Lewontin and White, 1960).

The kinds of behavior that we call *social* behavior have meaning only in the context of a population. One cannot talk about aggressive behavior or mating behavior of an individual insect without reference to other members of the population to which it belongs.

Populations also have genetic properties above and beyond those of the constituent individuals. These will be dealt with below.

Finally, but not exhaustively, populations evolve; i.e., they change in their genetic properties in the course of time. Individuals go through a process of *development.* Evolution, which results from the appearance of *new* individuals with *different* genetic characteristics, is a phenomenon of populations. When it is caused by genetic differences between individuals, variation of the sort discussed above is an essential element in the process of evolution.

But why should an entomologist be interested in the properties of insect populations? There are many reasons. Populations of insects compete with us for our food. Some are vectors of diseases. It is of great economic and medical importance to humanity that the factors which control the size of these pest populations be understood so that their numbers can be controlled. Thus entomologists are often concerned with the *dynamics* of populations, i.e., how they change through time. This concern leads automatically to a careful consideration of how the insects interact with one another and with their environments, i.e., *insect ecology.*

Insect populations influence their environment in many ways, e.g., by consuming a food resource. Tiny beetles consume substantial portions of humanity's stored grain; *Euphydryas* larvae feed on plants of the snapdragon and plantain families and may totally defoliate the plants. Hordes of migratory locusts often devour virtually all the green vegetation of areas of Africa. Similarly, the environment influences the population in many ways, as by altering its size or changing its genetic composition. For example, the food resource may be so depleted that the population is greatly reduced in size or goes extinct.

In addition to understanding population dynamics, if one wishes insight into the workings of an insect population it is necessary to understand its population genetics as well, especially since the dynamics and genetics are intimately related. If, for instance, one is attempting to suppress a population of *Culex* mosquitoes carrying the encephalitis virus, the choice of control measure may depend heavily on the genetics of the population—i.e., on the proportion of individuals that are genetically resistant to one or more pesticides.

It is only by understanding the population biology of economically important insects that the entomologist can hope to manage their populations over the long term. As you will learn in this chapter, populations are continuously changing in ways which make any pat "formula" for their control rapidly obsolete.

Although the details of population dynamics can be endlessly complex, the essence of analyzing changes of insect numbers is relatively straightforward. A number of points should be kept clearly in mind throughout the discussion.

1. The analysis basically amounts to keeping track of inputs and outputs.
2. The inputs are natality (births) and immigration.
3. The outputs are mortality (death) and emigration.
4. In most discussions in this section, migration will be ignored, and the analysis will focus on natality and mortality.
5. In populations with overlapping generations (such as those of people), the *age composition* of the population—i.e., the proportion of individuals in various age classes—has a substantial effect on the future course of population growth. In populations such as those of many, if not most, insects, where adults invariably die before their offspring mature, the generations are *non-overlapping* and the analysis is relatively simple.

CHANGE IN NUMBERS

Consider the problem of measuring the rate of change—i.e., increase or decrease—in an insect population *at a given moment.* What we would like to calculate is the *instantaneous rate of increase* (IRI) for a population—a quantity something like the [speed] readout on a speedometer, which tells how fast the car is going at a given instant. The IRI will be positive when the population is growing and negative when it is shrinking—a negative IRI being analogous to the speedometer reading when the car is backing up.

The rate of change of the size of a population (N) in the course of time (t) itself *depends* on the population size. (Or, more precisely, the rate of change of N is a *function* of N.) Whatever the average individual's contribution to population growth, it must be multiplied by the population size in order to determine the amount of change there will be in the population as a whole. These simple relations, expressed in the notation of calculus, yield the most basic equation of population dynamics:[1]

[Instantaneous rate of increase] equals

$$\frac{dN}{dt} =$$

[average individual contribution to population growth] multiplied by [population size]

$$r \times N = rN$$

Here dN/dt is the standard notation for "instantaneous rate of change of N with t." The average contribution, r, is the *instantaneous rate of increase per individual* (often called the *intrinsic rate of increase*). Thus, at a given moment, if a population of aphids contains a million individuals (N) and is growing at a moment at the rate of 10,000 individuals per week (dN/dt), then r would be 0.01. In other words, the growth rate amounts to one-hundredth of an aphid per aphid per week.

The factor r is more easily understood if it is split into its input and output components. If input is b, the instantaneous birth rate, and output is d, the instantaneous death rate, then r is seen to be simply b minus d. (We are ignoring migration. Note that d here is not the same d as in dN/dt.) If more individuals are hatching from eggs at a given instant than are dying, the population is growing and r is positive. If more are dying than are hatching, the population shrinks and r is negative. Very often we are concerned simply with the size of the adult population in insects, so that instead of natality, emergence from the pupa or molting from the preadult takes the place of birth or hatching from an egg in calculations.

[1]Some familiarity at least with the notation and fundamental approach of the calculus will be helpful for what follows. For an introduction or review, see E. Batschelet, *Introduction to Mathematics for Life Scientists,* 2d ed., Springer Verlag, New York, 1975, or any basic calculus text.

EXPONENTIAL GROWTH (Fig. 8.1a). The equation $dN/dt = rN$ relates the IRI to population size itself. It does not, however, show explicitly how the size of a population (growing according to this relationship) at one time is related to the size at another time. Consider for instance, a population whose size at an initial time, $t = 0$, is known and is denoted $N(0)$. How can the size at t units of time later, here denoted $N(t)$, be calculated? The answer can be derived by elementary calculus from the equation for the IRI. If r itself does not vary with time, one obtains:

[Size after t time units] equals

$$N(t) =$$

[initial size] multiplied by [e raised to the power of the IRI per capita multiplied by t time units]

$$N(0) \times e^{rt}$$

Here e is the base of the natural logarithms; it is roughly equal to 2.7183. When the quantity e is raised to a power that is a variable—e.g., the product rt where t changes—the result is called the *exponential function.* This function has unique mathematical properties that cause it to arise naturally in the mathematical representation of many physical and biological processes. Its presence in the growth equation is the basis for calling this particular process *exponential growth.*

A quantity growing exponentially increases in any given period of time by a fixed percentage of its size at the beginning of the period. A savings account at a bank where interest is compounded

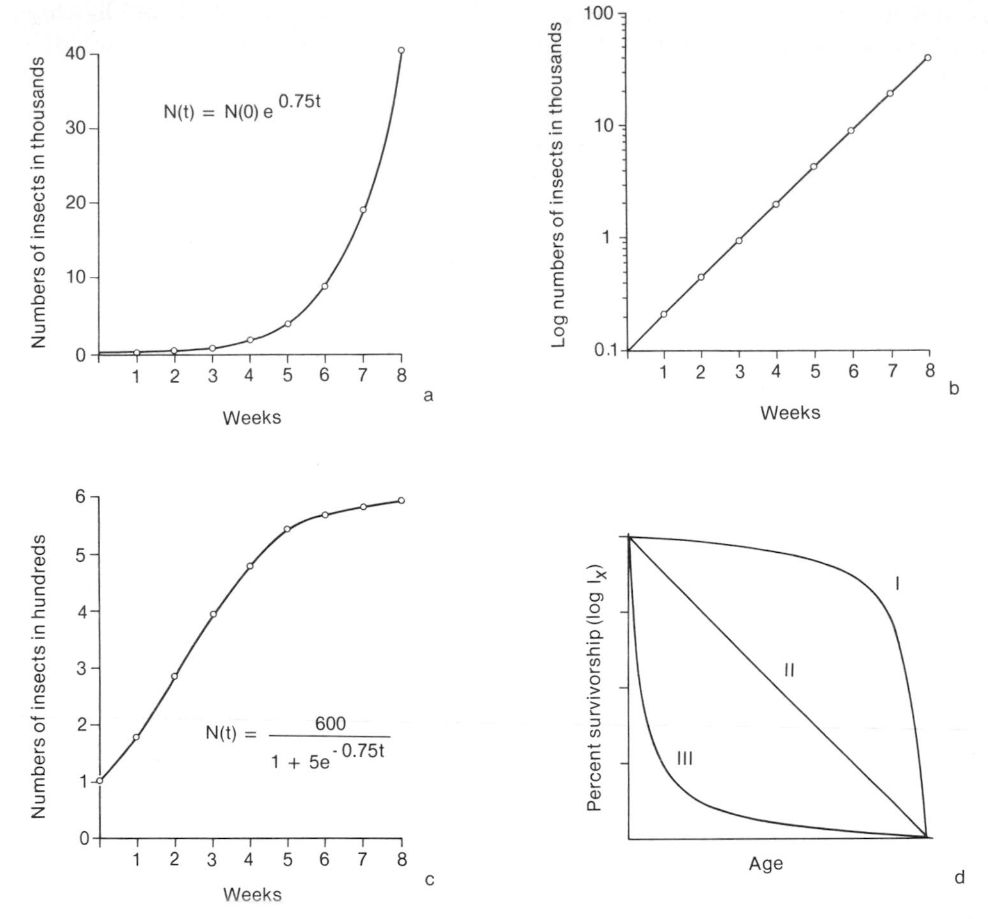

FIGURE 8.1 Population curves: *a*, exponential growth illustrated by text example where 100 beetles are placed in a bin and $r_m = 0.75$ per individual per week; *b*, the same example with population size plotted on a logarithmic scale; *c*, logistic growth of same example where $K = 600$; *d*, survivorship curves (diagramatic).

will grow exponentially. The interest becomes part of the balance and in turn earns more interest. As time goes on, the balance gets bigger and the additions in the form of interest get bigger in proportion. When growth is exponential, as it approximately is in the early stages of growth of many insect populations, each addition becomes a contributor of new additions.

Restraints on the growth of a population include predators, parasites, disease, and shortages of resources such as food and water. When all such restraints are removed—i.e., there are no predators or disease and there are superabundant resources—r reaches a theoretical maximum symbolized by r_m (*m* for Malthusian). Actually, r_m for a given species is not a single number but has a range of values depending on temperature, humidity, and other conditions. For a specified set of such conditions, then, the maximum possible exponential rate of growth is the corresponding maximum intrinsic rate of increase multiplied by the population size:

$$\left(\frac{dN}{dt}\right)_{max} = r_m N$$

Suppose, for instance, that a grain beetle has established a population of 100 individuals in a wheat bin in which the temperature is 29°C and the moisture content of the wheat is 14 percent. Assuming growth without significant restraint, and an $r_m = 0.75$ per individual per week, how many beetles would one expect to find after 3 weeks?

In the grain-beetle case, $N(0) = 100$, $r_m = 0.75$, $t = 3$, and $N(t) = N(3)$ is to be calculated.

After t weeks,

$$N(t) = N(0)e^{r_m t}$$

After 3 weeks,

$$N(3) = 100e^{(0.75)(3)} = 100e^{2.25} = 949$$

Thus one would expect to find 949 beetles after 3 weeks of unrestrained growth. After 10 weeks there would be 180,804 beetles, and in 20 weeks 326,901,737 beetles. Essentially unrestrained growth obviously cannot continue for 20 weeks unless the wheat bin is large, since at that point more than 300 million beetles would be competing for the remaining grain. Indeed, unrestrained growth for a year would lead to:

$$\begin{aligned} N(52) &= 100e^{(0.75)(52)} = 100e^{39} \\ &= 8{,}659{,}340{,}042{,}000{,}000{,}000 \\ &= 8.7 \times 10^{18} \end{aligned}$$

If each beetle weighed 10 mg on the average, the entire mass of beetles would weigh 8.7×10^{11} metric tons. After 82 weeks there would be 6.1×10^{28} beetles, and their weight of 6.1×10^{21} metric tons would be equal to that of Earth! A general rule about long-continued exponential increase is that it leads to preposterous numbers with surprising speed.[1]

Remember that in the more general case of exponential growth, where predators and disease are not eliminated, r is substituted for r_m. Hence growth may be exponential at a rate below the theoretical maximum. Exponential growth produces characteristic curves when numbers are plotted against time (Fig. 8.1*a*). These curves transform into a straight line if the population size is plotted on a logarithmic scale (Fig. 8.1*b*), which is why it is often called "logarithmic growth."

[1]The r_m conditions in this example approximate those calculated for the grain beetle *Sitophilus oryzae* (Curculionidae) by L. C. Birch (1953).

It is important to keep in mind both the properties of exponential growth curves and the assumptions underlying them. Remember especially that the growth rate, r, is assumed to be constant and that the rate of change of population size (derivative of size with respect to time, dN/dt) is this constant r, multiplied by the size at a particular time. Thus, when the beetle population was just 1000 individuals, the instantaneous growth rate was:

$$\begin{aligned} \frac{dN}{dt} &= rN = 0.75(1000) \\ &= 750 \text{ individuals per week} \end{aligned}$$

GROWTH WITH RESTRAINTS. Suppose that, instead of being a constant, an individual insect's reproductive contribution varied with population size. Now the basic equation becomes:

$$\begin{bmatrix}\text{Instantaneous}\\ \text{rate of}\\ \text{increase}\end{bmatrix} \text{ equals } \begin{bmatrix}\text{function of}\\ \text{population}\\ \text{size}\end{bmatrix} \text{ multiplied by } \begin{bmatrix}\text{population size}\end{bmatrix}$$

$$\frac{dN}{dt} = f(N) \times N$$

$$= r \times N$$

where $f(N)$ stands for "a function of N"—i.e., for r, which now varies with N.

The simplest mathematical model which can describe this situation is one in which r decreases linearly as N increases. In this model r is no longer a constant. Rather the expression for r is:

$$r = r_m \frac{K - N}{K}$$

Here K is the *carrying capacity*—the maximum number of individuals that can be supported in a given environment. According to this relation, at the beginning of the growth process (when

$N = 0$), r is equal to r_m, the maximum rate of increase. Later, as N approaches the carrying capacity, K, r approaches zero.

The differential equation for this simple model of growth is:

$$\frac{dN}{dt} = r_m \frac{K - N}{K} N$$

This equation can be integrated to obtain the following solution:

$$N(t) = \frac{K}{1 + ce^{-r_m t}} \qquad \text{when } c = \frac{K - N(0)}{N(0)}$$

This equation is that of the famous logistic curve (Fig. 8.1*c*)—the archetypal presentation of "density dependence" in ecology. Note in this figure that the growth rate of the population at each moment depends on its density (size). Growth starts off slowly and then accelerates rapidly [in our example, we already have 100 beetles at $N(0)$]. The first part of the curve is sometimes called the "log phase," because when plotted with N on a logarithmic scale (as in Fig. 8.1*b*), it approximates the straight line of exponential growth. Then the curve rapidly flattens out, approaching K asymptotically (i.e., the difference between N and K becomes infinitesimally small as t gets very large and the expression $ce^{-r_m t}$ gets very small).

There are many simplifying assumptions in the logistic equation. One is that all individuals are presumed to be alike ecologically—there is, for instance, assumed to be no change with age in the likelihood of giving birth, eating, or being eaten. This assumption is obviously violated in many insect populations, even if only adults are considered. For example, older females may lay many fewer eggs than younger ones, or fertility may differ among eggs laid at different ages. There is also assumed to be no time lag between a change in the environment and the reactions of the organisms—a very unrealistic assumption. Very low densities are presumed not to hinder mate finding. And, most importantly, there are the assumptions that the carrying capacity, K, is constant and that there is a linear relationship between $K-N$ and r. In spite of these simplifications, the logistic curve has proved to be remarkably close to observed patterns of population growth in laboratory cultures of organisms such as *Drosophila* (Drosophilidae), yeasts, and protozoa.

AGE COMPOSITION AND POPULATION GROWTH

The assumption that all individuals are ecologically identical—equally likely, for instance, to give birth or die, is not even approximated in most insect populations. Understanding the effects of the age composition of populations (and changes in that composition) on natality and mortality rates is critical to understanding their dynamics, especially when they have overlapping generations.

Consider the input side of the equation first. Given the requisite information, one can construct a schedule of age-specific birthrates for a population. This schedule is the maternity function, m_x (m for maternity, x for age). It is normally presented in terms of the number of female offspring per female in a given age class. To make statements about population growth, however, it is necessary to know more than that number of female offspring. One also needs to know how many females there will be in each age class, i.e., to know the probability that an individual will survive to any given age (l_x). This probability can be obtained from the age-specific death rates.

Conventionally, however, death-rate data are presented in terms of l_x, defined either as the number of individuals surviving or the proportion of individuals surviving. Most animal populations show survivorship (l_x) curves of one of three types (Fig. 8.1*d*). No insect populations show a type I curve, in which low early mortality is followed by a period of rapidly increasing death rates in middle age (this pattern is found, for example, in human populations in industrial countries). The type II curve, in which a constant proportion of the population dies in each age interval, is approximated by adults of some butterflies (which, however, suffer high mortality in their egg, larval, and pupal stages). Type III curves, in which there is a great die-off at early ages, are the most common in natural insect populations.

If age-specific death-rate and birthrate (l_x and m_x) data are available, it is possible to combine them in order to make some statements about population dynamics. Let us look at a single female cohort (group of females hatched at the same time) of a hypothetical insect which never lives more than 4.5 months. A version of a *life table* called an $l_x m_x$ table can be constructed. In Table 8.1, x is the designator of the age class $x - \frac{1}{2}$ to $x + \frac{1}{2}$ (if x is 3 months, all individuals $2\frac{1}{2}$ to $3\frac{1}{2}$ months old are in class 3). In the same table, l_x is the proportion of the cohort surviving to age x, and m_x is the age-specific female hatch rate (in the example, females first reproduce between 0.5 and 1.5 months of age, and the vast majority of eggs are devoured by predators and parasitoids!). The key column is the final one, the product of l_x and m_x. The sum of that column, $\Sigma_0^\infty l_x m_x$, tells the number of female offspring expected per female. This sum is designated R_0.

TABLE 8.1 Life Table for a Hypothetical Insect

x (age class, months)	l_x (proportion of survivors)	m_x (average no. of offspring)	$l_x m_x$
0	1.00	0.0	0.00
1	0.80	1.0	0.80
2	0.30	0.5	0.15
3	0.16	0.25	0.04
4	0.05	0.20	0.01
5	0.00	0.00	0.00
		$R_0 = \sum_0^\infty l_x m_x =$	1.00

If $R_0 = 1$, as it does in this example, it means that the female cohort is exactly replacing itself; for 1000 females in generation n, there will be 1000 females in generation $n + 1$. If $R_0 = 2$, then in generation $n + 1$ there will be 2000 females. If $R_0 = 0.5$, there will be 500 in generation $n + 1$. If generations do not overlap—i.e., if all the parents die before any of their offspring mature (as in an insect with one generation per year)—then R_0 indicates whether the population is growing, is shrinking, or is *stationary* (not changing in size). If R_0 is greater than 1, the population is growing; if R_0 is less than 1, the population is shrinking; if R_0 is exactly 1, the population is stationary.

If generations overlap, however, no statement about population growth at a given moment can be made on the basis of a calculation of R_0 alone. It is necessary also to know the age composition of the population. The reason for this is seen by considering how R_0 is calculated in a population with overlapping generations. Age-specific female birthrate and death-rate schedules are determined for a given point in time; and since there are no discrete generations of females in which reproduction and death can be observed, those schedules are applied mathematically to a hypothetical cohort in order to calculate R_0. The age composition of the actual population at the time the birthrate and death-rate schedules are determined is not taken into consideration.

Life tables have not been widely used in past analyses of insect populations, but they are being employed increasingly. Understanding the principles by which they are constructed is important for all entomologists. Those interested in actual examples and further references should refer to *Insect Ecology* by P. W. Price (1975).

DYNAMICS OF NATURAL INSECT POPULATIONS

The entomologist working on an insect population in the field is usually interested in two basic questions: (1) How are the numbers of insects in the populations changing? (2) What factors are causing those changes?

To answer the first it is necessary to census the population at several times. While workers on large animals or plants can sometimes simply count individuals, entomologists usually must resort to indirect methods. One of the most frequently employed is the technique of mark, release, and recapture (MRR). A sample of individuals is collected, given a distinctive mark, and released. Time is allowed for the released individuals to mix with the population, and then a second sample is taken and the proportion of recaptures noted. A simple population size estimate may then be obtained by the so-called Lincoln Index formula:

$$\frac{\text{No. marked individuals released}}{\text{Population size}} = \frac{\text{no. of marked individuals recaptured}}{\text{total in second sample}}$$

In other words one determines the population size by assuming that the ratio of recaptured marked individuals to the total of that sample is the same as the ratio of the number of marked individuals "at risk" to the total population size. The accuracy of MRR estimates depends on how well the various assumptions underlying it are met—for instance, how thoroughly the marked individuals mix in with the remainder of the population, and whether or not marked individuals change their behavior so that they are more or less readily captured than unmarked individuals. The MRR formula given also obviously includes the assumption that there will be neither recruitment (natality or immigration) nor loss (mortality, emigration) between the two samplings. Since this assumption is virtually never met in insect populations, more complex procedures of dealing with MRR data which attempt to account for recruitment and loss have been widely adopted, although there is considerable dispute over their accuracy.

One great advantage to the MRR approach to population censusing is that it also provides data on movements of individuals. Thus, among other things, it becomes possible to define population units. This, as we shall see, is extremely important to understanding the dynamics of populations.

Let us look at some actual insect populations in nature. Among the best known are three populations of the checkerspot butterfly *Euphydryas editha* at the Jasper Ridge Reserve near San Francisco. The colony of *E. editha* on the Ridge has been under intensive MRR study since 1960 (Ehrlich et al., 1975). In that year it was discovered that what had been thought to be a single population living in an island of grassland surrounded by a sea of chaparral was, in fact, three separate populations (*H,C,G*) which only rarely exchanged individuals. A glance at Fig. 8.2, which shows estimates of the sizes of those populations between 1960 and 1974, suffices to show why identification of population units is essential to understanding the factors controlling population size. Note, for example, that if the three populations had been considered a single dynamic unit, the total population size in 1965 would have been at a peak of over 4000, and the extinction (and subsequent reestablishment) of the *G* population would have been missed entirely. What if we had not realized that three populations were being sampled and had considered all Jasper Ridge *E. editha* to belong to one population? Then we would have been attempting to explain why a single population had reached a maximum instead of trying to explain why one of three had reached a maximum (*H*), while a nearby one continued to increase (*C*) and

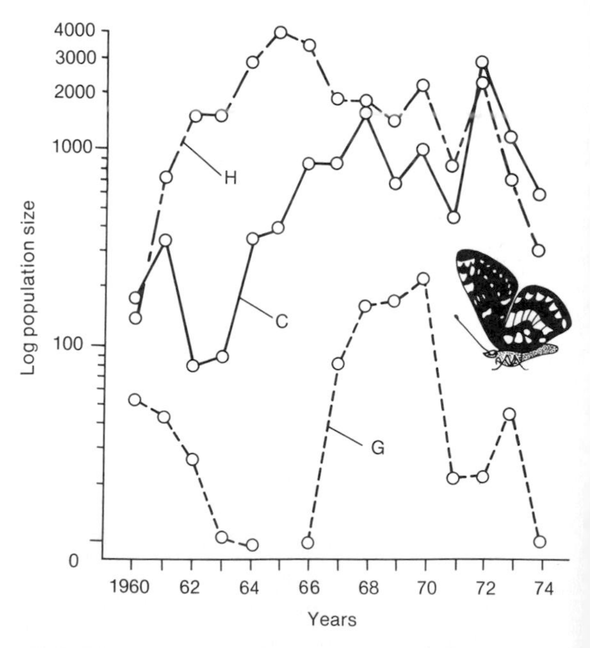

FIGURE 8.2 Sizes of three *Euphydryas editha* butterfly populations (*H, C, G*) on Jasper Ridge, Stanford, Calif. 1960–1974. (*Data from Ehrlich et al.,* 1975.)

a third (G) (which happens to be located in the grassland between the other two) declined to extinction!

The work of Michael Singer, Lawrence Gilbert, Raymond White, and others has slowly revealed the factors which affect the size of these and other *Euphydryas* populations. On Jasper Ridge the adults fly in March and April, laying their eggs primarily on or in the vicinity of a small annual plantain (*Plantago erecta*). Larvae must reach the appropriate size (mid-third instar) for diapause before the supply of food plant dries up with the approach of summer. Summer is spent in diapause, growth resumes with the fall rains, and pupation occurs in the early spring.

On Jasper Ridge it was determined that the overriding factor affecting numbers was changes in food resources for the prediapause larvae. In most cases the *Plantago* dries up before the larvae reach diapause size. Most of the larvae that survive do so by transferring to the flowers of the hemiparasitic annual plant *Orthocarpus densiflorus*. This plant remains in bloom long enough for the larvae to reach the necessary size. Occasionally *Plantago* stays green long enough, especially on soil tilled by gophers where the plant's roots can penetrate the soil further. But the size of a subsequent generation of *E. editha* adults can be quite successfully estimated by surveying the density of the *Orthocarpus* population in the spring (Fig. 8.3)—a density which is probably determined by microclimatic factors such as early spring rainfall. The "gopher effect" is of relatively little importance.

Changes in the size of the Jasper Ridge populations seem to show little relationship to the density of the population—i.e., population-control factors are largely *density-independent.* The amount of food plant consumed is negligible, much less than 5 percent year after year, even though 98 to 99 percent of the prediapause larvae starve annually because almost all the available food dries up.

At Del Puerto Canyon, some 84 km east southeast of Jasper Ridge, there is a population of *E. editha* which oviposits and feeds on *Pedicularis densiflora* (Scrophulariaceae). At Del Puerto intraspecific competition for food is severe—with up to 94 percent of the food-plant crop destroyed in a given year. When it occurs, such defoliation is accompanied by rates of larval starvation of 50 to 90 percent because of the influence of other larvae. Thus changes in size of the Del Puerto population are greatly influenced

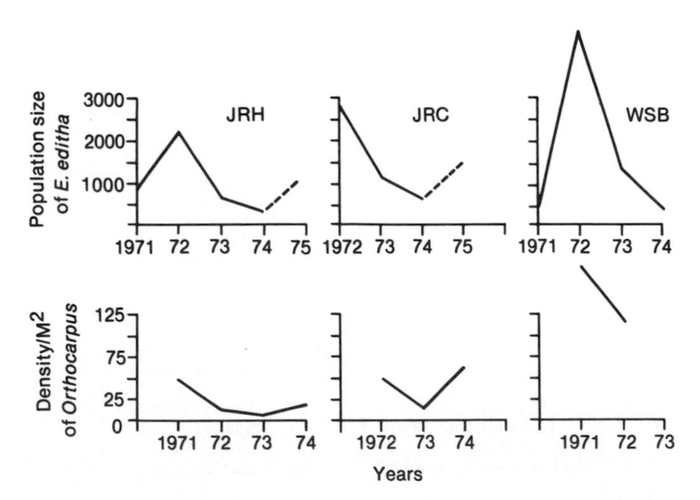

FIGURE 8.3 Sizes of *Euphydryas editha* butterfly populations compared to populations of the larval food plant, *Orthocarpus* sp., in the previous year. *JRH* and *JRC* populations are from Jasper Ridge, Stanford, Calif., and *WSB* is from 6.4 km away. Dashed lines were predictions of 1975 population sizes. These predictions proved accurate. (*Data from Ehrlich et al.,* 1975.)

by the size of that population. For instance, if the adult population is very large in year n, the chances are slight that it will again be large in year $n + 1$ because there will not be enough food plant to permit many larvae to mature. Population-size changes at Del Puerto show *density-dependence.*

The difference between density-dependent and density-independent factors has long fascinated entomologists interested in the control of insect numbers. Especially in agricultural ecosystems it seems desirable to introduce controls for pests which will function in a density-dependent manner—i.e., controls which will act more severely against each individual of the population as the size of population increases. A predator often functions in such a manner. At very low population sizes the chance that an individual pest will be consumed may be negligible. As the pest population increases, it may become increasingly profitable for the predator to attack the pest. If the predator is a bird, it may switch its "search image" to the pest; if the predator is an insect, the pest population may have become large enough to support a population of the predator. In either case the probability of an individual pest's being eaten increases, because the pest population is increasing.

As the pest population grows further, predator pressure becomes even more severe as the abundant food promotes population growth among the predators, and the prey population is decimated. This feedback pathway from the population being controlled to the controlling factor means that predation can act as a *reciprocal* density-dependent factor. When such a pathway is absent, as when, say, the number of old mouse nests available for bumblebee colonies is acting as a density-dependent factor, the number of old mouse nests can limit the size of the bumblebee population, but the bumblebee population cannot influence the number of mouse nests. Similarly food is a reciprocal factor in the case of *E. editha* at Del Puerto Canyon; food for a population of fruit flies living on fallen bananas in a plantation is nonreciprocal.

A useful schema of control factors is shown in Fig. 8.4. It should not be interpreted too rigidly, however. For instance although various "weather" factors are generally considered to operate in a density-independent manner, as Andrewartha and Birch (1954) pointed out,

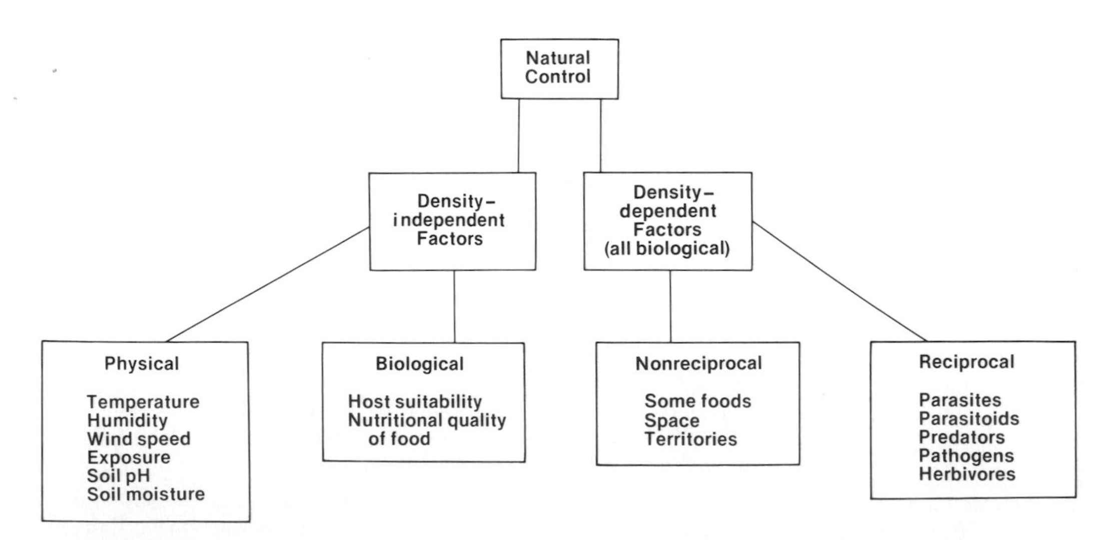

FIGURE 8.4 Major factors in the natural control of insect populations. (*Modified from van den Bosch and Messenger,* 1973.)

weather may act in a density-dependent fashion. For example, it may kill a greater proportion of individuals in a large population than in a small one, because in the large population individuals are forced to live in relatively exposed positions. One should also bear in mind in using any schema for classifying the factors affecting population size that those factors may change from place to place and from generation to generation. Thus we cannot say, "Intraspecific competition is a *key factor* [a mortality factor having value in predicting future population sizes—Morris, 1959] in the control of *E. editha*," because its significance varies from population to population.

There is a vast literature on the factors affecting population size in insects (for access see Price, 1975), but there are relatively few detailed studies of the sort done on *E. editha*. Nonetheless some generalization may be attempted. First of all, weather tends to provide a relatively density-independent control on insect populations in areas outside the Tropics and to a lesser degree in tropical areas with a severe dry season, which it does not in the humid Tropics. To put it another way, winter is a good friend of the farmer—it provides a free, annual pest-control treatment. Distinct seasons tend to have characteristics which permit the rapid building of population size when conditions become favorable. Specialization is for reproductive ability. An organism such as *E. editha* is said to be "*r*-selected" (*r* being the IRI per individual). Food reserves are accumulated by the larva, and the female adult emerges from the pupa ready to lay a thousand or more eggs in a few weeks and then die. Interspecific competition and predation play relatively small roles in controlling many (but not all) of its populations.

In contrast, a butterfly of humid Tropics such as *Heliconius ethilla* (Ehrlich and Gilbert, 1973) is not subject to catastrophic weather-related mortality, but it faces heavy losses in the larval stage from predatory ants. Females do not emerge ready to lay masses of eggs rapidly. Instead a minimum of growth occurs at the vulnerable larval stage, and the adults are long-lived (sometimes living more than 5 months). The females feed on pollen to obtain the protein necessary for egg production (Gilbert, 1972), and lay small numbers of eggs each day for extended periods. These butterflies come as close as any to the description of a "*K*-selected" organism (from *K*, the carrying capacity, in the logistic equation). They have evolved high competitive ability at the expense of the potential for rapid population growth, and have a quite constant population size, probably largely regulated by density-dependent predation and availability of nectar resources (Fig. 8.5).

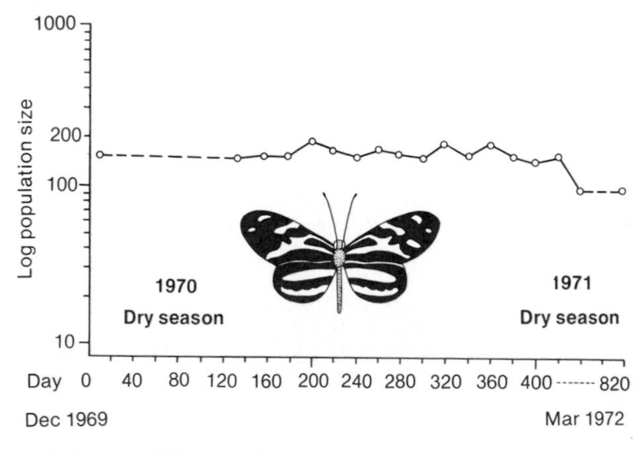

FIGURE 8.5 Size of *Heliconius ethilla* butterfly population at a locality in the forests of northern Trinidad, December 1969, to March, 1972. (*Data from Ehrlich and Gilbert,* 1973.)

One cannot, of course, conclude from these studies of butterflies that populations of Temperate-Zone insects are never controlled by predators or that those of tropical insects are never subject to density-independent control by the weather. We know, for example, that in California the cottony-cushion scale, *Icerya purchasi,* a pest originally imported from Australia, is controlled by predators and parasites. This can be stated with assurance because in Australia it is not a pest, but in California it was a pest until a predatory beetle *Rodolia cardinalis* and a parasitic fly *Cryptochaetum iceryae* were introduced. A year or so after the introductions of these natural enemies, the scale, which had threatened to wipe out the citrus industry, was no longer of economic importance.

At present one of the most promising challenges in entomology lies in the need to extend the samples of populations whose dynamics are being studied in detail. Only when this is done will it be possible to test generalities properly—e.g., the predominance of largely r-selected species in the Temperate Zones and K-selected species in the Tropics—and to induce new generalities.

In the terms r-selected and K-selected the word "selected" refers, of course, to natural selection. Natural selection, as we shall see, is the key process in evolution, and it is critical that entomologists thoroughly understand it and its relationship to other evolutionary processes. For not only are insects (like all other organisms) products of evolution, but they are also constantly evolving. Because insects often have short generation times, a population studied by an entomologist one year may be quite different from the one found in the same area the previous year. In order to come to grips with problems like the evolution of resistance to pesticides or the ways social behavior could have evolved it is necessary to understand basic population genetics—the topic of the following chapter.

SELECTED REFERENCES

For additional readings in population ecology consult Andrewartha and Birch (1954), Clark et al. (1967), Harcourt (1969), Odum (1971), Price (1975), Southwood (1968, 1975), Varley and Gradwell (1970), and Varley et al. (1973).

INHERITANCE IN POPULATIONS

9

Let us first examine those genetic properties of populations of insects that are not properties of individual insects themselves. These properties can be seen in the kinds and distribution of genetic information in the population. In the following discussion, a single locus with only two alleles will be considered. We shall call these alleles "A" and "a" but shall not assume automatically that A is dominant to a.

FREQUENCIES OF GENOTYPES AND GENES

Two of the basic parameters of population genetics are genotype frequency and gene frequency. Suppose that in a population of 200 insects each individual's genotype with respect to the A-a locus was determined. There are three possible types: AA, Aa, and aa. Suppose that all three genotypes are actually present in the population. The frequency of each is calculated simply by dividing the number of individuals with a particular genotype by the total number of individuals. In our hypothetical population, the results might be as given in Table 9.1.

TABLE 9.1

Genotype	*AA*	*Aa*	*aa*	Sum
No. of individual insects	120	40	40	200
Genotype frequency	0.60	0.20	0.20	1.00
Genotype frequency	*D*	*H*	*R*	

Note that the genotype frequencies are proportions and therefore sum to 1.00. Note also that the frequencies are symbolized by the letters D, H, and R, and that in this case $D = 0.6, H = 0.2$, and $R = 0.2$; and in all cases $D + H + R = 1$ (if the allele A were dominant over the allele a, then these would stand for Dominant, Heterozygote, and Recessive genotypes; they will be used here without regard for dominance relationships).

Genotype frequencies describe the proportions of different kinds of *individuals* in a population, and one could theoretically segregate the insects with different genotypes in different bottles. Gene frequencies, on the other hand, describe the proportions of *alleles* in a population without regard for their distribution among individuals. One can think of each AA individual as having two alleles A and each aa individual as having two alleles a. Each heterozygote would have one allele A and one allele a. If these genes could somehow be dissected out of the individuals and placed in a container, the proportion of A alleles in the container would be the gene frequency of A (the terms *allele* and *gene* are sometimes used interchangeably in population genetics). The proportion of a alleles would be the gene frequency of a.

Gene frequency thus may be easily calculated from the genotype frequencies. Let us calculate the gene frequency of allele A in our hypothetical population. Since the D individuals contain nothing but A genes, and one-half the genes in the H individuals are A genes, we get the gene frequency of A by simply adding one-half the genotype frequency of the heterozygotes to that of the AA homozygotes. The gene frequency of A, conventionally labeled p, is $D + \frac{1}{2}H$. Similarly the gene frequency of a, conventionally labeled q, is $R + \frac{1}{2}H$. In short, in this situation the frequency of the heterozygotes is simply divided and half is added to each homozygote genotype frequency in order to calculate the gene frequency. In the hypothetical population

$$p = D + \tfrac{1}{2}H = 0.60 + 0.10 = 0.70$$
$$q = R + \tfrac{1}{2}H = 0.20 + 0.10 = 0.30$$

Because they are proportions, $p + q = 1.00$ and $q = 1 - p$. Thus, in a population with 60 percent AA insects, 20 percent Aa insects, and 20 percent aa insects, 70 percent of the alleles are A and 30 percent are a.

Note that if the population consisted of only homozygous individuals ($H = 0$) the gene frequencies would be equal to the frequencies of the genotypes ($p = D, q = R$). This makes intuitive sense, for each genotype would donate two identical alleles to the mythical container. Heterozygotes, on the other hand, carry one allele of each type and thus make equal donations to the frequencies. Table 9.2 gives genotype and gene frequencies for a series of eight hypothetical populations. Be sure you understand the relationship between the two kinds of frequencies, and how gene frequencies are calculated from genotype frequencies, before you continue.

GENETIC EQUILIBRIUM AND THE HARDY-WEINBERG LAW

You will recall that the gametes of diploid organisms are normally haploid. The following statements can be made about gametes from an insect population:

If the gametes are drawn at random from the population, the proportion of sperm carrying A will be p, and of sperm carrying a will be q. Similarly, eggs carrying A and a will be produced in the proportions p and q. The production of gametes by the population may be viewed as a breaking down of the diploid genotypes, the gamete pool being analogous to the mythical container into which the genes were placed in the discussion above. Both alleles are present in the proportions in which they are present in the population of diploid individuals, but, except in the cases where either p or q is equal to 1.00, it would be impossible to say what the genotype frequencies were in the parental population if only the gamete frequencies were known. For instance, in Table 9.2 the gamete frequencies of populations 3, 4, 5, and 6 would be identical, but their genotype frequencies are very different.

If the gametes in the gamete pool combine at random, the chances of any particular combination will be the product of the frequencies of the respective kinds of gametes. If A sperm do not tend to combine more frequently with A eggs than with a, or vice versa, and if the a sperm are equally lacking in preference, then gamete union is at random. The chance of an A sperm uniting

TABLE 9.2

Population number	D(AA)	H(Aa)	R(aa)	Sum	p(a)	q(a)	Sum
1	1.00	0.00	0.00	1.00	1.00	0.00	1.00
2	0.35	0.00	0.65	1.00	0.35	0.65	1.00
3	0.50	0.00	0.50	1.00	0.50	0.50	1.00
4	0.00	1.00	0.00	1.00	0.50	0.50	1.00
5	0.25	0.50	0.25	1.00	0.50	0.50	1.00
6	0.01	0.98	0.01	1.00	0.50	0.50	1.00
7	0.20	0.20	0.60	1.00	0.30	0.70	1.00
8	0.02	0.02	0.96	1.00	0.03	0.97	1.00

with an A egg then is equal to the chance of taking a single sperm at random from the gamete pool and having it be an A sperm and then taking a random egg and getting an A egg. If the gene frequencies were $p = q = 0.50$, the chances would be 50-50 in each case. Since the probability of two independent events occurring together is the product of the probabilities of their separate occurrence, the chance of an A sperm and an A egg uniting would be $0.50 \times 0.50 = 0.25$ (the same as the chance of getting two "heads" in two independent flips of a coin). The results of gamete union in any particular gamete pool can be displayed in a checkerboard with gamete frequencies on the margins and zygote frequencies in the body of a table such as Table 9.3. In the table above, the gamete frequencies in the margins are those of our original hypothetical population (see Table 9.1). What are the genotype frequences of the resultant zygotes? Heterozygotes formed by the fusion of an A sperm with an a egg are indistinguishable from those formed by the fusion of an a sperm with an A egg. The new genotype frequencies (compared with the old) are

		New	Old
$D\ p^2$	**AA zygotes** =	0.49	0.60
$H\ 2pq$	**Aa zygotes** =	0.42	0.20
$R\ q^2$	**aa zygotes** =	0.09	0.20
	Sum =	1.00	1.00

The new *gene* frequencies are easily calculated from the genotype frequencies:

$$D + \tfrac{1}{2}H = 0.49 + 0.42/2 = 0.70 = p$$
$$R + \tfrac{1}{2}H = 0.09 + 0.42/2 = 0.30 = q$$

As you can see, the new gene frequencies are the same as the old gene frequencies (see Table 9.3); the genotype frequencies have changed, but the gene frequencies have remained constant. If the whole process is repeated, not only will the gene frequencies remain the same, but the genotype frequencies will not change again. In the absence of factors other than those mentioned, if the gene frequencies are p and q, then after one generation the genotype frequencies will be given by the expression

$$p^2(AA) + 2pq(Aa) + q^2(aa) = 1.00$$

In these circumstances, the gene frequencies do not change at all, and after one generation the genotype frequencies do not change. In other words, the population is in *genetic equilibrium* after one generation. This is the Hardy-Weinberg equilibrium, named after the scientists who independently described it early in this century.

The idea that populations do not change genetically *unless there is some factor causing them to change* has been called the Hardy-Weinberg law. This deceptively simple idea is central to the discipline of population genetics. It might be said that the essence of evolution is the changing of gene frequencies in populations. The importance of the Hardy-Weinberg law is that it defines a base-line situation in which there is no evolution. Evolution may be viewed as deviation from the base line, and the factors which may cause such deviations are those responsible for evolution. There are a number of such factors, discussed below, one or more of which occur in every population. Therefore Hardy-Weinberg equilibrium is a theoretical base-line condition and would, at best, only be approximated in a natural population, if it occurred at all.

TABLE 9.3

		Eggs	
		$p(A)$ 0.70	$q(a)$ 0.30
Sperm	$p(A)$ 0.70	$p^2(AA)$ 0.49	$pq\ (Aa)$ 0.21
	$q(a)$ 0.30	$pq\ (aA)$ 0.21	$q^2\ (aa)$ 0.09

FACTORS PREVENTING ESTABLISHMENT OF HARDY-WEINBERG EQUILIBRIUM

NONRANDOM MATING. One factor which can cause deviations from Hardy-Weinberg equilibrium is nonrandom mating. You will recall that one of the conditions for the equilibrium is that gametes must combine at random. This condition will not be met if mating is nonrandom. Imagine a population of dragonflies with both A and a alleles. If AA dragonflies prefer to mate with other AA dragonflies, then clearly the probability of an A gamete uniting with another A gamete will be higher than it would be if mating were at random.

MUTATION AND DIFFERENTIAL MIGRATION. Two other factors which can cause deviation from Hardy-Weinberg equilibrium are mutation and differential migration. Mutation changes one allele into another and therefore obviously can change gene frequencies. In an extreme example, if a dragonfly population consisted entirely of AA individuals, then $p = 1.00$ and $q = 0.00$. A single mutation of an A gene into an a gene will immediately change p to something less than 1.00 and q to something more than 0.00. Differential migration is the addition or subtraction of individuals to or from a population where the migrant individuals collectively have a *different* gene frequency from the population as a whole. In the example just given, the addition of a single heterozygote to the population would have roughly the same effect as a single mutation. Indeed the effects of mutation and of differential migration receive very similar treatment in the mathematics of population genetics.

SAMPLING ERROR (GENETIC DRIFT). Another factor which may be responsible for change in gene frequency is sampling error. Sampling error may be thought of as the kind of chance occurrences which prevent you from getting five "heads" and five "tails" each time you flip an "honest" coin ten times. Reproduction by any insect population can be thought of as a series of samplings. Gametes which are produced by the population are a sample of all the cells which could have become gametes. The gametes which actually unite into zygotes are a sample of all the gametes. The zygotes which mature are a sample of all the zygotes formed. If all these are random processes, gene frequencies will change from generation to generation because of sampling error. The magnitude of these changes is intimately related to the size of the population; *they are important only in quite small populations.*

When sampling error causes the frequency of a gene to change through a series of generations, that gene frequency seems to drift aimlessly from value to value. This has led to the description of these changes as *genetic drift.*

SELECTION. So far we have discussed four factors which can cause deviations from Hardy-Weinberg equilibrium: nonrandom mating, mutation, differential migration, and sampling error. We have left the most important factor, differential reproduction of genotypes, or *selection,* for last. Let us look at our three hypothetical genotypes and assign to each genotype a number indicating its relative reproductive success. This number is called the *adaptive value,* or *fitness,* of the genotype. The genotype which, on the average, has the highest number of its gametes represented in individuals of the next generation is assigned the adaptive value of 1, and others are assigned proportional values. Adaptive values are assigned without regard for the frequency of the genotype in the population; at a given instant the genotype which is the most successful reproducer may be quite rare in the population.

Suppose that in a growing population, *on the average,* each AA individual contributed 10 gametes to individuals of the next generation; each Aa individual, 6; and each aa individual, 4. Then the adaptive values of the genotypes would be

Genotype	*AA*	*Aa*	*aa*
Adaptive value	1	0.6	0.4

If you measure the differential reproduction of genotypes and you find that it exceeds the differential you could reasonably expect on the basis of sampling error, you can say that *selection* has occurred. Selection is now understood to be a matter of differential reproduction, rather than simply a matter of differential survival. The strength of selection is measured by the selection coefficient, s, which is simply 1 minus the adaptive value. Thus the above table is equivalent to the following:

Genotype	*AA*	*Aa*	*aa*
Selection coefficient(s)	0.0	0.4	0.6

Differential reproduction can, of course, run the gamut from a situation in which some individuals have a slightly decreased fertility to one in which some individuals are sterile. An individual that does not reproduce is selectively equivalent to one that is dead, except for rare exceptions such as the sterile castes of social insects. Of course, selection often involves selective death; one of the best ways to reduce the adaptive value of a genotype to zero is simply to kill all individuals of that genotype before they can reproduce.

SELECTION CHANGING GENE FREQUENCIES. It should be obvious that differential reproduction of genotypes may result in changes in gene frequency. Consider an extreme and hypothetical case. There is a population of tropical *Anopheles* mosquitoes which are segregating at the *A*-*a* locus. Assume further that *aa* individuals are especially effective vectors of malaria and are also susceptible to an insecticide which does not harm *AA* or *Aa* individuals. Suppose that a population of these mosquitoes in Hardy-Weinberg equilibrium is treated with the insecticide which kills the *aa* individuals before they mature, leaving the others untouched:

Genotypes	*AA*	*Aa*	*aa*
Frequency before selection	0.25	0.50	0.25
Adaptive value	1	1	0

After selection, 75 percent of the original population remains, and the genotype frequencies are now $AA = 0.33$, $Aa = 0.67$, $aa = 0.00$. The new gene frequencies are:

$$p = D + \tfrac{1}{2}H = 0.33 + 0.67/2 = 0.67$$
$$q = R + \tfrac{1}{2}H = 0.00 + 0.67/2 = 0.33$$

With some algebraic manipulation, it can be shown that in a situation where recessive homozygotes do not reproduce at all, the change in gene frequency per generation is given by the expression:

$$\Delta q = \frac{-q^2}{1 + q}$$

That is, Δq, the rate of removal of the a genes, is a function of the gene frequency of a. The more a genes there are in the population, the easier it is for their frequency to be reduced by failure of the homozygous recessives to breed. But, as the frequency of the a genes gets small, the rate of decrease gets lower and lower. What will happen if the selective regime is such that *aa* individuals have a fitness of 0, and the other two genotypes have a fitness of 1, generation after generation? At first q will decrease rapidly, but then the rate of removal of a genes will slow down dramatically.

SELECTION STABILIZING GENE FREQUENCIES. If the heterozygotes at a locus have a higher fitness than either homozygote, selection results in a gene-frequency equilibrium that is *stable.* Both alleles are retained in the population, and all three genotypes are present. The equilibrium value of p ($\hat{p}$, read "p hat" $= 1 - \hat{q}$) is determined solely by the relative fitness of the homozygotes. If the latter are equally inferior to the heterozygotes, $p = q = 0.50$, and half the genes in the population are one allele and the other half the alternate allele. This gene-frequency equilibrium is stable because if it is disturbed (say by immigration of a group of individuals homozygous for one allele), the gene frequencies will return to the equilibrium value if the fitnesses remain the same.

A situation in which two or more alleles at one locus persist in the same population, with the rarest of them in a frequency *too high to be accounted for by mutation alone,* is known as *genetic polymorphism.* When heterozygotes are favored over homozygotes, the establishment of a gene-frequency equilibrium creates a balanced polymorphism. This type of polymorphism is important in evolution, in part because it permits a certain amount of variability to be retained in the population. This means that the population may be able to react very rapidly to an environmental change and thus avoid extinction.

VARIATION AND SELECTION IN NATURAL POPULATIONS

How much genetic variation exists in natural populations has long been a question of central importance to population biology, and much of the work on this question has been done with insects. Experimental evidence was, for a long time, difficult to obtain, since most visible mutational changes were in characters whose expression is affected by many genes. *Drosophila* eye color, for instance, represents the final result of the action of multistep pigment pathways mediated by many genes. Nor was there any means of determining the incidence of genetic variants without visible effects. In the absence of direct experimental data, many population geneticists concluded on theoretical grounds that the total amount of genetic variation maintained in natural populations could not be large. They reasoned that different genetic variant types would have differing effects upon overall fitness, one type being optimal in a given environment. If the fitness of the optimal genotype is 1.0, then the more of the other suboptimal types present, the less the average fitness of the population (the frequency of the optimal type being less). They thus concluded that the maintenance of genetic variation in populations imposed a *genetic load* in terms of reduction in average population fitness. The amount of genetic variation maintained in natural populations was not thought to be large because too great a genetic load would so reduce average fitness that the population would be driven to extinction.

Within the last few years, the technique of gel electrophoresis has provided a direct experimental approach to this problem, and much of the important work has been done with insects. A tissue homogenate is placed in a supporting gel and subjected to an electric field. The soluble proteins of the homogenate migrate in this field, the rate of migration depending upon their charge (and to some degree upon their size). Most proteins have differing charge distributions and thus migrate at different rates. The value of this approach is that it allows investigation of discrete gene characters. A locus whose enzyme's substrate is known can be studied independently of other loci by simply placing a gel, after electrophoresis, in an assay mix containing the enzyme's substrate and reagents for rendering the reaction product visible. Only at that position on the gel to which the enzyme has migrated will the reaction occur, producing a band of visible reaction product. Because this technique is experimentally simple to carry out, it has rapidly become widespread. It should be noted, however, that electrophoresis characterizes variant types in terms of their mobilities in an electric field, and such measures of rate of movement depend not only upon protein charge but also upon time, temperature, current, and other variables. Comparability thus requires careful standardization. Allelic alternatives characterized by differences in electrophoretic mobility are called *allozymes.*

The first estimate of the amount of genetic variation in natural populations obtained by electrophoretic analysis was reported by Lewontin and Hubby in 1966 for natural populations of *Drosophila pseudoobscura.* From their studies they concluded that if all populations are considered, the average individual is heterozygous for 12.3 percent of its genes. Similar studies with other species of *Drosophila* and with butterflies, certain grasses, mice, and human beings have yielded essentially the same result. This is much more genetic variation than was predicted on the basis of the arguments for genetic load. Later treatments of genetic load have taken account of possible threshold effects in fitness as well as demonstrating that the interaction of epistasis and linkage between different loci also reduces load.

Characterizing the amounts of genetic variation by electrophoresis requires independent demonstration that the observed differences indeed reflect genetic differences rather than other factors (separate bands may also occur because of binding of ions or cofactors, enzymatic modification of the protein, etc.). A great strength of the initial report by Lewontin and Hubby was in its careful documentation of the genetic basis of each variation type examined.

It should be noted that estimates of the amount of genetic variation in natural populations based upon surveys of electrophoretic variation are almost certainly underestimates, as not all structural changes in an enzyme will affect its mobility in an electric field. Such estimates also assume that the sample of the genome is random, i.e., that the level of variation seen in the loci examined is typical of all loci. This assumption may not be valid if the levels of heterozygosity reflect the nature of the loci examined. The degree to which enzyme function and amount of genetic variation are related is not yet clear. Although some workers, e.g., Johnson (1973), argue for a strong correlation, little clear evidence exists.

It has been suggested by Kimura and others (e.g., Kimura and Ohta, 1971) that the large amounts of genetic variation detected by electrophoresis do not contradict the predictions of the genetic-load theory if one assumes that the different allozyme types do not produce differences in fitness. They argue that the electrophoretic variation simply reflects genetic "noise," being minor changes in protein structure sufficient to alter electrophoretic mobility but not enzyme activity. Such variants, being functionally equivalent, would respond similarly to selective forces; they are referred to by Kimura as being *selectively neutral.* This theory of selective neutrality of allozyme polymorphisms is highly controversial and has been a hotly debated issue in population genetics. Alternative arguments have been advanced strongly by Ewens, Johnson, and others that the existence of these polymorphisms has a selective basis. For example, Johnson (1973) finds that the occurrence of particular enzyme polymorphisms in butterflies is correlated with the pattern of certain environmental factors, transcending species boundaries.

One of the key problems in determining whether most observed patterns of allozyme variation are produced by selection or by drift is that there is insufficient knowledge of gene flow among the populations of most organisms. If, for instance, 10 or more isolated populations have similar frequencies of a given allele, we can be reasonably sure that this is not a random pattern—that selection is involved. If, however, there is even a relatively low level of gene flow among the populations, it would be sufficient, in the absence of selection, to keep gene frequencies similar.

Careful MRR studies of movements of individuals of *Euphydryas editha* have allowed gene flow to be eliminated as a force keeping most populations in genetic "lockstep." Populations in California and Oregon which are isolated from one another (Fig. 9.1) were found to have very similar frequencies of a major allele at the PGM locus (controlling production of the enzyme phosphoglucomutase) and systematic variation

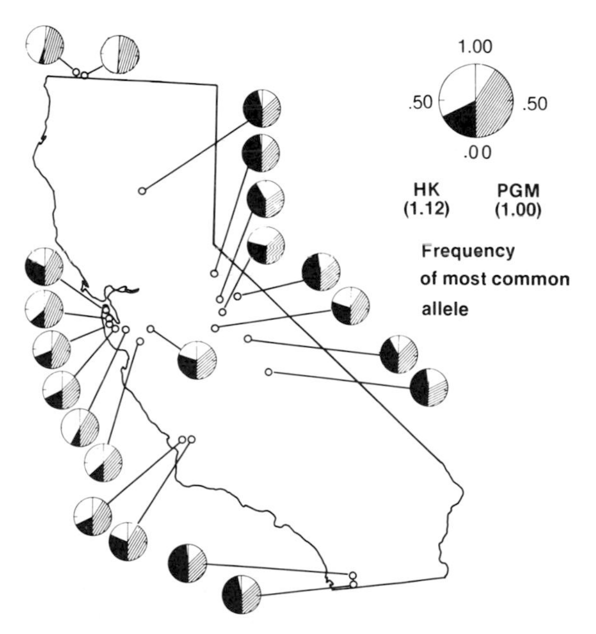

FIGURE 9.1 Genetic variation of two enzyme loci in populations of the butterfly *Euphydryas editha* in California and Oregon. Pie diagrams are connected to the localities where the populations were sampled. Left side of pie diagram indicates frequency of Hk-1.12 allele (in black); right side indicates frequency of Pgm-1.00 allele (hatched). Example diagram, at upper right indicates a frequency of 0.35 for Hk-1.12 and 0.82 for Pgm-1.00. (*Data from McKechnie et al.,* 1975.)

for one at the HK (hexokinase) locus (e.g., high frequencies in the South and in the Sierras, variable in the central coastal area, and low in the North). Such a pattern is strong evidence that, at least for these populations and loci, selection is playing a major role, either directly or by operating on linked loci (McKechnie et al., 1975).

ARTIFICIAL SELECTION IN INSECT POPULATIONS. Human beings affect evolutionary processes in insect populations both purposefully and inadvertently. Much of the purposeful intervention has been in the form of artificial selection experiments, performed in the laboratory in the hope of gaining understanding of natural evolutionary processes. *Artificial selection* may be defined as selection in which a human being determines the fitnesses of various genotypes—say by choosing the flies of each generation with most bristles or the cows that give the most milk as the parents for the next generation. However, even under the most carefully controlled laboratory conditions, *natural selection still is found in conjunction with artificial selection.* A great deal of work in artificial selection has been concerned with so-called quantitative characters. Quantitative characters are those influenced by many pairs of alleles at many different loci, such as height in humans, bristle number in *Drosophila,* or color pattern in ladybird beetles. Variation in such characters is continuous, rather than discontinuous as in classical Mendelian characters controlled by alleles at one or a few loci.

The experiments of Mather and Harrison (1949) are rather typical of laboratory work in artificial selection. They applied selection for the number of abdominal chaetae (bristles) in *Drosophila melanogaster.* Figure 9.2 summarizes the results of more than 100 generations of selection for higher bristle number. For 20 generations, progress was extremely rapid. At generation 20, reduced fertility and fecundity in the population made it necessary to discontinue selection (or "relax" selection, as it is often put) in order to prevent the strain from dying out. After several generations without selection the population reached an equilibrium bristle number higher than the original one but much lower than that achieved at the peak in generation 20. From the line in which selection was relaxed, a new selected line (line S) was extracted at generation 24, and progress was rapid to a point near that achieved with the first selected line. When selection in this line was relaxed, there was only a slight return toward the original bristle number. In addition,

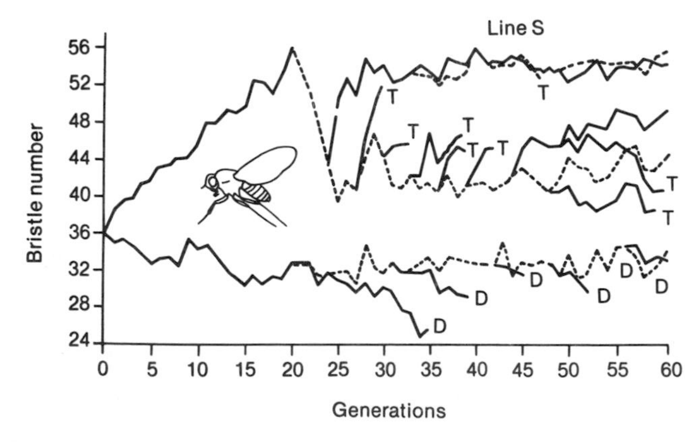

FIGURE 9.2 Artificial selection for abdominal bristle number in *Drosophila melanogaster.* Solid lines indicate cultures under selection for high (upper line) or low (lower line) numbers of bristles; dashed lines are cultures not under selection; *T* indicates deliberate termination of line; *D* indicates that line died out through sterility. (*Data from Mather and Harrison,* 1949.)

after some 85 generations in the continuously selected line, further increase in bristle number was achieved.

Artificial selection often produces rapid change at first. Then a plateau is reached at which further progress is difficult or impossible, or the viability of the line reaches such a low ebb that either selection must be discontinued or the line dies out. Generally, if selection is relaxed before a plateau is achieved, the lines return toward the control level. If selection ends after the population has reached a plateau, there may be little or no tendency to regress toward the original state. Continuous selection of a population that has achieved a plateau often will not produce appreciable results for long periods. If selection is continued long enough, however, progress once again may be made.

One reason for these phenomena presumably is the balance between artificial and natural selection. It seems reasonable that natural selection must lead to a balanced system in which the best possible relationship of characteristics determining fitness is produced; i.e., *fitness must be maximized.* The available evidence seems to indicate that, especially in animals, a high degree of heterozygosity in the genotype produces a high degree of physiological fitness. It also seems likely that extremes of quantitative characters often are produced by a high degree of homozygosity at the loci concerned. Therefore artificial selection for high or low bristle number may well be countered by natural selection for fitness if the bristle-number extremes are produced by homozygosity at a series of loci.

One might make a crude analogy to an airplane. Trying to improve the airplane by making the engine more powerful will do little good if increased speed would tear off the wings. This problem might be solved by strengthening the fusilage or the structural members of the wings, but this would not help if it made the airplane too heavy to get off the ground. In an insect, as in an airplane, a viable balance of all the various factors that ensure successful functioning must be attained. There is a limit to how much one characteristic alone can be changed before the "working combination" is seriously disrupted.

THE LESSONS OF ARTIFICIAL SELECTION. The importance of what has been learned from artificial-selection experiments becomes clear when one considers inadvertent human intrusion into the evolutionary processes of insect populations. Most critical, of course, has been the evolutionary response of pest insects to the routine broadcast use of pesticides. The pesticides constitute a powerful selective force, and most pest insects have the large population sizes and genetic variability which permit the rapid evolution of resistance. The all-too-frequent human response to the emergence of resistance is to increase the dose of the pesticide until resistance is so complete that the pesticide becomes useless, and then to hope to find another pesticide which will work.

A more intelligent approach would be to retard the development of resistance by using a pesticide only when the pest population has reached a level which threatens economically important damage, and when other control measures fail. Then the same pesticide should not be used for more than a few consecutive sprayings—after which a "moratorium" should be declared in which nonchemical controls (or an unrelated pesticide) are used if the pests still threaten serious damage. During the "moratorium" the level of resistance that was building in the pest population will, hopefully, decline under the influence of natural selection.

Interestingly some plants attacked by insects in natural situations appear to employ a sort of "selective moratorium" to retard the evolution of resistance to their defenses. For instance the populations of lupines (plants of the pea family) in Colorado which are most successful in reducing the damage caused by larvae of *Glaucopsyche lygdamus* (a small lycaenid butterfly) are those populations which are most variable in the kinds and amounts of alkaloids they contain (Dolinger et al., 1973). The alkaloids are thought to be defensive chemicals—natural pesticides (Ehrlich and Raven, 1964). Each butterfly matures on a single plant, and in most cases its offspring will mature on a plant containing an entirely different pesticidal array. The variation in the plant population may be maintained by a *frequency-dependent* selection. That is, as one alkaloidal type becomes more common, resistance to that type starts to evolve and it is more readily attacked. A system in which the commonest morph is most vulnerable to predation will remain polymorphic. When two different kinds of organisms are so intimate ecologically that evolutionary changes in one are likely to induce changes in the other, we say that *coevolution* is occurring. In Part III various coevolutionary systems (e.g., plant-herbivore, predator-prey, host-parasite) involving insects will be examined.

It is important to pay close attention to population genetics when dealing with biological controls also. One of the most successful insect-control programs ever carried out has been the use of the "sterile-male" technique against the screwworm fly, *Cochliomyia hominivorax* (Calliphoridae), a serious pest which infests wounds of cattle and other animals, often causing death. Masses of male flies were reared and then sterilized with radiation and released. Since the female flies frequently mate only once, it was possible in the late 1950s, by the release of some 3 billion sterile males, to swamp the population in 64,000 square miles of Florida and eradicate the screwworm there.

A similar program was started in Texas, where the task of control was more difficult. Screwworm flies could invade continually from Mexico, in contrast to the isolated Florida situation. Initial success in the mid-1960s was followed by a resurgence of the fly populations in the early 1970s. One of the major problems was that in the sustained effort necessary to prevent reinfestation not enough attention was paid to maintaining the genetic quality of the stocks being reared for release. Natural selection began to change them into stocks more suitable for the hatchery than for the field, inbreeding effects occurred, and the released males apparently became unsatisfactory mates for the wild females (Bush and Neck; 1976; Bush et al., 1976). New genetic programs have been started which it is hoped will overcome this difficulty (Bushland, 1975).

SELECTED REFERENCES

For additional readings in population genetics consult Dobzhansky (1970), Ehrlich et al. (1974), Johnson (1973), Kimura and Ohta (1971), and Lewontin (1974). Control of pests by genetic techniques is treated by Davidson (1974) and Pal and Whitten (1974).

DIFFERENTIATION OF POPULATIONS AND SPECIATION

10

It is obvious to anyone observing animals and plants that there is not a continuum of variation in nature. There are, for example, no living intermediates between monarch and queen butterflies. Within a genus or family of organisms variants may sometimes be arranged in a continuum, but there are gaps in the variation from continuum to continuum. Inscets, viewed by our usual techniques of studying organisms, seem to be aggregated into discrete or nearly discrete clusters or kinds, usually called *species.* Species are kinds of organisms sufficiently different from one another to be considered distinct by taxonomists, kinds that are usually not observed interbreeding with each other. There is a continuum of differentiation from that of a single panmictic (random mating) population through the degrees of difference formally recognized by taxonomists by the designation of different species, genera, families, orders, etc. A fundamental process of evolution, then, would seem to be the differentiation of populations, and an important question is how a single, supposedly interbreeding population can evolve into two or more discrete populations with different characteristics. This process of differentiation is often called *speciation,* but since species represent only one level of taxonomic distinctness we will discuss differentiation in more general terms. Working out the mechanisms of differentiation has been one of the central problems of evolutionary theory.

GEOGRAPHIC VARIATION

There are clearly many different levels of similarity or difference among groups of organisms. In many instances, within a single species variation is sufficiently continuous so that no dividing lines between segregates are obvious. Variation in some characters may occur in gradients. These gradients in single characters are called *clines,* and the variation is then called *clinal.* Geographic variation in color, pattern, size, and genetic characteristics is one of the most widely studied of all biological phenomena. This variation is often of the sort already described for allozymes of *Euphydryas editha* (Fig. 9.1) in which populations differ primarily in the frequency of different types of individuals present. But in many cases, variation is not in frequency of types; individuals within a given population may all be similar (little intrapopulational variation), but average color, leg length, chirp frequency, or some other character may change from population to population in broad geographic trends.

ENVIRONMENTAL HETEROGENEITY AND SELECTION. That there is ubiquitous geographical variation among insects (and other organisms) should not be surprising. After all no two points on Earth are identical, so that organisms in different places are subject to different environments and selection pressures. This means that the organisms in different places will not be the same, and this, in turn, will make the environments even more diverse, increasing the amount of differentiation. It is in fact more unusual (and more difficult to explain) when populations are phenetically (phenotypically) or genetically similar while living in disparate habitats. For example the small lycaenid butterfly, *Lycaena phlaeas,* has populations high in the Sierra Nevada of California which are phenetically very similar to those in Eastern North America and Europe (nothing is known of their genetic similarities). In such cases one can only assume, rather lamely, that the most important aspects of the environment are quite similar in all three places (unlikely) or that *L. phlaeas* has evolved a "jack-of-all-trades" genotype which interacts with a wide variety of environments to produce a successful phenotype (more likely).

Sticking with butterflies (in which geographic variation has been more thoroughly studied than in any other group of insects) it has been relatively easy to explain the selective bases for some patterns of differentiation. For example, populations of the eastern tiger swallowtail *Papilio glaucus* have high frequencies of dark females in certain areas, presumably because these dark-brown females resemble the Aristolochia (pipevine) swallowtail *Battus philenor,* also found in these areas. *Battus philenor* is distasteful to birds, and birds have been observed feeding on adult *P. glaucus* in the field. Selection apparently favored the development and maintenance of the mimetic form of *P. glaucus* in these areas.

In the high Sierra Nevada of California, two forms of the butterfly *Oeneis chryxus* are found, a light form in areas of granite rock and a dark form in areas of basaltic rock. The selection pressure involved has not been discovered, but the correlation suggests the work of an unknown visual predator. Similar examples of geographic variation in color which can be related to variation in the habitat have been reported in many groups of insects, as well as in numerous other animals. To reiterate, geographic variation is ubiquitous, and selection has been shown to play a major role in differentiating most populations that have been studied thoroughly. It is a truism that populations of organisms in different places will, under most circumstances, be genetically different.

THE PROCESS OF SPECIATION

While it is easy to explain why *E. editha* or *D. pseudoobscura* populations are each slightly different from one another, how does one explain the existence of different species of *Euphydryas* or *Drosophila?* With geographic variation as a starting point of our explanation, how can we explain two different kinds of organisms living together and not interbreeding (as *E. editha* and the closely related checkerspot butterfly *E. chalcedona* do on Jasper Ridge)?

THE CLASSICAL MODEL OF ALLOPATRIC SPECIATION. This model explains speciation in terms of geographic isolation. In its simplest form a barrier of some sort appears in the middle of the distribution of a geographically varying species (Fig. 10.1). Perhaps the larval food plant of the species is exterminated in the middle of the range, or perhaps gradual geographical changes create a water or mountain barrier. Now it is no longer possible for populations on either side of the barrier to exchange genes, and the gene pools of the two isolates begin to evolve on separate courses in response to differing environmental pressures (Fig. 10.1). Eventually the isolated populations become sufficiently distinct so that if the barrier disappears and the two isolates reunite, they will not successfully interbreed—i.e., they will be distinct species (Mayr, 1963).

The lack of interbreeding of reunited differentiated isolates may be due to premating isolating mechanisms (e.g., termites swarming at different times, crickets having different songs, butterflies not recognizing each other's mating pheromones) or postmating isolating mechanisms (e.g., failure of sperm to fertilize egg, death of zygote, hybrid of reduced viability or fertility). This process of differentiation in isolation is called *allopatric speciation* (allopatric organisms are those with non-overlapping ranges; sympatric organisms live together).

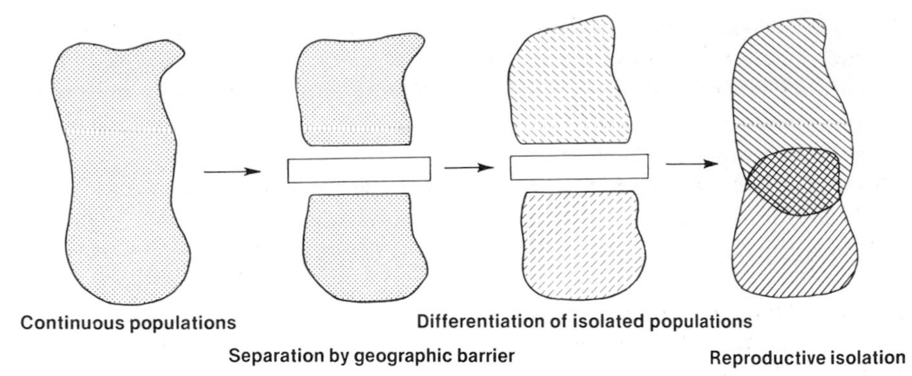

FIGURE 10.1 Diagrams of steps in allopatric speciation. See text for explanation.

SELECTION AGAINST HYBRIDS. When there are post-mating isolating mechanisms, some individuals of the reunited isolates may interbreed, but these matings are less successful than those within isolates. In these cases selection is operating against individuals that hybridize (they leave fewer offspring than those that mate with their own kind). Such selection will reinforce factors tending to prevent hybridization. The exchange of genetic information between the isolates then becomes negligible. It is entirely possible that when more is known about the processes of differentiation, it will be discovered that hybridization between individuals of rejoining segregates is almost universal, in other words, that mechanisms preventing exchange of genetic material between differentiated forms usually arise only through relatively unsuccessful hybridization after sympatry has been reestablished. Selection against hybrids and the subsequent evolution of mechanisms which prevent hybridization is the basis of *character displacement,* i.e., where closely similar species have partly overlapping ranges they tend to be more differentiated in the zone of overlap than where they occur alone.

An interesting experimental demonstration of this mechanism was obtained by Koopman (1950), who synthesized artificial mixed populations of *Drosophila pseudoobscura* and *D. persimilis* and held them at low temperatures (16°C) at which sexual isolation between the two is at a low level: hybrids are formed more readily at low than at high temperatures. Under the experimental conditions, the hybrids were extremely unsuccessful, but Koopman intervened to produce complete failure of hybridization by removing all hybrid individuals before they could reproduce (hybrids were identified by visible genetic traits).

Over a period of several generations the proportion of hybrids formed showed a marked decrease, indicating a reinforcement of whatever factors were operating to prevent hybridization. Koopman was able to show that the isolating mechanism was at least in part sexual; i.e., males "preferred" to mate with females of their own kind.

In nature, *D. pseudoobscura* and *D. persimilis,* although occurring in the same geographic areas, presumably do not hybridize for two reasons. First, there is considerable ecological isolation, *D. persimilis* usually occurring higher in the mountains and preferring cooler, shadier spots than *D. pseudoobscura.* Sexual isolation must also play a part, for except at low temperatures newly captured flies show little tendency to hybridize. It is suspected that other undetected factors also help to keep the two entities apart in nature. In the experiments just described, the two known factors were removed by crowding the flies together at low temperatures. In a very short time the action of natural and artificial selection established a barrier that was at least partly sexual where one had not existed previously.

THE EVIDENCE FOR ALLOPATRIC SPECIATION. An abundance of evidence suggests that allopatric speciation is an important cause of organic diversity. It is the splitting mechanism which, coupled with extinction, has been employed to explain the large numbers of relatively distinct kinds of animals, plants, and microorganisms that populate the earth—a diversity of species which reaches full expression in the insects.

Little is known about the time required for populations to differentiate. In any given case, many variables would affect the required time span, including population size, magnitudes of selection pressures, degree of isolation, and the genetic system of the organism. Usually speciation seems to be a much more drawn-out process than could be conveniently observed by the evolutionist. However, some work, e.g., that on industrial melanic moths (see "Industrial Melanism," in Chap. 15) and on polymorphic land snails of the genus *Cepaea* (Cain and Sheppard, 1950) and common water snakes *Natrix sipedon* (Camin and Ehrlich, 1958), indicates that selection pressures in nature may be generally higher than once thought; if this is the case, speciation may also occur more rapidly than has been assumed in the past. At any rate, the evidence for the view that allopatric speciation is a primary splitting mechanism in evolution is not direct observation. The presence in nature of patterns of variation that seemingly represent every conceivable stage in the postulated process is the actual basis for this view.

It has been observed, for instance, in organisms as diverse as frogs, salamanders, butterflies, and *Drosophila*, that populations may become sufficiently differentiated so that they can live together without interbreeding, and yet be able to exchange genetic information through gene flow via intermediate "bridging" populations. For instance in the neotropical fruit fly, *Drosophila paulistorum*, there is a series of strains of semispecies (i.e., populations that have acquired some, but not all, the attributes of separate species; Dobzhansky, 1970) which generally do not interbreed when two or more occur together. For example, the Amazonian and Orinocan semispecies often occur together, as do the Amazonian and Andean-Brazilian (Fig. 10.2). In laboratory tests there is greater behavioral isolation between individuals of different semispecies where they come from sympatric

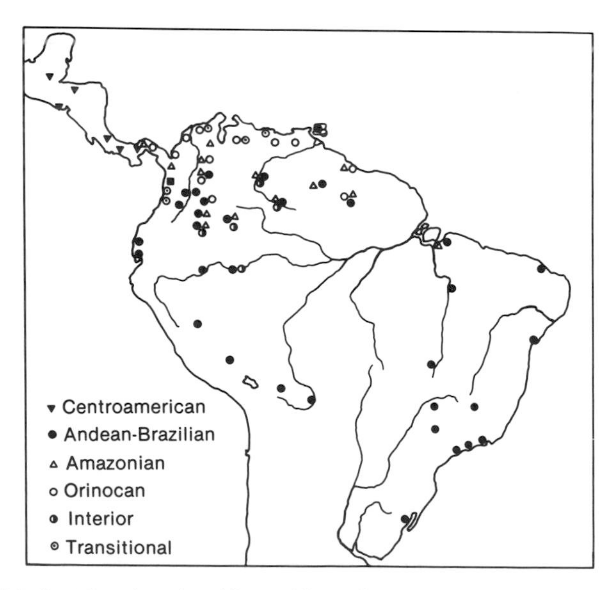

FIGURE 10.2 Distribution of semispecies of *Drosophila paulistorum* in South America. (*Data from Dobzhansky, 1970.*)

populations than when they come from allopatric populations—an example of character displacement.

In northwestern South America there occur "transitional" populations which show diverse genetic patterns. Some, for instance, produce fertile hybrids with Centro-American, Andean-Brazilian, or other transitional strains. Through these transitional populations it is, at least in theory, possible for the semispecies to exchange genes. Dobzhansky describes *D. paulistorum* as "a group of species still in *statu nascendi*" (in the process of being born). It could be thought of as a snapshot showing one stage in the process of speciation.

THE NEOCLASSICAL MODEL OF ALLOPATRIC SPECIATION. There has recently been some shift away from the classic model of allopatric speciation. The degree to which gene exchange (gene flow) limits the amount of differentiation between populations is unknown. However, data from plant and animal populations suggest that gene flow is not as important as it was once thought to be (Ehrlich and Raven, 1969). Studies of migratory behavior in various animal populations, as well as pollination and seed-dispersal studies in plant populations, indicate that on the whole the exchange of genetic information among populations is *quite limited.* Many cases are known in which populations have been totally isolated for thousands of generations and yet show little differentiation. This, for instance, is the case with *Euphydryas editha* populations in California and in Colorado, or the California and eastern populations of *Lycaena phlaeas.* In other cases gene flow between distant populations may be so slow as to be in practice indistinguishable from mutation as a source of variability.

In contrast there also are situations where differentiation has occurred in spite of gene flow—as in some characters of *E. editha* in different Jasper Ridge populations. Ehrlich and Raven (1969) have argued that it is selection itself that is both the primary cohesive force and the primary disruptive force in evolution, and that the selective regime determines what influence gene flow has on observed patterns of differentiation. Populations will differentiate if they are subjected to different selective forces and will tend to remain similar if they are not. In most circumstances differential selection pressures will be generated primarily by geographic heterogeneity of the environment, and speciation will be largely allopatric. But there exists also the possibility of populations splitting without geographic separation.

SYMPATRIC SPECIATION. Evolutionary theorists disagree on whether or not speciation without geographic isolation (sympatric speciation) can occur in sexual organisms. One argument against sympatric speciation has been that hybridization will swamp out any differences produced by disruptive selection—i.e., selection pressures operating in different directions, as when both large and small individuals are more reproductively successful than medium-sized individuals. Some work, however, indicates that this is not necessarily so. Laboratory experiments with *Drosophila* have shown that disruptive selection can produce divergence in the absence of isolation. For instance Thoday and Gibson (1962) subjected a wild-type population to disruptive selection for numbers of chaetae, with both high- and low-selected individuals being placed in a common vial for mating. At the end of 12 generations the original population had split into two populations, which produced few hybrids. Attempts to repeat this work, however, have not been successful.

Bush (1969) has concluded that the widespread (Northeastern and Central United States) race of the apple maggot *Rhagoletis pomonella,* a trypetid fly, was derived from the hawthorn-infesting race of this species. The new race most likely originated in the Hudson River Valley, where it was first reported and where both apples and hawthorns commonly occur. There is a positive relationship between host plant and mate selection in these races. It is believed that host selection has a genetic basis and represents minor genetic changes. Bush has argued that such initial differences have led to the sympatric evolution of these races. Reproductive isolation was reinforced by factors such as different times of emergence on plants with different fruiting times, disruptive selection, conditioning, and semigeographic isolation.

There is also evidence that temporal factors may lead to sympatric speciation. Among the *Neodiprion* sawflies, each species of which tends to be associated with one typical host tree species, there are several pairs of sawfly species on the same host. One species mates in the fall and spends the winter in the egg stage. The other overwinters in the pupal stage and then emerges, mates, and oviposits in the spring. Here a simple change—possibly of a single gene—affecting the onset of diapause can produce complete temporal isolation of two groups. The larvae of these species pairs are often found feeding together in common clusters.

It is, unfortunately, very difficult to demonstrate the occurrence of sympatric speciation. The actual physical separation of a population into two segments is a positive event which can be observed. To prove that two gene pools have never been so separated is a logical impossibility. Field studies will inevitably uncover positive evidence of events much more frequently than they will ever find any kind of convincing evidence that something never happened, i.e., that two sympatric species never were allopatric. There can be little question that additional data are needed to resolve the sympatric-speciation controversy, data especially from field and laboratory experiments.

THE TAXONOMIC DESCRIPTION OF POPULATIONS

THE BIOLOGICAL-SPECIES CONCEPT. Biologists have attempted to integrate their ideas of how populations became differentiated with their taxonomic descriptions of the populations. Thus in the classical model the isolates evolved first into subspecies and then into full species, i.e., reproductively isolated entities. Biological species are defined as *groups of actually or potentially interbreeding natural populations, which are reproductively isolated from other such groups* (Mayr, 1963). In practice taxonomists have always considered as different species those populations of sexually reproducing organisms that lived together and remained distinct. Whether or not allopatric populations are classified as the same or different species has normally been determined by a subjective judgment as to whether they were sufficiently different to be considered different kinds of organisms. Because of the difficulty of applying the biological species concept to allopatric populations there has been much controversy over the utility of the concept. The interested reader may consult Ehrlich (1961), Mayr (1963), and Sokal and Crovello (1970).

The key point to remember is that the taxonomic system is a relatively rigid communications device that permits us to talk about the products of evolution but which can never adequately describe the intricate phenetic, genetic, phylogenetic, and ecological relationships of those products. One should not be disturbed by this imprecision, for when doing detailed evolutionary studies other methods of description are substituted for the formal taxonomic descriptions. For example, in the *Euphydryas* work, population dynamic and genetic units were simply given locality codes. Above all, it must never be assumed that because the taxonomic system is neat and hierarchical, nature must conform to it!

SELECTED REFERENCES

For additional readings on speciation consult Dobzhansky (1970), Ehrlich et al. (1974), and Mayr (1963).

PART THREE

INSECTS IN RELATION TO ENVIRONMENT

BIOTIC AND PHYSICAL FACTORS OF THE ENVIRONMENT

11

The study of insects in relation to their environment includes interactions of three kinds: (1) among members of the same species; (2) among members of different species, including plants, animals, and microbes; and (3) interactions with the physical environment. Phase polymorphism, mating behavior, social behavior, and population ecology are examples of the first category and have been discussed in Parts I and II. In Part III we focus on the second and third categories, beginning with the physical factors of the environment and then exploring the immense variety of relationships that insects have with other insects and other organisms.

LIMITING FACTORS. The details of these relationships not only are fascinating to discover but, taken together, also serve to circumscribe the existence of an insect in its *habitat,* or the place in which it lives. For example, temperature and humidity can be accurately measured. Each presents a gradient within which we can find a zone habitable by the species under study. The maximum and minimum values mark the *limits of tolerance.* Some insects have a wide tolerance, others a narrow tolerance; e.g., terrestrial species generally tolerate a wider range of temperatures than aquatic species.

This principle can be extended to other aspects of an insect's chemical and physical environment. A space in which to live, oxygen, essential nutrients in food, energy sources, and water are necessary in certain minimal amounts to sustain development and reproduction. On closer analysis we find that each stage of the life of an insect has its own limits of tolerance beyond which certain physiological processes fail, sometimes irreparably. If the maximum or minimum limit is approached or exceeded by any one factor, that factor becomes the *limiting factor* controlling the survival of the population.

The effects of an extreme condition in a limiting factor are usually ameliorated by a behavioral response to move to another, less stressful, place. If exposure is gradual, then some degree of physiological acclimation may occur. Over a number of generations, the recurrence or persistence of a factor or combination of factors in a limiting condition will let selection favor individuals that are genetically resistant or tolerant.

Limiting factors must be considered in the context of other factors, because not only are the factors intercorrelated to some degree, but also the joint effects differ. At high temperatures, for example, a high humidity may reduce the rate of desiccation. On the other hand, a low humidity will permit a brief period of evaporative cooling that may spell the difference in survival. An insect under stress from one limiting factor may be more susceptible to unfavorable conditions in other factors, especially if energy reserves are expended in order to maintain a normal state. For example, a continual search for moisture reduces the energy available to search for scarce food or shelter.

TROPHIC RELATIONSHIPS. Within a favorable physical environment, insects interact in a variety of ways with other organisms living in the same place. Among the interactions, the *trophic,* or feeding, relationships are most important. Energy and nutrients in green plants are consumed by *phytophagous,* or plant-feeding, insects. These in turn are eaten by *predators* and *parasitoids.* The latter feed parasitically at first, not harming vital tissues, then kill their victims. The true *parasites* feed on blood, tissues, or the stored food provisions of their hosts without killing the host. *Scavengers* devour microbe-rich, decaying plant or animal matter, thus serving as decomposers that help break down tissues and return nutrients to the soil.

Closely correlated with finding and consuming food are features of the insect's sense organs, feeding apparatus, and intestine, as well as characteristics of nutrition, composition of hemolymph, excretory products, etc. Likewise, when insects serve as food or as hosts for other organisms, we can detect structural and functional adaptations that reduce mortality. Insect populations are largely regulated in numbers by the supply of food available to them and the losses to predators, parasitoids, and pathogens that feed on them. Trophic relationships, therefore, have pervasive effects on the population biology, morphology, and physiology of a species.

COMPETITION. A second major type of interaction is *competition* for food and space. Competition is most severe when members of the same species compete with one another. Population growth becomes self-limiting, and a tendency exists for individuals to disperse into less favorable, but also less crowded, habitats. When two or more species have closely similar requirements and compete for resources that are in limited supply, one species tends to displace the others. This is known as the *competitive exclusion principle* (Hardin, 1960). Each species ultimately dominates a separate, allopatric area that is usually adjacent to but not overlapping the areas occupied by the other species.

When two or more related species coexist sympatrically, i.e., in the same geographic area, two inferences are possible: (1) the coexistence is recent in origin and competitive exclusion has not yet occurred, or (2) the species have evolved different requirements and no longer directly compete.

BENEFICIAL RELATIONSHIPS. The third and last major type of interaction involves the regular and often close associations of two or more species or organisms, at least one of which benefits by the association and the others are either not affected or also benefit. A *commensal* is an organism that benefits by association with a host organism, but the host is neither benefited nor harmed. Examples are mites and insects that regularly scavenge refuse and obtain shelter in the nests of other animals such as ants or rodents. A *mutualistic* relationship exists when all participants benefit. The pollination of flowers by bees is an example. Sexual reproduction of the plants is achieved, and the bees obtain nectar and surplus pollen as food.

ECOTYPES AND THE ECOLOGICAL NICHE. Taking a broad viewpoint, an insect species will not occupy all the habitats in the world that provide a suitable physical environment. It will be prevented from colonizing every such habitat because at least some barriers to dispersal, i.e., oceans or mountains, will be absolute. Furthermore, the insect will be excluded from some areas by competition, intolerable mortality from predators or disease, or absence of a favored food plant or mutualistic associate.

In the area occupied by a species, the characteristics of the habitat will vary from place to place, depending on the climate, nature of the soil or water, and the distribution of other plants and animals with which the insect interacts. Populations of widespread species tend to evolve adaptations to local environments. Such locally adapted populations are called *ecotypes.*

We have now briefly reviewed the boundaries of an insect's world. The combination of all aspects—physical space, environmental gradients, trophic relations, and interactions with other organisms—is unique for each species and constitutes the *ecological niche* of the species.

PHYSICAL ENVIRONMENT

Considerable attention has been given to the influence of weather on the distribution and numbers of insects. By weather we mean the hourly and daily changes in temperature, humidity, sunlight, atmospheric pressure, wind currents, and amounts of various kinds of precipitation. The correlation of weather with insect activity offers tantalizing possibilities to the physiologist, ecologist, and pest manager.

Insects are generally *poikilothermic,* or cold-blooded, animals, whose body temperature varies directly with environmental temperature. In the Temperate Zone, however, an insect's physiological response to temperature is often suspended in autumn. At that time day length decreases, triggering a change in the brain's neurosecretory cells that results in a period of arrested development, or diapause. After diapause terminates during winter, the insect is again responsive to temperature.

Within certain limits as temperature rises, metabolism rises, leading to accelerated development, increased reproductive activity, and, ultimately, increases in population number. Growth of food plants and the numbers and activity of natural enemies also respond to weather. To the extent that we understand the responses of insect populations to physical and biotic factors, and to the extent that we can forecast the weather, we should be able to anticipate the timing of certain events of an insect's life cycle and estimate the numbers of the population at a series of points over the growing season. The practical advantages in agriculture are immense; economic damage can be estimated and control measures, if needed, can be applied at the most effective times. Unfortunately the relationship of the physiology of insects to physical factors is neither simple nor well understood. Furthermore the most sophisticated procedures for prediction carry the risk of the unexpected: for example, a violent storm or killing frost may have catastrophic effects on the pest population, natural enemies, and crop plants.

LIGHT

Sunlight plays a key role in the orientation of insects in their environment and in the timing of events in their life cycles. Momentary responses to light, perception of images, color vision, and behavioral reaction to light and polarized light are discussed in Chap. 6. Of interest here is the influence of the natural alternation of light and darkness in the rhythmic activity of insects.

ACTIVITY AND SUNLIGHT. As a consequence of the earth's inclined rotation, physical factors important to the life of terrestrial organisms fluctuate daily and annually on a regular basis. Recall that the earth's axis is tilted 23.5° from a plane vertical to the plane of the earth's annual orbit around the sun. Any given area of the earth's surface receives sunlight for only a part of the 24-hour rotation. With the appearance of the sun's heat each day, temperature usually rises and relative humidity falls. The tilted axis produces an annual variation in the quantity of heat reaching the nonequatorial regions, resulting in the seasons. Furthermore, the earth's rotation around the tilted axis precisely changes the proportion of light and darkness each 24 hours. This results in long summer days and short winter days in the Temperate and Polar Zones.

The daily activities of insects usually take place during specific periods of the 24-hour cycle. Such synchronization, for example, permits the sexes to find each other, limits the search for food to periods of peak abundance, and allows insects to avoid enemies or to take advantage of a favorably high humidity for ecdysis or oviposition. Insects usually spend a part of each day inactive. For example, at night some bees and wasps (Fig. 11.1) clasp the stems of plants with their mandibles or crawl into flowers and seem to "sleep."

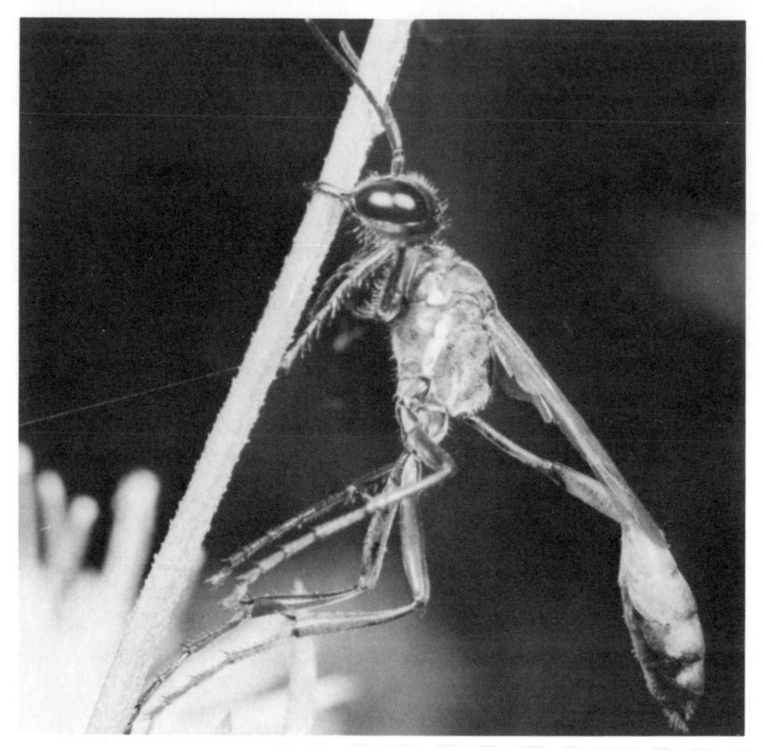

FIGURE 11.1 Sphecid wasp, *Ammophila* sp., asleep and clinging to stem by its mandibles. (*Photo courtesy of and with permission of Carl W. Rettenmeyer.*)

An activity during the day is termed *diurnal;* at dawn or dusk, *crepuscular;* at night, *nocturnal;* and at dawn only, *matinal.* These terms should be used in reference to specific activities or events rather than loosely to overall activity: the emergence of adult *Drosophila pseudoobscura* (Drosophilidae) from pupae is matinal; attraction to food baits, mating, and oviposition are crepuscular; and pupation is mainly crepuscular or nocturnal.

One might interpret a daily rhythm in activity as simply a response to a certain level of light or other physical factor that occurs regularly each day. In other words, the insect's activity is a response to *exogenous* factors. This is true for oviposition behavior in the walkingstick *Carausius morosus.* Most eggs are laid at or near dusk, or at the time the lights are switched off in laboratory experiments. After *C. morosus* has been kept only a day in constant light or darkness, the laying of eggs no longer occurs at 24-hour intervals but is distributed over the 24-hour period (Lavialle and Dumortier, 1975). A direct response to an environmental signal may also be delayed a specific interval as if timed by a "sand hourglass."

CIRCADIAN RHYTHMS. In most insects activity rhythms have a physiological, or *endogenous,* basis that is partly independent of environmental signals. In a light/dark cycle of 12:12 hours, mature male crickets, *Teleogryllus commodus,* begin to stridulate their mating song 1 to 2 hours before the onset of darkness. Singing continues during darkness and is terminated 2 to 3 hours before the lights switch on. When these crickets are placed in constant darkness, singing begins at intervals of 23.5 hours and lasts the normal period. In constant light, singing occurs at intervals of about 25.3 hours (Fig. 11.2, days 1 to 12). A return to the 12:12 hour cycle of light/dark results in a return to the 24-hour rhythm, beginning 2 hours before darkness (Fig. 11.2, days 13 to 31; Loher, 1972).

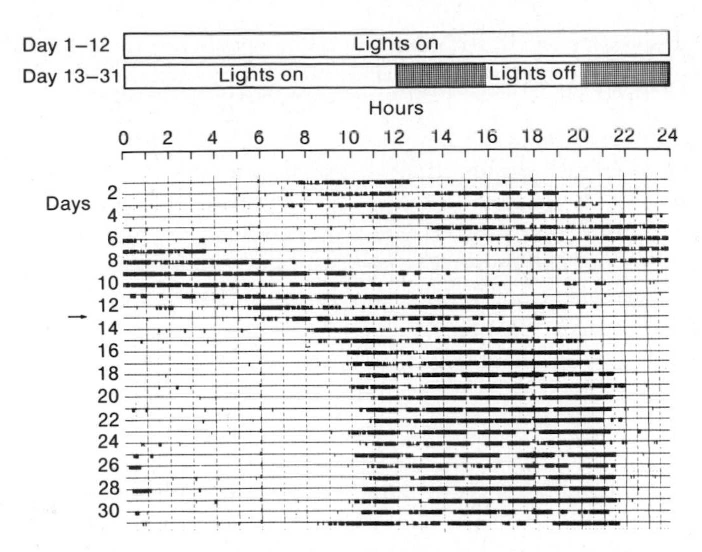

FIGURE 11.2 Circadian rhythm of stridulation in a male cricket, *Teleogryllus commodus*. Marks in rows are tic marks from automatic recorder that is actuated by sound of stridulation. Each row records stridulation during 1 day. When the record began, the cricket had been in constant light and the stridulation rhythm was free-running. On day 13 the lights were switched off at 1200 and remained off for 12 hours. Thereafter, the cricket was exposed to 12 hours of light and 12 hours of dark. Note that stridulation is temporarily interrupted when the lights are switched off at 1200 each day. (*Data courtesy of Werner Loher.*)

When a given activity or event can be demonstrated experimentally to recur at about 24-hour intervals after all environmental signals have been excluded, it is said to occur in a *circadian rhythm* (meaning "approximately daily"). Under constant conditions the time interval is nearly always slightly more or less than 24 hours, resulting in a progressive drift each day in the time of initiation (Fig. 11.2, days 1 to 12). The presence of the drift is evidence that other possible signals with 24-hour periodicities are not influencing the insect. Cosmic radiation and geomagnetic forces, for example, fluctuate on an exact 24-hour cycle with the earth's rotation.

The physiological basis for circadian rhythm is largely a mystery. Individual cells exhibit rhythmic changes in nuclear volume and protein synthesis. Specific organs function cyclically; e.g., endocrine glands synthesize hormones cyclically. Susceptibility to insecticide poisoning is rhythmic, as is general oxygen consumption. Clearly, rhythms are evident in a variety of physiological processes. The rhythm itself, however, is not determined by the metabolic rate of the process. Unlike nearly all physiological processes, circadian rhythms are largely independent of temperature, within normal limits of tolerance.

The underlying mechanism is conceptualized as a "biological clock" or an "oscillator." The clocks of circadian rhythms are reset daily or entrained by brief environmental signals. The most common cues are dawn or dusk, when light intensity varies rapidly. The location of the photoreceptors varies with the species. Compound eyes or photosensitive tissues in the brain have been shown to be essential for entrainment. Even insects inside fruits or leaves respond to light changes if the surrounding media are translucent.

An entrained circadian rhythm is called a *diel rhythm*. It is possible that more than one clock controls the different rhythms of cells, development, general physiology, and behavior. The clock or clocks serve to coordinate life processes within the insect while at the same time scheduling the insect's activity in a changing environment. Locomotor activity, feeding, mating, and the insect's varying responsiveness to stimuli are programmed to take place at opportune times in the daily cycle.

A different type of diel rhythm involves populations and "once-in-a-lifetime" events. If *Drosophila melanogaster* is reared in constant darkness, the adults emerge from pupae at any time. When the pupae are given as brief as 1-minute exposure to light, emergence becomes synchronized to occur during one period each 24 hours. The period of emergence has been compared to a developmental "gate" that opens or closes. If the open gate is missed by an individual, then its emergence is delayed 24 hours. Gated phenomena are known for egg hatching in the pink bollworm *Pectinophora gossypiella* (Gelechiidae), pupation of the mosquito *Aëdes taeniorhynchus* (Culicidae), and emergence of the intertidal midge *Clunio marinus* (Chironomidae) in Europe.

The genetics of circadian rhythms has been studied to a limited extent. The rhythmic emergence of *Drosophila* from pupae has been observed to persist in constant dim light for over 200 generations. In other experiments with *Drosophila* the time period can be changed by mutations of a single gene. The same clock seems to control the pupal emergence gate and the adult's locomotor rhythm.

The complexity and geographic variation of rhythms is illustrated by the intertidal midge, *Clunio marinus*. This species probably has the shortest adult life of any insect. Adults emerge, mate, oviposit, and die during the 2 hours of low tide when their breeding sites are exposed. Along the coasts of France and Spain the midge larvae live in the lowest intertidal zone. Their breeding areas are exposed only twice monthly during the greater-than-average tides associated with the new and full moons. On the days of these spring tides, low water occurs in the morning and again in the evening. *Clunio* adults emerge only in the evening during days of the spring tide. The winged males emerge first, followed by the wingless females. On the exposed tidal flats, the male must find the female, assist her in emergence, copulate, and carry her to the larval habitat. Here she places the eggs, enclosed in a sticky jelly, on the exposed substrate. The eggs will float away if not attached.

In the laboratory, Neumann (1976) discovered the circadian and semilunar rhythms that make this closely timed emergence possible. His experiments demonstrated that a circadian rhythm controls the evening emergence. Furthermore, by simulating the light of a full moon for 4 days, an emergence gate could be instituted that recurred at 15-day intervals for more than three semilunar cycles. The moonlight synchronizes the beginning of pupation, not the actual emergence. Pupation can begin at any time of day. When the adult is fully developed 2 to 5 days later, the circadian rhythm limits the emergence to the evening only. In Neumann's experiment, the emergence continued at semilunar intervals until all the larvae initially exposed to moonlight had emerged. Thus the full moon can synchronize emergences during the new moon as well.

Although the low tides occur on the same day along the coast, the exact times differ. Accordingly, the emergence times of *Clunio* must also differ from place to place. In genetic experiments, Neumann demonstrated that the time of emergence was inherited. Crosses between populations gave offspring with intermediate times of emergence. Further north, at the latitude of the British Isles, the summer nights remain brighter and moonlight is weaker. The entraining effect of moonlight diminishes, and semilunar emergence is timed by a combination of the circadian rhythm and a response to tidal action. North of the Arctic Circle, emergence occurs 10 hours after the previous ebb tide and not in a semilunar rhythm. Neumann theorizes that tidal action initiates a 10-hour timer that functions like an hourglass.

The genetic basis of circadian rhythms is also illustrated by the mating behavior of moths. Most moths have specific periods for mating activity. Certain closely related species have identical sex pheromones. Consequently, the males can be attracted to females of different species. Some of these species are allopatric or mate at different seasons. In the sympatric species, mating occurs at different times of day under circadian control, thus preventing cross attraction.

INSECT SEASONALITY. In addition to daily rhythmic activity, the life cycles of insects are also synchronized on a long-term or seasonal basis. Insects survive annual periods of winter cold, drought, summer heat, or food shortage by undergoing a state of dormancy. A temporary dormancy that lasts only until favorable conditions return is *quiescence.* In the context of our present discussion, however, we are interested in the prolonged dormancy, or *diapause,* that usually results from an interruption in the hormonal stimulation of growth in immature insects or reproduction by adults.

Diapause is induced in most insects by a specific proportion of day length to night length, called the *critical day length.* This is, strictly speaking, insect *photoperiodism,* in contrast to the brief cues that entrain circadian rhythms. It is not yet clear whether the underlying mechanism of photoperiodism also resembles an oscillator or is better explained as a physiological hourglass, "tipped over" by light each day, and with an additional mechanism for accumulating the products of daily cycles (Lees, 1971). Some insects respond to long days/short nights, others to short days/long nights, and some require more complex schedules in which long days are followed by short days, or vice versa. Near the equator only slight differences in day length may be sufficient. The stage of the insect that experiences the critical day length is not necessarily the stage that will diapause. For example, whether or not eggs will diapause is determined by the female parent.

The length of time that an insect remains in diapause is the result of a complex interaction involving (1) seasonal changes in photoperiod and temperature; (2) the insect's changing responses to these physical factors; and (3) the rate of *diapause development* (Chap. 4). Ordinarily diapause development takes place only at temperatures below the minimum temperature for growth. A period of chilling, therefore, is often required for diapause development.

In autumn, the diapause of overwintering insects is initiated by the progressively shorter day lengths. Warm temperatures also have the effect of delaying diapause development. The insect is thus doubly protected against resuming growth during temporary periods of favorable weather, later to be caught in a susceptible state when winter arrives. By midwinter, however, diapausing insects are usually no longer responsive to photoperiod. A few insects do remain responsive, and their diapause is terminated by a return to long days. Some evidence suggests that the temperature threshold below which diapause development takes place also changes. With cooler winter temperatures, the threshold rises.

At some point in the winter, diapause development is completed. The diapause spontaneously terminates in most insects without specific environmental stimulus. In the exceptional insects, some respond to photoperiod as mentioned above. Others appear to require food or moisture; parasitoids require a hormonal interaction with the host. The actual time at which diapause terminates is difficult to establish precisely. In some the transition is gradual. Increased oxygen consumption, neuroendocrine activity, response to photoperiod, feeding behavior, and gametogenesis have all been used as signs of the beginning of the postdiapause period.

In nature, diapause of individual insects of the same species may terminate at different times during midwinter. Yet growth does not resume throughout the population until the minimum temperature threshold has been reached. Other factors, such as the availability of water, are also important. The local population is consequently synchronized in growth by environmental conditions during spring.

Summer diapause has been less well analyzed. In the insects studied, it can be induced and maintained by photoperiod. Diapause development is favored by cooler conditions during summer. The termination of summer diapause, in contrast to winter diapause, usually requires a specific stimulus such as food.

In those insects with more than one generation each year, the insects that develop in different seasons may exhibit different characteristics. Seasonal forms are known in Lepidoptera, Gryllidae, and Cicadellidae. Early taxonomists frequently described the different forms as separate species. For example, the European nymphalid butterfly *Araschnia levana* has two forms. Caterpillars reared under short-day photoperiods, diapause and emerge as the spring, or levana, form (Fig. 11.3*a*). When reared under 16 or more hours of light, the caterpillars do not diapause and the summer, or prorsa, form develops (Fig. 11.3*b*).

As a final example, the complex annual cycles of aphids (Fig. 4.8) are partly controlled by photoperiod (Hille Ris Lambers, 1966; Lees, 1966). Production of the oviparous forms in autumn is stimulated by short day length and lower temperature. In the spring, the first aphids (fundatrices) hatch from eggs that have overwintered in diapause. The female offspring of the fundatrix are produced parthenogenetically. Their offspring also reproduce parthenogenetically, even though days are short, apparently because they are inhibited in some way from responding to photoperiod. The inhibition lasts a definite time, depending on temperature, and spans several subsequent generations. The effect has been compared to an hourglass timer that must run its course, regardless of the number of generations. During the same time the days are increasing in length so that no sexuals are produced. Once the inhibitory effect has vanished, the parent aphids become sensitive to photoperiod. Under long days they continue to produce offspring that will reproduce parthenogenetically. Under short days, the parent alters her influence on the development of her embryos such that they develop into oviparous females. By carefully illuminating minute areas of parent aphids reared under short days, Lees (1966) increased the "day length" for specific internal organs. He revealed that the brain, probably the pars intercerebralis, is the location of the photoperiodic receptor.

HUMIDITY

The moisture content of air affects the rate at which water is lost or gained by terrestrial insects. Air at higher temperatures holds more water vapor than air at lower temperatures. Consequently, an amount of water that will saturate air at a lower temperature will not be sufficient at

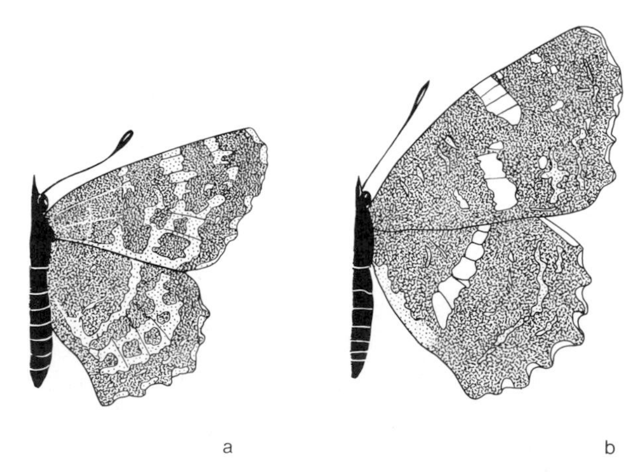

FIGURE 11.3 Seasonal forms of the European nymphalid butterfly *Araschnia levana: a*, spring, or levana, form; *b*, summer, or prorsa, form.

a higher temperature. The ratio between the actual water vapor present at a given temperature and the water vapor necessary for saturation is the familiar "relative humidity."

In saturated air, or 100 percent relative humidity (RH), an insect will not lose water by evaporation. In air of the same temperature and increasingly lower RH (or in air of constant water content at increasingly higher temperatures), a point will be reached where the water content of the insect exceeds that of the air, resulting in the chance for evaporative loss. Because their surface/volume ratio is so high, insects are continually subject to water loss even under moderate conditions.

To maintain water content within well-defined limits, insects acquire water by drinking, by metabolism, and, in some cases, by absorption from moist surfaces or air. Eggs of many insects absorb moisture. Certain insects that normally live in rather dry environments have been demonstrated to absorb water directly from air at 80 to 90 percent RH, or in fleas as low as 50 percent RH. The site of absorption is not known except in the mealworm *Tenebrio,* in which the cryptonephric system around the rectum is responsible (Fig. 5.7*b*). The various structural devices for water conservation have been discussed previously: the waxy layer of the epicuticle that prevents loss through the integument (Fig. 2.1), spiracular closing mechanisms that regulate moisture loss from the tracheal system (Fig. 5.4*c*), and excretory organs that reclaim water from urine and feces (Fig. 5.7*a*). Total loss by transpiration tends to be lowest in eggs and pupae, the immobile stages that are least able to acquire additional water. If protected against excessive loss and if they have access to water to restore the normal content, insects are able to exist under a broad range of humidities. At either extreme, harmful effects become conspicuous.

Humidities near 100 percent RH are prevalent deep in soil, under objects on soil, in decaying organic matter, and at the transpiring surfaces of plants and mammals. Many insects live in such places. Drowning in condensation is a distinct hazard, avoided partly by the waxy epicuticle, spiracular closing devices, and, in the eggs of some insects, by a plastron (Fig. 5.5). Fungal pathogens may thrive under high humidity. Certain insects also seem unable to eliminate excess metabolic water under these conditions.

Dry environments, on the other hand, place maximum stress on the retention of water. Common insects of the desert include Lepidoptera, Coleoptera, Formicidae and other Hymenoptera, Diptera, and Acrididae (Orthoptera). The pests of dry stored products are mainly Lepidoptera and Coleoptera. Studies indicate that insects normally living in dry habitats avoid desiccation by behavioral responses to humidity gradients, nocturnal activity to take advantage of higher humidities, and the increased effectiveness of water-conserving devices. The positive response to higher humidity increases as the insect loses water. It has been shown that the cuticle of desert insects usually is less permeable to water than is the cuticle of comparable insects living under higher humidity. The elytra of Coleoptera cover the abdominal spiracles and reduce water loss by transpiration.

Even at moderate humidities, the physiology of insects is influenced by different levels. Rate of oviposition and duration of development in *Locusta* are optimal at 70 percent RH. Oviposition, development, and longevity among insects are not favored by low humidity. Consequently, periodic droughts may greatly reduce populations of insects normally accustomed to moderate humidity.

TEMPERATURE AND DEVELOPMENT

The effect of temperature on insects is best understood if we recall certain relationships between temperature and chemical reactions. Temperature describes the intensity of heat and gives a measure of the average motion of molecules. The addition of a quantity of heat sufficient to raise the temperature of an object also increases the kinetic energy, or motion of molecules. At a certain temperature, the collisions between reactive molecules in a mixture of chemicals will be strong enough to cause a chemical reaction. The rate, or frequency, of reactive collisions will increase a specific amount with further increases in temperature.

Accordingly, the reactions of the enzymes that govern metabolic processes increase with temperature up to a certain maximum temperature, beyond which the catalytic activity declines as the protein molecules become altered or denatured. In general, the overall metabolic rates of organisms approximately double with each rise of 10°C in body temperature. The upper limit of tolerance to temperature for an organism is determined by the harmful effects of high temperature on the physiological properties of membranes and possibly the denaturation of certain enzymes.

TEMPERATURE AND GROWTH RATE. We can anticipate, then, that an increase in temperature within a favorable range will speed up the metabolism of an insect and consequently increase its rate of development. The relationship will not be the simple linear relationship seen in a single chemical reaction, because many chain reactions are involved in growth. Each stage in the life history will develop at its own rate in relation to temperature. How can this information be used to anticipate events in the life of insects in nature? A simplified example will illustrate the general approach:

The first step is to establish in the laboratory the average time required for each stage to develop at different constant temperatures. When egg development of *Drosophila* is plotted directly on a graph we see that a curve, not a straight line, is formed (Fig. 11.4*a*) This reflects the complexity of the metabolic response to temperature. The data can be expressed in another way: how much of the total development takes place during 1 hour at a given temperature? This can be expressed as the reciprocal of the total hours at a given temperature or as a percentage. If 16.94 hours are required at 30.8°C, then $1/16.94 = 0.059$, or 5.9 percent of the total development occurs in 1 hour at 30.8°C. By plotting the reciprocals (or percentages) against temperature we create a sigmoid curve (Fig. 11.4*b*). This can be interpreted as follows: developmental rate increases linearly over the middle part of the curve but is relatively slower at the

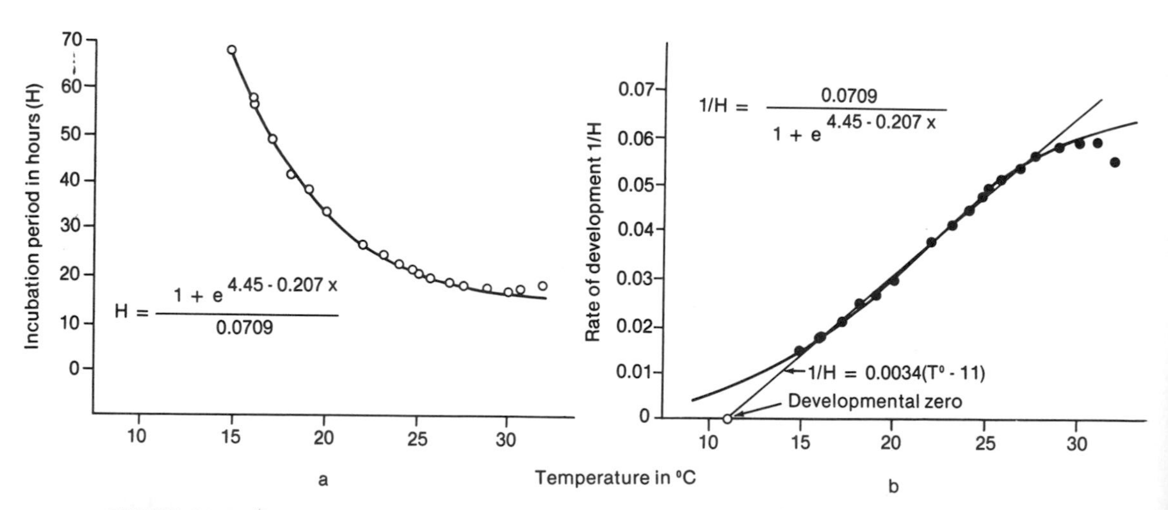

FIGURE 11.4 Temperature and egg development of *Drosophila melanogaster: a*, total development period in hours for eggs incubated at various temperatures; *b*, same data expressed as percent development in 1 hour at different temperatures. (*Data from Davidson,* 1944.)

lowest and highest temperatures. For our purposes, we can fit a straight line through the flat middle portion to estimate the slope or amount in hours by which developmental rates are increased for each degree of increased temperature. In our example, this has been calculated to be 0.0034 hour-degrees (Varley et al., 1973).

Below a certain temperature, development does not occur. This is called the *growth threshold* or minimum threshold of development. The threshold can be determined in various ways, but here we will estimate the temperature at which no growth occurs. This is done by projecting the straight line to the abscissa (zero hours of development) and noting that the line crosses at 11°C.

Restating the results thus far, we have learned that to develop completely, *Drosophila* eggs require 1/0.0034, or 294.12, hour-degrees, *above* 11°C. This provides a measure of the *physiological time* required to complete development.

Under natural conditions, air temperature undergoes a daily fluctuation, rising with heat gained from the sun during the day and decreasing at night as heat is lost. During the growing season, the temperature at night may fall below the growth threshold of the insect, and during the day it varies above the threshold.

As the second step in our example, we place freshly laid eggs of *Drosophila* outdoors and record the temperature each hour. To estimate the time of emergence, we discard all readings below the 11°C growth threshold. From all other readings we subtract 11°C, giving an *effective temperature* each hour. By adding the degrees of effective temperature for each hour after the eggs were laid, we accumulate the hour-degrees required for complete development and expect hatching when 294 hour-degrees are reached. Alternatively, we may take the maximum and minimum temperatures recorded for each 24-hour period, divide by 2 to give an average for the day, then subtract the growth threshold temperature. This approximates the effective temperature in day-degrees. In the example of *Drosophila,* we expect hatching when a little over 12 day-degrees is reached.

The general method can be applied to development of larvae and pupae and to events in the adult life such as the beginning of mating or oviposition. Each stage has its own characteristic physiological time.

The estimate of physiological time in nature is complicated by several circumstances. Genetic differences may exist between the laboratory and natural populations, and among natural populations. Fluctuating temperatures may have accelerating or retarding effects on growth when compared to the same number of day-degrees under constant temperature. Circadian rhythms may limit activities to certain times of the day or night. Furthermore, the daily temperature regime actually experienced by the insect in soil, under bark on the shaded side of a tree, or among leaves exposed to full sun in day and chilling night breezes, will differ from the temperatures registered in a standard weather shelter. Geiger (1965) has documented the greatly varied conditions of microclimate (the climates of small spaces) in contrast to the climates recorded by meteorological stations. Finally, elaborate mathematical procedures are usually needed for accurately relating development to the joint effects of temperature and other factors.

In practice, a population of insects in nature is monitored by identifying certain events that will serve as signals to us to begin accumulating the day-degrees required for certain stages. In the spring one can assume that diapause is broken and growth has resumed after a certain day length is reached and the specific growth-threshold temperature is exceeded.

Direct evidence of the seasonal progress of the population is obtained by periodic sampling. For example, traps baited with sex pheromone are useful in determining when the males first seek mates. With such information from the field, the forecasts can then be corrected.

THERMOREGULATION

Insects are known to be active within definite ranges of environmental temperature. The snow scorpion flies *Boreus* (Boreidae), snow fly *Chionea* (Tipulidae), snow "fleas" *Achorutes* (Poduridae), and winter stone flies *Allocapnia* (Nemouridae) are active at low temperatures on snow. *Boreus* dies if held for long in the human hand. At the other extreme, the firebrat *Lepismodes inquilinus* (Lepismatidae; = *Thermobia domestica*) prefers 12 to 50°C and dies at 51.3°C. Although insects can usually not lower their temperature below that of ambient, many can actively regulate their temperature above that of the environment and benefit from a higher metabolism rate. This is in spite of the distinct disadvantage of small size. The high surface/volume ratio results in rapid transfer of body heat to the environment. At rest, the body temperature of an insect soon becomes nearly the same as the surrounding medium.

At unfavorably high ambient temperatures, an insect is severely limited in the extent to which it can cool itself. Cooling by the evaporation of water quickly leads to desiccation unless water is freely accessible. Heat is readily regained because of the surface/volume relationship. The tsetse fly *Glossina morsitans* (Muscidae) feeds while exposed to sun on the hot skin of African mammals (Edney and Barrass, 1962). The opening of the spiracles at temperatures above 39°C permits evaporative cooling within the tracheal system to lower the body temperature about 2°C. This is possible because the meal contains ample liquids to restore lost water. Honeybees transport water to their hives, deposit it in the combs, and vigorously fan with their wings so that evaporative cooling takes place. Most insects, however, escape heat by seeking shade or cooler places.

ECTOTHERMY. On the other hand, insects are able to increase their body temperature well above ambient by absorbing heat from the sun. Organisms that derive heat almost entirely from their environment are called *ectothermic*. Butterflies assume distinctive postures while basking in the sun. "Dorsal baskers" spread their wings and point their heads away from the sun. The skippers (Hesperiidae) commonly spread only the hind wings, leaving the forewings vertical. "Lateral baskers" keep both wings vertical over the body and orient to present one side to the sun's rays.

Watt (1968) carefully analyzed the thermal regulation of species of *Colias* butterflies in warm and cold climates. In cold climates the butterflies are much darker, especially in a basal area on the hind wings. *Colias* is a lateral basker. Below 34 to 35°C, the insect orients to present the lateral surface at right angles to the sun, achieving maximum heating; at about 37 to 38°C, the body is oriented parallel to the sun's rays to avoid heating, or the insects seek shade. When basking, the dark area of the hindwings covers the abdomen and part of the thorax, increasing the absorption of solar heat by the body. Thus, in cold climates, dark *Colias* are able to reach the optimum activity range of 35 to 38°C before the pale *Colias*. This leads to greater feeding, mating, and oviposition activity and ultimately to greater reproductive success. In warm climates, the dark form is at a disadvantage because of overheating in sun and forced inactivity.

ENDOTHERMY. A second source of heat is that derived from energy metabolism. The flight muscles of insects are metabolically the most active tissues known. A substantial part of the energy conversion is degraded to heat.

Insects with insulating hairs or scales, or with free access to high-energy food, or both, are able to raise and maintain their bodies above ambient temperature by metabolic heat for varying periods. Such insects are called *endothermic*. Essentially all of the heat comes from muscular activity.

The flight muscles of larger insects require higher temperatures, in some species near or above 40°C, to generate wingbeat frequencies sufficient to lift the insect. Various species, but not all, in Odonata, Hemiptera, Orthoptera, Lepidoptera, Coleoptera, Hymenoptera, and Diptera have been observed warming up before flight by contracting the flight muscles. Those with indirect flight muscles of the synchronous type visibly quiver or "shiver" the wings as the muscles contract against one another. Flight mechanisms with asynchronous muscles may become uncoupled from the wings so that no movement is seen. Basking can be combined with shivering to bring the thorax up to the required temperature.

Once in flight the muscles continue to produce heat. Thoracic temperatures of flying butterflies and locusts are 5 to 10°C above ambient, while well-insulated moths and bumblebees fly at 20 to 30°C above ambient. Heinrich (1974) recorded 36°C in the thorax of a large queen bumblebee that was flying in air at 3°C.

Thoracic temperatures of flying endothermic insects are regulated within narrow limits even though a wide range of ambient temperatures may be experienced. Temperatures in other parts of the body vary. The study by Heinrich and Bartholomew (1972) of the thermoregulation in the sphinx moth *Manduca sexta* reveals how this is accomplished. Overheating is avoided in the thorax by transferring heat to the abdomen. The abdomen is thermally isolated from the thorax by a large air sac and is less well insulated by scales. Hemolymph flows posteriorly from the hot thorax to the abdomen, where the heat is dissipated by convection. The cooler hemolymph is pumped forward by the heart to the thorax and travels a sinuous course in the aorta between the flight muscles, thus cooling them. The rate of heartbeat is apparently controlled by the nervous system in the thorax or head. During warm-up the flow of hemolymph is restricted between the thorax and abdomen, allowing heat to accumulate. When overheated, the heart pumps strongly.

Queen bumblebees "brood" their first offspring, sometimes for several weeks or more. The queen presses the underside of the thorax and abdomen against the wax cups that hold the eggs, larvae, and pupae. Again, Heinrich (1974) has elucidated the mechanism. Heat is generated by the thoracic flight muscles, where temperatures reach 35 to 38°C. At the option of the queen, heat is transferred by hemolymph to the abdomen. Heat is conducted by close contact from the abdomen at 31 to 36°C to the brood clump. At ambient temperatures of 3 to 33°C, the brood is maintained at 24 to 34°C. Worker bumblebees are also able to transfer heat to their abdomens.

INSECT COLD-HARDINESS

Insects suffer varying consequences when temperatures fall below their growth and activity thresholds. They may become immobilized at temperatures above freezing and, if kept in this state, will die from desiccation because they are unable to restore lost water by drinking. Insects living in temperate or colder regions, such as the Arctic or high mountains, risk freezing when air temperatures drop below 0°C. Some insects, i.e., northern populations of the screwworm *Cochliomyia hominivorax* (Calliphoridae), annually die in winter and are repopulated again by migrants from warmer climates to the south. Yet some insects do survive subzero weather, and others live normally after being frozen. The latter are the most complex and largest animals to tolerate freezing.

Some insects are able to avoid low air temperatures by virtue of their habitat or behavior. Aquatic insects and those in deep, moist soil escape because freezing usually occurs only at the water or soil surface. In winter, honeybees cluster together in a mass in the combs of their hive. Even when air temperatures outside are well below 0°C, they maintain 20 to 30°C inside the cluster. Heat is derived from the metabolism of stored honey. Winged adult insects that migrate, such as the monarch butterfly, fly south in autumn and return northward in spring.

Immature stages, however, are often on the soil surface, on plants or other objects, and must endure exposure to cold weather. Diapause was discussed earlier as a physiological state that is associated with overwintering. In this section we will describe other processes that are usually combined with diapause to ensure survival even in the Arctic. These processes are called cold-hardening.

SUSCEPTIBILITY AND TOLERANCE TO FREEZING. When water is cooled to 0°C and below, freezing does not necessarily take place even though 0°C is ordinarily considered the freezing point. Before ice can be formed, some molecules of water must be arranged in a certain pattern to provide a nucleus for growth of the crystal. Until by chance the pattern occurs, the liquid can be further cooled. In careful experiments with small droplets of pure water, −40°C can be reached before crystallization. Water that remains liquid below its freezing point is said to be *supercooled.* The temperature at which a crystal begins is the supercooling point. Usually crystallization is instantaneous and only one crystal is formed. During the process of crystallization heat is released, raising the temperature toward the freezing point. Once the crystal is complete, the temperature can be further lowered.

Mineral particles, especially dust, supply surfaces on which ice nuclei readily form. In other words, the particles serve as "nucleators." The presence of dust nucleators in the gut of an insect or of other nucleator substances in the hemolymph raises the supercooling point by increasing the likelihood of freezing. On the other hand, the addition of dissolved substances in water lowers the points of freezing and supercooling. Even without special antifreeze substances, the body fluids of overwintering insects can often be supercooled to −20°C. Thus many insects survive subzero weather by evacuating their guts in autumn and supercooling without harm, to temperatures above the supercooling point. When the supercooling point of the insect is reached, an ice crystal will form internally around a nucleator.

Insects that die when frozen are called *freezing-susceptible.* Those that can survive freezing during winter are called *freezing-tolerant.* The nature of the fatal damage to susceptible insects is not clear. Possibly crystals mechanically rupture cells, or the removal of water by ice formation causes an imbalance in the concentration of chemicals in the hemolymph and cell fluids.

Freezing generally begins in the hemolymph and may or may not include the cells. Extracellular freezing is not as severe as intracellular freezing, yet the cells may become dehydrated. Shorter periods of freezing are less damaging than longer periods. Upon thawing some tissues may revive, but the fate of the insect depends on the tissues most susceptible to injury. Tissues involved in development are the most sensitive. Death may not ensue for several days or more. Pupae, for example, may fail to complete metamorphosis after being frozen.

GLYCEROL AND FREEZING TOLERANCE. Glucose, trehalose, low-molecular-weight lipids, and sorbitol in the hemolymph confer additional protection against freezing. No natural chemical, however, equals glycerol as an antifreeze. Wyatt and Kalf (1958) discovered glycerol in the hemolymph of diapausing pupae of the cecropia moth, revealing for the first time that this compound might aid insects to survive freezing. Subsequently, glycerol has been found in most overwintering, freezing-tolerant insects, especially larvae and pupae. R. W. Salt (1961) records that up to 25 percent of the fresh weight of the wasp larva *Bracon cephi* (Braconidae) is glycerol, an amount sufficient to lower the freezing point to −15°C and the supercooling point to −47°C.

Some confusion surrounds the role of glycerol and freezing tolerance: not all tolerant insects seem to possess glycerol, and the presence of glycerol does not necessarily confer tolerance. Part of the conflicting evidence may be due to inadequate analytical procedures. Recent studies by Baust and Miller (1970) demonstrate that glycerol concentrations may vary rapidly. The collection of insects in nature and their storage in the laboratory may alter the cold-hardened condition before tests are performed. Furthermore, glycerol might become localized in tissues and not detectable in hemolymph.

Nevertheless, glycerol synthesis in insects is closely associated with lower environmental temperatures. Its presence in hemolymph is correlated with lowered freezing and supercooling points. High concentrations are found in extremely tolerant insects. This does not exclude other possible protective compounds or mechanisms of freezing tolerance. The action of glycerol is to extend the temperature range of supercooling without freezing, to retard the rate of freezing, and to reduce the size of crystals. When the insect is indeed frozen, glycerol presumably reduces the deleterious osmotic effects and prevents the intracellular freezing that is fatal.

WINTER SURVIVAL OF AN ARCTIC BEETLE. The studies by Baust (1972, 1976), Miller (1969), Baust and Miller (1970, 1972), and Kaufmann (1971) of overwintering adults of the carabid *Pterostichus brevicornis* provide an example of the ecology, behavior, and physiology of an insect in a cold environment. Adult insects are generally not known to be freezing-tolerant. In this respect the carabid is unique, but the annual cycle of glycerol production and loss (Fig. 11.5) resembles that known in less detail for overwintering larvae and pupae of other insects. It is also interesting to note that the carabid is able to digest food, develop eggs, and exhibit other signs of physiological activity while in a cold-hardened and frozen state. Although only the adults have thus far been analyzed, the beetles also overwinter as larvae or pupae. Some individuals overwinter twice and live up to 36 months.

Pterostichus brevicornis is a scavenger that lives in forest litter in circumpolar regions. Near Fairbanks, Alaska, summer is spent in the upper 2 to 3 cm of surface litter that is warmed to 10 to 14°C by air and intermittent sunshine. Only a few centimeters below the surface, the temperature of litter drops to 0 to 1°C because of the underlying permafrost. After the first frost in autumn and independently of photoperiod, the adult beetles migrate by the thousands into the old galleries of wood-boring beetles in decaying tree stumps. The beetles continue to move toward the interior of the stump, favoring regions of −1 to 2°C, until immobilized and finally frozen by the approaching winter.

In winter the stumps may be exposed to −40°C for several weeks and may reach −60°C for shorter periods. Baust (1976) transferred a stump from 20 to −42°C and measured the decline in temperature at various distances inside the stump. After 27 hours, the interior of the stump at 5 to 9 cm depth had not yet reached −42°C. Cooling is slow because the moist core of the stump combines the insulating qualities of wood with the thermal properties of water. At 7 cm depth, the cooling rate was 25°C per hour. Stumps also are often covered by an insulating

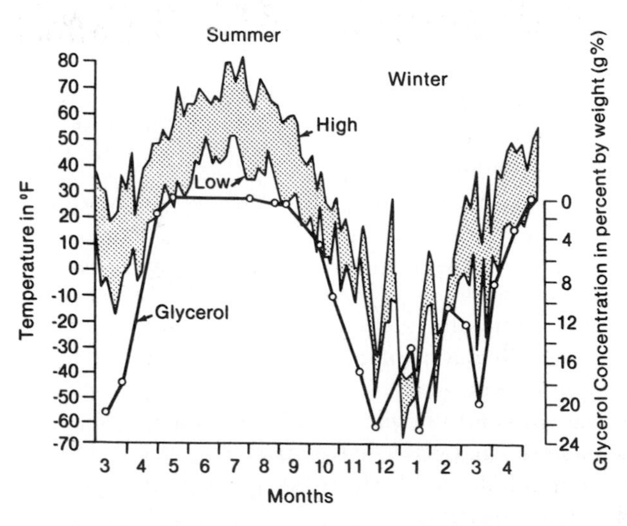

FIGURE 11.5 Seasonal variations in the average glycerol concentration in hemolymph of the arctic carabid, *Pterostichus brevicomis,* and high-low ambient temperatures near Fairbanks, Alaska. Note that the glycerol scale is inverted. (*Data from Baust and Miller,* 1970.)

blanket of snow. Thus the hibernating beetles are buffered against rapid temperature changes.

When beetles are kept at 5°C, no glycerol is detectable in the hemolymph. Without glycerol, the freezing point of hemolymph is −2°C and the supercooling point is −7.3°C. A sudden drop in temperature to −15°C or more is fatal. Beetles inside stumps, however, are protected against such rapid changes. If given time to synthesize glycerol, the beetles become freezing-tolerant.

Synthesis of glycerol is stimulated by exposure to 0°C for more than 24 hours. Glycerol then continues to accumulate at temperatures below freezing. Between 0 and −15°C, the lower the temperature, the more rapid is the buildup. Between 0 and −10°C, concentrations may peak or "overshoot" during the first week. Excessive amounts produced at this time would be beneficial in the event of a further decline in temperature. From November to April, the glycerol concentration fluctuates between about 10 to 23 g per 100 ml of hemolymph (percent by weight). The freezing point of hemolymph is in the range of −5.0 to −4.2°C, and the supercooling point is −18.3 to −11.5°C. In this condition, beetles can survive down to −87°C for 5 hours with little mortality.

Kaufmann (1971) found that beetles start hibernation with a water content at about 62 percent of the body weight, abundant fat, and the guts filled with the summer diet of small insects and other arthropods. During the first few months of hibernation, digestion and excretion continue until the gut is emptied. Body weight declines as the gut is cleared, fat is consumed, and water content drops about 10 percent. Then in January and February whenever temperatures exceed −3 or −4°C, she noted that the beetles began to crawl about and feed. The diet is now rotten wood containing bacteria and fungi. The ovaries become active and eggs start to develop. Fat droplets in fat cells fluctuate in abundance as a result of the renewed intake of food (increasing fat) and the growth of eggs (decreasing fat). Water content of the beetle's body begins to rise as water in the stump thaws and becomes available.

In the spring, at temperatures of 0 to 14°C, the glycerol concentration declines steadily at about 0.3 g per 100 ml per hour. The drop is more rapid at higher temperatures. At 23°C, glycerol at 19.8 g per 100 ml drops to 2.4 g per 100 ml in 24 hours and to 0 g per 100 ml in 36 hours, a loss rate of 0.7 g per 100 ml per hour. After one day at 21°C, the beetles have completely lost their tolerance to freezing; all die if returned to below-freezing temperatures.

SELECTED REFERENCES

General ecology is treated by Andrewartha and Birch (1954) and Odum (1971), and insect ecology by Chauvin (1967) and Price (1975). Reviews of specific aspects of biotic and physical factors in insect biology are: trophic relations by Brues (1946) and Frost (1942); competitive exclusion by De Bach (1966); circadian rhythms by Beck (1968), Brady (1974), Lees (1966, 1971), and Saunders (1974, 1976); insect seasonality by Tauber and Tauber (1976); microclimate by Geiger (1965); humidity by Bursell (1974b), Cloudsley-Thompson (1975), and Edney (1974); temperature effects on physiology by Bursell (1974*a*) and Heinrich (1974); and cold-hardiness by Asahina (1966, 1969), Downes (1965), and Salt (1961, 1969).

INSECTS OF SOIL AND WATER

12

Insects that spend part or all of their lives in soil or water exhibit special structural and behavioral adaptations to the physical, chemical, and biotic conditions found in each. In this chapter we shall examine some of the distinctive features of these media as habitats for insect life and discuss how these features are related to insect biology. We shall also note the important role of insects as decomposers in the renewal of these vital natural resources. The chapter concludes with a discussion of the early evolution of insects in soil and water.

SOIL INSECTS

Soil is the medium that connects all habitats of the land, from the intertidal to the alpine. It is also a distinct realm and one on which green plants and ultimately all forms of terrestrial life depend. The soil surface and overlying litter of plant debris are inhabited by at least some members of nearly all insect orders, as well as by other terrestrial arthropods.

Beneath the surface the variety of species markedly declines, but millipedes, mites, springtails, beetles, termites, ants, and fly larvae are still represented, sometimes abundantly. Relatively few kinds of insects inhabit caves, but they are nonetheless interesting because of their unique adaptations.

SOIL AS A HABITAT FOR INSECTS. The process of soil formation leads to horizontal layers that are readily visible in a cross section such as a road cut. On the surface is usually a duff, or litter of whole or partly decayed leaves and other plant parts. The nature of the litter varies with moisture and the covering vegetation, i.e., grassland, coniferous forest, deciduous forest, or lichens and mosses. The soil's uppermost layer, or topsoil, is darkened by a complex, organic substance called *humus.* Topsoil is the place where most plant roots and soil organisms occur.

The transformation of the litter into humus is accomplished in stages. Arthropods contribute significantly to the process by ingesting organic matter at the surface, mechanically breaking it down, and defecating deeper in the soil. Their excavations also mix the surface layer between litter and soil. Bacteria and fungi in the moister regions below then act on the fragments, now with greater surface area, to produce humus. The deeper subsoil, if present, is paler, being only slightly enriched from above, and grades below into the unaltered rock of the earth's crust.

Cracks, old root channels, and tiny cavities between soil particles create pore spaces in soil. The total pore space of topsoil is usually about a third of the soil volume, but in some places may exceed half the volume. At greater depths, the soil is more compact and has smaller pore spaces. Depending on the nature of the soil particles, rainfall, groundwater, and drainage, the pore spaces are filled partly by air and partly by water held by capillary forces. Soil air is usually saturated with water vapor except near the soil surface. In well-drained topsoil, oxygen in soil air is replenished by gaseous diffusion from the atmosphere. Soil water derives its oxygen from soil air. Oxygen, however, diffuses 10,000 times slower in water. The consumption of oxygen in soil water by bacterial decomposition of organic matter can create anaerobic conditions in as short a distance as 1 mm from a water-air interface. Deeper in the soil gaseous diffusion from the atmosphere decreases. Here the respiration of plants and animals reduces oxygen and increases the carbon dioxide content of soil air.

Providing the covering vegetation is not too dense, sunlight penetrates the litter, open burrows, and superficial soil crevices. Below the surface, at depths of only a few centimeters at most, is permanent darkness. In deserts, grasslands, alpine regions, and deciduous forests after the leaves have fallen, the surface is largely exposed to the sun's radiation. Thus heated by the sun, the surface temperature of soil rapidly climbs well above that of either the air or the soil at slight depths beneath. Surface temperatures above 50°C are not infrequent even in the Temperate Zone. Owing to the low thermal conductivity of soil, the heating and cooling of deeper layers lags behind the fluctuations in temperature of the air, litter, and the soil surface. Heat acquired during day penetrates into the soil quite slowly. While the surface cools in the evening, the deeper layers of soil are still rising in temperature. In the morning, the deeper layers have cooled but the surface is now being heated by the sun, creating a steep, vertical gradient in temperature. A high moisture content in soil will reduce the amplitude of these daily fluctuations in temperature. "Cover objects," such as large rocks or logs resting on the surface, shelter the soil beneath from extreme temperatures and maintain a high humidity by preventing evaporation. The spaces beneath cover objects provide insects a place to escape from predators and environmental extremes, especially in grasslands.

LITTER INSECTS. The most numerous insects in the litter and superficial layers of topsoil, in terms of both species and individuals, are colonies of Isoptera, colonies of Formicidae, all stages of Coleoptera, immature Diptera, and immature Lepidoptera, especially pupae. The most numerous arthropods of all here are the Acarina and Collembola. Each of the other insect orders is represented by at least one species, except the Ephemeroptera, Plecoptera, and Phthiraptera. The insects of certain orders spend their entire lives in litter, often in association with cover objects (Archeognatha, Thysanura, Grylloblattodea, Zoraptera, and some species of Blattodea, Dermaptera, Embioptera, and Orthoptera). Other orders are represented at least by the immature stages which feed as scavengers or predators or which feed on algae, fungi, or mosses (Mantodea, Thysanoptera, Psocoptera, Neuroptera, Raphidioptera, Mecoptera, and Siphonaptera). Parasitoids and parasites, including Strepsiptera, attack other litter insects.

Remarkably few aquatic insects are also active in moist litter, even though the habitat would seem suitable and, in most instances, the aquatic forms probably evolved from litter inhabitants. Exceptions are the terrestrial naiads of *Megalagrion* (Odonata) in Hawaii, and the terrestrial larvae of *Enoicyla pusilla* (Trichoptera) in forest litter of Europe. Megaloptera and aquatic Coleoptera, however, regularly pupate out of water in soil. The transition between aquatic and terrestrial life is often made by fly larvae. For example, crane-fly larvae (Tipulidae) readily burrow from beneath streams into moist soil.

Immobile stages of insects are commonly spent on the ground. Eggs are often deposited here: Phasmatodea drop eggs in litter, and Acrididae lay eggs in soil. Lepidoptera pupae and the resting "pupae" of Thysanoptera are in litter. Periods of quiescence or diapause also may be spent among dead leaves. Aggregations of diapausing adults of *Hippodamia convergens* (Coccinellidae) are found on the ground in Sierra Nevada of California from June to February.

The active stages of litter insects are greatly varied in shape. Most are somewhat dorsoventrally compressed, and often parallel-sided or anteriorly narrow, a form that facilitates movement among varied obstacles. Most are agile and possess well-developed means for locomotion on ground. Escape by jumping is often a characteristic (Archeognatha, Collembola, Orthoptera, Schizopteridae (Hemiptera), alticine Chrysomelidae, and rhynchaenine Curculionidae). The antennae and tactile sense organs are usually well developed. Exposed, membranous wings are apparently an encumbrance in narrow passageways. Except for mating and dispersal, flight is an unnecessary investment of energy and structure. The Embioptera spin silken runways to move out from cover objects into the litter. The wings of male web spinners, if present, are flexible and bend freely inside the tunnels. The Collembola, Diplura, Protura, and Apterygota are, of course, wingless, and the immature pterygotes have small wing pads at most. Among adult pterygotes some have protective forewings (Orthoptera, Dermaptera, and Coleoptera); others fail to develop wings (some Psocoptera, Grylloblattodea, and the sterile castes of Isoptera and Formicidae); or discard them after mating and dispersal (Zoraptera, reproductive castes of Isoptera and Formicidae).

Certain features of the body are evidently correlated with daily illumination and hazard of predation by visual hunters such as birds: the compound eyes are well developed; circadian rhythms in behavior are present, often with nocturnal activity; and the body is often pigmented with concealing patterns. Litter insects are generally resistant to desiccation. Respiration is by tracheae, and the spiracles are usually fitted with closing devices.

SOIL INSECTS. Below the litter layer and soil surface the variety of arthropod taxa drops abruptly, yet some groups are found here only. Collembola and Acarina are the most numerous of the permanent residents. All species of Pauropoda, Symphyla, Diplura, and Protura are entirely confined to soils, moist litter, or crevices under cover objects. No entire order of insects is similary restricted, but some groups in various orders spend part or all of their lives underground.

The most common are Isoptera, Hymenoptera (Formicidae and certain other aculcates), Diptera, Coleoptera, Hemiptera, and Orthoptera. Of these, the Coleoptera are the most numerous in species. All stages of the life cycle of Coleoptera can be found in soil (Fig. 12.1*c*, *d*),

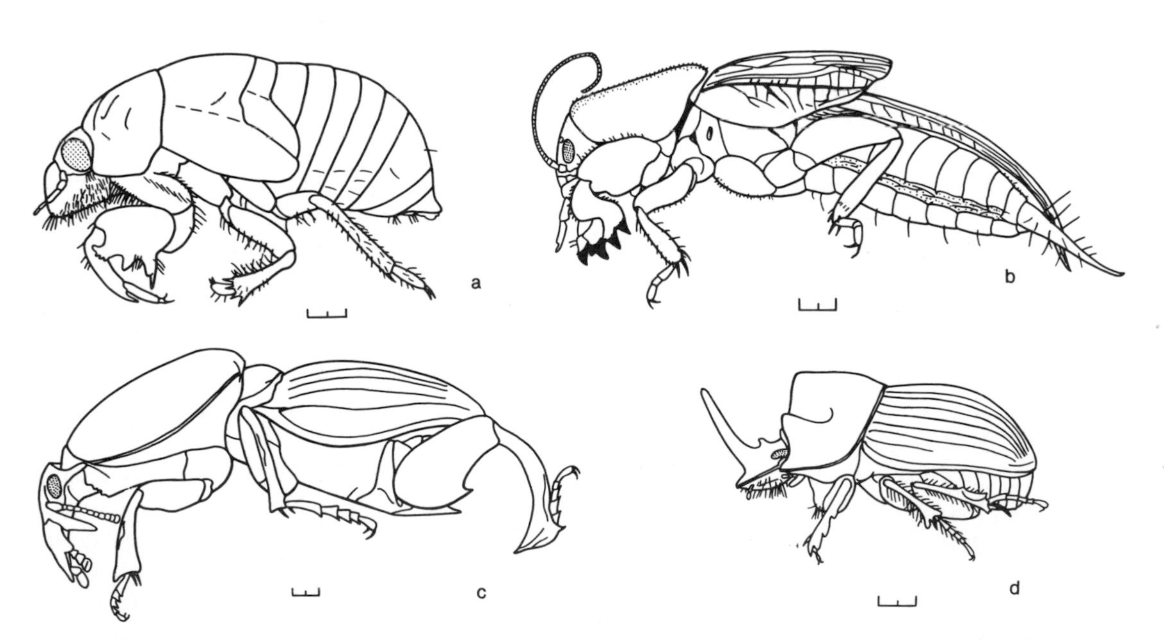

FIGURE 12.1 Examples of soil insects: *a*, cicada nymph, *Okanagana* sp. (Cicadidae); *b*, mole cricket, *Gryllotalpa* sp. (Gryllotalpidae); *c*, unusual soil-dwelling cerambycid beetle from Brazil, *Hypocephalus armatus;* *d*, dung beetle, *Copris lugubris* (Scarabaeidae). Scales equal 2 mm.

but generally only the immature stages of Diptera are present. Termites and ants make their nests in soil. Their excavations and the plant matter brought into the nest contribute to soil development. Galleries of *Atta* ants may be enormous and reach 3 to 6 m in depth. Many species of bees and predatory wasps excavate simple burrows in earth that are stocked with provisions for their young. Roots are the food of nymphal Cicadidae (Fig. 12.1*a*) and certain other subterranean Hemiptera. Mole crickets (Gryllotalpidae, Fig. 12.1*b*) feed on roots but also take insect larvae. These orthopterans and the omnivorous Stenopelmatinae (Gryllacrididae) are true inhabitants of the soil.

The high humidity of soil air, films of soil water, and moderate temperature of soil create conditions approaching an aquatic environment. Many invertebrate phyla are represented in soil by minute aquatic forms that are found elsewhere in freshwater. Protozoa, for example, are the most numerous soil animals. In the soil air of pore spaces live the smaller Collembola, Acarina, Pauropoda, Symphyla, Diplura, and Protura. When the number and size of organisms are compared to the surface area of the walls of pore spaces, it is evident that the cavities are greatly underpopulated (Christiansen, 1964).

An important factor limiting insects in soil is the obvious difficulty in moving about. Unlike the small arthropods listed above, insects are generally not small enough to negotiate interconnecting pore spaces for any distance. They must either tunnel by pushing aside particles and squeezing through, or excavate the soil in front and deposit it behind. The physical resistance of the soil, therefore, raises the energetic cost of locomotion and reduces the distance to which an individual can hunt for mates or food. Sight and wind-borne odors are also eliminated as aids in the search. For this reason, mating of insects is usually accomplished above ground. Except for the roots of green plants, insects must find their food in the form of other soil organisms, dead or alive. Active excavators such as ants, termites, silphid beetles, larval cicindelids, wasps, and bees obtain their food from richer sources above ground.

Reduction in wings is a characteristic of soil insects as well as of litter insects. Soil insects are generally round in cross section but are greatly varied in shape otherwise. "Tunnelers" have reduced or no legs and are of two kinds. Larval Elateridae and Therevidae are smooth, stiff-bodied, and slender. By sinuous movements of the trunk, they force their hard, tapered heads forward through the substrate. Larval Tipulidae and Bibionidae are soft-bodied and use peristaltic movements to penetrate soil. "Excavators" may be fairly thick-bodied and often exhibit conspicuous molelike modifications of the head and forelegs for digging. Such fossorial adaptations are seen in Gryllotalpidae, Cydnidae, nymphal Cicadidae, and Scarabaeidae (*Copris, Geotrupes*), to list only a few (Fig. 12.1). Worker ants and termites are not so clearly equipped for digging, but they effectively use their mandibles to remove and carry away particles.

In the absence of light, soil insects and the other hexapodous arthropods tend to have reduced or no compound eyes; antennae variably developed, even absent (Protura); well-developed tactile sense organs; and pale pigmentation of the integument without patterns. They are usually repelled by light and higher temperatures. The upper limits of tolerance to temperature are relatively low, but some soil insects can remain active a few degrees above 0°C. Rhythmic behavior has been poorly studied. Vertical migration in response to daily temperature fluctuations may occur if the soil is exposed to solar radiation. Resistance to desiccation is low, probably because the epicuticle becomes abraded by soil particles or is naturally permeable to water. Some respire cutaneously. Those that normally inhabit the deeper subsoil are tolerant of higher levels of carbon dioxide, at least for several days (some Collembola, larval Elateridae, Scarabaeidae).

Many insects that inhabit deserts are frequently dependent on the sandy substrate for protection against extreme temperatures and desiccation. Those running on the surface during the day may have long legs that lift the body well above the hot surface (Mutillidae, Tenebrionidae). During the hot season many forage openly only at night, spending the day in burrows or simply immersed in sand. Edney (1974) has shown that surface temperatures up to 72°C and humidity of 10 percent are avoided by the desert cockroach *Arenivaga investigata.* At depths of 30 to 45 cm in sand, the roach inhabits a zone where daily temperatures are well below the upper lethal limit (about 45.5 to 48.5°C) and humidity may reach above 82 percent, a level at which the roach can actually absorb water vapor. Like some tenebrionid beetles, the roach literally swims in sand. The body is biconvex in cross section, oval, and smooth, and has sharp lateral margins, resembling that of an aquatic beetle.

CAVE INSECTS. Insects found in caves may be the same species that are normally found in litter or soil outside. Certain flying insects, especially flies (Culicidae, Mycetophilidae), seek shelter in caves during the day. The darkness, high humidity, even temperature, and absence of green plants are conditions similar to those of deeper soil, but without the physical restraint. Soil insects, therefore, already have many of the modifications required for continuous life in caves.

The absence of plants except for their roots, however, largely limits cave-dwelling insects to scavenging, parasitism, or predation. Phytophagous insects are usually absent. New organic matter must be washed in or be deposited as waste by organisms that enter and leave. Bats attract several taxa of parasites (Polyctenidae, Streblidae, and Nycteribiidae; some Siphonaptera and Cimicidae) and the accumulated feces of bats (called guano) attracts scavengers (Arixeniina, Dermaptera; Blattodea, Diptera, etc.).

Animals which are adapted to continuous life in caves and which reproduce there are called *troglobites.* Insects and other hexapodous arthropods in this category are almost entirely limited to Collembola (Entomobryidae), Diplura (Campodeidae), Orthoptera (cave crickets, Rhaphidophoridae, Fig. 12.2*a*), and Coleoptera (Pselaphidae; trechine Carabidae, Fig. 12.2*b*; and additionally in Europe, bathysciine Leiodidae). Aquatic insects are absent, but some cave Collembola and beetles have been observed submerged and alive in bodies of water in caves.

The restriction of troglobite insects to caves has stimulated research into their origin and geographic distribution (Mitchell, 1969). Most troglobites inhabit caves in the Northern Hemisphere, where colder climates and glaciation occurred in the Pleistocene. Apparently, insects became adapted to cold, humid, forest litter near the edges of glaciers. At that time glaciers extended to lower latitudes and altitudes. When the glaciers receded, some cool-adapted insects found refuge in caves, while the remainder of the fauna moved to higher latitudes and altitudes or became extinct. The scarcity of

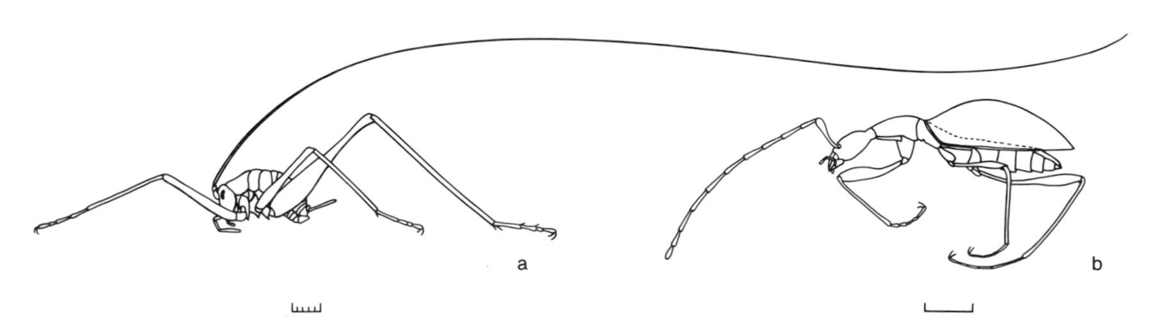

FIGURE 12.2 Examples of cave insects: *a,* cave cricket, *Tropidischia xanthostoma* (Gryllacrididae; scale equals 5 mm); *b,* blind cave beetle, *Glacicavicola bathyscioides* (Carabidae; scale equals 1 mm).

food and lower temperatures of caves in the Temperate Zone place strong selection pressure on efficient use of food at a low metabolic rate. In separate, isolated cave systems, insects independently evolved economizing adaptations: reductions of eyes, wings, flight muscles, pigmentation, and, in some beetles, number of offspring. Troglobites are rare in tropical regions. Presumably this is because food entering the caves is more abundant, and less selection pressure exists for energy-economizing adaptations.

Troglobite insects and Collembola exhibit various degrees of these reductions. On the average, individuals of cave species tend to be larger than their close relatives outside. Cave crickets and trechine Carabidae have more slender bodies and longer appendages. Antennae are generally longer, sometimes exceeding the length of the body in Collembola. Beetles may have erect, long hairs scattered over the body that are probably tactile. Most troglobites are repelled by light. The tracheal system and spiracles of trechine carabids are rudimentary; respiration probably takes place through the membranous abdominal terga. Resistance to desiccation and higher temperature is low. Long periods without food are tolerated, and growth is slow. Rhythmic activity and rest in bathysciine beetles are independent of days but are associated with temperature fluctuations.

Some bathysciine beetles that live far from the entrance where food is rare have larvae that do not feed. The female lays one egg at a time; each is large and full of yolk. Shortly after hatching the larva builds a clay capsule and remains inside 5 to 6 months, then pupates. The adult searches widely for food, in a manner not possible for a young larva. Adults live for about 3 years.

AQUATIC INSECTS

In this section we discuss insects that spend at least the immature stages of their lives floating or submerged in water. Aquatic insects are a minor fraction of all insects, probably numbering no more than 3 to 5 percent of all species. The limited number is probably the result of the limited amount of freshwater habitat in comparison with the land surface. Yet these insects are taxonomically diverse (Table 12.1), are fascinating in structure and biology, and some of them, such as mosquitoes, are of extreme importance in public health.

A few insects are truly marine. Some others live in the intertidal zone, in brackish coastal waters or saline lakes. Most aquatic insects, however, are restricted to freshwater, where they are among the most important organisms in the aquatic ecosystem. As herbivores, predators, scavengers, and even parasitoids, insects function in the trophic structure and, in turn, are food for fish, birds, and amphibians.

Aquatic insects exhibit conspicuous adaptations for respiration and locomotion within the constraints of a liquid medium (Fig. 12.3). Similar adaptations, especially for respiration, can be observed among insects in semiliquid media: scavengers in decaying plant or animal bodies or animal waste; phytophagous insects that burrow in juicy fruits or other plant tissues; and endophagous parasites and parasitoids immersed in the hosts' blood.

MARINE INSECTS. Whereas insects constitute nearly three-quarters of the earth's animal species, they are greatly reduced in number of species in the earth's largest habitat, the ocean (Usinger, 1957). By no means are insects scarce as individuals on the coasts or even far out at sea. Some are exceedingly abundant. Each marine species has overcome certain physical barriers: tidal submergence (the collembolan *Anurida maritima;* insects in several orders, including the hemipteran *Aepophilus bonairei* of England and Europe); wave action [especially midges of the genus *Clunio* (Chap. 11) and the barnacle-eating larvae of the dryomyzid fly *Oedoparena glauca*]; salinity (especially salt-marsh mosquitoes, ephydrid flies, and the water boatman *Trichocorixa*); depth (the midge *Chironomus oceanicus,* dredged from 20 fathoms); and life away from land (five species of the pelagic water striders, *Halobates,* that occur hundreds of miles from land).

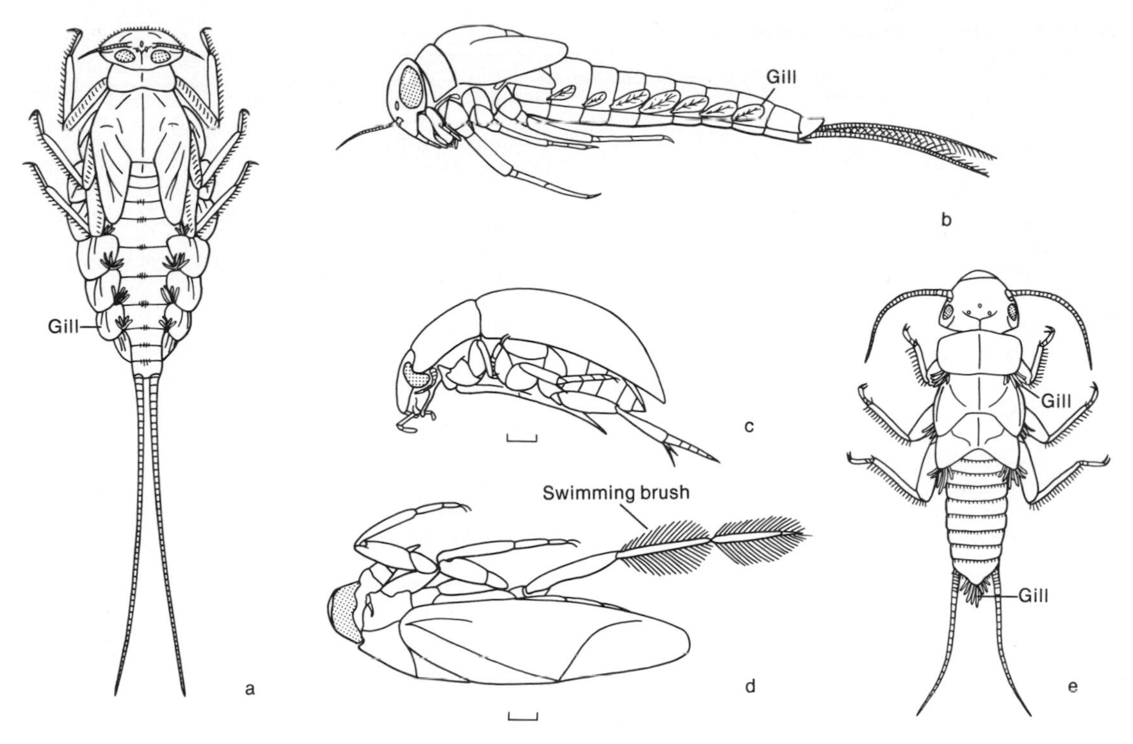

FIGURE 12.3 Examples of aquatic insects: *a*, mayfly naiad, *Epeorus* sp. (Heptageniidae); *b*, mayfly naiad, *Ameletus* sp. (Siphlonuridae); *c*, water beetle, *Tropisternus ellipticus* (Hydrophilidae); *d*, back swimmer, *Notonecta undulata* (Notonectidae); *e*, stone-fly, naiad, *Acroneuria pacifica* (Perlidae). Scales equal 1 mm.

Of these barriers perhaps the most limiting is the inability to respire for long beneath the surface. No marine insect is known to spend its entire life under water. Usinger (1957) has pointed out that insects are no more successful in colonizing deep freshwater lakes. Only certain midge larvae (Chironomidae, Diptera) and perhaps the peculiar stone fly *Capnia* spp. (Jewett, 1963) can exist indefinitely in deep, still water.

Added to the physical barriers is the intense biological competition from marine arthropods and other organisms, plus the threat of predatory fish. During the Paleozoic Era while insects flourished on land, the marine fauna was diversifying and occupying the habitats of the sea. The oceans now seem effectively closed to insects.

EXTENT OF FRESHWATER HABITATS. On a global scale the total extent of fresh water is remarkably small. Fresh water lakes and rivers constitute less than 0.01 percent of the total volume of water in the biosphere (Wetzel, 1975). Even including the inland seas such as the Black and Caspian, all inland bodies of still water occupy only 1.8 percent of the land surface. Rivers or running waters occupy about 0.001 percent. Much fresh water is concentrated in a few large lakes in the Temperate and Subarctic regions of the Northern Hemisphere. Lake Baikal in Siberia and the Laurentian Great Lakes of North America each contain about 20 percent of all fresh water.

Fresh water has been available to terrestrial life because of the hydrological cycle. Actual geological evidence of ancient freshwater habitats is scarce. The sediments of lakes and rivers, like other terrestrial deposits, seldom escape destruction by erosion. At any instant lakes contain about 100 times more water than rivers, but rivers are continually renewed. Over a year rivers

TABLE 12.1 Taxa of aquatic insects[7]

Immature stages in freshwater

- Ephemeroptera (all)
- Odonata (nearly all)
- Plecoptera (nearly all)
- Megaloptera (all)[5]
- Neuroptera
  - Sisyridae[5]
- Coleoptera
  - Helodidae
  - Limnichidae (some)
  - Psephenidae[6]
  - Ptilodactylidae
- Diptera[6]
  - Anthomyiidae (*Limnophora*)
  - Blephariceridae
  - Ceratopogonidae
  - Chironomidae
  - Culicidae
  - Deuterophlebiidae
  - Dixidae
  - Dolichopodidae (*Dolichopus*)[2]
  - Empididae (Clinocerinae, Hemerodromiinae)
  - Ephydridae
  - Ptychopteridae[2]
  - Phoridae (*Diploneura*)
  - Psychodidae[2]
  - Rhagionidae (*Atherix*)
  - Sarcophagidae (*Sarcophaga*)
  - Sciomyzidae
  - Scopeumatidae (*Hydromyza*)
  - Simuliidae
  - Stratiomyidae (Stratiomyiinae, Adoxomyiinae)
  - Syrphidae (Sericomyiinae, Eristalinae)
  - Tabanidae[2]
  - Tanyderidae (*Protoplasa*)[2]
  - Thaumaleidae
  - Tipulidae (Tipulinae, Limoniinae)
- Lepidoptera
  - Pyralidae (Nymphulinae, *Acentropus*)[6]

Immature stages in freshwater

- Trichoptera (nearly all)[6]
- Hymenoptera[6]
  - Braconidae (*Chorebus*)
  - Diapriidae (*Trichopria*)
  - Ichneumonidae (*Asilops*)
  - Mymaridae (*Caraphractus*)
  - Scelionidae (*Tiphodytes*)
  - Trichogrammatidae (*Hydrophylita*)

Immature and adult stages in freshwater

- Collembola[3]
  - Entomobryidae (*Sinella*)
  - Hypogastruridae (*Xenylla*)
  - Isotomidae (*Isotomurus*)
  - Onychiuridae (*Onychiurus*)
  - Poduridae (*Podura aquatica*)
  - Sminthuridae (*Sminthurinus*)
- Orthoptera
  - Acrididae (*Marellia remipes*)[3]
  - Tridactylidae[4]
- Hemiptera (Heteroptera)
  - Aepophilidae
  - Belostomatidae
  - Corixidae
  - Dipsocoridae[4]
  - Gelastocoridae[4]
  - Gerridae[3]
  - Hebridae[4]
  - Hydrometridae[3]
  - Macroveliidae[4]
  - Mesoveliidae[3,4]
  - Naucoridae
  - Nepidae
  - Notonectidae
  - Ochteridae[4]

carry a volume equal to perhaps a third of all freshwater lakes (Hynes, 1970*a*).

INSECTS OF RUNNING WATER. Although the total volume of running water is less than still water, continental drainages exist for periods of a million years or more. This provides a continuously habitable environment for aquatic organisms over a wide geographic area. Because major drainages are more or less permanent and are otherwise favorable, insects have evolved the greatest variety of taxa in running waters. Indeed most kinds of aquatic insects are represented in running waters, and some occur only there. This is especially true in the *rhithron*, or rapidly flowing, shallow, rocky streams of the headwaters where oxygen concentration is high and the

TABLE 12.1 Taxa of aquatic Insects[7] *(continued)*

Immature and adult stages in freshwater

- Pleidae
- Saldidae[4]
- Veliidae[3]

- Coleoptera[5]
  - Amphizoidae
  - Curculionidae (Hydronomini, Erirrhinini, Bagouini)
  - Chrysomelidae (Donaciinae)[6]
  - Dytiscidae
  - Elmidae
  - Gyrinidae (adults[3])
  - Haliplidae
  - Heteroceridae (*Heterocerus*)[2]
  - Hydraenidae (larvae[4])
  - Hydrophilidae (most)
  - Hydroscaphidae
  - Limnichidae (some)
  - Noteridae[6]
  - Sphaeriidae[2]

Adult stage in freshwater

- Coleoptera
  - Dryopidae

Intertidal marine insects

- Collembola
  - Entomobryidae (*Entomobrya*)
  - Hypograstruridae (*Anurida maritima*)
  - Isotomidae (*Isotoma*)
- Hemiptera (Heteroptera)
  - Corixidae (*Trichocorixa*)
  - Gerridae (*Rheumatobates*)
  - Omaniidae
  - Saldidae (*Aepophilus*)
- Hemiptera (Homoptera)
  - Aphididae (*Pemphigus*)
  - Psyllidae (*Aphalara pulchella*)

Intertidal marine insects

- Coleoptera
  - Carabidae (*Thalassotrechus*)
  - Chrysomelidae (*Haemonius*)
  - Curculionidae (*Emphyastes*)
  - Heteroceridae (*Heterocerus*)[2]
  - Hydraenidae (*Ochthebius*)
  - Hydrophilidae (*Cercyon*)
  - Limnichidae (*Hyphalus, Mexico*)
  - Melyridae (*Endeodes*)
  - Salpingidae (*Aegialites*)
  - Staphylinidae (Aleocharinae, *Bledius, Thinopinus*)
- Diptera
  - Anthomyiidae (*Fucellia*)
  - Canaceidae
  - Ceratopogonidae
  - Chironomidae (Clunioninae)
  - Coelopidae
  - Culicidae
  - Dolichopodidae (*Aphrosylus*)
  - Dryomyzidae (*Oedoparena*)
  - Ephydridae
  - Helcomyzidae
  - Heleomyzidae (*Anorostoma*)
  - Tabanidae
  - Tipulidae (*Limonia*)
- Trichoptera
  - Limnephilidae (*Limnephilus affinus*)
  - Philanisidae (*Philanisus*)
- Hymenoptera
  - Encyrtidae (*Psyllaephagus*)
  - Eulophidae (*Tetrastichus*)

Insects of the open ocean

- Hemiptera (Heteroptera)
  - Gerridae (*Halobates*)[3]
- Diptera
  - Chironomidae (*Pontomyia natans* adults[3])

[1]Adults enter water to oviposit on aquatic hosts.

[2]In saturated soil or semiliquid media.

[3]On water surface.

[4]On shore near water, occasionally enter water.

[5]Pupae terrestrial.

[6]Pupae aquatic.

[7]This list of taxa is intended to be representative, not exhaustive. The listing of a family without qualification does not necessarily mean that all species in the family have the same habits. Well known or exceptional taxa are given in parenthesis.

Source: Summarized from Usinger, 1956*a*, Smith and Carlton, 1975, and Cheng, 1976.

mean monthly temperature seldom exceeds 20°C.

Water seeping into a stream from the ground or issuing from springs is commonly cooler than air in summer and warmer than air in winter. This difference may be sustained for some distance if the stream is shaded in summer or insulated by a snow cover in winter. Although subject to daily and seasonal fluctuations, the range of variation is usually less than that of the shallow regions of lakes. Furthermore, cooler water absorbs much more oxygen than warmer water, e.g., water saturated at 4°C holds 1.5 times the amount at 22.5°C. Turbulence over the rocky bed maintains dissolved oxygen at the point of saturation and distributes it to the depth of the stream. Consequently, insects are able to feed and grow throughout the year, even in winter in Temperate regions.

Species of Ephemeroptera, Plecoptera, Trichoptera, and Megaloptera are the most numerous in streams. Larvae of the fly families Blephariceridae, Simuliidae, and Deuterophlebiidae, and the beetles Helminthidae, Dryopidae, Elmidae, Ptilodactylidae, and Psephenidae occur among rocks in swift, cool waters. Naiads of the odonates Agrionidae, Cordulegastridae, and Gomphidae also favor running water.

An ever-present hazard to current dwellers is to be swept off the bottom and drift downstream exposed to predators. The number of insects drifting usually rises on darker nights, because many are nocturnally active. Drifting may involve only short distances and not be entirely deleterious. Competition for space is relieved, and the population is more evenly distributed downstream. Evidence is inconclusive that adults tend to fly upstream to lay their eggs and recolonize the uppermost waters. The eggs of mayflies, stone flies, and some dragonflies are equipped to sink rapidly, and firmly attach to the bottom, thus remaining near the ovipositional site.

Insects that inhabit swift water are generally prevented from visiting the surface for air. Their respiratory structures are tracheal gills or plastrons (Chap. 5) that function while bathed in oxygen-rich water. Naiads of some mayflies and stone flies that normally reside in still pools create their own currents. The abdominal gills of mayflies have an undulating beat that circulates the surrounding water. Similarly, stone flies pump up and down on their legs at a rate proportional to the oxygen need. Insects living in the swiftest waters do not exhibit such respiratory movements and depend entirely on the current to bring oxygen. They die quickly even when placed in still water of the same temperature and oxygen saturation because the oxygen supply becomes depleted in the proximity of the gills and is not renewed by currents.

A number of similar morphological adaptations has been evolved independently by various insects in torrential waters (Hynes, 1970*a*). The body may be dorsoventrally compressed so that the insect can creep in the thin boundary layer of 2 to 3 mm next to the substrate where the current declines to zero. Naiads of the mayflies Heptageniidae (Figs. 12.3*a*, 20.4*c*), and larval beetles of Psephenidae are thus flattened (Fig. 39.8*c*). Small size also permits an insect to utilize the boundary layer. Nearly all adult beetles in swift water are small.

Various attachment devices allow insects to cling to the substrate. The beetles Elmidae and Dryopidae have the last tarsal segment and claws enlarged for anchoring themselves. In addition, the larvae of Elmidae have prehensile hooks near the anus. Stout tarsal claws and anal hooks also are found on the free-living caddis larvae of Rhyacophilidae (Fig. 45.2*a*) and the hellgrammites or larvae of Corydalidae (Megaloptera). Lacking true legs, the aquatic larvae of the fly Families Simuliidae, Deuterophlebiidae, Blephariceridae, Chironomidae, Rhagionidae, Empididae, Syrphidae, Ephydridae, and *Limnophora* (Anthomyiidae) have developed leglike prolegs for attachment. The prolegs are usually fitted with fine hooks. Suckers are developed near the anus of larval simuliids and on the ventral surface of *Maruina* (Psychodidae) and Blephariceridae. Frictional resistance to detachment is developed by ventral gills or hairy pads on the mayfly naiads *Rhithrogena* (Heptageniidae),

Ephemerella doddsi and *E. pelosa* (Ephemerellidae) the African *Dicercomyzan* (Tricorythidae), and the larval beetles of Psephenidae (Fig. 39.8*c*). Silk and sticky secretions are used for attachment by some Chironomidae, and the caterpillars of Pyralidae (Lepidoptera). Larval simuliids move about with their prolegs on silken mats. *Hydropsyche* (Trichoptera) spin nets in the current and harvest waterborne food. Most Trichoptera attach their pupal cases to rocks with silk. The stone cases built by Trichoptera in streams are usually composed of heavier grains and a few heavy pebbles (Fig. 45.3). This provides ballast so that the caddis larva quickly sinks into still water if dislodged.

Few insects actively swim against the current. The streamlined, minnowlike mayflies *Ameletus* (Fig. 12.3*b*), *Isonychia, Centroptilus,* and some *Baetis* move in and out of the current with powerful, up-and-down movements of their caudal filaments. On the surface of riffles are the bugs *Rhagovelia* (Veliidae, Fig. 34.7).

Insects in crevices can utilize the rich supply of oxygen and food of streams without the immediate danger of the current. Flattened mayflies are also found in the dead water beneath rocks. Tiny, young instars of insects, the narrow-bodied stone flies and mayflies, and crane-fly larvae, *Antocha* and *Hexatoma* (Tipulidae), normally are found amongst the bottom gravel. Leaves packed against obstructions have a special fauna of mayflies, caddis flies, and stone flies. The stout-limbed *Ephemerella* (Ephemerellidae, Fig. 20.2*a*) is commonly found here.

The lower region of a drainage, or *potamon,* is characterized by slower flow, greater depth, occasional low oxygen concentrations, and monthly mean temperatures close to that of air and rising above 20°C. The bed of the river is formed by deposits of sand and mud.

Fewer kinds of insects are specialized for life in the potamon, but some species are extremely abundant. Naiads of the burrowing mayflies Ephemeridae (Fig. 20.4*b*) excavate U-shaped burrows in silt. Undulating movements of their feathery gills circulate water in the burrow, bringing in organic matter, which is eaten, and oxygen (Fremling, 1960). Other mayflies, such as Caenidae, Tricorythidae, and some Ephemerellidae, inhabit silty bottoms of streams and rivers. These have the functional gills kept clean of particles by the operculate uppermost gill. The body surfaces of some bottom-dwelling mayflies, stone flies, and dragonflies are covered by a protective layer of fine hairs. Larvae of chironomid midges are often abundant in soft, mucky bottoms where they build tubes.

INSECTS OF STILL WATERS. Lakes are relatively short-lived in comparison to rivers. Most lakes existing today originated in the Pleistocene and are less than 25,000 years old. Only a few large, deep lakes are much older. Lake Baikal and Lake Tanganyika in Africa probably existed in the Tertiary. Bodies of still water normally fill with sediments or evaporate during prolonged dry periods.

In contrast to running water, the insect life of still water is influenced more by temperature and oxygen supply than by water movement. Temperatures in shallow, marginal waters of lakes or small bodies of water such as ponds tend to fluctuate with air temperature. Large bodies of still water warm slowly in summer, becoming stratified into an upper, warmer layer, or *epilimnion;* a zone of transition, or *thermocline,* where temperature decreases 1°C or more per meter of depth; and a deep layer at 4°C called the *hypolimnion.* In winter, lakes cool slowly. Water reaches maximum density at 4°C and is less dense at lower or higher temperatures. Surface water cooled to 4°C sinks. Ice forms at the surface after deeper water is uniformly 4°C and the colder, lighter water is at the top. Insects in lakes are thus protected against freezing. Oxygen is absorbed at the water's surface and is produced by photosynthetic plants. In summer the oxygen supply of the epilimnion may exceed saturation. Decomposition in the hypolimnion reduces the supply.

Lakes that are poor in nutrients are called *oligotrophic.* Young, deep lakes are characteristically of this type. Precipitation and runoff from the watershed carry dissolved substances into the lake, promoting the growth of free-floating algae in sunny, open water. The gradual enrichment of water with nutrients, especially nitrogen and phosphorus, is called *eutrophication.* Continued addition leads to the production of organic matter in excess of decomposition. The decomposed organisms in deep water deplete the water of its oxygen, leaving the organic matter to accumulate with other sediments. Older lakes, rich in nutrients and shallow, are called *eutrophic.* When lakes receive large amounts of acid humic substances from the watershed, the water becomes stained brown and poor in variety of aquatic life. Organic matter accumulates from plants in the shallow marginal waters. Such lakes or bogs are called *dystrophic.* Ultimately both eutrophic and dystrophic lakes fill completely with organic matter, are overgrown by vegetation, and disappear.

Slender fly larvae dominate the insect life of the deeper, open waters and the bottom beneath. The thin sediments of oligotrophic lakes are characterized by colorless midge larvae of *Tanytarsus* (Chironomidae). On the surface of the bottom ooze of eutrophic lakes, where oxygen is often depleted, the predominate or only insects are red, hemoglobin-containing larvae of *Chironomus.* These are able to survive the irregular oxygen supply and continue to function as important decomposers. The open waters of eutrophic lakes are inhabited by the phantom midge larvae, *Chaoborus* (Culicidae). As one of the few free-floating insects, the predatory *Chaoborus* spend the day near the bottom and swim toward the surface at night. The phantom larva is nearly transparent—hence the name. Gas sacs are visible at each end and act as organs of equilibrium. The volume of gas is controlled by the nervous system. When the sacs are compressed, the larva sinks; when they are expanded, it rises.

Adult midges emerge synchronously in enormous numbers from productive lakes throughout the world. *Chaoborus edulis* forms spectacular clouds seen at great distances over Lake Malawi in Africa. The emergence of *C. astictopus,* the Clear Lake gnat of Clear Lake, California, creates an annual nuisance to resort owners. When carried over land by winds, the masses of chironomids and *Chaoborus* remove significant quantities of organic matter from the lake ecosystem.

The shallow water of lakes and ponds contains more different kinds of insects than the deeper, open waters. Certain areas approximate the physical and chemical conditions of running water. On the wave-washed, rocky shore of oligotrophic lakes we find some of the same mayflies, stone flies, and caddis flies that are in streams. Stone flies are sensitive to low oxygen and are rare in lakes except in such places. The numbers and variety of insects in shallow water increase with eutrophication. Mayflies, caddis flies, dragonflies, and damselflies live among submerged plants or on the bottom. These insects have gills and do not rise to the surface for air. Also confined to quiet water are the many Hemiptera, Coleoptera, and Diptera that breath air, either directly or in air bubbles. On the surface are the water striders, Gerridae, Veliidae, and marsh treaders, Hydrometridae.

Shallow lakes may be created each year by rain and snow melt. These last but a few weeks or months before the water is lost through evaporation or absorption into the soil. Such bodies of water formed in the spring are called *vernal lakes* or *ponds.* Insect life ordinarily teems in these temporary habitats because predatory fish are absent. Even the small volumes held in tree holes, cavities at the bases of leaves of plants such as the tropical bromeliads, and hoofprints are sufficient for some insects to complete their life. Growth is usually quite rapid. For example, less than 2 weeks is needed for the yellow-fever mosquito, *Aëdes aegypti,* to complete development.

INSECT SCAVENGERS

Many, but by no means all, insects that live in soil and water are scavengers. Perhaps the word "scavenger" is poorly chosen. Does it indicate an animal that eats whatever comes along, one that eats mostly dead things, or one whose diet consists of items so small and miscellaneous that an accounting is difficult? Here we intend all these meanings when we call certain insects scavengers. The term is broad enough to convey the diversity of food that some insects are known to eat while also concealing our ignorance of the exact diets of others.

A complex terminology has developed to describe the diets of scavengers. Some may be *omnivorous,* eating plants or animals whether they are dead or alive. *Saprophagous* scavengers feed on dead organisms and have been subdivided into *xylophagous* forms, boring into and feeding on sound or decaying wood; *phytosaprophagous* forms, feeding on decaying vegetable matter; *scatophagous* or *coprophagous* forms, feeding on feces or dung; and *zoosaprophagous* or *necrophagous* forms, feeding on dead animals. *Detritivores* feed on small bits of animal or vegetable trash. The food of saprophagous insects, however, is not all dead matter, because microbes such as bacteria, fungi, and yeast flourish in such places and are nutritious.

Small living food is taken by *microphytic* insects that selectively eat bacteria, yeasts, fungi, algae, diatoms, lichens, spores, and loose pollen. Animal foods include Protozoa, and small invertebrates and their eggs. *Fungivorous* insects are adapted to feeding exclusively on living and dead fungi. Thus among the insects loosely called scavengers are those with specific choices of food. Major taxa of scavenging and fungivorous insects are listed in Table 12.2.

Saprophagous insects have unity in an ecological sense because they function in the complex world of the decomposers. These should be counted among our most beneficial insects. Intact bodies of dead vascular plants and vertebrates, smaller dead organisms, and excrement are progressively disintegrated. The nutrients in streams and soils are derived from plant material that is decomposed initially by insects and other arthropods. Scavenging species are usually specific to either plant or animal matter. In terrestrial environments the species appear in a predictable sequence as characteristic chemicals are emitted from the decaying organism at each stage of disintegration. Tissue is ingested and partially or completely digested, often with the aid of symbionts or unusual enzymes such as cellulase, keratinase, or collagenase. The feces of scavengers are further decomposed by other scavengers or microbes, thus aiding in nutrient recycling.

EARLY EVOLUTION OF INSECTS IN SOIL AND WATER

It is generally agreed that insects evolved from some form of marine arthropod. Opinions vary, however, as to whether the intermediate steps in this evolution took place in water or on land. In other words, were the first insects orginally aquatic and secondarily terrestrial or vice versa? The weight of evidence favors a terrestrial origin. The tracheal system of insects and their allies seems to have evolved for the transport of air in a terrestrial environment. The aquatic stages of insects possess the same gas-filled tracheae as terrestrial insects. Only by special modifications can the tracheal system function under water. Even then, many insects must return frequently to the surface for air (Chap. 5).

In most groups of aquatic insects the immature stages alone are aquatic. The adults are aerial or terrestrial, returning near the water or reentering it for oviposition. The aquatic Hemiptera and Coleoptera are the only large groups of insects that live submerged in water as adults. Neither has entirely relinquished its terrestrial adaptations: both usually retain functional wings, and most beetles pupate in soil away from the water. Furthermore, only the specialized Aphelocheiridae (Hemiptera) of Europe, and possibly some *Capnia* spp. (Nemouridae, Plecoptera) are able to remain submerged throughout their lives. For these reasons, one must conclude that insects were originally terrestrial and are incompletely adapted to life under water.

During the evolutionary history of insects, aquatic forms have evolved repeatedly, perhaps hundreds of times, from terrestrial ancestors. Often, but not always, the more primitive orders of a group of orders, or the more primitive

TABLE 12.2 Major taxa of scavenging and fungivorous insects[1]

Omnivorous, generally saprophagous, or microphytic insects

- Collembola
- Diplura (Campodeidae)
- Protura (?)
- Archeognatha
- Thysanura
- Ephemeroptera[2,5]
- Blattodea
- Isoptera (some)
- Grylloblattodea
- Dermaptera (some)
- Plecoptera[2,5]
- Orthoptera (Gryllidae)
- Zoraptera
- Psocoptera
- Coleoptera
  - Cantharidae
  - Carabidae (some)
  - Chrysomelidae[2]
  - Cleroidea (many)
  - Colydiidae (many)
  - Cryptophagidae
  - Cucujidae
  - Dryopidae
  - Elmidae
  - Helodidae[2,5]
  - Hydrophiloidea
  - Hydroscaphidae[5]
  - Lathridiidae
  - Limnichidae
  - Nitidulidae
  - Psephenidae[2,5]
  - Staphylinoidea (many)
  - Tenebrionidae (many)
- Mecoptera[2]
  - Bittacidae
  - Choristidae
  - Panorpidae
- Diptera[2]
  - Anthomyiidae (some)
  - Nematocera (many)[5]
  - Stratiomyiidae
  - Syrphidae (*Volucella*)
- Siphonaptera[2]
- Trichoptera (many)[2,5]
- Hymenoptera (Formicidae)

Phytosaprophagous insects

- Isoptera (some)
- Embioptera
- Coleoptera
  - Anthicidae
  - Cerylonidae
  - Corylophidae
  - Hydrophilidae
  - Monommidae
  - Nitidulidae
  - Nosodendridae
  - Ptiliidae
  - Ptinidae
  - Scarabaeidae
  - Silphidae (some)
  - Staphylinidae (some)
  - Tenebrionidae
  - Throscidae
- Diptera[2]
  - Ceratopogonidae (*Forcipomyia*)
  - Chloropidae
  - Coelopidae
  - Dolichopodidae
  - Dryomyzidae
  - Helcomyzidae
  - Heleomyzidae
  - Lauxaniidae
  - Lonchopteridae
  - Muscidae
  - Mycetophilidae
  - Sciaridae
  - Stratiomyidae
  - Syrphidae
  - Tipulidae
- Trichoptera (many)[2,5]

TABLE 12.2 Major taxa of scavenging and fungivorous insects[1] (*continued*)

Scatophagous and zoosaprophagous insects	Xylophagous insects in dead wood
Thysanura (Lepismatidae)[4]	Blattodea (Cryptocercidae)
Isoptera (some)	Isoptera
Hemiptera (Heteroptera)[4]	Coleoptera
Gerridae	Anobiidae
Mesoveliidae	Archostemata
Coleoptera	Bostrichidae
Cleridae	Brentidae
Dermestidae	Buprestidae
Gyrinidae[3,4]	Cephaloidae[2]
Hydrophilidae	Cerambycidae[2]
Nitidulidae	Corylophidae (a few)
Scarabaeidae	Cucujidae
Silphidae	Curculionidae (*Cossonus*)
Staphylinidae	Elateridae (some)
Tenebrionidae (rare)[3]	Eucnemidae[2]
Trogidae	Lucanidae[2]
Mecoptera (Panorpidae)[3,4]	Lyctidae (all)
Diptera[2]	Lymexylidae[2]
Anthomyiidae	Mordellidae[2]
Bibionidae	Oedemeridae[2]
Calliphoridae	Passalidae
Dolichopodidae (some)	Pyrochroidae[2]
Heleomyzidae	Rhysodidae
Lauxaniidae	Scarabaeidae
Muscidae	Scolytidae
Psychodidae	Scraptiidae
Sarcophagidae	Tenebrionidae[2]
Scatopsidae	Diptera[2]
Sepsidae	Tipulidae
Sphaeroceridae	Xylophagidae
Syrphidae	
Lepidoptera[2]	
Pyralidae (*Bradypodicola*)	
Tineidae	
Hymenoptera (Formicidae)	

TABLE 12.2 Major taxa of scavenging and fungivorous insects[1] (*continued*)

Insects infesting stored plant products: grains, nuts, seeds, dried fruits, flour, and other foods

- Thysanura (Lepismatidae)
- Psocoptera (Liposcelidae)
- Coleoptera (Polyphaga)
  - Anobiidae
  - Anthicidae[6]
  - Bostrichidae
  - Bruchidae
  - Cleridae
  - Colydiidae
  - Cryptophagidae[6]
  - Cucujidae
  - Curculionidae
  - Dermestidae
  - Erotylidae[6]
  - Lathridiidae[6]
  - Mycetophagidae[6]
  - Nitidulidae[6]
  - Trogositidae
  - Ptinidae
  - Tenebrionidae
- Diptera
  - Piophilidae[2]
- Lepidoptera[2]
  - Cosmopterygidae
  - Gelechiidae
  - Oecophoridae
  - Pyralidae
  - Tineidae
  - Tortricidae
- Hymenoptera
  - Eurytomidae[2]
  - Formicidae

Fungivorous insects

- Isoptera (Macrotermitinae)
- Zoraptera
- Hemiptera
  - Aradidae
  - Derbidae
- Thysanoptera (some)
- Coleoptera (Polyphaga)
  - Anthribidae
  - Cucujoidea (many)
  - Derodontidae
  - Leiodidae
  - Nosodendridae
  - Trogositidae
  - Ptiliidae
  - Scaphidiidae
  - Scarabaeidae (a few)
  - Scolytidae
- Diptera[2]
  - Cecidomyiidae (some)
  - Ceratopogonidae
  - Drosophilidae
  - Mycetophilidae
  - Platypezidae
  - Sciaridae
- Lepidoptera[2]
  - Lycaenidae (a few)
  - Lyonetiidae
  - Oecophoridae
  - Tineidae
- Hymenoptera
  - Formicidae (*Atta*)
  - Siricidae[2]

[1] This list of taxa is intended to be representative, not exhaustive. The listing of a family without qualification does not necessarily mean that all species in the family have the same food habits. Unless otherwise indicated, both the immature and adult stages are believed to have the same food habits. Well-known or exceptional genera are given in parenthesis.

[2] Feeding on this food is mainly by immature stages.

[3] Feeding on this food is mainly by adults.

[4] Food is mainly dead insects.

[5] Algae are probably important in diet.

[6] Fungi are important in diet.

families in an order, are aquatic. Life in water offers certain advantages, especially for the immature stages. They are sheltered from desiccation and freezing and protected from certain natural enemies such as parasitoids. The aquatic habitat was apparently colonized early during each new evolutionary radiation of insects, and the aquatic forms tended to outlast their terrestrial progenitors. Hence the more primitive forms are well represented in the aquatic environment because they survived there, not because insects originated in water.

Turning now to the early terrestrial environment, the first episode of plant evolution on land began in the late Silurian Period (Table 17.3) and ended with the close of the Devonian Period. During this relatively short span of time, a rapid succession of vascular plants appeared: the Subdivisions Psilopsida, Lycopsida, Sphenopsida, and the ferns or Pteropsida. The diversity of this early flora can be estimated by the 83 "genera" of spores that have been described (Chaloner, 1970). When the sporophyte plant was preserved, it was sometimes treelike. The Devonian *Pseudosporochnus* reached 2 m or more in height. All these plants were confined to wet places because the flagellated sperm of the gametophyte generation required water in which to swim to the egg cells.

The fossil record indicates that a variety of aquatic mollusks, worms, and arthropods could have served as decomposers when plants fell in the water. But on land, few animal fossils exist to reconstruct the ancient food webs of the early Paleozoic Era. The moist, decaying plant matter would have been a suitable place for arthropods to make the transition from aquatic to terrestrial life. Here was an uncontested food supply, including spores and microbial life as well as protection from desiccation.

The fossils that do exist from this time support this line of reasoning. Predatory scorpions are represented in both the Silurian and Devonian Periods. The earlier forms may have been aquatic, but the later scorpions were terrestrial. A myriapod, *Archidesmus,* is known from the Silurian and may have been a terrestrial scavenger. In the Devonian a mite, *Protacarus,* and a few small arachnids have been found. The mites were probably microphytic and the arachnids predatory. Thus some evidence exists that arthropod scavengers and predators colonized the land as early as the Silurian and Devonian. Furthermore, they probably lived in litter. Certain protective characteristics of the spores of terrestrial plants and fungi that lived at this time indicate that they may have been fed upon and dispersed by arthropods. Fragments of arthropods have also been found inside a fossil stem fragment and inside fossil sporangia, further supporting the idea that arthropods probably thrived in litter (Kevan et al., 1975).

Definite insect fossils are not known from the Devonian, yet about 30 million years later an abundance of varied insects are found in strata of the Upper Carboniferous Period. It seems most probable that insects evolved as scavengers in plant litter and made their appearance at least by the close of the Devonian. The Apterygota are probably representative of the early insects, and it is not coincidental that they are scavengers in litter today. Other survivors of the same early radiation of terrestrial arthropods include the Pauropoda, Symphyla, Collembola, Diplura, Protura, Chilopoda, Diplopoda, and the arachnids Pseudoscorpionida, Scorpionida, Palpigrada, Podogona (Ricinulei), and the more primitive Acarina and Phalangida. These groups are also found today mostly in moist soil and litter.

EVOLUTION OF SCAVENGERS. As the diet of Paleozoic insects became predominantly living prey, blood, fungi, or green plants, scavengers became predators, parasites, fungivores, or herbivores respectively. Most evolutionary lineages in insects seem to be from scavenging to other feeding habits. Except for the mouthparts of larval and adult Diptera, the mouthparts of the scavenging taxa are of the chewing type. This versatile design is readily modified into more specialized feeding structures.

In the Carboniferous Period, the amount of food for scavengers increased enormously. A rise in the carbon dioxide content of the atmosphere is thought to have created a "greenhouse effect" and produced a warm, moist climate. The increased availability of carbon dioxide plus a favorable climate resulted in an increase in photosynthetic activity and the explosive growth of land plants. The dominant lycopod, *Lepidodendron,* reached over 40 m in height. The rapid accumulation of dead plants created the great coal deposits upon which we now depend for fuel. The animal decomposers preserved as fossils were insects, mites, millipeds, and giant arthropods (now extinct) of the Class Arthropleurida. The last were shaped like millipeds and measured up to 2 m in length. They were probably the world's largest exclusively land arthropods and a dramatic measure of the amount of food available to scavengers. Other decomposers, such as pulmonate snails and terrestrial isopods, were apparently absent.

At this time insects evolved wings and the Paleoptera made their appearance (Chap. 17). Adult Paleoptera, with their outstretched wings, would have been prevented from returning to life among the crevices in the litter. The nymphs were probably less restricted, and those with chewing mouthparts probably continued as scavengers. The scavengers and herbivores were prey to increasing numbers of predators. These now included amphibians and the first reptiles, as well as the first spiders and other arachnids, predatory insects, and possibly chilopods. The pressure to escape terrestrial predation may be responsible for the origins of the aquatic nymphs in the Odonata and Ephemeroptera. The latter continued their scavenging habits in their new environment.

The evolution of the wing-folding mechanism permitted the entire life cycle of the early Neoptera to be completed in or near the litter. The success of this adaptation can be judged by the prevalence of scavengers among the early Neoptera that survive today (Table 12.2). Furthermore, the first scavenging Neoptera to be fossilized were the Blattodea. They were so abundant that the period is often called the "Age of Cockroaches."

The Permian Period was characterized by geologic and climatic disturbances, often violent. Some areas were glaciated, some became windblown deserts, and in other regions thick salt deposits were laid down from drying seas. One can imagine the stresses placed on the Carboniferous insect fauna that had become adapted to moist, warm forests over a period of 80 million years. Besides struggling to survive under new climatic regimes and amidst the competition caused by shrinking forests, they were confronted with an increasing variety of reptiles which doubtless included insectivores.

During this period of changing and adverse conditions, the holometabolous life cycle was evolved. The larvae of scavengers were able to bore into decaying tissues, or into soil, thus gaining increased protection from desiccation and predators, as well as access to new food sources. Among the first scavenging endopterygotes to be preserved as fossils are Coleoptera of the Family Cupesidae. This family survives today (Fig. 39.3*a*), and the larvae bore in decaying wood.

Through the Mesozoic and into the Tertiary, the scavenging endopterygotes became specialists on larger, dead organisms and their wastes. Wood and decaying foliage of the gymnosperms and angiosperms were penetrated. Alone among the exopterygote orders, the Isoptera were also able to penetrate sound wood. The dead bodies of reptiles and later mammals, and their excrement, were invaded by insects able to feed and respire in semiliquid media. The higher metabolic activity of mammals was accompanied by higher food intake and increased quantities of feces. Dead animals and mammalian feces as habitats are exclusively occupied by endopterygotes, especially Coleoptera and Diptera.

Scavengers in the nests of birds and mammals became scavenging ectoparasites such as the Phthiraptera (Suborder Mallophaga), and some became obligate bloodsuckers. Certain dung-feeding maggots became myiasis-producing parasites (discussed in Chap. 15). Insects feeding on seeds stored by mammals or the provisions of presocial and eusocial insects later became pests of human granaries and pantries.

SELECTED REFERENCES

Soil insects are treated by Kevan (1962), Kühnelt (1961), R. F. Lawrence (1953), Lee and Wood (1971), and Wallwork (1970). Cloudsley-Thompson (1975) and Edney (1974) discuss adaptations of insects to deserts. For ecology and geography of cave insects, see Barr (1967), Mitchell (1969), and Vandel (1965). The ecology of Collembola, a dominant group in soil and caves, is reviewed by Butcher et al. (1971) and Christiansen (1964). Texts on aquatic ecology and insects include Amos (1967), Cheng (1976), Hynes (1970*a*), Niering (1966), and Usinger (1956*a* and *b*, 1967). Reviews of aquatic ecology and physiological adaptations are provided by Bay (1974), Cummins (1973), Hynes (1970*b*), Miller (1974*b*), Nachtigall (1974*b*), Neumann (1976), Waters (1972), and Wilhm (1972). For references on specific soil and aquatic taxa, consult the appropriate chapters in Part IV.

13  INSECTS AND VASCULAR PLANTS

Approximately half the living species of insects are *phytophagous,* i.e., they feed on living, green plants. Although primitive kinds of vascular plants figured significantly in the early evolution of insects and will be discussed, we will be concerned mainly with the seed plants, or Spermopsida. It is here that we find the most diverse and intimate insect/plant relationships. Here also are the conifers and hardwoods that supply wood products, and the flowering plants that yield food and fibers for human use.

Insects and plants have evolved together from near the beginning of terrestrial life. The dominant plants today, the Angiosperms or flowering plants, owe their origin to the feeding behavior of Mesozoic insects that functioned as pollinators. Insects now are the most important primary consumers of land plants, easily exceeding vertebrate herbivores and competing directly with humans. The impact of insect feeding is checked by natural enemies and ameliorated by varied plant defenses.

Phytophagy has evolved repeatedly from other food habits, usually scavenging. Some whole orders and many families are almost exclusively phytophagous. Insects in the six largest orders (Fig. 1.1) derive much or most of their food from plants. In the following sections we will examine in turn the evolution of phytophagous insects and details of their relationships with their plant hosts. These relationships, both beneficial and detrimental to plants, are envisioned as coevolutionary, i.e., they involve the interaction of two or more kinds of organisms whose survival is intertwined to an extent that evolutionary changes in one affect the evolution of the others.

PHYTOPHAGOUS INSECTS IN THE PALEOZOIC ERA

In the previous chapter we explained that the first insects were probably scavengers in the moist debris that accumulated beneath early land plants. Living plant tissues, i.e., reclining stems and rhizomes along the ground, were also within range of the litter dwellers. As plants evolved greater height and tree-like forms in the Devonian Period, (Table 17.3) a new habitat was created for terrestrial animals. The spores of these plants may have been an important item in the diet of insect scavengers. Measuring less than 200 μm in diameter, spores of the Lower Devonian plants could have been easily ingested when found singly or in windrows on the ground. At the tops of the sporophyte plants spores also could be found fresh and concentrated in exposed sacs or sporangia.

An arboreal insect faces greater risk of desiccation than an insect that lives in moist litter; furthermore it must cling, while walking, to smooth and sometimes vertical surfaces, and it risks greater exposure to predators and parasitoids. Exposure to aboreal or aerial predators, however, was not a problem for Devonian insects because none existed. Nor were any other organisms in competition for food borne high on erect plants. It is likely, therefore, that some of the insect scavengers acquired the resistance to water loss and the tarsal modifications necessary to climb Devonian plants. The shift from scavenging to feeding on vegetative parts of plants also probably required physiological adjustments to the new diet.

The Carboniferous swamp forests would have provided an abundant food supply for phytophagous insects. Among the fossils preserved in the Upper Carboniferous Period are insects already highly specialized for external plant feeding. The paleopterous orders Palaeodictyoptera, Megasecoptera, and Diaphanopterodea had mouthparts prolonged in a beak (Fig. 17.2*b*). The beak seems most suited for feeding on plants, but it is a matter of speculation whether the insects took tissue fluids by piercing and sucking (Carpenter, 1970), or probed for spores, pollen, and seeds in reproductive cones (Smart and Hughes, 1973). The abrupt decline of the plant taxa that dominated the Carboniferous Period may explain the extinction of these beaked Paleoptera.

Direct feeding on vascular phloem probably began with the evolutionary appearance of leaves. Feeding on stems probably did not take place until after the appearance of the arborescent gymnosperms known as the Cordaitales. The cambium and phloem of these plants were close to the surface and accessible by piercing mouthparts. The stems of other common plants had a thick cortex around the vascular tissue. The evolution of hemipterous piercing-sucking mouthparts, therefore, is correlated with the increasing availability of phloem tissue in the Carboniferous. Numerous fossils of Hemiptera (Suborder Homoptera) are known in the next period, the Permian. These are the oldest surviving insects that are exclusively phytophagous.

When the numbers of species of beaked Paleoptera and Homoptera are compared with the number of other fossil insects, Carpenter (1970) estimates that nearly half the Paleozoic insects had piercing-sucking mouthparts. This is some measure of the amount of plant food available and the extent of its utilization by insects. The proportion of insects that are phytophagous has apparently remained approximately the same up to the present.

The feeding habits of Paleozoic insects with chewing mouthparts are less easily associated with plant feeding. Some may have been predators or scavengers. The jumping Orthoptera were present in the Permian Period and probably were mostly phytophagous, as they are today. The first possible evidence of insect damage to leaves was found in early Permian rocks in South Africa. Leaves of the ancient fern *Glossopteris* were discovered with marginal scallops resembling the notches made by edge-feeding, chewing insects today.

Fleshy fruitlike or berrylike reproductive structures and nutlike seeds of gymnosperms have been found in the Lower Permian. The fleshy fruits were probably eaten and dispersed by reptiles which then dominated the vertebrate scene. These fruits were probably also a source of food for insects long before the origin of the fruits and nuts of angiosperms.

Recall that the first endopterygote insects appear in the Permian. Among other advantages, larval insects were able to penetrate the tissues of plants and fleshy fruits for the first time. We do not know when mining and boring in living plants began. The earliest insects capable of such activity may have been Coleoptera, which were present in the Permian and which today exhibit these habits. The miners and borers may have evolved repeatedly from scavengers that bored in dead wood or decaying vegetable matter, or from external phytophagous forms that extended their feeding into the plant from the surface. Today the endopterygotes are the most numerous of the phytophagous insects.

THE FLOWER-VISITING INSECTS

Angiosperms are the dominant land plants today. It is generally agreed that they owe their origin and much of their diversification directly to the behavior of insects. The earliest flowering plants are known only from pollen grains. Like the fossil spores of the Paleozoic, pollen is more readily preserved than plant fragments. The pollen of gymnosperms and that of early angiosperms are so much alike that no sharp distinction can be made. The first fossil grains with predominantly angiospermous features occur near the end of the Lower Cretaceous Period. Later, at the close of the Cretaceous, angiosperm pollen exceeds that of gymnosperms and fern spores. The evolution of angiosperms was so rapid that an astonishing 67 families are represented at this time. What are some of the events which led to the origin of this successful group?

During the evolution of the gymnosperms, fertilization by swimming sperm was replaced by the growth of a pollen tube. Although the need for moisture or special fluids at fertilization was thus eliminated, and drier regions became habitable, the pollen grains must lodge in contact with the ovule. Sexual outcrossing among gymnosperms is assured because individual plants are unisexual, producing either pollen or ovules, but not both. Large quantities of wind-borne pollen are needed in order for a small fraction to land by chance in the correct spot for fertilization. The transfer of pollen from male to female structures is called *pollination.*

Large pollen sacs, seeds, and other edible tissues probably attracted insects to the cones of Paleozoic gymnosperms. A pollen-eating insect that moved only among male cones would not bring about pollination. Bisexual cones having both sexual organs, however, would be suited to benefit from such an insect. Attractive food would be combined with the receptive ovules. An insect would be able to transport pollen from the male organs to the female organs, increasing the likelihood of correct placement of pollen of the same species next to the ovules. The total amount of pollen needed would then be greatly reduced. Such an insect is called a *pollen vector.*

The first pollen vectors may have included terrestrial or flying reptiles or early birds, but the most important were probably the flying insects. Among these, the Coleoptera are thought to be the most significant. They were probably well diversified in the Mesozoic Era. Interest in them is heightened by the fact that a number of primitive angiosperms today are beetle-pollinated. However, many beetles consume the ovules of the plants that they pollinate. For this reason, certain flower structures such as carpels can be explained as defensive measures initially evolved against the powerful, chewing jaws of these insects (Takhtajan, 1969).

Other features seem designed to aid pollinators in locating and recognizing flowers. The colored petals aid visual recognition and orientation. Odors are emitted which are attractive to insects at a distance. The first floral odors may have imitated odors of fruits or decay that were attractive to scavenging beetles. Thus the early flowers presumably had both pollen and ovules, showy petals and odors, and their pollinators came mainly for pollen.

The addition of nectaries, (glands that secrete nectar) probably came after beetles had established insect pollination as a regular part of angiosperm reproduction. Primitive beetle flowers generally lack floral nectar. Nectaries are lacking among all gymnosperms, but are present on new fronds of the fern *Pteridium.* A small amount of sugary fluid is secreted from the ovule of certain gymnosperms as a part of the pollination process, but this may or may not be significant in the evolution of angiosperm nectaries.

Nectar is an aqueous fluid rich in sugars. Recently Baker and Baker (1975) demonstrated the presence of other nutrients of value to pollinators: amino acids, proteins, and lipids. Other substances include ascorbic acid, possibly serving as an antioxidant, and alkaloids, which might be toxic to certain unwanted flower visitors. Nectaries are associated with the vascular phloem system of plants.

The first nectaries of angiosperms may have been outside the flower, i.e., extrafloral nectaries, and they may have served a role different from that of floral nectaries. Although nectaries were rare before the appearance of angiosperms, insects had access to a fluid of comparable composition beginning at least in the Permian, if not earlier. This is the honeydew excreted by the phloem-feeding Homoptera. The Homoptera living today are wasteful feeders, ejecting the phloem sap largely unaltered and in quantity. Among the insects attracted to honeydew are natural enemies of phytophagous insects such as ants, predatory and parasitoid wasps, and lacewings, as well as other insects such as bees and moths. In some cases this is a regular or major part of their diet.

Fossil ants are known from the Cretaceous, and we conjecture that they were as fond of honeydew as are their descendants (Endpaper 1*b*). Ants and certain Homoptera have evolved mutualistic associations in which the ants protect the Homoptera from natural enemies in return for honeydew. The host plants also benefit by the feeding of ants on phytophagous insects (Janzen, 1967*b*; Bentley, 1977). Even vertebrate herbivores are discouraged by ants on foliage. The first nectaries, therefore, may have been the plants' device for supplying imitation honeydew as an attractant for beneficial ants and other predaceous insects, without the injury of phloem-feeding Homoptera.

The secretion of nectar inside the flower was an incentive to actively flying insects in need of carbohydrate fuel such as Lepidoptera (Endpaper 1*a*), Diptera, and Hymenoptera. Floral nectar differs in composition from both honeydew and extra-floral nectar in ways that suggest it is secreted expressly for the food needs of favored pollinators. Advanced families of plants have higher concentrations of amino acids than primitive families, and butterfly-pollinated flowers have higher concentrations than bee-pollinated flowers. This is associated with the inability of most butterflies to ingest protein-rich pollen, whereas bees obtain their amino acids from pollen. *Heliconius* butterflies, however, collect pollen on their galeae and digest it there, taking up free amino acids (Gilbert, 1972). Flies which feed on protein-rich dung are attracted to the nectars of fly-pollinated flowers that are high in amino acids (Baker and Baker, 1973).

Now let us turn our attention to the insect visitors of flowers. The most common pollinators are Coleoptera, Lepidoptera, Diptera, and Hymenoptera (Table 13.1). They have several features in common: all are actively flying adults of neopterous, endopterygote insects. Their search for mates, oviposition sites, and plant or animal food is aided by a strong flight apparatus, highly developed senses, and, in some groups, learning ability. These same attributes aid pollinators as they search for and remember flowers.

Individual insects which visit flowers of the same plant species during a single flight or longer period are said to be *flower-constant.* When all individuals of an insect species are restricted to visiting a single species of plant for food (nectar, pollen, other substances), the insect is said to be *monotrophic.* If several, possibly related, plant species are visited the insect species is *oligotrophic,* and if many are visited, the term *polytrophic* applies. An individual bee may be flower-constant to each of a series of plant species during successive time periods and be a member of a polytrophic species. When visits are for pollen, the terms *monolectic, oligolectic,* and *polylectic* are used.

Flower constancy is beneficial to both plants and insects. It is advantageous for a plant to attract flower-constant visitors because they are the most effective cross-pollinators. It is advantageous to insects to become temporary or permanent specialists because they reduce competition for food and forage more efficiently, learning to recognize a given flower and operate its floral mechanism.

In a general way, the flower size, shape, position of reproductive parts, color patterns, odor, nectar composition, and time of flowering can be matched with the sizes, anatomies, diets, sensory physiologies, rhythmic activity, and foraging behaviors of its pollinators. Even among related species of plants, different species may depend on quite different kinds of pollinators. This specificity attracts effective pollinators and tends to reduce losses of pollen and nectar to nonpollinating visitors. The reward given to pollinators is thereby more closely regulated to promote cross-pollination.

Commonly nectar is situated deep within a floral tube so that casual visitors are unable to

TABLE 13.1 Major taxa of terrestrial phytophagous insects[1]

External feeders on foliage, stems, roots, fruits, and/or seeds

Exposed feeders
- Isoptera (some)
- Dermaptera
- Plecoptera (some)[3]
- Orthoptera
  - Acrididae
  - Gryllidae
  - Gryllotalpidae
  - Tettigoniidae
- Phasmatodea (all)
- Hemiptera (Heteroptera)
  - Coreidae
  - Largidae
  - Lygaeidae[4]
  - Miridae
  - Pentatomidae
  - Piesmatidae
  - Pyrrhocoridae[4]
  - Tingidae
- Hemiptera (all Homoptera)
- Thysanoptera (most)
- Coleoptera (Adephaga)
  - Carabidae (some)
- Coleoptera (Polyphaga)
  - Anthicidae (some)[3]
  - Anthribidae (some)[3]
  - Byrrhidae (some)[3]
  - Byturidae
  - Cantharidae (some)[3]
  - Cerambycidae (some)[3]
  - Chrysomelidae
  - Coccinellidae (*Epilachna*)
  - Curculionidae
  - Elateridae (some)[3]
  - Meloidae (some)[3]
  - Scarabaeidae (Melolonthinae, Rutelinae)[3]
- Lepidoptera[2]
  - Bombycoidea
  - Geometroidea
  - Hesperioidea
  - Noctuoidea
  - Notodontoidea
  - Papilionoidea
  - Sphingoidea
- Hymenoptera (Symphyta)[2]
  - Tenthredinoidea

Leaf rollers and shelter makers[2]
- Coleoptera
  - Attelabidae

External feeders on foliage, stems, roots, fruits, and/or seeds

- Lepidoptera
  - Gelechiidae
  - Gracilariidae
  - Lasiocampidae
  - Pyralidae
  - Tortricidae (Tortricinae)
  - Yponomeutidae
- Hymenoptera (Symphyta)
  - Megalodontidae
  - Pamphiliidae

Case bearers[2]
- Coleoptera (Polyphaga)
  - Chrysomelidae (Clytrinae, Cryptocephalinae)
- Lepidoptera
  - Coleophoridae
  - Incurvariidae
  - Psychidae
  - Tineidae

Open galls
- Thysanoptera (some)
- Hemiptera (Homoptera)
  - Aphididae (some)

Internal feeders on foliage, stems, and/or roots[2]

Borers
- Coleoptera (Polyphaga)
  - Brentidae
  - Buprestidae
  - Cerambycidae (some)
  - Curculionidae
  - Languriidae
  - Platypodidae
  - Scolytidae
- Diptera
  - Agromyzidae
  - Anthomyiidae
  - Chloropidae
  - Ephydridae
- Lepidoptera
  - Cossidae
  - Hepialidae
  - Noctuidae
  - Pyralidae
  - Sesiidae
  - Tortricidae (Olethreutinae)
- Hymenoptera (Symphyta)
  - Cephidae
  - Siricidae (some)
  - Syntexidae
  - Xiphydriidae

TABLE 13.1 Major taxa of terrestrial phytophagous insects[1] (*continued*)

Internal feeders on foliage, stems, and/or roots[2]

Leaf miners
- Coleoptera
  - Buprestidae
  - Chrysomelidae
  - Curculionidae
- Diptera
  - Agromyzidae
  - Anthomyiidae
  - Cecidomyiidae
  - Chironomidae
  - Drosophilidae
  - Ephydridae
  - Lauxaniidae
  - Psilidae
  - Sciaridae
  - Syrphidae
  - Tephritidae
- Lepidoptera
  - Coleophoridae
  - Cosmopterygidae
  - Cycnodiidae
  - Elachistidae (some)
  - Eriocraniidae
  - Gracilariidae
  - Heliodinidae
  - Heliozelidae
  - Incurvariidae
  - Lyonetiidae
  - Nepticulidae
  - Noctuidae
  - Opostegidae
  - Pyralidae
  - Tischeriidae
  - Tortricidae (a few Oleuthreutinae)
  - Yponomeutidae (Argyresthiinae)
- Hymenoptera
  - Argidae
  - Tenthredinidae

Closed galls
- Coleoptera
  - Buprestidae
  - Cerambycidae
  - Curculionidae
- Lepidoptera
  - Cosmopterygidae
  - Gelechiidae
  - Tortricidae (Olethreutinae)
- Diptera
  - Agromyzidae
  - Cecidomyiidae

Internal feeders on foliage, stems, and/or roots[2]

- Hymenoptera
  - Cynipidae
  - Eurytomidae
  - Tenthredinidae

Flower feeders

Flower-tissue feeders[3]
- Coleoptera (Polyphaga)
  - Anthribidae
  - Buprestidae
  - Cantharidae
  - Cerambycidae
  - Chrysomelidae
  - Curculionidae
  - Elateridae
  - Meloidae
  - Melyridae
  - Nitidulidae
  - Scarabaeidae
- Lepidoptera
  - Lycaenidae (some)[2]

Pollen feeders
- Collembola (some)
- Blattodea (some)
- Dermaptera (some)
- Plecoptera (some)[3]
- Hemiptera (Heteroptera)
  - Anthocoridae
  - Miridae
- Thysanoptera (some)
- Coleoptera (Polyphaga)
  - Cephaloidae[3]
  - Meloidae[3]
  - Mordellidae[3]
  - Nitidulidae
  - Oedemeridae[3]
  - Phalacridae[3]
- Diptera (probably many)[3]
  - Anthomyiidae
  - Bibionidae
  - Bombyliidae
  - Calliphoridae
  - Mycetophilidae
  - Muscidae
  - Scatopsidae
  - Syrphidae
  - Tachinidae
- Lepidoptera[3]
  - Micropterygidae
  - Nymphalidae (*Heliconius*)[5]

TABLE 13.1 Major taxa of terrestrial phytophagous insects[1] (*continued*)

Flower feeders

- Hymenoptera
  - Apoidea[6]
  - Vespidae (Masarinae)[6]
  - Xyelidae[2]

Nectar feeders[3]

- Neuroptera
  - Chrysopidae
- Mecoptera
  - Panorpidae (*Panorpa*)
- Diptera (many)
- Lepidoptera (most)
- Trichoptera (some)
- Hymenoptera (most)[6]

Internal feeders in fruits and/or seeds on living plants[4]

- Coleoptera (Polyphaga)
  - Bruchidae
  - Byturidae[2]
  - Chrysomelidae
  - Curculionidae
  - Nitidulidae
  - Scarabaeidae[3]
- Diptera[2]
  - Cecidomyiidae (*Contarinia*)
  - Tephritidae (*Ceratitis, Rhagoletis*)
- Lepidoptera[2]
  - Gelechiidae (*Pectinophora*)
  - Incurvariidae (*Tegeticula*)
  - Lycaenidae (*Strymon*)
  - Noctuidae (*Heliothis*)
  - Nolidae (*Celama*)
  - Pyralidae (*Ostrinia*)
  - Tortricidae (*Laspeyresia*)
- Hymenoptera[2]
  - Agaonidae (*Blastophaga*)
  - Eurytomidae (*Bruchophagus*)

[1]This list of taxa is intended to be representative, not exhaustive. The listing of a family without qualification does not necessarily mean that all species in the family have the same food habits. Unless otherwise indicated, both the immature and adult stages are believed to have the same food habits. Well-known or exceptional genera are given in parenthesis.

[2]Feeding is mainly by larvae.

[3]Feeding is mainly by adults.

[4]Feeding is mainly on seeds.

[5]*Heliconus* adults ingest nutrients from pollen that have been dissolved in nectar.

[6]Adult females of Apoidea and Vespidae (Masarinae) also store pollen and nectar in their nests as food for their larvae.

reach it. Elongation of the mouthparts into a sucking tube is consequently a frequent adaptation among specialized flower-visiting insects. The elongation is achieved in many ways and often independently (Fig. 13.1). Curiously, the Hemiptera, with their long sucking beaks, never became regular flower visitors for nectar.

The largest group of efficient pollinators are the bees, or Apoidea. About 20,000 species are known in the world. Bees evolved from visually hunting, predaceous wasps that frequently visit flowers for nectar. Both sexes of bees take nectar as flight fuel. Females eat pollen as a source of the protein used in producing their eggs. All females, except the cleptoparasites and social queens, also collect pollen and nectar for their larvae. Either this is stored in the nest as provisions, or, in certain social species, the pollen is fed directly to the larvae.

The adaptations of bees that are associated with flower visitation are plumose or featherlike hairs; special pollen-transporting devices; modifications of the tongue, or glossa, for extracting nectar; a diet of pollen and nectar; and, in the honeybees, a highly developed system for communication. A few species of bees are monotrophic, but most bees are oligotrophic or polytrophic. Among the latter are honeybees which visit an enormous variety of flowers, including those designed to attract insects other than bees. During times of food scarcity, honeybees collect honeydew or fruit juices as nectar substitutes. When pollen is scarce, honeybees have been observed to collect flour or even inert dust.

Certain primitive bees, such as *Hylaeus* (Colletidae), are relatively hairless, like wasps. They eat pollen and later regurgitate it with nectar while preparing the nest provisions. But most bees have abundant, plumose hairs which retain pollen grains brushed on the body during flower visits. Female bees methodically groom themselves and pack the pollen into special devices for transport to the nest. These are of two kinds: (1) pollen brushes, or *scopae,* of long, dense hairs on the hind legs of most bees or on the underside of the abdomen in Megachilidae;

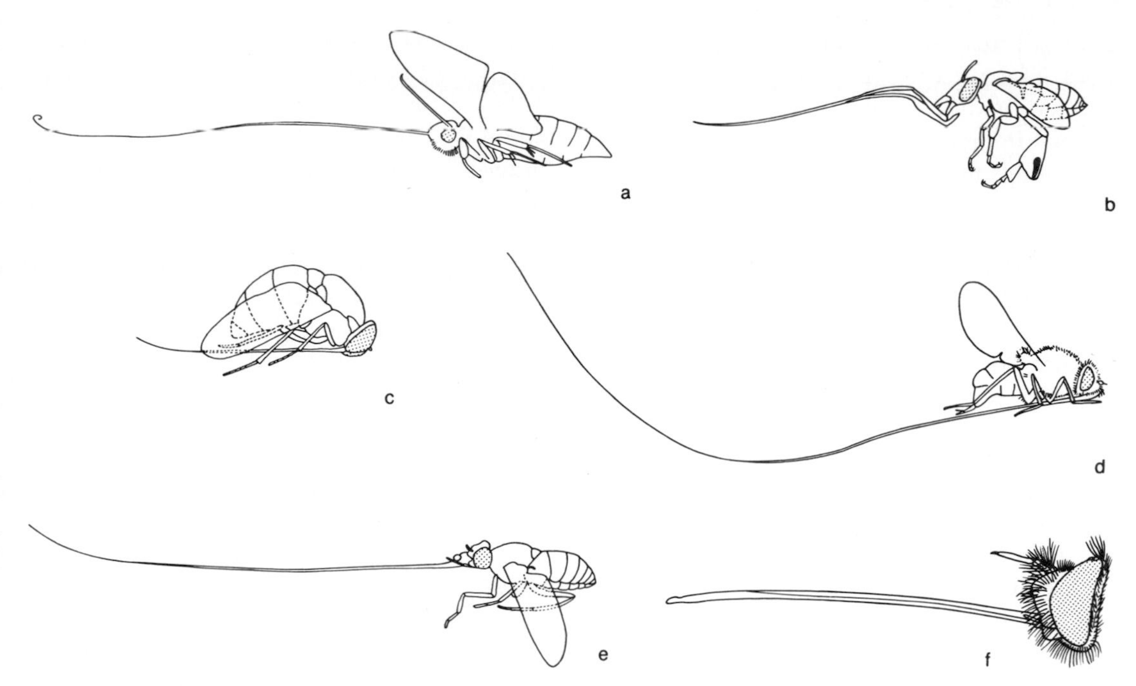

FIGURE 13.1 Extreme modifications of mouthparts of flower-visiting insects (length of proboscis in parentheses): *a,* sphingid moth, *Manduca quinquemaculata*—proboscis (12 cm) is coiled under head when not in use; *b,* male apid bee, *Euglossa asarophora*—proboscis (3 cm) is folded between legs when not in use; *c,* acrocerid fly, *Lasia kletti*—proboscis (2.4 cm) is swung forward in use; *d,* nemestrinid fly, *Megistorrhynchus longirostris*—proboscis (8.4 cm) is swung forward in use; *e,* tabanid fly, *Philoliche longirostris*—proboscis (5.8 cm) is carried in flight as shown; *f,* head of bombyliid fly, *Bombylius lancifer*—proboscis (9.7 mm) is carried as shown.

and (2) pollen baskets, or *corbiculae,* which are created by a circle of stiff hairs on the outer surface of the hind tibiae of honeybees, bumblebees, and their relatives. Andrenidae, Colletidae, and Halictidae have short tongues suited to take nectar from exposed nectaries or flowers in which the bee can bodily enter. These are considered less specialized than the long-tongued Megachilidae, Anthophoridae, and Apidae, which can reach nectar hidden in the inner recesses of specialized flowers. The expansible crop carries nectar back to the nest and also functions on outbound flights as a fuel tank.

"Bee flowers" characteristically open at certain times during the day, emitting sweet or aromatic odors and presenting their pollen and nectar. Petals of bright blue, purple, yellow and other colors within the bee's range of color vision are common. Recall that red is invisible to bees; it is an uncommon color of bee flowers. Patterns reflecting ultraviolet are seen by the bee but are invisible to us. Separate petals which create a broken outline are suited to detection as a mosaic image by the bee's compound eyes. The two-lipped form of certain bee flowers, such as those of legumes or mints, provides a landing platform. This places the bee in a position favoring access to the food and pollination. Distinctive stripes or spots serve as nectar guides which orient the bee to the food.

Two other large groups of pollinating insects show conspicuous modification for flower visiting: certain Diptera and most Lepidoptera. In both orders, the special modifications are mainly elongation of the mouthparts to reach hidden nectar. The mouthparts of butterflies and moths are suited only to sucking fluids. Most species take nectar, but some moths do not feed at all as adults. The longest tongue of any insect is the 22.5-cm proboscis of the Madagascar hawkmoth,

Xanthopan morgani praedicta (Sphingidae). The moth is apparently the sole pollinator of the orchid *Angraecum sesquipedale,* a plant with nectar situated in an equally long tube and accessible only to this moth.

Although the mouthparts of flies are variously modified for blood-sucking or sponging, the general ability seems to be retained to ingest nectar and small particles such as pollen grains. Certain species in each of several families of flies have developed exceptionally long mouthparts for probing deep flowers. Some species with long mouthparts are found in Bombyliidae, Apioceridae, Nemestrinidae, Acroceridae and Tabanidae (Fig. 13.1). Shorter, but distinctly specialized, mouthparts are also seen on species of *Rhingia* (Syrphidae), Conopidae, and Tachinidae.

The long-tongued flies visit the same kinds of flowers that bees visit. The so-called "fly flowers" are mainly adapted to less specialized, short-tongued insects which normally feed on fluids from dead animals, feces, or plant juices. The flowers depend more on odor than their appearance to attract these insects. The shallow flowers are often white or dull-colored; the nectar is exposed; and the smell is often musty or rank.

Flowers attractive to butterflies open during the day, have sweet odors, and often have nectar at the base of a deep tube. The flowers are erect and have a horizontal surface for landing. Red is visible to butterflies and is a common color of "butterfly flowers," such as carnations or the butterfly weed, *Asclepias tuberosa.* "Moth flowers" bloom in the evening or night; most of them have heavy, sweet scents; and they are often a highly visible pale or white color. Hawkmoths (Sphingidae) characteristically hover in front of a flower and take nectar by extending their long probosces. Flowers visited by hawkmoths are horizontal or drooping, with the reproductive parts situated to contact the hovering moth.

The Coleoptera which visit flowers are active fliers that frequent open, sunny places, in contrast to their terrestrial, cryptic relatives. Beetles tend to linger in flowers, feeding on pollen and flower parts with their powerful mouthparts, and sometimes taking nectar. Elongation of the mouthparts has occurred in only a few cases. The heads of the cerambycids *Strangalia* spp. and *Cyphonotida laevicollis* (Fig. 13.2*b*) are prognathous, with the anterior portions somewhat elongated.

The most spectacular adaptations of this sort in the Coleoptera are the greatly elongated galeae of *Nemognatha* (Meloidae), which form a sucking tube (Fig. 13.2*a*).

Turning now to certain examples of extreme interdependence between plants and their polli-

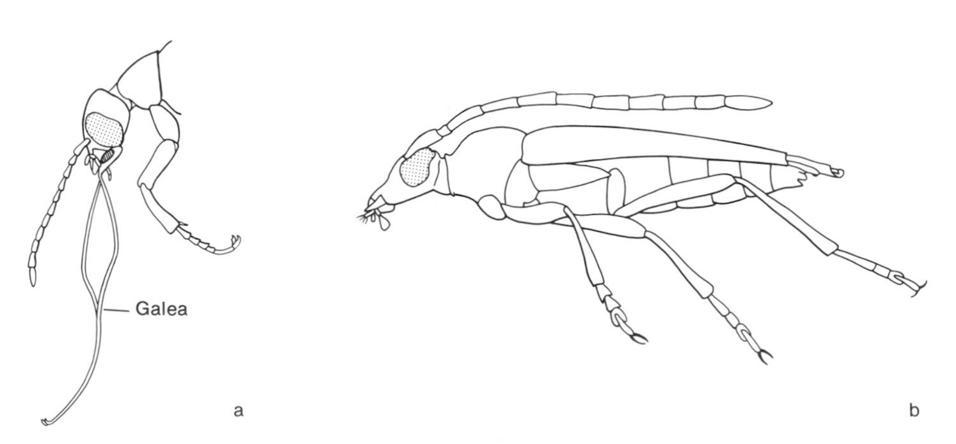

FIGURE 13.2 Adaptations of flower-visiting Coleoptera: *a,* meloid beetle, *Nemognatha* sp., with elongate galeae 7.5 mm long; *b,* cerambycid beetle, *Cyphonotida laevicollis,* with prognathous elongate head.

nators, we find in these associations that pollen and nectar are not the prime incentives for the insects. The host specificity is compounded by the specificity of the larvae which feed on the plant tissues: the fig wasps and yucca moths lay their eggs in the ovaries of their hosts. Initially they probably were only accidental pollinators. Seed production, however, is advantageous to both the plant and the seed-infesting insects. The relationships now are completely mutualistic.

The largest group of highly specialized pollinators are the tiny wasps of the Family Agaonidae which pollinate figs. The genus *Ficus* (plant Family Moraceae), found in tropical regions, includes about 800 species. Virtually every species of wasp is confined to a single species of fig. The fig that we eat is actually an inflorescence composed of an enlarged receptacle which encloses many small flowers.

In general, the pollination of the caprifig, *Ficus carica,* by the wasp *Blastophaga psenes* takes place in the following manner (Fig. 13.3). The fig has flowers of two kinds: pollen-producing male florets and female florets with short styles. The female wasp flies to a fig in the proper state of maturation and forcibly enters it through the narrow opening, or ostiole, which is guarded by scales (Fig. 13.3*a*). The restricted opening presumably excludes nonpollinating insects. In the process she loses her wings and antennal flagella. She penetrates deeply in the fig to reach the female florets. There she inserts her ovipositor

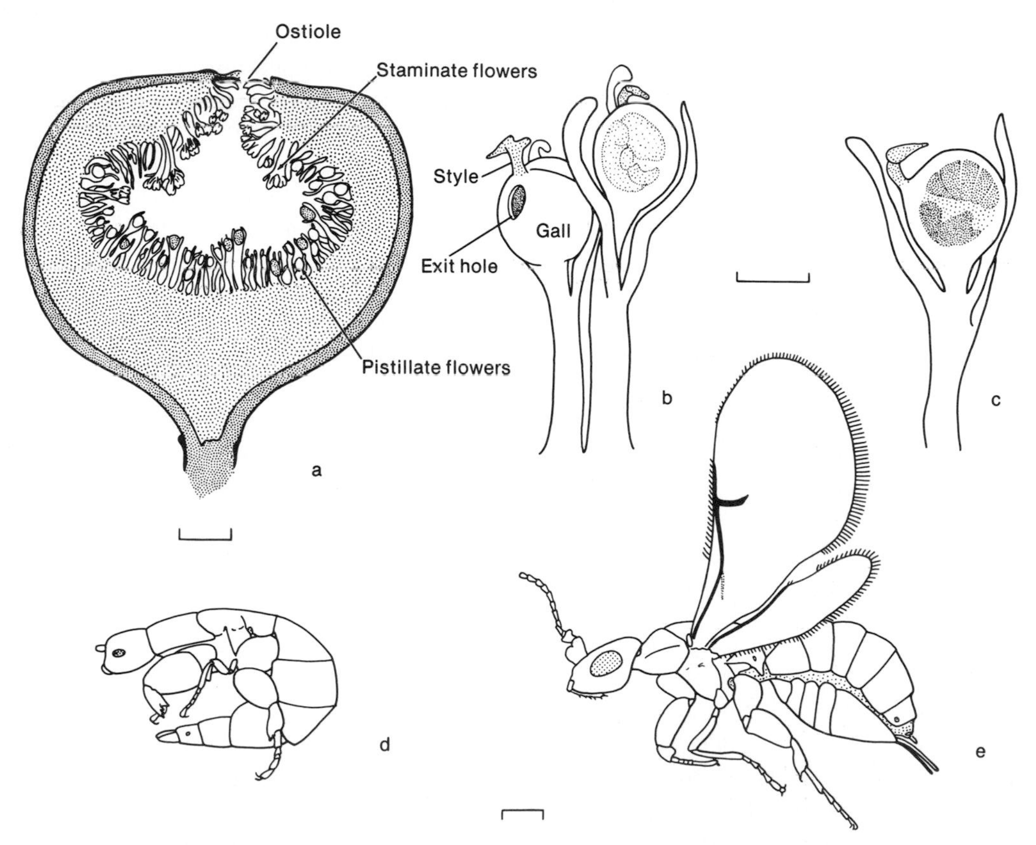

FIGURE 13.3 Pollination of caprifig by *Blastophaga* wasps: *a,* caprifig at time of emergence of wasps (scale equals 5 mm); *b,* gall flower with male wasp inside and another with exit hole; *c,* gall flower with female wasp inside (scale for *b* and *c* equals 1 mm), *d,* male *Blastophaga, e,* female *Blastophaga* (scale for *d* and *e* equals 0.25 mm).

through the short style of a floret to the ovarian region, where she lays an egg. She also is seen to remove some fig pollen with her front legs from special pouches on her body and brush it on adjacent stigmas, thus pollinating the female florets. After laying her eggs, she dies, still in the fig. Each wasp larva feeds inside a floret, causing a tiny gall. At maturity, the adult males are flightless and have reduced legs, eyes, and antennae (Fig. 13.3*d*). They chew their way out of their galls first (Fig. 13.3*b*) and seek the galls containing adult female wasps. A hole is bored in the gall by the male, and he copulates with the female by means of an extensible abdomen. The male wasp then dies. The inseminated female emerges from her gall (Fig. 13.3*c,e*) and seeks the pollen contained in the male florets near the ostiole. She packs pollen into special cavities on her body, leaves the fig, flies to another fig in the proper state of development, enters it, and the process is repeated. The exact cycle varies with the species of wasp and fig.

When edible figs were introduced in California in the late 1800s, the Smyrna variety failed to produce fruit. An enterprising grower traveled to Turkey to learn the secrets of successful fig culture. After some difficulty, including an epidemic of plague, he returned with the knowledge that special pollination is necessary. The Smyrna variety has only female florets with long styles. The wasps and suitable pollen must be obtained from the caprifig, in which the normal cycle can be completed. Caprifigs fruits, containing pollen-laden females of *Blastophaga*, are hung in perforated bags among the limbs of the Smyrna fig trees. The *Blastophaga* emerge and enter the Smyrna figs. They are unable to lay eggs because their ovipositor is too short for the long-styled florets. The fig is nevertheless pollinated and normally ripens with seeds into an edible fruit.

Another group of host-specific pollinators are the Yucca moths, *Tegeticula* and *Parategeticula* sp. (Incurvariidae). All species of *Yucca* (plant Family Agavaceae) are American in origin, but they have been introduced elsewhere in the world. More than two dozen species of *Yucca* east of the Rockies and the Mojave Desert are pollinated by one moth species, *T. yuccasella*. In the West, *Yucca brevifolia* is pollinated by *T. paradoxa; Y. whipplei* by *T. maculata* (see Powell and Mackie, 1966); and *Y. schottii* by *Parategeticula polleniferae,* as well as *Tegeticula yuccasella* (Davis, 1967).

The eastern moths are active at night, the time at which flower scent is also strongest. The female *T. yuccasella* enters the white flower, climbs up the stamens to the anthers, and gathers the pollen in a ball (Fig. 13.4*a*). The pollen mass from one to four anthers is carried under her head, clasped by a prehensile elongation of the maxillary palpi and the bases of the forelegs. She then flies to another flower in the proper state of ovarian development. After inspecting the ovary, she bores in with her sclerotized, elongate ovipositor and lays an egg (Fig. 13.4*b*). Climbing the style, she packs some pollen on the stigma. This behavior is often repeated after each egg is laid. On the average one egg is inserted into each of three compartments of the ovary. A few of the many seeds in pollinated flowers serve as food for the moth larvae. Unpollinated flowers do not develop seeds. When fully fed, the larvae leave the seed pod (Fig. 13.4*c*) and pupate in the ground. Emergence of the adults is timed to coincide with the flowering season.

A final example is provided by the male euglossine bees of the American Tropics, which visit orchids. Males take nectar as flight fuel from various flowers, but they visit orchids to obtain odors. The floral odors are created by species-specific blends of volatile compounds such as benzyl acetate, cineole, eugenol, methyl salicylate, and methyl cinnamate. Males of species of *Euglossa* (Fig. 13.1*b*) and *Eulaema* are each attracted to certain odors. The males brush the odor-producing surfaces and apparently store the fragrance in special cavities in their large hind tibiae. The bees' use of the odor is not clear, but they seem dependent on an adequate supply. Possibly it is metabolized or converted to attractants for either or both males or females. During their contact with the orchid, the male bees become intoxicated. When disabled, they fall into a trap device. During their escape, a packet of pollen is attached to a specific place on their

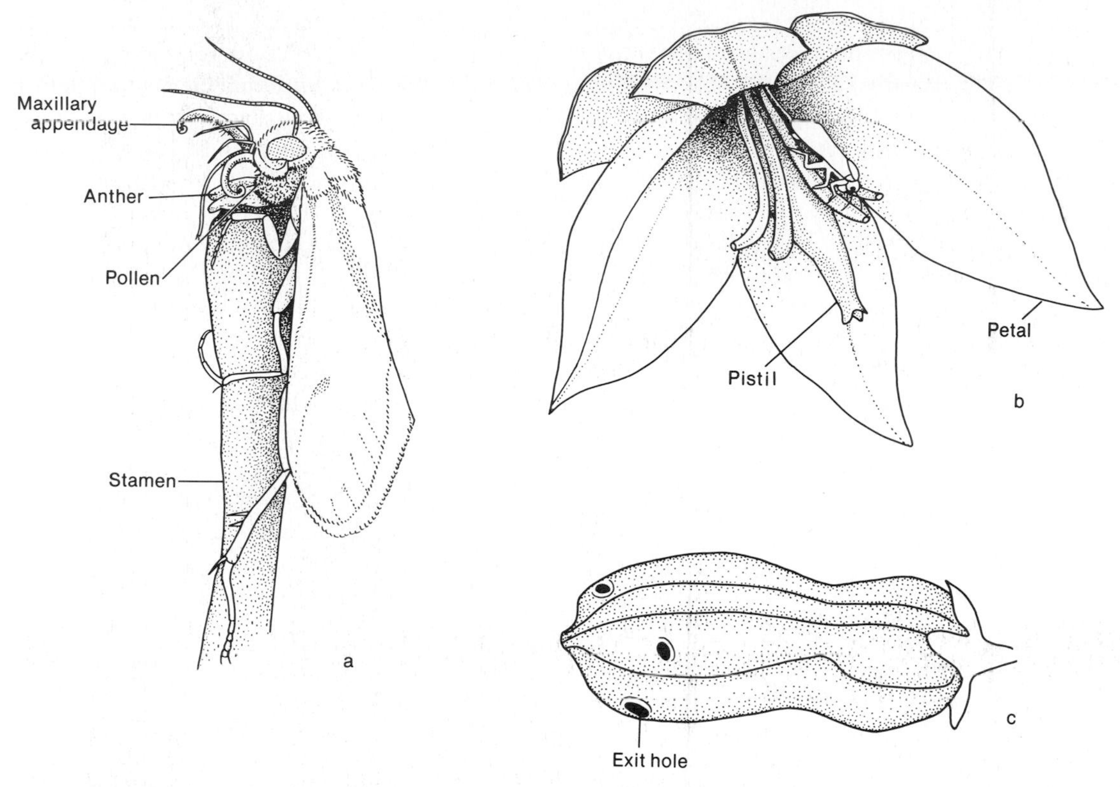

FIGURE 13.4 Pollination of *Yucca* sp. by *Tegeticula* moths: *a*, female of *Tegeticula yuccasella* collecting pollen from anther; *b*, position of female moth during oviposition in ovary of *Yucca; c*, mature pod of *Yucca* showing emergence holes. (*Redrawn from Riley,* 1892.)

body. At their next visit to an orchid of the same species, the packet is removed by the orchid, thus achieving pollination.

MODERN PHYTOPHAGOUS INSECTS AND THEIR MODES OF FEEDING

An adequate classification of phytophagous feeding habits would be overly complex for our purposes. Here we shall divide insects by their taxonomic groupings and by their general mode of feeding (Table 13.1).

The mouthparts and other characteristics of the immature and adult insects should be kept in mind when considering each group and its food habits. Orthoptera and Phasmatodea have chewing mouthparts in all stages and feed externally on plants. The phytophagous species among the endopterygote orders Hymenoptera, Coleoptera, and larval Lepidoptera also have chewing mouthparts. Some of these feed externally, and some bore bodily into the plant tissues. The rasping mouthparts of all stages of Thysanoptera are applied externally to disrupt cells and suck cell fluids. The piercing-sucking mouthparts of all stages of Hemiptera and the fruit-piercing moths take vascular or tissue fluids by penetrating the plant while the insect remains outside. Mouthparts of larval Diptera vary from the normal chewing type to hooklike structures which tear loose and ingest plant tissues and fluids. Larval flies are able to feed internally by boring. Adult Diptera, whether of the biting or sponging type, ingest only liquid food and particles in suspension. They feed externally and probably take fluids which are freely available without further damage to the plant. Nectar and pollen, as well as saps and juices from previously injured tissues or fruits, are eaten. The sucking mouthparts of adult Lepidoptera pass fluids of the same kinds, but not particles.

Insects which eat the vegetative parts of plants can be broadly divided into external (Fig. 13.5) and internal feeders (Fig. 13.6*b*). During the life history of some species, especially small moths, the larvae may feed first inside then outside. External feeders may be freely visible to predators and parasitoids. Such *exposed feeders* are usually protectively colored and patterned if they are large enough to be edible by vertebrates. The manner of feeding is characteristic of the species. Many caterpillars and sawfly larvae feed along the leaf edge. Surface feeders with chewing mouthparts may ingest whole pieces of leaf, leaving holes or notches. Others remove most of the photosynthetic tissues, leaving a delicate skeletonized vascular network. The cellulose-rich feces are called *frass*. Exposed feeders simply drop frass from their feeding stations.

The feeding injuries of leaf-feeding Hemiptera are often recognized as small, discolored spots where the inner palisade and spongy mesophyll cells are broken down and emptied. The insect's saliva sometimes has physiological effects on the leaf which are also injurious. Thysanoptera also remove the contents of the inner leaf cells, resulting in silvery air spaces inside the leaf. Damaged leaves and stems may wither, discolor, or fall prematurely.

Some external feeders gain protection by feeding on roots or within enfolding leaves. Those that are able to spin silk may bind together leaves in a sheltering cluster. The *leaf rollers* are caterpillars and sawfly larvae which roll or fold leaves to create a tube. The larvae hide in the tube, feeding at its edge or on nearby leaves. Some moth species pupate in the roll. Adult attelabrid weevils cut the leaf bearing their egg, and the leaf forms a roll naturally. The weevil larva feeds inside. Parenthetically, we should note that the predaceous gryllacridid cricket *Camptonotus carolinensis* also rolls leaves and fastens them with an oral secretion. The leaf roll is used as a retreat.

Insects that feed inside the plant, completely surrounded by living tissues, are exclusively endopterygotes and usually the larval stages of the life history. *Leaf miners* are small larvae that eat some or all of the mesophyll tissues between the outer layers of leaf blades or needles. Female insects select the host and lay their eggs on the leaf surface, or inserted in the mesophyll, or nearby on the plant. Usually mature leaves are mined. The larvae feed in a manner characteristic of the species, leaving a "signature" which is sometimes beautifully serpentine. Larvae of Diptera are cylindrical but soft-bodied and adaptable to the restricted mine. The other larval miners are highly flattened, usually colorless, legless, and with a flattened, wedgelike head which slopes forward. No adult insects are leaf miners. Linear mines begin as a tiny channel and

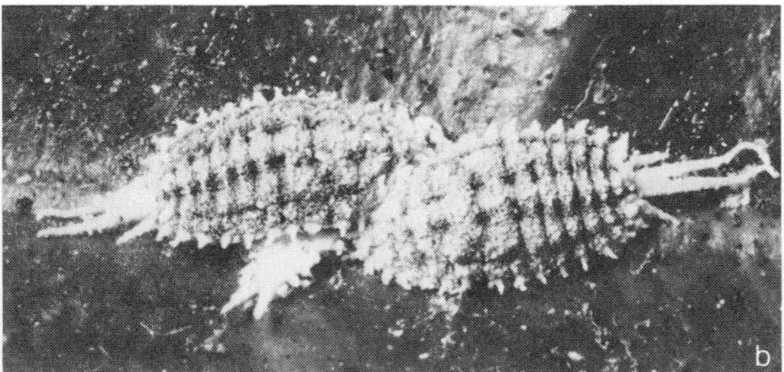

FIGURE 13.5 Phytophagous exopterygotes: *a*, cicadellid leafhopper, *Graphocephala atropunctata*; *b*, pseudococcid mealybug, *Pseudococcus fragilis*. (*Photos by Howell V. Daly.*)

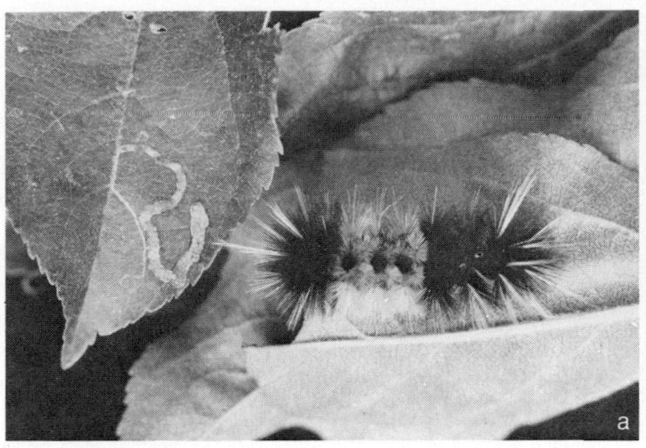

FIGURE 13.6 Phytophagous endopterygotes on apple: *a*, leaf mine of *Nepticula* sp. (Nepticulidae) on leaf at left, caterpillar of *Halisidota maculata* (Arctiidae) at right; *b*, gallery of codling moth larva, *Laspeyresia pomonella* (Tortricidae). (*Photos by Howell V. Daly.*)

progressively widen as the larva grows in size and appetite (Fig. 13.6*a*). The direction of the mine may be altered when vascular bundles are encountered. Blotch mines are created when the larva feeds in various directions, eating both vascular and mesophyll tissues. The frass is often deposited in the mine or ejected through an opening. Lepidoptera may lay frass inside in a continuous line, or pack it at one end or plaster it randomly on the minefloor, while Diptera lay two rows. Hymenoptera scatter the frass about the cavity or in piles. Some miners pupate in the mine, while other species leave to form a cocoon elsewhere.

Plant *borers* are insects that burrow in living or dead plant tissues other than leaves. The xylophagous scavengers discussed in the previous section are borers in dead plants. Their life-span is usually long because of the lower nutrient value of the food, variable and sometimes low moisture content, and the stable environment. We are concerned here primarily with larvae that bore into living buds, stems, roots, fruits (Fig. 13.6*b*), seeds, nuts, or grains. Life-spans of these borers may be short, not only because the food value and moisture are higher, but also because the living plant tissues change.

Although this is a heterogeneous group, the larvae share several similarities arising from their sheltered mode of life. The bodies are often cylindrical, pale, legless, and with bumps or rough areas of skin which provide traction against the burrow walls. The mouthparts, mounted on a retractable head capsule, excavate tissue bit by bit. Antennae are short and retractable into grooves. These specializations give borers access to food rich in nutrients: newly developing leaves, cambium tissue, the flesh of fruits, and endosperm of seeds. Frass is often packed behind the larva as it advances.

Certain secretions of insects and mites stimulate abnormal development of growing plant tissue, resulting in misshapen leaves or in swellings called *galls*. These are usually initiated in the spring and early summer when the meristematic tissue is active. Aphids feeding on the underside of a new leaf can cause the leaf to curl around them and enclose a favorably high humidity. Homoptera and mites create galls by their feeding secretions. The gall is typically a hollow cavity which remains open to the exterior and consequently is called an *open gall*. A group of insects or mites or several generations may live inside the gall, feeding and taking shelter.

The galls of endopterygote insects are usually, but not always, occupied by a single larva. Either the ovipositing female or the larva or both seem to be responsible for the biochemical stimulants. The galls do not have a permanent opening and are called *closed galls*. The host plant responds with a growth of specific color and design such that the insect can be identified by the gall it makes. The larva is surrounded by moisture and food and is largely protected from natural enemies. Some wasp parasitoids, however, are able to lay their eggs in or near the gall-making larvae so that several kinds of insects may emerge from a single gall.

HOST SPECIFICITY OF PHYTOPHAGOUS INSECTS

How specific are insects in their choice of food? Some exposed feeders accept a great variety of plants as food. The larvae of the gypsy moth, *Porthetria dispar* (Lymantriidae), are known to feed on 458 species of plants in the United States (Leonard, 1974). Such insects that accept a wide variety of plants for food are called *polyphagous*. *Oligophagous* insects feed on a few species of plants, often related to one another or having certain similar biochemical constitutents. *Monophagous* insects feed on only one species. Insects with the more restricted range of hosts are usually the leaf miners, borers, and gall makers which are surrounded by plant tissues, or insects adapted to toxic plants. The early plant-feeding Neoptera are assumed to have been polyphagous, exposed feeders, that are exemplified now by Orthoptera. The oligophagous and monophagous taxa evolved as specialists, with occasional reversions to polyphagous habits (Dethier, 1954).

What factors favor the evolution of host specificity? To answer this question, we must consider in detail the coevolutionary nature of insect/plant relationships (Chap. 9). Part of the explanation lies, on the one hand, in the defense strategies of plants against attack. We have seen that all parts of a plant may be eaten: roots, stems, leaves, sap, flower parts, fruits, and seeds. Plant growth is affected by loss of food when photosynthetic tissue is eaten or phloem sap is drained. Damage to roots and xylem deprives the plant of water and minerals. Chemicals in insect's saliva or secreted by the ovipositor may have general or specific effects on the host's physiology. Injured cambium or meristematic tissue results in abnormal growth. These influences lead to stunting, lowered production of seeds, or death. Damage to flowers, fruits, and seeds directly affects the population growth of plants. Phytophagous insects also transmit plant diseases.

In the coevolution of insects and plants, the plant evolves a protective measure, then insects evolve effective means to overcome the plant's defense which in turn selects for new plant defenses, and so on. In the context of this interaction, a plant which is less damaged because of heritable characteristics is said to be *resistant*. Three basic kinds of resistance have been identified by Painter (1958): (1) Resistant plants may be *nonpreferred* for oviposition, shelter, or food; i.e., some chemical or physical feature of the plant either is lacking and fails to attract insects *or* some feature is present which is repellent. (2) Resistant plants may adversely affect the biology of the insect. Physical or chemical properties of the plant may result in the insect's early death, abnormal development, decreased fecundity, or other deleterious conditions. For example, toxins, repellents, copious sap or pitch, or tissue which is nutritionally inadequate for insects could reduce or eliminate attack. This kind of resistance is called *antibiosis*. (3) Resistant plants may be *tolerant* and survive even when infested by insects at levels that kill or injure susceptible plants. Rapid replacement of lost parts, excess production of seeds, rapid wound healing, and detoxification of insect salivary toxins are some of the possible ways a plant might survive damage.

The nonpreference and antibiotic properties prevent or reduce attack. These properties vary (as do other genetic traits) from plant species to species, geographically with the ranges of species, locally depending on ecology, and from individual to individual (see discussion of alkaloids in lupines, Chap. 9). For example, the oleoresins or pitch of conifers traps, expels, or is toxic to boring insects in needles or bark. The variation in physical and chemical composition of oleoresins gives almost every tree a unique individuality

(Hanover, 1975). Angiosperms have so-called secondary plant substances such as essential oils, alkaloids, and glycosides whose primary function seems to be in chemical defense, although some may be otherwise integrated into the plant's metabolism (Ehrlich and Raven, 1967; Whittaker and Feeny, 1971; Schoonhoven, 1972). This toxic, biochemical shield has partially protected flowering plants against herbivores, both insect and vertebrate. These chemicals, incidentally, also give us spices and flavorings.

The plant-feeding insects, on the other hand, compete for food. The resistance factors evolved by plants further limit the number of kinds of plants available to eat. Insects which are able to survive the antibiotic properties and develop a preference for a nonpreferred host will acquire not only food but also some relief from interspecific competition. Those able to bore into plant tissues must tolerate immersion in the host's chemical environment and physical constraints, but here also are food and relief from competition, plus escape from certain enemies and protection against desiccation, freezing, etc. Some insects store toxic plant substances and thereby acquire a defense for themselves against vertebrate predators (Rothschild, 1973; Chap. 15). Thus a number of advantages accrue to the monophagous or oligophagous insect that is adapted to the special conditions of life associated with one or a few kinds of plant hosts.

Some insects may be secondarily polyphagous because they are able to tolerate a variety of resistance factors in plants. Their adaptation to biochemical stresses also may permit them to resist or detoxify insecticides manufactured by humans.

The defensive mechanisms of plants and host specificity of insects have evolved in response to essentially two periods of contact (Beck, 1965): (1) the period of oviposition, during which the female seeks suitable host plants on which to lay her eggs; and (2) the period of feeding, during which the immature and sometimes adult insects eat the plant. Oviposition behavior involves recognition and orientation to a host plant at some distance, the search for specific sites in the plant, and, finally, the deposition of eggs, followed by dispersal. The behavior is a series of complex events and involves many of the insect's sensory receptors. Any heritable physical or chemical feature of the plant which reduces the numbers of eggs, and thus ultimately the number of feeding insects, will be selectively favored. This is resistance of the nonpreference type and is achieved either by failing to attract ovipositing females or by providing some inhibition.

Feeding behavior of insects on plants is similarly complex. Beck (1965) identified four steps: (1) host recognition and orientation; (2) initial biting or piercing of the plant; (3) maintenance of feeding; and (4) cessation of feeding, usually followed by dispersal. The behavioral response at each step depends on releasing stimuli provided by the plant and on the insect's response thresholds, which vary with its physiological state. Resistance of the nonpreference type would be given a plant which lacked the appropriate releasing stimuli or which discouraged feeding at some step.

Physical and chemical stimuli have been classified according to the response they elicit from insects (Dethier et al., 1960; Beck, 1965). During orientation to a host plant at a distance, certain stimuli may act positively as *attractants* or negatively as *repellents.* When in close contact with the plant, positive stimuli may stop further locomotion, i.e., be *arrestants,* or act as repellents to hasten the insect's departure. At the initiation of feeding, positive stimuli are called *incitants* and negative stimuli are *suppressants.* Feeding is maintained by *stimulants* or terminated by *deterrents.*

The sensory apparatus and orientation behavior of insects are finely tuned to the characteristics of the desired host plants. Secondary plant substances which are repellent to most insects are, in fact, often the feeding stimulants for the appropriate monophagous insects. Nutrients, including sugars, amino acids, phospholipids, and ascorbic acid, can also be stimulants to certain insects.

The sensory receptors on the antennae and maxillae of caterpillars are chiefly involved in discrimination. When such receptors are experimentally removed, oligophagous caterpillars often accept as food a broader range of plant species. Although the female usually selects the host plant when she lays her eggs, the caterpillar must select the parts of the plant to eat, avoiding concentrations of toxins and finding the richest food.

SELECTED REFERENCES

Insect/plant relationships are treated by van Emden (1972*b*), Felt (1940), Graham and Knight (1965), Hering (1951), Jermy (1974), Kulman (1971), and Needham et al. (1928). Insect/plant relationships in the Paleozoic are discussed by Carpenter (1970), Hughes and Smart (1967), Plumstead (1963), and Smart and Hughes (1973). Pollination is reviewed by Baker and Hurd (1968), Fraegri and van der Pijl (1971), Free (1970), McGregor (1976), Percival (1965), and Proctor and Yeo (1973). Plant resistance to insect attack is discussed by Beck (1965), Gallun et al. (1975), Hanover (1975), Maxwell et al. (1972), Painter (1958), and Schoonhoven (1972, 1973). Insect/plant coevolution is discussed by Dolinger et al. (1973) and Ehrlich and Raven (1964, 1967).

ENTOMOPHAGOUS INSECTS

14

In this chapter we discuss those insects that kill or injure one or more other invertebrates before completing their life cycle. Most of these carnivorous insects feed on other insects and are said to be *entomophagous.* Snails, earthworms, millipeds, mites, and other terrestrial and freshwater invertebrates are also eaten by insects. Victims are called *prey* if killed directly, *hosts* if fed upon while still living. Entomophagous insects are divided into three major groups according to their mode of feeding: (1) *predators* kill and consume more than one prey organism to reach maturity; (2) *parasitoids* require only one host to reach maturity, but ultimately kill the host; and (3) *parasites* feed on one or more hosts, but do not normally kill the host.

Parasitoid insects are intermediate between predators and parasites in the sense that they live at first parasitically at the host's expense, but ultimately kill the host. In entomological literature the word "parasite," alone and in combinations (e.g., hyperparasite), is often used loosely for an insect that lives at a host's expense, regardless of the host's fate. From an ecological viewpoint, predators and parasitoids both act to eliminate individuals in the prey or host population, whereas true parasites permit the host to continue functioning in the community, though often at a subnormal level. This is an important distinction, especially when predators and parasitoids are used to control pests. Whenever possible in this text we have used the word "parasite" only for those insects that normally do not cause host mortality. However, we use the term "endoparasite" indiscriminantly for both parasites and parasitoids that are physically inside the body of the host.

Although the three modes of feeding defined above can be easily distinguished among most insects, the spectrum of entomophagous habits is actually continuous. For example, within a species an individual of a normally predaceous insect, such as a ladybird beetle, might complete its life cycle by eating only one large host and thus be technically, according to number of victims, a parasitoid. Alternatively, the host of a parasitoid might survive attack, or the host of a parasite might be fatally injured. These abnormal relationships make distinction somewhat arbitrary, but are instructive by indicating how one mode of feeding might evolve into another.

Before further discussion of predators, parasitoids, and parasites, some additional kinds of behavior should be mentioned in which an insect consumes food collected at the energetic expense of the host. Although the insect may not be directly entomophagous, the net result is a reduction in the number of host offspring and an increase in the offspring of the insect that benefits. *Cleptoparasites,* or "cuckoo" parasites, lay their eggs in the nests of other species, in the manner of cuckoo birds. The cleptoparasitic parent or the larva may kill the egg or larva of the host immediately, or the host larva dies of starvation after the cleptoparasite larva eats the nest provisions. In the former instance, the young cleptoparasites often possess enlarged mandibles suited for attack. A *social parasite* is a female that enters the nest of a social host and takes over the role of the queen. The social parasite's offspring are fed by host workers at the expense of host offspring. The host is usually a closely related species. Parasites that spend much of their life in their host's nests are known as *inquilines* (Fig. 14.1). Slavery is practiced by certain ants (Wilson, 1975*a*). Worker pupae of another species are taken by slave-making workers. The resulting adult slaves then become members of the colony and do most of the work. Nest robbing also involves social species. The robbing species enters the nest of another colony and removes the stored food.

Phoresy is the transport of an insect on or physically inside the body of another insect. The insect that provides the transportation is usually not harmed. When the passenger is an adult female parasitoid or predator, however, the usual result is that the passenger remains aboard until the host lays its eggs. The passenger then oviposits on the host's eggs. This habit occurs frequently in the wasp family Scelionidae. The first instar larvae of meloid and rhipiphorid beetles, stylopid parasites, and eucharitid wasps are commonly phoretic on adults of their favored hosts. By this means the larvae gain entrance to the nests of the hosts.

RESISTANCE OF HOSTS OR PREY TO ATTACK. Hosts or prey insects may avoid or actively prevent attack by entomophagous insects. Enemies may be excluded by tough cocoons, hard puparia, nests of mud or leaves, or by the inaccessible position of the insect in burrows in soil or plant tissues. Some still succumb to parasitoids and predators that are suitably equipped with strong mandibles or long, powerful ovipositors.

Exposed insects may violently resist attack or quickly escape. When approached by an enemy, an aphid may drop from the plant, walk away, or kick and secrete oily droplets from its siphunculi. The odor of the siphuncular secretion alarms other aphids, and they drop off or walk away. Larvae and pupae of Lepidoptera or Coleoptera thrash about when contacted. Larvae may bite, or regurgitate or secrete defensive fluids. One of the advantages of social behavior is mutual defense against insect enemies.

It seems possible that some of the defensive strategies to avoid predation by vertebrates (Chap. 15) are also effective against insect enemies. Certain aphids are distasteful and are avoided by insect predators after an initial contact. The gustatory senses and learning ability of the predator are therefore important to the success of this defense. The role of protective

FIGURE 14.1 Inquilines in army ant colony: *a*, ant-mimicking staphylinid beetles (three beetles are visible on the food, a grasshopper's femur; ant hosts are in foreground and background); *b*, a thysanuran, *Trichatela manni*, in an army ant column. (*Photos courtesy of and with permission of Carl W. Rettenmeyer.*)

coloration and behavior, however, is generally less effective because many insect predators, unlike vertebrate predators, do not depend on vision to find prey.

Insect hosts are able to respond defensively to endoparasites by a reaction called *encapsulation.* G. Salt (1963, 1968, 1970) has investigated this kind of reaction in detail. Hemocytes collect around those foreign bodies that are too large to be engulfed by a single cell. A capsule forms as the inner layer of cells flatten over the object's surface and new cells are added outside. The inner layer secretes an envelope of connective tissue. The capsule may remain clear or become darkened by melanization. If the foreign body is an endoparasite, it is killed by the encapsulation. The process may take only a day to complete.

Fourteen orders of insects are known to encapsulate foreign bodies, endoparasitic worms, and parasitoids. Among the parasitoids, larvae of certain species definitely stimulate encapsulation in hosts that are not the normal hosts: maggots of Tachinidae and some, but not all, wasp larvae of Ichneumonidae, Braconidae, Chalcidoidea (Encyrtidae, Eulophidae), Proctotrupoidea, and Cynipoidea. Yet in their normal hosts, these parasitoids usually escape encapsulation. How do they succeed?

Among wasps, some attack only eggs. By rapidly completing their growth before the host's hemocytes develop, they avoid encapsulation altogether. Rapid destruction of a larval host also avoids encapsulation by simply killing the host quickly. If the parasitoid lingers for a while and allows the host larva to grow, a larger food supply becomes available, but at a greater risk of encapsulation. Certain parasitoids insert their eggs precisely into an organ of the host where the endoparasite will be separated from the host's hemolymph by a layer of connective tissue. This prevents encapsulation. Others first invade the alimentary canal and are sheltered from the host's hemocytes. When the host is suitably larger, the parasitoid feeds voraciously and so debilitates the host that it is unable to react in time. Some parasitoids are initially invested in cellular membranes the cells of which become dissociated, enlarged, and circulate in the host's hemolymph. Apparently the cells absorb nutrients and similarly act to reduce the host's ability to muster an encapsulation.

Salt has shown that the eggs and first instar larvae of the wasp *Venturia canescens* (Ichneumonidae) are coated with a particulate layer that inhibits encapsulation by the hemocytes of its moth host, *Anagasta kuehniella* (Pyralidae). The layer is first deposited on the egg by a specialized region of the wasp's oviduct and later is somehow transferred from the egg shell to the wasp larva once it is inside the moth larva. If the coating is disrupted, encapsulation ensues.

EVOLUTION OF ENTOMOPHAGY

Among the earliest insect fossils in the Upper Carboniferous Period are Odonata. Today the dragonflies are exclusively predaceous, and we assume they were also in ancient times. Predatory habits have apparently evolved repeatedly throughout the orders from scavenging and plant-feeding ancestors (Table 14.1). Parasitoids and entomophagous parasites did not appear until the holometabolous life cycle was evolved. Thus we can conclude that predation is the oldest mode of the entomophagous habits.

Other living orders present as fossils in the Paleozoic Era (Table 17.3) that were mostly predatory are the Neuroptera and Raphidioptera. Certain extinct orders have been suspected to be predatory. The *Mischoptera* of the Megasecoptera had stout, raptorial forelegs. Were these used to capture prey or to cling to vegetation? The giant *Meganeura* of the Meganisoptera had biting mandibles and large eyes. These insects resembled dragonflies and were doubtless predatory. Many fossils are merely wings without the body parts from which we could infer food habits.

The early evolution of parasitoids and parasites is not well documented by fossils. Most parasitoids are Hymenoptera and Diptera. These orders must have evolved diverse forms during the Mesozoic Era following the first appearance

TABLE 14.1 Major taxa of entomophagous insects[1]

Predaceous as both immatures and adults

- Collembola
  - Isotomidae (*Isotoma*)
- Diplura
  - Japygidae
- Odonata (all)
- Mantodea (all)
- Dermaptera
  - Arixeniina (probably)
  - Forficulina (most)
- Orthoptera
  - Gryllacrididae
  - Gryllidae (*Oecanthus*)
  - Tettigoniidae (*Conocephalus*)
- Psocoptera
  - Caeciliidae (*Caecilius*)
- Hemiptera (Heteroptera)
  - Anthocoridae (*Anthocoris*)
  - Belostomatidae
  - Enicocephalidae
  - Gelastocoridae
  - Gerridae (occasionally)
  - Hebridae
  - Largidae (*Euryopthalmus*)
  - Lygaeidae (*Geocoris*)
  - Miridae (some)
  - Nabidae
  - Naucoridae
  - Nepidae
  - Notonectidae
  - Ochteridae
  - Pentatomidae (some)
  - Phymatidae
  - Pyrrhocoridae (*Dindymus*)
  - Reduviidae (except Triatominae)
  - Saldidae (occasionally)
  - Veliidae (occasionally)
- Thysanoptera
  - Aeolothripidae (*Aeolothrips*)
  - Phlaeothripidae (*Leptothrips*)
  - Thripidae (*Scolothrips*)
- Neuroptera (most)
- Raphidioptera (all)
- Coleoptera (Adephaga)
  - Amphizoidae
  - Carabidae (most)
  - Cicindellidae
  - Dytiscidae
  - Gyrinidae

Predaceous as both immatures and adults

- Coleoptera (Polyphaga)
  - Anthicidae (*Anthicus*)
  - Anthribidae (*Brachytarsus*)
  - Cantharidae (many)
  - Cleridae
  - Coccinellidae (most)
  - Cucujidae (*Cryptolestes*)
  - Elateridae (some)
  - Histeridae
  - Lampyridae
  - Malachiidae
  - Melyridae (many)
  - Nitidulidae (*Cybocephalus*)
  - Nosodendridae
  - Orthoperidae
  - Phengodidae
  - Silphidae (*Xylodrepa*)
  - Staphylinidae (many)
- Diptera (Brachycera)
  - Asilidae
  - Dolichopodidae
  - Empididae
- Diptera (Cyclorrhapha)
  - Anthomyiidae (*Hylemya*)
- Hymenoptera (Apocrita)
  - Chalcidoidea
  - Chrysididae
  - Formicidae
  - Ichneumonoidea
  - Vespidae

Predaceous primarily as immatures

- Ephemeroptera
  - Siphlonuridae (*Isonychia*)
- Plecoptera
  - Setipalpia (most)
- Megaloptera (all)
- Neuroptera
  - Chrysopidae (*Chrysopa*)
- Coleoptera (Polyphaga)
  - Dermestidae (*Thaumaglossa*)
  - Drilidae
  - Hydrophilidae (some)
  - Lycidae (most)
  - Meloidae
  - Rhipiphoridae (most)

TABLE 14.1 Major taxa of entomophagous insects[1] (*continued*)

Predaceous primarily as immatures

- Diptera (Nematocera)
  - Cecidomyiidae (*Aphidoletes*)
  - Ceratopogonidae
  - Culicidae (*Chaoborus*)
  - Chironomidae (Tanypodinae)
  - Culicidae (*Megarhinus*)
  - Mycetophilidae (*Platyura*)
  - Tipulidae (Hexatomiini)
- Diptera (Brachycera)
  - Bombyliidae (most)
  - Mydaidae
  - Rhagionidae
  - Tabanidae (*Tabanus*)
  - Therevidae
  - Xylophagidae
- Diptera (Cyclorrhapha)
  - Anthomyiidae (some)
  - Calliphoridae (*Stomorhina*)
  - Chamaemyiidae
  - Chloropidae (*Siphonella*)
  - Drosophilidae (*Gitonides*)
  - Lonchaeidae (*Lonchaea*)
  - Otitidae (*Elassogaster*)
  - Phoridae (*Syneura*)
  - Sarcophagidae (*Sarcophaga*)
  - Sciomyzidae
  - Syrphidae (most)
- Lepidoptera
  - Blastobasidae (*Holcocera*)
  - Cosmopteryidae (*Stathmopoda*)
  - Cyclotornidae
  - Heliodinidae
  - Lycaenidae (some)
  - Noctuidae (*Eublemma*)
  - Lyonetiidae (*Ereunetis*)
  - Psychidae (*Platoeceticus*)
  - Pyralidae (*Laetilia, Macrotheca*)
  - Sesiidae (*Synanthedon*)
  - Tineidae (*Dicymolomia*)
  - Tortricidae (some *Tortrix*)
- Trichoptera
  - Hydropsychidae (*Hydropsyche*)
  - Polycentropodidae
  - Rhyacophilidae
- Hymenoptera (Apocrita)
  - Proctotrupoidea
  - Sphecoidea (some)

Predaceous primarily as adults

- Coleoptera (Polyphaga)
  - Cerambycidae (*Elytroleptus*)
  - Scarabaeidae (*Cremastocheilus*)
- Mecoptera
  - Bittacidae (some)
- Diptera
  - Anthomyiidae (some)
  - Blephariceridae
  - Calliphoridae (*Bengalia*)
  - Ceratopogonidae (some)
- Hymenoptera (most Symphyta)
- Hymenoptera (Apocrita)
  - Bethylidae
  - Dryinidae
  - Mutillidae
  - Tiphiidae

Parasitoids

- Coleoptera
  - Carabidae (*Lebia, Brachinus*)
  - Colydiidae (*Deretaphrus*)
  - Meloidae
  - Rhipiceridae
  - Rhipiphoridae
  - Staphylinidae (some Aleocharinae)
- Diptera
  - Acroceridae
  - Agromyzidae (*Cryptochaetum*)
  - Anthomyiidae
  - Conopidae
  - Nemestrinidae
  - Phoridae (*Megaselia*)
  - Pipunculidae
  - Pyrgotidae
  - Tachinidae
- Lepidoptera
  - Cyclotornidae
  - Epipyropidae
- Hymenoptera
  - Bethyloidea
  - Chalcidoidea
  - Chrysidoidea
  - Evanioidea
  - Figitidae
  - Ibaliidae
  - Ichneumonoidea
  - Megalyroidea
  - Pompiloidea
  - Proctotrupoidea
  - Scolioidea
  - Sphecoidea (some)
  - Trigonaloidea

TABLE 14.1 Major taxa of entomophagous insects[1] (*continued*)

Insect ectoparasites of insects
Diptera
Ceratopogonidae (some)
Chironomidae (*Symbiocladius*)
Hymenoptera
Scelionidae (*Rielia*)
True insect endoparasites of insects
Strepsiptera (all)
Hypermetamorphic taxa
Neuroptera
Mantispidae
Coleoptera
Carabidae (*Lebia, Brachinus*)
Colydiidae (a few)
Drilidae
Meloidae
Rhipiceridae (*Sandalus*)
Rhipiphoridae
Staphylinidae (Aleocharinae)
Strepsiptera
Diptera
Acroceridae
Bombyliidae
Calliphoridae (some)
Nemestrinidae
Tachinidae (some)
Lepidoptera
Cyclotornidae
Epipyropidae
Hymenoptera
Chalcidoidea (some)
Ichneumonoidea (some)
Proctotrupoidea
Trigonaloidea

[1]This list of taxa is intended to be representative, not exhaustive. The listing of a family without qualification does not necessarily mean that all species in the family have the same food habits. Well-known or exceptional genera are given in parenthesis.

Source: Modified from Askew, 1971; Balduf, 1935; Clausen, 1962; and Hagen et al., 1976.

of fossils of both orders in the Triassic Period. The parasitoids were probably part of this radiation. Fossil chalcidoid wasps are known from the Cretaceous Period. In contrast to most predators, parasitoids are host-specific. Presumably as a consequence of coevolutionary interactions with their hosts, parasitoid taxa also tend to have many more species than predatory taxa. The parasitic Strepsiptera are known from primitive forms preserved in Baltic amber. They may have existed no earlier than the Tertiary Period.

PREDACEOUS INSECTS

Predators are usually larger in body size than other entomophagous insects, and more than one and often many prey are eaten. The greater need for prey also makes predators dependent on prey populations of higher density than other entomophagous insects. Prey are usually smaller than the predator but proportional to the predator's size. In other words, larger predators take larger prey and smaller predators take smaller prey. Wasps with stings and ants in groups can kill prey larger than themselves. Victims are rapidly subdued, and, except when the predators are wasps that store prey for later consumption by their larvae, the prey are eaten immediately.

The same terminology can be used here as in describing the choice of foods by phytophagous insects. Monophagous predators feed on one species of prey; oligophagous predators take several prey species; and polyphagous predators take many kinds of prey. The last type are the "generalists." They take individuals of prey species largely in proportion to their relative abundance. Thus polyphagous predators, by continually shifting to the most abundant prey, tend to stabilize populations of prey in the community. Furthermore, if mobile, they are able to thrive in disturbed communities where prey species vary in abundance as ecological succession proceeds. At the other extreme, monophagous predators are density-dependent on one species of prey, and may regulate the prey at lower levels than polyphagous predators. Monophagous predators tend to be associated with undisturbed communities where their host maintains continuous, stable populations.

Perusal of Table 14.1 leads to the conclusion that a majority of predaceous taxa are carnivorous in all feeding stages. This does not mean they are always exclusively carnivorous; exact diets vary species by species. Adults of some of these visit flowers for nectar, as do certain adults of taxa that are predaceous only as immatures. Scavenging, honeydew, symbionts, and plant foods also may supplement a predator's diet.

Some of the complexities of diet are revealed in the following example. Hagen and his associates (1970) have carefully examined the diets of lacewings of the genus *Chrysopa* (Chrysopidae). Adults of about half the species have larger mandibles and are predaceous like the larvae. Adults of other species have shorter mandibles and feed only on honeydew and pollen. The foreguts of the second group contain yeast symbionts and are supplied with larger tracheal trunks for increased respiration. By this means the essential amino acids are acquired even though the adult diet lacks animal prey. Predaceous adults of the first group lack the yeast and tracheal modifications. Females of both groups of *Chrysopa* require substantial amounts of food before eggs are produced. As a consequence, females lay eggs only near abundant food sources that will later supply the lacewing larvae. An advantage to the honeydew-feeding species is that prey of the larvae may include not only aphids but a variety of other insects attracted to honeydew.

In contrast to other entomophagous insects, predators of both sexes must repeatedly find and subdue prey. They may be active during day and night. The eggs of predaceous insects are usually deposited by the females in close proximity to or at least in the vicinity of suitable prey. In this way the adult's well-developed compound eyes, chemoreceptive organs, and ability to fly are used to search for prey-rich habitats within which the less well-equipped immatures can forage. Plants of a certain height, odors of honeydew, odors of decay, and even the pheromones of prey are attractive to searching predators. The adult may eat prey or other attractive foods when they are found or only lay eggs near the prey.

STRATEGIES OF INSECT PREDATORS. Several kinds of strategies are used by insect predators in finding and capturing prey: (1) *random searching,* (2) *hunting,* (3) *ambush,* and (4) *trapping.*

Insects of the first type roam in the appropriate microhabitat and seize prey after physical contact. Their orientation to objects in the microhabitat and their movements may increase the probability of encountering prey. For example, the predator may patrol leaf edges, veins, or stems and eat insect eggs or sedentary Homoptera. Random searchers may have either monophagous or polyphagous habits. The prey is accepted if proper incitants are detected by receptors on the forelegs, mouthparts, or antennae. The victim is then devoured by use of mouthparts that are usually specially adapted for predation. After an initial contact or meal, further searching often involves more frequent turns so that the predator stays in an area of previous success. While seemingly inefficient, the random searchers are probably the most common type of insect predator and include species highly effective in regulating prey populations.

Most predatory Coleoptera are random searchers. Predatory carabids are distinguished by long, sharply hooked mandibles that contrast with the broad, blunt jaws of plant-feeding relatives. Coccinellid beetles, syrphid larvae (Endpaper 1*b*), and neuropteran predators that feed on aphids also search at random. The mandibles of predaceous coccinellids may be incisors with one or two apical teeth and a basal tooth or they may be small with ducts for sucking prey juices. Neuropteran larvae have sickle-shaped jaws each formed by the mandible and maxilla locked together to create a tube through which body fluids of prey are sucked (Fig. 38.1*b*).

The Lepidoptera are almost exclusively phytophagous. Caterpillars of certain moths retain their close association with plants but attack other phytophagous insects such as coccids and leafhoppers as well as mites. Caterpillars of the lycanid butterflies may be phytophagous or partly or wholly predaceous on ant larvae and pupae or aphids.

The relatively small size and high nutritive value of insect eggs make them vulnerable to predation by insects of varied food habits. Scavenging, phytophagous, predatory, or parasitoid insects may consume eggs. Collectively they are called *egg predators* when more than one egg is consumed. These predators are of special ecological importance because the prey never function in the community. For example, the large clustered eggs of Orthoptera are subject to frequent attack by many predatory larvae: clerids, meloids, bombyliids, rhagionids, anthomyiids, calliphorids, otitids, phorids, sarcophagids, and eurytomids. The adult female of the predatory species finds the eggs, lays her own eggs, and the larvae feed essentially as random searchers.

The hunting insects differ from random searchers by utilizing sight or other stimuli to orient to prey at a distance. Visual hunters have enlarged compound eyes with overlapping fields of vision that permit distance perception. Mandibles may be sharply toothed, and the legs may be strong and spiny for seizing elusive prey. Strong fliers often carry their prey to perching places or nests. Conspicuous examples of such aerial hunters are the adults of dragonflies, asilid flies, and the aculeate or stinging wasps. Stinging wasps are able to subdue prey such as katydids or spiders that are physically larger than themselves by injection of paralyzing chemicals with their sting. Dragonflies such as *Anax* (Aeschnidae) fly almost continuously, "hawking" for prey, while *Libellula* (Libellulidae) remain perched until prey approaches, then quickly dash in pursuit. Asilids commonly perch and await flying prey, then return to the perch after the victim is caught (Fig. 14.2*b*). Adult cicindelid beetles pursue their prey by running fast on open ground.

Stimuli other than visual may be used by hunting predators. Although blind, the famous army ants (genus *Eciton*) of the New World Tropics are enormously successful predators (Fig. 7.2*b*). Odor and movement of prey are detected by chemical and tactile receptors while the colony is on one of its periodic raids. Prey many times larger than the individual worker ants are attacked by massive swarms (Wilson, 1971). Cooperative foraging by "packs" of other kinds of ants also subdues large prey.

The back swimmers (Notonectidae) are voracious aquatic predators. They flush and stalk prey by sight and are able to detect the vibrations of prey movements with receptors situated along the forelegs. The water striders (Veliidae) also perceive prey by sight and vibrations (Bay, 1974).

Insects that ambush prey conserve energy by simply waiting for prey to approach within striking distance. Reduviid and phymatid bugs often remain motionless on flowers awaiting flower visitors. Such predators are exposed to predation by vertebrates and have either con-

FIGURE 14.2 Insect predators: *a*, praying mantis; *b*, asilid fly perching with beetle prey. (a, *photo courtesy of and with permission of Carl W. Rettenmeyer;* b, *photo courtesy of and with permission of Kenneth Lorenzen.*)

cealing or warning coloration (Chap. 15). Concealing coloration may also prevent detection by alert prey. Raptorial or clasping forelegs are frequently characteristic of insects that ambush, especially among the Hemiptera (Fig. 34.2*a* - *c*). Some reduviids aid prey capture by smearing sticky secretions or plant resins on their legs. The beaks of predatory bugs are stout and inject saliva that contains proteolytic enzymes. This is why their bites are painful to humans in contrast to the mild bites of bugs that are parasites of humans. Some reduviids also have a potent toxin that quickly renders prey helpless (Miles, 1972). Once the tissues of the victims are digested, the resultant fluid is sucked by the bug.

The praying mantis is a familiar predator that strikes prey from ambush (Fig. 14.2*a*). The coordination of the depth-perceiving vision, mobile head and prothorax, and toothed, raptorial forelegs has been carefully analyzed by Mittelstaedt (1962). The complete strike, timed by high-speed photography, takes 50 to 70 milliseconds. According to Roeder (1967), a fly or cockroach would require about 45 to 65 milliseconds to respond if startled by the first movement of the mantis; alas, too slow to escape the strike already in motion. The dragonfly naiad similarly grasps prey with quick strikes of its extensible labium. Cicindelid larvae seize prey that pass near the entrance to their burrows in soil.

Only a few kinds of insects trap prey. The "ant lion" larvae of Myrmeleontidae (Neuroptera) excavate conical pits in fine, loose sand (Fig. 38.2*a*). The larva waits motionless and buried at the bottom until a small insect ventures into the pit. The ant lion flicks sand toward the victim to cause the unstable sand to slide down. Once the prey is seized, the ant lion extracts the body fluids with its sickle-shaped jaws. Larvae of the fly *Vermileo* (Rhagionidae) also construct pits and trap prey. Wheeler (1930) aptly describes these insects as "demons of the dust."

Larvae of the New Zealand glowworm, *Arachnocampa luminosa* (Mycetophilidae) live in moist caves. They spin slimy webs that dangle from the ceiling. The larvae glow in the darkness, attracting small flies that become entangled in the sticky webs. Larval mycetophilids of the genus *Platyura* also secrete webs that entangle prey and contain a toxic fluid. Even though silk is secreted by various insects, webs are surprisingly rarely used to trap prey in the manner of spiders. The aquatic caddis fly larvae of the Hydropsychidae spin webs that filter small prey and other food from passing currents. Though they do not spin their own webs, one group of slender reduviid bugs, the emesines, wait in ambush at spider webs to attack trapped prey.

PARASITOID INSECTS

Only the larvae of holometabolous insects, especially Diptera and Hymenoptera, are parasitoids. Their unique attributes combine certain features of true parasites and predators. Parasitoids differ from predators in the following features: (1) only one host is required, (2) the host is larger than the parasitoid, (3) parasitoids are frequently host-specific, attacking one or several related host species, (4) a lower density of host population will sustain a parasitoid population, and (5) the victim is usually searched for and selected by the diurnally active, adult female (Endpaper 1*c*). A species of parasitoid may be specific not only in the choice of host species, but also in the choice of the life stages of the host to be attacked. Eggs or young or both are most frequently attacked, but pupae and sometimes adult hosts are also eaten.

Parasitoids may feed externally as ectoparasitoids or internally as endoparasitoids. Exposed hosts are usually attacked by endoparasitoids, whereas hosts in protected situations such as leaf mines, galls, or nests, are attacked by either endoparasitoids or ectoparasitoids. When the larva of a species characteristically develops in the ratio of one to a host, the species is termed a *solitary parasitoid.* When several larvae of the same species normally develop in a single host, the species is called a *gregarious parasitoid.* If the host is a phytophagous insect, the parasitoid is a *primary parasitoid.* Parasitoids that attack other parasitoids are called *hyperparasitoids* or secondary parasitoids. *Multiple parasitoidism* occurs when two or more species of primary parasitoids attack one host individual. *Superparasitoidism* occurs when a host is attacked by more larvae of the same species than can reach maturity in the one host.

Adult parasitoids are free-living, and most of them are winged except for female velvet ants (Mutillidae) and certain other scolioid wasps. Energy for their activity and egg production is derived from food stored by the larva or by feeding as an adult. In the latter case, nectar and honeydew are frequently consumed. Some parasitoid wasps feed on the body fluids that are released from the host's body when it is punctured by the parasitoid's ovipositor. This is called *host feeding*. In some instances, the host is inside a cocoon or for some other reason can be reached only by the parasitoid's ovipositor. A tube of coagulated host's hemolymph, or possibly of a secretion produced by the parasitoid, forms around the ovipositor. When the ovipositor is withdrawn, the parasitoid is able to suck the host's hemolymph as it wells up in the tube.

The adult parasitoid searches first for the habitat in which the appropriate hosts exist. Primary parasitoids are often attracted mainly by the plants that shelter favored insects hosts. In this behavior the parasitoids respond like phytophagous insects to shapes, colors, and odors of plants, yet they do not feed on plant tissues. Once near the host, female Hymenoptera usually search for suitable individuals on which to oviposit, using their tactile and olfactory senses. Their movements may be essentially random or somewhat systematic in response to stimuli detected at a distance. Female Diptera lack the well-developed antennae of wasps and the needlelike ovipositor for precisely inserting eggs in hosts. Consequently, the flies commonly depend on first instar larvae to actually locate hosts after the eggs are laid in the proper habitat.

Insects that have two or more successive larval instars specialized for different modes of life are *hypermetamorphic* (see taxa listed in Table 14.1). Among the hypermetamorphic parasitoids, the first instar larva is specialized for active host finding and the subsequent instars specialized for feeding.

The ovipositor of Hymenoptera plays an important role in selecting the host, sometimes in paralyzing it, and in delivering the egg to the host (Figs. 46.9, 46.11*d*). Dethier (1947) demonstrated that chemoreceptors on the ovipositor tip responded to a variety of chlorides, aliphatic alcohols, and hydrochloric acid in much the same way as taste receptors on the anterior appendages. These receptors function in the final discrimination of hosts before the egg is laid. Hosts inside tough cocoons or sclerotized puparia, in leaf mines, or even deep in burrows in solid wood can be reached by parasitoids equipped with ovipositors of suitable length and strength. By alternate movements of the sharply ridged valvulae, parasitoids can drill through resistant substrates. Receptors at the tip sense the presence and suitability of the host. The highly elastic eggs then pass down the minute channel inside the ovipositor by becoming greatly elongated. After deposition in the host, eggs of some species increase in size by as much as 1000 times. Some wasps temporarily or permanently paralyze or kill the host before ovipositing. For example, the venom of *Bracon hebetor* (Braconidae) causes permanent paralysis of host caterpillars when diluted up to 1 in 200 million parts of host hemolymph.

Superparasitoidism and multiple parasitoidism may be advantageous to the parasitoids in that the host's encapsulation reaction is dissipated. The resulting competition among larvae also may be disadvantageous. First instar larvae of certain wasps are specialized to compete successfully in such situations. Some have large sickle-shaped mandibles and kill competitors; others release inhibitory toxins or, if larger, deprive smaller competitors of oxygen. Competition is also commonly reduced by the behavior of the female wasp. For example, *Trichogramma* (Trichogrammatidae, Chalcidoidea) reject hosts on which other females have walked. Some wasps seem to be able, by means of sensilla on their ovipositors, to select healthy host individuals from among those already parasitized.

Endoparasitoids are immersed in the body tissues and fluids of their hosts. Many synchronize their development with that of the host by responding to the host's hormones. Some obtain food in the normal manner via the mouth, but others, especially those in eggs, are greatly simplified and absorb food through the outer integument.

Endoparasitoid wasp larvae commonly exchange gases through their outer integument, with or without a tracheal system and rarely with open spiracles. Respiration may be aided by caudal filaments, anal vesicles, or an everted hindgut. Endoparasitoid fly larvae usually have at least a metapneustic tracheal system (Chap. 5). Some perforate the host's trachea or integument and obtain atmospheric air. Others take advantage of the host's encapsulation reaction and mold a respiratory "funnel" or tube of host tissue connected to a trachea or the integument.

Among the parasitoids, Hymenoptera have several features of reproductive biology that are noteworthy. Recall that sex is determined in this order by a haplodiploid mechanism. The ratio of sexes is often unbalanced in favor of females. A peculiarity of Aphelinidae (Chalcidoidea) is that the female may select the host according to the sex of the egg to be laid. Furthermore, the sex of the larva may influence its feeding behavior; males may be regular hyperparasitoids. Some wasps are exclusively parthenogenetic, in contrast to other entomophagous insects that are rarely so. The unusual phenomenon of polyembryony (Chap. 4) is found only among parasitoids (Fig. 14.3*b*) and parasites; it occurs in some endoparasitoid Hymenoptera and a few Strepsiptera. As many as 1500 embryos in a single caterpillar have been counted after oviposition by one female of *Litomastix* (Encyrtidae).

PARASITES

True parasitic relationships between one insect and another are surprisingly rare. As we have seen in the previous sections, the usual result of attack is the death of the victim. Entomophagous parasites may be divided into ectoparasites and endoparasites.

Adult biting midges of the genus *Forcipomyia* (Ceratopogonidae) and several other genera are ectoparasites that take blood from both vertebrates and insects. When feeding on the latter, the flies puncture the wing veins or intersegmental membranes (Fig. 14.3*a*). The flies visit the host only during feeding.

The small wasps of the family Scelionidae lay their eggs among the freshly laid eggs of host

FIGURE 14.3 Parasitic and parasitoid insects: *a*, ceratopogonid fly sucking hemolymph from leg of a phasmid; *b*, emergence of most of 789 brood produced by the polyembryonic wasp parasitoid, *Pentalitomastrix plethroicus* (Encyrtidae), from caterpillar of navel orangeworm, *Paramyelois transitella* (Pyralidae). (a, *photo courtesy of and with permission of Carl W. Rettenmeyer;* b, *photo courtesy of and with permission of Frank E. Skinner.*)

insects. Females of several genera have been observed attached phoretically to the bodies of female hosts. After the host's eggs are laid, the wasp immediately oviposits. Young winged females of the French *Rielia manticida* (Scelionidae) search for and attach themselves to mantids of both sexes. Female hosts are more frequently selected. Once attached by their mandibles, the wasps shed their wings and await the host's oviposition, which may not take place for several months. In the interim, the wasp feeds as an ectoparasite on the mantid's hemolymph. When the mantid's ootheca has been deposited and is still soft, the wasp leaves the mantid to lay its own eggs. The wasp is said then to attempt to return to its host's body.

An unusual instance of ectoparasitic behavior is provided by the aquatic larvae of the midge *Symbiocladius* (Chironomidae). The larvae attach themselves behind the wing buds of mayfly naiads and feed.

Strepsiptera are the only entomophagous insects that are true endoparasites. Numbering about 300 species, they probably evolved in the early Tertiary Period from some group, as yet unknown, of parasitoid Coleoptera. Fossils have been found in Baltic amber. Hosts include insects in the orders Thysanura, Blattodea, Mantodea, Orthoptera, Hemiptera, and Diptera, but most of them are aculeate Hymenoptera. The infested hosts are said to be "stylopized," because the name of a common genus is *Stylops* (Fig. 40.2*d*).

The life history of *Stylops pacifica* (Stylopidae) was studied by Linsley and MacSwain (1957) near Berkeley, California. The hosts are two species of bees in the genus *Andrena* (Andrenidae). On bright warm days in February and early March, adult bees emerge from their burrows in soil and visit flowers of the buttercup, *Ranunculus* (Fig. 14.4*a,c*). As many as 16 percent are parasitized. The visible evidences are the puparia that protrude from the body, commonly between the fourth and fifth terga of the abdomen (Fig. 14.4*a*). The male *Stylops* is winged and less than 3 mm long (Fig. 14.4*b*). On emergence from the puparium, it immediately begins a rapid, vibrating flight in search of a female. The latter is reduced to virtually a sac of reproductive organs inside the puparium. Females release a sex attractant while the host flies from flower to flower. Males fly upwind, tracking the odor, until the female is located on the dorsum of a feeding bee. The male *Stylops* lands and inserts the aedeagus by puncturing the female's puparium (Fig. 14.4*c*). Once inseminated the female no longer releases the attractant. Males probably die the same day as they emerge and mate but once. Except during copulation their intense flight activity never ceases and they rarely feed. The tarsi even lack claws for clinging to objects. Linsley and MacSwain were able to capture the rare males by putting bees with virgin female parasites in cages and placing the cages among flowers.

The first-instar larvae are triungulins, and they develop from the fertilized eggs while still inside the female's body. During the next 30 to 40 days, the female dies and the larvae move into a median brood passage in preparation for their exit. A large female *Stylops* probably produces 9000 to 10,000 triungulins. On warm days while the host is visiting flowers, the active larvae emerge through the ruptured puparium (Fig. 14.4*d*). Parasitized hosts move more rapidly among flowers than normal hosts. The triungulins are brushed singly or several at a time onto the flowers.

In contrast to the triungulins of Meloidae, Rhipiphoridae, and possibly other Strepsiptera, those of *Stylops pacifica* do not readily attach themselves to new hosts, but are ingested with nectar by bees (Fig. 14.4*e*). The bee returns to its nest and prepares a ball of pollen mixed with regurgitated nectar. In this way the triungulin is deposited in a nest cell with the bee's egg (Fig. 14.4*f*). The triungulin penetrates the egg, transforming as it does into the second instar, and becomes an endoparasite (Fig. 14.4*g*).

As the host feeds and grows, the *Stylops* larva feeds on the host's nonvital tissues and passes through an unknown number of instars (Fig. 14.4*h*). When the bee pupates, the *Stylops* protrudes the anterior portion of its body through the intertergal membrane and also pupates (Fig.

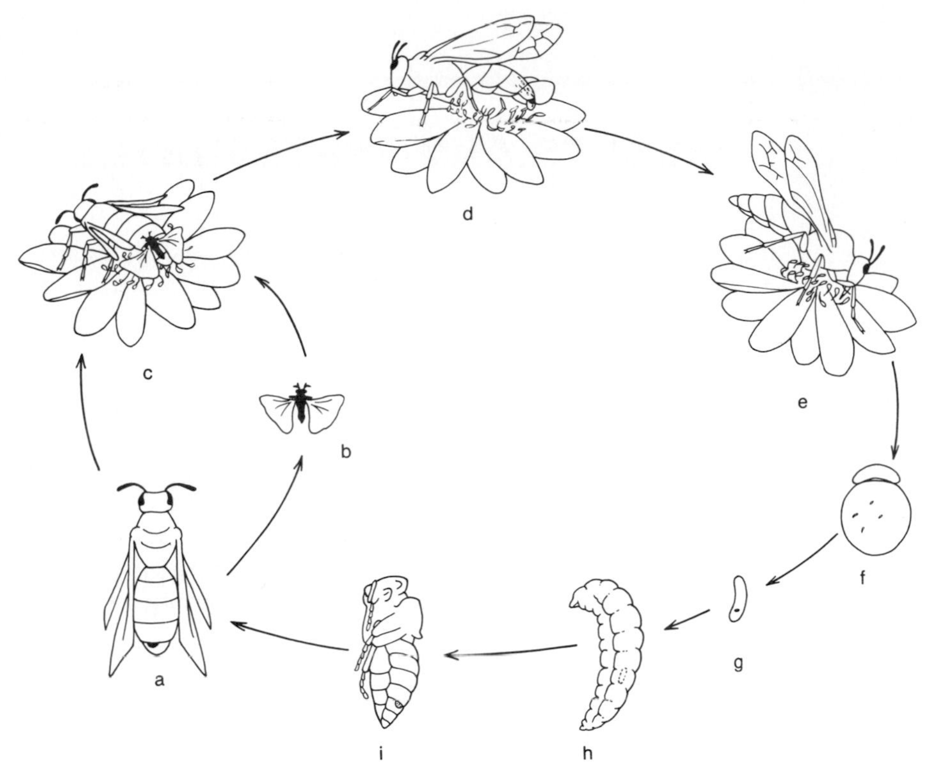

FIGURE 14.4 Life history of *Stylops pacifica* (Strepsiptera). See text for explanation. (*Redrawn from Linsley and MacSwain,* 1957; *published* 1957 *by The Regents of the University of California; reprinted by permission of the University of California Press.*)

14.4*i*). The last larval skin is not shed, but forms a tanned puparium.

Strepsiptera have no known natural enemies. Major causes of mortality are the initial losses of triungulins that fail to find a female host of the proper species, and superparasitism. Linsley and MacSwain noted that usually only the larger female bees with a solitary female parasite lived long enough for the triungulins to escape. Smaller male hosts, those with several *Stylops,* or those with *Stylops* plus nematode parasites, usually died before the cycle was completed.

Bees that survive parasitism are variously affected. The genitalia of both sexes may be reduced. Secondary sexual characters, especially in the female, may be shifted toward the opposite sex. Stylopized females may have more malelike yellow color and reduced pollen-collecting hairs on the legs.

SELECTED REFERENCES

Comprehensive treatments of entomophagous insects include Askew (1971), Balduf (1935, 1939), Clausen (1940), Hagen et al. (1976), and Price (1974). Control of insect pests by use of natural enemies is treated by DeBach (1964, 1974), Huffaker (1971), Huffaker and Messenger (1976), Metcalf and Luckman (1975), and van den Bosch and Messenger (1973). Reviews are given for cellular defense reactions by G. Salt (1970) and Whitcomb et al. (1974); host selection by Vinson (1976); and phoresy by Clausen (1976). For references on specific entomophagous taxa, consult the appropriate chapters in Part IV.

15  INSECTS AND VERTEBRATES

Insects have coexisted with freshwater and terrestrial vertebrates throughout most of their evolutionary history. In the course of this long association, trophic relationships have evolved such that certain insects are parasitic on vertebrates and certain vertebrates are predatory on insects. The orders Phthiraptera and Siphonaptera, and various families and species of Dermaptera, Hemiptera (Suborder Heteroptera), Lepidoptera, Coleoptera, and Diptera are dependent on the tissues or blood of vertebrates as food during part or all of their lives. As you will learn in the next chapter, this parasitic behavior also has made possible the transmission of certain microbes and helminths among vertebrate hosts, including some of the most important diseases of human beings.

In the geological past, the abundance of insects as food for vertebrates played a key role in the early evolution of reptiles, birds, mammals, and the primates. Judging by the structure of fossils, the early forms of each group appear to have been mainly insectivores. Today insects continue to be an important source of food for vertebrates. Many familiar birds in urban and rural areas feed almost exclusively on insects and in some instances significantly reduce pest populations. Likewise some freshwater fish include insects as a regular part of their diet. The mosquito fish, *Gambusia affinis,* is an ideal agent for biological control of mosquitoes in ponds and irrigated rice fields. By at least 200 A.D., human anglers had learned to imitate insects by making "flies" of feathers or hair. Thus the popular recreation of fly-fishing takes advantage of the predatory behavior of fish on insects.

INSECT PREDATORS OF VERTEBRATES

The difference in size between vertebrates and insects protects most vertebrates from insect predators. Some large insect predators, however, are able to catch and kill small vertebrates. Large dragonfly naiads, dytiscid larvae (Fig. 39.4*b*), and belostomatids capture small fish or tadpoles. These aquatic predators are unwelcome in home aquaria or in fish hatcheries. Large praying mantises may kill young birds. Chicks and young birds are also killed by ants as small as Argentine ants. Ants probably initially seek moisture, but soon also take blood. Poultry farms have reported losses of chicks killed by ants.

INSECT PARASITES OF VERTEBRATES

The larger size of vertebrates prevents predation on them by most insects, but does not prevent parasitism. All terrestrial vertebrates are subject to attack. Only vertebrates that remain partly or completely submerged in water throughout their lives—fish, whales, and seacows—are exempt. Other marine vertebrates, such as seals and marine birds, including penguins, periodically leave the water and are hosts to lice and occasionally other kinds of parasites. For example, mosquitoes transmit filarial worms to sea lions and have even been seen biting fish at low tide (Downes, 1971).

Blood is the usual food and often is ingested with the aid of anticoagulants in the saliva. Other parasites subsist on sebaceous (oil) secretions, mucus, pus, lachrymal (tear) secretions, epidermis, skin debris, hair or feathers, or internal tissues. In some instances, the eating of foreign matter, fungi, or microbes on the host's skin is probably beneficial, but most parasites are detrimental to their hosts in some degree. Nevertheless, the health and survival of the host are essential to the parasite.

As a rule, the host's reaction to a new parasite is severe, while long-established parasites have evolved in such a way as to minimize irritation. On its normal host the bite of a bloodsucking insect is often apparently painless, but on other hosts the reaction may be swift. Occasionally hosts die as a direct result of parasitism. Excessive numbers of bloodsucking insects may kill a host by exsanguination or toxic effects, or internal parasites may destroy vital organs. Debilitating or fatal pathogens may be transmitted, or secondary infections may develop in wounds.

Some terms will be needed for our discussion of insect parasitism (James and Harwood, 1969). Parasites located on the outside of the host's body are *ectoparasites,* those inside the body are *endoparasites. Continuous* parasites remain on the host's body throughout the parasite's life. *Transitory* parasites spend a part of their life on the host and part elsewhere as a free-living insect. *Intermittent,* or *temporary,* parasites visit the host only to feed. *Facultative* parasites are insects which normally complete their life without parasitism, but are able to survive as parasites under certain circumstances. *Obligatory* parasites require a host to complete their life.

Parasites have evolved repeatedly from different lineages during the long association of insects and vertebrates as the codominant terrestrial animals. For our purposes, we divide parasites into three broad categories based on their general adaptation to parasitism: (1) *crawling ectoparasites* are apterous or sometimes winged and cling to their host or crawl on their hosts at least during feeding; (2) *flying ectoparasites* are winged, active fliers and alight at or near the site of feeding; and (3) *myiasis-producing Diptera* are tissue- or blood-feeding larvae.

TAXA OF PARASITES AND THEIR CHARACTERISTICS

CRAWLING ECTOPARASITES. Names of taxonomic groups of *continuous or apparently continuous ectoparasites* follow. (1) Suborder Hemimerina (Dermaptera), 10 species, eat epidermis and skin debris of southern African rats of the genus *Cricetomys* (Fig. 15.2*b*). (2) Polyctenidae (Hemiptera), or bat bugs, 6 species in the Old World and 12 in the New World, suck blood of bats (Fig. 15.1*c*). (3) Phthiraptera, or lice are divided into four suborders: Amblycera and Ischnocera, the biting lice (Fig. 33*a,b*), include 2800 species, most of which feed on feathers, skin debris, and sometimes blood of birds, but some species in each suborder parasitize mammals. Rhynchophthirina, or elephant lice, include only *Haematomyzus elephantis* on African and Asian elephants, and *H. hopkinsi* on African warthogs. Anoplura, or sucking lice, include over 400 species which suck blood of many mammals. Species attacking humans are the human body louse, *Pediculus humanus* (Fig. 15.1*a*), and the pubic louse, *Pthirius pubis* (Fig. 33.3*d*). (4) Hippoboscidae (Diptera), or louse flies, include about 100 species which suck blood from birds and large mammals. (5) Streblidae (Diptera), or bat flies, suck blood of bats in both the New and Old Worlds (Fig. 15.2*c*). (6) Nycteribiidae (Diptera), or spiderlike bat flies, 195 species, suck blood of bats, mostly in the Malaysian region (Fig. 15.2*e*). (7) *Platypsylla castoris* (Platypsyllidae), or beaver beetles (Fig. 15.2*a*) feed on skin debris of beavers in Europe and North America, and the relict rodent, *Aplodontia rufa,* of coastal Northwestern United States. (8) Beetles of the tribe Amblyopinini (Staphylinidae, Coleoptera), 41 species, feed mostly on rodents and also on marsupials in

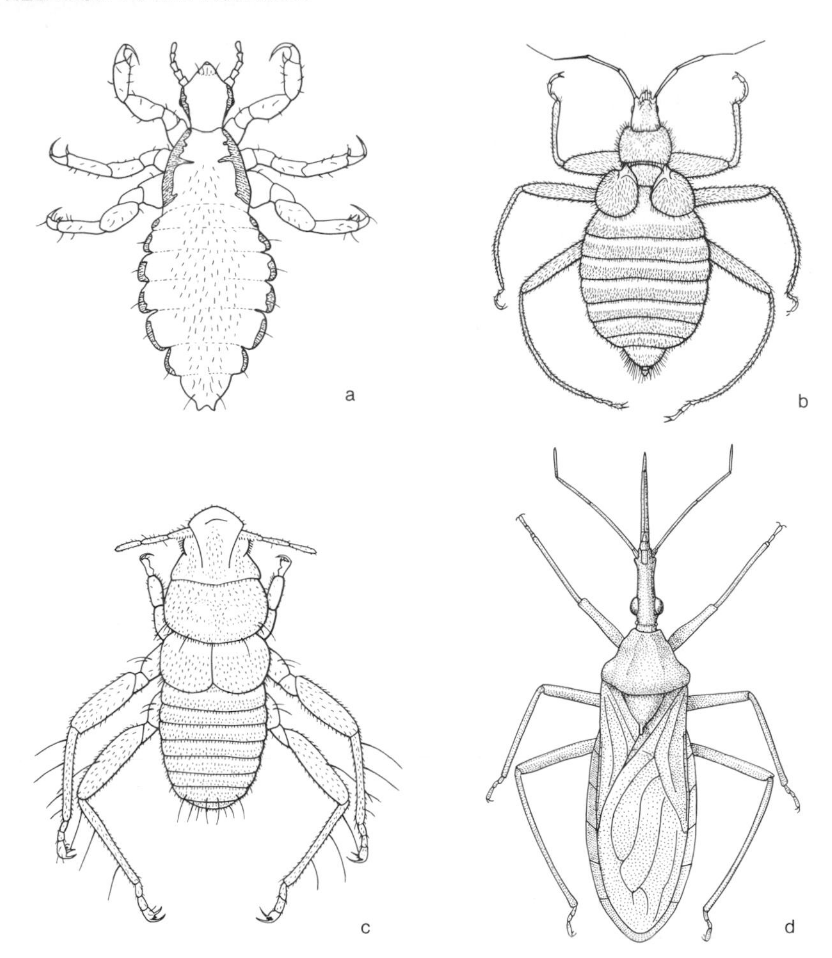

FIGURE 15.1 Parasitic hemipteroids: *a*, human body louse, *Pediculus humanus* (Pediculidae); *b*, primitive cimicid bat bug from Texas cave, *Primicimex cavernis; c*, polyctenid bat bug, *Hesperoctenes fumarius; d*, blood-sucking reduviid, *Rhodnius prolixus.*

Central and South America; one species feeds on a rat from Tasmania. Parasites become embedded in skin where they may suck blood, but specimens have also been found in bat guano. (9) Sloth moths are three species in the large Family Pyralidae (Lepidoptera): *Bradypodicola hahneli* on the three-toed sloth, *Bradypus* spp.; *Cryptoses choloepi* on the two-toed sloth, *Choloepus hoffmanni;* and *Bradypophila garbei* on *Bradypus marmoratus.* Adults are phoretic on sloths. It is not known whether they feed on sebaceous secretions or on the algae that grow in the grooves of the sloth's hair. When the sloth descends to the forest floor to defecate, the female moth leaves the host and oviposits in the feces. The larvae develop in the feces and later, as adults, find their way to a sloth.

Transitory ectoparasites are as follows. (1) Siphonaptera, or fleas, including nearly 1800 species, are ectoparasitic bloodsuckers in the adult stage only. The larvae live as scavengers in the host's nest or in ground litter of the host's habitat. Hosts are mostly mammals that nest in dens, holes, or caves. Fleas are not known to parasitize flying lemurs, primates [the so-called human flea, *Pulex irritans,* normally attacks pigs; the burrowing flea, *Tunga penetrans,* burrows under toenails of humans; the oriental rat flea, *Xenopsylla cheopis* (Fig. 15.2*d*), and others also bite

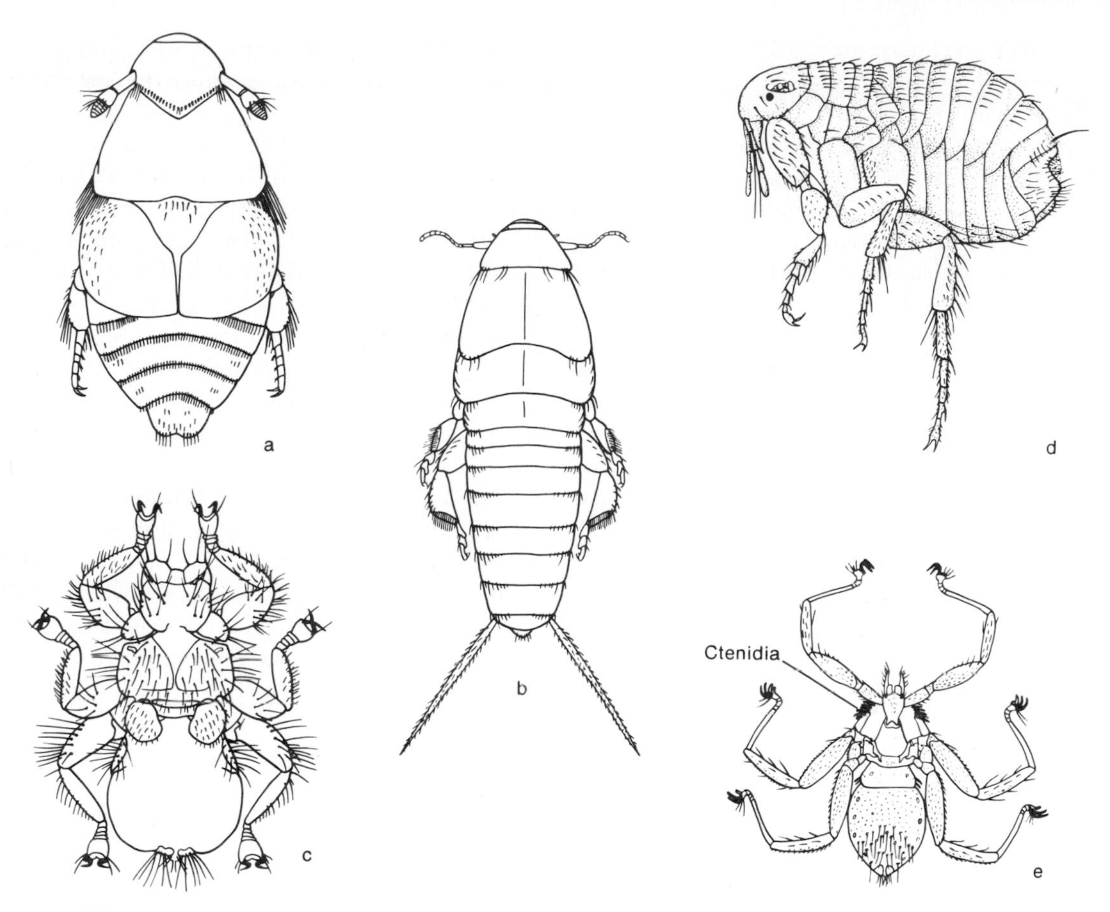

FIGURE 15.2 Various parasites of vertebrates: *a*, beaver beetle, *Platypsylla castoris; b*, parasitic dermapteran, *Hemimerus talpoides; c*, streblid bat fly; *d*, oriental rat flea, *Xenopsylla cheopis*, a vector of plague; *e*, nycteribiid bat fly.

humans among other hosts], aardvarks, elephants, or most horses and their allies. Birds are also hosts, mostly seabirds and perchingbirds, especially swallows. (2) Ectoparasitic beetles of the Scarabaeidae include three genera: adults of *Uroxys gorgon* and *Trichillum brachyporum* live in the hair of the South American sloth, *Bradypus,* and the larvae live in dung. Adults of species of *Onthophagus* (= *Macropocopris*) cling to hairs near the anus of kangaroos and wallabies, feeding on secretions and awaiting the host's feces, which are buried by the adult beetle as food for the larvae.

Intermittent, or temporary, ectoparasites are as follows. (1) Cimicidae (Hemiptera) or bedbugs (Fig. 15.1*b*), including 74 species, suck blood from hosts in all stages but visit the host only to feed. Twelve genera feed only on bats, nine genera feed only on birds, and only *Cimex* has some species feeding on bats and others on birds. *Cimex hemipterus* and *C. lectularius* (Fig. 34.6*b*) are human bedbugs which also feed on chickens and bats. (2) Triatominae (Reduviidae, Hemiptera), or cone-nose bugs, including about 80 species in the New World and 7 in the Old World, feed on blood in all stages and parasitize lizards, birds, and mammals, including humans (Fig. 15.1*d*)

The Arixeniina (Dermaptera) are sometimes regarded as ectoparasites because of their association with bat caves and bats, but actual parasitism has not been demonstrated.

Crawling ectoparasites exhibit a number of convergent adaptations associated with life in the hair or feathers of their hosts. In most species both sexes are parasitic and adults are more or less flattened. Siphonaptera are laterally compressed for movement through hair, the others are dorsoventrally flattened and manage equally well in hair or feathers. The adults of continuous parasites and Siphonaptera have generally reduced sensory and locomotory apparatus: (1) short antennae; (2) compound eyes small in some, absent in Hemimerina, Polyctenidae, some Anoplura, and Siphonaptera; (3) ocelli absent in endopterygote parasites, except some Hippoboscidae; (4) wings absent except in some Hippoboscidae and Streblidae, in which the wings may be well developed, vestigal, or absent; (5) legs usually short (Nycteribiidae have long legs), more or less stout, and with well-developed tarsal claws which may be single on mammalian hosts; (6) stiff, posteriorly directed hairs or scales, or a comblike ctenidium in Polyctenidae, Siphonaptera (Fig. 43.1*a*), Nycteribiidae, and *Platypsylla;* and (7) where the life history is known, mating and oviposition usually takes place on or near the host. The temporary parasites Cimicidae and Triatominae move to and from their hosts. Consequently they retain long antennae, compound eyes, and walking legs. The Cimicidae are otherwise apterous, and have small eyes and no ocelli. The Triatominae are winged and have well-developed eyes and ocelli.

FLYING ECTOPARASITES. All flying ectoparasites are intermittent parasites in the adult stage. Three groups are (1) *mandibulate biting flies,* which include species of Psychodidae (or sand flies), Ceratopogonidae (or biting midges), Simuliidae (or blackflies), Culicidae (or mosquitoes), Tabanidae (or horse- and deerflies), and Rhagionidae (or snipe flies). The parasites are females with bloodsucking mouthparts which include functional mandibles (Fig. 42.3*b*). A wide range of vertebrate hosts, both warm- and cold-blooded, are attacked. Humans are attacked by species in all families. Blood meals are usually required for egg maturation and production. Some species are able to produce the first clutch of eggs without a blood meal and are termed *autogenous.* The immature stages are free-living and usually aquatic or associated with damp soil. Larvae of tabanids and snipe flies are also found in dry litter or soil. (2) *Bloodsucking muscoid flies* include about 20 species of *Glossina,* or tsetse flies; *Stomoxys calcitrans* (Muscidae), or stable flies and their allies; *Haematobia irritans* (Muscidae), or hornfly; *H. exigua,* the buffalo fly of Australia; and *H. stimulans,* the cattle fly of Europe. Both sexes suck blood. Most feed on mammals, but *Glossina palpalis* prefers crocodiles and monitor lizards. Humans are bitten by various species of *Glossina* and *Stomoxys.* Larvae of *Stomoxys* and *Haematobia* live in manure, but *Glossina* is viviparous, producing a mature larva which soon pupates. (3) *Eye gnats* include species of *Hippelates* (Chloropidae) and certain muscids throughout the world. (4) *Eye moths* include 23 species of Noctuidae, Pyralidae, and Geometridae in Africa and Southeast Asia. These insects are attracted to the eyes of mammals and humans, where they feed on lachrymal and sebaceous secretions, pus, and sometimes blood. The fruit-puncturing noctuid moth, *Calpe eustrigata,* is able to puncture mammalian skin and take blood.

MYIASIS-PRODUCING DIPTERA. Larvae of certain flies infest the bodies of animals, including humans, and produce a disease called myiasis (Zumpt, 1965). Pupation is usually in the soil, and adults of the flies are free-living. Four groups of myiasis flies have been recognized (James and Harwood, 1969): (1) *Enteric myiasis and pseudomyiasis*-producing larvae of many species, mostly in Muscidae, Calliphoridae, Sarcophagidae, and Gasterophilidae, have been reported in alimentary canals of vertebrates. Pseudomyiasis involves fly larvae which are ingested accidentally or enter through the anus and are able to survive for varying periods in the intestine. The larvae may eat the ingested food of the host or, less frequently, tissues of the intestine. True enteric myiasis is created by the larvae of horse bot flies, *Gasterophilus* (Gasterophilidae), which live attached to the stomach wall while feeding on the wall and absorbing the host's food. (2) *Facultative myiasis*–producing larvae of many species, mostly in Muscidae, Calliphoridae, and Sarcophagidae, normally live as scavengers in dead animals. Adult females, attracted by odors of decomposition, lay eggs on the carcass. Wounds or deep, soiled wool of sheep also attract ovipositing females. The resulting larvae become parasites in such places on living animals. (3) *Obligatory myiasis*–producing larvae regularly infest living hosts. The primary screwworm, *Cochliomyia hominivorax* (Calliphoridae), of the New World; Old World screwworm, *Chrysomyia bezziana* (Calliphoridae); and *Wohlfahrtia magnifica* (Sarcophagidae), also of the Old World, enter the body at wounds or through mucous membranes and create local infestations. Larvae of the cattle grubs, *Hypoderma lineatum* and *H. bovis* (Oestridae) penetrate unbroken skin, migrate extensively in the host's body, and return to complete their development in tumors beneath the skin before dropping from the host to pupate. The female of the human bot, *Dermatobia hominis* (Cuterebridae), attaches her eggs to another fly, such as a blowfly, tabanid, or mosquito. While the carrier insect is feeding, the bot larva hatches and drops to the host's skin, burrows into the subcutaneous tissues, and remains in the same place until fully fed. Mature larvae drop from the host to pupate. These flies occur in Latin America. (4) *Blood-sucking larvae* that are temporary feeders include the Congo floor maggot, *Auchmeromyia luteola* (Calliphoridae), which lives in huts and feeds on sleeping persons; and various species of *Protocalliphora* (Calliphoridae) in the Old and New Worlds which feed on nestling birds.

EVOLUTION OF INSECT PARASITES

When insects first appear abundantly as fossils in the Carboniferous Period (Table 17.3), cold-blooded vertebrates were already present. Amphibians appeared earlier near the end of the Devonian Period, but the first reptiles coincide with the sudden appearance of insects. In the Upper Carboniferous and Permian Periods, reptiles diversified and replaced amphibians as the most abundant land vertebrates. By the close of the Paleozoic Era, insects and vertebrates had coexisted for more than 80 million years, or nearly the first quarter of their association. Did parasites evolve during the first radiation of insects?

Recalling that piercing-sucking mouthparts were well developed at this time in both the Paleoptera and Neoptera, it would be surprising if some insects so equipped did not occasionally suck vertebrate blood or become regular parasites. Yet no Paleozoic insects are believed to be parasites, and parasites are almost entirely lacking among the more primitive insects which survive today: Apterygota, Paleoptera, and Orthopteroidea (except the Suborder Hemimerina, which probably evolved much later). Furthermore, no major groups of modern insect parasites are exclusively adapted to cold-blooded vertebrates, but some, such as sand flies, do attack lizards, snakes, and toads. To answer our question, we must conclude that no positive evidence exists of early insect parasitism.

The possibility remains, of course, that parasites of the first vertebrates perished with the progressive extinction of their hosts and left no record. In contrast, other arthropods, such as mites and ticks, did evolve specialized parasites of cold-blooded vertebrates. The Phylum Pentastomida or Linguatulida are sometimes considered arthropods. These are internal parasites of snakes, crocodiles, birds, and mammals.

In the Mesozoic Era the reptiles dominated the land by virtue of their numbers and variety, plus the immense size of some species. The separate reptilian lineages leading to the mammals and to the birds are now thought to have been warm-blooded as early as the Permian and to have evolved insulating coats of hair or feathers during the Mesozoic. True mammals are recorded from the Triassic Period. If the Jurassic *Archaeopteryx* is considered really a dinosaur, then the first birds are known in the Cretaceous Period. Insects of the Mesozoic, therefore, had cold- and warm-blooded reptiles, mammals, and birds as potential hosts. In this era we would expect definite evidence or inferences of parasitism, and we are not disappointed.

The earliest parasites were probably the mandibulate biting flies. The reasoning behind this inference is as follows: Four major subgroups of Diptera are recorded in the Triassic, indicating that considerable evolution had already occurred before the fossils were preserved. Among the modern flies in these groups, only the bloodsucking parasites and closely allied predators retain functional mandibles. Piercing mouthparts, with only slight modifications, serve either parasitic or predatory functions. Sucking blood, whether of insects or vertebrates, must have been the original habit or was developed early in the history of the Diptera before mandibles were lost in the main lineages (Downes, 1958, 1971). Flying ectoparasites are not dependent on the host's body covering for shelter. Biting flies today are noted for their broad spectrum of hosts, including cold-blooded vertebrates. Thus, the Diptera could have arisen in the Permian as the first parasites of vertebrates and could later have expanded their choice of hosts to include warm-blooded vertebrates. The latter offer additional cues for hostfinding, such as warmth, water vapor, and increased carbon dioxide from the higher metabolism. Carbon dioxide is a common attractant for parasites of mammals.

Some of the major groups of crawling ectoparasites also could have evolved at this time after some vertebrates acquired a coat of hair or feathers. In contrast to the skin of cold-blooded vertebrates, this covering provides a uniformly warm, humid environment with something to clasp while the host is in motion, and protection against the host's efforts to rid itself of small passengers. When coupled with the nest-making habit of terrestrial vertebrates, the stage was set for repeated, independent evolution of crawling ectoparasites. Most of these probably evolved from insects that lived in the host's nest or den as scavengers or as predators on other insects. The scavengers probably crawled on the host and fed on hair, feathers, skin debris, or secretions. Later they could take blood from wounds made by their biting. This is probably the origin for most of the continuous and transitory crawling ectoparasites listed above. Bedbugs and conenose bugs probably arose from a predatory ancestor because the closest relatives of each are predators.

The Siphonaptera are believed to have evolved first on hairy hosts, because the laterally compressed body seems suited for passage through a pelt. The ancestor was probably a mecopteran which lived in nests of vertebrates as a scavenger and became parasitic in the adult stage. Fleas retain the scavenging habit in the larval stage. Fossil Mecoptera are well represented in the Permian, and what may be fossil fleas have been described from the Lower Cretaceous (Riek, 1970*b*).

The origins of the Phthiraptera are more difficult to unravel. The ancestor was doubtless a scavenging insect related to the Psocoptera. These insects commonly occur in or near the nests of vertebrates (Hicks, 1959) and freely crawl on the bodies of captive reptiles. Once the insects became flightless, continuous parasites, and dependent on their hosts, their evolution was strongly influenced by the evolution and habits of their hosts. Transfer among host individuals of the same species can occur when the hosts accidentally contact each other, copulate, or care for their young. Rare instances of phoresy are known, in which lice attach to flying insects and are possibly transported to other hosts. Opportunities to transfer to new species of hosts are presented when the normal host is killed and the lice seek the warmth of the vertebrate predator—or when cuckoo birds lay their eggs in the nests of other species. Host specificity, however, is highly developed in the lice. In fact, only rarely are lice species on two or more hosts that are not closely related phylogenetically. Yet the evolution of the higher categories of lice remains clouded. Both suborders of biting lice, Amblycera and Ischnocera, have bird and mammalian hosts, indicating that host transfer did occur at times in the past. Possibly the first hosts were warm-blooded reptiles whose extinction leaves the early history of lice unknown. The Rhynchophthirina and Anoplura, both exclusively on mammals, are thought to have evolved from the Ischnocera (Clay, 1970).

Other parasites that are closely associated with certain birds or mammals probably evolved in the Tertiary Period, when these vertebrates became dominant. Bats, for example, are exclusive hosts for Nycteribiidae, Streblidae, Polyctenidae, and many Cimicidae, including the most primitive species *Primicimex cavernis* (Fig. 15.1*b*). The first bat fossils are from the Lower Eocene Epoch, and this would seem to be the earliest origin for this peculiar assemblage of parasites. Several circumstances favor bats as hosts, including the resting of bats in sheltered caves, the accumulations of guano that attract insect scavengers and their predators, and the inability of bats to groom themselves.

Bats are oddly not now hosts for Phthiraptera, but possibly lice were displaced by other parasites at an earlier geologic time. Competition is apparently keen among individuals of several parasitic species on a single host. The nature of the competition is not well studied, but ordinarily one species predominates or excludes others from certain areas on the host or excludes them entirely from the host's body. This may provide an explanation for the curious absences of certain parasites from major vertebrate groups which are suitable hosts for other parasites.

Many insects seek moisture and proteinaceous substances such as mucus from the skin surfaces of vertebrates. The eye gnats and eye moths undoubtedly evolved their association with vertebrates in this manner. The sponging labella of muscoid flies is constructed for ingesting superficial fluids. One can easily imagine that the bloodsucking muscoids evolved from sweat- and mucus-seeking flies. As the labella was progressively sharpened and a rasping device developed at the tip, the flies were able to create wounds while taking moisture. Thus the bloodsucking habit was reacquired independently by a second group of flies, probably long after the evolution of mandibulate biting flies. Their later appearance is inferred not only because the muscoids are advanced phylogenetically, but also because the hosts are mainly perrisodactyls and artiodactyls which date from the Lower Eocene.

Even within the bloodsucking muscoids, there is evidence that the evolution of a piercing labella occurred more than once. The tsetse flies, *Glossina*, resemble the stable flies, *Stomoxys*, but Pollock (1971) has shown that tsetse flies actually are closer to Gasterophilidae and Hippoboscidae. The mouthparts of the tsetse must have evolved convergently to resemble those of *Stomoxys*. *Glossina* is found as a fossil in the Florrisant shales of Colorado, which date back to the Oligocene Epoch. Apparently the flies became extinct in North America with the extinction of their hosts, and survive now only in Africa.

A recent event gives insight into the evolution of myiasis-producing Diptera. Blowflies (Calliphoridae) are normally scavengers as larvae on dead animals and animal wastes. Several circumstances in Australia led to the rapid evolution of parasitic behavior by the sheep blowfly, *Lucilia cuprina*. As a part of a range management plan, rabbits were killed in massive quantities. Large populations of flies were produced from the dead rabbits. Poisons used to kill rabbits and dingos also killed carrion birds, which ordinarily reduce maggot populations by tearing open the carcasses. At the same time, sheep breeders developed animals with thick wool on wrinkled skin and with increased oily content. Bacteria thrive in deep folds where the oily wool is soiled and moist with sweat. The increasingly abundant blowflies began to lay their eggs in such areas. When the larvae bore into the sheep's flesh, a myiasis is created which may lead to death of the host (Zumpt, 1965).

EVOLUTIONARY EFFECTS OF PREDATION BY INSECTIVOROUS VERTEBRATES

Insects are a regular part of the diet of many vertebrates: freshwater fish, amphibians, reptiles, birds, and small mammals, including bats. Even humans are known to vary their diet with insects. The predatory effectiveness of these insectivores is enhanced by one of the following or a combination of (1) physical agility; (2) keen senses of hearing, smell, taste, and/or vision (including color perception by some fish, some reptiles, most birds, and primates); (3) hunting strategies; and (4) the capacity to learn from experience. Insects are subject to predation in aquatic, terrestrial, and aerial habitats, during both day and night. To survive, exposed insects have evolved special patterns of coloration, behavior, and defensive properties to take advantage of what the predator cannot detect, cannot capture, or will not eat.

The protective strategies in each case have evolved and are maintained by selection pressures from one or more kinds of insectivores in a given area. It follows that a widespread insect species may be attacked by different combinations of predators at different stages of its life, in different seasons, at different times of day, and accordingly may vary in response over its geographic range. In spite of the unique circumstances surrounding each instance of protective coloration and behavior, certain strategies recur again and again in different insect taxa. This is because within each dominant group of predators (e.g., birds) the habits, learning ability, and senses are so much alike that closely similar strategies to escape capture are evolved convergently by insects. These convergences provide some of the best opportunities to study natural selection in action. Much speculation has been made, but too few instances of the protective strategies have been analyzed with critical experiments (Cott, 1957).

For our discussion, we first divide insect prey into those that are *palatable* (edible or acceptable) to a given kind of predator and those that are *unpalatable* (inedible or rejected). This distinction must be made with qualifications. Some individual insects will be relatively more or less palatable than others, and individual predators will vary in accepting prey depending on hunger and previous experience. Hard or spiny bodies, sticky secretions, stings, stinging hairs, and distasteful or toxic chemicals are defenses that tend to render an insect unpalatable (Rettenmeyer, 1970). Even a low frequency of unpalatable individuals however, will be an advantage favored by selection in the prey population. Unfortunately we do not know the palatability of most insects relative to their main predators, but observations indicate that most species of insects are edible by a wide variety of insectivores.

PALATABLE INSECTS. It is assumed that relatively palatable species evolve either an effective means of *escape,* or a disguise, or both. Two kinds of disguises are common: (1) *camouflage,* or *concealing coloration,* to avoid detection, and (2) *Batesian mimicry* of an unpalatable species to avoid attack. Batesian mimicry will be discussed later.

Escape by jumping or flying is often aided by coloration that diverts the predator's strike or attention. *Eyespots* may be of two main types: (1) small, dark spots, with or without a light mark simulating a reflection from the "eye"; and (2) large spots with concentric rings depicting a colored iris and dark pupil marked with a reflected highlight. The first are thought to imitate an insect's eye, and the second are unmistakable representations of a terrestrial vertebrate's eye. The small eyespots may be situated at the posterior end of the insect in conjunction with antennalike processes, to imitate a head. Examples are hairstreak butterflies (Lycaenidae), which have tails that resemble antennae on the caudal part of their wings, and, at the bases of the tails, spots that resemble eyes. When perched, the butterfly moves its wings to wiggle the false antennae. This presumably directs a predator's attention to the wrong end of the body and permits the insect to escape in the opposite direction. Small spots near the outer wing margin of other butterflies, such as the satyrs (Satyrinae, Fig. 44.11*i*), presumably attract the bird's peck away from the vital body and give the insect a chance to free itself. Evidence that such butterflies do escape is provided by collected specimens that have notched margins or V-shaped bill marks on their wings.

Larger eyespots have been shown to frighten vertebrate predators (Blest, 1957). Cryptically colored insects may hide eyespots and suddenly display them when contacted by a predator. The nymphalid butterfly *Caligo* (Fig. 15.3) and the Io moth, *Automeris* (Saturniidae), have conspicuous eyespots. Unexpected displays of bright *flash colors* are similarly designed to startle enemies at the moment of escape. Underwing moths, *Catocala* (Noctuidae, Fig. 44.13*b*), have conspicuous red or yellow bands on the hind wings. The colors are suddenly revealed and visible during the moth's erratic flight, but abruptly disappear when the moth alights and resumes a cryptic posture.

Insects with concealing coloration are overlooked by predators unless the predator learns to seek them. A low density of the insect or insects with similar protective strategies therefore prevents accidental detection and reinforcement of the predator's search image. The coloration may be one or a combination of the following: (1) *color and pattern resemblance* or *cryptic coloration* such that the insect either closely matches the background against which it is seen, or resembles an inanimate object (Endpaper 2); (2) *obliterative shading,* in which the rounded shape of the insect, such as a caterpillar, is obscured by having the illuminated surface of the body more darkly pigmented and the shaded surface lighter in color; (3) *shadow elimination* by checkered borders or marginal fringes of hairs which reduce the contrast along shaded edges next to the substrate (Endpaper 2*c*), and (4) *disruptive coloration* of contrasting patterns and colors to alter the perception of the true body outline or contour (Endpaper 2*c*). These patterns are usually accompanied by one or a combination of special behavioral traits in which the insect (1) selects an appropriate background and orients its body in a definite position relative to the background; (2) adopts a special posture to eliminate shadows or more closely resemble the object imitated; (3) rests motionless even when closely approached; and (4) escapes quickly, silently, and erratically.

Concealing coloration is the rule among palatable insects exposed to diurnal, visually hunting predators. Aquatic nymphs and larvae may be complexly patterned to resemble aquatic vegetation, detritus, or the variegated gravels or other substrates of the bottom of the body of water. Immature and adult insects that inhabit plants, whether as predators or herbivores (Endpaper 2*a*), are often green (commonly created

FIGURE 15.3 Neotropical nymphalid, the owl butterfly *Caligo* sp., illustrating large eye spots on the hind wings. (*Photo courtesy of and with permission of Carl W. Rettenmeyer.*)

by the pigment insectoverdin, less often by chlorophyll) or colored or sculptured to resemble leaves, stems, or bark. The larger insects of litter, soil, sand, or gravel (Endpaper 2*d*) may closely match their surroundings. The color patterns are usually characteristic of the instar, not changing rapidly or in the manner of chameleons by physiological processes. The walkingstick, *Carausius* (Phasmatidae), however, does become darker or paler depending on various stimuli.

INDUSTRIAL MELANISM. The best-documented study of the evolution of concealing coloration concerns *industrial melanism,* i.e., the evolution of dark or melanic populations of insects in industrially polluted areas (Kettlewell, 1959, 1961, 1973). Prior to the Industrial Revolution in England, trunks of trees such as oak were naturally light in color and often encrusted with light-colored lichens. Resting against this background, the pepper moth, *Biston betularia* (Geometridae), is virtually invisible. With the development of smoke-producing industries in the latter half of the eighteenth century, soot accumulated on vegetation in the areas surrounding and downwind from the factories. Lichens were killed and the bark of trees was stained black. The light, or *typica,* form of the pepper moth was conspicuous against this darkened background.

In 1848 a dark, or melanic, form of the moth, called *carbonaria,* was first recorded at Manchester in the polluted district. At that time the melanic form probably was not more than 1 percent of the population, but by 1898 it was estimated to be 95 percent.

Melanic forms of over a 100 of the 780 species of larger moths in England have also increased in frequency during the last 130 years. Such dark forms of moths are becoming evident elsewhere in industrial areas throughout the world's Temperate Zone, including North America. Industrial melanics are not known in tropical regions, possibly because the heat-absorbing dark color is a disadvantage in warmer regions. The melanics are always members of species that are concealingly colored. They rest during the day on such objects as lichen-encrusted bark, rocks, or fallen logs. In the majority of species, the switch from predominantly light forms to dark forms does not involve intermediate shades and is usually controlled genetically by a single dominant gene.

Kettlewell carefully investigated various explanations for the rapid spread of the *carbonaria* form of the pepper moth. Mutagenic effects of the pollutants were discounted by experiments; the melanic forms must arise in populations by normal, low rates of mutation. Except where the melanics are favored by selection, the frequency remains quite low. In certain unpolluted environments such as heavily shaded forests, melanic forms of moths may increase in frequency because they are less visible when flying. In industrially blackened areas, the resting melanic forms are less visible to insectivorous birds. Observation in the field and experiments verified that birds of several species selectively eat the conspicuous form (*typica* against dark trunks in polluted districts and *carbonaria* against light trunks in unpolluted districts) and overlook the concealed forms. The rate of spread of melanics in polluted areas can be accounted for by predation alone. Analysis of other possible advantages, such as superior larval vigor in polluted areas, remains inconclusive.

Thus in a relatively brief period, the cryptic patterns of many species of moths over wide areas shifted to a dark, patternless design as a result of bird predation in polluted areas. It now seems that this dynamic relationship is a sensitive monitor of pollution. With improved pollution controls now in England, nonmelanic moths are reappearing in districts where only melanic forms occur (Bishop and Cook, 1975).

UNPALATABLE INSECTS. The coloration and behavior of unpalatable species contrast strikingly with those of cryptic palatable species. The strategy here is twofold: (1) to create an unpleasant, but not lethal, ordeal for the vertebrate predator by means of poisonous or nauseous secretions, repulsive odors, effective stings or bites, or spiny surfaces; and (2) to be easily recognizable so that an experienced predator probably will not attack again. Unpalatable species usually have *aposematic,* or *warning,* coloration and behavior, including one or more of the following: (*a*) bold, simple patterns of contrasting black and red, orange or yellow which are highly visible to vertebrates with color vision (Endpapers 3*c*, *d*; 4) and which also contrast against the predominantly green color of the landscape (green is the complementary color of red); (*b*) gregarious behavior that leads to local, dense aggregations (Endpaper 3*c*); (*c*) free exposuure when walking or resting, and slow, conspicuous flight in full view of predators; (*d*) characteristic movements, sounds, or smells that warn against attack; and when attacked, (*e*) only sluggish efforts to escape, repulsive taste and a durable body permitting survival without fatal injury. For example when disturbed, caterpillars of Papilionidae release defensive odors from brightly colored *osmeteria* (Endpaper 3*d*).

A naive predator has ample opportunity to sample an aposematic species and, even without color vision, has a number of cues by which to remember the experience. Bird predators are also known to learn vicariously by witnessing another predator's response. Learning can be reinforced by merely seeing an aposematic insect again without touching it. Not all the cues may be necessary to elicit an avoidance reaction. It is not known to what extent a predator's response includes some innate avoidance of aposematic colors or behavior. In any event, the sensory and learning abilities of relatively long-lived vertebrate predators have allowed aposematic species to evolve. Furthermore, evidence exists that experienced predators may generalize their learning to avoid similar species after experience with only one.

Aposematic coloration is oddly missing among most insects that live in water, even though some fish are known to have color vision and learning ability.

MIMICRY. This last aspect of vertebrate psychology would give some advantage to insects that are in the process of evolving aposematic coloration and behavior but have not acquired the full set of advertising characteristics. The predator's ability to generalize (or failure to discriminate) is also the basis for the evolution of two types of mimicry: (1) *Müllerian mimicry,* named for Fritz Müller (1879), involving two or more unpalatable and aposematic species; and (2) the previously mentioned *Batesian mimicry,* named for Henry W. Bates (1862), involving one or more unpalatable species that serve as models and one or more palatable species that mimic the model's coloration and behavior.

The mutual resemblance of Müllerian mimics confers protection on all members of the association in a given area. Because all are unpalatable, and predators can generalize, the resemblance among the species need not be exact, although it is often astonishingly precise. Naive predators presumably attempt to eat mimics in proportion to the relative abundance of each species, and each species suffers fewer losses in the process of educating predators when several mimics are together. Examples are found among the Neotropical butterflies, especially the genus *Heliconius* (Nymphalidae) and the nymphalid Subfamily Ithomiinae; the monarch and queen butterflies (Fig. 15.5); netwinged beetles (Lycidae) and certain other insects (Endpaper 4*c*,*d*); and the convergent similarity of many ants, wasps, and bees, the females of which are equipped with stings.

Batesian mimics are avoided by predators familiar with the unpalatable model. An argument can be made that the populations of mimics must remain less than the model's population. Otherwise a naive predator will be more likely to eat palatable mimics. Generally, Batesian mimics are rarer than their models (in some, only the female is the mimic), but recent studies show that unpalatable models may confer protection even when outnumbered. Apparently tasting a model is truly a memorable experience for the predator! The resemblance of Batesian mimics to their models is often quite close. Examples include Neotropical moths and katydids that resemble wasps (Fig. 15.4); the syrphid flies (Endpaper 4*b*) and other insects that

FIGURE 15.4 Batesian wasp mimics: *a*, moth, probably *Pseudosphex* sp. (Ctenuchidae), taking liquid with proboscis (note narrow wings, petiolate abdomen, and banded pattern on abdomen); *b*, katydid (note thick antennae, white marks on body that produce illusion of slender dark wasp). (*Photos courtesy of and with permission of Carl W. Rettenmeyer.*)

resemble wasps and bees; and long-horned beetles (Cerambycidae) and moths that resemble lycid beetles. In some cases, however, the model and mimic share only a general resemblance, as in the viceroy butterfly mimic (Fig. 15.5*c*) and monarch model (Fig. 15.5*a*).

Batesian mimicry may be thought of as involving coevolutionary interactions similar to those occurring in host-parasite complexes (Ehrlich, 1970). The mimic, of course, plays the role of the parasite. Its strategy is to take advantage of the model without destroying it. The model gains nothing—and faces the danger of a "credibility gap" developing in its potential predators. For at some point, if the mimics get too common, most predators will associate only pleasant experiences with the aposematic pattern of the model. Such a development, of course, ruins the game for both model and mimic, since a conspicuous pattern now means "tastes good" to the local insectivores. One would expect that the model would evolve a pattern different from that of the mimic at a maximum rate, everything else being equal. It is to the mimic's advantage to maintain a maximum of resemblance to the model, until that critical point mentioned above is reached. Then the advantage becomes a disadvantage—the mimic is conspicuously patterned, but predators now associate that pattern with tastefulness. As a result, selection would tend to move the mimic away from the model into a more cryptic pattern. It is not inconceivable that imperfect resemblances, such as that of the monarch and viceroy butterflies (Fig. 15.5*a*,*c*), often attributed to mimicry in the process of being perfected, are quite the opposite. They may represent mimics moving away from the model of mimics in an equilibrium situation between perfect mimicry and cryptic coloration.

In all cases it obviously is of advantage to the Batesian mimic to become distasteful if it can do so without sacrificing too much physiologically. In butterflies, at least, it appears that the usual source of noxious compounds is plant biochemicals, so that food-plant relationships must play a large role in the evolutionary dynamics of any given situation. A butterfly has several different routes to distastefulness open to it. If it eats a food plant which does not produce an appropriate compound, it may switch food plants. If it is feeding on a plant with an appropriate compound or its precursors, the butterfly may evolve the ability to use the compound or synthesize a noxious compound from precursors. Finally the food plant of the butterfly may evolve an appropriate compound, which then may be picked up by the butterfly. In the latter case the mimetic butterfly would be involved in a complex of "selection races" involving the model, the food plant, and predators. As the food plant becomes more and more toxic, the butterfly must find ways of "breaking even" by avoiding poisoning or "winning" by turning the poison to its own advantage. Predators may simultaneously be undergoing selection for ability to discriminate between model and mimic, and for "resistance" to the noxious properties of the model. Of course the presence or strength of such selection will depend on many variables. For instance, in some cases butterflies in a single population may make up such a small proportion of the targets of a single predator that selective influence on the predator will be negligible.

On careful analysis of mimetic associations, the distinction between Müllerian and Batesian becomes more difficult to make. Lincoln P. Brower and his associates (1964, 1968) have studied the mimetic relationships of the monarch butterfly, *Danaus plexippus* (Nymphalidae) (Fig. 15.5*a*), and queen butterfly, *Danaus gilippus* (Fig. 15.5*b*). The larval food plants of both species are milkweeds, mostly in the genus *Asclepias.* Two species of milkweeds in the Southeastern United States, *A. curassavica* and *A. humistrata,* contain cardiac glycosides. These chemical compounds cause vomiting or, in larger amounts, death in birds and cattle. In Costa Rica, cattle learn to avoid eating *A. curassavica;* hence the plant is protected. Monarch larvae feed on these milkweeds and acquire various amounts of cardiac glycosides, which are retained in the body of the adult butterfly. Insectivorous birds vomit after eating a sufficiently toxic adult monarch and will avoid pursuing other monarchs when sighted. Thus, not only are emetic monarchs (those that cause vomiting) effectively unpalatable and aposematic, but also the larvae on the poisonous milkweed are undisturbed by grazing cattle.

Not all milkweeds, however, contain enough cardiac glycosides to cause emesis, and some lack the drugs altogether. Monarchs reared on these plants are palatable in laboratory tests with naive birds, but are rejected by birds previously fed emetic monarchs. The natural population of monarchs in an area may include both palatable and unpalatable individuals, yet the predators learn to avoid most of both types. Even as low as 25 percent unpalatable monarchs theoretically will gain protection for 75 percent of the population. Brower has called this kind of intraspecific relationship *automimicry.* The same concept would apply, for example, to aposematic male bees and wasps that lack the stings of the females.

The queen butterfly also feeds on toxic milkweeds and resembles the monarch. Where they occur together, they are usually considered Müllerian mimics. In Trinidad, however, Brower found both monarch and queen larvae feeding

FIGURE 15.5 Multiple mimicry: *a*, monarch butterfly, *Danaus plexippus; b*, queen butterfly, *Danaus gilippus; c*, viceroy butterfly, *Limenitis archippus.* See text for explanation. (*Photos by University of California Scientific Photography Laboratory.*)

on both poisonous and nonpoisonous species of milkweed. Adult monarchs were 65 percent emetic, but only 15 percent of queen adults were emetic. The mimetic relationships here include automimicry between palatable and unpalatable members of each species, Batesian mimicry between palatable members of one species and unpalatable members of the other, and Müllerian mimicry between unpalatable members of both species.

In many ways mimetic assemblages make ideal subjects for the study of coevolution—as has been amply demonstrated by the Browers, Phillip Sheppard, and others. A great deal is understood about them, and yet many questions remain to be answered. For instance, detailed studies of supposed Müllerian complexes are needed to solve a variety of problems. One would expect, for example, that the various members of the complex would have different effects on predators, since they presumably are picking up poisons from different sources. As an example, one Müllerian butterfly complex consists of a *Lycorea* species (Danainae), presumably feeding on Asclepiadaceae or Apocynaceae, several ithomiines on Solanaceae, two *Heliconius* (Nymphalinae) on Passifloraceae, and a *Perrhybris* (Pieridae) with an unknown food plant. Ideally, of course, each member of the complex would give strong and similar reinforcement to all local predators, so that multivalent noxiousness might evolve in various members. It would be particularly interesting if biochemical mimicry could be detected in some of these organisms—i.e., two quite different chemical compounds obviously selected to give similar effects in the same predator. Rothschild (1961) has suggested that this occurs with defensive odors.

Although it is clear that, in general, Batesian complexes should evolve toward Müllerian complexes, the fate of Müllerian complexes is less obvious. It would probably be unwise to think of them as stable "end points" of evolutionary sequences. If this were the case, one might picture all the diurnal Lepidoptera (and perhaps many other herbivores and small predators) in an area eventually being recruited into one large complex. It would really save the memories of the birds, but the birds would not have to remember for long because they would starve to death. Obviously, the larger a Müllerian complex gets, and the more similar the defenses of its members become, the more "profit" accrues to a predator which devises a way of consuming the Müllerian mimics. Thus a large selective premium is placed on a strong stomach, and one would expect predators evolving rapidly to deal with the entire complex. If this happens, the advantages of belonging to the Müllerian complex are reduced, and one might expect it to break up.

SELECTED REFERENCES

Behavior and ecology of insect parasites of vertebrates are discussed by Downes (1958, 1971), Friend and Smith (1977), Hocking (1971), James and Harwood (1969), Rothschild (1965), and Rothschild and Clay (1952). For references on specific parasitic taxa, consult the appropriate chapters in Part IV. Protective coloration and mimicry are treated by Cott (1957), Kettlewell (1973), Rettenmeyer (1970), Rothschild (1973), and Wickler (1968).

INSECTS, MICROBES, AND HELMINTHS

16

An individual insect carries with it a small community of other organisms. Some are merely phoretic passengers, but others live more or less intimately in or outside their hosts. Extraordinary examples of such relationships have been described by Gressitt and his colleagues (1965). In the high mountains of New Guinea, large flightless weevils live amidst mosses in humid forests. The backs of the long-lived, slow-moving weevils are camouflaged with living fungi, algae, lichens, and liverworts (Endpaper 1*d*). The integumental surface and setae, plus a secretion, encourage plant growth. Mites, nematodes, rotifers, diatoms, and psocid insects live and feed among the plants, causing no harm to the weevils.

Less visible but no less interesting or important are the microbes and occasional helminth worms that are associated with all other insects. Lower forms of life were present before insects evolved. We can speculate that as scavengers in decaying vegetable matter the early insects would have been immersed in a microbe-rich environment and would have ingested and excreted saprophytic forms. Inside the insect's gut, some were digested and absorbed, and some persisted harmlessly or as injurious parasites. In time, some microbes evolved a beneficial partnership such that their hosts could exploit foods that were otherwise nutritionally deficient.

The mobility of insects and their resistance to desiccation allow their less hardy associates to survive and be dispersed in the terrestrial environment. When insects began feeding on the vascular fluids of higher plants and vertebrates, certain viruses and microbes were introduced into the wounds. Some proved to be highly dangerous to their new hosts. Insects also became involved as intermediaries in the natural cycles of other disease agents. Although insects alone rarely cause death of plants and vertebrates, as vectors of disease they can have catastrophic effects on entire populations.

SYMBIOTIC RELATIONSHIPS

Broadly defined, symbiosis is the living together of dissimilar organisms, regardless of the possibly injurious or beneficial interactions. In our present discussion, the microbes are called *symbionts* and insects are the *hosts*.

Symbionts may live parasitically at the host's expense. If the host's vitality is impaired, the symbiont is called a *pathogen*. Agents causing fatal diseases of insects include certain viruses, bacteria, protozoa, fungi, and probably most parasitic helminths. Some of these pathogens can be used effectively in controlling pest insects (Burges and Hussey, 1971). Insect hemocytes defensively phagocytize the smaller agents and encapsulate the larger ones (Chap. 14). Insects do not produce specific antibodies against pathogens, but lysozymes in the gut, fat body, and hemolymph can destroy microbes by enzymatic action.

A coevolutionary tendency exists for the hosts and symbionts to become mutually adjusted. Parasites that kill their host also die, whereas those causing less harm to their host survive.

Hosts that are susceptible to parasitic damage are selectively eliminated, leaving resistant hosts in the majority. Thus the relationship evolves toward benign parasites and resistant hosts.

Symbionts that cause no harm to their hosts are *commensals,* such as the fungi Laboulbeniales. *Mutualistic* symbionts have evolved a reciprocally beneficial relationship with their host. When the symbionts are outside the insect's body, e.g., the fungus gardens of insects, they are called *ectosymbionts. Endosymbionts* are mutualistic microbes sheltered inside the insect's body. Buchner (1965) and Koch (1967) give detailed treatments.

Buchner recognized that certain groups of insects were dependent on symbionts because the insects' diet lacked essential items. For example, the vascular fluids of both higher plants and vertebrates, and wood are deficient as food for insects.

Phloem-sucking Homoptera obtain carbohydrate in surplus, but the diet is inadequate in protein; all possess symbionts that supply missing nutrients. Predatory Hemiptera lack symbionts but phytophagous Pentatomidae have them. Larvae of *Dasyhelea* flies (Ceratopogonidae) and *Nosodendron* beetles (Nosodendridae) live in sap and have symbionts.

Vertebrate blood is notably lacking in B vitamins. Phthiraptera, Cimicidae, *Triatoma* (Reduviidae), *Glossina* (Muscidae), and crawling ectoparasitic flies are parasites that feed exclusively on blood in all feeding stages and have symbionts. Adult parasites with scavenging larvae, i.e., Siphonaptera and biting flies, lack symbionts. The larvae of these insects obtain B vitamins sufficient for the rest of the life cycle from bacteria-rich food.

Wood-feeding insects depend on ectosymbionts such as the fungus gardens mentioned above or on endosymbionts in the intestine. Scarab grubs contain bacteria; anobiids and certain cerambycids contain yeasts; and certain termites contain protozoa or bacteria. Certain pests of seeds and stored grain have symbionts: *Nysius* (Lygaeidae), *Lasioderma* (Anobiidae), *Sitophilus* (Curculionidae), *Rhizopertha* (Bostrychidae), *Oryzaephilus* (Cucujidae), and *Coccotrypes* (Scolytidae). Oddly, the Bruchidae apparently lack symbionts.

Symbionts also occur among insects with a variety of other food habits. Many phytophagous weevils harbor bacteria. *Bromius, Cassida,* and *Donacia* are chrysomelid beetles whose larvae feed on green plants and have bacterial symbionts; other chrysomelids lack them. Tephritid flies and lagriine tenebrionid beetles feed in fresh or decayed plants. Both have bacterial symbionts. Many species of scavenging cockroaches have bacterial symbionts, as do certain ants of the genera *Camponotus* and *Formica.* The yeasts of nonpredatory *Chrysopa* adults have been described previously (Chap. 14).

Microbes commonly gain entrance to the insect's body through the mouth as a part of the food. It is not surprising to find most endosymbionts still reside in the gut, having evolved the ability to live in the intestinal environment. They are often transmitted from generation to generation, or among individuals simply by ingestion. For example, the cellulose-digesting protozoa and bacteria of termites are passed among the colony members by anal feeding. Insects without overlapping generations infect the eggs. Phytophagous Pentatomidae defecate on the eggs and the nymphs acquire symbionts after hatching. Some insects have special organs for smearing bacteria on the egg as it is deposited. The female *Donacia* covers the egg with a secretion that encloses a mass of bacteria near the larva's head. On hatching, the larva eats through the mass and ingests the bacteria.

The host's intestine may be modified to accommodate endosymbionts. The hindgut of certain scarab beetle larvae is greatly enlarged as a fermentation chamber containing bacteria and wood particles. Lateral saclike chambers, or caeca, are characteristic of insects with intestinal endosymbionts. Crane-fly larvae may have a large diverticulum on the hindgut. Phytophagous Hemiptera-Heteroptera characteristically have many bacteria-filled caeca along the posterior portion of the hindgut. Malpighian tubules may also serve as bacterial crypts. Presumably the localization of endosymbionts in blind chambers reduces their loss in feces, allows a longer time for them to act on food, and increases the insect's absorption of nutrients derived from the microbes.

A significant coevolutionary step is seen in those endosymbionts that become intracellular. In addition to other possible advantages, the microbes thus sheltered are protected from phagocytosis. Intermediate stages have been observed in intestinal caeca and Malpighian tubules where symbionts are both in the lumen and inside the insect's cells. Cells containing yeast or bacterial symbionts are called *mycetocytes.* These may be part of the epithelium or may be clustered in discrete organs in the hemocoel called *mycetomes.* Mycetocytes and mycetomes containing one to several kinds of symbionts may occur in various places within a single insect. The numbers of symbionts are regulated by the insect host. Often the symbiont-containing tissues are associated with or modified from the gut, Malpighian tubules, gonads and genital ducts, or the fat body. Some mycetomes are easily visible, even from outside. Early microscopists Robert Hooke and Jan Swammerdam saw the yellow mycetome on the gut of the human louse.

In contrast to those symbionts associated with the gut and transmitted to the next generation via feces, the symbionts residing in internal organs are moved to the ovaries by their mycetocytes or by the hemolymph. Depending on the structure of the ovary, the symbionts are transferred into the egg and later included in the embryonic mycetocytes.

The identities of intracellular symbionts in terms of microbial classification remain controversial. Earlier they were sometimes considered waste products or normal cell organelles, such as mitochondria. It is now certain that many intracellular symbionts of insects are actually microorganisms. Now we are uncertain whether mitochondria are derived from ancient symbionts!

For critical analysis, the symbionts must be cultured, but unfortunately this has rarely been achieved. Judging by other evidence and observations of their anatomy with the electron microscope, insect symbionts include yeastlike, bacteriumlike, and rickettsialike organisms.

INSECTS AS VECTORS OF DISEASE

Among the most devastating diseases of crop plants, domestic animals, and humans are those carried by insects. Insects that transmit disease agents are called *disease vectors.* The life history, host selection, and feeding behavior of the vector largely determine the epidemiology of the disease. By controlling the insect, it is sometimes possible to break the link in transmission and to stop the spread of disease.

Microbes causing disease in plants and animals rarely also cause disease in the vector even when they live and multiply in the vector. An exception is Western X mycoplasma, which reduces the longevity of its leafhopper vector. Yet certain agents of plant disease actually seem to be beneficial to the vector by favorably changing the host plant's structure or functioning. Experiments have shown that vectors may feed or ovipost more frequently on diseased plants. A comparable relationship between animal diseases and their vectors is not known.

The nature of the transmission of pathogenic organisms by insect vectors to plant hosts may be *mechanical* or *biological.* Mechanical transmission involves the contamination of the body or mouthparts, or at most the foregut, after feeding on an infected host. The pathogen is transferred to a new host when the vector next feeds or makes contact. Infectivity is lost when the supply of pathogen is depleted or at ecdysis if the vector is immature. For example, mosquitoes function literally as flying pins in the transmission of the myxoma virus among rabbits. Similarly, the nonpersistent plant viruses are rapidly spread by aphids during their brief and repeated probing behavior in search of suitable plant hosts.

Biologically transmitted pathogens, acquired after prolonged feeding, enter the vector's gut, hemocoel, or various organs. Infectivity is not immediate, but occurs only after a *latent period* during which the pathogen moves about in the host or further develops or reproduces. Once infective, the vector tends to remain infective for life.

The persistent circulative plant viruses, such as the aphid-transmitted pea enation mosaic, do not replicate in the vector, but circulate in the body. On the other hand, helminths causing animal diseases undergo developmental changes but do not multiply in their intermediate insect hosts. This is called *cyclodevelopmental* transmission of animal pathogens.

When the pathogen multiplies in the vector, the transmission is called *propagative.* Persistent propagative plant pathogens, such as the aphid-transmitted lettuce necrotic yellows, and plague bacteria in fleas, are examples. If the multiplication is accompanied by cyclic development, as in the case of malaria plasmodia in anopheline mosquitoes, the term is *cyclopropagative.*

Certain insects and acarines acquire infections as immatures and remain infective after one or more ecdyses. This is called *transstadial* transmission. Some transmit pathogens to their offspring by *transovarial* transmission, a feature already described above for endosymbionts. Certain arboviruses (see below) and rickettsiae are transovarially transmitted by ticks, but few insects have been shown to transmit animal pathogens similarly to their offspring. Plant pathogens, however, can be transovarially transmitted. Some persistent viruses, rickettsialike pathogens, and mycoplasmas are transmitted in this way by leafhoppers.

INSECTS IN RELATION TO VIRUSES AND MYCOPLASMALIKE DISEASE AGENTS

Viruses are highly variable, submicroseopic particles consisting of either deoxyribonucleic acid (DNA) or ribonucleic acid (RNA) and usually, but not always, with a protein cover. Viruses multiply or replicate only in living cells, primarily those of bacteria, plants, insects, and vertebrates. The best-known viruses associated with insects are those that are pathogenic to insects, higher plants, or vertebrates. David (1975) and Vaughn (1974) have reviewed the viruses affecting insects, and Falcon (1976) discusses their use in controlling pests. From the standpoint of numbers of species of insects, the most important pathogens are the DNA-containing nuclear polyhedrosis viruses (NPV), DNA granulosis viruses (GV), and RNA cytoplasmic polyhedrosis viruses (CPV). Lepidoptera caterpillars and larval sawflies are most susceptible to these viruses. The polyhedrosis viruses are characterized by polyhedral crystals.

Carter (1973) lists 261 viruses that are transmitted to plants by insect vectors. Aphids transmit the largest number of viruses, mostly those producing the mottled "mosaic" diseases. Aphids also rank first in numbers of species of vectors. Some species may vector numerous viruses, for example, *Myzus persicae* is the vector of more than 100 viruses.

Leafhoppers rank next in importance and are noted as vectors of persistent propagative viruses. Recently, one group of diseases called the "yellows diseases" has been discovered to be caused not by viruses, but by organisms called mycoplasmas in the phloem that resemble minute bacteria (Whitcomb, 1973; Whitcomb and Davis, 1970). Like bacteria, mycoplasmas are procaryotes, i.e., the nucleus lacks a membrane. The cells are highly variable in shape and uniquely lack a cell wall. Mycoplasmas were previously known as pathogens of mammals and birds, but not associated with arthropods. It has been estimated that as many as 350 species of plants may be involved in the natural cycle of aster yellows disease and its vector, the aster leafhopper, *Macrosteles fuscifrons* (Cicadellidae).

Of historical interest is the tulip "breaking" virus that causes the petals of red or purple flowers to be beautifully striped. The first record was published in 1576, making this the oldest known plant disease (McKay and Warner, 1933). Broken tulips became greatly admired in Europe. Paintings of the sixteenth and seventeenth centuries often depict the attractive blossoms. Demand was heightened because the bulb of a strikingly patterned flower might degenerate or rapidly die. Various methods of cultivation were alleged to increase breaking, but the capricious cause was unknown. A colossal gambling craze developed in Holland in 1634–1637 called "tulipomania." Fortunes were traded for single prized bulbs. The episode was finally stopped by the government (Frylink, 1954). We now know that the virus is transmitted mechanically by several aphids, including *Myzus persicae.*

Viruses pathogenic to vertebrates that are transmitted by arthropods are collectively called "arboviruses," an abbreviation of "arthropod-borne viruses." These contain RNA as the nucleic acid. Out of more than 200 arboviruses, 50 infect humans (Horsefall and Tamm, 1965). Yellow fever, dengue, and various encephalitis viruses are the most important. Mosquitoes are the vectors for the great majority of arboviruses, followed by ixodid ticks, *Culicoides* flies (Ceratopogonidae), *Phlebotomus* flies (Psychodidae), and possibly laelapid mites.

Yellow fever is an arbovirus transmitted by the mosquito *Aëdes aegypti.* This disease may be fatal in up to 10 percent of human cases. Hundreds of thousands have died, especially in tropical Africa and South America. The yellow fever mosquito was originally from Africa, but now occurs worldwide as a vector in urban environments. Although controlled in cities and towns, the virus persists in tropical forests. Here it is transmitted among monkeys by forest mosquitoes such as *Haemagogus* in the New World and various *Aëdes* in Africa. Jungle yellow fever is transmitted to humans living in forest clearings and working in forests.

INSECTS IN RELATION TO BACTERIA, SPIROCHETES, AND RICKETTSIAE

The classification of procaryotic organisms is continually revised as new information is obtained (Buchanan and Gibbons, 1974). Among procaryotes broadly called the bacteria are distinctive groups closely associated with arthropods: certain true or eubacteria, spirochetes, rickettsias, and mycoplasmas. The last group was mentioned in the preceding discussion of viruses because, until recently, insect-borne mycoplasmas have been confused with certain persistent plant viruses. The procaryotes in this section all have cell walls.

Eubacteria are commonly found on the integument, in the gut, and in mycetocytes. Endosymbiotic bacteria occur in Orthoptera, Isoptera, Phthiraptera, Hemiptera, Coleoptera, Hymenoptera, and Diptera. Certain larval calliphorid flies, known as "surgical maggots," are exceptional in that the gut is sterile. A bactericidal substance, allatonin, kills bacteria in the midgut. Napoleon's surgeon, Larrey, noted that healing was actually enhanced when neglected battle wounds were infected with maggots. Maggots were later used medically because they ingest bacteria and necrotic tissues.

Bacteria pathogenic to insects have been reviewed by Faust (1974). Infection is usually through ingestion. Important examples of bacterial diseases are milky disease of Japanese beetles, caused by *Bacillus popilliae* and *B. lentimorbus;* and the honeybee diseases, American foulbrood, caused by *Bacillus larvae,* and European foulbrood, caused by *Streptococcus pluton. Bacillus cereus* and its variant, *B. thuringiensis,* produce crystals toxic to more than 182 species of pest insects, mostly Lepidoptera caterpillars. Commercial preparations of bacteria and crystals are applied with regular spray equipment to trees and crop plants in the control of pests.

Bacterial diseases of plants are mostly transmitted mechanically by insects, but some seem to be endosymbionts of the gut and may be transovarially transmitted. Fire blight of pears, apples, and some 90 kinds of other trees is caused by *Erwinia amylovora* and mechanically transmitted by many insects, especially bee pollinators.

Human diseases caused by bacteria and mechanically transmitted by insects are dysentery bacteria, *Shigella* and *Salmonella,* carried by muscid filth flies, and tularemia, caused by *Francisella tularensis* and transmitted by the bite of tabanid deerflies.

The organism causing plague, or Black Death, *Yersinia pestis,* is transmitted biologically by fleas. Plague bacilli are naturally transmitted among wild rodents by rodent fleas. While hunting or camping human beings may acquire infections by handling diseased rodents or rabbits. Urban populations are endangered when the pathogen infects domestic rodents such as the black rat, *Rattus rattus,* and its ectoparasite, the oriental rat flea, *Xenopsylla cheopis.* Bacteria multiply in the flea's gut, forming a gelatinous plug in the proventriculus. A "blocked flea" is unable to pass blood to the midgut and repeatedly attempts to suck. The result is that plague bacilli are regurgitated into the wounds. As the disease spreads among rats, the susceptible rats die. Fleas leave cold rats and transfer to live rats or humans. Following infection of a human by flea bites, the lymph nodes become inflamed—hence the name "bubonic plague," referring to the swollen buboes. In the most deadly phase of an epidemic, bacilli are rapidly spread from human to human by inhalation of infected respiratory droplets.

Plague has devastated the populations of entire countries since ancient times and has played a decisive role in history. In addition to the staggering statistics of human deaths, the impact of this disease is reflected in works of art and literature. In *The Decameron,* Giovanni Boccaccio (1313–1375) describes plague in Florence in 1348. This was at the beginning of the great epidemic that swept away 25 million Europeans and contributed to social change in the fourteenth century.

The terror of plague is shown in art beginning in 1348 (Brossollet, 1971). Reproduced here is an engraving, "The Pest," by Pierre Mignard (1610–1695) (Fig. 16.1). The scene was created from Biblical accounts of the plague of David (II

FIGURE 16.1 *The Pest* by Pierre Mignard, 1610–1695. See text for explanation of this scene of plague. (*Reproduced by permission of M. A. S., Ampliaciones y Reproducciones, Barcelona, Spain.*)

Samuel 24 and I Chronicles 21) and Thucydides's description of the plague in Athens. Accurately illustrated are symptoms of unquenchable thirst (fountains in center background), insanity (delirious patient at left being restrained), and axillary buboes ruptured or lanced (armpit of woman in center foreground). The angel in the sky is pouring sulfurous fumes as a disinfectant. Smoke from torch and brazier were also intended to be purifying.

Spirochetes are slender, motile bacteria of spiral configuration. Some are associated with insects. Termites of the advanced Family Termitidae may depend on spirochetes as intestinal endosymbionts. In the same genus as the spirochetes of syphilis, *Treponema,* are pathogens causing human skin diseases called pinta and yaws. These are transmitted mechanically from lesions by *Hippelates* (Chloropidae) eye gnats.

Rickettsiae are small, nonmotile bacteria that are spherical or rod-shaped. All are associated with arthropods at some point in their natural cycle. Diptera, Hymenoptera, and ticks possess commensalistic or mutualistic forms. The best-known rickettsiae are pathogens of vertebrates that are transmitted by bloodsucking fleas, lice, ticks, or mites. Epidemic, or louse-borne, typhus fever is caused by *Rickettsia prowazeki* and transmitted by human body lice, *Pediculus humanus* (Pediculidae). The microbes multiply in the louse's gut, passing out with feces. Human infection results when infected feces are inhaled or lice are crushed and the pathogens are introduced by fingers into skin abrasions or the eyes. Louse-borne typhus spreads rapidly during times of social strife. Crowding and poor sanitation, as in war or famine, favor louse infestations. Typhus was partly responsible for the collapse of Napoleon's army in 1812.

Murine, or flea-borne, typhus is caused by *Rickettsia typhi.* The disease is transmitted among rats by rat fleas such as *Xenopsylla cheopis* (Pulicidae) and others. Humans acquire infection in the same manner as in louse-borne typhus, i.e., through infected flea feces and crushed fleas. Other important rickettsial diseases are transmitted by acarines: Rocky Mountain spotted fever caused by *Rickettsia rickettsii* and transmitted by ticks; and scrub typhus, caused by *Rickettsia tsutsugamushi* and transmitted by chiggers.

Until recently rickettsias have been considered confined to insect or vertebrate hosts. Several diseases of plants, previously thought to be persistent viruses, are now associated with rickettsialike organisms: Pierce's disease of grapes, phony peach disease, and clover club leaf (Whitcomb, 1973). All are transmitted by leafhoppers. Clover club leaf has been shown to multiply in both the plant and insect.

INSECTS IN RELATION TO PROTOZOA

Protozoa are endosymbionts in nearly all orders of insects but are best known in Orthoptera and Isoptera. Extraordinary quantities are found in termites, where flagellates constitute up to one-third of a nymph's weight. Protozoa enter insects via the mouth, and many types remain associated with the gut, but others become intracellular parasites or live in the hemocoel. They may be pathogenic, commensalistic, or mutualistic in relation to the insect host. Pathogens causing several major diseases of vertebrates are transmitted by insects, as are a few minor protozoan pathogens of plants.

The geologic record gives no clues to the antiquity of insect-symbiont relationships. That these associations can be quite old is inferred from the distribution of flagellates among the primitive termites and *Cryptocercus* roaches (Chap. 7). Cleveland and his associates (1934) found 25 species of flagellates in *Cryptocercus.* Several species of the flagellate genus *Trichonympha* are found in both the roach and various termites. Because few opportunities for exchange of endosymbionts now seem to exist between the two orders of insects, we infer that both probably derive their *Trichonympha* from a common ancestor in the Mesozoic Era.

Protozoa pathogenic to insects have been reviewed by Brooks (1974). The most important are the microsporidians, which are parasites of many, if not most insects, especially of Lepidoptera and Diptera. The pebrine disease of silkworms, caused by *Nosema bombycis,* and nosema disease of honeybees, caused by *N. apis,* are examples. The coccidian *Adelina* is pathogenic in Coleoptera and Diptera (Fig. 16.2*a*). Relatively harmless are the eugregarines that are also known from many insects, especially Coleoptera, Orthoptera, and Diptera. Some are easily seen with the unaided eye because they may be up to 16 mm long. *Mattesia grandis* is, however, highly pathogenic to the boll weevil *Anthonomis grandis,* and may prove useful in its control.

At one time malaria was considered the most important disease of humans. It is caused by four species of sporozoans in the genus *Plasmodium* and transmitted by mosquitoes of the genus *Anopheles.* Persistent chemical insecticides, especially DDT, and antimalarial drugs have substantially reduced mortality. The prospect was so encouraging that in 1955, the World Health Organization resolved to undertake a worldwide eradication program. The use of residual insecticides continues to be the main strategy, but a variety of other methods are being developed to control resistant populations of the vectors. Unfortu-

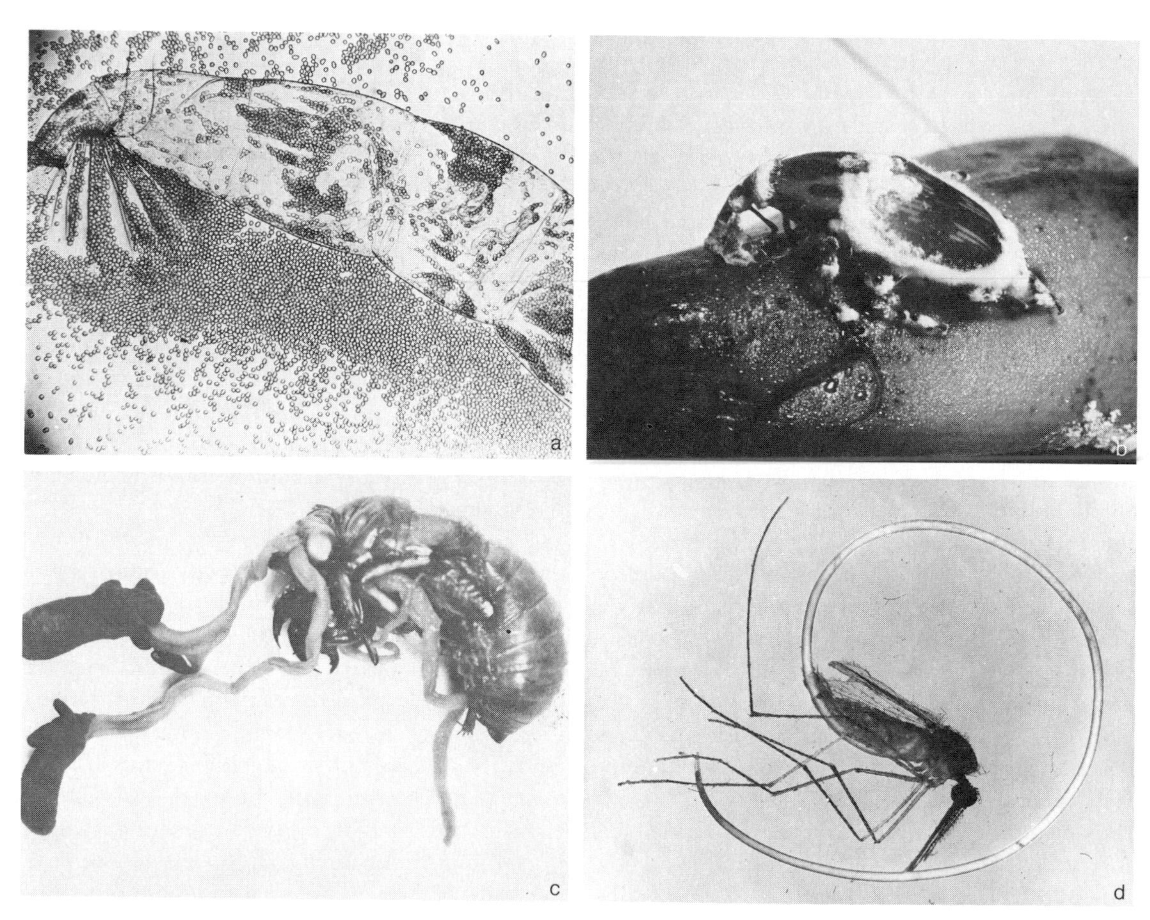

FIGURE 16.2 Insect pathogens: *a*, *Chaoborus astictopus* (Chaoboridae) larva ruptured by infection of *Adelina* sp. (a coccidian protozoan); *b*, *Metamasius hemipterus* (Curculionidae) infected with *Beauveria bassiana* (fungus)—mycelia are emerging from between segments; *c*, *Diceroprocta apache* (Cicadidae) nymph infected with *Cordyceps sobolifera* (fungus)—note mature spore-producing structures; *d*, *Anopheles funestus* (Culicidae) with nematode *Empidomeris cozii* emerging from body. (*Photos courtesy of and with permission of Gerard M. Thomas and George O. Poinar, Jr.*)

nately, malaria is now increasing in several parts of the world.

The biological transmission of plasmodia is of the cyclopropagative type. Female *Anopheles* ingest microgametocytes and macrogametocytes along with blood from infected humans. In the mosquito's midgut, the microgametocytes produce male microgametes that seek and penetrate the female cell, now enlarged and called a macrogamete. The resulting zygote develops into a motile ookinete. This enters the midgut epithelium and encysts as an oocyst between the epithelium and the outer tissues of the gut. Within the oocyst the cell undergoes mitotic and meiotic divisions, ultimately producing thousands of haploid sporozoites. The oocyst bursts, releasing the sporozoites in the hemocoel. The sporozoites invade various tissues, but those reaching the salivary glands are passed into another human's blood when the mosquito feeds. In the human, the sporozoites first enter cells of the liver, then produce merozoites that enter red blood cells. Merozoites develop into trophozoites that destroy the blood cell and release either more merozoites or gametocytes. The release of merozoites and their toxins occurs at regular periods according to the species of *Plasmodium.* This is the cause of the recurring fevers and chills that characterize malarial attacks.

Other important diseases of humans in tropical regions are caused by flagellates of the genera *Trypanosoma* and *Leishmania.* Several kinds of diseases are produced by the latter, including kala azar and espundia. *Phlebotomus* flies (Psychodidae) are known as vectors. Trypanosomes cause trypanosomiasis, or sleeping sickness. In Africa, both sexes of tsetse flies, *Glossina* (Muscidae), feed on blood and ingest flagellates. Some mechanical transmission may occur immediately by infected mouthparts, but cyclopropagative transmission follows. The flagellates multiply in the gut, and moving anteriorly to the salivary glands, they are passed to new hosts when the fly feeds. Trypanosomiasis in America is called Chagas' disease. The vectors are bloodsucking reduviid bugs of the Subfamily Triatominae. The transmission is also cyclopropagative in the insect's gut, but infection of new hosts is by feces. The bug defecates after feeding. Humans accidentally rub the contaminated feces into the wound or into the eyes or mouth.

Flagellates of the genus *Phytomonas* infest plants, especially those with latex. Vectors are various phytophagous Hemiptera-Heteroptera.

INSECTS IN RELATION TO FUNGI

Fungi have many interesting relationships with insects as commensals, mutuals, and pathogens (Fig. 16.2*b*, *c*). Many fungal diseases of plants are transmitted by insects, but apparently no pathogens of vertebrates.

The Laboulbeniales are fungi that live almost exclusively on the integument of Coleoptera, Diptera, and Neuroptera. As commensalistic ectosymbionts, they are transmitted from insect to insect by contact. The insect hosts are rarely killed by the fungi.

A mutualistic relationship exists between *Septobasidium* fungi and various scale insects in the Southeastern United States. Patches of fungus grow in concentric, annual rings on the bark of trees. Inside are tunnels and chambers enclosing scale insects. Thus embedded and protected, the scales feed on the plant with their long sucking mouthparts. The fungus apparently derives nourishment parasitically from the living insects and may kill the insect (Couch, 1938).

Mutualistic ectosymbionts among the fungi are frequently cultivated in "fungus gardens." Batra and Batra (1967), Francke-Grosmann (1967), Graham (1967), and Weber (1966) provide many details of the relationships. The insects are gall midges (*Lasioptera,* Cecidomyiidae), wood wasps (Siricidae and Xiphydriidae), ambrosia beetles (various Scolytidae, Platypodidae, and Lymexylidae), fungus ants (*Atta,* Formicidae), and fungus termites (Macrotermitinae, Termitidae, Fig. 24.3). Nutritious food is obtained from fungal saprophytes that grow on cellulose-rich wood or vegetable matter. Some fungus is carried by the adult insects to new locations. Queen *Atta* ants carry a pellet of fungus in a pouch in the mouth when they fly forth to establish new colonies. Wood-boring ambrosia beetles and wood wasps have special integumental cavities called *mycetangia.* Here the fungus is protected from desiccation and is nourished by a secretion. The mycetangia are filled with fungus before the insect leaves its infected gallery. When the beetles bore into a new host or the wasps oviposit, the fungus is inoculated in the area where the progeny will develop.

Blue-stain fungi of the genus *Ceratocystis* are associated with certain scolytid bark beetles. The spores are carried on the body, in mycetangia, or passed with feces. Sapwood moisture of the infected tree is reduced, making the tree more favorable for beetles. The stain lowers the commercial value of the wood.

Dutch elm disease is caused by *Ceratocystis ulmi.* First discovered in Ohio in 1930, this European disease has killed or has made necessary the destruction of elm trees in many urban areas. It has been recently recognized in California. *Scolytus multistriatus* and *Hylurgopinus rufipes* are vectors. The former feeds on twigs of healthy trees after emergence from infected trees or logs. The beetle then seeks sick trees or freshly cut elm wood to construct its breeding galleries. The healthy trees are infected with spores that cling to the insect. The fungus blocks water-conducting tissues, causing limbs or the whole tree to wilt. The afflicted parts then become suitable breeding sites for more beetles.

Endosymbiotic fungi and yeasts are found in the mycetomes of Hemiptera-Homoptera and intestines of anobiid and cerambycid beetles.

Fungi that are pathogenic to insects normally invade through the integument, but a few enter via the gut. Warm humid environments favor fungal attack. The fungal filaments and reproductive structures are often evident on the body surface (Fig. 16.2*b*, *c*). Species of *Entomophthora* and *Beauveria* are common pathogens. In the past, the latter caused serious losses of silkworms in China and Europe. *Beauveria globulifera* was early noted as causing a contagious disease of the chinch bug *Blissus leucopterus* (Lygaeidae). Near the end of the last century, the fungus was mass-produced and distributed to farmers for control of the pest. *Entomophthora* infects all life stages of insects, especially Diptera, Hemiptera, Lepidoptera, Coleoptera, and Orthoptera. In the fall, *Entomophthora muscae* causes epidemics among flies, leaving them attached in lifelike postures inside houses or on plants.

Insects mechanically transmit fungal pathogens of plants, but this does not preclude intimate biological relationships. The Dutch elm disease mentioned above is an example. Spores of *Fusarium moniliforme,* causing the disease endosepsis of figs, are transmitted when the fig wasp pollinates the flowers (Fig. 13.3). A honeydew-like secretion is caused by the ergot fungus *Claviceps,* when it attacks grasses and cereals. The secretion, infected with spores, is rich in amino acids and attractive to Diptera. The insect visitors spread disease by carrying the spores externally or by passing them in feces.

INSECTS IN RELATION TO HELMINTHS

Wormlike animals are called *helminths.* We are concerned here with the phyla Platyhelminthes, Aschelminthes, and Acanthocephala. You may wish to renew your knowledge of these phyla by consulting a recent text in zoology (Storer, et al., 1972) or animal parasitology (Olsen, 1974; Dunn, 1969).

Among the flatworms, or Platyhelminthes, are two classes that involve insects as intermediate hosts. The digenean flukes of the class Trematoda usually depend on snails as the first intermediate host, then enter a second intermediate host (snails, crustaceans, or fish) or encyst on aquatic plants before being ingested by the definitive vertebrate host.

The fluke *Prosthogonimus* spp., however, seeks dragonfly naiads as the second intermediate host after parasitizing snails. The parasites enter the anus while the naiad takes in water during respiration. Birds eat the infected naiads or adult dragonflies and become the definitive hosts.

One of the most remarkable effects of a parasite on its vector has been observed in the case of the fluke *Dicrocoelium* spp. Eggs are released by the adult flukes in the definitive host and excreted. The eggs are then ingested by land snails and the parasites later expelled from the snails' respiratory pore in a ball of slime. Hundreds of parasites may be protected in each ball. The balls are collected by several species of ants in the genera *Formica* and *Proformica.* In the nest, the slime balls are eaten by the colony, resulting in a high rate of infection. Most parasites encyst in the ants' abdomen, but one or two often migrate to the brain, where a lesion develops. In the evening, ants usually descend from vegetation, but those with brain lesions remain on the vegetation until warmed by the next day's sun. Ruminants grazing early in the day ingest the infected ants, thus completing the cycle.

The tapeworms, or Cestodea, depend primarily on arthropods as intermediate hosts. Crustaceans are the usual hosts because most tapeworms are marine. The adults occupy the intestines of vertebrates. Some tapeworms in the terrestrial environment utilize insects as intermediate hosts. Eggs of *Dipylidium caninum,* the dog tapeworm, are ingested by larval dog fleas, *Ctenocephalides canis* (Pulicidae); cat fleas, *C. felis;* human fleas; *Pulex irritans* (Pulicidae); and the dog louse, *Trichodectes canis* (Trichodectidae). The larval fleas and the biting lice have chewing mouthparts. They are able to ingest eggs, whereas the adult fleas cannot. Dogs and cats and their wild relatives acquire tapeworms by eating adult fleas or lice that were infected in earlier instars. Humans are occasionally hosts.

Various species of *Hymenolepis* infect aquatic crustaceans, earwigs, dung and scavenger beetles, millipeds, meal moths, and fleas. *Hymenolepis nana,* the dwarf tapeworm of humans, mice, and rats, and *H. diminuta,* another tapeworm of mice and rats, both depend on insects as intermediate hosts. Coprophagic scavengers, including larval fleas and the mealworm beetle, *Tenebrio molitor,* ingest infected rodent feces. The insect is eaten in turn by the definitive vertebrate hosts. Humans may be parasitized also. A number of other genera of tapeworms involve beetles, grasshoppers, larval and adult flies, and ants in their life cycles (Dunn, 1969).

The spiny-headed worms, or Acanthocephala, also live as endoparasitic adults in the intestines of terrestrial and aquatic vertebrates. On land, the eggs are ingested by insects, and in water probably by both insects and crustaceans. The intermediate hosts are later eaten by the vertebrate. *Macracanthorhynchus hirudinaceus* is a parasite of pigs and their wild relatives, and occasionally humans. The insect hosts are coprophagic beetle larvae of the Families Scarabaeidae, Tenebrionidae, and Hydrophilidae. They acquire eggs of the parasite by eating pig feces. The pigs then eat the infected larvae in the course of rooting in the soil.

The Aschelminthes include two classes of worms that infect insects: roundworms or Nematoda, and hairworms or Nematomorpha. The immature stages of the latter are endoparasitic in insects and crustaceans. Adult hairworms are free-living in saltwater or freshwater. The complete cycle in terrestrial and freshwater habitats is not well known. The adult worms copulate, lay eggs, and die in water. Larval worms emerge and may either penetrate soft-bodied aquatic invertebrates or be ingested by aquatic and possibly by terrestrial insects. Some hosts are not suitable for further development. In these the larvae encyst. Development proceeds, however, in larger terrestrial insects that acquire infection directly from water or by eating intermediate hosts infested with encysted larvae. The parasite enters the hemocoel, feeds, and matures. When the host is near or in water, the adult worm emerges from the host quickly and becomes free-living. Heavily parasitized hosts may have the reproductive organs reduced in size. Emergence of the worm is fatal to the host.

May (1919) found up to 20 percent of tettigoniid grasshoppers near Urbana, Illinois, infected with *Gordius robustus. Paragordius varius* was found in gryllid crickets. May believed the parasites were obtained directly from water. Inoue (1962) traced part of the life cycle of *Chordodes* in Japan. Mayfly naiads ingest the larvae, and these encyst. The infected adult mayflies are eaten in turn by larger predators such as the praying mantis. The cyst is digested, freeing the larva to penetrate the host's gut. Development is completed in the hemocoel.

The most numerous helminths associated with insects are nematodes (Fig. 16.2*d*). Much remains to be learned about the relationships between the largest group of animals, i.e. the insects, and what is possibly the second largest group, the nematodes (Welch, 1965; Poinar, 1975). Nematodes have essentially five kinds of relationships to insects: (1) phoresy, (2) facultative parasitism, (3) obligatory parasitism that depends only on insects as hosts, (4) parasitism involving both insects and plants as hosts, and (5) obligatory parasitism with insects as intermediate hosts and vertebrates as definitive hosts. Insects are vectors of important nematode diseases of humans and other animals, but they are curiously not vectors of nematodes causing plant diseases.

Insects regularly serve to transport nematodes in a phoretic relationship. Juvenile stages of rhabditoid nematodes in decaying matter or beetle galleries in wood attach themselves externally to insect scavengers or borers and are carried to new habitats. Some harmlessly enter the insect's body, effectively avoiding desiccation.

The internal phoretic relationship becomes one of facultative parasitism when the nematode feeds on the host's tissues, but is also able to complete its life cycle without an insect host (Poinar, 1972). The effect on the host varies from no apparent harm to death. Species of *Neoaplectana* have an odd relationship to their insect hosts that defies classification. The juvenile worm penetrates the insect's body like a true endoparasite. Specific bacteria are released by the worm after it enters the hemocoel. The bacteria rapidly kill the host, and the nematode completes up to several generations, feeding on the dead host's tissues and bacteria. *Neoaplectana* is probably obligatorily associated with insects in nature, yet can be reared on artificial media such as dog food. Is it a facultative parasite, predator, or saprophyte? Regardless of label, the nematodes are among the most promising organisms for use in the control of insect pests.

Obligate parasites feed only on insects. The juvenile stages of obligate nematode parasites of insects develop in the host's hemocoel, intestine, or reproductive organs. Adults may be free-living or remain in the host. One important family are the mermithids that infest aquatic insects and terrestrial insects in moist habitats. The host is killed when the nematode emerges. Infection is sufficiently high in some areas to reduce significantly or to eradicate the host populations. Nematodes influence the host's development by destroying organs, removing nutritional reserves, or possibly producing toxic effects on the corpora allata. Altered behavior, reduced fecundity, sterilization, intersexes, and intercastes of social insects may result from parasitism (Welch, 1965).

Juvenile nematodes of the genus *Deladenus* infest the hemocoel of siricid wood wasps. The adult parasites mature outside the host. Two types of females are produced. One type does not feed and after mating, penetrates a new host. The other type feeds on a fungal symbiote of the wood wasp. Female wood wasps inoculate the fungus where they oviposit. Both sexes of the wood wasps are sterilized by the parasites, but the female continues to fly to new trees for oviposition. The fungus-feeding nematodes are deposited during the futile oviposition in new places where other wood wasps are present (see Poinar, 1972). A related nematode is parasitic on weevils and mustard plants.

The last group of nematodes involve species whose definitive hosts are vertebrates. These are mainly the filarial worms, some of which are host specific to humans. The adult worms feed on lymph and tissue fluids. The young or microfilariae are produced viviparously. Blood-sucking Diptera ingest the microfilariae. These penetrate the insect's gut and become intracellular parasites in fat, muscle, or Malpighian tubule cells. In 1 to 2 weeks, they metamorphose into an infective stage that migrates to the proboscis. When the insect feeds, the parasites migrate into the wound.

The grossly swollen legs, breasts, or genitalia characteristic of elephantiasis are caused by *Wuchereria bancrofti. Culex pipiens quinquefasciatus* (Culicidae) is the principal vector, but various species in other mosquito genera are also found infected. Species of *Simulium* (Simuliidae) transmit *Onchocerca volvulus,* a blinding disease of the Tropics. Another eye disease in Africa, caused by *Loa loa,* is carried by deerflies of the genus *Chrysops* (Tabanidae).

Other nematodes combining insects and vertebrates as hosts include the ascaridoid *Subulura* spp. that infects beetles and cockroaches, and then galliform birds. Many genera of spiruroid nematodes utilize coprophagous scarab beetles, cockroaches, termites, and anthomyiid flies as intermediate hosts. Definitive hosts are domestic animals and birds and their wild allies (Dunn, 1969).

SELECTED REFERENCES

The general relationships of insects, microbes, and helminths are treated by Buchner (1965), Francke-Grosmann (1967), Poinar (1972, 1975), and Steinhaus (1946, 1949). Diseases of insects are discussed by Cantwell (1974), Stairs (1972), and Weiser (1970). Insects as vectors of plant diseases are dealt with by Carter (1973), Garrett (1973), Watson and Plumb (1972), Whitcomb (1973), and Whitcomb and Davis (1970). Insects as vectors of human and animal diseases are treated in the texts by Dunn (1969), Horsfall and Tamm (1965), James and Harwood (1969), and Olsen (1974). Control of insects by pathogens is treated by Burges and Hussey (1971).

PART FOUR
INSECT DIVERSITY

AN EVOLUTIONARY PERSPECTIVE OF THE INSECTA

17

The Phylum Arthropoda is split into major subdivisions that reflect basic differences in structural organization (Table 17.1). The insects belong to the Subphylum Uniramia, which includes all those arthropods with the limbs primitively uniramous, or one-branched. In contrast, in Crustacea, Trilobita, and Chelicerata, the limbs are primitively biramous, usually with a ventral walking part and a dorsal gill or swimming part. The body plans of these subphyla are quite distinct, so that Crustacea and Chelicerata which have secondarily uniramous legs (e.g., spiders, terrestrial isopods) are easily separated from the Uniramia.

TABLE 17.1 Conspectus of the Higher Classification of the Phylum Arthropoda

Subphylum Trilobita—trilobites
Subphylum Chelicerata
  Class Merostomata—horseshoe crabs and eurypterids
  Class Pycnogonida—sea spiders
  Class Arachnida—spiders, scorpions, mites, etc.
Subphylum Crustacea—crabs, shrimp, lobsters, etc; usually divided into about 8 classes
Subphylum Uniramia
  Superclass Onycophora—onycophorans
  Superclass Myriapoda
    Class Diplopoda—millipeds
    Class Chilopoda—centipedes
    Class Pauropoda—pauropods
    Class Symphyla—garden centipedes
  Superclass Hexapoda
    Class Entognatha—proturans, collembolans, diplurans
    Class Insecta—insects

Important features shared by all Uniramia include a long tubular alimentary canal without diverticula, and a common pattern of embryogenesis. In Crustacea and Chelicerata the gut is usually short and is produced into diverticula where digestion occurs. Embryogenesis in these classes is fundamentally different from that in Uniramia. Finally, the Uniramia and Biramia differ in the structure of the mandibles. In the Uniramia the structure of the mandibles is so highly modified that its morphological composition is difficult to interpret. Manton (1964, 1973) believes that mandibles of Uniramia are constituted from entire, whole limbs with the biting portion at the tip. Snodgrass (1958) and others argue that the mandibles of both Uniramia and Biramia are fundamentally similar and represent the basal segments (gnathobase) of modified limbs. In the Biramia, the mandibles are clearly derived from basal portions of limbs. For example, in some Crustacea the mandibles have functional palps.

The Uniramia includes three major lineages, differing in body organization and cranial structure. Onycophora, containing a few species restricted to tropical regions, have a multilegged, unsclerotized trunk without a differentiated head capsule. The mandibles articulate with the second head segment. Myriapoda have a multilegged, sclerotized trunk with the head developed as a distinct body region, or *tagma.* The genital opening is situated anteriorly on about the fourth trunk segment. In the Hexapoda the body is divided into three distinct tagmata—head, thorax, and abdomen, with walking legs on the three thoracic segments. The genital opening is situated posteriorly on about the eighth to ninth trunk segment. In Myriapoda and Hexapoda the mandibles articulate with the third head segment.

Tagmosis is the differentiation of the body into regions specialized for different functions. This is one of the most important developments in the evolution of the Arthropoda, and has produced the distinctive body plans of the various arthropod classes. Particularly significant in the Uniramia was the specialization of the thorax for locomotion, leading first to shortening of the body and the characteristic hexapod gait, and eventually to the evolution of wings in pterygote insects.

Until recently all six-legged arthropods have been included in the Insecta. However, the hexapodous condition represents an *evolutionary grade* and may have been evolved more than once. In other words, six-leggedness is a level of structural organization which could easily have evolved in different groups of myriapods which lost all but three pairs of legs. Common grades of body organization are frequently shared by distantly related organisms adapted for similar modes of life. For example, in both birds and bats the forelimbs are modified as wings (though in different manners). In arthropods the six-legged body probably evolved at least twice from different multilegged ancestors—once in the Insecta, and once in the Entognatha (Fig. 17.1; Table 17.1).

Independent evolution of the Entognatha and Insecta is strongly indicated by their fundamental differences in head structure. The Entognatha have the mandibles and maxillae deeply retracted into pouches in the head capsule (Fig 18.1*c*), the tips being protruded during feeding. This condition is termed *entognathy* or endognathy. The entognathous hexapods have *monocondylic* mandibles, adapted for externally triturating food particles into minute pieces before ingestion. In the Insecta, mandibles are primitively *dicondylic,* articulated with the head capsule at two points. These mandibles move transversely and are adapted for biting off and grinding food particles. The elongate styliform or bladelike mandibles of the Entognatha lack the basal grinding lobe of Insecta. In addition, all antennal segments of Entognatha are musculate, while in the Insecta only the basal two segments bear muscle insertions.

Ecologically, as well as in some morphological features, the Entognatha resemble myriapods more than insects. Like myriapods, the entognathous orders are primarily inhabitants of litter and the upper soil layers, requiring high relative humidity or sources of free water. Direct evidence of their former multilegged condition is provided by their styliform, rudimentary abdominal appendages, which function as skids in supporting the abdomen. Sperm transfer is indirect, by means of a spermatophore which is attached to the substrate and later picked up by the female, with or without direct contact with the male. Some of these primitive features are also characteristic of apterygote insects, as discussed below.

Among the Entognatha, Diplura are most similar to Insecta in general body plan, and also bear cerci on the last abdominal segment, as in primitive insects. Protura and Collembola differ from insects as well as Diplura in basic features. Protura lack antennae, and they grow by a process of *anamorphosis,* whereby abdominal segments are added at the time of molting. The newly eclosed nymphs have nine segments, while mature individuals have twelve segments. Anamorphosis is characteristic of myriapods. In contrast, insects (and other hexapods) grow by *epimorphosis,* in which the number of segments remains constant. Collembola are unique among hexapods in their six-segmented abdomen, also found in early instars of some myriapods. Eggs of Collembola are *microlecithal* (small-yolked) and develop by complete or *holoblastic cleavage.* Eggs of other hexapods divide by *meroblastic,* or superficial, cleavage because of the large yolk mass.

The great differences among the various entognathous orders indicate a very early divergence. No undisputed fossils older than Mesozoic have been discovered. This is probably because the small, soft bodies of Entognatha do not facilitate preservation. The extant orders of the primitive hexapods perhaps represent remnants of a much more diverse Paleozoic fauna.

ORIGIN OF WINGS

Wings have contributed more to the success of insects than any other anatomical structures; yet the historical origin of wings remains largely a mystery. The earliest insect fossils that have been discovered, from the Upper Carboniferous Period, were already winged. The primitively wingless hexapods, from which we might have gained insight into the origin of flight, are poorly represented in the living fauna. Thus the body structures that developed into wings, the steps in the evolution, and the ecological circumstances that favored wings, are largely matters of speculation. Despite the vigor of opposing arguments, unequivocal evidence is simply missing.

PREADAPTATIONS FOR FLIGHT. Before discussing two theories of wing origin, let us examine certain general deductions about the origin of pterygotes and their early environment. Wings evidently evolved only once, because the veins and articulatory sclerites at the wing base can be homologized among most Pterygotes. Furthermore, wings probably evolved after the apterygotes were resistant to desiccation, breathed air by means of tracheae, and were six-legged. These attributes may be considered preadaptations for flight. In other words, these are features that evolved in response to one set of environmental conditions and have provided the basis for the evolution of new features under another set of conditions. Here adaptations to terrestrial life and locomotion have provided the preadaptations for aerial life and locomotion.

Protection against desiccation is obviously needed by small organisms that are freely exposed to the drying effect of air currents. The assumption that the tracheal system preceded wing evolution is supported by the association of that system with wing ontogeny. At one time tracheae were even thought to influence the pattern of venation, but this is incorrect. The veins arise as blood-filled extensions of the hemocoel between the layers of wing epidermis. Nevertheless, tracheae supply oxygen within the wing and presumably antedate the wing in evolutionary origin.

The six-legged condition probably also evolved before the wings. The reasoning is that wings are restricted to the thoracic segments because only those segments had the skeletal and muscular arrangements associated with walking legs. Certain flight muscles were originally leg muscles, and the enlarged pleura that first evolved to support the legs were modified as a stiff fulcrum for the wings. Of the three segments, the mesothorax and metathorax supported the more powerful legs and possibly were also in the most favorable position aerodynamically to evolve the flight mechanism.

The early pterygotes may have molted more than once after the wings became functional, thus perpetuating the apterygote pattern of life history in which adults molt repeatedly. Today only the Ephemeroptera have two instars with wings. All other Pterygotes possess wings only in the final instar. Judging by the occasional failure of living insects properly to shed the cuticle enclosing the wings, one suspects that repeated molting of wings large enough to fly would involve a decided risk during growth. Flight is also expensive energetically and would divert nutritional resources away from growth in the immature stages. The strategy of delaying reproduction and dispersal until the last instar has been successful in the insects. This style of life history combines the advantages of flight and offsets the hazards of wing molting. Flight is advantageous in promoting sexual outcrossing among unrelated mates, thus reducing inbreeding; in dispersing to new habitats; in escaping enemies; and in locating specific feeding and oviposition sites.

ECOLOGICAL CONSIDERATIONS. Turning to the probable circumstances under which flight evolved, we should note that by the middle of the Devonian Period some plants reached 6 m or more in height. Later, in the Carboniferous Period, trees of the coal swamps reached 41 m in height. Food in the form of spores, pollen, seeds, or vegetative parts might have provided incentives for early apterygotes to adopt arboreal habits. Arboreal herbivores also may have been joined by predatory relatives and arachnids. Therefore, wings may have arisen first as a device to escape predators or to move from plant to plant.

The distribution of vegetation and bodies of fresh water was probably patchy and seasonal in the Devonian. Dispersal by air would have been a distinct asset, both for terrestrial forms as well as for those that had become secondarily aquatic. It has been emphasized by Rainey (1965) that dispersal by wind is especially advantageous in arid regions. Thermal updrafts over heated ground lift objects aloft. The prevailing winds tend to converge at low-pressure areas where rainfall is likely. Thus wind movement is toward areas of renewed plant growth and fresh water.

Flight is not essential for passive airborne dispersal, but organisms must be quite small to benefit from passive dispersal. Glick (1939) collected wingless arthropods by airplane at heights up to 4500 m. Included in his catch were mites, spiders, Thysanura, Collembola, Siphonaptera, ants, and immature instars of Hemiptera, Orthoptera, Coleoptera, Lepidoptera, and Diptera. The possession of wings, however, permits insects both large and small to reach altitudes of favorable wind currents, and to remain there longer, thus greatly increasing the distances traveled. Wings are, of course, important for local movements because insects can search for mates and select suitable feeding or breeding sites.

An obstacle that all theories of wing origin must overcome is to explain how selection would favor intermediate steps of wing development before the wing could function in flight. To lift the insect in air and provide a forward thrust, the wings must be thin, stiff, and large relative to the body; articulated at the bases; and moved by muscles in a complex propulsive stroke.

The time from the appearance of early land plants to the first fossils of winged insects spans the Devonian and Lower Carboniferous Periods, or 66 million years. Perhaps it is not surprising that arthropods could become fully terrestrial and then aerial during that time. A period of equivalent length in the Cenozoic Era was sufficient to evolve humans from shrewlike prosimians.

PARANOTAL ORIGIN OF WINGS. According to this theory, flight first evolved in an arboreal insect. If dislodged by wind or while escaping predators, a wingless insect would fall to the ground unchecked. Although probably not harmed, it would be necessary for the insect to regain its footing before escaping further. The addition of thin, lateral expansions on the terga, called paranotal lobes, might create a stable orientation or attitude during falling so that the insect would land on its feet for a quick getaway. Calculations have shown that some degree of attitudinal control is given when even small, fixed wings are added to models, provided the models are no smaller than 1 cm in length. Paranotal lobes also might have functioned in other ways, such as giving increased lateral protection from predators when the insect was flattened against the substrate.

Larger paranotal lobes would facilitate gliding to the ground or from plant to plant. The development of a suitable basal hinge on the thoracic nota would permit the paranotal lobes to flap when the thorax was deformed by muscular contractions. The final refinements would include modification of the basal hinge, muscle attachments, and nervous system so that the wing angle could be varied during each stroke.

The anatomical evidence for this theory is based mainly on the presence of broad thoracic nota in Thysanura and Archeognatha, and winglike prothoracic lobes on some fossil Palaeodictyoptera, Ephemeroptera, and Protorthoptera. The prothoracic lobes of some fossils appear to be articulated and to have well-developed venation (Fig. 17.2*b*). Presumably they indicate what the wings were like in an incipient stage of evolution. Alexander and Brown (1963) proposed that wings first developed as articulated notal flaps used by male insects in courtship display, then became winglike organs for gliding and were shared by both sexes.

GILL ORIGIN OF WINGS. This theory proposes that flight originated in an insect with an aquatic, gill-bearing nymph. The lateral abdominal gills of certain Ephemeroptera bear some resemblance to wings, being movable, thin, and membranous and having a ramifying pattern of tracheae. The similarity is quite striking in nymphs of Permian mayflies that have been illustrated by Kukalova-Peck (1968) (Fig. 17.2*a*). These fossils have wings on the mesothorax and metathorax that are too small for flight, but are curiously held outstretched and curved obliquely backward from the body. The winglets have fairly distinct venation and what appears to be a movable basal articulation.

The similarity of the winglets and abdominal gills in the Permian mayfly nymph reinforces the idea that the gills and wings are serially homologous. By beating the gills, living mayfly nymphs create currents of water over the body to replenish oxygen. Sclerotized gills or gill plates also function to some extent in locomotion. The articulated winglets of Permian nymphs may have moved in the same fashion and for the same purposes. More powerful strokes might have propelled the nymph forward. Thus the winglets might have functioned like fins, stiffened by venation, and moved in a propulsive stroke by muscles. The final step from aquatic fin to aerial wing would involve mainly the enlargement of the winglet and the thoracic musculature.

If wings and gills are serially homologous, what familiar structures of the arthropod body do they represent? Opponents of the gill theory, such as Snodgrass (1958), believed that the abdominal gills were serially homologous to the thoracic legs. Since legs are already present on the thoracic segments, the wings must have evolved from something else, such as paranotal lobes. Wigglesworth (1976), however, argued in favor of the gill theory by homologizing gills with basal lobes of the arthropod limb, called *exites*. If derived from exites, both wings and legs could be present in the thorax. The mesothoracic and metathoracic legs of Archeognatha, for example, have small styli on the coxae that could be interpreted as exite lobes, but morphologists have usually considered the styli to be endite lobes.

CLASS INSECTA

As restricted here, the Class Insecta may be defined as those hexapodous arthropods which at some time in their life show the following features: head, five-segmented with ectognathous mouthparts comprising mandibles without palps; maxillae and labium both with palps; antennae with intrinsic muscles in first two segments; thorax, three-segmented, with one pair of legs per segment; legs, six-segmented (coxa, trochanter, femur, tibia, tarsus, pretarsus); abdomen, primitively eleven-segmented with cerci; respiration through tracheal system with spiracles on last two thoracic and first eight abdominal segments; excretion by Malpighian tubules; postembryonic development epimorphic. In addition, the vast majority of insects are characterized by the possession of compound eyes and wings in the adult stage, though these are absent in some primitive forms and secondarily lost in many specialized ones.

By conservative estimates the Class Insecta encompasses at least 702,170 extant species, classified in 29 orders and about 750 families (Table 1.1). These numbers dwarf those in any other class of animals. Indeed, the four largest orders of insects (Coleoptera, Lepidoptera, Hymenoptera, Diptera) each exceed the next largest phylum of animals (Mollusca) by a considerable margin. The Family Curculionidae (weevils), with an estimated 65,000 species, is by far the largest family of animals, larger than all nonarthropod phyla except Mollusca. Species known only from fossils represent at least 10 additional orders which arose and radiated in the late Paleozoic and early Mesozoic Eras.

The morphological and ecological diversity within the Insecta is enormous. As detailed elsewhere in this book, insects occupy essentially every conceivable terrestrial habitat and have extensively colonized freshwaters. Only marine situations are relatively devoid of insects, although several thousand species inhabit the sea-land interface presented by intertidal regions, and a few Hemiptera live on the surface of open seas far from land. Insects include scavengers, herbivores, predators, and parasites. Many species are narrowly specialized for highly restricted foods, substrates, or activity periods, with consequent morphological modifications. The immensity of diversity within the Insecta and the attendant complexity of their phylogenetic relationships have resisted attempts to develop an entirely consistent and comprehensive evolutionary history. Nevertheless, the broad outlines of insect evolution have been repeatedly confirmed by independent lines of investigation. Areas of uncertainty or dispute are indicated where appropriate in the following account.

The classification proposed by Handlirsch (1908, 1926) has formed the framework for most modern classifications (Brues et al., 1954; O. W. Richards and Davies, 1957; MacKerras, 1970; Wille, 1960; Hennig, 1953; Rohdendorf, 1969), although many minor differences are evident. Kristensen (1975) provides an excellent review of ideas regarding hexapod phylogeny.

Perhaps the most significant advance over Handlirsch's classification is the recognition of the major differences in flight mechanism between the paleopterous and neopterous orders, as first set forth by Martynov (1925). More recently Tiegs and Manton (1958), Manton (1964, 1972, 1973), and Anderson (1973), among others, have shown conclusively that the entognathous hexapods are fundamentally different from Insecta, as detailed above. The classification adopted here is shown in Table 1.1 and Fig. 17.1.

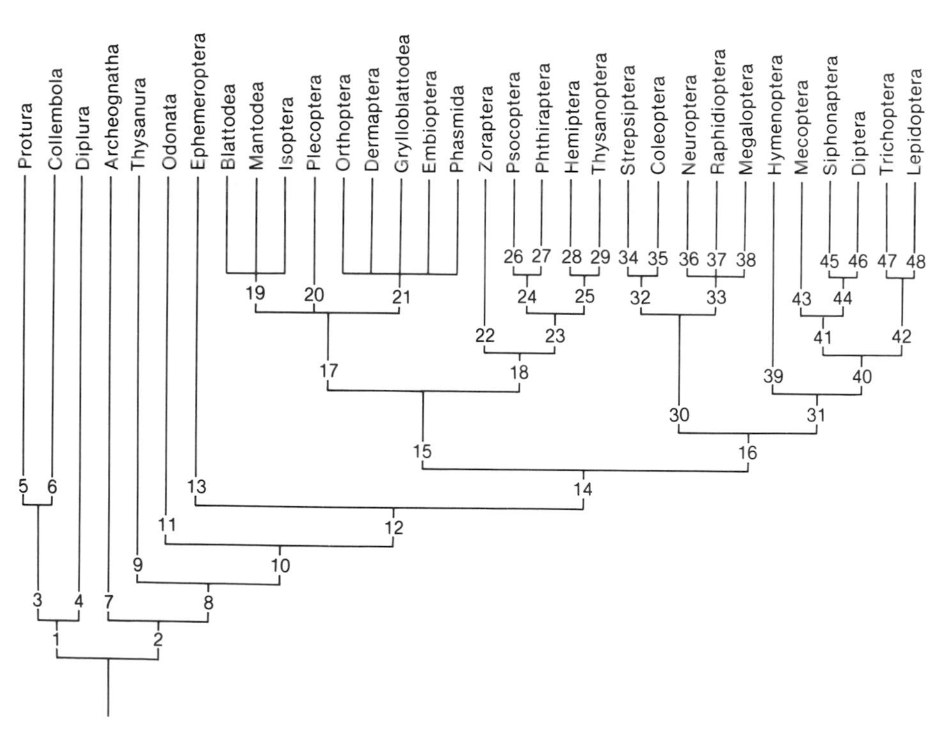

FIGURE 17.1 Diagram of relationships of the insect orders.

Important characteristics of insects and closely related hexapodous arthropods. Numbers refer to lineages in Fig. 17.1.

1: mouthparts entognathous
2: mouthparts ectognathous
3: cerci lost
4: cerci retained
5: antennae lost; embryonic development anamorphic
6: abdomen with 6 segments; abdominal appendages modified as jumping apparatus
7: mandibles elongate, with incisor and molar lobes isolated (fig. 19.1*b*); single articulation
8: mandibles shorter (fig. 19.2*b*), with incisor and molar lobes approximate
9: wingless; abdominal legs present
10: wings present (or secondarily lost); adults without abdominal legs
11: flight musculature indirect; thoracic segments fused as synthorax
12: flight musculature direct; thoracic segments separate
13: paleopterous; winged subimago and imago
14: neopterous; single winged imaginal stage
15: exopterygote postembryonic development
16: endopterygote postembryonic development
17: mouthparts mandibulate (adapted for chewing); central nervous system diffuse (with many segmental ganglia); numerous malpighian tubules
18: mouthparts usually adapted for sucking; central nervous system concentrated (with fused ganglia); 4-6 malpighian tubules
19: dorsal longitudinal muscles (wing depressors) weak or absent; corporotentorium perforated; immatures terrestrial
20: dorsal longitudinal muscles large; corporotentorium not perforated; immatures aquatic
21: dorsal longitudinal muscles large; corporotentorium not perforated; immatures terrestrial
22: cerci present; nymphs with ocelli
23: cerci absent; nymphs without ocelli
24: mandibles retained; laciniae modified as rods
25: mandibles and laciniae modified as stylets
26: free living
27: ectoparasitic
28: both mandibles modified as stylets; labium modified as sheath
29: right mandible lost; labium not modified
30: head with gula or pregular bridge; larvae campodeiform or highly modified; abdominal legs absent in larvae
31: head without gula (pregular bridge may be present); larvae usually eruciform or apodous; larvae frequently with abdominal legs
32: forewings modified as elytra
33: forewings membranous
34: legs without trochantins; immatures endoparasitic in insects
35: legs with trochantins; immatures never endoparasitic in insects
36: larvae with incomplete intestine and suctorial mouthparts
37: larvae with complete intestine; terrestrial
38: larvae with complete intestine; aquatic
39: adult with mandibulate (chewing) mouthparts; malpighian tubules numerous
40: adult with rostrate (modified as rostrum or proboscis) mouthparts; 4-6 malpighian tubules
41: larvae usually apodous (exception, Mecoptera); labial glands seldom producing silk; wing membranes bare or with few unmodified setae
42: larvae eruciform, frequently with abdominal legs; silk production by labial glands; adults with wing membrane covered with hairs or scales
43: adults with four wings; larvae with thoracic legs and abdominal pseudopods
44: adults with two wings, or wings absent; larvae apodous
45: adults without wings; ectoparasites living on mammals and birds
46: adults almost always with 1 pair wings; sometimes ectoparasitic, but seldom living on hosts
47: adults with mouthparts atrophied, wings covered with hairs; larvae aquatic
48: adults with mouthparts developed as coiled proboscis (occasionally atrophied); wings covered with scales; larvae almost always terrestrial

SUBCLASS APTERYGOTA. These insects are similar in body organization to pterygote (winged) insects, but primitively lack wings. Apterygotes have ectognathous mouthparts and intrinsic musculature in only the basal two antennal segments, and they have compound eyes and ocelli. The long, slender cerci and median filament recur in Ephemeroptera (mayflies), and the body form of Apterygota strongly resembles that of Ephemeroptera nymphs. Like the Entognatha, apterygotes have rudimentary abdominal appendages, employ indirect insemination, and molt throughout life. These are primitive features which have been retained in both hexapod lines of evolution; they do not indicate close relationship.

Two levels of organization are evident within the Apterygota. In the Archeognatha the elongate monocondylic mandible has the incisor lobe distinct from the molar lobe (Fig 19.1*b*), a primitive condition similar to that in many Crustacea. In Thysanura the dicondylic mandible has the incisor and molar lobes much closer, as in pterygote insects. However, Archeognatha have large, compound eyes, as in Pterygota, while Thysanura have the eyes reduced to a few lateral facets, and it is not clear which order is more nearly related to winged insects. The Archeognatha and Thysanura also show differences in the endocrine system (Watson, 1965), structure of the spermatozoa (Wingstrand, 1973), and abdominal musculature (Birket-Smith, 1974). The Order Monura, known from Paleozoic fossils, appears to be more primitive than extant Apterygota, and has been interpreted as ancestral to winged insects and other Apterygota.

SUBCLASS PTERYGOTA. The Pterygota are primitively winged insects with the mesothoracic and metathoracic segments enlarged and bearing wings, or secondarily wingless; mandibles primitively dicondylic, adapted for chewing, or highly modified; abdomen primitively with 11 segments; the anterior 10 segments without appendages and the eleventh segment frequently with cerci.

Pterygota differ biologically from Apterygota in employing *direct insemination* via copulation and in molting only until sexual maturity. No Pterygota possess abdominal appendages similar to those of apterygotes, but the abdominal gills of mayflies are probably homologous (Riek, 1970*a*). The abdominal appendages found on some pterygote larvae may be secondarily derived structures.

Infraclass Paleoptera. Pterygota comprise two major divisions, differing in their wing hinging mechanism and flight musculature. In the Paleoptera the wings cannot be flexed over the back at rest, whereas in the Neoptera wing flexing is allowed by the configuration of axillary sclerites and folds in the wing base. Wing flexing was necessary to permit winged insects to reenter the protected environments provided by litter and by subcortical (under-bark) and other restricted situations. Some paleopterous insects, such as the Suborder Zygoptera (Odonata), simulate wing flexing through the steep backward tilt of the pterothorax (Chap. 21).

The extant orders of Paleoptera exemplify many primitive features. In Odonata (dragonflies) the male stores the spermatozoa in a secondary copulatory organ on the second abdominal sternite. The female is responsible for completing sperm transfer, as in Apterygota. Possibly this behavior represents a modified indirect fertilization. Furthermore, in Odonata the flight musculature is entirely *direct* (attached directly to the wing bases) rather than primarily *indirect* (attached to the thoracic terga) as in Ephemeroptera and nearly all Neoptera. Notably primitive features of Ephemeroptera include the monocondylic mandibles of naiads, and the paired male external genitalia of some species. The winged but sexually immature subimaginal instar of mayflies is unique, but suggests the indeterminate molting of Apterygota.

Extant Paleoptera are aquatic as naiads, leading to the idea that the Pterygota may have arisen from aquatic ancestors whose gills evolved into wings, as discussed above. However, during the Paleozoic Era a much more diverse paleopterous fauna existed than at present, including many forms with apparently terrestrial nymphs.

Infraclass Neoptera. The Neoptera comprise about 99 percent of all insects. They present an exceedingly diverse assemblage but share in common the ability to flex the wings over the back by means of a pleural muscle inserting on the third axillary sclerite. For almost every other character the Neoptera are variable, or at least subject to exceptions. Neopterous insects may be classified in two major divisions, the Exopterygota and Endopterygota, based on modes of growth and development. In Endopterygota the wings and other presumptive adult structures develop as internal buds, or *anlagen,* in the immature (larva), which usually differs from the adult in many features. In Exopterygota the immature (nymph) is generally similar to the adult except in the size of the wings, which develop as external pads. Some Hemiptera and Thysanoptera are physiologically holometabolous, with a "pupal" instar in which occurs histolysis of muscles, alimentary canal, and other structures. However, the ontogeny of the wings is exclusively external in these species, which are highly evolved members of their evolutionary lineages. Thus, holometabolism represents an evolutionary grade which has developed independently in Exopterygota and Endopterygota. For these reasons, Exopterygota and Endopterygota are preferable to Hemimetabola and Holometabola as taxonomic names.

TABLE 17.2 Major Evolutionary Trends in Exopterygote Insects

Orthopteroid
Nymphs with ocelli
Chewing mouthparts; gular region well developed
Antennae long, multiarticulate
Hind wings dominate in flight ("posteromotoria")
Tarsomere number variable
Cerci usually large,multiarticulate
Malpighian tubules numerous
Central nervous system with separate ganglia in thorax and abdomen
Herbivores, predators, very few parasites; some social, subsocial (termites, cockroaches)
Hemipteroid
Nymphs lacking ocelli
Chewing mouthparts in most primitive groups; gradual development of piercing-sucking mouthparts in advanced groups
Tendency toward reduction in number of antennal segments
Forewings dominate in flight ("anteromotoria")
With 3 or fewer tarsomeres
Cerci absent (except in Zoraptera)
Malpighian tubules 4 or 6 in number
Central nervous system strongly concentrated in thorax
Herbivores, predators, parasites; none social, few subsocial

DIVISION EXOPTERYGOTA. The Exopterygota have diverged as two major lineages, designated as the Superorders Orthopteroidea and Hemipteroidea. These differ in numerous major morphological characteristics (Table 17.2), the most important involving the structure of the central nervous system, the number of Malpighian tubules, and wing venation. In all these features, as well as in the structure of the mouthparts, the Hemipteroidea are specialized compared to the more primitive Orthopteroidea. Hemipteroidea also appear much later in the fossil record, suggesting derivation from some orthopteroid ancestor.

SUPERORDER ORTHOPTEROIDEA. The orthopteroid orders are characterized by chewing mouthparts, long multiarticulate antennae, hind wings with a large anal lobe, multiarticulate cerci, and numerous Malpighian tubules, as well as other primitive features. Undoubtedly the orthopteroid insects represent the most primitive Neoptera. Yet the interrelationships of the orthopteroid orders are poorly understood because of the mosaic distribution of primitive and specialized characteristics. For example, Dermaptera are the only Neoptera which possess paired external genitalia (paired penes of some species), but they display highly specialized hind-wing venation (to permit folding beneath the specialized forewings) and strongly sclerotized cerci modified as forceps. Morphological features of Embioptera are strongly specialized for life within self-constructed silken tubes, and Phasmatodea for camouflage. In contrast, nearly all features of the Grylloblattodea and Plecoptera are so generalized that their relationships with other orders are difficult to determine.

Within the Orthopteroidea, the Blattodea (cockroaches), Mantodea (mantids), and Isoptera (termites) form a group of clearly related orders, sometimes combined as the single Order Dictyoptera. All Dictyoptera share several unique morphological features (proventricular armature, wing venation pattern, structure of female genitalia, perforated corporotentorium), and Blattodea and Isoptera share additional structural and biological features (see Chaps. 22, 24; Table 24.1). Phylogenetically the termites and mantids might be considered specialized cockroaches.

Among the remaining Orthopteroidea, Plecoptera are noteworthy for their primitive morphological features, including tracheal gills in nymphs, distinct anapleurite and coxopleurite in some nymphs, and multiarticulate cerci. They have sometimes been separated as the Polyneoptera, a separate division of Neoptera. Orthoptera show a long, separate evolution, with recognizable fossils as early as the Carboniferous Period. Their closest affinities are probably with the Phasmatodea. In the Zoraptera are combined features of both the Orthopteroidea (chewing mouthparts, cerci) and the Hemipteroidea (six Malpighian tubules, concentrated central nervous system, two-segmented tarsi, and reduced wing venation). The balance of characters indicates that Zoraptera should probably be regarded as the most primitive hemipteroid insects.

SUPERORDER HEMIPTEROIDEA. The most important feature in interrelating the hemipteroid orders is the configuration of the mouthparts. With the exception of Zoraptera, all Hemipteroidea have the lacinia modified as a sclerotized, styliform organ. In other mouthparts much variation exists. In one evolutionary line leading through Psocoptera (book lice) to Phthiraptera (lice) the lacinia is a sturdy rod which is applied to the substrate as a brace to stabilize the head while the mandibles scrape food particles. The hypopharynx and pharynx are modified as a mortar-and-pestle—like crushing device. Some Phthiraptera (Suborder Anoplura) have the mouthparts further modified to suck blood, but their general similarity in numerous characters to more primitive lice (Suborders Amblycera and Ischnocera) and Psocoptera leaves little doubt of their evolutionary origin.

In the evolutionary line leading to Thysanoptera (thrips) and Hemiptera (bugs), both mandibles and maxillae are modified as styliform piercing organs. The more primitive arrangement occurs in Thysanoptera, where the maxilla and single mandible are relatively short, thick blades which are guided by the conical labrum and labium. As in Hemiptera the maxillary stylets are attached to the stipes by a special lever arm (Chaps. 34, 35). In Hemiptera the two mandibular and two maxillary stylets interlock to form a highly efficient piercing-sucking tube which is enclosed by the troughlike labium. A few Thysanoptera and Hemiptera are physiologically holometabolous, as mentioned above. Their highly specialized mouthparts, concentrated central nervous system, and other features preclude them from being ancestors of the Endopterygota.

The exopterygote orders display limited biological diversity, compared to the Endopterygota. Most Orthopteroidea live on the surface of ground, on vegetation or in litter, while the great majority of Hemipteroidea inhabit foliage. Specialized modes of life, such as ectoparasitism in Phthiraptera and Hemiptera, or social behavior in Isoptera, are exceptional. There are no parasitoids, no endoparasites (see Chaps. 14, 15), and very few aerial species. Exopterygota scarcely participate in exploiting pollen and nectar, very important factors in the diversification of the Endopterygota. In general, exopterygotes utilize the same food resources in all instars (exceptions include cicadas) and are almost all external feeders. Inability to feed internally has barred them from a multitude of niches occupied by endopterygote larvae. For example, except for termites (Isoptera), almost no exopterygotes bore or mine living or dead wood, foliage, fruits, or seeds. By external feeding only Hemiptera have exploited the vascular region of woody plants, an important habitat for endopterygotes, especially Coleoptera.

DIVISION ENDOPTERYGOTA. Included here are about 85 percent of the extant species of insects, representing immense taxonomic and biological diversity. While the evolution of holometabolism is not entirely clear its chief adaptive value seems to be in allowing adults and larvae to utilize different resources. Resource division is usually accompanied by morphological divergence. This divergence is extreme in Diptera and most Hymenoptera, where the legless, grublike larvae frequently lack eyes and have the antennae and mouthparts reduced to papillae. Differences between adults and larvae are least in Megaloptera and Raphidioptera, which are among the most primitive endopterygote orders. Besides sharing holometabolous development, the Endopterygota also flex the wings along the jugal fold, in contrast to flexion along the anal fold in Exopterygota.

The origin of the Endopterygota is obscure. As mentioned above, holometabolous development has independently evolved in the Hemipteroidea, but there are no other Exopterygota which suggest a tendency toward holometabolism. The most primitive endopterygotes—Mecoptera, Megaloptera, and Raphidioptera—are extremely generalized, with little hint of what their ancestors may have been. Their wing venation has been compared to that of Palaeodictyoptera, but the resemblance is apparently convergent. Venational similarities to Plecoptera (Hamilton, 1972) are probably more indicative of the true relationships of the Endopterygota.

The endopterygote orders form three evolutionary lineages, recognized as the Superorders teroidea (Table 1.1). The more specialized orders, such as Coleoptera, Diptera, or Lepidoptera, clearly belong to one lineage or the other, but the primitive orders (Megaloptera, Mecoptera) share a similar body plan and evidently arose from a common ancestor.

SUPERORDER NEUROPTEROIDEA. The most primitive Neuropteroidea are certainly the Megaloptera and Raphidioptera. The pupae of both groups are capable of limited movements and have functional (decticous) mandibles which are used in defense. Although larvae of Megaloptera are aquatic, those of Raphidioptera terrestrial, their morphological similarity is so great that they are frequently classified as a single order.

Adult Neuroptera are similar to Megaloptera, but the larvae are highly specialized in having the mandibles and maxillae adapted for piercing and sucking, and in having the midgut end blindly. Fossil evidence shows a long period of evolution independent of Megaloptera, with great diversification in the Permian and Cretaceous Periods.

Coleoptera and Strepsiptera are so highly specialized that their exact relationships are uncertain. Larvae of some aquatic Coleoptera are exceedingly similar to immature Megaloptera, and fossils such as *Tshekardocoleus* may represent intermediates between Megaloptera and Coleoptera. Mickoleit (1973) has shown that the coleopteran ovipositor is homologous with that in other neuropteroids.

The Strepsiptera are narrowly specialized for endoparasitism, and are morphologically distinct from other neuropteroids. Like Coleoptera they fly using the hind wings (posteromotoria). First-instar larvae (triungulins) of Strepsiptera are similar to triungulins of the parasitoid beetle Families Meloidae and Rhipiphoridae.

SUPERORDER MECOPTEROIDEA. Extant Mecoptera number no more than a few hundred species, but fossils record a large, diverse fauna in the late Paleozoic and Mesozoic Periods. Like Megaloptera, larvae of Mecoptera are decticious. They share compound eyes with larvae of some primitive Lepidoptera. Most modern Mecoptera are characterized by the elongation of the lower face and mouthparts, producing a strong resemblance to primitive Diptera (Figs 41.1*b*, 42.3*a*). Likewise, wing venation in primitive flies is similar to that of Mecoptera, and some fossils are difficult to place in either order. Modern Diptera are characterized by the reduction of the hind wings to halteres, but metathoracic wing pads are visible in some *Drosophila* pupae, and certain fossil diptera may have been four-winged.

Siphonapteran (flea) adults are extremely specialized as ectoparasites of mammals and birds, but the larvae are similar to larvae of primitive Diptera, from which they presumably evolved. The former winged condition of Siphonaptera is revealed by pupal wing pads in some species.

Lepidoptera and Trichoptera are generally similar in wing venation, mouthpart morphology, and larval body form, though their larvae are terrestrial and aquatic, respectively. The most primitive Lepidoptera (Suborder Zeugloptera) and all Trichoptera have decticous pupal mandibles, and Zeugloptera have compound larval eyes, as in Mecoptera. Lepidoptera are generally characterized by the elongation of the apposed maxillary galeae as a sucking proboscis. In Zeugloptera the maxillae are unmodified, and in Eriocraniidae the proboscis is formed by relatively short galeae, apposed only during feeding.

SUPERORDER HYMENOPTEROIDEA. Relationships of Hymenoptera are uncertain. They possess numerous Malpighian tubules, differentiating them from all other Endopterygota, which typically have four or six tubules. Hymenopteran wing venation is highly distinctive, with fusions of many veins to produce a few large cells. The eruciform larvae of primitive Hymenoptera share many characteristics with larvae of Lepidoptera and Mecoptera, including abdominal pseudopodial legs, a single tarsal claw, and labial silk glands. Probably the Hymenoptera diverged relatively early from the mecopteroid lineage.

THE GEOLOGICAL RECORD OF INSECTS

Our knowledge of insects that lived in the past is based on traces or remains preserved as fossils in the earth's sedimentary rocks. The hard parts of insects are resistant to decay. If the insect is quickly immersed in a protective, finegrained medium such as mud or volcanic ash, the chances are great that at least its wings will be preserved. Unfortunately, through the ages terrestrial deposits on the surface are usually destroyed by the erosive action of water or wind. Other deposits become too deeply buried to be discovered. Today the places where insect fossils can be found are limited in number. From these localities we are able to obtain brief glimpses of the insect life of millions of years ago.

Fossil evidence can be used to establish the first appearance of a taxon and, in the case of extinct taxa, the last appearance. When contemplating the duration of their existence, remember that the organisms must have existed for some period prior to the earliest record we have found. Likewise, taxa now extinct undoubtedly persisted for some time after the last fossils.

The chronological sequence in which the insect orders appear in the geological record is of limited value in determining their phylogenetic relationships. For example, the Archeognatha and Collembola are obviously derived from ancient lineages, yet unquestioned fossils are no older than the Mesozoic Era. The extant Orders Protura, Grylloblattodea, and Zoraptera and the Suborder Mallophaga of the Phthiraptera are not represented at all in the fossil record.

Unfortunately evidence of the crucial steps leading to the origin of insects have not yet been found in the fossil record. The oldest supposed remains of insects or their allies are found in rocks of the Devonian Period (Table 17.3). *Eopterum devonicum* and *Eopteridium striatum* in these strata were first thought to be the oldest winged insects. They have now been identified not as insects but as the winglike tails of crustaceans.

Another organic trace in Devonian rock was named *Rhyniella praecursor*. The remains are clearly those of a collembolan, possibly belonging to a family existing today. Crowson (1970) has noted fossil-like traces of modern insects in the same rocks and suspects the insects died in crevices at more recent dates, after the rocks were formed. Consequently, *Rhyniella* may not be the ancient fossil it was once thought to be.

EXTINCT ORDERS. The most common fossils are wings or fragments of wings. Many extinct orders have been described on the basis of unusual

TABLE 17.3 Geological Time Scale in Millions of Years

Era	Period	Epoch	Began	Duration
Cenozoic	Quaternary	Recent		
		Pleistocene	1	1
	Tertiary	Pliocene	10	9
		Miocene	25	15
		Oligocene	40	15
		Eocene	60	20
		Paleocene	70	10
Mesozoic	Cretaceous		130	60
	Jurassic		180	50
	Triassic		230	50
Paleozoic	Permian		270	40
	Upper Carboniferous		334	64
	Lower Carboniferous		350	16
	Devonian		400	50
	Silurian		440	40
	Ordovician		500	60
	Cambrian		600	100

Source: From Braziunas, 1975.

patterns of venation exhibited by fossil wings. Here we will follow the practice of Carpenter (1977) and recognize only those 10 orders where the head, mouthparts, and, in the case of Pterygota, both the fore- and hindwings are known. Fossil evidence indicates that these orders flourished during at least the last 104 million years of the Paleozoic Era (Upper Carboniferous and Permian Periods). Most became extinct at the end of the Permian, perhaps because profound changes took place in the climate and vegetation. Of the 10, only the Protodonata and Glosselytrodea are known to have survived into the Triassic Period of the next era, the Mesozoic. During this same time at the end of the Paleozoic many orders existing today made their first appearance. Consequently, life of the Permian may have been more diverse in insect orders than even the present.

SUBCLASS APTERYGOTA

Order Monura. Insects resembling Archeognatha, with ventral abdominal appendages and a long, caudal style, but without lateral cerci. Some specimens exceed 30 mm in body length excluding the style (Upper Carboniferous-Permian).

SUBCLASS PTERYGOTA

Infraclass Paleoptera

Order Protodonata (= Meganisoptera). Closely related to Odonata, but lacking the nodus, pterostigma, and arculus in the venation. *Meganeura* reached 75 cm in wing spread, but this was exceptional. Most were the size of large dragonflies. Nymphs are unknown; presumably they were aquatic. Adults were predatory (Upper Carboniferous-Triassic).

Order Palaeodictyoptera. This order (Fig. 17.2*b*) and the next two are closely related. All possessed cerci (no caudal styles), external ovipositors, and long, piercing beaks of five stylets [according to Kukalova-Peck (pers. comm.), the paired mandibles and maxillae and the hypopharynx are supported in an elongate, troughlike labium]. The beaks were held vertically beneath the head or slightly thrust forward, not directed to the rear as in Hemiptera. The postclypeus was swollen and may have housed a sucking pump. Nymphs of this order and Megasecoptera were evidently terrestrial. The wing pads were free and curved obliquely backward. Nymphs had beaks like the adults, and all stages presumably fed on plant

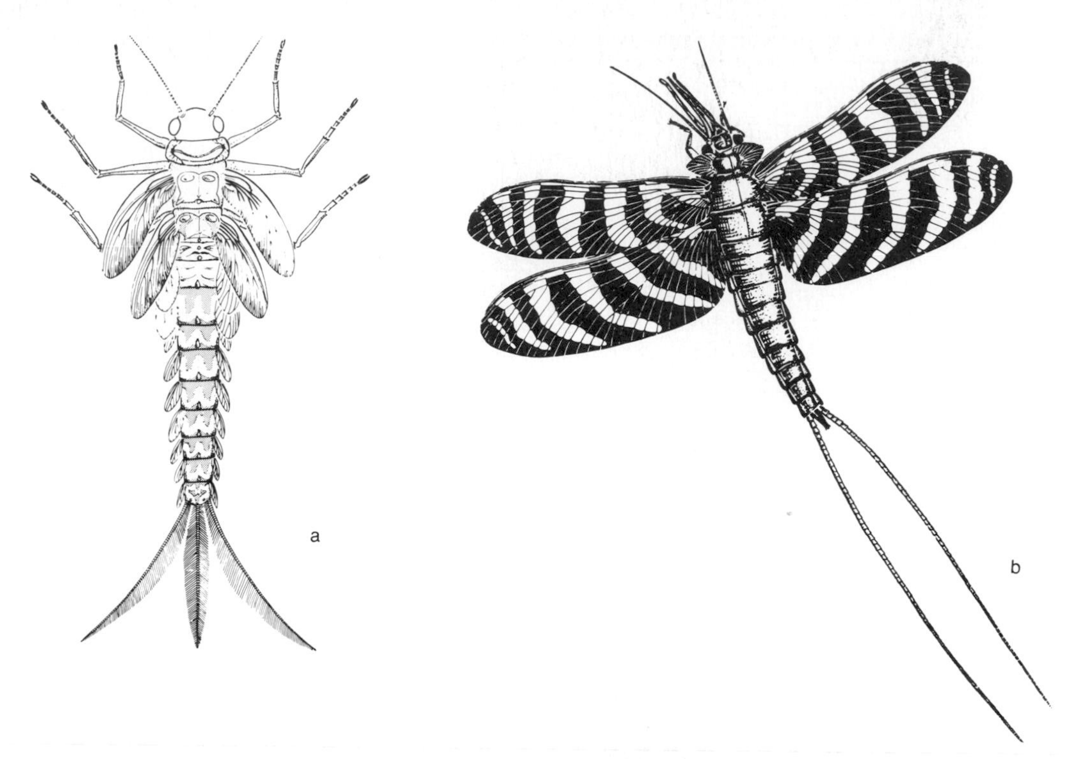

FIGURE 17.2 Reconstruction of Paleozoic insects by J. Kukalova-Peck: *a,* naiad of *Protereisma* sp., a Permian mayfly; *b, Homaloneura joannae,* a palaeodictyopteran from the Upper Carboniferous Period. (*Courtesy of and with permission of J. Kukalova-Peck.*)

juices. Adults of this order reached 50 cm in wing spread, but many were smaller. Adults may have continued molting. The wings were often dark in color with patterns of light spots or transverse bands. The hind wings were variable in size and shape, being larger than, equal to, or smaller than, the forewings. The pronotal lobes were large and sometimes exhibited a veined membrane. The abdominal terga had large lateral lobes. Cerci were about twice as long as the abdomen (Upper Carboniferous-Permian).

Order Megasecoptera. Distinguished by long cerci and by distinctive wings that were long, nearly equal in size and shape, frequently elongate and petiolate at the base. All Megasecoptera and some Palaeodictyoptera had simple or branched integumental processes on the body that were densely covered with setae. Some processes were quite long, even longer than the body, and formed fringelike rows on the tergites (Upper Carboniferous-Permian).

Order Diaphanopterodea. Distinguished by a wing-folding mechanism superficially like that of Neoptera. Nymphs unknown (Upper Carboniferous-Permian.)

Infraclass Neoptera

DIVISION EXOPTERYGOTA

Order Protorthoptera. The most diverse of the extinct orders. Characterized by coriaceous or membranous forewings, more or less expanded anal area of the hind wing, cerci, and mandibulate mouthparts. Some had raptorial forelegs. Presumably the earliest of the Neoptera to radiate into various habitats (Upper Carboniferous-Permian).

Order Caloneurodea. Small to large insects with long antennae, mandibulate mouthparts, subequal wings, slender abdomen, and short cerci (Upper Carboniferous-Permian).

Order Miomoptera. Small insects with mandibulate mouthparts, short cerci, and wings of equal size with simplified venation. In some Permian deposits they are more numerous than all other orders (Upper Carboniferous-Permian).

Order Protelytroptera. Small, robust insects superficially resembling beetles, but related to Blattodea and Dermaptera. Forewings formed into elytra, the broad hind wings folded beneath. Short cerci were present (Permian).

DIVISION ENDOPTERYGOTA

Order Glosselytrodea. Small insects that are related to Neuroptera. The wing is elytriform and has specialized venation (Permian-Triassic).

FOSSIL HISTORY OF INSECTS. The earliest fossils that are unquestionably ancient and insectan are found in the Upper Carboniferous Period of the Paleozoic Era. A few isolated wings have been found near the base or oldest strata, but the best fossils begin to appear near the middle of the period. Vast tropical swamps with luxuriant forests characterize the period. These have yielded extensive coal deposits. Two of the most important fossil localities are associated with such beds of coal. Fossiliferous shales at Commentry, France, contain diverse and well-preserved fossils. Insects from nearby coal forests evidently were buried in the fine sediments of stream deltas along the shore of a freshwater lake. Near Morris, Illinois, iron-bearing concretions or nodules that enclose fossil insects can be found along Mazon Creek. The ancient environment was a low, coastal plain with coal-producing forests.

Fossils from all localities indicate that at least 11 orders were present in the Upper Carboniferous: Monura, Palaeodictyoptera, Megasecoptera, Diaphanopterodea, Protodonata, Ephemeroptera, Protorthoptera, Orthoptera, Blattodea, Miomoptera, and Caloneurodea. Evidently, considerable insect evolution had already taken place before the Upper Carboniferous because all the higher taxa (above the category of order) are represented except the Endopterygota. From this fauna, however, only the Ephemeroptera, Orthoptera, and Blattodea survived until the present. Fossil wings of Blattodea are so common that the period is often called the "Age of Cockroaches." These insects were similar to modern cockroaches except that the wings were different in some features, their ovipositors were much longer, and they probably did not deposit oothecae. The Orthoptera had jumping hind legs and stridulatory areas on the wings.

The climates of the world changed drastically during the Permian Period. In the Northern Hemisphere increasing aridity is evidenced by great wind-blown deserts and thick salt deposits left by drying seas. Colder climates in the southern Hemisphere were marked by several periods of glaciation.

Permian limestones of the Wellington Formation in Kansas and Oklahoma have abundant insect fossils. The strata measure some 213 m and include fossils of terrestrial plants, marine arthropods, and salt deposits. This is interpreted as a coastal swamp with intermittent freshwater habitats.

Associated with the changing climate and landscape of the Permian, the insect fauna also underwent major changes in composition. Ten new orders that have survived until the present made their appearance: Odonata, Plecoptera, Hemiptera (Homoptera), Psocoptera, Neuroptera, Raphidioptera, Megaloptera (including a larva), Mecoptera, Trichoptera, and Coleoptera. The last six are the earliest definite endopterygotes. Two additional orders are also new: Glosselytrodea and Protelytroptera, but the latter as well as six of the orders from the Upper Carboniferous became extinct by the end of the Permian. The Glosselytrodea and Protodonata survived into the Triassic. Permian Ephemeroptera had wings of equal size, and some had conspicuous mandibles. Nymphs are also found in Permian deposits (Fig. 17.2*a*). The mandibles, known from one specimen, are large and have well-developed teeth. The small wings have distinct veins and curve obliquely backward from the body. Gills were present on the first nine abdominal segments.

Unfortunately fossils are scarce from the Mesozoic Era. This was a crucial time when modern insects were evolving in association with flowering plants and mammals. Archeognatha, Hymenoptera, Phasmatodea, and Diptera are first represented in the Triassic Period. During the Jurassic Period, fossils of Dermaptera, Thysanoptera, and Hemiptera (Heteroptera) first appear. Also at this time, existing families of Odonata, Orthoptera, Diptera, and Hymenoptera can be recognized.

In the Cretaceous Period, many of the insects are preserved in hardened lumps of fossil plant resin called amber. The brownish, translucent substance often contains organic matter such as pollen, hair, leaves, or insects. Polished pieces are prized as jewelry. Once entrapped in the resin, the external features of insects are perfectly preserved in microscopic detail, but the internal structures are lost. The resins do not decay or dissolve, but harden with time. Because of their light weight, lumps are carried by stream water to be deposited, eroded, and redeposited several times. Consequently, pieces may be much older than the rock in which they are finally embedded. When found with lignite or coal seams, however, it is likely that the amber has not been carried far from the trees that produced the resin.

Cretaceous amber has been found in Alaska, Canada, Siberia, and Lebanon. The insects thus preserved include the earliest Lepidoptera, worker ants, and authentic Collembola, plus Aphididae, parasitic Hymenoptera, and Chironomidae. Isoptera are also known from the Cretaceous, but not in amber. Siphonaptera have been described, but the identification is questionable.

Of the several epochs of the Tertiary Period in the Cenozoic Era, the Oligocene is best known. Amber from the shores of the Baltic Seas is dated from the Oligocene. Here for the first time are Diplura, Thysanura, Mantodea, Embioptera, Siphonaptera, and Strepsiptera. The species do not exist now, but the insects belong to modern types. The silverfish Family Lepidotrichidae was thought extinct, but living relatives of those entrapped in Baltic amber were discovered in 1959 in the coastal forests of northern California. Oligocene and Miocene amber from Chiapas, Mexico, has yielded many insects (Fig. 17.3). Included are stingless bees of the genus *Trigona* that differ only slightly from species living today in Central America.

Another important locality of Oligocene age is the Florissant Fossil Beds National Monument, near Colorado Springs, Colorado. A creek flowing through the area was dammed by lava and mudflows emanating from volcanoes about 24 km away. Fine ash from the volcanoes settled into the lake and buried a great variety of plants and animals that fell into the still water. Over a hundred species of higher plants and thousands of insect species have been described from these deposits. Half or more belong to genera existing today. Some, however, are no longer present in North America. *Glossina,* the tsetse fly, is present as a fossil at Florissant, but today is found only in tropical Africa.

Calcareous nodules containing insects have been found in the Mojave Desert, California, from deposits of the Miocene Epoch. Apparently a Miocene freshwater lake was surrounded by volcanoes. Many orders of insects are represented, but none for the first time. Some are beautifully preserved by replacement of the original organic matter by collodial silica.

Asphalt or tar pools in southern California, dating from the Pleistocene Epoch of the Quaternary Period, have yielded insect remains. Inside the skulls of the sabre-toothed cat, *Smilodon,* are puparial shells of dipterous larvae that presumably were scavengers on animals trapped in the tar. Adult aquatic insects, such as dragonflies and water beetles, are attracted today to the waterlike reflections of the liquid tar, and entombed.

The oldest record of the Suborder Anoplura of the Phthiraptera was obtained when lice were found on the carcass of a rodent frozen during the Pleistocene in Siberia.

USE OF DICHOTOMOUS KEYS

The keys in this book consist of series of numbered dichotomies or two-way choices. Each pair

FIGURE 17.3 Insects preserved in Oligocene and Miocene amber from Chiapas, Mexico: *a* (*left*), fulgorid bug; *b*, (*right*), stingless bee, *Trigona silacea* (Apidae). Note sensilla on antennae. (*Photos courtesy of Joseph H. Peck, Jr.;* b *reprinted by permission of the Journal of Paleontology.*)

of choices is called a *couplet*. Each half of a couplet either leads to a subsequent couplet (as indicated by the appropriate number), or name of a taxon (order, family), which has then been identified.

Beginning with the first couplet, one works through the key, comparing the characteristics of the specimen at hand with the dichotomous choices. If a couplet lists more than one feature, the primary, or most diagnostic, character is compared first. Secondary characters which follow should be used to confirm identifications made with the primary character, or as alternatives when primary characters are missing or damaged. Numbers in parentheses indicate the couplet immediately preceding, so that the keys can be worked backward or forward. It should be emphasized that any large group of organisms, such as the Insecta, contains exceptional species which do not fit keys. The keys in this book should identify nearly all the insects encountered during general collecting.

KEY TO THE CLASSES AND ORDERS OF COMMON HEXAPODOUS ARTHROPODS

Body without wings, or with rudimentary or vestigial wings less than half body length **Key A, p. 279**

Body with 1 or 2 pairs of wings at least half as long as body (wings may be modified as rigid covers over abdomen, or folded) **Key B, p. 283**

KEY A
Wings Absent or Rudimentary

1. Legs absent or reduced to unsegmented papillae shorter than one-fifth body width 2

   Legs with 4 to 5 distinct segments, almost always terminated by 1 or 2 claws 3

2(1). Mouthparts enclosed in a slender, tubular rostrum (Fig 17.4*a*); antennae and eyes usually absent; body segmentation indistinct or absent; sessile plant feeders, frequently covered by a waxy or cottony shell (Fig 34.12*d*) **Hemiptera** (Chap. 34)
Mouthparts mandibulate or internal, never enclosed in a tubular rostrum; antennae and eyes present or absent; body segmentation usually distinct; seldom with protective covering **Legless endopterygote larvae, not keyed further**

3(1). Legs terminated by a single claw or without claws 4
At least middle legs terminated by 2 claws 16

4(3). Head with large compound eyes almost always present laterally; ocelli frequently present on vertex 13
Head with compound eyes absent or vestigial, stemmata (lateral ocelli) often present (Fig. 17.4*b,d*); ocelli absent from vertex 5

5(4). Mouthparts enclosed in a slender, tubular rostrum (Fig. 17.4*a*); antennae and eyes usually absent; body segmentation frequently indistinct or absent; plant feeders, frequently with body covered by a waxy or cottony shell (Fig. 34.12*d*) **Hemiptera** (Chap. 34)
Mouthparts mandibulate or concealed in head capsule; body segmentation rarely indistinct 6

6(5). Antennae with 2 or more segments 8
Antennae absent 7

7(6). Tarsus terminating in single claw; lateral ocelli absent; body cylindrical, elongate (Fig. 17.4*c*); minute, pale arthropods found in soil **Protura** (Chap. 18)
Tarsus without claw; stemmata (lateral ocelli) large, usually set in pigmented patches (Fig. 17.4*d*); body fusiform (Fig. 17.4*d*); minute insects usually found on flowers, foliage. **Strepsiptera (First instars,** Chap. 40)

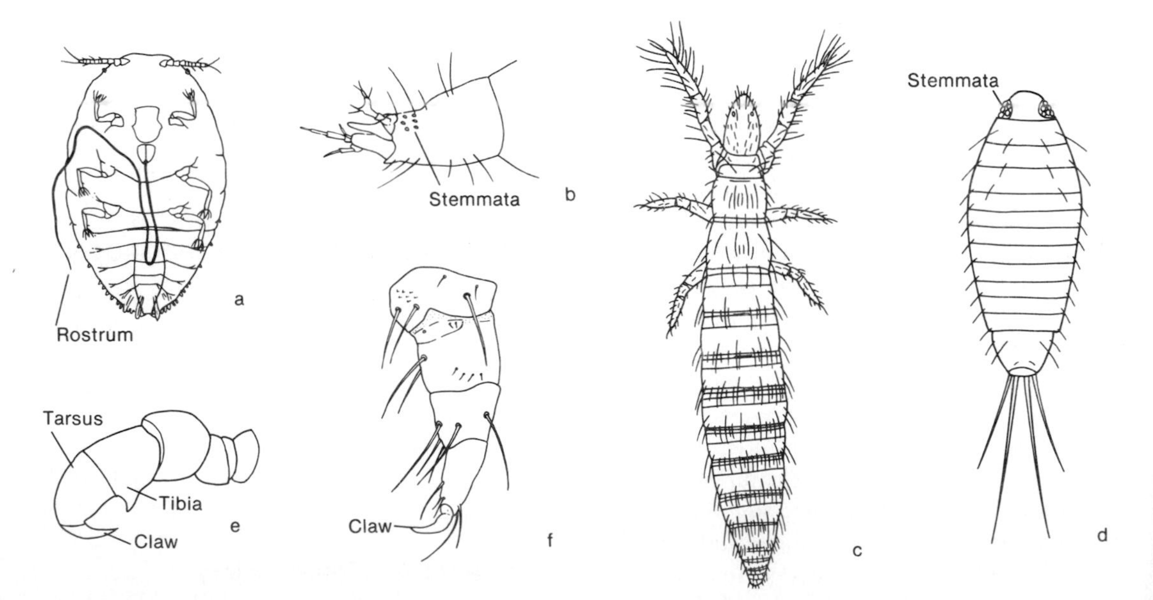

FIGURE 17.4 *a*, A scale insect, Hemiptera (*Parlatorea*); *b*, head of larval carabid beetle, showing stemmata (lateral ocelli); *c*, Protura (Acerentomidae); *d*, triungulin larva of Strepsiptera (Mengeidae); *e*, chelate tarsus of Phthiraptera (Anoplura); *f*, leg of endopterygote larva (Lepidoptera, Noctuidae). (a *from University of California Extension;* b *from Melis,* 1936; c *from Berlese,* 1909.)

8(6). Abdomen with 6 segments; segments 1, 3, and 4 usually with medial unpaired appendages (Fig. 18.4) **Collembola** (Chap. 18)
Abdomen with 8 to 11 segments; without appendages or with paired unsegmented appendages on some segments **9**

9(8). Tarsus and claw chelate (Fig. 17.4*e*); abdomen without appendages; ectoparasites with flattened body, thick, short, 3- to 5-segmented antennae **Phthiraptera** (Chap. 33)
Tarsus and claw usually fused (Fig. 17.4*f*), very rarely chelate; abdomen with or without appendages; body shape extremely variable **(Endopterygote orders, larvae) 10**

10(9). Abdomen with unsegmented, paired walking appendages on some preterminal segments; body caterpillar-shaped (Fig. 17.5*a*) **11**
Abdomen without walking appendages on preterminal segments; not usually caterpillarlike **12**

11(10). Abdominal appendages bearing rows, circles, or patches of short curved spines (crochets, Fig. 17.5*b*) **Lepidoptera** (Chap. 44)
Abdominal appendages without crochets **Hymenoptera** (Chap. 46)

12(10). Abdomen with ventrally directed hooked prolegs on last segment (Fig. 45.2); thoracic legs with trochanters 2-segmented; aquatic larvae usually dwelling in tubular cases **Trichoptera** (Chap. 45)
Abdomen without appendages or with dorsal or lateral appendages without terminal hooks; if prolegs are ventrally directed, then terminal hooks are absent **Coleoptera** (Chap. 39)

13(4). Mouthparts enclosed in tubular rostrum (Fig. 17.4*a*); cerci absent **Hemiptera** (Chap. 34)
Mouthparts mandibulate, never enclosed in a rostrum; cerci present or absent **14**

14(13). Terminal abdominal segment with 2 or 3 long filaments; antennae multiarticulate, slender; long-legged aquatic insects, usually with leaflike abdominal gills **Ephemeroptera (naiads)** (Chap. 20)
Terminal abdominal segment without filaments; terrestrial, gills absent **15**

15(14). Body caterpillar-shaped (Fig. 17.5*a*) or grublike (Fig. 17.5*c*), with short, thick legs; tarsi with 1 segment **Mecoptera (larvae)** (Chap. 41)
Body slender, elongate, with long, slender legs; tarsi with 4 to 5 segments **Mecoptera (adults)** (Chap. 41)

16(3). Anterior legs with first tarsal segment globular, at least 3 times as thick and long as second segment **Embioptera** (Chap. 28)
Anterior legs with tarsal segments about equal in thickness and length **17**

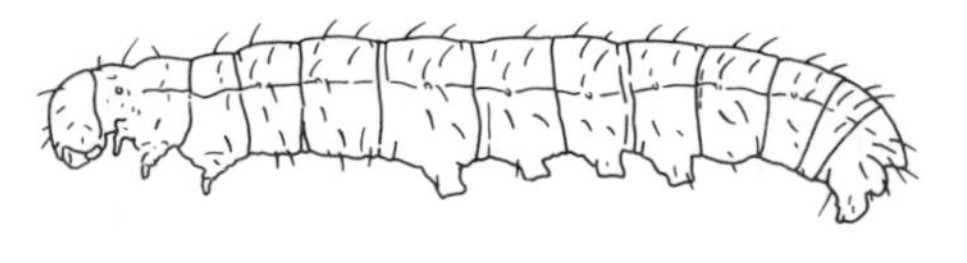

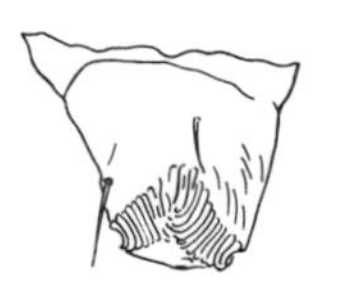

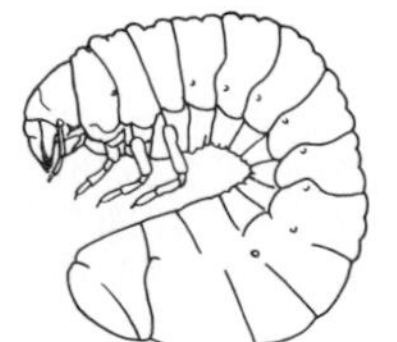

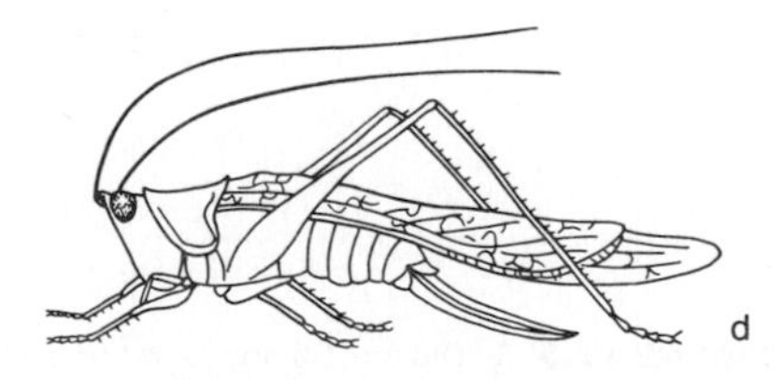

FIGURE 17.5 *a*, Larva of Lepidoptera (Noctuidae); *b*, abdominal leg of larva of Lepidoptera (Noctuidae); *c*, scarabaeiform grub of Endopterygota (Coleoptera, Scarabaeidae); *d*, a grasshopper (Orthoptera). (a, b *from Melis*, 1936; c *from Ritcher*, 1944; d *from University of California Extension*.)

17(16). Posterior legs modified for jumping, with femur greatly thickened (Fig. 17.5*d*) **Orthoptera** (Chap. 29)
Posterior legs similar to middle pair **18**

18(17). Last abdominal segment bearing cerci, either single-segmented (Figs. 17.6*b*, 17.7*a*) or multiarticulate (Fig. 17.6*a*) **19**
Last abdominal segment without cerci **32**

19(18). Last abdominal segment with median, multiarticulate filament; short, 1-segmented appendages present on at least abdominal segments 7 to 9 **20**
Last abdominal segment without median filament; appendages absent from segments 7 to 9 **21**

20(19). Head with large, contiguous compound eyes; maxillary palp with 7 segments; body cylindrical, arched **Archeognatha** (Chap. 19)
Head with widely separated lateral ocelli or without eyes; maxillary palp with 5 segments; body usually flattened **Thysanura** (Chap. 19)

21(19). Mouthparts modified as a rostrum or haustellum (Fig. 17.6*c*) or projecting beaklike below the head (Figs. 17.4*a*; 34.1*b*); labial palp with 2 segments **22**
Mouthparts short, mandibulate (Fig. 17.6*d*), never forming a beak or haustellum; labial palp with 3 segments **23**

22(21). Mouthparts modified as a rostrum or haustellum (Fig. 17.6*c*); metathorax frequently with halteres; antennae frequently with 5 or fewer segments **Diptera** (Chap. 42)
Mouthparts with elongate mandibles and maxillae projecting beaklike (Fig. 41.1*b*); halteres absent; antennae with at least 12 segments **Mecoptera** (Chap. 41)

23(21). Tarsi with 5 segments **24**
Tarsi with 1 to 4 segments **27**

24(23). Anterior legs raptorial (Fig. 17.6*e*) **Mantodea** (Chap. 23)
Anterior legs not raptorial, fitted for walking **25**

25(24). Head prognathous (mouthparts directed anteriorly) **26**

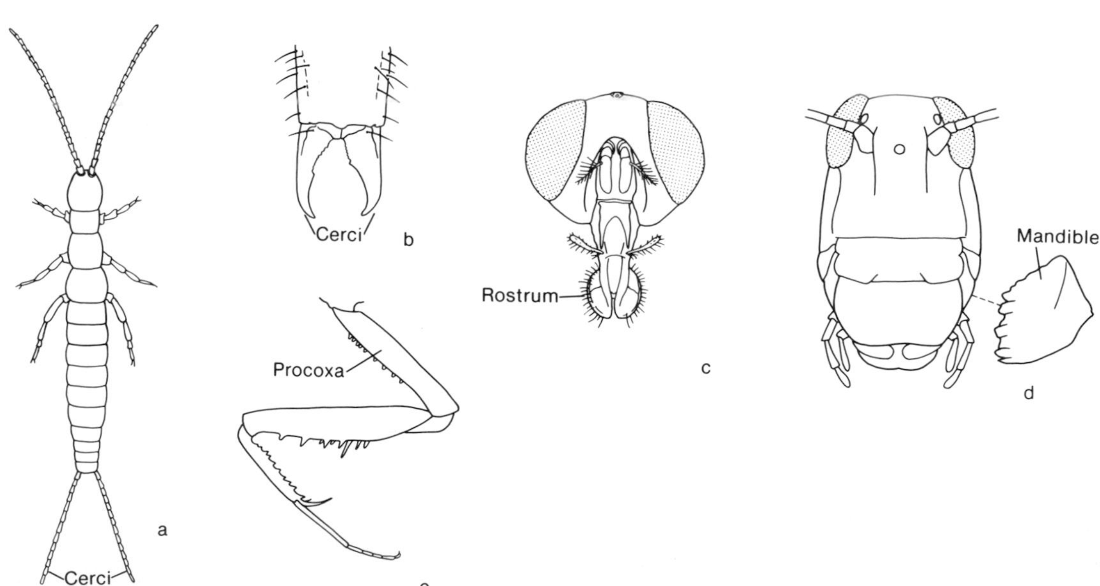

FIGURE 17.6 *a,* Diplura, Campodeidae; *b,* forceps of Diplura, Japygidae; *c,* haustellum of Diptera (Muscidae); *d,* chewing mouthparts of a grasshopper (Orthoptera); *e,* raptorial leg of a mantid (Mantodea). (a, d *from University of California Extension;* c *from Snodgrass,* 1953.)

Head hypognathous (mouthparts directed ventrally) **Blattodea** (Chap. 22)

26(25). Cerci with 5 to 9 segments, long, flexible; body not modified for camouflage **Grylloblattodea** (Chap. 25)
Cerci with 1 segment, usually short; body almost always sticklike or leaflike **Phasmatodea** (Chap. 30)

27(23). Tarsi with 1 to 2 segments **28**
Tarsi with 3 to 4 segments **29**

28(27). Cerci either multiarticulate (Fig. 17.6*a*) or 1-segmented, forceps-shaped (Fig. 17.6*b*); tarsi with 1-segment; soil or litter dwellers with long, parallel-sided body (Fig. 17.6*a*) **Diplura** (Chap. 18)
Cerci with 1 segment, short, not forceps-shaped, tarsi with 2 segments; body stouter, not parallel-sided **Zoraptera** (Chap. 31)

29(27). Cerci forceps-shaped, strongly sclerotized (Fig. 17.7*a*) **Dermaptera** (Chap. 26)
Cerci not forceps-shaped **30**

30(29). Tarsi with 3 segments **31**
Tarsi with 4 segments **Isoptera** (Chap. 24)

31(30). Antennae more than half as long as body; labium small, without movable, apical teeth **41**
Antennae much less than half as long as body; labium jointed, with large, movable apical teeth (Fig. 21.4). **Odonata** (Chap. 21)

32(18). Tarsi with 5 segments **36**
Tarsi with 1 to 3 segments **33**

33(32). Mouthparts enclosed in a long, slender rostrum projecting beneath the head (Figs. 17.4*a*, 34.1*b*); maxillary and labial palps absent **Hemiptera** (Chap. 34)
Mouthparts not in the form of a rostrum; maxillary and labial palps usually present **34**

34(33). Antennae longer than head, with at least 5 segments, usually with more than 10 segments **35**
Antennae shorter than head, with 3-7 segments **40**

35(34). Head cone-shaped, directed ventrally or posteriorly (Fig. 17.7*c*); antennae with 4 to 9 segments; body elongate, slender **Thysanoptera** (Chap. 35)
Head not cone-shaped; antennae almost always with more than 12 segments; body stout (Fig. 17.7*d*) **Psocoptera** (Chap. 32)

36(32). Abdomen strongly constricted at base (Fig. 17.8*a*); antennae frequently elbowed (Fig. 17.8*a*) **Hymenoptera** (Chap. 46)
Abdomen not constricted at base; antennae not elbowed **37**

37(36). Body densely covered with scales or long hairs; mouthparts usually a coiled proboscis (Fig. 17.8*e*) (sometimes vestigial) **Lepidoptera** (Chap. 44)
Body bare or sparsely covered with hairs, rarely scaled; mouthparts not a coiled proboscis **38**

38(37). Mouthparts a slender tube-shaped rostrum Figs. 43.1*b*, 42.3*b* or a haustellum (Fig. 17.6*c*); antennae usually with 3 or fewer segments **39**
Mouthparts mandibulate (Fig. 17.6*d*), never forming a rostrum or haustellum; antennae almost always with 9 to 11 segments **Coleoptera** (Chap. 39)

39(38). Body strongly flattened laterally; thorax and head usually bearing large flattened, backwardly directed spines **Siphonaptera** (Chap. 43)
Body cylindrical or flattened dorsoventrally; head and thorax not bearing special spines **Diptera** (Chap. 42)

40(34). Antennae usually concealed in grooves; ectoparasites on birds, mammals, with flattened bodies, reduced eyes and pigmentation (Fig. 17.7*b*). **Phthiraptera** (Chap. 33)
Antennae free; free living aquatic naiads with long legs, large compound eyes and darkly pigmented bodies (Fig. 21.5) **Odonata** (Chap. 21)

41(31). Cerci with numerous segments; aquatic naiads with gills usually present on legs, thorax or abdomen **Plecoptera** (Chap. 27)
Cerci with 1 segment; terrestrial, without gills **Phasmatodea** (Chap. 30)

KEY B

Wings Present, Functional

1. Mesothoracic wings thick, strongly sclerotized or parchmentlike at least at base, or vestigial 2
   Mesothoracic wings membranous, sometimes covered with scales; never vestigial 10

2(1). Front wings vestigial, scalelike or club-shaped; hind wings large, fan-shaped 9
   Front wings covering about one half or more of abdomen; never club-shaped or scalelike 3

3(2). Abdomen with large, strongly sclerotized forceps-shaped cerci (Fig. 17.7*a*); forewings short, leaving at least 3 abdominal segments exposed **Dermaptera** (Chap. 26)
   Abdomen with cerci absent or not forceps-shaped; forewings usually covering entire abdomen 4

4(3). Mouthparts a slender, elongate rostrum projecting beneath head (Figs. 17.4*a*, 34.1*b*) **Hemiptera** (Chap. 34)
   Mouthparts short, mandibulate, never forming a rostrum (Fig. 17.6*d*) 5

5(4). Forewings without venation, usually strongly sclerotized and meeting at midline at rest (Fig. 17.8*b*); antennae rarely with more than 11 segments, frequently clubbed **Coleoptera** (Chap. 39)
   Forewings with extensive, reticulate venation; antennae usually with more than 12 segments, filiform, never clubbed 6

6(5). Hind legs with femora greatly enlarged for jumping (Fig. 17.5*d*) **Orthoptera** (Chap. 29)
   Hind legs not modified, similar to middle legs 7

7(6). Anterior legs raptorial (Fig. 17.6*e*) **Mantodea** (Chap. 23)
   Anterior legs not raptorial, fitted for walking 8

8(7). Head prognathous (mouthparts directed anteriorly); body sticklike or leaflike **Phasmatodea** (Chap. 30)
   Head hypognathous (mouthparts directed ventrally); body not sticklike or leaflike **Blattodea** (Chap. 22)

9(2). Hind legs with femora enlarged for jumping (Fig. 17.5*d*); prothorax projecting posteriorly over wings and abdomen **Orthoptera** (Chap. 29)
   Hind legs not modified for jumping, similar to middle legs; prothorax small, not projecting over hindbody **Strepsiptera** (Chap. 40)

10(1). One pair of wings present 11
   Two pairs of wings 13

11(10). Abdomen bearing 1 to 3 long filaments on terminal segment; mouthparts vestigial **12**

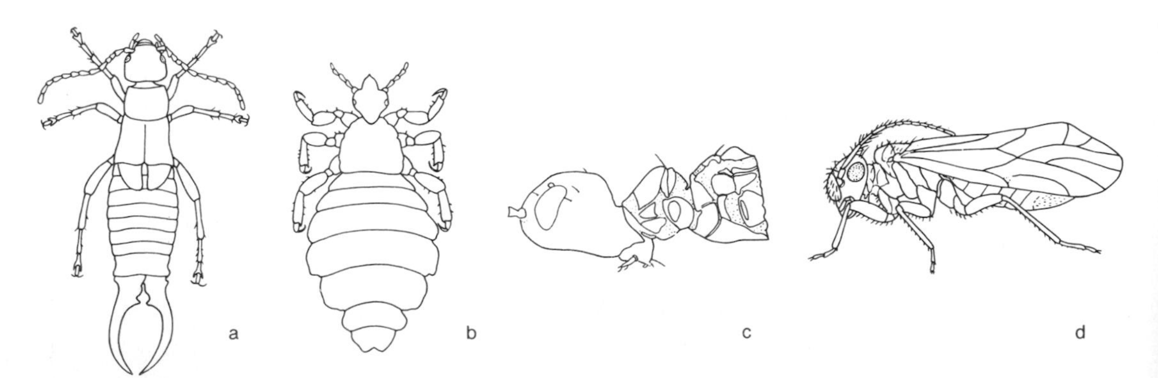

FIGURE 17.7 *a*, An earwig (Dermaptera); *b*, a louse (Phthiraptera); *c*, lateral aspect of the anterior end of a thrips (Thysanoptera); *d*, a book louse (Psocoptera). (a, b *from University of California Extension;* c *from Faure,* 1949.)

Abdomen not bearing long filaments; mouthparts rarely vestigial **Diptera** (Chap. 42)

12(11). Antennae long, filiform; wing with a single vein, without cells **Hemiptera** (Chap. 34)
Antennae short, bristle-shaped; wing with closed cells, usually with veins very numerous **Ephemeroptera** (Chap. 20)

13(10). Abdomen bearing 2 to 3 long filaments; mouthparts vestigial **Ephemeroptera** (Chap. 20)
Abdomen without filaments; mouthparts rarely vestigial **14**

14(13). Tarsi with 1 to 4 segments **15**
Tarsi with 5 segments **22**

15(14). Mouthparts a slender, tube-shaped rostrum projecting below the head (Figs. 17.4*a*, 34.1*b*) **Hemiptera** (Chap. 34)
Mouthparts short, mandibulate, never an elongate projecting rostrum (Fig. 17.6*d*) **16**

16(15). Forelegs with basal segment of tarsi globular, at least twice as thick as second segment **Embioptera** (Chap. 28)
Forelegs with tarsal segments about equal in size **17**

17(16). Antennae shorter than head, bristle-shaped; insects at least 1.5 cm long; abdomen long, slender; wings with extensive network of veins **Odonata** (Chap. 21)
Antennae longer than head, filiform **18**

18(17). Tarsi with 4 segments **Isoptera** (Chap. 24)
Tarsi with 2 to 3 segments **19**

19(18). Last segment of abdomen bearing cerci **21**
Cerci absent **20**

20(19). Wings linear, narrow, with no more than 2 veins; head cone-shaped, directed ventrally or posteriorly (Fig. 35.1*a*) **Thysanoptera** (Chap. 35)
Wings oval with at least 4 longitudinal veins; head not cone-shaped **Psocoptera** (Chap. 32)

21(19). Forewings with 3 longitudinal veins, no closed cells; hind wings smaller than forewings; minute insects occurring in decaying wood **Zoraptera** (Chap. 31)
Wings with numerous longitudinal veins and many closed cells; hind wings larger than forewings **Plecoptera** (Chap. 27)

22(14). Forewings densely covered with hairs or scales **23**

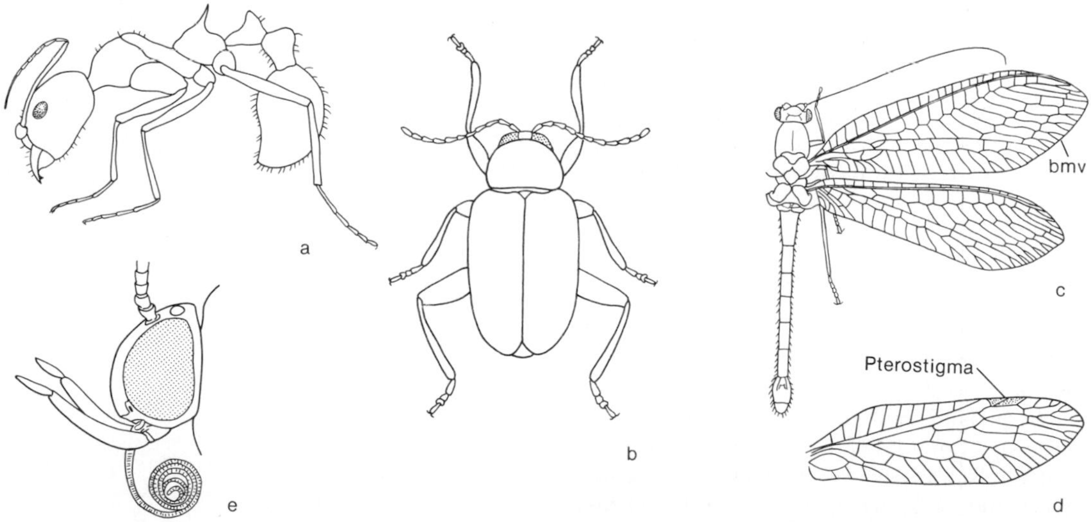

FIGURE 17.8 *a,* An ant (Hymenoptera, Formicidae); *b,* a beetle (Coleoptera); *c,* wings of Neuroptera, showing bifurcating marginal veins (bmv); *d,* wing of a snakefly (Raphidioptera), showing pterostigma; *e,* head of a moth, showing coiled proboscis. (a, b *from University of California Extension;* e *from Snodgrass,* 1935.)

Wings bare, or with fringe of marginal hairs 24

23(22). Wings covered with scales; mouthparts usually a coiled proboscis (Fig. 17.8*e*) **Lepidoptera** (Chap. 44)
Forewings covered with hairs; mouthparts mandibulate **Trichoptera** (Chap. 45)

24(22). Forewings about 1.5 times longer than hind wings; fore- and hind wings usually markedly different in shape and venation; abdomen usually strongly constricted at base **Hymenoptera** (Chap. 46)
Forewings and hind wings similar in size, shape, and venation; abdomen not constricted at base 25

25(24). Head not prolonged ventrally; wings usually with numerous crossveins in costal margin (exception, Coniopterygidae) 26
Head prolonged ventrally, beaklike (Fig. 41.1*b*); wings with 1 to 3 crossveins in costal margin **Mecoptera** (Chap. 41)

26(25). Hind wings broader at base than forewings; veins not bifurcating near wing margin (Fig. 36.1*c*) 27
Hind-wing breadth at base less than or equal to forewing breadth; at least radial veins bifurcating just before wing margin (Fig. 17.8*c*) (exception Coniopterygidae) **Neuroptera** (Chap. 38)

27(26). Pronotum quadrate or nearly so; wings without pterostigma **Megaloptera** (Chap. 36)
Pronotum at least 3 times longer than wide; wings with pterostigma (Fig. 17.8*d*) **Raphidioptera** (Chap. 37)

SELECTED REFERENCES

PHYLOGENETIC RELATIONSHIPS

Anderson, 1973 (embryology and phylogeny of arthropods); Hamilton, 1972 (phylogeny of winged orders); Hinton, 1955*a*, 1958, 1963*a* (panorpoid orders); Istock, 1966 (evolution of metamorphosis); Kristensen, 1975 (comprehensive review of insect phylogeny); Manton, 1973; Remington, 1956 (entognathous orders); Sharov, 1966; E. L. Smith, 1969 (evolutionary development of external genitalia); Snodgrass, 1950, 1958; Tiegs and Manton, 1958; Tuxen, 1959, 1970*b* (entognathous orders); Wille, 1960.

ORIGIN OF WINGS

Paranotal lobe theory: Müller, 1879; Hinton, 1963*b*; Flower, 1964; Hamilton, 1971. Tracheal gill theory: Oken, 1831; Wigglesworth, 1976. Fossil evidence of wing origin: Kukalova-Peck, 1977; Wootton, 1976. Additional theories or general discussions: Alexander and Brown, 1963; Sharov, 1966; Snodgrass, 1958.

GEOLOGICAL RECORD OF HEXAPODA

Braziunas, 1975 (revised ages for geological time scale); Carpenter, 1953, 1970, 1977 (reviews of insect fossil record); Kukalova-Peck, 1977 (morphological studies of Paleozoic insects); Martynova, 1961 (review); Plumstead, 1963 (insect paleoecology); Smart and Hughes, 1973 (insect paleoecology); Smart and Wootton, 1967 (critical dates for insect fossils).

THE ENTOGNATHA

18

ORDER PROTURA

Very small to minute entognathous hexapods. Eyes, antennae absent; mandibles, maxillae styliform, retracted into invaginations in cranium; 1 or 2-segmented, nonambulatory appendages present on abdominal segments 1 to 3; cerci absent; postembryonic development anamorphic, with 9 abdominal segments in first instar, 12 segments in adult.

Protura are weakly sclerotized, elongate, fusiform arthropods which occur exclusively in moist decaying leaf litter, moss, soil, and similar situations. About 175 species are known; all are easily recognized by the absence of antennae and the characteristic body shape with conical head. Because of their small size and cryptic habits, Protura are seldom encountered, but soil or litter extracts frequently contain large numbers of individuals. The biology of Protura is very poorly known, and their embryology is unknown. First-instar larvae eclose with 9 abdominal segments, with one additional segment added between the last two abdominal segments after each of the first three molts. One or two additional molts follow without increased segmentation.

Among hexapodous arthropods, Protura are unique in lacking antennae in all instars, but *pseudoculi* (Fig. 18.1*a,b*), present on the anterior cranium, may represent rudimentary antennae. The forelegs are usually held in front of the body, apparently functioning as sensory organs. Mouthparts of Protura are quite unlike those of insects but resemble those of Collembola. Mandibles and maxillae are in the form of elongate, simple blades, apparently used to triturate food particles. At rest the mouthparts are deeply retracted into invaginations in the head (Fig. 18.1*c,d*), with only the labrum visible externally. The Protura have been monographed by Tuxen (1964).

KEY TO THE FAMILIES OF
Protura

1. Spiracles present on mesothorax and metathorax; tracheae present **Eosentomidae**
   Tracheal system and spiracles absent 2
2. Abdominal terga with transverse sutures and laterotergites **Acerentomidae**
   Abdominal terga without transverse sutures or laterotergites **Protentomidae**

ORDER DIPLURA

Entognathous hexapods with elongate body, short legs, slightly differentiated thoracic and abdominal segments. Eyes absent, antennae multiarticulate, flagellum musculate. Abdomen with 10 to 11 segments, appendages on segments 1 or 2 to 7; cerci developed as multiarticulate, moniliform appendages or as short, sclerotized forceps, bearing terminal gland openings.

Like the other entognathous orders, Diplura inhabit leaf litter, soil interstices, moss, and other moist, cool situations. Typically, they are unpigmented and weakly sclerotized organisms, 5 to 15 mm long, but *Anajapyx* (Japygidae) reaches 50

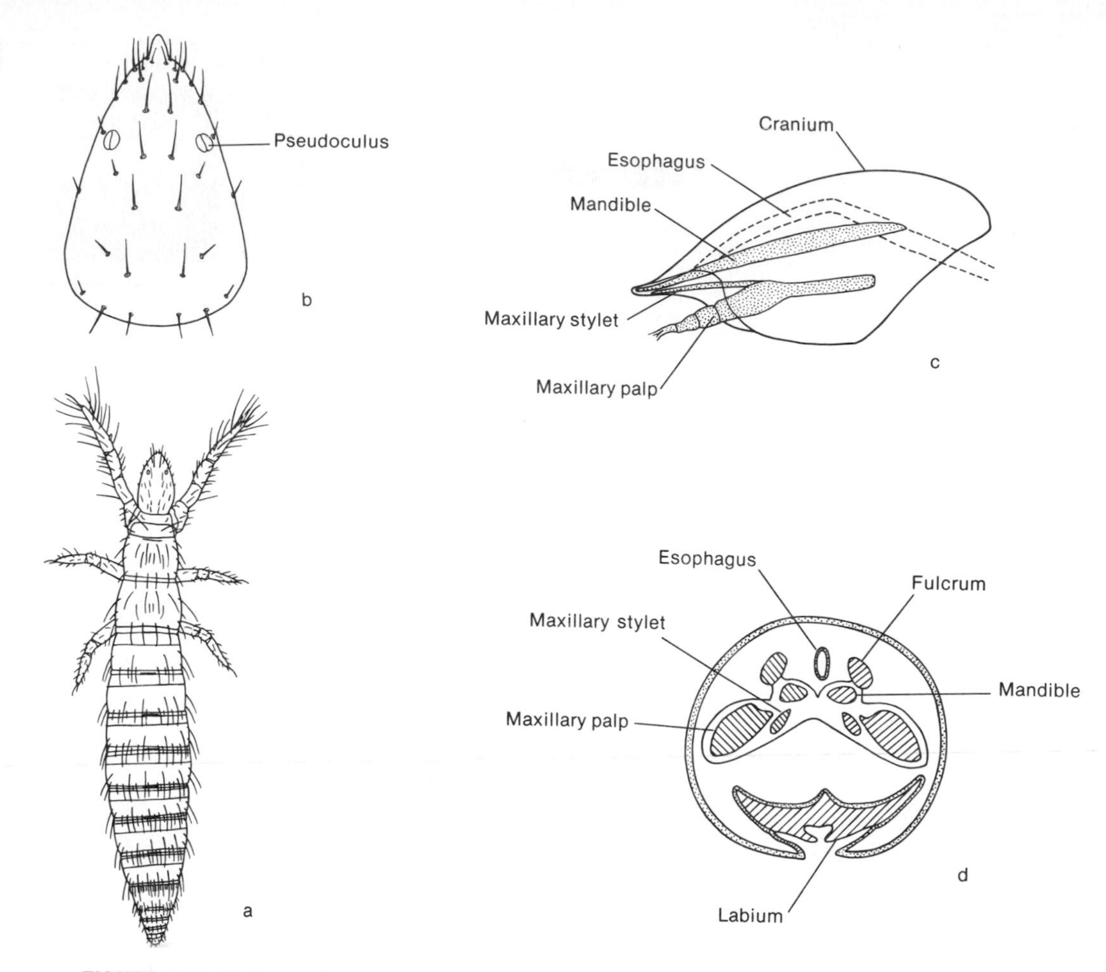

FIGURE 18.1 Protura: *a,* Acerentomidae, dorsal (*Acerentus barberi*); *b,* Acerentomidae, dorsal aspect of head (*Acerentomon*); *c,* longitudinal section of head, diagramatic; *d,* transverse section of head, diagramatic. (a *from Ewing,* 1940; b *from Price,* 1959.)

mm in length. Diplura are cosmopolitan, with the Campodeidae best represented in temperate regions, especially in the Northern Hemisphere. The Japygidae are predominantly tropical, subtropical, and south temperate.

Known species are bisexual. Insemination is indirect, by means of a stalked spermatophore which the males attach to the substrate. Early instars are similar to adults except for chaetotaxy, and metamorphosis is essentially absent. Molting continues throughout life, as in the Thysanura, but sexual maturity is attained when the setation is complete. Campodeidae apparently feed on decaying vegetation, while Japygidae are predatory, using the forceps to capture prey.

Diplura were formerly placed in a single taxon with Thysanura, which are also primitively wingless with multiarticulate antennae and cerci. However, the structure of dipluran mouthparts, with very elongate mandibles and maxillae partly retracted into the head capsule at rest (Fig. 18.2*b,c,d*), is very different than in the Thysanura. Imms (1936) has shown that the Diplura are

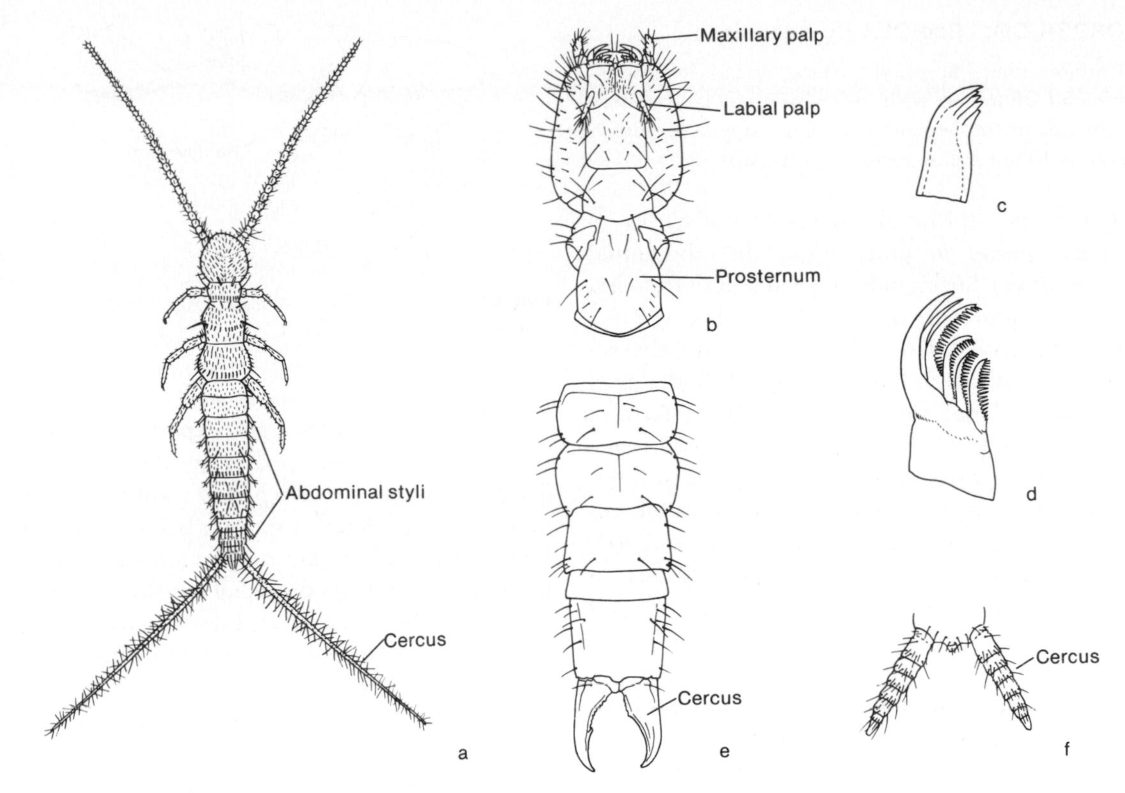

FIGURE 18.2 Diplura; *a*, Campodeidae (*Campodea montis*); *b*, Japygidae (*Japyx kofoidi*), ventral aspect of head; *c*, same, apical portion of mandible; *d*, same, apical portion of lacinia; *e*, Japygidae (*Metajapyx steevsi*), abdomen; *f*, Projapygidae (*Anajapyx menkei*), cerci. (a *from Gardner,* 1914; b, c, d *from Silvestri,* 1928; e *from Smith and Bolton,* 1964; f *from Smith,* 1960.)

generally similar to the Symphyla in body organization. The Japygidae are unique among hexapodous arthropods in possessing four pairs of thoracic spiracles, with two pairs on each of the mesothorax and metathorax.

KEY TO THE FAMILIES OF
Diplura

1. Cerci multiarticulate, not sclerotized — 2
   Cerci with 1 segment, sclerotized, forcipulate (Fig. 18.2*e*) — **Japygidae**
2. Cerci less than half length of antennae (Fig. 18.2*f*), strongly moniliform, and bearing terminal gland openings — **Projapygidae**
   Cerci about as long as antennae, filiform or weakly moniliform, without terminal gland openings (Fig. 18.2*a*) — **Campodeidae**

Campodeidae (Fig. 18.2*a*) White, delicate, active species, mostly about 3 to 5 mm long. Abundant in leaf litter and soil throughout North America.

Japygidae Body more compact than in Campodeidae, with terminal abdominal segment and forceps sclerotized. Commonly ranging from 5 to 15 mm in length; in North America most abundant in Southwest and West.

Projapygidae Body compact, as in Japygidae, but cerci multiarticulate. In North America occurring only in California.

ORDER COLLEMBOLA (Springtails)

Entognathous hexapods characterized by 6-segmented abdomen with specialized, nonambulatory appendages on segments 1, 3, 4; antennae with 4 (rarely 6) musculate segments; cerci absent.

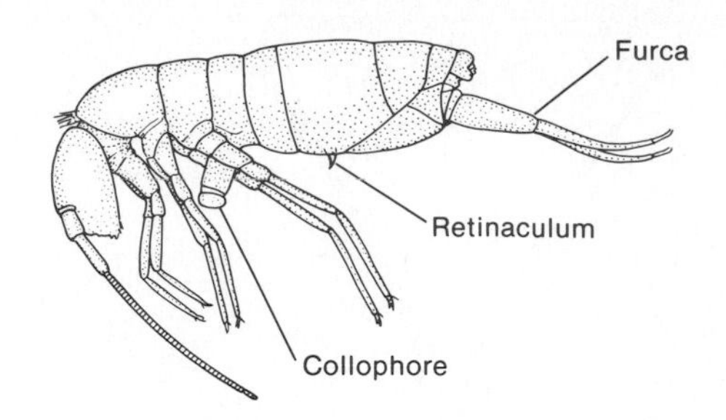

FIGURE 18.3 Collembola: Entomobryidae, *Tomocerus flavescens.* (*From Folsom,* 1913.)

The name "springtail" refers to the ability of many species to jump using the abdominal appendages. Springtails range in length from less than 1 mm to more than 10 mm, averaging 2 to 3 mm. Several different body plans are characteristic of different families (Fig. 18.4), the most distinctive being the compact, globular form in Sminthuridae. Collembola occur in a wide variety of moist situations, especially soil and litter, where they are frequently exceedingly abundant (densities of 53,000 per m^2 have been estimated) and are among the most important consumers in many soil ecosystems. Collembola also inhabit fungi, lower levels of vegetation, the surface of slowly moving or stagnant water, and occasionally buildings. Several species are abundant on seashores, ranging into the upper intertidal zone; a few forms occur in ant or termite nests. Collembola are readily wind-borne, and many species are extremely widespread, appearing even on coral atolls and oceanic islands.

Collembola share a six-segmented abdomen with early instars of myriapods, which add segments after each molt, until the adult segmentation is attained. The arrangement and structure of abdominal appendages (Fig. 18.3) are unique among arthropods, but the structures are believed to be homologous to the abdominal legs of myriapods. The *collophore,* used in imbibing free water, is almost always present on the first abdominal segment. The *furca,* usually three-segmented, is held under tension beneath the abdomen by the *retinaculum* when at rest. When released it propels the animal up to 0.2 m through the air.

Most Collembola are bisexual, but parthenogenetic forms are known. Insemination is indirect, males depositing stalked spermatophores on the substrate. The spermatophores are later picked up by the females, usually without direct participation by males. Eggs, deposited singly or in clutches, are dropped onto the substrate. Those oviposited during unfavorable seasons may diapause for many months before hatching. Eggs of Collembola are unique among those of hexapodous arthropods, being *microlecithal* (lacking large yolk reserves), and developing by *holoblastic* or complete cleavage. Immature instars are essentially similar to adults except in size. Four to eight molts have been recorded in various species; but sexual maturity is acquired before the ultimate molt. Most springtails probably feed on decaying vegetation, but spores, pollen, fungal mycelia, or animal material are included in the diet of various species. Some species attack living plant material, occasionally becoming economically important.

KEY TO THE FAMILIES OF

Collembola

1. Body elongate, cylindrical or subcylindrical; first 6 postcephalic segments distinct (Figs. 18.3; 18.4*b*) — 2
   Body globular; first 6 postcephalic segments fused (Fig. 18.4*a*) — **Sminthuridae**

2. Prothorax developed as distinct segment, bearing short setae (Fig. 18.4*b*) — 3
   Prothorax reduced, indistinct, without setae (Fig. 18.3) — 4

3. Eyes present (as in Fig. 18.4*b*); third antennal segment with large sensilla — **Onychiuridae**
   Eyes absent; third antennal segment without unusually large sensilla — **Poduridae**

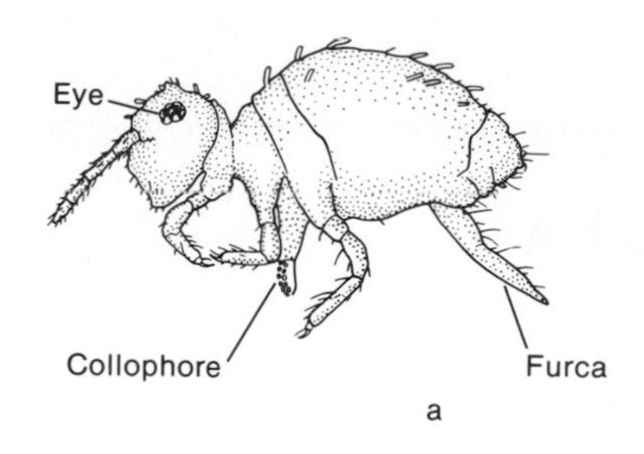

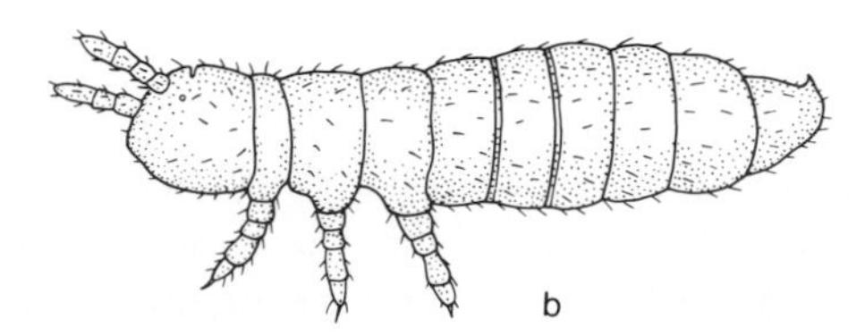

FIGURE 18.4 Collembola: *a*, Sminthuridae (*Neosminthurus curvisetis*); *b*, Onychiuridae (*Onychiurus subtenuis*). (a *from Mills*, 1934; b *from Folsom*, 1917).

4. Body with scales or clavate setae; fourth abdominal segment usually much longer than third (Fig. 18.3) **Entomobryidae**
   Body with simple setae only; third and fourth abdominal segments subequal **Isotomidae**

Sminthuridae (Fig. 18.4*a*). Easily distinguished from other Collembola by globose body, frequently covered with brightly colored scales. The collophore, retinaculum, and furca are usually present. The few agriculturally important Collembola are sminthurids.

Onychiuridae (Fig. 18.4*b*). Usually depigmented, lacking furca and retinaculum. Mostly inhabitants of soil interstices.

Poduridae. Body form variable; small, active species with furca and retinaculum present are similar to Onychiuridae; large (to 1 cm long), sluggish species superficially resemble immature Coccinellidae (Coleoptera).

Entomobryidae (Fig. 18.3). A large family, frequently occurring under bark, on the lower leaves of plants, or occasionally in buildings and other relatively dry situations, as well as in leaf litter and soil. The dominant group of Collembola, easily separated from other families by the enlarged fourth abdominal segment.

Isotomidae. A large group of common species which occur in leaf litter, garden soil, or rotting vegetation.

SELECTED REFERENCES

Protura

Ewing, 1940 (systematics); Tuxen, 1964 (systematics); Kevan, 1962 (biology); Kühnelt, 1961 (biology); Wallwork, 1970 (biology).

Diplura

BIOLOGY

Kevan, 1962; Kühnelt, 1961; Shaller, 1970 (reproductive behavior); Wallwork, 1970.

SYSTEMATICS AND EVOLUTION

Imms, 1936; Remington, 1956.

Collembola

BIOLOGY AND ECOLOGY

Butcher et al., 1971; Christiansen, 1964; Hale, 1965; Joose, 1976 (excellent general discussions); Kevan, 1962; Kühnelt, 1961; Shaller, 1970 (reproductive behavior); Waldorf, 1974 (sex pheromone); Wallwork, 1970.

SYSTEMATICS

Maynard, 1951; Mills, 1934; Richards, 1968 (generic classification, world); Salmon, 1956*a*,*b*; Scott, 1961 (pictorial key, nearctic genera).

19

THE APTERYGOTA

ORDER ARCHEOGNATHA (Machilids, Bristletails)

Elongate, subcylindrical, with arched trunk and 3 multiarticulate terminal filaments; antennae multiarticulate; 3 ocelli and large, dorsally contiguous compound eyes (Fig. 19.1*c*); mouthparts partly retracted into head, mandibles (Fig. 19.1*b*) elongate, monocondylic; maxillary palp elongate, 7-segmented; 1-segmented appendages present on abdominal segments 1 to 9 (Fig. 19.1*a*).

Superficially the Archeognatha are similar to the Thysanura, and the two are sometimes classified as a single order. Structurally the Archeognatha share certain features, especially mouthpart structures, with the entognathous hexapods, especially the Diplura, as well as showing similarities to the Thysanura and Pterygota (Chaps. 17, 18). As Sharov (1966) points out, the occurrence of archeognathan-like fossils such as the Monura (Chap. 17) seems to indicate that a much larger fauna of primitive, flightless insects existed during the Paleozoic Era. The Archeognatha probably represent a relictual stock of such forms.

In contrast to the entognathous hexapods, the Archeognatha have successfully adapted to a variety of habitats besides soil and litter. Machilids commonly occur in chaparral or dry coniferous woodlands, where they are dependent on localized sources of free water, which they imbibe through the eversible membranous vesicles adjacent to their abdominal appendages. Meinertellids inhabit extremely xeric habitats, includ-

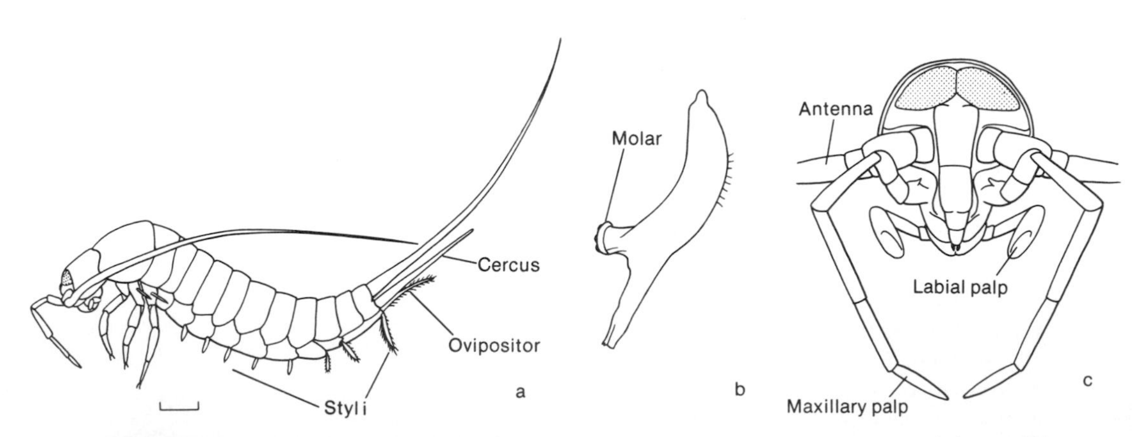

FIGURE 19.1 Apterygota, Archeognatha, Machilidae: *a, Machilis,* female (scale equals 1 mm); *b,* mandible of *Machilis; c,* head of *Machilis.* (b *from Snodgrass,* 1950.)

ing desert sand dunes, and require little if any free water.

Biology and life histories are known in detail for only a few species. Fertilization is indirect, as in the entognathous groups. During courtship the male spins a fine thread from the gonapophyses. One end of the thread is attached to the substrate, the other end held at the tip of the abdomen. Several spermatophores are deposited on the thread as it is formed. These are later picked up by the female after a series of complicated courtship maneuvers. Females deposit clutches of up to 30 eggs in crevices.

Immatures resemble adults except for chaetotaxy and the lack of styli on the thorax. Food consists of lichens, algae, vegetable detritus, and probably dead arthropods. Sexual maturity is attained after about 8 to 10 molts over a period of up to 2 years. These are active, fast-running insects, capable of random escape jumps by suddenly arching the body and slamming down the abdomen. The abdominal appendages are used as skids to support the abdomen.

Meinertellidae. Abdominal sterna small; antennae without scales. Mostly small (<5 mm) diurnal species found in dry regions.

Machilidae. Abdominal sterna relatively large; antennae with scape and pedicel usually scaled. Larger (5 to 15 mm), mostly crepuscular species. Cosmopolitan.

ORDER THYSANURA (Silverfish)

Elongate to ovate, flattened insects with 3 multiarticulate terminal abdominal filaments; antennae multiarticulate; eyes of compound type but small, rudimentary, or absent; mouthparts mandibulate; mandibles short, semidicondylic (Fig. 19.2*b*), maxillary palps 5-segmented; 1-segmented appendages (Fig. 19.2*c*) absent or present on variable number of abdominal segments.

In the morphology of the mouthparts the Thysanura are comparable to primitive pterygote insects, such as mayflies, rather than the superficially similar Archeognatha. Structurally they are more variable than the Archeognatha, ranging from slender, elongate, free-living forms, to ovate, short-tailed types which frequent mammal burrows or ant nests. Thysanura are rapid runners but lack the jumping ability of archeognathans. Biologically Thysanura are similar to Archeognatha, with indirect insemination, ametabolous metamorphosis, and surprisingly slow

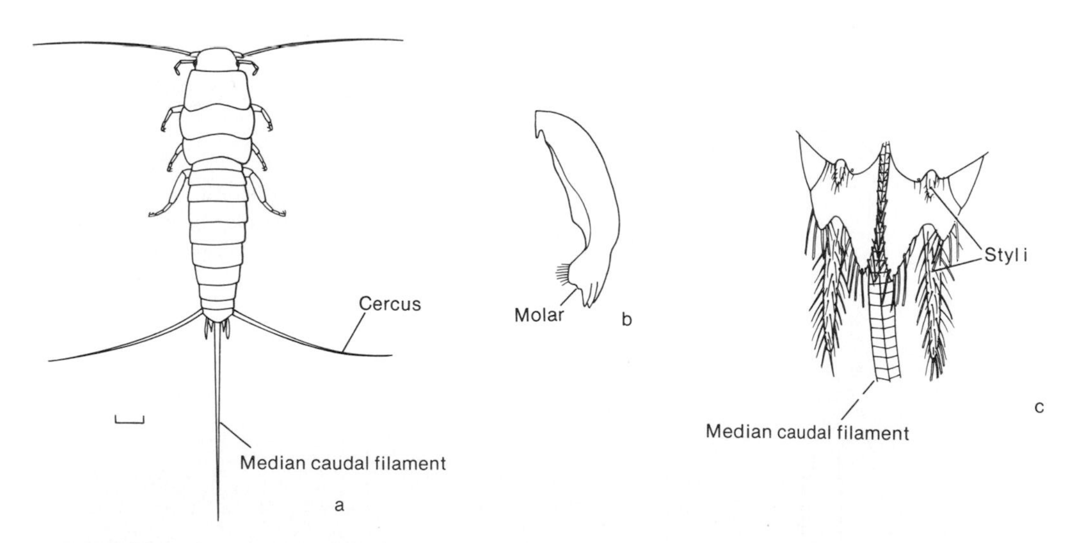

FIGURE 19.2 Apterygota, Thysanura, Lepismatidae: *a, Ctenolepisma lineata* (scale equals 1 mm); *b,* mandible of *Ctenolepisma; c,* abdominal apex of *Heterolepisma.* (b *from Snodgrass,* 1950; c *from Folsom,* 1924.)

development and long life-spans. *Tricholepidion gertschi* (Lepidotrichidae) (Fig. 19.3), the most primitive species, has a 2-year cycle, as do many Archeognatha. Survival of lepismatids over periods of 3 to 4 years is recorded, during which molting and reproduction continue. Most Thysanura are probably omnivorous.

KEY TO THE FAMILIES OF **Thysanura**

1. Compound eyes present — 2
   Compound eyes absent — **Nicoletiidae**
2. Body hairy; ocelli present; tarsi with 5 segments; exertile vesicles present on abdominal segments 2 to 9 — **Lepidotrichidae**
   Body scaly; ocelli absent; tarsi with 3 or 4 segments; exertile vesicles absent — **Lepismatidae**

Nicoletiidae. Mostly commensal forms, occurring in ant or termite nests or mammal burrows, or cavernicolous or subterranean. Eyes and ocelli are absent; body either elongate, slender or broad, oval, with short terminal filaments. In the United States, restricted to the South and Southwest.

Lepidotrichidae. Body slender, elongate, moderate in size (12 to 15 mm); lateral and median ocelli present. *Tricholepidion* (Fig. 19.3), the only genus, has the body clothed with simple setae. The family was known only from Oligocene fossils until living insects were discovered in northwestern California in 1959.

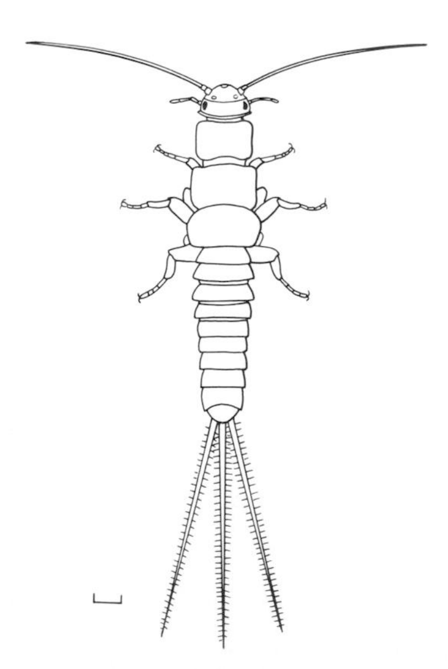

FIGURE 19.3 Apterygota, Thysanura, Lepidotrichidae (*Tricholepidion gertschi*, scale equals 1 mm).

Lepismatidae. Body slender, elongate, small to moderate in size; lateral ocelli absent, median ocellus reduced. The body of lepismatids is always clothed in scales, differentiating them from Lepidotrichidae and most Nicoletiidae. *Lepismodes inquilinus* (the firebrat) and *Lepisma saccharina* (the silverfish), as well as other common household species, are lepismatids.

SELECTED REFERENCES

Archeognatha

BIOLOGY AND ECOLOGY

Benedetti, 1973; Joose, 1976 (excellent general discussion); Shaller, 1971 (reproductive behavior); E. L. Smith, 1970.

SYSTEMATICS AND EVOLUTION

Remington, 1954, 1956; Sharov, 1966.

Thysanura

MORPHOLOGY

Barnhart, 1961; Smith, 1970; Wygodzinsky, 1961.

BIOLOGY AND ECOLOGY

Delaney, 1957, 1959; Joose, 1976; Shaller, 1971; E. L. Smith, 1970.

SYSTEMATICS AND EVOLUTION

Remington, 1954 (generic classification); Slabaugh, 1940 (domestic species); Wygodzinsky, 1961 (*Tricholepidion*), 1972 (revision, North American species).

ORDER EPHEMEROPTERA (Mayflies)

20

Adult. Delicate paleopterous insects with elongate, subcylindrical body (Fig. 20.1*a*). Head with short, multiarticulate, filiform or setaceous antennae, 3 ocelli, and large, compound eyes; mouthparts mandibulate, vestigial. Mesothorax enlarged, supporting large, triangular forewings; hind wings small, rounded, or absent; venation extensive, with numerous crossveins (Fig. 20.3). Abdomen slender, elongate with 10 segments; tenth segment with long cerci and usually with median caudal filament. *Naiads* extremely variable in body form (Fig. 20.4) but almost always with large compound eyes, short multiarticulate antennae; mouthparts mandibulate; tarsi 1-segmented, with single claw; 10-segmented abdomen with cerci and usually median caudal filament; gills present on at least some of anterior 7 abdominal segments. *Subimago:* similar to adult.

Adult and subimaginal mayflies are easily recognized by the large, triangular forewings which are held vertically above the body at rest, and by the long abdominal filaments. Naiads are usually distinguished by the combination of three terminal filaments and abdominal gills. Forms with two terminal filaments (Fig.20.4*c*) may superficially resemble nymphs of Plecoptera. Mayflies are exceedingly abundant in nearly all permanent freshwater habitats, where they constitute a basic food item for nearly all

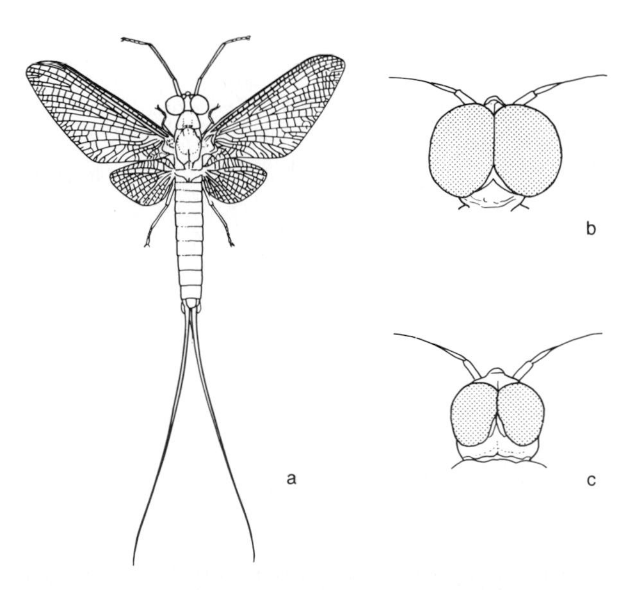

FIGURE 20.1 *a,* Adult of *Hexagenia bilineata* (Hexageniidae); *b, Homoneura dolani,* dorsal aspect of head of male; *c,* same of female. (a *from Needham,* 1917; b, c *from Edmunds et al.,* 1958.)

predators. Adults of many species, especially in temperate climates, tend to emerge synchronously, sometimes in extremely large numbers, and may be temporary nuisances near lakes or large streams.

The Ephemeroptera, together with the Odonata, are the only extant insects with a paleopterous flight mechanism. These two relatively small orders, sometimes classified as the Infraclass Paleoptera, constitute the remnants of an extensive fauna which mostly became extinct at the end of the Mesozoic Era. In mayflies, although wing-flexing capability is lacking, the dorsal longitudinal thoracic muscles are relatively large, and flight is partly powered by these indirect muscles. In Odonata, flight is powered entirely by direct musculature. The major differences in flight mechanism, as well as the grossly different body plans and biological adaptations, indicate the extreme antiquity of the paleopterous orders and their long period of independent evolution.

Mayflies are remarkable in possessing paired penes and oviducts opening through separate gonopores. Among other insects, the external genitalia are paired only in male earwigs (Dermaptera). Mandibles of nymphs (Fig. 20.2*b*) are notable in their general similarity to those of Archeognatha (Fig. 19.1*b*), and in having a single cephalic articulation. Musculature and development of the nymphal gills show similarities to legs, suggesting possible homology with the abdominal appendages of the entognathous hexapods (see Chap. 17). In most other morphological features, mayflies are highly specialized, as described below.

Mayfly naiads molt to *subimagos* similar to the adults but sexually immature. Duration of the subimaginal instar is generally correlated with the length of adult life. In species which mate and reproduce the same night, the subimago usually lasts only a few minutes; in other species the adult may live several days, and the subimago usually persists for about 12 to 24 hours. In a few species the imaginal molt is partly or completely suppressed. Adults engage in mating flights, which may involve large, dense swarms, especially in nocturnally active, Temperate Zone species. Nuptial flights or dances frequently entail characteristic movements, especially by males, which rapidly fly upwards several meters, then slowly drift downward. These patterns are assisted by the caudal filaments, which act as a counterbalance during the downward segment of

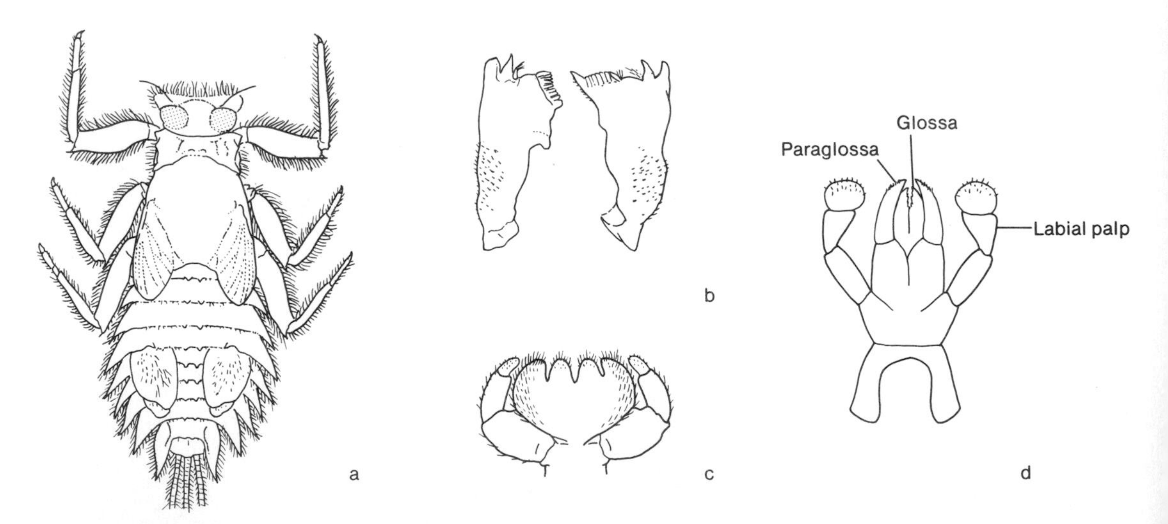

FIGURE 20.2 Structures of mayfly naiads: *a, Ephemerella hecuba,* dorsal, caudal filaments removed (Ephemerellidae); *b,* mandibles, *E. hecuba; c,* labium, *E. hecuba; d,* labium, *Baetis* (Baetidae). (a, b, c *from Allen and Edmunds,* 1959; d *from Murphy,* 1922.)

the flight. Adults and subimagos have vestigial mouthparts and never feed. The midgut is enlarged as an air sac, increasing the efficiency of flight. The esophagus and hindgut are modified as valves, apparently for regulating and maintaining the amount of air in the midgut.

Before copulation, the male approaches the female from below, grasping her mesothorax with his elongate forelegs, which are provided with a reversible joint at the first tarsal segment. Also correlated with courtship is the enlargement of the eyes in males (Fig. 20.1*b,c*). In many genera the dorsal facets are larger than the ventral ones; in some the eyes are divided into distinct dorsal and ventral lobes.

Several hundred to several thousand minute eggs are deposited, either in small batches or en masse, depending upon the species. Females may oviposit while in flight or may descend below the water surface. Eggs are quite variable in shape and sculpturing among different genera and families, and are of some taxonomic importance. Eggs may hatch after one to several weeks or may diapause for several months before developing. Development of immatures requires one to several years, with up to 27 instars. Most species are scavengers or herbivores; a few predaceous forms are known. Naiad morphology is generally correlated with habitat. Free-swimming, campodeiform or shrimplike nymphs (Figs. 12.3*b*, 20.4*d*) usually occupy small, standing bodies of water or sluggish streams. Flattened bodies with widely separated, laterally directed legs (Fig. 20.4*a,c*) are adaptations for clinging to stones or logs in fast-flowing streams. In burrowing forms, the legs or mandibles (Fig. 20.4*b*) may be specialized for digging, and the plumose gills for creating a current in the burrow, as in *Ephemera* and *Polymitarcys*.

Classification of mayflies has been subject to numerous rearrangements, with great differences in the number of families recognized. The classification adopted here follows Edmunds (1959) and Edmunds et al. (1976). Adult mayflies, probably because of their high degree of specialization for a short adult life and aerial copulation, show only minor morphological differences among families. The key to adults is based largely on wing venation, and must be used carefully. Nymphs are highly specialized for living on or in different substrates, as reflected in their great morphological divergence, and usually can be easily keyed to family. The general geographic and habitat distribution of nymphs and the numbers of North American species in each family are listed in Table 20.1

KEY TO THE NORTH AMERICAN FAMILIES OF **Ephemeroptera**[1]

ADULTS

1. Forewings with veins MP_1 and CuA strongly divergent at base; vein MP_2 abruptly bent toward MP_1 at base (Fig. 20.3*a,b*) **2**
   Forewings with veins MP_1 and CuA gradually divergent at base; vein MP_2 gradually bent toward MP_1 at base (Fig. 20.3*d,e,h*) **6**

2(1). Basal portion of forewing with costal crossveins faint or incomplete; hind wing with veins MP_1 and MP_2 separating distally to middle of wing (as in Fig. 20.3*f*) **Neoephemeridae**
   Forewing with costal crossveins strong, never incomplete; hind wing with veins MP_1 and MP_2 separating proximally to middle of wing (Fig. 20.3*a*) **3**

3(2). Forewings with veins Sc and R_1 curving posteriorly around apical margin of wing (Fig. 20.3*a*); middle and hind legs atrophied to trochanter **Polymitarcidae**
   Forewings with veins Sc and R_1 straight to apex, or slightly curved, ending anterior to apex (Fig. 20.3*b*); middle and hind legs with femur, tibia normal **4**

4(3). Apical anterior portion of forewing with dense, interconnecting network of marginal crossveins (Fig. 20.3*a*); cubital intercalary veins straight, not connected with vein CuA (Fig. 20.3*a*) **Polymitarcidae**

[1]Names of the veins (Fig. 20.3*b*) are those of Edmunds et al. (1976); MA, medius anterior; MP, medius posterior; CuA, cubitus anterior (not cubitoanal); CuP, cubitus posterior.

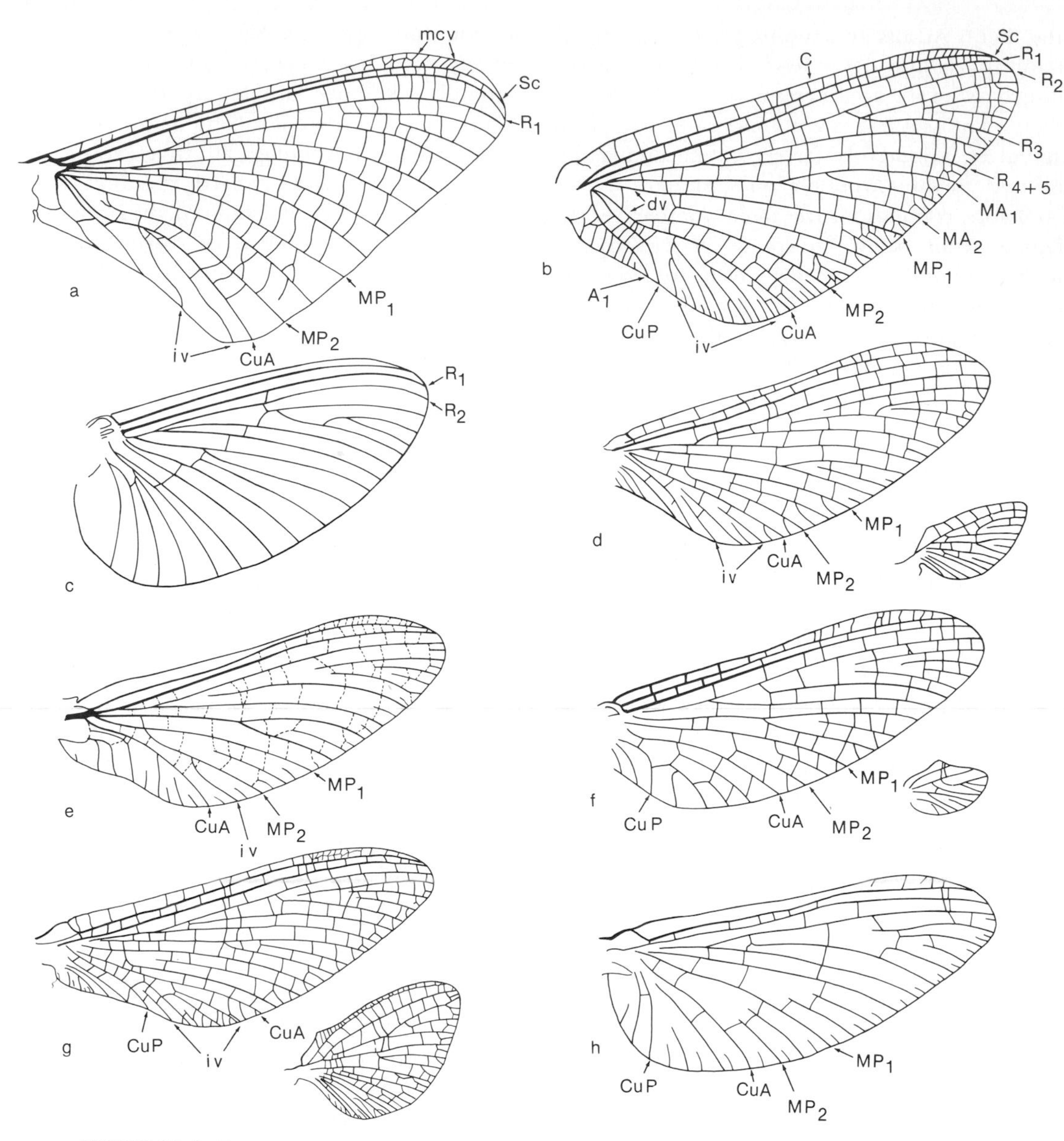

FIGURE 20.3 Representative mayfly wings: *a,* Polymitarcidae (*Ephoron leukon*); *b,* Ephemeridae (*Ephemera simulans*); *c,* Caenidae (*Brachycercus lacustris*); *d,* Heptageniidae (*Heptagenia maculipennis*); *e,* Ephemerellidae (*Ephemerella lutulenta*); *f,* Leptophlebiidae (*Thraulodes speciosus*); *g,* Ametropodidae (*Siphloplectron basale*); *h,* Baetidae (*Callibaetis fluctuans*). *iv,* intercalary veins; *mcv,* marginal crossveins; *dv,* divergent veins; (*From Burke,* 1909.)

Radial portion of forewing with marginal crossveins sparsely interconnecting, not forming network; cubital intercalary veins sinuous, joining vein CuA (Fig. 20.3*b*) 5

5(4). Vein A_1 forked **Potamanthidae**
Vein A_1 not forked (Fig. 20.3*b*) **Ephemeridae**

6(1). Lateral ocelli about one-tenth to one-fourth as large as compound eye 8
Lateral ocelli about one-half as large as compound eye 7

7(6). Forewing with numerous crossveins between veins R_1 and R_2 **Tricorythidae**

Forewing with 1 or very few crossveins between R_1 and R_2 (Fig. 20.3*c*) **Caenidae**

8(6). Forewing with venation reduced to 7 to 9 longitudinal veins; veins Sc and R_1 fused **Oligoneuriidae**

Forewing with at least 15 longitudinal veins; veins Sc and R_1 separate from base to apex (Fig. 20.3*d* to *h*) 9

9(8). Forewing with cubital intercalary veins present (Fig. 20.3*d*) 10

Forewing with cubital intercalary veins absent **Baetiscidae**

10(9). Hind leg with tarsus 5-segmented **Heptageniidae**

Hind leg with tarsus 3- or 4-segmented; basal segment(s) fused with tibia 11

11(10). Forewing with 1 or 2 long intercalary veins between veins MP_2 and CuA (Fig. 20.3*e*) **Ephemerellidae**

Forewing without long intercalaries between veins MP_2 and CuA (Fig. 20.3*f,g,h*) 12

12(11). Forewing with vein CuP abruptly angled near midpoint (Fig. 20.3*f*) **Leptophlebiidae**

Forewing with vein CuP straight or evenly curved 13

13(12). Forewing with 1 to 2 pairs of long, parallel cubital intercalary veins; marginal veinlets never free (Fig. 20.3*g*) **Ametropodidae**

Forewing with several short cubital intercalary veins, or with marginal veinlets free, detached (Fig. 20.3*h*) 14

14(13). Hind wing very small or absent; forewing with vein MP_2 detached from MP_1 (Fig. 20.3*h*) **Baetidae**

Hind wing large; forewing with vein MP_2 basally connected to MP_1 **Siphlonuridae**

NAIADS

1. Thorax with notum enlarged posteriorly, leaving only 4 or 5 abdominal tergites exposed, and concealing gills (Fig. 20.4*a*) **Baetiscidae**

Thorax with notum smaller, leaving at least 7 abdominal tergites exposed; gills exposed (Fig. 20.4*b,c,d*) 2

2(1). Mandibles produced as long tusks projecting forward beyond the head (Fig. 20.4*b*) 3

Mandibles short, without tusks 6

3(2). Gills with fringed margins (Fig. 20.4*b*) 4

Gills bifurcate, with margins straight, unfringed **Leptophlebiidae**

4(3). Gills dorsal, held above abdomen; fore tibiae broad, flattened 5

Gills lateral, held at sides of abdomen; fore tibiae slender, cylindrical **Potamanthidae**

5(4). Hind tibiae with apex produced into a sharp, angulate point, with a row of spines on the apical margin. **Ephemeridae**

Hind tibiae with apex rounded or truncate. **Polymitarcidae**

6(2). Head with dense tufts of setae on anterior corners. **Behningiidae**[1]

Head without dense tufts of setae 7

7(6). Forelegs with dense row of long setae on inner margin; maxillae with tuft of gills at base 8

Forelegs without setae arranged in dense row on inner margin; maxillae without gills 9

8(7). Abdomen with dorsal gills on first segment; forelegs with gills on coxae **Siphlonuridae**

Abdomen with ventral gills on first segment; forelegs without gills on coxae **Oligoncuriidae**

9(7). Abdominal segment 2 with gills subquadrate, meeting at midline of abdomen 10

Abdominal segment 2 with gills rounded or triangular, not meeting at midline of abdomen 11

10(9). Mesonotum with rounded lobes projecting laterally from anterior corners; metathorax with distinct wing pads **Neoephemeridae**

Mesonotum without projecting lateral lobes; metathorax without wing pads **Caenidae**

11(9). Gills absent from abdominal segment 1, or vestigial, threadlike; sometimes absent from segments 2 or 3 12

Gills present on at least abdominal segments 1 to 5 13

12(11). Gills present on abdominal segment 2, triangular or oval **Tricorythidae**

[1]Known only from naiads in North America.

TABLE 20.1 Characteristics of Families of Mayflies in North America

Family	Approximate no. of species in North America	Distribution	Habitat	Substrate	Habits
Siphlonuridae	75	Widespread	Streams, rivers, lakes, ponds, swamps, temporary pools	Sandy or rocky or on vegetation or debris	Active swimmers; filter feeders
Oligoneuriidae	5	Southeast to southwest	Moving water, small to large streams, rapids	Shifting sand or under sticks, stones	Burrowing in sand or clinging to substrates; poor swimmers; filter feeders
Heptageniidae	130	Widespread	Mostly in moving water; also in ponds, rock pools, lake margins	Usually on stones; sometimes in vegetation or debris, or under stones, logs	Clinging to substrate with gill holdfast; poor swimmers
Ametropodidae	8	Northeast to southwest	Slow-moving streams	On vegetation, stones, or sand	Crawling and swimming
Baetidae	140	Widespread	Streams, rivers, ditches, lakes, ponds	Usually in vegetation; also in debris, under stones	Most actively swimming or climbing about on vegetation; negatively phototactic; herbivorous
Leptophlebiidae	60	Widespread	Mostly slowly moving streams, backwashes, lake margins	Under stones, logs; in accumulations of leaves or other debris	Mostly crawling; negatively phototactic; herbivorous or omnivorous
Ephemerellidae	80	Widespread	Variable	On stones, logs, other objects	Climbing about substrate; poor swimmers
Tricorythidae	5	Widespread	Streams, rivers	On sticks, logs, branches, or on sand, gravel	Crawling about substrate; poor swimmers; herbivorous

Caenidae	20	Widespread	Slowly moving, quiet, or stagnant water	Sand or silt bottoms	Resting on bottom, partly covered by fine sand or silt; herbivorous or omnivorous
Neoephemeridae	4	Southeast to central	Slow- to fast-moving streams	On or in debris anchored in current; in vegetation	Crawling about substrate; poor swimmers
Potamanthidae	8	East	Swift streams	On sandy or silty bottoms or under stones	Unknown
Behningiidae	1	Southeast	Large rivers	Sand bottoms	Burrowing in sand
Ephemeridae	23	Widespread	Streams, rivers, lakes	Sand, silt, or mud	Burrowing in substrate; probably ingest mud
Polymitarcidae	7	Widespread	Large rivers; swift streams	Sand, silt, or mud substrates	In burrows under partly embedded stones, or in clay banks; filter feeders
Baetiscidae	10	Widespread	Small to moderate streams; lake margins	Sandy or gravelly substrates	Resting partially buried in sand or silt

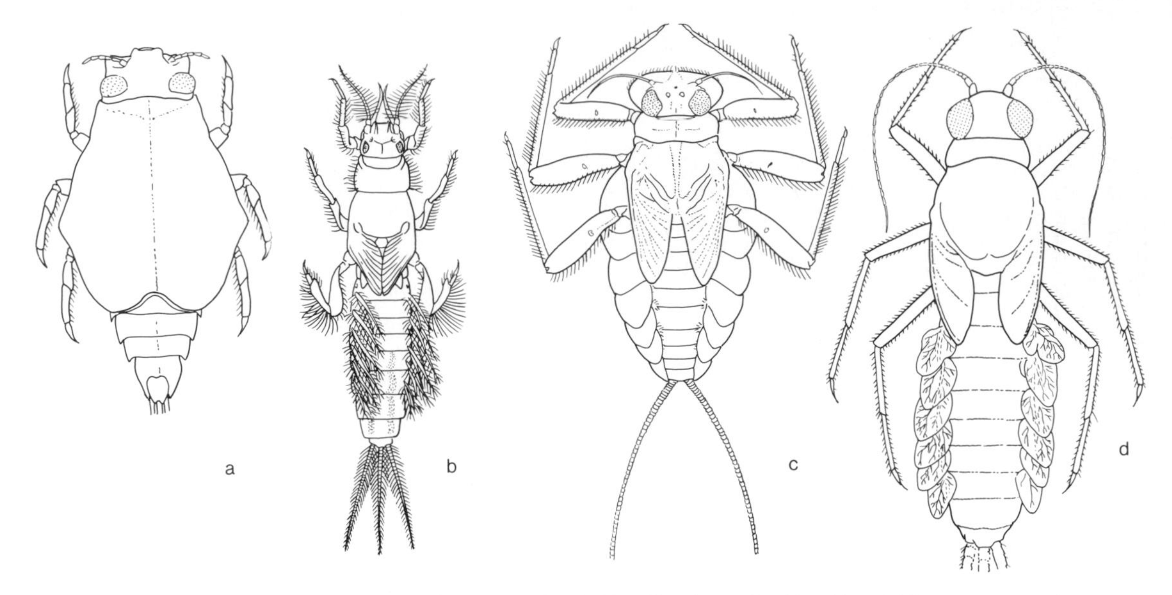

FIGURE 20.4 Representative mayfly naiads: *a,* Baetiscidae (*Baetisca columbiana*), caudal filaments removed; *b,* Ephemeridae (*Ephemera varia*); *c,* Heptageniidae (*Epeorus longimanus*); *d,* Baetidae (*Callibaetis coloradensis*), caudal filaments removed. (a *from Edmunds,* 1960; *b from Needham,* 1920; c, d *from Jensen,* 1966.)

Gills absent on abdominal segment 2, sometimes absent from segment 3 **Ephemerellidae**

13(11). Body cylindrical; eyes or antennae located anteriorly or laterally (Fig. 20.4*d*) 15
Body dorsoventrally flattened; both eyes and antennae located dorsally (Fig. 20.4*c*) 14

14(13). Gills with 2 equal filaments; labial palps with 3 segments **Leptophlebiidae**
Gills unbranched or with slender filament attached near base; labial palps with 2 segments **Heptageniidae**

15(13). Claws on middle and hind legs long, slender and without spines; claws on forelegs bifid or with several stout spines **Ametropodidae**
Claws on all legs similar in structure 16

16(15). Labium with glossae and paraglossae long, slender (Fig. 20.2*d*); abdomen with posterolateral angles of segments 8 and 9 prolonged as flattened spines **Siphlonuridae**
Labium with glossae and paraglossae short, broad (Fig. 20.2*c*); abdomen seldom with segments 8 and 9 produced posteriorly (Fig. 20.4*d*) **Baetidae**

SELECTED REFERENCES

GENERAL

Eatonia, 1954 to date (international journal); Needham et al., 1935; Peters and Peters, 1973.

MORPHOLOGY

Edmunds and Traver, 1954; Rick, 1970*a*.

BIOLOGY AND ECOLOGY

Berner, 1959 (nymphs); Brinck, 1957 (mating, reproduction); Britt, 1962 (detailed life histories of two species); Elliott, 1968 (activity patterns of nymphs); Eriksen, 1966 (burrowing forms); Fremling, 1960 (biology of *Hexagenia*); Hunt, 1950 (biology of *Hexagenia*); Ide, 1935 (effects of temperature); Lyman, 1955 (seasonal distribution).

SYSTEMATICS AND EVOLUTION

Berner, 1950 (Florida species); Burks, 1953 (Illinois species); Day, 1956 (California species); Edmunds, 1959 (North America); Edmunds et al., 1963 (family classification of nymphs); Edmunds et al., 1976 (North and Central America); Edmunds and Traver, 1954; Leonard and Leonard, 1962 (Michigan species); Koss, 1968 (eggs); Merritt and Cummins, 1978 (North America).

ORDER ODONATA (Dragonflies and Damselflies)

21

Adult. Medium to large, paleopterous insects with large mobile head, enlarged thorax, two pairs of long, narrow, membranous wings, and slender, elongate abdomen (Fig. 21.1*a,c*). Head globular, hypognathous, constricted behind into a petiolate neck; compound eyes large, multifaceted; median, and two lateral ocelli present; antennae short, setaceous with 3 to 7 segments; mouthparts mandibulate. Prothorax reduced, mobile; posterior thoracic segments fused into rigid pterothorax with enlarged pleura, reduced sterna, terga; wings similar, with extensive network of veins, apical pterostigma near anterior margin. Abdomen of male with sterna 2 to 3 modified as complex copulatory organs; tergite 10 with unsegmented appendages. *Naiad.* Body more robust than in adult, with less mobile head, smaller eyes, and longer antennae; mouthparts mandibulate, with labium modified as jointed, extensile grasping organ (Fig. 21.4); abdomen 10-segmented; respiration by tracheate lamellae in rectum; Zygoptera also with 3 external caudal gills or lamellae.

Their distinctive body configuration makes adult Odonata one of the most readily distinguished groups of insects. From the superficially

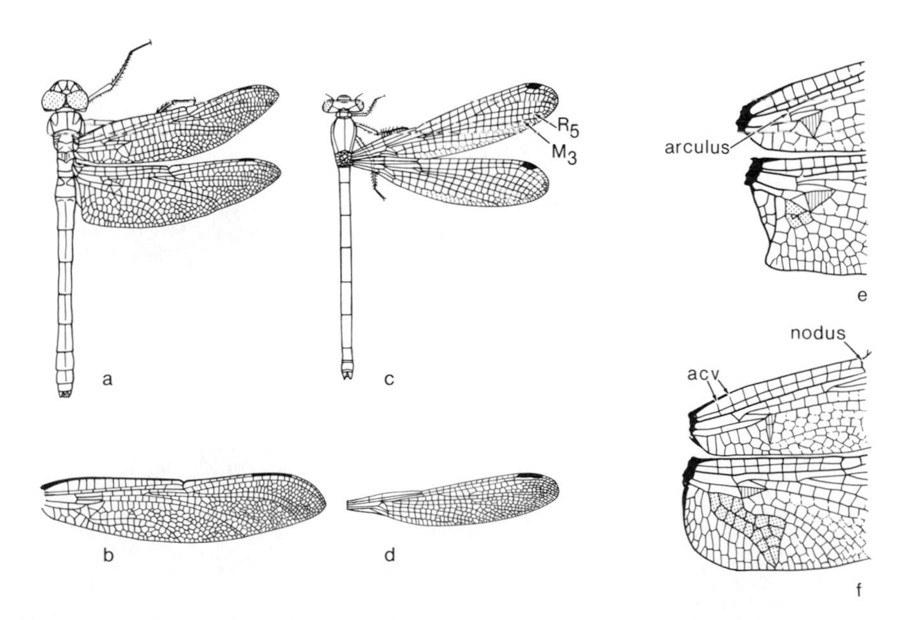

FIGURE 21.1 Adult dragonflies: *a,* Anisoptera, Macromiidae (*Macromia magnifica*); *b,* Petaluridae (*Tanypteryx hageni*); *c,* Zygoptera, Coenagrionidae (*Argia*); *d,* Lestidae (*Archilestes*); *e,* Gomphidae (*Ophiogomphus*); *f,* Libellulidae (*Erythemis*). Stippled pattern indicates anal loops; vertical lines indicate triangles. (a–d, *from Kennedy,* 1915, 1917; e, f *from Needham and Westfall,* 1955.)

similar Myrmeleontidae (Neuroptera) they differ in having setaceous antennae (clubbed in Myrmeleontidae) and in numerous other features. The unique prehensile labium differentiates the naiads from all other insects. All Odonata are predatory in all instars, mostly on organisms much smaller than themselves, and are probably important natural control agents of mosquitoes and other aquatic insects. Odonata are commonly associated with aquatic habitats, but many dragonflies are highly vagile, straying many miles into deserts or other arid regions, where they may colonize animal watering troughs or other very limited sources of water. Two broad adaptive modes exist within the order. The more robust dragonflies (Anisoptera) are powerful, active, searching predators. The more frail damselflies (Zygoptera) spend more time perching and darting after suitable prey.

The flight mechanism is more primitive in Odonata than in any other insects. Not only is wing-flexing capability lacking, as in Ephemeroptera, but the flight muscles attach directly to the bases of the wings, with a consequent reduction in the size of the thoracic terga and sterna. Aerial efficiency is increased by synchronization of the two pairs of wings (which move slightly out of phase), and dragonflies are rapid, agile fliers despite their "primitive" thoracic structure. In the Zygoptera, wing flexing is simulated by the extreme backward tilt of the pterothorax.

Odonata are unique in the development of elaborate secondary male copulatory organs (Fig. 21.2*b,c*). Prior to courtship sperm are transferred from the primary genitalia on the ninth abdominal segment to the genital fossa on sternite 3. The cerci are modified as strong clasping organs, used to hold the female by corresponding grooves on the prothorax (Zygoptera) or occipital region of the cranium (Anisoptera). While the courting partners are joined in tandem (Fig. 21.2*a*), the female bends her abdomen forward to the male genital fossa, where copulation and insemination ensue.

Other adaptive features facilitate efficient predation. In Anisoptera the nearly hemispherical eyes allow almost 360° vision. In Zygoptera

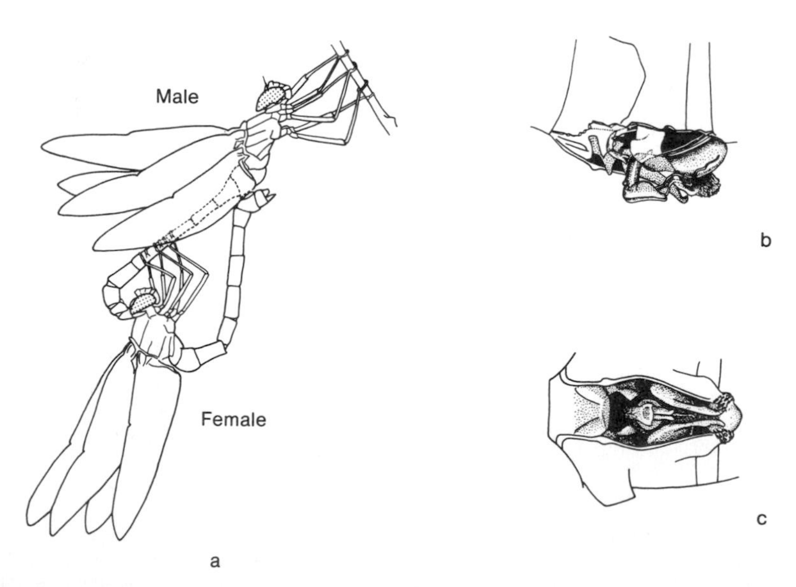

FIGURE 21.2 Copulation in Odonata: *a*, tandem pair of *Macromia magnifica* (Macromiidae) in "wheel position"; *b, c*, secondary male genitalia of *Macromia* (*b*, lateral aspect; *c*, ventral aspect). (*From Kennedy*, 1915.)

the eyes are smaller but are placed at the lateral extremities of the transversely elongate head, maximizing the field of vision. The arrangement of the legs, with the posterior pairs successively longer, creates a basket. The spinose femora and tibiae are useful in holding prey items, and the forward shift of leg attachments allows easy transfer of prey items to the mouth while in flight.

Naiads emerge from the water before the final molt. Dispersal flights away from the place of emergence succeed metamorphosis in many species. The first few weeks of adult life, during which sexual maturity and complete coloration develop, may be spent well away from aquatic habitats, but courtship and mating usually occur around water. Male Anisoptera frequently defend territories, either by patrolling elongate areas over open water, or by protecting more compact areas from a central perch. Territoriality is less obvious but also important in Zygoptera (Paulson, 1975).

Both sexes of Odonata are polygamous. The tandem position may be maintained for long periods before or after insemination and often during oviposition. Zygoptera and some Anisoptera (Aeshnidae) oviposit in stems of vegetation; most Anisoptera lack an ovipositor and deposit eggs freely or attach them to submerged vegetation. Naiad development ranges from a few months in species which colonize temporary ponds, to several years in larger species inhabiting permanent bodies of water. Naiad morphology is broadly correlated with microhabitat. Naiads of Zygoptera are typically slender and shrimplike. They usually frequent aquatic vegetation. Body form in the Anisoptera is variable. Relatively elongate, slender naiads (Aeshnidae, Petaluridae) mostly frequent vegetation or surfaces of substrates. Burrowing forms tend to be shorter, broader, and often flattened (Gomphidae, Petaluridae). Long-legged, flattened naiads (Libellulidae) usually sprawl on soft bottoms.

KEY TO THE SUBORDERS OF **Odonata**

ADULTS

1. Hind wings with base broader than in forewings; venation dissimilar (Fig. 21.1*a*) **Anisoptera**

   Fore- and hind wings similar in shape and venation (Fig. 21.1*c,d*) 2

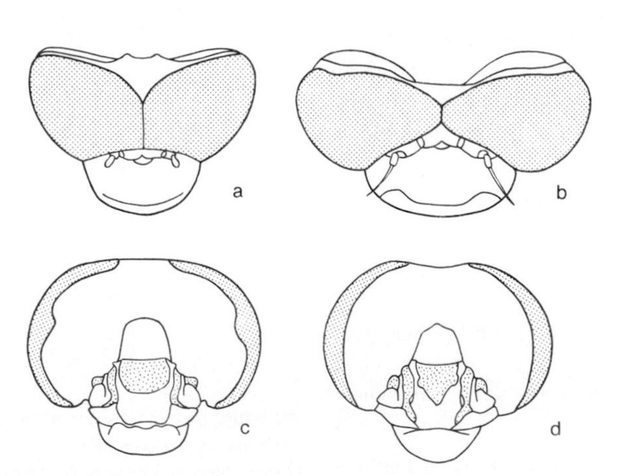

FIGURE 21.3 Anisoptera heads: *a*, Aeshnidae (*Anax junius*); *b*, Cordulegastridae (*Cordulegaster drosalis*); *c*, Corduliidae (*Cordulia shurtleffi*); *d*, Libellulidae (*Libellula modistictus*); *a*, *b*, dorsal views; *c*, *d*, posterior views.

2. Eyes separated by more than their width in dorsal view (Fig. 21.1*c*) **Zygoptera**
   Eyes separated by less than their width in dorsal view (Oriental region) **Anisozygoptera**

NYMPHS

1. Body slender with 3 caudal tracheal gills (Fig. 21.5*b*) **Zygoptera**
   Body stout, without external gills (Fig. 21.5*a*) 2

2. Antennae with 4 or 6 to 7 segments **Anisoptera**
   Antennae with 5 segments (Oriental region) **Anisozygoptera**

KEY TO NORTH AMERICAN FAMILIES OF

Zygoptera

ADULTS

1. Wings with 10 or more antenodal crossveins (as in Fig. 21.1*b*) **Calopterygidae**
   Wings with 3 or fewer antenodal crossveins 2

2. Veins M_3 and R_5 arising closer to arculus than nodus (Fig. 21.1*d*) **Lestidae**
   M_3 and R_5 arising closer to nodus than to arculus (Fig. 21.1*c*) **Coenagrionidae**[1]

NYMPHS

1. Median and lateral gills similar 2
   Median gill bladelike, flattened; lateral gills triangular in cross section **Calopterygidae**

2. Median lobe of labium grooved (Fig. 21.4*c*) **Lestidae**
   Median lobe of labium projecting, not grooved (Fig. 21.4*d*) **Coenagrionidae**

KEY TO NORTH AMERICAN FAMILIES OF

Anisoptera

ADULTS

1. Triangles similar in fore- and hind wings (Fig. 21.1*e*) 2
   Triangles dissimilar in fore- and hind wings (Fig. 21.1*f*) 5

[1]Protoneuridae, with two uncommon species in southern Texas, will key here.

2(1). Eyes broadly contiguous dorsally (Fig. 21.3*a*) **Aeshnidae**
   Eyes separated or barely touching dorsally (Fig. 21.3*b*) 3

3(2). Labium with median lobe deeply notched 4
   Labium with median lobe not notched **Gomphidae**

4(3). Pterostigma moderate, intersecting 4 to 5 crossveins **Cordulegastridae**
   Pterostigma elongate, intersecting 6 to 7 crossveins (Fig. 21.1*b*) **Petaluridae**

5(1). Eyes with posterior margin lobed (Fig. 21.3*c*) 6
   Eyes with posterior margin straight (Fig. 21.3*d*) **Libellulidae**

6(5). Hind wing with anal loop elongate, with two parallel rows of cells (Fig. 21.1*f*) **Corduliidae**
   Hind wing with anal loop short, rounded, with cells not arranged in 2 parallel rows (Figs. 21.1*a,e*) **Macromiidae**

NAIADS

1. Labium strongly convex, projecting anteroventrally 2
   Labium flat 5

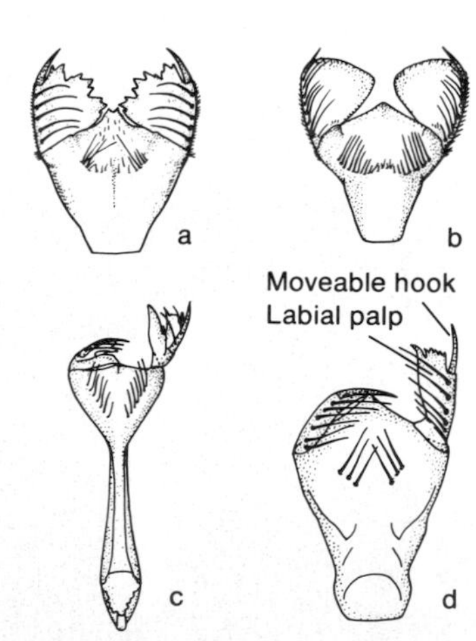

FIGURE 21.4 Ventral aspects of labia of naiads of Odonata: *a*, Cordulegastridae (*Cordulegaster*); *b*, Libellulidae (*Tarnetrum*); *c*, Lestidae (*Lestes*); *d*, Coenagrionidae (*Argia*). (a, b *from Musser*, 1962; c, d, *from Kennedy*, 1915.)

2(1). Lateral lobes of labium with large, irregular, interlocking teeth (Fig. 21.4*a*) **Cordulegastridae**

Lateral lobes of labium with regular dentition; teeth usually small, not interlocking (Fig. 21.4*b*) 3

3(2). Head with prominent, pyramidal horn or tubercle; squat naiads with very long legs **Macromiidae**

Head bare or sometimes with small tubercle; body form variable 4

4(3). Cerci nearly as long as median anal spine (epiproct); lateral spines on abdominal segment 9 less than one-fourth length of segment **Corduliidae**

Cerci usually about one-half to three-fourths length of epiproct; if longer, lateral spines of segment 9 more than one-fourth length of segment **Libellulidae**

5(1). Antennae with 4 segments; fore and middle tarsi with 2 segments **Gomphidae**

Antennae with 6 or 7 segments; tarsi with 3 segments 6

6(5). Antennae with segments hairy, wider than long **Petaluridae**

Antennae with segments bare, longer than wide **Aeshnidae**

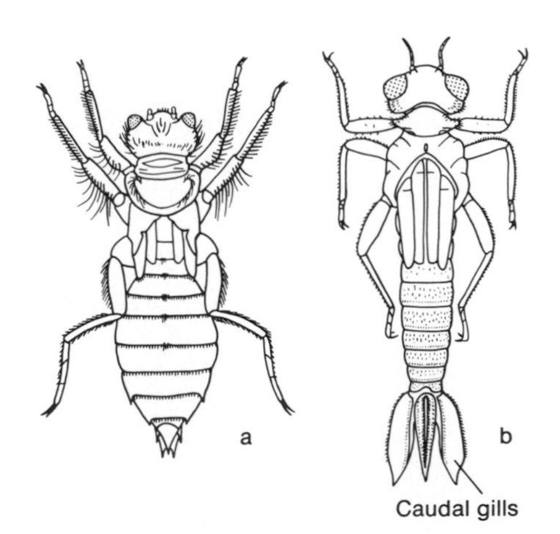

FIGURE 21.5 Naiads of Odonata: *a*, Anisoptera, Libellulidae (*Libellula*); *b*. Zygoptera, Coenagrionidae (*Argia*).

Suborder Zygoptera

Coenagrionidae. The dominant family of Zygoptera, especially in temperate regions. Naiads inhabit a variety of situations, especially sluggish or static waters, where they clamber about on vegetation.

Lestidae. Relatively large damselflies which frequent marshlands and swamps; bodies usually metallic blue, green, or bronze, wings transparent. Eastern North America to the Pacific Coast.

Calopterygidae. Primarily a tropical family, with only two genera in North America; bodies variably colored, wings usually pigmented, at least basally; naiads inhabit riffles in streams and creeks, eastern North America to Pacific Coast.

Suborder Anisoptera

Aeshnidae. Cosmopolitan; possibly the most familiar dragonflies to nonentomologists; adults strong-flying, sometimes migratory, and found far from water. Naiads occupy a variety of habitats, including brackish water.

Gomphidae. Cosmopolitan, medium to large in size, found most abundantly about moving water; the enlarged abdominal apex is a useful diagnostic character for adults; naiads usually burrow or hide in debris.

Petaluridae. Large, dull-colored, clear-winged species which mostly frequent bogs (where the naiads burrow) or, less frequently, streams; a relictual family with only about a dozen species, distributed mostly in temperate regions. Two species in North America.

Cordulegastridae. A small family of medium to large dragonflies, dark brown to black with yellow striping; adults frequent brooks and streams, where the naiads inhabit bottom debris. Holarctic and Indomalaysian, 25 species.

Corduliidae and Macromiidae. Widespread and abundant, especially in north temperate regions; adults mostly metallic blue or green with spots, bands; naiads usually occur in vegetation in a variety of aquatic habitats, including streams, ponds, and swamps.

Libellulidae. A dominant family, especially in the Tropics; in cross section the abdomen is triangular; cuticle usually nonmetallic, but often powdered with silvery pruinosity; wings frequently banded or clouded; nymphs stout, usually inhabiting vegetation or bottom debris in still water.

SELECTED REFERENCES

GENERAL

Odonatologica, 1972 to date (international journal).

MORPHOLOGY

Ando, 1962 (embryology); Snodgrass, 1954 (naiad).

BIOLOGY AND ECOLOGY

Corbet et al., 1960 (general); Corbet, 1962 (general); Kormondy, 1961 (territoriality and dispersal); Paulson, 1974 (reproductive isolation).

SYSTEMATICS AND EVOLUTION

Fraser, 1957 (higher classification; Garman, 1927 (Connecticut species); Gloyd and Wright, 1959 (North America); Needham and Westfall, 1955 (North America); Robert, 1962 (Quebec); Walker, 1953, 1958, and Walker and Corbet, 1975 (Canada and Alaska); Wright and Petersen, 1944 (nymphs).

ORDER BLATTODEA (Cockroaches)

22

Small to large Exopterygota with broad, flattened body, strongly hypognathous head (Figs. 22.1, 22.2). Mouthparts mandibulate, generalized; compound eyes large, ocelli developed as two "ocelliform spots"; antennae long, multiarticulate. Prothorax large, mobile, with shieldlike notum; meso- and metathorax similar, rectangular, slightly smaller than prothorax. Wings with longitudinal veins usually much branched, connected by numerous crossveins; forewing sclerotized as a protective *tegmen,* with anal lobe undifferentiated; hind wing membranous, with large, fan-shaped anal lobe which is folded by longitudinal pleats at rest. Legs robust, long, often spiny; coxae very large, tarsi 5-segmented. Abdomen with 10 distinguishable segments; tergite 10 bearing cerci of 1 to many segments; tergites 5 to 6 and sometimes sternites 6 to 7 bearing openings of scent glands; ovipositor of 3 pairs of small internal valves; male genitalia complex, strongly asymmetrical (Fig. 22.3*b*). Nymphs similar to adults except in development of wings, genitalia, and sometimes color and integumental texture.

About 30 species of cockroaches are cosmopolitan inhabitants of man's dwellings, placing them among the most familiar insects to nonentomologists. Domiciliary species are commonly regarded with disgust because of their invasions of cupboards and pantries, which they spot with their feces and, depending on the species, may permeate with the musty-smelling secretions from their abdominal glands. Cockroaches have not been definitely implicated in transmission of

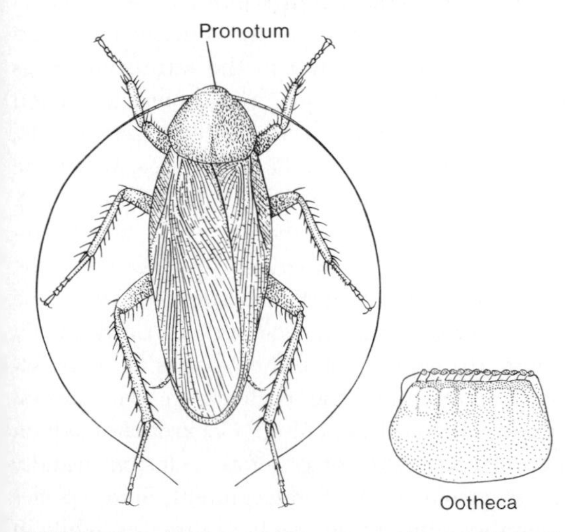

FIGURE 22.1 Blattidae, *Periplaneta americana:* adult female and ootheca. (*From Patton,* 1931.)

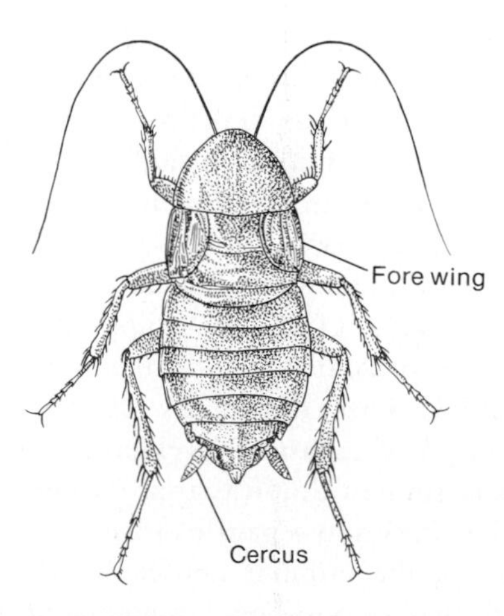

FIGURE 22.2 Blattidae, *Blatta orientalis.* (*From Patton,* 1931.)

human diseases, and they neither bite nor sting. Yet they must be considered of potential medical importance because of the large number of pathogenic microorganisms, including bacteria, viruses, protozoa, and parasitic worms, which have been isolated from their bodies or excrement. It may be pointed out that the same characteristics which adapt species such as *Periplaneta americana* (Blattidae) to life in dwellings also produce a hardy laboratory animal. Cockroaches have been important subjects for investigation of hormonal control of insect growth and development, and especially for the study of insecticidal mode of action and insecticide resistance.

Cockroaches are among the first neopterous insects to appear in the fossil record, and are extremely generalized in most morphological features. The Blattodea are sometimes classified as a suborder of Orthoptera, with which they share many features, such as mandibulate mouthparts, a common pattern of wing venation, and the presence of cerci. However these groups differ in several profound characters, the most important probably being the mode of oviposition and the type of flight musculature. All cockroaches produce characteristic oothecae (Fig. 21.1), consisting of a double-layered wrapper protecting two parallel rows of eggs. This habit extends to the Mantodea, and the egg arrangement occurs in the primitive termite *Mastotermes* (Chap. 24), although the oothecal covering is absent. In contrast, Orthoptera lay eggs singly or in pods, never with an ootheca. In most neopterous insects, including Orthoptera, wing movements are powered almost entirely by indirect flight muscles. In Blattodea, Mantodea, and Isoptera the indirect muscles are very small, the downstroke of the wing apparently being effected by basalar and subalar muscles (Snodgrass, 1958). These differences, together with the long period of separate development indicated by fossils, suggest that it is appropriate to recognize Blattodea as a separate order.

Most of the familiar household cockroaches are nocturnal, positively thigmotactic crevice dwellers. Tropical species, comprising the bulk of the world fauna, include diurnal species, often brightly colored, and sometimes arboreal. North American species of *Arenivaga* (Polyphagidae) are burrowers in sand or litter, and *Attaphila fungicola* (Polyphagidae), less than 3 mm long when mature, is a symbiont of the leaf-cutting ant *Atta fervens* in Texas. Most species are solitary or simply gregarious, but *Cryptocercus punctulatus* (Cryptocercidae), the wood roach, occurs subsocially in galleries in punky, rotting wood. In this species, as in termites, wood is digested with the aid of symbiotic flagellate protozoa (Chap. 16), which are lost after each molt. The defaunated nymphs become reinfested by consuming fecal pellets of adults or older nymphs.

Sex pheromones are produced by some cockroaches, such as *Periplaneta americana.* Courtship by males, involving stridulation or posturing, normally precedes copulation. Males also produce a pheromone from the abdominal tergites, which induces the female to mount the male just before copulation. After genitalic contact the pair turn tail to tail. Most species are bisexual, but some, including *Periplaneta,* may reproduce parthenogenetically.

Oothecae are quite variable in structure, enabling recognition of those of household species. Species also differ in the length of time the ootheca is retained by the mother, in its placement upon deposition, in the number produced by a single female, and in the number of eggs contained. For example, the American roach deposits an average of 50 oothecae per female, each containing about 12 to 14 eggs, while the German roach, *Blattella germanica* (Blattellidae), averages only five oothecae, each containing about 40 eggs. The American roach cements the oothecae to objects in the environment, while the German roach retains each ootheca until shortly before the eggs hatch. Other species are ovoviviparous, retaining the ootheca in an internal brood sac until hatching. Cockroaches which deposit oothecae long before eclosion usually attempt some form of concealment. Some species tuck the ootheca into cracks or crevices, while in others dirt or debris may adhere to its outer surface.

Embryonic and nymphal development is relatively slow, ranging from a few months in smaller species, to over a year in larger ones. Food of domestic species is extremely generalized, but many wild species are difficult to rear, especially in early instars. Adult roaches are often long-lived, some species surviving 4 years or more under laboratory conditions.

Approximately 3500 species of cockroaches are recognized, principally from tropical regions. The North American fauna, numbering about 50 species, includes about a dozen important domiciliary species. Only *Parcoblatta* (Blattellidae) among native roaches is of any economic significance. *Periplaneta americana*, the so-called American roach, apparently originated in Africa (Cornwell, 1968).

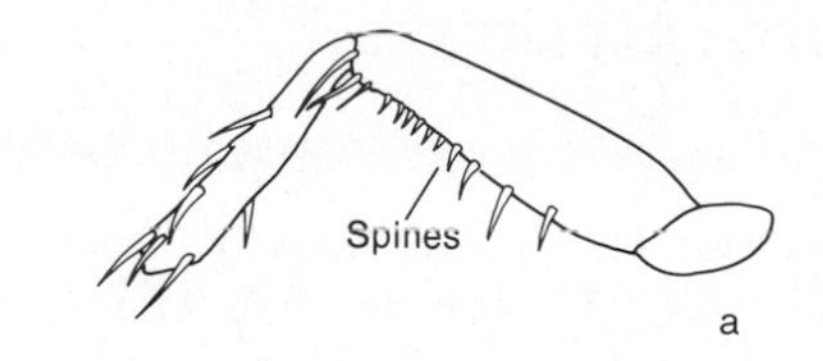

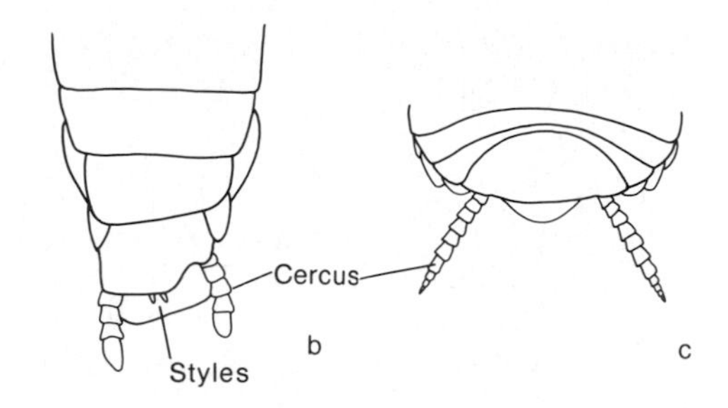

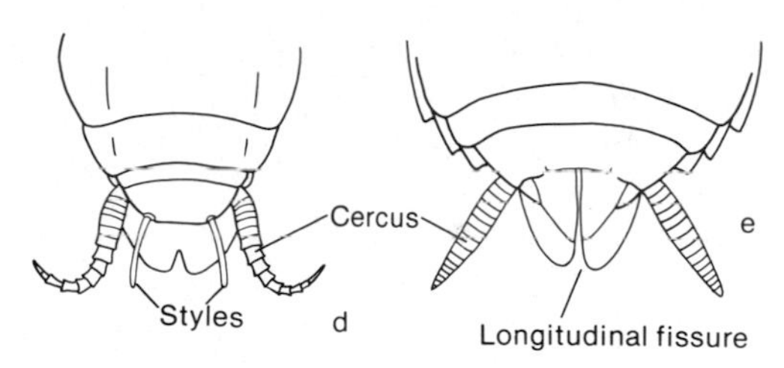

FIGURE 22.3 Taxonomic characters of Blattodea: *a*, anterior leg, Blattellidae (*Blattella*); *b* to *e*, ventral aspects of external genitalia; *b*, Blattellidae, male (*Blattella*); *c*, Blattellidae, female; *d*, Blattidae, male (*Periplaneta*); *e*, Blattidae, female.

KEY TO COMMON NORTH AMERICAN FAMILIES OF **Blattodea**

1. Body length less than 3 mm; eyes very small; symbiotic in nests of *Atta* (Formicidae) **Polyphagidae** (*Attaphila*)
   Body length greater than 3 mm; eyes almost always large; not in ant nests 2

2(1). Middle and hind femora with numerous stout spines on posterior margin 3
   Middle and hind femora with one or two apical spines or lacking spines (hairs may be present) 5

3(2). Fore femora with posterior row of similar spines 4
   Fore femora with spines long and stout proximally, shorter and slender distally (Fig. 22.3*a*) **Blattellidae**

4(3). Female with last visible sternal plate divided by a longitudinal fissure (Fig. 22.3*e*); male styli slender, long, straight (Fig. 22.3*d*) **Blattidae**
   Female with last visible sternite not divided (Fig. 22.3*c*); male with last sternite and styli usually asymmetrical or unequal in size (Fig. 22.3*b*) **Blattellidae**

5(2). Abdomen with seventh tergite and sixth sternite expanded as plates which conceal abdominal apex, including cerci; body elongate, parallel-sided **Cryptocercidae**
   Abdomen with eighth to tenth segments and cerci exposed; body oval or broadly oval 6

6(5). Fore femora with 1 to 3 spines on posterior margin **Blaberidae**
   Fore femora without spines on posterior margin 7

7(6). Frons flat; wings with anal region folded in fanlike pleats at rest; at least 16 mm long; pale green **Blaberidae**
   Frons bulging, swollen; wings with anal region folded flat against preanal portion at rest (females frequently wingless); less than 16 mm long; brown, tan, or gray **Polyphagidae**

SELECTED REFERENCES

GENERAL

Cameron, 1961 (*Periplaneta*); Cornwell, 1968; Guthrie and Tindall, 1968; Scharrer, 1951 (wood roach, *Cryptocercus*).

MORPHOLOGY

Guthrie and Tindall, 1968.

BIOLOGY AND ECOLOGY

Cleveland et al., 1934 (wood roach, *Cryptocercus*); Guthrie and Tindall, 1968; Roth, 1970 (reproduction); Roth and Hartman, 1967 (sound production); Roth and Willis, 1960 (biotic associations).

SYSTEMATICS AND EVOLUTION

McKittrick, 1964; Rehn, 1950 (North American genera); Rehn, 1951 (higher classification); Roth, 1970; Roth and Hartman, 1967.

ORDER MANTODEA (Mantids)

23

Moderate-sized to very large Exopterygota adapted for predation. Head small, triangular, freely mobile on slender neck; mouthparts mandibulate; eyes large, lateral; ocelli usually 3 or none; antennae slender, filiform, multiarticulate. Prothorax narrow, elongate, strongly sclerotized; meso- and metathorax short, subquadrate; forewings narrow, usually cornified with reduced anal region; hind wings much broader, membranous, with large anal region; venation of largely unbranched longitudinal veins, usually with numerous crossveins; wing folding of longitudinal pleats; wings frequently reduced, especially in females; forelegs raptorial, with very long coxae; middle and hind legs long, slender, adapted for walking; tarsi 5-segmented. Abdomen with 10 visible segments, weakly sclerotized, slightly flattened dorsoventrally; segment 10 with variably segmented cerci; ovipositor short, mostly internal; male genitalia bilaterally asymmetrical. Nymphs similar to adults except in development of wings and genitalia.

Mantids are primarily tropical in distribution but are among the more familiar insects to nonentomologists because of the large size and striking appearance of many species. The common mantid pose, with the anterior part of the body elevated and the forelegs held together as if in prayer, has inspired folk beliefs since ancient times. In reality, this is the posture assumed for ambushing prey.

Mantids show a high morphological resemblance to cockroaches (Blattodea). Important similarities include the strong development of the direct pleurosternal flight muscles (*see* Chap. 22) and relatively small size of the indirect longitudinal muscles, five-segmented tarsi, multisegmented cerci, and strongly asymmetrical male genitalia. The eggs are always enclosed in an ootheca, which differs in details but is generally similar to that of cockroaches. These features strongly differentiate roaches and mantids from Orthoptera, with which they are sometimes classified as a single order. The features which distinguish mantids from roaches are nearly all related to their predatory mode of life. Most obvious is the modification of the forelegs as strong grappling hooks for snaring arthropod prey (Fig. 23.1*a*). The elongate thorax and coxae increase the reach, while the sharp femoral and tibial teeth secure struggling victims. The long, slender posterior pairs of legs elevate the body, allowing unimpeded vision. The large eyes are situated far apart, as in other visually orienting predators such as damselflies, and the mobile head can be cocked in any direction. The cryptic coloration, usually in shades of green, brown, or gray, is a further adaptation for the sedentary mode of predation practiced by these insects.

In temperate regions adult mantids appear late in summer, when they are seen on vegetation, tree trunks, and other perches. As predators they are generalists, consuming any arthropod small enough to subdue, including smaller mantids. The larger species may be quite pugnacious, readily unfurling the hind wings, which may be brightly colored, and adopting threatening postures. All mantids are "sit-and-wait" predators which rely on camouflage to get within reach of prey, as mentioned above. Tropical species may be gaudily colored with foliate appendages and specially textured or sculptured cuticle, creating

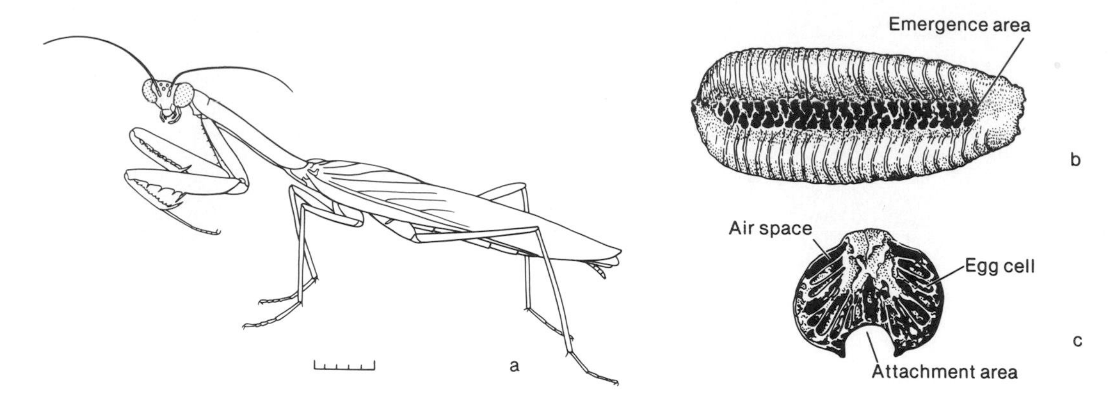

FIGURE 23.1 Mantodea: *a, Stagmomantis carolina,* male (scale equals 5mm); *b,* ootheca, dorsal (diagramatic); *c,* ootheca, transverse section (diagramatic). (b, c *from Breland and Dobson,* 1947.)

a strong overall resemblance to flowers, leaves, bark, or other plant parts (Endpaper 2).

Copulation may be preceded by courtship, allowing the male to reduce the chances of attack as he approaches the larger female. During copulation the male sits astride the female with his abdomen curled ventrally. Sperm is transferred by a spermatophore. In many species the female attacks and partly or wholly consumes the male during or immediately after copulation.

Gravid females become heavily swollen by the large mass of developing eggs. At the time of oviposition the female constructs a characteristic ootheca (Fig. 23.1*b,c*) from the frothy secretion from the accessory glands. This secretion quickly hardens into a protective envelope around the eggs, which are aligned in rows, four abreast. Structural details of oothecae, as well as site of attachment, vary greatly among species. Each female normally lays several to many egg masses. Females of a few species protect the ootheca until the nymphs have emerged. In temperate regions overwintering usually occurs in the egg stage.

As in cockroaches, the first-instar nymph is a nonfeeding stage which quickly molts. Succeeding instars vary in number among different species from 3 or 4 to 12. There is usually a single generation per year.

North American species are included in a single family, MANTIDAE. The Carolina mantid, *Stagmomantis carolina,* is the most familiar native species. Other conspicuous forms include the Chinese mantid, *Tenodera aridifolia,* and the praying mantid, *Mantis religiosa,* both introduced into Eastern United States. In the Western states the small, wingless gray or brown species of *Litaneutria* are not infrequently encountered on the ground or on low vegetation. The world fauna comprises approximately 2000 species in eight families.

SELECTED REFERENCES

Chopard, 1949 (general); Crane, 1952 (defensive behavior); Gurney, 1948 (North American species); Roeder, 1935 (sexual behavior).

ORDER ISOPTERA (Termites)

24

Neopterous Exopterygota living socially in small to very large colonies. Individuals small to moderate in size, with subcylindrical, weakly sclerotized body (Figs. 24.1, 24.2). Head hypognathous or prognathous with mandibulate mouthparts, 10 to 32 segmented moniliform antennae; compound eyes and sometimes 2 ocelli present in reproductives, usually absent in soldiers and workers; frons almost always bearing conspicuous *frontal pore* or fontanelle, through which discharges a *frontal gland.* Thorax with subequal segments, usually with lightly sclerotized tergites, membranous sterna; reproductives with 2 pairs of similar wings with simple, longitudinal venation or extensive network of crossveins; wings broken at *humeral suture* and lost after dispersal flight; soldiers, workers and nymphs without wings; legs short, adapted for walking. Abdomen with 10 similar segments and reduced eleventh segment; terminal segment bearing 1 to 8 segmented cerci.

Termites constitute a relatively small order of insects, with about 1900 described species. All are of retiring habits, seldom venturing out of their concealed galleries, with the exception of the reproductives, which disperse aerially prior to founding new colonies. Yet most nonentomologists are familiar with termites because of their nearly worldwide distribution and especially because of their proclivity to invade and consume manmade wooden or paper objects. Termites are economically important everywhere except extremely cold regions, and in parts of the Tropics and Subtropics may be among the most destructive arthropods. In northern Australia *Mastotermes* (Mastotermitidae) is considered the single most important insect pest. Aside from their economic importance, termites are of general biological interest because of their highly developed social behavior (see below and Chap. 7).

Termites are highly adapted for living in densely populated colonies, sociality apparently having evolved by the early Mesozoic Era or even prior. Yet many characteristics, especially of the more primitive families, are shared with cockroaches (Blattodea). In particular, the primitive termite family Mastotermitidae, represented by a single extant species in northern Australia, shares many similarities with the roach family Cryptocercidae (Table 24.1). Most convincing, perhaps, are the similarities in wing structure, egg deposition, and intestinal symbionts. The woodroach, *Cryptocercus,* harbors 21 to 22 species of hypermastigote and flagellate protozoa. All of these are members of families which occur in primitive families of termites, including the shared genus *Trichonympha.* The nymphs of *Cryptocercus,* which are not unlike termite workers in general appearance, occur gregariously with adults in galleries in rotting wood. Another similarity, not listed in Table 24.1, is the presence of relatively large direct pleurosternal muscles and only small, poorly developed indirect dorsal longitudinal flight muscles in both roaches and termites (see also Chap. 22).

Termites, as in social Hymenoptera, are characterized by the development of morphologically differentiated *castes* which perform different biological functions. Least modified morphologically are the *reproductives* (Fig. 24.1) which have

TABLE 24.1 Comparison of Selected Characters of Isoptera and Blattodea

	Isoptera		Blattodea	
	Isoptera except Mastotermitidae	**Mastotermitidae**	**Cryptocercidae, Panesthiidae**	**Blattodea except Cryptocercidae**
No. of antennal segments	10–22	20–32	>32	>32
No. of tarsal segments	3–4	5	5	5
Wing venation	Reticulate in Kalotermitidae, Hodotermitidae; reduced in Termitidae	Reticulate	Reticulate	Reticulate
Hind wing anal lobe	Absent	Present	Present	Present
Humeral suture	Present in both wings	Present in forewings; absent in hind wings	Wings absent in Cryptocercidae; sutures present in Panesthiidae	Absent in both wings
Female external genitalia	Absent or vestigial	Short ovipositor with 3 pairs of valves	Short ovipositor with 3 pairs of valves	Short ovipositor with 3 pairs of valves
Egg deposition	Laid separately	Laid in pods	Contained in oothecae	Contained in oothecae
Intestinal symbiotes	Flagellate protozoa (spirochaetes in Termitidae)	Flagellate protozoa;	Flagellate protozoa in Cryptocercidae; amoebae in Panesthiidae	Bacteria
Bacteriocytes	Absent	Present	Present	Present
Social organization	Eusocial	Eusocial	Primitively subsocial	Nonsocial

relatively large eyes, unmodified mandibulate mouthparts, and a sclerotized body with distinct thoracic and abdominal tergites. Primary reproductives always bear wings initially, later breaking them at a basal fracture zone (humeral suture) by pressing the extended wings firmly against the substrate. Males do not change in external appearance after shedding their wings, but the abdomens of females gradually become swollen, or *physogastric,* from the enlargement of the ovaries. Physogastry is especially evident in the more specialized termites of the Families Termitidae and Rhinotermitidae, in which the abdominal sclerites are eventually isolated as small islands on the bloated, membranous abdomen (Fig. 7.1). Many species, especially of the more primitive families, have the capacity to develop substitute kings and queens. These are neotenic individuals which never develop wings, do not become so physogastric as the primary reproductives, and differ in other minor morphological features.

The worker caste (Fig. 24.2*a*) is distinguished mostly by regressive characteristics. Compound eyes are usually absent, the number of antennal segments is usually smaller than in other castes, and the body is very lightly sclerotized except in a few species which actively forage outside their galleries. In Kalotermitidae and Hodotermitidae (such as *Zootermopsis* in North America) the worker caste is absent, being functionally replaced by nymphal reproductives, or *pseudergates.*

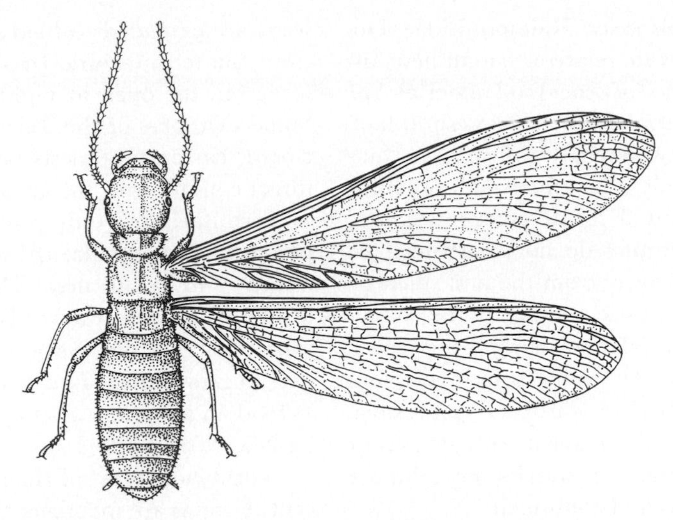

FIGURE 24.1 Adult female termite, Hodotermitidae (*Zootermopsis nevadensis*). (*From Banks,* 1920.)

Workers or pseudergates are the most abundant caste, performing foraging, care of young nymphs, feeding of soldiers and reproductives, and nest construction and maintenance.

Soldiers (Figs. 24.2*b*, *c*) are characterized by specialization of the head for defense of the colony, especially against ants. The mandibles are usually enlarged as long, slender, scissorlike blades, sometimes with grotesque serrations, ridging, or twisting. In some Termitidae the mandibles of soldiers are atrophied; the head is drawn into a nozzle-shaped projection bearing the opening of the frontal gland. These *nasute* soldiers deter invaders by spraying or exuding repellent or entangling substances. The proportion of soldiers varies from none in some species of Termitidae and Kalotermitidae to over 15 percent in *Nasutitermes* (Termitidae). In some species two classes of soldiers may occur, differing in structure as well as size.

Social organization in Isoptera is extremely complex and variable (see Chap. 7). In general,

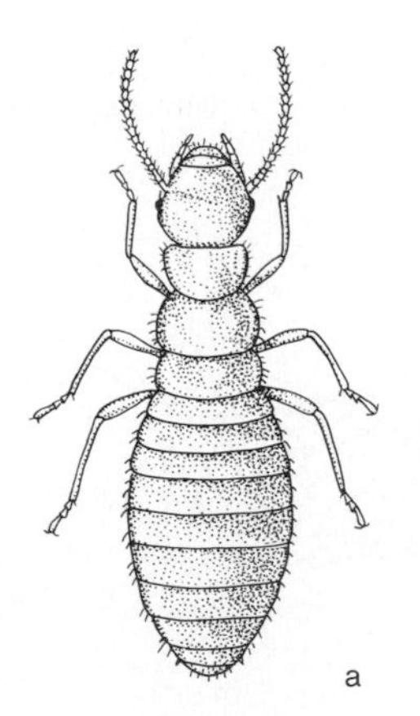

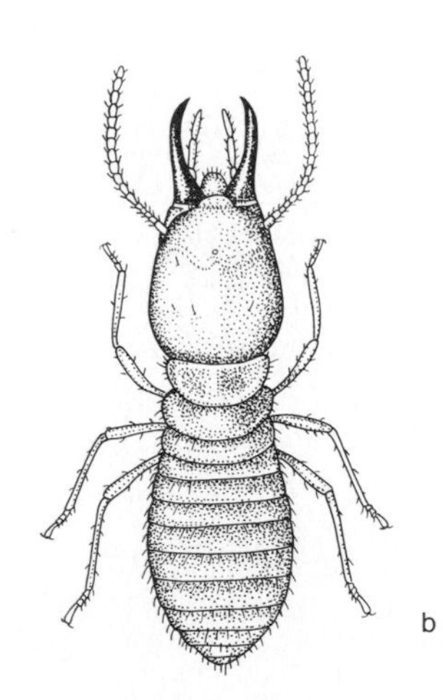

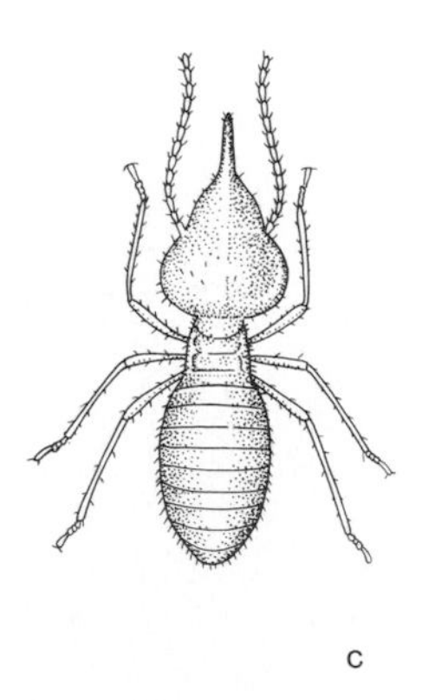

FIGURE 24.2 Variation in body form of termites: *a,* worker, Rhinotermitidae (*Prorhinotermes simplex*); *b,* soldier, Rhinotermitidae (*Prorhinotermes simplex*); *c,* soldier, Termitidae (*Nasutitermes costaricensis*). (*From Banks,* 1920.)

the more primitive species (Kalotermitidae, Hodotermitidae) excavate relatively small nests directly in logs, timbers, or other food material. For example, in *Incisitermes minor* (Kalotermitidae), which infests dry wood in California, mature colonies number only a few thousand individuals, including about 8 to 10 percent soldiers. These primitive termites do not construct passageways or forage away from the nest, thereby limiting the potential size and age of the colony by the size of the initial food source. *Mastotermes*, generally considered the most primitive living termite, is exceptional in constructing massive, densely populated subterranean colonies, from which a wide variety of starchy or cellulose materials is invaded and consumed.

Most members of the Families Rhinotermitidae and Termitidae construct subterranean galleries from which the workers forage for food. The size of such colonies varies considerably, but frequently they are very large, containing millions of individuals. In many species the mature *termitarium* includes a portion which extends above the ground surface as a dome or spire which may reach 7 m in height. Such structures consist of an ensheathing hard layer of soil particles cemented together with excrement or glandular secretions, enclosing a honeycomb of living chambers (Fig. 24.3). Covered passageways are extended to food sources up to 100 m from the termitarium. In a few species workers forage in the open at night or on cloudy days. Some members of the Termitidae construct enclosed, carton-type nests on trees, without any direct contact with the ground.

Termitaria not only provide protection from predators but stabilize the temperature and humidity within the nest. The Old World Subfamily Macrotermitinae (Termitidae) has capitalized on the homeostatic climate within the nest to grow fungi (*Termitomyces*), which are used as food, in a manner analogous to fungus culture by New World attine ants.

Worldwide, most of the economically important termites are members of the Termitidae. In parts of Africa and Asia damage to neglected structures occurs with remarkable rapidity. In temperate regions, where Termitidae are rare (14 species in North America), Rhinotermitidae cause more damage, principally by invading foundation timbers from their subterranean nests.

As in social Hymenoptera, the life history of termites involves dispersal of reproductives from established colonies. Dispersal occurs at characteristic times for each species. For example, *Incisitermes minor*, discussed earlier, flies during warm dry periods, whereas *Reticulitermes hesperus*

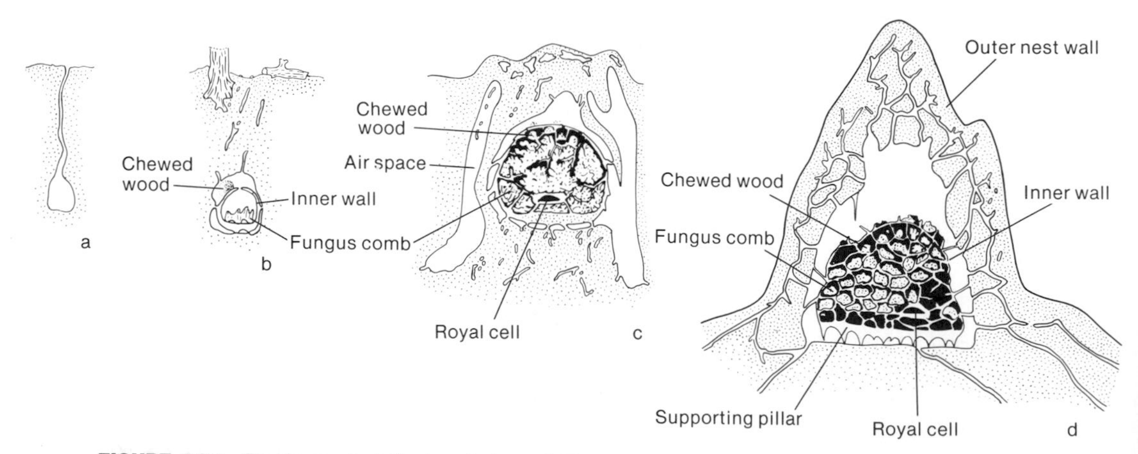

FIGURE 24.3 Development of the termitarium of *Macrotermes bellicosus* (Termitidae) of Africa; *a*, initial chamber excavated by founding couple; *b*, *c*, intermediate stages of development; *d*, structure of mature nest. (*From Grassé and Noirot*, 1958.)

(Rhinotermitidae) flies after the first fall rains in California. Dispersal flights cover variable distances. A few species are efficient colonizers of islands and other isolated habitats, but sexuals of most species probably fly no more than a few hundred meters. After returning to earth the unmated males and females dehisce their wings, then join in pairs which almost always exhibit tandem running, the male following the female (Chap. 7). Founding pairs of reproductives initiate colonies in suitable locations in soil, logs, or other substrates. Colony growth is slow at first. After the first year a typical colony may number only a dozen workers and the primary reproductives. The first soldiers appear in 2 or 3 years, and winged dispersal castes are not produced for 4 years or more. Individuals of all castes develop slowly and survive for considerable periods. The ultimate instar is usually achieved after 7 to 10 molts, requiring a year or more. Workers and soldiers live up to 4 years; reproductives may survive 15 years or more.

As in social Hymenoptera, termite colonies are exploited by a variety of other organisms. Vertebrate predators include anteaters, aardvarks, marsupials, reptiles, and frogs. Staphylinid beetles are particularly well represented among insect associates or *termitophiles*. Some termitophiles consume eggs or young nymphs of their host, while others are nest scavengers. The latter frequently show considerable structural modification, with special secretory areas which are attractive to the termite, or with the body modified to resemble the contour or texture of the host.

Classification of Isoptera has attained a relatively mature state. New genera are infrequently discovered, and no new species have been described from North America since 1941. The Family Termitidae is dominant, with about 1400 species, mostly tropical in distribution. Only 14 Termitidae occur in North America, all restricted to the Southwest. The Kalotermitidae contains 16 North American members, including several economically important species. Hodotermitidae are represented by three species and Rhinotermitidae by eight species, including *Reticulitermes*, the most destructive genus in North America. In most of the Northern and Central United States only a single species of *Reticulitermes* is present.

Workers are extremely similar and difficult to distinguish, and are not keyed here. All termites should be stored in liquid preservatives to prevent shriveling of their bodies.

KEY TO NORTH AMERICAN FAMILIES OF

Isoptera

REPRODUCTIVES (WINGS OR WING BASES PRESENT)

1. Head with median dorsal pore or fontanelle; wings with 2 strong and several faint veins, without crossveining — 2
   Head without fontanelle; wings with at least 3 strong veins, usually with extensive crossveining — 3

2(1). Pronotum flat; basal portion of dehisced wing longer than pronotum — **Rhinotermitidae**
   Pronotum concave, saddle-shaped; basal portion of dehisced wing shorter than pronotum — **Termitidae**

3(1). Head with ocelli — **Kalotermitidae**
   Head without ocelli — **Hodotermitidae**

SOLDIERS

1. Head produced anteriorly into a hornlike projection (Fig. 24.2*c*); mandibles vestigial — **Termitidae**
   Head without hornlike projection; mandibles large — 2

2(1). Mandibles with one or more large, marginal teeth — 4
   Mandibles without marginal teeth — 3

3(2). Head longer than broad, with median dorsal pore or fontanelle — **Rhinotermitidae**
   Head about as long as broad, without fontanelle — **Kalotermitidae**

4(2). Mandibles with 1 prominent marginal tooth; fontanelle present; head narrowed anteriorly — **Termitidae**
   Mandibles with at least 2 prominent marginal teeth; fontanelle absent; head not narrowed anteriorly — 5

5(4). Antennae with third segment enlarged, 2 to 4 times length of fourth segment **Kalotermitidae**
Antennae with third segment subequal to fourth **6**

6(5). Hind legs with femora swollen; antennae with at least 23 segments **Hodotermitidae**
Hind legs with femora slender; antennae with fewer than 23 segments **Kalotermitidae**

SELECTED REFERENCES

GENERAL

Ebeling, 1968 (economic); Harris, 1961 (economic); Krishna and Weesner, 1969*a,b*; Skaife, 1961; Snyder, 1956, 1968 (bibliography).

BIOLOGY AND ECOLOGY

Howse, 1970 (social behavior); Krishna and Weesner, 1969*a,b*; Lüscher, 1961 (social behavior); Wilson, 1971 (social behavior and biology).

SYSTEMATICS AND EVOLUTION

Emerson, 1952, 1955 (biogeography); Roonwal, 1962 (higher classification); Snyder, 1954 (North American species).

ORDER GRYLLOBLATTODEA (Grylloblattids)

25

Elongate, cylindrical or slightly flattened Exopterygota of moderate size. Antennae multiarticulate, filiform, about as long as thorax; compound eyes reduced or absent, ocelli absent; mouthparts mandibulate. Prothorax subquadrate, larger than mesothorax and metathorax; wings absent; legs long, slender, adapted for walking, tarsi 5-segmented. Abdomen 10-segmented, with filiform, 8- or 9-segmented cerci; female with straight, short ovipositor. Nymphs similar to adults.

Grylloblattids comprise one of the smallest orders of insects, with about half a dozen species in the Family GRYLLOBLATTIDAE occurring in North America and eastern Asia. Collections have been made from as low as 500-m elevation in northern California, but most specimens are found at high altitudes under cool, moist conditions, frequently near snowfields from Central California to western Canada and Wyoming. Activity occurs at low temperatures, and grylloblattids scavenge on the surface of snow or ice, as well as about snowmelt or moist surfaces in or near rockslides or under rotting logs. Food includes both plant and animal material. In *Grylloblatta campodeiformis* (Fig. 25.1) the eggs are deposited in moist soil or moss, apparently hatching only after a prolonged diapause of about a year. Development is slow, with the eight recorded nymphal stadia requiring several years for completion.

These generalized insects are of special interest because they combine features of the Orthoptera (external ovipositor; structure of the tentorium) and the Blattodea (five-segmented tarsi; multiarticulate cerci; asymmetrical male genitalia). It has been suggested that the presently reduced fauna represents the remnants of an evolutionary line which produced both the blattoid and orthopteroid orders.

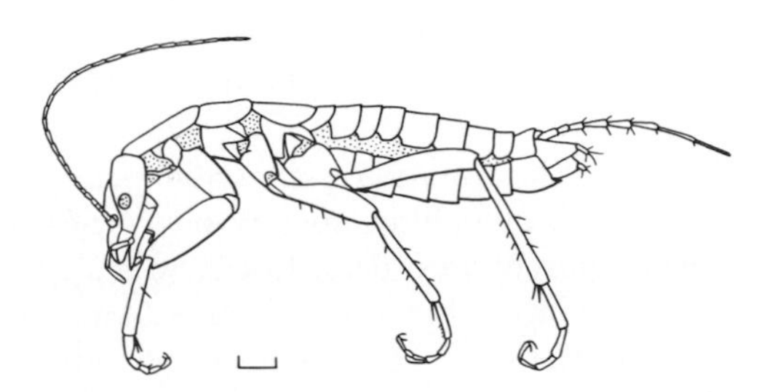

FIGURE 25.1 Grylloblattidae: *Grylloblatta campodeiformis*, male (scale equals 1 mm).

SELECTED REFERENCES

Ford, 1926 (behavior); Gurney, 1948 (systematics); Kamp, 1963, 1973 (systematics); Walker, 1949 (morphology).

26

ORDER DERMAPTERA (Earwigs)

Neopterous Exopterygota of medium size; body elongate, slightly flattened in dorsoventral plane, with leathery cuticle (Figs. 26.1*a,c*). Head prognathous; compound eyes moderate or absent, ocelli absent; antennae multiarticulate, filiform; mouthparts mandibulate, adapted for chewing. Prothorax mobile, with large notum (except in wingless forms); cervical region of several separate sclerites; mesothorax small, mesothoracic wings short, elytriform, or absent; metathorax large, wings semicircular with distinctive radiating venation (Fig. 26.1*b*), or absent; legs adapted for walking, tarsi with 3 segments. Abdomen with 8 segments in female, 10 in male; cerci modified as strongly sclerotized forceps, except in parasitic forms.

The forcipulate cerci, terminating the flexible, telescoping abdomen, distinguish earwigs from all insects except Japygidae (Diplura), which are entognathous and blind and which differ in numerous other features. In the elytriform forewings Dermaptera resemble staphylinid beetles (Coleoptera), which, however, never have forceps. About 1200 species occur in all parts of the world but are most diverse in tropical and subtropical regions. In most temperate regions the familiar earwigs represent a few cosmopolitan, synanthropic species. Although they frequently become extremely abundant, most are of no more than nuisance importance. Earwigs have unmodified, mandibulate mouthparts and unspecialized legs, and are generalized in most aspects of their internal anatomy. The retention of three pairs of cervical sclerites is probably a primitive feature. Male genitalia are distally paired, as in Ephemeroptera, the only other insects with paired external genitalia.

With a few exceptions the Dermaptera are quite uniform in most anatomical and biological features, being rather narrowly adapted for a cryptic life. For example, entry into crevices, or cavities under stones, or logs, where nearly all Dermaptera take refuge by day, is facilitated by the specialized wing structure. In the flying wings, the sclerotized, platelike preanal veins are crowded into the wing base. The membranous anal portion, with numerous, radiating veins, folds fanwise and transversely beneath the abbreviated mesothoracic wings, with only a sclerotized apical portion exposed. The mesothoracic wings are sclerotized, without defined veins. At rest they are held closed by rows of bristles on the metanotum.

Forceps are used in defense, prey capture, and courtship. In most Dermaptera forceps are single-segmented throughout life, but are usually straighter, less strongly sclerotized, and setose in early instars. In *Diplatys* and *Bormansia* (Forficulina) the cerci are multiarticulate until the last nymphal instar, in which they are unsegmented but straight, becoming forcipulate only after the ultimate molt.

Biological information is available primarily for a few cosmopolitan species but probably applies to most of the order. Most earwigs appear to be omnivorous, but some, such as *Anisolabis*, are reported to be predatory. Nearly all are nocturnal, retreating to protected shelters during the day. Nymphs and adults differ little in

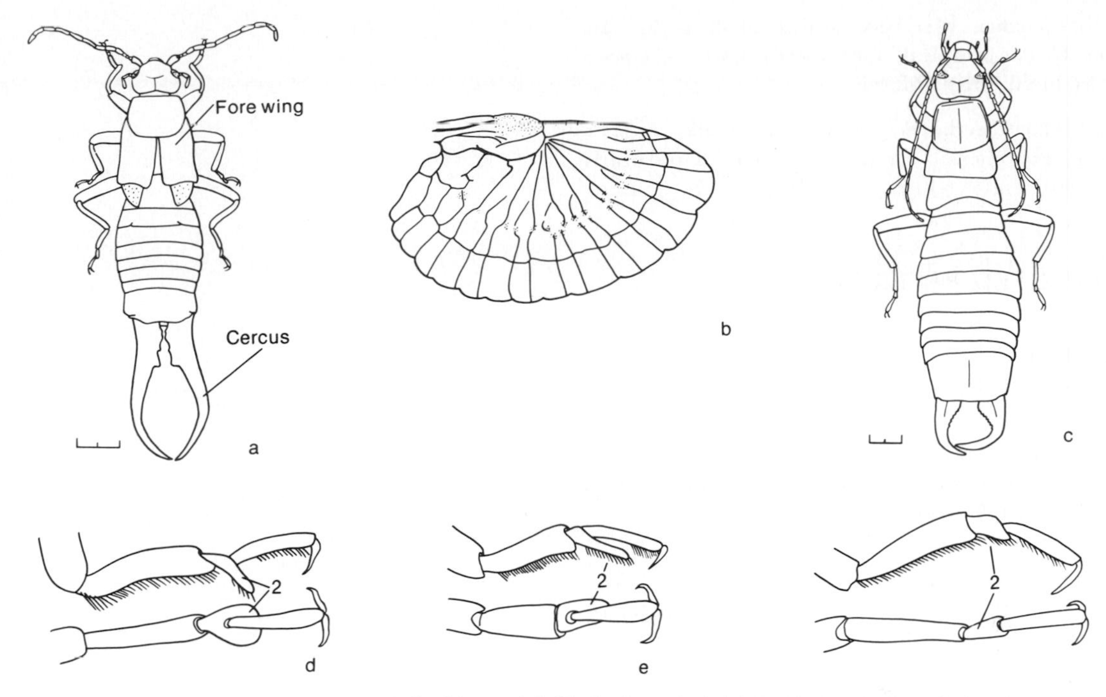

FIGURE 26.1 Dermaptera: *a,* Forficulidae, adult (*Forficula auricularia*) (scale equals 2 mm); *b,* wing, Forficulidae (*Doru linearis*); *c,* Labiduridae (*Anisolabis maritimus,* male (scale equals 2 mm)); *d,* tarsus, Forficulidae (*Forficula*); *e,* tarsus, Chelisochidae (*Chelisoches*); *f,* tarsus, Labiduridae (*Labidura*).

biological requirements. The order is notable for the development of maternal care of the young. Females oviposit clutches of ovoid, delicate eggs in self-constructed burrows or retreats, then protect the eggs and first-instar nymphs for several weeks. Nymphs pass through four or five instars in various species. *Arixenia* and *Hemimerus* are parasites or nest associates of mammals. Both show reduction of sense organs, flattening of the body, and other morphological modifications associated with parasitism. In both genera embryonic development is internal, with first-instar nymphs appearing viviparously.

KEY TO THE SUBORDERS AND NORTH AMERICAN FAMILIES OF

Dermaptera

1. Eyes vestigial or absent; cerci not forcipulate — Suborders **Arixeniina** and **Hemimerina**
   Eyes well developed; cerci forcipulate, at least in adults — (Suborder **Forficulina**) 2

2. Second tarsal segment produced longitudinally or laterally (Fig. 26.1*d, e*) — 3
   Second tarsal segment cylindrical, not produced (Fig. 26.1*f*) — 4

3. Second tarsal segment expanded laterally (Fig. 26.1*d*) — **Forficulidae**
   Second tarsal segment produced longitudinally below third (Fig. 26.1*e*), not expanded laterally — **Chelisochidae**

4. Antennae with segments 4 to 6 together longer than first segment — **Labiidae**
   Antennae with segments 4 to 6 together subequal to or shorter than first segment — **Labiduridae**

Labiduridae. A large, cosmopolitan family with many species spread by commerce; antennae with 15 to 25 segments; nymphs with forcipulate cerci. Many species are apterous. *Anisolabis* (Fig. 26.1*c*), *Euborellia*, and *Labidura* are introduced in North America.

Labiidae. A large, heterogeneous assemblage; antennae usually with 15 to 20 (25) segments; usually winged. *Labia* and *Prolabia* are common introduced members of this cosmopolitan group.

Chelisochidae. Members of this small family are mostly paleotropical. One species *Chelisoches morio* introduced into California.

Forficulidae. A large, cosmopolitan family. *Forficula auricularia* (Fig. 26.1*a*) has been introduced into nearly all temperate regions, including North America.

SELECTED REFERENCES

GENERAL

Chopard, 1949.

BIOLOGY AND ECOLOGY

Eisner, 1960 (defensive behavior); Fulton, 1924; Guppy, 1950 (biology of *Anisolabis*).

SYSTEMATICS AND EVOLUTION

Blatchley, 1920 (species of Northeastern states); Giles, 1963 (higher classification); Hebard, 1934 (Illinois species); Langston and Powell, 1975 (California species); Popham, 1965 (families and subfamilies, world).

ORDER PLECOPTERA (Stoneflies)

27

Neopterous Exopterygota with aquatic naiads, terrestrial adults. *Adults* (Fig. 27.1*a*) with head hypognathous or prognathous with mandibulate mouthparts, bulging, lateral compound eyes, 2 to 3 ocelli, long, filamentous antennae with numerous segments. Thorax with 3 subequal segments composed of unmodified tergal, pleural, and sternal sclerites; wings membranous; forewing narrower than hind wing, with anal region much reduced; venation of longitudinal veins with complex network of crossveins, or with crossveining absent, wing folding of longitudinal pleats; legs robust, long, with 3-segmented tarsi bearing 2 terminal claws. Abdomen weakly sclerotized, cylindrical or flattened, with 10 similar segments and cerci of 1 to many segments. Thorax and/or abdomen often with remnants of nymphal gills. *Naiads* (Fig. 27.1*b*) similar to adults, but with gills usually larger or more extensive, cerci longer, wings and genitalia undeveloped.

Plecoptera are rather obscure as adults, though some species are present throughout most of the year. The nymphs, which are familiar to fishermen and anyone who has turned stones in cool brooks, are important in food chains in aquatic ecosystems.

Stoneflies are extremely generalized in most features of their morphology. Their unmodified, mandibulate mouthparts (Fig. 27.2), with two-lobed maxilla and four-lobed labium, each with well-developed palps, are similar to those of Orthoptera. The multiarticulate antennae and cerci, unspecialized cursorial legs, and (in primitive forms) complex network of wing veins are shared, at least in part, with primitive insects such as Ephemeroptera and Grylloblattodea. Plecoptera are notable for the primitive structure of the thoracic pleura, which are not divided into episternum and epimeron in the prothorax. The single, crescentic pleurite is suggestive of the subcoxal arc of primitive insects such as silverfish. Venation is most primitive in the Family Eustheniidae, presently restricted to Australia, New Zealand, and South America, but Pteronarcidae of the North American fauna show a relatively complete venation (Fig. 27.3*a*). In other families, especially Nemouridae, venation is variably reduced (Fig. 27.3*b* to *d*).

Nymphs (Fig. 27.1*b*) of most stoneflies respire through the general body surface, but many bear external tracheated gills. The gills vary from straplike, unbranched filaments to profusely branched tufts, and may occur on the head, thorax, legs, or abdomen, or in the rectum. Differences in the distribution and structure of gills are of uncertain functional significance but are extremely useful in identification.

Adult stoneflies usually remain near the streams or lakes from which they emerged. Many species rest exposed on stones or vegetation, but others hide in cracks and crevices, and may be encountered in such unexpected places as beneath slightly loose bark. Most species are weak, fluttery fliers, and many will run rather than fly to elude predators. Most stoneflies are diurnal, but some members of the Suborder Setipalpia are nocturnal. Food of adults consists of algae, lichen, and foliage, though many species appear not to feed.

Emergence is highly seasonal for most species, and since the adult life-span is usually only a few

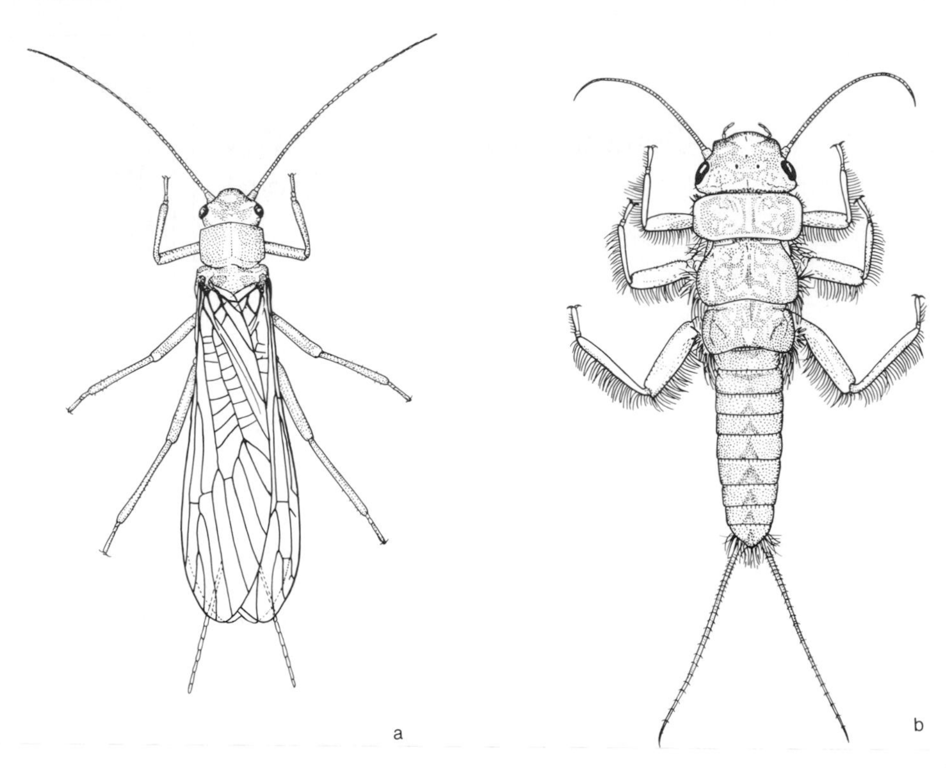

FIGURE 27.1 Representative adult and nymphal stoneflies: *a*, adult of *Isoperla confusa* (Perlodidae); *b*, naiad of *Acroneuria pacifica* (Perlidae). (a *from Frison*, 1935; b *from Claassen*, 1931.)

weeks, the composition of stonefly faunas is highly characteristic of time and location. Many species emerge during the coldest months, retaining normal abilities to crawl or fly, copulate and oviposit, even during periods of prolonged subfreezing temperatures (Sailer, 1950). Brachyptery, or winglessness, is not uncommon among these "winter stoneflies."

In some species mating is preceded by courtship involving drumming of the male abdomen against the substrate, sometimes producing an audible sound. Copulation, which may occur several times, takes place on substrates, with the male astride the female. Males of some species seek female nymphs, then immediately copulate with the freshly emerged adult before the exoskeleton has hardened.

Eggs, numbering from several hundred to several thousand per individual, are extruded into a ball temporarily carried on the venter of the abdomen, which is curled up so that the egg mass appears to be dorsal. Eggs are later dropped into water during flight, washed off by dipping the abdomen just beneath the surface, or deposited when the female crawls into the water, as in wingless species. Eclosion usually occurs after 2 to 3 weeks, or even several months, but in some species eggs are not deposited until they are ready to hatch.

Nymphs develop slowly, requiring about a year to reach maturity in many species, and up to 3 to 4 years in others. Full growth requires numerous molts. From 22 to 33 instars, differing in the number of gills and segments in the antennae and cerci, are recorded for North American species. Stoneflies are most abundant in cool streams with moderate or swift current, but also occur around lake margins and in warm, sluggish streams. Stony, irregular bottoms support the most varied and densest fauna. In general, stonefly nymphs require well-oxygenated water and are among the first aquatic life to disappear in polluted streams.

Most stonefly nymphs are primarily herbivorous, consuming submerged leaves, algae, diatoms, and similar items. Nymphs of Perlidae and Chloroperlidae are commonly predaceous, favoring mayfly nymphs as prey. These feeding differences are reflected in mouthpart structure. Herbivorous species have short, stout mandibles with a pronounced molar lobe (Fig. 27.2*b*); predatory species have elongate, sickle-shaped mandibles with the molar lobe reduced or absent (Fig. 27.2*e*). A few days before adult emergence feeding ceases, and shortly before the last molt the nymph crawls onto an emergent stone, log, or vegetation, often several feet from the water surface.

About 1500 species of Plecoptera have been described worldwide, with about 300 species in North America north of Mexico. The order is most abundant and diverse in cool temperate regions, and in North America is best represented in the north and in the western mountains, although other species are characteristic of the Great Plains, midwest, and southeast. The classification adopted here follows Ricker (1952, 1959). Nymphs and adults should be preserved in 70 percent alcohol. The exuviae of the last nymphal instars may be softened in warm water to relax the gills, and are nearly as useful as nymphs for identification. Nymphs are superficially similar to those of Ephemeroptera, but differ by having no median caudal filament, two tarsal claws, and gills, if present, on the head, thorax, and abdomen, but not along the sides of the abdomen (compare Figs. 27.1*b* and 20.4).

KEY TO NORTH AMERICAN FAMILIES OF **Plecoptera**

ADULTS

1. Labium with glossae and paraglossae of approximately equal length (Fig. 27.2*a*) (Suborder **Filipalpia**) 2

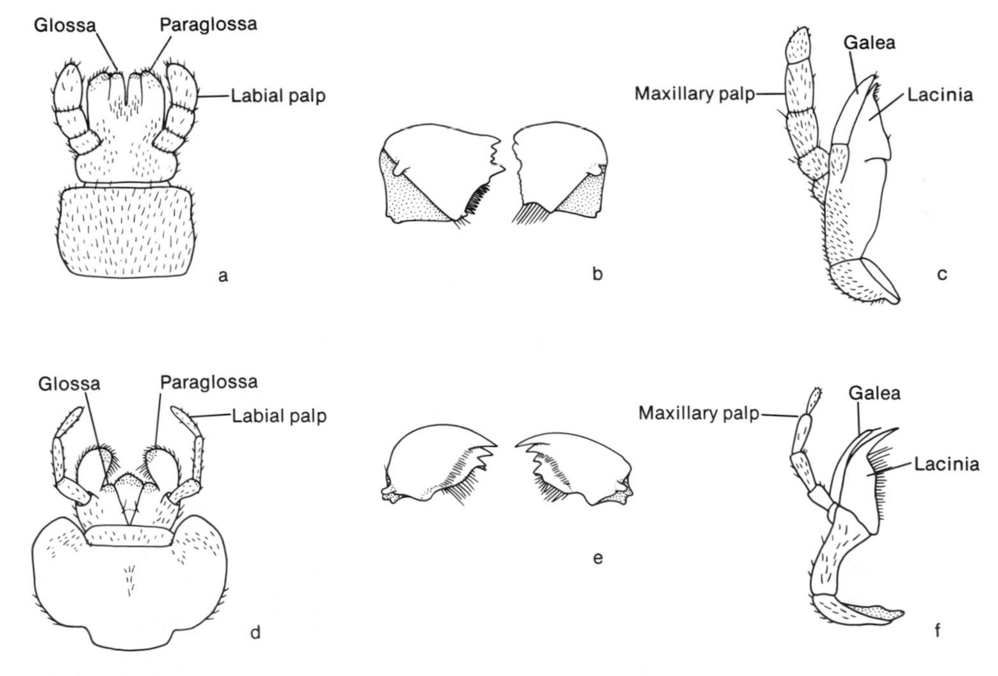

FIGURE 27.2 Taxonomic characters of stonefly naiads. (*Left to right*) *a, d,* labia; *b, e,* mandibles; *c, f,* maxillae. *a* to *c,* Filipalpia (Nemouridae, *Nemoura venosa*); *d* to *f,* Setipalpia (Perlidae, *Perlinella drymo*). (*From Claassen,* 1931.)

Labium with glossae much shorter than paraglossae (Fig. 27.2*d*) (Suborder **Setipalpia**) 4

2(1). Abdomen with branched gill remnants on venter; fore coxae contiguous or nearly so; forewing with at least 2 rows of anal crossveins (Fig. 27.3*a*) **Pteronarcidae**

Abdomen without gills or with unbranched gills; fore coxae widely separated; forewing with no more than 1 row of anal crossveins (Fig. 27.3*b,c,d*) 3

3(2). Head with 2 ocelli; forewing with at least 10 costal crossveins (Fig. 27.3*b*); broad, flattened, cockroach-like in form **Peltoperlidae**

Head with 3 ocelli; forewing with fewer than 10 costal crossveins, except in *Isocapnia;* slender, elongate in form **Nemouridae**

4(1). Thoracic sterna or pleura with profusely branched gill remnants at corners; forewing with cubitoanal crossvein intersecting anal cell (Fig. 27.3*b*) or distad from it by no more than its own length **Perlidae**

Thoracic segments without gills or with unbranched, straplike gills; forewing with cubitoanal crossvein distad from anal cell by its own length or more (Fig. 27.3*c*) or absent 5

5(4). Forewing with veins 2A and 3A arising from anal cell separately (Fig. 27.3*c*); gill remnants sometimes present on thorax or abdomen **Perlodidae**

Forewing with veins 2A and 3A arising from anal cell as single vein (Fig. 27.3*d*); gill remnants absent **Chloroperlidae**

NYMPHS

1. Labium with glossae and paraglossae of approximately equal length (Fig. 27.2*a*); maxilla with lacinia bearing blunt teeth (Fig. 27.2*c*) (Suborder **Filipalpia**) 2

Labium with glossae much shorter than paraglossae (Fig. 27.2*d*); maxilla with lacinia bearing sharp spines (Fig. 27.2*f*) (Suborder **Setipalpia**) 4

2(1). Gills finely dissected; present on the venter of the thorax and of the first 2 or 3 abdominal segments **Pteronarcidae**

Gills absent on abdominal segments 1 to 3; when present on thorax, not finely dissected 3

3(2). Head with 3 ocelli; thoracic segments with sterna short, not produced posteriorly **Nemouridae**

Head with 2 ocelli; thoracic segments with sterna produced posteriorly over following segment **Peltoperlidae**

4(1). Thoracic sterna with profusely branched gills at corners and usually above coxae; paraglossae rounded **Perlidae**

Thoracic gills as single or double filaments or straps; paraglossae pointed 5

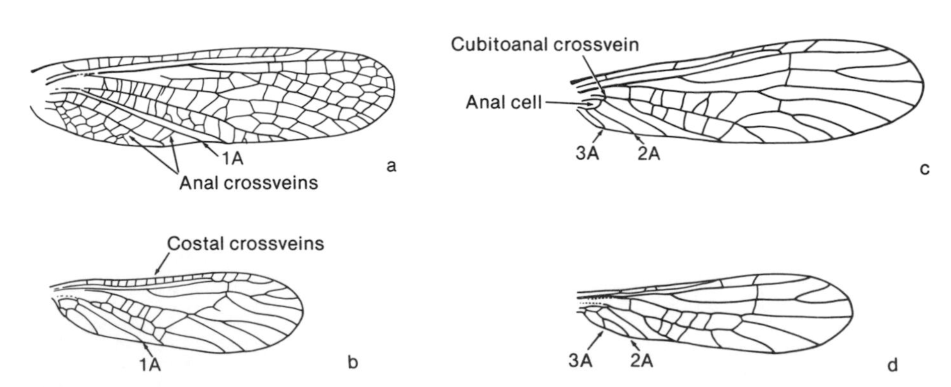

FIGURE 27.3 Wings of stoneflies: *a,* Pteronarcidae (*Pteronarcys proteus*); *b,* Peltoperlidae (*Peltoperla mariana*); *c,* Perlodidae (*Isoperla bilineata*); *d,* Chloroperlidae (*Utaperla sopladora*). (a, c *from Needham and Claassen,* 1925; b, d *from Ricker,* 1943, 1952.)

5(4). Hind wing pads with lateral margin parallel to long axis of body, or nearly so; cerci not more than three-fourths as long as abdomen **Chloroperlidae**

Hind wing pads with lateral margin diverging from axis of body; cerci usually at least as long as abdomen **Perlodidae**

SELECTED REFERENCES

GENERAL

Needham and Claassen, 1925

MORPHOLOGY

Wu, 1923.

BIOLOGY AND ECOLOGY

Holdworth, 1941 (biology of *Pteronarcys*); Hynes, 1976 (general); Newcomer, 1918 (economic); Sailer, 1950 (low-temperature activity).

SYSTEMATICS AND EVOLUTION

Claassen, 1931 (naiads); Frison, 1935, 1942 (Illinois); Gaufin et al., 1966 (Utah); Harden and Mickel, 1952 (Minnesota); Illies, 1965 (phylogeny, biogeography); Jewett, 1956 (California), and 1959 (Northwest); Merritt and Cummins, 1978 (North America); Needham and Claassen, 1925 (general); Ricker, 1959 (general).

28

ORDER EMBIOPTERA (Web Spinners, Embiids)

Small to moderate-sized Exopterygota, with subcylindrical, elongate body modified for living in tubular galleries (Fig. 28.1*a*). Head prognathous with mandibulate mouthparts, 12 to 32 segmented antennae, kidney-shaped eyes; ocelli absent. Prothorax small, narrower than head; mesothorax and metathorax broad, flattened, and usually bearing wings in male but narrower, elongate, without wings in female; wings long, narrow, with similar, simple venation (Fig. 28.1*c* to *e*); legs short, robust, tarsi 3-segmented (Fig. 28.1*b*), forelegs with basal tarsomere enlarged, globular; hind legs with femur enlarged. Abdomen with 10 subequal segments, 2-segmented cerci on segment 10. Nymphs similar to adults except in development of wings and genitalia.

Embioptera are monotonously similar in body form but vary in size and coloration. Males of some species are attracted to lights, but most are obscure because of their small size, dull color, and retiring habits. This predominantly tropical order is poorly represented in the Temperate Zones, and except for introduced species is absent from islands, presumably because of the lack of wings in females. Although the world fauna included fewer than 200 named species in 1970, it seems likely that as many as 1800 species remain undescribed (E. S. Ross, 1970). The United States fauna comprises seven native species and three others introduced from the Old World.

With the exception of dispersal by adult males, all instars of all species of Embioptera exclusively inhabit self-constructed silken galleries. This narrow specialization for a single niche is strongly reflected in the overall morphological uniformity of embiids. The most obvious modification involves the fore tarsi (Fig. 28.1*b*), which contain the silk-producing glands. Each swollen basitarsomere may contain up to about 75 gland cells, each connected to a hollow spinning bristle by a minute cuticular tubule. Silk production is involuntary, and since each tarsal bristle extrudes an individual fiber, a broad swath of silk is available continuously. The forelegs are rapidly shuttled back and forth, and the body is rotated as galleries are extended onto new food sources. Additional silk is continually added to old galleries, which eventually become opaque from multiple laminations of silk. As individuals grow they progressively construct larger tunnels, so that an active embiid colony contains a maze of passageways of various sizes and ages (see Endpaper 3*a*).

Rapid backward movement within the narrow passageways is aided by the highly tactile cerci and strongly developed tibial depressor muscles of the swollen hind femora. In alate individuals, the wings are soft and flexible, allowing them to flex quickly over the head during backward motion. During flight the wings are stiffened by inflation of a blood sinus formed by the radial vein. The remaining veins are relatively weak, with few branches, and with almost no venational differences between fore- and hind wings. In most other morphological features embiids are somewhat similar to earwigs (Dermaptera) and stoneflies (Plecoptera), indicating a very early divergence from the orthopteroid lineage.

Embioptera are strongly gregarious, a typical colony consisting of one or more adult females

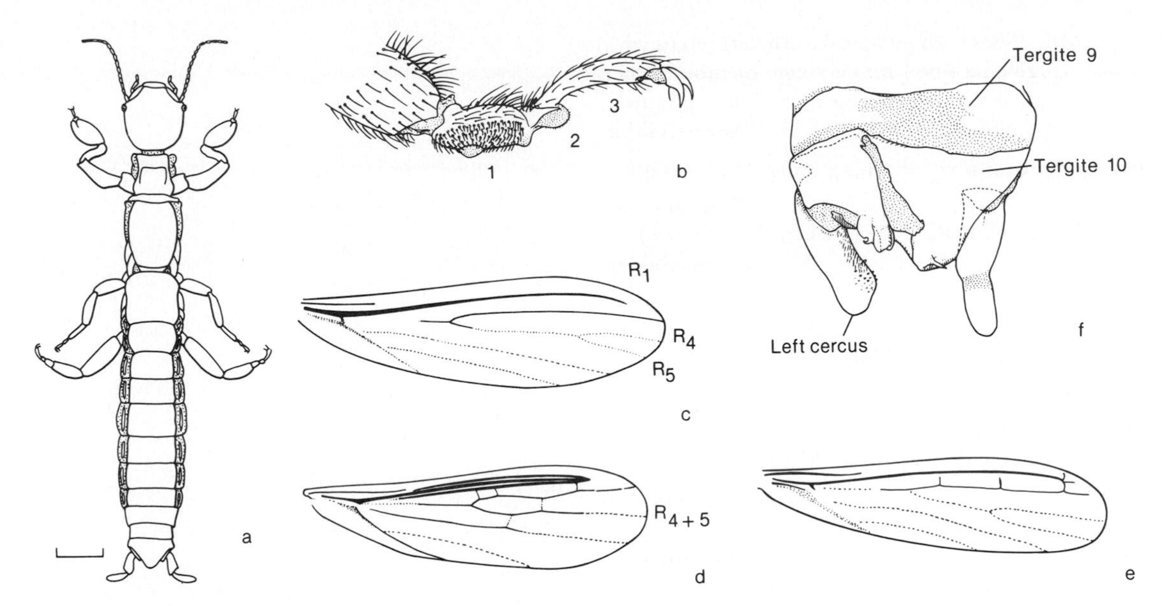

FIGURE 28.1 Embioptera: *a*, adult female of *Haploembia solieri* (Oligotomidae; scale equals 1 mm); *b*, fore tarsus of *Gynembia tarsalis*; *c*, forewing of Teratembiidae (*Oligembia brevicauda*); *d*, forewing of Anisembiidae (*Anisembia schwarzi*); *e*, forewing of Teratembiidae (*Teratembia geniculata*); *f*, cerci of Anisembiidae (*Chelicera galapagensis*). (b to f *from Ross*, 1940, 1952, 1966.)

with offspring of various numbers and sizes. A colony is usually centered around a secure retreat, such as a crevice beneath a stone or a crack in a log. Passageways are extended to close-by leaf litter, bark, lichen, moss, or other vegetable material. In humid tropical habitats webways may run exposed over bark or other irregular surfaces, sometimes covering considerable areas. In North America, colonies are normally relatively small and concealed beneath stones, logs, or other shelters. In arid regions the galleries extend down cracks into cooler soil regions.

Eggs are laid within the galleries and are sometimes covered by masticated cellulose. Eggs and newly emerged nymphs are briefly tended by the mother. The sexes are similar until wing buds develop in males. At maturity males leave the home colony for a short dispersal flight before entering another colony. They do not feed, and after mating they live only a short time, frequently being consumed by the females. The elongate, prognathous mandibles of male embiids are used to grasp the head of the female during mating. The highly complex, asymmetrical male cerci (Fig. 28.1*f*) of some species function as clasping organs during copulation. Parthenogenesis is recorded in a few species, and wingless males occur in others, especially those inhabiting arid regions.

In North America the order is restricted to the Southern states, including California. The best differentiating characters are found in the wings, genitalia, and mouthparts of adult males. The following key is based on the work of E. S. Ross (1940, 1944) and Davis (1940).

KEY TO THE NORTH AMERICAN FAMILIES OF

Embioptera

ADULT MALES[1]

1. Mandibles with apical teeth; left cercus with 2 segments; inner apical surface of cercus smooth 2

[1] *Haploembia solieri* (Oligotomidae), common in the Southwest, is parthenogenetic. In all instars the hind basitarsi have two ventral papillae. One papilla is present in all other North American species.

Mandibles without apical teeth; left cercus with segments fused into a single member; inner apical surface of cercus lobed and minutely spiculate (Fig. 28.1*f*) **Anisembiidae**

2(1). Wings with R_{4+5} unbranched (as in Fig. 28.1*d*) **Oligotomidae**

Wings with R_{4+5} 2-branched (Fig. 28.1*c,e*) **Teratembiidae**

SELECTED REFERENCES

MORPHOLOGY

Barth, 1954 (tarsal silk glands); Mukerji, 1927.

BIOLOGY AND ECOLOGY

Mills, 1932; Mukerji, 1927; E. S. Ross, 1944, 1970.

SYSTEMATICS

Davis, 1940; E. S. Ross, 1940 (North American species), 1944 (New World), 1957 (California species), and 1970 (general).

ORDER ORTHOPTERA (Grasshoppers, Crickets, etc.)

29

Usually medium to large Exopterygota with subcylindrical, elongate body; hind legs enlarged for jumping (Fig. 29.1). Head hypognathous, with compound eyes; ocelli present or absent; antennae multiarticulate; mouthparts mandibulate. Prothorax large, with shieldlike pronotum curving ventrally over pleural region; mesothorax small, with wings narrow, cornified; metathorax large, wings broad, with straight, longitudinal veins, numerous crossveins; wing folding by longitudinal pleats between veins; tarsi with 1 to 4 segments. Abdomen with first 8 or 9 segments annular, terminal 2 to 3 segments reduced; cerci with 1 segment. Nymphs similar to adults except in development of wings and genitalia.

Many Orthoptera, especially tropical species, are bizarrely modified to mimic living or dead vegetation or other insects (Fig. 15.4*b*). In subterranean or myrmecophilous forms wing loss may be accompanied by lack of differentiation of thoracic and abdominal segments, and body proportions may be quite different than in surface dwelling types. Yet nearly all Orthoptera are readily distinguished from other insects by the enlarged hind femora and characteristic grasshopperlike shape. The Orthoptera are easily the dominant group of mandibulate hemimetabolous insects, with about 20,000 species distributed throughout the world. Most species are terrestrial herbivores, but various adaptations for predation or for subaquatic or subterranean life characterize many genera or tribes.

A majority of Orthoptera conform to two broad adaptive modes, which correspond taxonomically to suborders. The **Caelifera** include mostly alert, diurnal forms with acute vision and hearing, which rely on jumping to escape from predators. Antennae and legs are relatively short, allowing unimpeded, rapid movement. These species, exemplified by the Acrididae (grasshoppers and locusts), are predominantly terrestrial (as opposed to arboreal). They are especially adapted to exposed situations in open habitats, such as savannah woodlands, steppes, and deserts. The great majority are herbivores; many

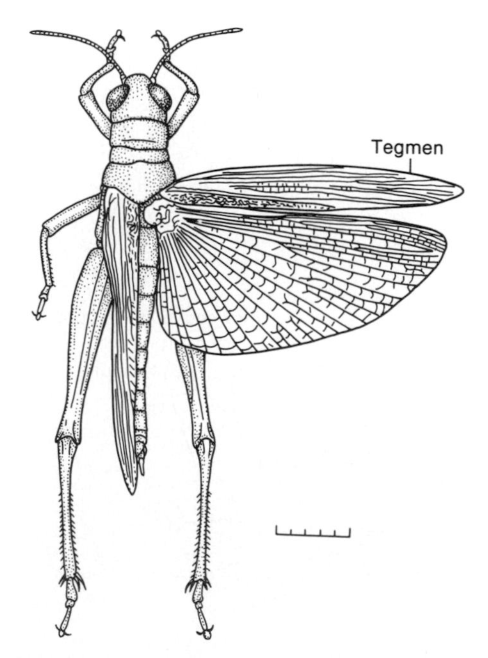

FIGURE 29.1 A grasshopper, *Melanoplus cinereus*, Acrididae, dorsal aspect. Scale equals 5 mm.

have specialized dentition for consuming the tough foliage of monocotyledonous plants.

In contrast to the Caelifera, the **Ensifera** (katydids, crickets) tend to rely on crypsis to avoid predators. Most species are nocturnal, either retreating to protected niches during the day, or escaping notice through camouflage or mimicry. Antennae and legs are long, sometimes extremely so; tactile responses are highly developed, though vision and hearing are also important. Movements tend to be slow and careful, but become abrupt and vigorous upon physical contact. Many Ensifera are omnivorous or predatory. Phytophagous species usually feed on stems or foliage of dicotyledonous plants, rather than grasses.

Hearing is acutely developed in most Orthoptera, and auditory stimuli are important in courtship in many species. In the Caelifera, auditory organs (if present) are on the first abdominal tergite, beneath the folded wings (Fig. 29.2*b*). Sound production is by stridulatory structures located on the lateral, vertical part of the forewings. In the Ensifera, auditory organs are located on the anterior tibiae (Fig. 29.2*a*). Stridulation is accomplished by vibrating the horizontal, overlapping parts of the forewings over one another (Fig. 6.7*d,e*).

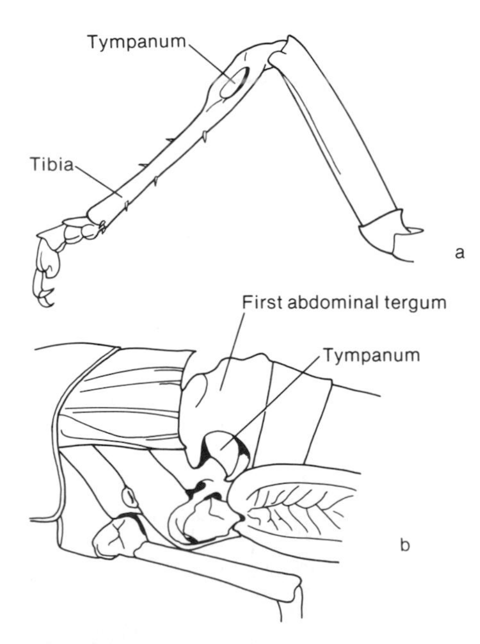

FIGURE 29.2 Auditory organs: *a*, Tettigoniidae (*Microcentrum californicum*); *b*, Acrididae (*Dissosteira carolina*).

Intraspecific polymorphism is common in Orthoptera, especially in the Caelifera. In the extreme case, distinct differences in behavior and physiology, as well as in color pattern and body proportions, may occur in response to population fluctuations and ecological conditions. Such phase transformation is very pronounced in periodically migratory grasshoppers such as the desert locust (*Schistocerca gregaria*) and occurs in species from all geographic regions (see Chap. 4). More typically, polymorphism involves only color, cuticular sculpturing, or wing length. The function in many aspects is obscure.

Nymphs are very similar to adults, except in the less-developed wings and genitalia. Immatures of apterous species may be very difficult to distinguish from adults. A peculiarity shared only with Odonata is that the orientation of nymphal wing pads changes during development. In early instars the costal margin is lateral. Later, usually at the penultimate molt, the costal margin is turned inward; subsequent reversal to the lateral position occurs at the adult molt.

Sound communication is a frequent adjunct to courtship (see Chap. 6) and stridulatory mechanisms are frequently restricted to males. Copulation occurs with the male astride the female and may endure for several hours or longer. Orthopteran eggs are typically elongate or oval. Eggs are usually inserted singly into plant tissue or soil substrates in the Ensifera. In the Caelifera clutches of 10 to several hundred eggs are buried in shallow burrows excavated by the ovipositor valves. Egg diapause is widespread in Orthoptera.

KEY TO THE NORTH AMERICAN FAMILIES OF
Orthoptera

1. Antennae usually with many more than 30 segments (Suborder **Ensifera**) 2
   Antennae usually with many fewer than 30 segments (Suborder **Caelifera**) 4

2(1). Tarsi with 4 segments, at least on middle and posterior legs 3
Tarsi with 3 segments **Gryllidae**

3(2). Anterior tibiae with auditory organs (Fig. 29.2*a*), without ventral, articulated spines **Tettigoniidae**
Anterior tibiae without auditory organs, usually with ventral, articulated spines **Gryllacrididae**

4(1). Tarsi with 3 segments 5
Tarsi with 1 or 2 segments **Tridactylidae**

5(4). Pronotum extended backwards to cover wings **Tetrigidae**
Pronotum not prolonged backwards over wings and abdomen **Acrididae**

Suborder Ensifera

Tettigoniidae (katydids). This variable family comprises approximately 5000 species predominantly from the Tropics. Most katydids are arboreal, winged insects, but aptery is common. Many Tettigoniidae are strikingly mimetic. Different species mimic ants and wasps as well as living or dead leaves, flowers, or lichen. Those resembling other insects often display unusual, mimetic behavior. Most familar North American species are relatively unmodified, green or brown katydids (Fig. 29.3*d*).

Gryllidae (crickets; Fig. 29.3*a*). Approximately 1200 species, distributed worldwide; includes both winged and wingless forms. Most species live on the ground under logs or stones or in dense vegetation, but others are arboreal or subterranean. The Subfamily Gryllotalpinae, with the antennae much shorter than the body and the forelegs flat and platelike, burrow in moist soil. The Myrmecophilinae, with reduced eyes and coxae almost contiguous ventrally, inhabit ant nests. Both subfamilies are sometimes accorded family status.

Gryllacrididae (camel crickets, etc.) North American representatives are apterous (Fig. 29.3*c*), but winged species occur in other areas. Most species are at least partly subterranean, and many are restricted to caves. One North American species constructs shelters of rolled leaves tied with silk. *Stenopelmatus,* of Pacific North America, lacking an obvious ovipositor, is also distinguished by the bicolored, banded abdomen. This genus is often classified as a separate family, the Stenopelmatidae. The Gryllacrididae include approximately 1000 species worldwide.

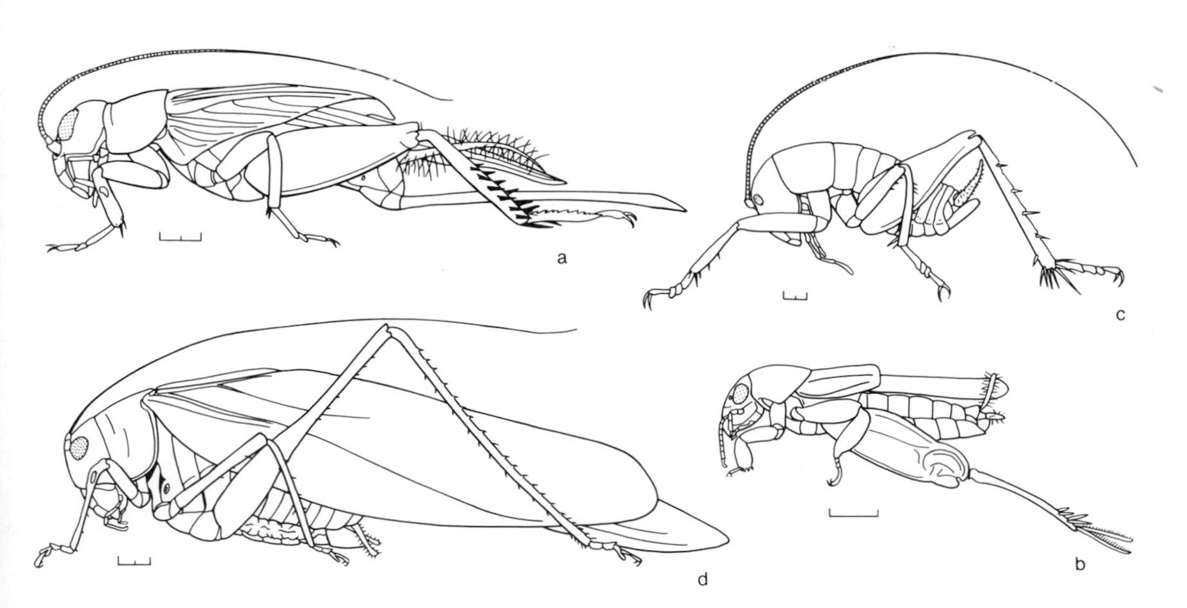

FIGURE 29.3 Representative Orthoptera: *a,* Gryllidae (*Acheta assimilis,* female); *b,* Tridactylidae (*Tridactylus minutus*); *c,* Gryllacrididae (*Ceuthophilus californicus*); *d,* Tettigoniidae (*Microcentrum californicum,* male). Scales equal 2 mm, except that for *b* equals 1 mm.

Suborder Caelifera

Acrididae (grasshoppers). The largest family of Orthoptera, with over 10,000 species. The great majority are typical "grasshoppers," with minor structural variation from the acridid body plan. The Tanaocerinae and Eumastacinae, sometimes recognized as families, are wingless forms confined in North America to the southwest. The latter are recognized by the greatly elongate antennae; the former have the antennae shorter than the anterior femora.

Tetrigidae (pygmy locusts). A well-defined group of about 1000 species, mostly tropical or subtropical. Few are more than 15 mm long. Polymorphisms involving wing and pronotal development occur in many species, and many are grotesquely flattened and contorted. Most are associated with moist situations and are proficient swimmers. They feed on algae or detritus.

Tridactylidae (Fig. 29.3*b*). At least 75 species of this specialized family occur worldwide. All frequent margins of ponds or streams. Eggs are laid in burrows in moist sand. Tridactylids are excellent swimmers. Food is apparently algae.

SELECTED REFERENCES

GENERAL

Acridia, 1972 to date (articles on grasshoppers and locusts); Beier, 1955; Chopard, 1938, 1949; Uvarov, 1966.

MORPHOLOGY

Albrecht, 1953; Judd, 1948.

BIOLOGY AND ECOLOGY

Alexander, 1957, 1962 (sound production); E. J. Clark, 1948 (general); Iseley, 1944 (feeding specificity); Mulkern, 1967 (food selection).

SYSTEMATICS AND EVOLUTION

Alexander, 1962; Ball et al., 1942 (Arizona species); Blatchley, 1920 (Northeastern states); Brooks, 1958 (Canadian species); Dirsh, 1975 (higher classification); Froeschner, 1954 (Iowa species); Hebard, 1934 (Illinois species); Helfer, 1953; Rehn and Grant, 1961 (North American species); Sharov, 1971 (higher classification); Strohecker et al., 1968 (California species); T. J. Walker, 1963 (calling songs).

ORDER PHASMATODEA (Stick Insects)

30

Cylindrical, elongate or flattened, leaflike Exopterygota, mostly of moderate to large size. Antennae multiarticulate, filiform or moniliform; compound eyes present, ocelli present in some winged males; mouthparts mandibulate, prognathous. Prothorax short, mesothorax and metathorax often elongate (winged species) or short (wingless species); pronotum without ventrolateral extensions covering pleural region; mesothoracic wings narrow, cornified; metathoracic wings broad, with straight longitudinal veins, numerous crossveins, anterior margin cornified, wing folding longitudinal, between veins; wing reduction or aptery common. Legs usually elongate, slender; tarsi with 5 segments. Abdomen with 10 (occasionally 11) apparent segments; terminal 2 or 3 sternites concealed beneath subgenital plate; cerci with 1 segment, external or concealed. Nymphs similar to adults except in development of wings and terminalia.

Approximately 2500 species of phasmatids are known, predominantly from the Tropics, especially in the Oriental region. Walkingsticks appear to be relatively rare insects, even considering the extraordinary difficulty of finding them, and seldom become economically important. Superficially they are similar to some Orthoptera, but differ in having asymmetrical male genitalia (symmetrical in Orthoptera), in lacking wingpad reversal in the nymphs, and in proventricular structure.

Nearly all phasmatids are modified to mimic vegetation of various sorts. Many species strikingly resemble either leaves or twigs, but others are generally similar to foliage without clearly conforming to either morphological type. Integumental sculpturing is diversely modified as thorns, tubercles, ridges, and other processes or textures which heighten the similarity to vegetation. Body form ranges from extremely slender to foliate, or only the appendages may be foliate. Resemblance to plant parts is further enhanced by posturing, frequently in grotesque, asymmetrical positions. Such postures may be maintained, even when the insects are disturbed, or they may drop from their purchase and feign death. A few species, such as *Carausius morosus,* are able to effect color changes by altering the dispersal of pigment granules in the epidermis. A few others, such as *Anisomorpha buprestoides* (Fig. 30.1*b*) of the Southeastern United States, are aposematically colored. These active, agile insects secrete large quantities of a defensive substance (anisomorphal) in the prothoracic glands, readily spraying or oozing the milky product upon very slight disturbance. Yet another common defensive mechanism is limb autotomy, which takes place between the femur and trochanter. Partial leg regeneration may occur in nymphs.

All Phasmatodea are herbivores, typically leading solitary, sluggish lives on vegetation, especially trees and shrubs. Males are usually distinctly smaller than females. During copulation, which may be very prolonged, the male sits astride the female. Eggs numbering about 100 to 1200 in different species, are deposited singly, either glued to vegetation or dropped to the ground. Phasmatid eggs are distinctive, with the thick, tough chorion often sculptured or patterned. Egg diapause is common and may endure more than a single season. Newly hatched

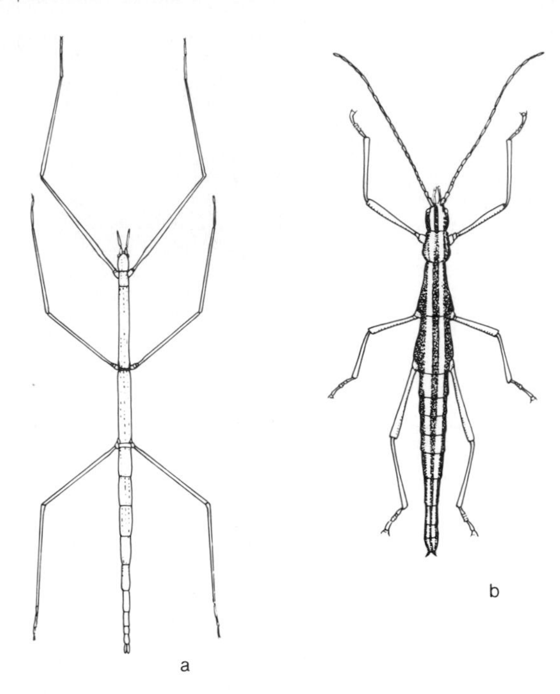

FIGURE 30.1 Representative Phasmatodea, Phasmatidae: *a, Parabacillus coloradus; b, Anisomorpha buprestoides.* (a *from Henderson and Levi,* 1938; b *from Caudell,* 1903.)

nymphs climb onto convenient vegetation. They are generally similar to adults, both in appearance and behavior.

Two families, both widespread, are recognized here. The PHYLLIIDAE include many leaflike species. There are no North American representatives. The PHASMATIDAE include species of diverse body form. Most North American species are sticklike, but *Anisomorpha* (Southeastern states) and *Timema* (California) are fusiform with relatively short legs. *Timema* is distinguished from all other genera in having the tarsi three-segmented.

SELECTED REFERENCES

Chopard, 1938, 1949 (general); Craighead, 1950 (biology of North American species); Littig, 1942 (morphology); Severin, 1911 (life history of *Diaphemeromera*).

ORDER ZORAPTERA

31

Minute to small Exopterygota with mandibulate, chewing mouthparts, 9-segmented moniliform antennae, compound eyes, and ocelli present or absent. Prothorax large, subquadrate; mesothorax and metathorax subquadrate or transverse; wings with large pterostigma and reduced venation or absent; legs adapted for walking, with 2-segmented tarsi. Abdomen with 11 similar segments, unsegmented cerci on segment 11. Nymphs similar to adults except in development of wings and/or genitalia.

These obscure insects resemble termites in general appearance, and frequently occur about decaying wood. The order is known from the single family ZOROTYPIDAE, which includes the 22 species of the genus *Zorotypus*. The species are distributed sporadically in most parts of the world except Australia.

Their unspecialized, chewing mouthparts and the presence of cerci indicate a relationship of the Zoraptera with the orthopteroid orders. The male genitalia, which are asymmetrical, suggest affinities with cockroaches and termites. However, the Malpighian tubules number only six, and the central nervous system is concentrated into two large thoracic and two abdominal ganglia, both features of the hemipteroid orders. The wings, when present, have the venation extremely reduced, somewhat as in Psocoptera or Thysanoptera, and the hind wings are smaller than the forewings, as in hemipteroids (Fig. 31.1*b*).

Zoraptera occur in moist leaf litter, in rotting wood or sawdust, and in or about termite colonies. Species which have been studied are polymorphic, one adult form being largely unpigmented and lacking wings, compound eyes, and ocelli (Fig. 31.1*c*). Winged forms are more darkly pigmented, with small compound eyes and three ocelli (Fig. 31.1*a*). The significance of polymorphism in Zoraptera is uncertain, but the winged forms probably function in dispersal and may develop when the substrate becomes unsuitable because of drying or other deterioration. Food apparently consists mostly of fungi, though arthropod fragments have also been found in the alimentary canal.

Eggs of Zoraptera are ovoid, without remarkable sculpturing or micropylar structures. Eclosion is by means of an egg burster on the head of the embryonic nymph. Polymorphism is evident in the nymphs (Fig. 31.1*d*), which develop external wing pads and have eyes and ocelli if destined to produce winged adults. The number of nymphal instars and other details of the life history and biology are unknown.

Two species occur in North America. *Zorotypus hubbardi* is widespread in the Southeastern United States. *Zorotypus snyderi* is known from southernmost Florida and Jamaica.

SELECTED REFERENCES

Caudell, 1920 (biology); Delamare-Deboutteville, 1948 (biology); Gurney, 1938 (systematics); Riegel, 1963 (distribution).

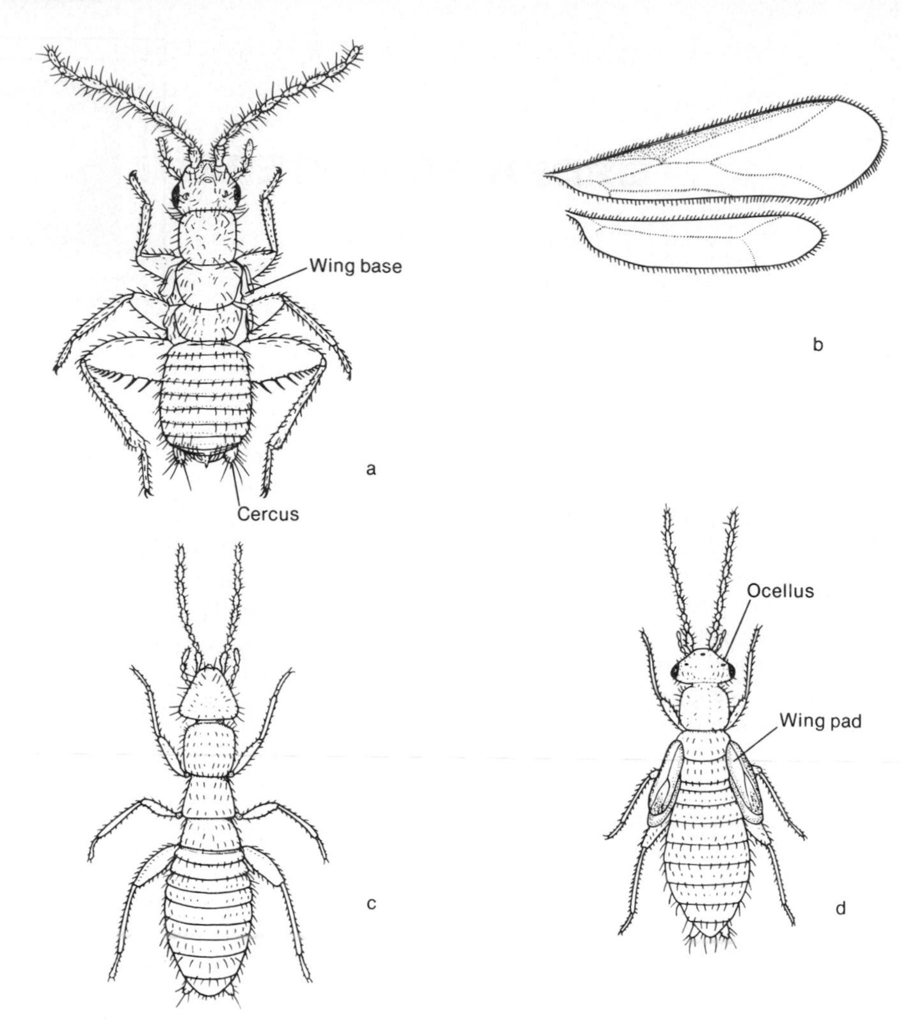

FIGURE 31.1 Zoraptera, Zorotypidae: *a,* dealate adult of *Zorotypus mexicanus; b,* wings of *Zorotypus hubbardi; c, d,* apterous adult and nymph of alate form of *Zorotypus hubbardi.* (a *from Bolivar y Peltain;* b, c, d *from Caudell,* 1920.)

ORDER PSOCOPTERA (Psocids, Book Lice)

32

Minute to small Exopterygota with stout, soft body and large mobile head with swollen clypeus (Fig. 32.1). Mandibles of chewing type, maxillae with lacinia modified as a slender rod, detached from stipes (Fig. 32.2*b*); compound eyes usually large, convex; 3 ocelli present in winged, absent in wingless forms; antennae filiform, usually with 13 segments; prothorax small, collarlike, especially in winged forms; mesothorax and metathorax subequal, sometimes with fused terga in wingless forms; forewings about 1.3 to 1.5 times longer than hind wings, with simplified venation; few crossveins or closed cells, pterostigma present; legs slender, adapted for walking, tarsi with 2 to 3 segments. Abdomen 9-segmented, without cerci. Nymphs similar to adults, except in the smaller number of antennal segments and ocelli, and undeveloped wings.

Because of their small size and retiring habits, Psocoptera frequently escape notice. However, they are exceedingly common animals in many habitats, and number over 1800 species worldwide. Most psocids are scavengers, and they often occur in buildings, where they occasionally cause minor damage to stored materials. For example, *Liposcelis*, the common book louse, sometimes infests insect collections or herbarium specimens, as well as invading dry, starchy foodstuffs.

Psocids are among the more generalized of hemipteroid insects, with chewing mandibles bearing well-developed molar and incisor lobes. The maxilla is specialized in having the detached lacinia modified as a stout, heavily sclerotized rod (Fig. 32.2*b*). Other maxillary structures are unmodified. The hypopharynx bears dorsally a concave *sitophore sclerite*, which is directly opposed by a cuticular knoblike structure on the dorsal wall of the cibarium (Fig. 32.2*a*). During feeding the rodlike laciniae are apparently shoved against the substrate as a brace to steady the head while lichen, fungal hyphae, algae, or detritus is scraped up with the mandibles. The sitophore apparatus provides an auxiliary grinding and crushing device. It may be pointed out here that the lacinia and sitophore in biting lice

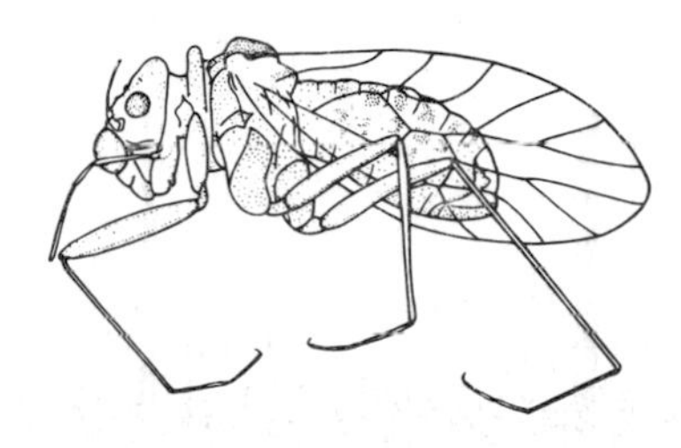

a

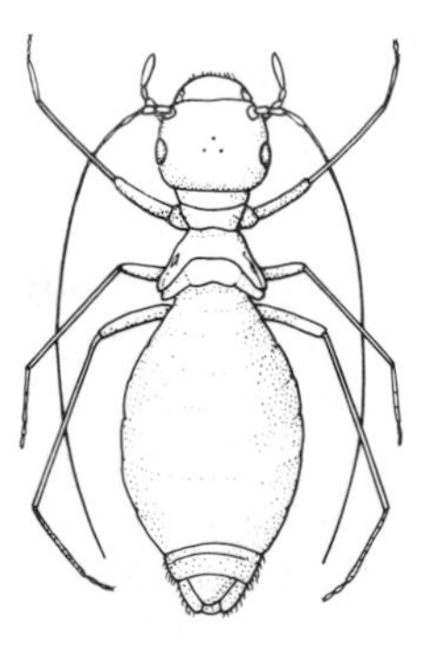

b

FIGURE 32.1 Representative adult Psocoptera: *a, Speleketor flocki;* b. *Psyllipsocus ramburi.* (*From Gurney,* 1943.)

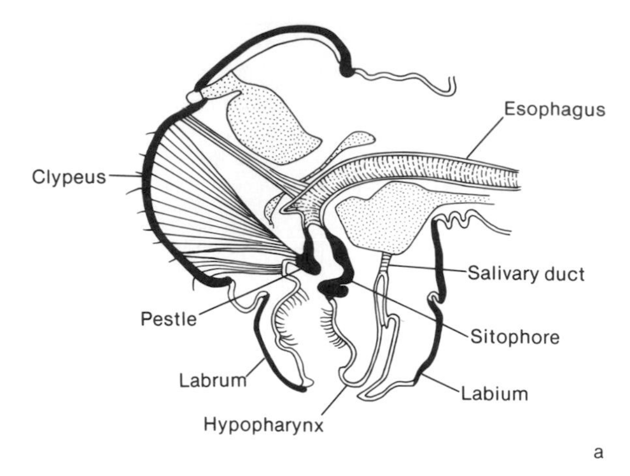

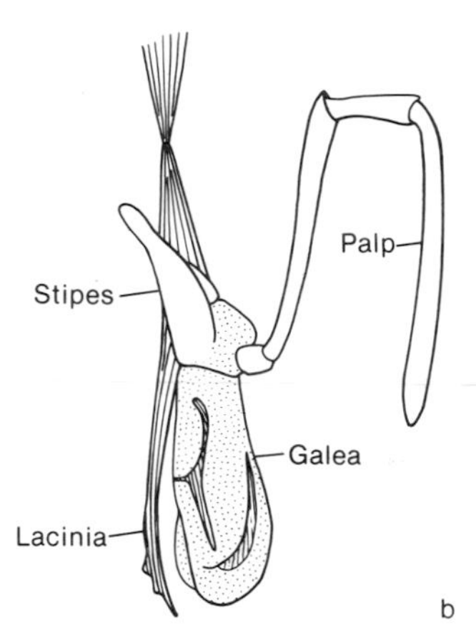

FIGURE 32.2 Oral structures: *a*, median section of head of *Ectopsocus briggsi*; *b*, maxilla of *Speleketor flocki*. (*From Snodgrass*, 1944.)

(Phthiraptera) are highly similar to the structures described above, and provide the best evidence of the origin of those highly specialized insects (see Chap. 33).

Psocids are frequently gregarious, mixed groups of winged or wingless adults and immatures of various sizes congregating in areas that are damp but not wet. Caves, hollow stumps, lichen-covered rock faces, and similar protected situations are favored by some species, and these frequently enter man-made structures. Other species frequent the foliage, twigs, and branches of trees and shrubs—hence the name "barklice." Fungi, especially bracket fungi on trees, leaf litter, and bird, mammal, wasp, and ant nests, provide suitable habitats for certain forms. Many species are rapid, nimble runners, often moving backward or forward with equal ease. At least some psocids have the ability to spin silk, which may be used to construct small canopies beneath which colonies shelter. When the insects are numerous, entire branches or trees may be fairly covered with unsightly but harmless webs.

Polymorphism is of frequent occurrence, involving either or both sexes. Wingless forms usually lack ocelli and may differ greatly in thoracic structure from winged individuals. Minor differences in development of various body setae and specialized sense organs have also been described. Unlike the seasonal changes of Homoptera, polymorphism in most Psocoptera is not obviously related to environmental factors and is of uncertain significance. Parthenogenesis is frequent, males being extremely rare or unknown in some species; in others both asexual and sexual races are known.

Mating is typically preceded by courtship dances by the male, which later copulates from beneath the female. Eggs are deposited singly or in small clusters in crevices in irregular substrates, or on foliage. A few species are viviparous. Six nymphal instars are typical. Species which breed out of doors usually have one to three generations per year, depending on climate, while those inhabiting man-made structures have continuous generations.

Identification of Psocoptera is difficult because of their small size, and slide-mounted specimens may be needed for examination of some of the important structures. Various authorities recognize different families, ranging from about 15 to 25 in number, and many genera have been repeatedly transferred among families. However, the great majority of commonly encountered species belong to a few families. The common book louse, *Liposcelis divinatorius*, is a member of the Liposcelidae (Suborder Troctomorpha). The Suborder Trogiomorpha contains many short-winged or wingless forms, many frequenting caves. A few species are common in granaries, barns, etc. Most of the common, winged species encountered outdoors are members of the Suborder Eupsocida.

KEY TO THE SUBORDERS OF
Psocoptera

ADULTS

1. Antennae with at least 20 segments; tarsi with 3 segments **Trogiomorpha**
   Antennae with 12 to 17 segments; tarsi 2- or 3-segmented 2

2(1). Tarsi 3-segmented; antennae with some segments secondarily ringed; wing with pterostigma not thickened **Troctomorpha**
   Tarsi 2-segmented, or if 3-segmented, antennal segments not ringed; pterostigma thickened **Eupsocida**

SELECTED REFERENCES

Cope, 1940 (morphology); Gurney, 1950 (general); Mockford, 1951 (Indiana species), and 1957 (biology); Mockford and Gurney, 1956 (Texas species); Pearman, 1928 (sound production); Sommerman, 1943 (biology).

33 ORDER PHTHIRAPTERA (Lice)

Small, wingless Exopterygota with the body strongly modified for ectoparasitism on vertebrates. Eyes reduced or absent, ocelli absent; antennae short, thick, 3- to 5-segmented; mouthparts mandibulate or highly modified for piercing and sucking. Thorax with segments poorly separated or fused, or with prothorax free, mobile; legs usually short, stout, often with tarsi modified, cheliform for clinging to host's pelage or plumage. Abdomen with 8 to 10 distinct segments, cerci absent. Nymphs similar to adults.

All members of this specialized order are permanent ectoparasites of birds and mammals, including domestic stock and fowls. Most species are scavengers of dried skin or other cutaneous debris. A relatively small number of species feed on blood, sebum, and other tissue fluids. All species can cause general irritation and may prove debilitating in severe infestations. The Anoplura are of special importance because of their ability to transmit epidemic typhus and other rickettsial diseases to humans; fortunately the prevalence of these diseases has decreased greatly since the advent of insecticides of low mammalian toxicity.

Lice are highly adapted to survive on the bodies of their hosts. The cuticle is tough and elastic, providing hydrostatic protection for the internal organs. Sensory structures are reduced or absent; antennae are short and can be tucked against the head; the body is usually streamlined or flattened. The tarsi are frequently shaped to cling to the pelage or plumage of the host, and in mammal-infesting species the tarsus and its single claw form a clamp for grasping hairs (Fig. 33.3*b* to *d*). Most of these modifications are characteristic of ectoparasitic insects from diverse orders and are evident in parasitic crustaceans as well.

The biting lice (Suborders **Amblycera** and **Ischnocera**) have clearly evolved from psocid ancestors, which they closely resemble in mouthpart configuration. As in psocids, the maxillary laciniae are sclerotized rods, detached from the stipes, and the hypopharynx and cibarium are specialized as a sitophore apparatus (Fig. 33.1*c*, *d*). These species have biting mandibles which are used to clip off the host's skin debris or feathers, or in some species to pierce or lacerate the host integument. In the sucking lice (Suborder **Anoplura**), the functional mouthparts consist of three slender stylets of uncertain homology (Fig. 33.2*a* to *c*). At rest these are withdrawn into a pouch in the head capsule. The oral opening bears small denticles which are anchored in the host's integument while the stylets are inserted. In the small Suborder **Rhynchophthirina,** whose members are parasites of elephants and warthogs, the very small mandibles are attached apically on an elongate rostrum.

Phthiraptera are dependent on their hosts for food, shelter, and oviposition sites; most of them survive only a short time if isolated from hosts. Many species are further restricted to limited parts of the host's body. For example, some bird lice occupy the hollow quills of the large primary wing feathers, while others infest the chin, nape, or other parts of the body that are inaccessible to preening. Many mammal-infesting lice range

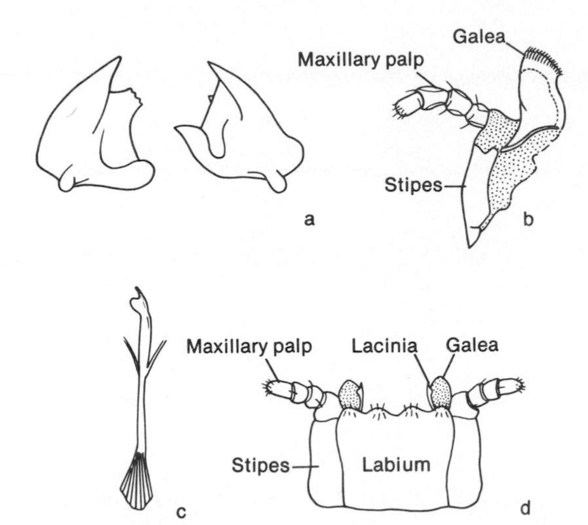

FIGURE 33.1 Mouthparts of Amblycera: *a,* mandibles of *Laemobothrion; b,* maxilla of *L. gypsis; c,* maxillary fork (lacinia) of *Ancistrona vagelli; d,* labium and maxillae of same. (*From Snodgrass,* 1944.)

widely over their hosts, whereas certain others are restricted to localized sites such as legs, tail, head, flippers, or even to the interdigital fossae of the feet, where their hosts have difficulty grooming. The species inhabiting humans occupy head or body hair, including pubic hair. Anoplura associated with marine mammals are unusual in being capable of surviving on land for long periods while their hosts remain at sea.

Most lice are bisexual, but males are frequently less common than females; a few species are known to be parthenogenetic. The number of eggs per individual varies from a few dozen to over 300 in different species, but usually only a few eggs are deposited each day. In most species the eggs are glued to hairs or feathers, or left in places inaccessible to preening and grooming, such as the interior of feather shafts. There are three nymphal instars, each usually molting after about 1 week.

No single classification of lice is generally accepted. The Suborders Amblycera, Ischnocera, and Rhynchophthirina are sometimes considered a single order, Mallophaga, coordinate with the Order Anoplura, which classification has the disadvantage of concealing the structural diversity within the Mallophaga. The most recent authoritative classifications do not use the term Mallophaga. Identification of families is frequently difficult without properly prepared slides. The key below includes suborders. Keys to families and genera appear in most medical entomology textbooks.

KEY TO THE SUBORDERS OF **Phthiraptera**

1. Head prolonged anteriorly as a rostrum, with mandibles at apex
Suborder Rhynchophthirina
Head not prolonged as a rostrum; small mandibles articulated ventrally on head capsule, or absent — 2

2(1). Head relatively small, narrower than prothorax, sometimes fused with thorax; mandibles absent **Suborder Anoplura** (Fig. 33.3*c,d*)
Head relatively large, wider than prothorax, free, mobile; mandibles adapted for chewing — 3

3(2). Antennae capitate, 4-segmented; head with grooves for receiving antennae; mandibles horizontal; maxillary palpi present
Suborder Amblycera (Fig. 33.3*b*)

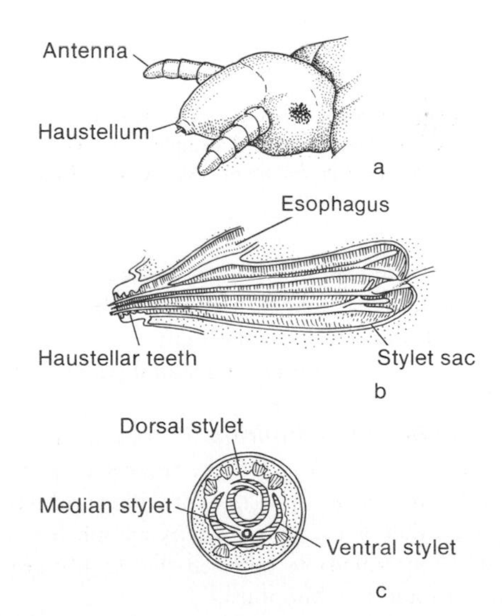

FIGURE 33.2 Mouthparts of Anoplura, diagramatic; *a,* head with sytlets protruding from oral opening; *b,* median section through head; *c,* transverse section near oral opening. (*From Furman and Catts,* 1961.)

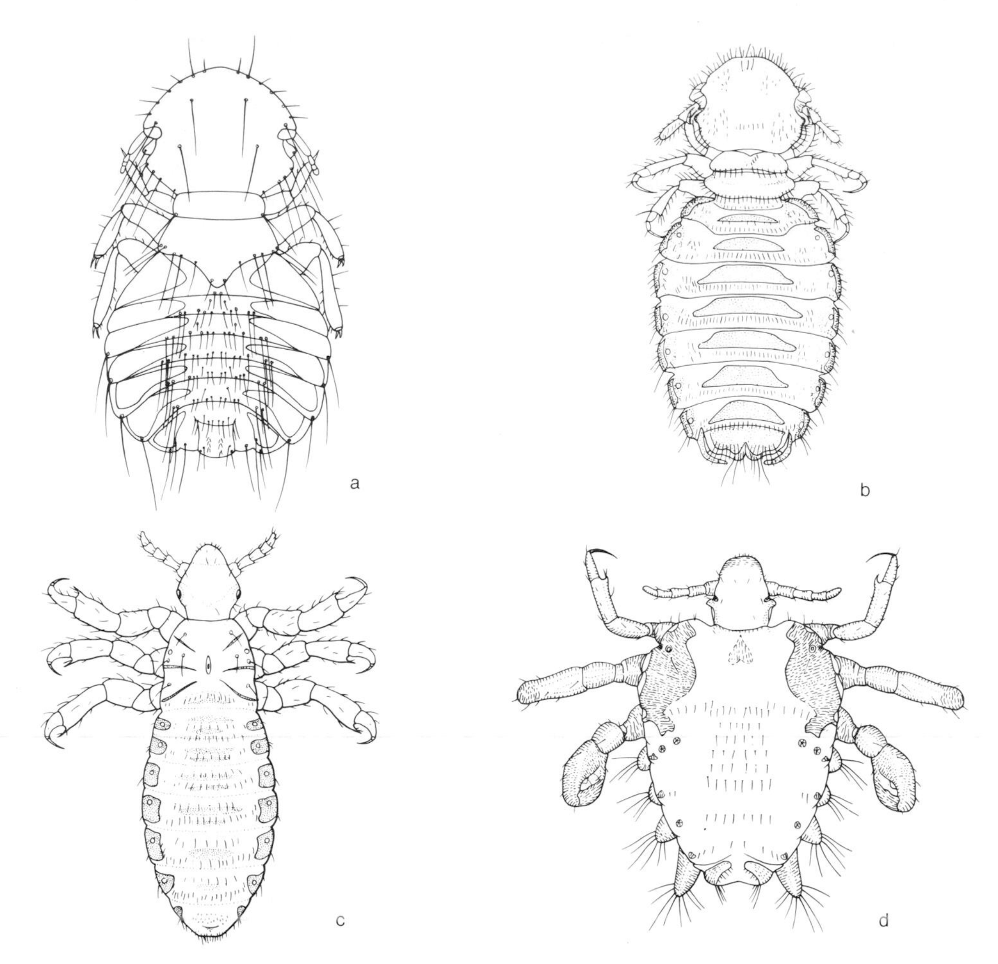

FIGURE 33.3 Representative lice: *a,* Ischnocera, *Goniodes gigas,* from poultry; *b,* Amblycera, *Trichodectes caprae,* from goat; *c, d,* Anoplura (*c,* the human body louse, *Pediculus humanus; d,* the pubic louse, *Pthirus pubis*). (a *from Emerson,* 1956; b *from Patton,* 1931; c *courtesy of D. Furman;* d *from Ferris,* 1951.)

Antennae filiform, 3- to 5-segmented; head without antennal grooves; mandibles vertical; maxillary palpi absent
Suborder Ischnocera (Fig. 33.3*a*)

Suborder Rhynchophthirina. Two species restricted to elephants and the warthog, respectively. In the reduced mandible and vestigial maxillae and labium, as well as other characteristics, Rhynchophthirina are somewhat intermediate between the Amblycera and Ischnocera and the Anoplura.

Suborders Amblycera and Ischnocera. The largest and most diverse group of lice, with about 2500 species, mostly associated with birds. The two suborders differ in biological as well as morphological characteristics. The Amblycera have horizontally biting mandibles, which in most species are used to scrape off feathers, loose skin, or other cutaneous debris. In a few genera the mandibles are modified for piercing integument. The mammal-infesting species are almost entirely confined to marsupials and rodents from the Neotropics. The Ischnocera have vertically biting mandibles. The bird-infesting lice of this suborder feed largely on feathers, and mammal-infesting species feed on dead or living skin. The mammal-infesting species occupy placental mammals.

Suborder Anoplura. About 250 bloodsucking species confined to placental mammals. The Family PEDI-

CULIDAE specializes on primates. Included are two species which infest humans—*Pediculus humanus,* the body and head louse (Fig. 33.3*c*), and *Pthirus pubis,* the pubic or crab louse (Fig. 33.3*d*).

SELECTIVE REFERENCES

GENERAL

Askew, 1971; Buxton, 1947 (species which infest humans); Furman and Catts, 1961 (medically important species); Rothschild and Clay, 1952.

MORPHOLOGY

Snodgrass, 1944 (mouthparts); Symmons, 1952 (head).

BIOLOGY AND ECOLOGY

Busvine, 1948; Emerson, 1956 (ecology of species on chicken); Hopkins, 1949 (host associations); Martin, 1934 (life history of *Columbicola*); Matthysse, 1946 (biology of cattle lice).

SYSTEMATICS AND EVOLUTION

Busvine, 1948 (races of body louse); Clay, 1970; Ferris, 1931 and 1951.

34 ORDER HEMIPTERA (Bugs, Leafhoppers, etc.)

Minute to large Exopterygota with styliform suctorial mouthparts (Fig. 34.1). Head capsule strongly sclerotized, seldom with separate sclerites delimited by sutures; compound eyes usually large, ocelli present or absent; antennae short to long, filiform or setaceous, with 10 or fewer segments (exception, numerous segments in male Coccoidea); mandibles and maxillae extremely elongate, needlelike piercing stylets, ensheathed in elongate, tubular labium, which is jointed or unjointed; palps absent. Prothorax large, distinct, except in wingless forms; mesothorax represented dorsally by scutellum, usually large, triangular; metathorax small. Forewings larger than hind wings, frequently coriaceous, at least basally; flexed flat or obliquely over abdomen, usually without folds in wing membrane; legs usually adapted for walking, forelegs raptorial in many predatory taxa (Fig. 34.2*a* to *c*); tarsi with 3 or fewer segments, frequently with arolia, pulvilli, or other pretarsal structures (Fig. 34.2*d*). Abdomen commonly with anterior 1 to 2 segments reduced or absent, posterior 1 to 2 fused, reduced, associated with genitalia; cerci always absent.

The Hemiptera, with approximately 55,000 species, are the fifth largest insect order. Nu-

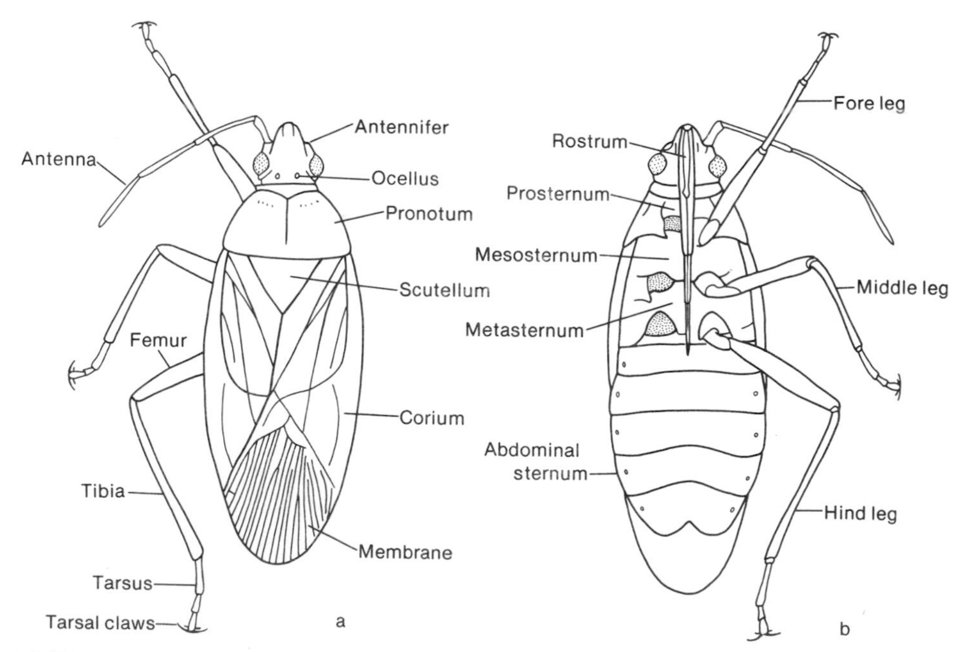

FIGURE 34.1 Major morphological features of Hemiptera: *a*, dorsal aspect, and *b*, ventral aspect of *Apateticus cynicus* (Miridae). (*From Britton, et al.*, 1923.)

merous fossils ascribed to the Hemiptera are known from the Permian Period, and it seems certain that the order arose earlier, in the Carboniferous Period. Presumably these early Hemiptera had evolved the suctorial mouthparts characteristic of the order, and were admirably preadapted to feeding on the vascular tissue of higher plants, which diversified during the Mesozoic Era. Among the present hemipteran fauna, members of one suborder, the **Homoptera,** feed almost exclusively as phloem feeders; the other suborder, the **Heteroptera,** includes a minority of predatory and ectoparasitic species. But, most species of the larger, more diverse families are principally phytophagous. Suctorial phloem feeding is perhaps the major feeding niche that endopterygote insects have failed to invade. It seems likely that their complete exclusion has been due to the prior occupation by the Hemiptera, which differentiated before the large orders of Endopterygota.

Specialized, styleform mouthparts are the most obvious feature shared by all Hemiptera. The mandibles and maxillae have become highly specialized as a two-channeled piercing tube for delivering salivary secretions and taking up food (Fig. 34.2*e* to *g*). Proximally, the maxillary and mandibular stylets are hinged to a short arm (as in Fig. 35.1*c*), on which are inserted the protractor and retractor muscles. This specialized structure differentiates the Hemiptera and Thysanoptera from the other hemipteroid insects (see Chap. 17), in which the muscles insert directly on the stylets. The labium is developed as a stout, usually jointed sheath for the stylets and is normally the only portion of the rostrum visible externally. Most Hemiptera probe only a short distance into the host tissue, and the stylets are not greatly longer than the rostrum. Penetration is accomplished by repeated contraction and relaxation of the protractor muscles, necessitated by the small scope of movement allowed by the

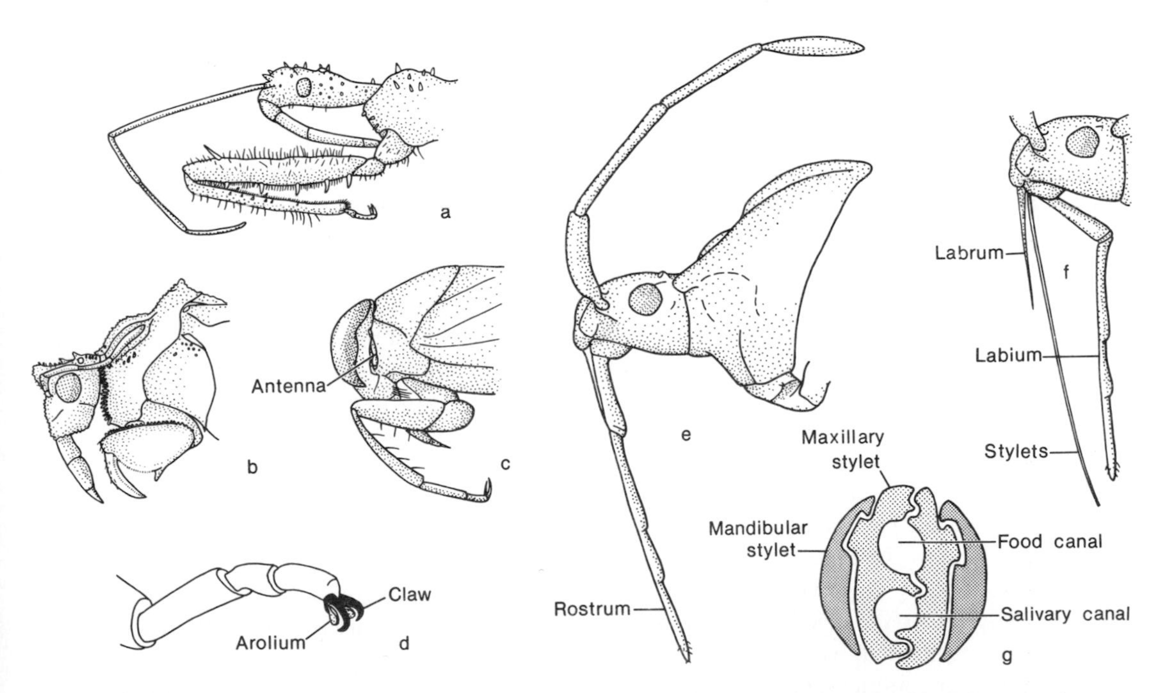

FIGURE 34.2 Legs and mouthparts of Heteroptera: *a*, raptorial forelegs of *Sinea* (Reduviidae); *b*, same of *Phymata granulosa* (Reduviidae); *c*, same of *Notonecta* (Notonectidae); *d*, arolium of pretarsus; *e*, rostrum of *Anasa tristis* (Coreidae); *f*, rostrum with labium removed to show feeding stylets; *g*, cross section of feeding stylets. (g *redrawn from Snodgrass,* 1935, *by permission of McGraw-Hill Book Company.*)

protractor levers. In Aradidae, which feed on fungal mycelia, which are usually long and sinuous, the stylets are commonly longer than the body and are coiled in pouches within the head capsule. In predatory forms such as Reduviidae, where quick penetration of a potentially dangerous prey is important, the protractor muscles are very long, allowing the stylets to be inserted with a single muscle contraction.

The alimentary canal of Hemiptera is modified for uptake of liquid food. Salivary glands are universally present. Extraoral digestion is apparently widespread, and relatively large quantities of saliva are injected into the host's tissue, causing local necrosis in plants, or, more rarely, systemic disturbances. Injection of saliva is also a major factor in transmission of microorganisms, especially plant viruses. In predatory species, saliva is highly toxic and paralytic, enabling relatively large prey to be quickly subdued. In many phytophagous Heteroptera, especially Lygaeidae, Pyrrhocoridae, and Pentatomidae, the midgut is differentiated as several distinct regions. These may include a distensible, anterior crop and a variable number of posterior caecae. The number and configuration of the caecae, which contain symbiotic bacteria, are characteristic of some families or genera.

Most Homoptera, with the major exceptions of the Fulgoroidea and some Cicadellidae, have the tubular posterior midgut coiled closely adjacent to the dilated, croplike anterior midgut. The Malpighian tubules may also be involved, forming a complex organ called a *filter chamber* (Chap. 5; Fig. 5.7*c*), whose function is apparently to convey excessive amounts of water to the hindgut. In Diaspididae (Coccoidea) no direct connection between midgut and hindgut exists, excess liquid being resorbed by the enlarged Malpighian tubules. In many Homoptera and some Heteroptera (Tingidae), surplus fluids, containing sugars, amino acids, and other small molecules, are excreted as honeydew, which forms the basis for commensal relationships with ants.

Flight in Hemiptera is powered primarily by the mesothoracic wings, which are usually longer than the metathoracic pair. This condition differentiates the Hemiptera from the orthopteroid orders, in which the anterior pair of wings functions primarily as a protective cover. Venation in Hemiptera is frequently reduced, especially in the smaller Homoptera, where only a few longitudinal veins support the wing membrane. In Heteroptera the wings usually conform closely to the outline of the body, and wing structure is frequently highly specialized. The anterior basal region, or *corium,* which covers the abdomen laterally (Fig. 34.3), is usually more heavily sclerotized. The corium may bear a distal line of weakness, the *cuneal fracture,* which allows the apex (membrane) of the wing to bend downward over the abdomen in repose. The arrangement of veins and closed cells in the corium and membrane is characteristic of families and is of great taxonomic utility.

Hemiptera commonly have pretarsal structures (Fig 34.2*d*), important in maintaining a purchase on smooth, exposed surfaces of plants, or in holding prey. It is notable that in several other phytophagous orders special holdfast organs have evolved (e.g., crochets in larval Lepidoptera, Chap. 44). Most Heteroptera secrete aromatic compounds, apparently defensive in nature. Gland openings in nymphs are on the dorsum of the abdomen; at the last molt, new openings appear on the metapleural region, below the wings.

Most Hemiptera are hemimetabolous. Nymphs differ from adults in such details as numbers of antennal segments, presence of ocelli, and chaetotaxy, as well as wing development. Reproduction is usually bisexual, with one to several generations per year. In several families of Homoptera, however, differences among instars may be extreme, and cyclical or permanent parthenogenesis is common. These complications in reproductive strategy are discussed below. Most Hemiptera are oviparous, ovipositing either on or in plant tissue or on soil, stones, or other substrates. Eggs of Homoptera are usually of simple, ovoid structure, whereas those of

Heteroptera are greatly differentiated among various families, frequently with opercula, spines, or appendages (Cobben, 1968).

In some classifications Homoptera and Hemiptera are considered separate orders. However, the differences between them are no greater than the differences between suborders of Coleoptera, Lepidoptera, or Hymenoptera. Furthermore, the superfamily Coleorrhyncha (Homoptera) is intermediate in several important characters, and in the past has been placed in the Heteroptera. Most recent classifications place Homoptera and Heteroptera as suborders of Hemiptera, and this arrangement is followed here.

KEY TO SUBORDERS OF Hemiptera

1. Mesothoracic wings almost always held rooflike over abdomen, uniformly membranous; if wings absent, rostrum arising near posterior margin of head, without gula **Homoptera** (p. 359)

   Mesothoracic wings folded flat over abdomen; if held rooflike, basal portion thickened, apical portion membranous; if wings absent, rostrum arising anteriorly on head (Fig. 34.2*a,b*), with sclerotized gula **Heteroptera** (p. 351)

Suborder Heteroptera (True Bugs)

Head horizontal, with base of rostrum clearly distinct from prosternum, although the distal segments may extend between the anterior coxae; antennae with 4 or 5 (occasionally fewer) segments; prothorax usually large, trapezoidal or rounded, frequently longer than broad; tarsi usually with 3 segments.

Whereas the Homoptera are exclusively terrestrial herbivores, the Heteroptera have become diversified as predators, ectoparasites, scavengers, and herbivores, with numerous amphibious and aquatic representatives. Heteroptera include a few viviparous species, but apparently none is parthenogenetic, in sharp contrast to the Homoptera. Phytophagous Heteroptera include many species which are general phloem feeders on higher plants (Miridae, Coreidae, Pentatomidae, Tingidae, etc.), and some which are restricted to particular plant parts. For example, many Lygaeidae, Pyrrhocoridae, and Largidae feed preferentially or exclusively on seeds, sometimes from a very narrow host range. Many Cydnidae are subterranean, apparently feeding on roots. Other Heteroptera limit their feeding to certain plant taxa. For example, Aradidae feed exclusively on wood-rotting fungi.

Among terrestrial forms, the Reduviidae, Nabidae, and Anthocoridae are almost entirely predaceous, and many Miridae, Pentatomidae, and Lygaeidae, as well as occasional members of other families, are predatory. In addition, many more species are facultatively predatory. Modes of predation range from "sit-and-wait" types (many Reduviidae) to active searchers, such as *Geocoris* (Lygaeidae). Some species, especially in the Pentatomidae, are host-specific, but most heteropteran predators prey on all suitably sized organisms in a given habitat. All aquatic Heteroptera are predators except Corixidae, which feed on plankton. The numerous amphibious families are either scavengers or predators, mostly of much smaller, weaker organisms.

Two families of Heteroptera are exclusively ectoparasitic. Cimicidae inhabit bird and mammal nests, including houses, and suck blood from the occupants. Polyctenidae suck blood from bats. In addition, all members of the reduviid subfamily Triatominae are ectoparasites of mammals and birds.

Heteroptera are among the most successful insect colonizers of marine habitats. *Halobates* (Gerridae) occurs in all tropical seas, commonly hundreds or thousands of kilometers from land. *Rheumatobates* occurs in brackish mangrove swamps and *Hermatobates* around coral reefs. Many Saldidae frequent seacoasts, and some, such as *Aepophilus,* are intertidal, enduring long periods of submergence during high water.

Approximately 40 families of Heteroptera are recognized. These are arranged in about 15 superfamilies, but no classification is universally accepted, and several of the superfamilies are very heterogeneous and difficult to characterize. The suborder is frequently treated as three broad divisions, the Hydrocorisae, Amphibicorisae, and Geocorisae. As suggested by their names, these categories are ecologically distinct, but their morphological separation is difficult. The following key includes North American families.

KEY TO THE FAMILIES OF THE SUBORDER Heteroptera

1. Antennae very short, inserted ventrally, with no more than 1 or 2 segments visible from above (aquatic or subaquatic species) 2
   Antennae much longer than head, usually inserted dorsally, and with at least 3 segments visible from above 9

2(1). Anterior tarsus with a single flattened, paddle-shaped segment (Fig. 34.8*c*); rostrum short, with 1 or 2 segments **Corixidae** (p. 358)
   Anterior tarsus usually with 2 or 3 segments, never paddle-shaped; rostrum long, with 3 or 4 segments 3

3(2). Ocelli present 4
   Ocelli absent 5

4(3). Antennae retractable into grooves beneath eyes; anterior femora thickened, raptorial **Gelastocoridae** (p. 359)
   Head without grooves for antennae; anterior femora slender **Ochteridae** (p. 359)

5(3). Abdomen with slender caudal respiratory appendages at least one-fourth length of body (Fig. 34.8*b*); tarsi with 1 segment **Nepidae** (p. 359)
   Abdomen with respiratory appendages short or absent; tarsi with 2 segments 6

6(5). Body oval, flattened; anterior legs raptorial, with femora thickened and grooved to receive tibiae 7
   Body strongly convex, arched dorsally, usually elongate; anterior femora not raptorial 8

7(6). Eyes bulging, prominent; wing membrane with reticulate veins; abdomen with short, flat apical breathing appendages **Belostomatidae** (p. 359)
   Eyes molded to outline of head and thorax, not prominent; wing membrane without veins; abdomen without breathing appendages **Naucoridae** (p. 359)

8(6). Hind legs with tibia and tarsus flattened, oarlike, without tarsal claws; body length greater than 5 mm **Notonectidae** (p. 358)
   Hind legs with tibia and tarsus cylindrical, with 2 tarsal claws; body length less than 4 mm **Pleidae** (p. 358)

9(1). Eyes present 10
   Eyes absent; body flattened; forewings short, hind wings absent **Polyctenidae** (p. 357)

10(9). Anterior legs with claws attached before apex of tarsi; apical tarsal segment longitudinally cleft 11
   Anterior legs with claws attached apically; apical tarsal segment not cleft 12

11(10). Middle coxae close to hind coxae, remote from fore coxae; hind femora much longer than abdomen **Gerridae** (p. 358)
   Middle coxae about equidistant from fore and hind coxae; hind femora about as long as abdomen **Veliidae** (p. 358)

12(11). Head linear, as long as thorax, including scutellum; eyes distant from base of head **Hydrometridae** (p. 358)
   Head shorter (usually much shorter) than thorax including scutellum; eyes usually near base of head 13

13(12). Scutellum concealed by pronotum; ocelli absent; tarsi with 2 segments; hemelytra frequently with raised, reticulate venation (Fig. 34.3*d*) **Tingidae** (p. 356)
   Scutellum usually visible, frequently large; if concealed, ocelli present or tarsi with 3 segments 14

14(13). Body greatly flattened dorsoventrally, with texture and appearance of tree bark; forewings in repose usually much narrower than abdomen (Fig. 34.6*c*) **Aradidae** (p. 356)
   Body not textured like bark, usually not greatly flattened; forewings in repose usually about same width as abdomen 15

15(14). Antennae with basal 2 segments short, thick; apical segments slender, filamentous 16

Antennae with segments of subequal length and diameter or gradually tapering, or with only first segment short, thickened **17**

16(14). Pronotum, wings, and head, including eyes, with long erect spines **Leptopodidae** (p. 358)
Body not spiney **Dipsocoridae, Schizopteridae** (p. 357)

17(16). Wings reduced to leathery flaps no longer than first abdominal segment, without veins; body ovoid, flattened, leathery (Fig. 34.6*b*) **Cimicidae** (p. 357)
Wings usually large; if reduced, usually extending beyond first abdominal segment and with veins; body elongate, thick **18**

18(17). Forewings with cuneus present (Fig. 34.4*a,b*) **19**
Forewings without cuneus **21**

19(18). Ocelli present; forewing with 0 to 1 basal cells in membrane (Fig. 34.4*b*) **20**
Ocelli absent; forewing with 2 to 3 closed cells in membrane near cuneus (Fig. 34.4*a*) **Miridae** (p. 356)

20(19). Rostrum with 4 segments; tarsi with 2 segments **Microphysidae** (p. 357)
Rostrum and tarsi each with 3 segments **Anthocoridae** (p. 356)

21(18). Antennae with 4 segments **25**
Antennae with 5 segments **22**

22(21). Scutellum large, triangular or U-shaped, reaching wing membrane (Fig. 34.6*d*) **23**
Scutellum smaller, not reaching wing membrane (as in Fig. 34.5*a* to *d*) **24**

23(22). Tibiae set with stout spines; apices of hind coxae fringed with stiff setae **Cydnidae** (p. 354)
Tibiae not set with spines; hind coxae without fringe of stiff setae **Pentatomidae** (p. 355)

24(22). Anterior femora thickened, raptorial (Fig. 34.2*a*); antennae with second segment about one-fourth length of third **Nabidae** (p. 357)
Anterior femora slender; antennae with second segments about one-half length of third **Hebridae** (p. 358)

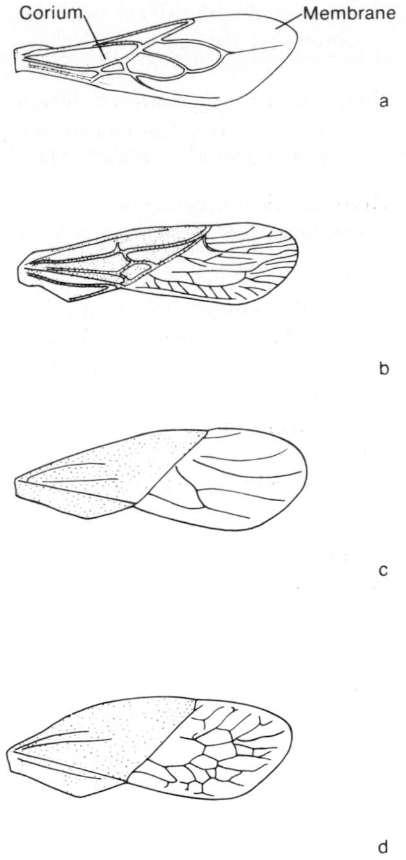

FIGURE 34.3 Wings of Heteroptera: *a*, Reduviidae (*Sinea diadema*); *b*, Nabidae (*Nabus alternatus*); *c*, Lygaeidae (*Lygaeus kalmi*); *d*, Largidae (*Largus*).

25(21). Ocelli present **26**
Ocelli absent **35**

26(25). Middle and hind tarsi with 1 or 2 segments; size minute to small **27**
Middle and hind tarsi with 3 segments; size small to large **30**

27(26). Forewings entirely membranous; forelegs thickened, raptorial; body slender, elongate, linear **Enicocephalidae** (p. 356)
Forewings with base thickened, coriaceous; forelegs slender; body not linear **28**

28(27). Rostrum with 4 segments; corium shorter than wing **29**
Rostrum with 3 segments; wing with corium extending to apex (1 species, Florida) **Thaumastocoridae** (p. 357)

29(28). Body covered with velvety hairs; femora extending far beyond body margins **Hebridae** (p. 358)

Body coarsely punctuate on dorsum; never with velvety hair; femora extending about to body margins **Piesmatidae** (p. 356)

30(26). Rostrum with 3 segments **31**

Rostrum with 4 segments **32**

31(30). Ocelli set between eyes; forewings with 4 to 5 elongate, parallel cells in membrane (Fig. 34.4*c*); head broader than long **Saldidae** (p. 358)

Ocelli set behind eyes; forewings usually with 1 or 2 large cells in membrane (Fig. 34.3*a*); head longer than broad **Reduviidae** (incl. **Phymatidae,** p. 356)

Cuneal fracture
Cuneus
Corium
Clavus
Membrane
a

Cuneus
b

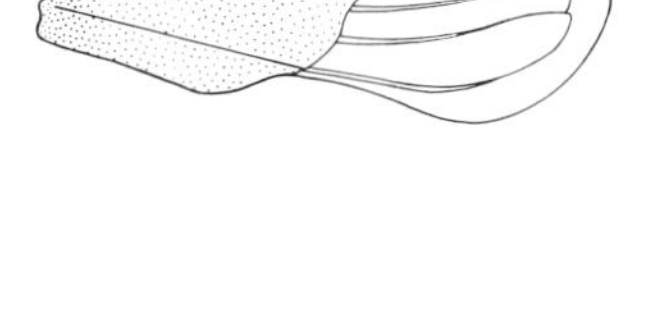
c

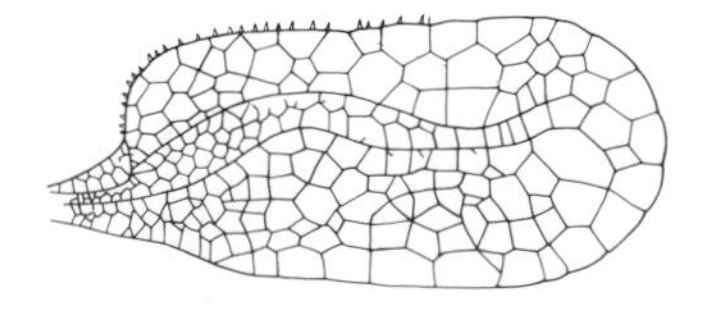
d

FIGURE 34.4 Wings of Heteroptera: *a*, Miridae (*Irbisia mollipes*); *b*, *Anthocoridae* (*Tetraphelps*); *c*, Saldidae (*Salda abdominalis*); *d*, Tingidae (*Corythuca obliqua*).

32(30). Rostrum with basal segment projecting, not capable of being apposed to ventral surface of head (Fig. 34.2*a*); tarsi without arolia **Nabidae** (p. 357)

Rostrum capable of being apposed to ventral surface of head and pronotum; tarsi with arolia (Fig. 34.2*d*) **33**

33(32). Forewings with 6 or fewer main longitudinal veins; if wings abbreviated, antennifers (tubercles on which antennae are inserted) located ventrolaterally **34**

Forewings with at least 7 (usually many) main longitudinal veins (Fig. 34.5*d*); antennifers located dorsolaterally **Coreidae** (incl. **Alydidae, Rhopalidae**) (p. 355)

34(33). Body elongate, slender, linear; antennae and legs extremely long, threadlike, with tips of femora and antennal scape slightly swollen (Fig. 34.6*a*) **Berytidae** (p. 356)

Body usually stout; legs and antennae not extremely long and slender **Lygaeidae** (p. 355)

35(25). Tarsi with 3 segments **36**

Tarsi with 1 or 2 segments **37**

36(35). Tarsi with 2 segments; rostrum retractable into groove between fore coxae **Reduviidae** (p. 356)

Tarsi with 1 segment; rostrum not retractable into groove **Nabidae** (p. 357)

37(35). Anterior femora swollen, bearing several teeth; forewing with 4 to 5 longitudinal veins in membrane (Fig. 34.3*c*) **Lygaeidae** (p. 355)

Anterior femora slender, usually without teeth; forewing with numerous longitudinal veins in membrane (Fig. 34.3*d*) **38**

38(37). Pronotum with upturned lateral margins **Pyrrhocoridae** (p. 356)

Pronotum with lateral margins rounded **Largidae** (p. 356)

Cydnidae (burrowing bugs). Broadly oval, brown or black bugs 3 to 10 mm long, sometimes with orange bars on the forewings. Ocelli present, antennae with five segments, tarsi with three segments; forelegs frequently modified for digging. Most species are probably phytophagous, feeding on either roots or foliage. As in pentatomids, thoracic repugnatorial glands are present. Worldwide in distribution.

Pentatomidae (shield bugs; Fig. 34.6*d*). Broad, usually triangular or trapezoidal in shape. Ocelli present, antennae with five segments; tarsi with three segments and arolia. Most species are phytophagous, including economically important pests on rice and crucifers; the Subfamily Asopinae are predators, mostly of lepidopterous larvae. All instars produce aromatic repugnatorial secretions released through openings in the thorax. Pentatomids include a variety of brightly metallic-colored or grotesquely sculptured members, particularly in the Tropics, where the majority of the approximately 3000 described species occur.

Coreidae (squash bugs, etc., Fig. 34.5*d*). Small to large, robust, elongate bugs with trapezoidal prothorax. Ocelli present, antennae with four segments, inserted dorsolaterally on head, tarsi with three segments and arolia; wing membrane with numerous veins (Fig. 34.5*d*). All species are phytophagous, mostly attacking growing shoots. A few, such as *Anasa tristis* in North America, are economically important. Most produce repugnatorial secretions, which some large species are capable of spraying short distances. Approximately 2000 species, worldwide in distribution.

Lygaeidae (seed bugs). Oval or elongate bugs, mostly 2 to 15 mm in length. Ocelli present, antennae with four segments (with few exceptions), inserted ventrolaterally on head; tarsi with three segments and arolia; wing with 4 to 5 veins in membrane (Fig. 34.3*c*). *Lygaeus, Oncopeltus,* and other large, bright orange or red and black "milkweed bugs" are the most commonly noticed members of this family. The great majority of species are small, usually dull-colored, frequently ground- or litter-dwelling insects. Most feed on seeds, which are grasped by the swollen,

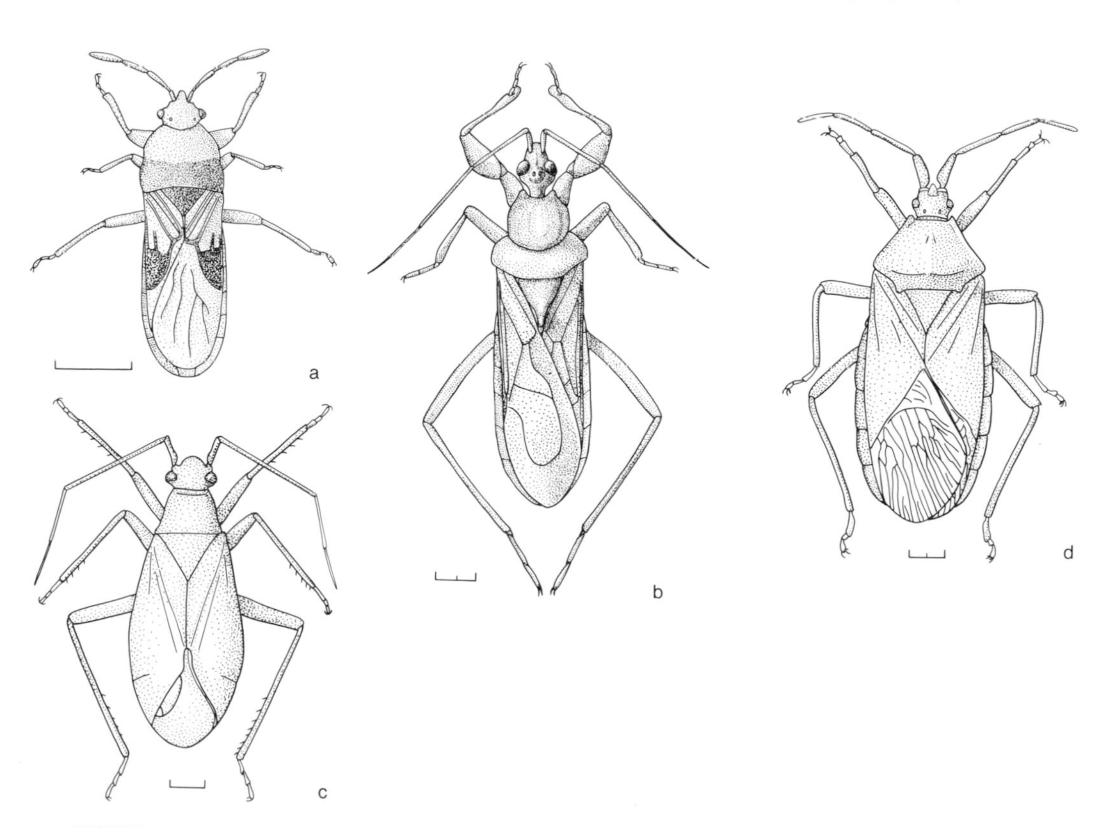

FIGURE 34.5 Representative Heteroptera: *a,* Lygaeidae (*Blissus leucopterus;* scale equals 1 mm); *b,* Reduviidae (*Triatoma protracta;* scale equals 2 mm); *c,* Miridae (*Irbisia mollipes;* scale equals 1 mm); *d,* Coreidae (*Anasa tristis;* scale equals 2 mm).

toothed forelegs; some species occur in mammal nests, apparently feeding on stored seeds. *Geocoris* is a common predator in low vegetation, and chinch bugs (*Blissus,* Fig. 34.5*a*) feed on stems and roots of grasses, including grain crops. Loss or shortening of wings is frequent in ground-dwelling lygaeids, which may mimic ants. There are over 2000 species, worldwide.

Pyrrhocoridae and Largidae (cotton stainers). Oval or elongate bugs, mostly more than 5 mm in length. Ocelli absent, antennae with four segments, tarsi with three segments and arolia; wing with anastomosing veins in membrane (Fig. 34.3*d*). Most species are bright orange, red, or yellow and black. Feeding is predominantly on fruiting structures of plants, and several species are economically important. Worldwide in distribution, but mostly tropical.

Piesmatidae. Small, oval insects resembling Tingidae in the lacelike texture of the dorsal surface, but with the scutellum exposed. Ocelli present, antennae with four segments; tarsi with two segments and pulvilli. Phytophagous, widespread but uncommon.

Berytidae (stilt bugs; Fig. 34.6*a*). Slender, elongate, parallel-sided bugs, similar to lygaeids in most characteristics. All are apparently phytophagous, mostly with broad host ranges. About 100 species, widespread.

Aradidae (flat bugs; Fig. 34.6*c*). Strongly flattened, oval in dorsal silhouette, with rough gray, black, or brown cuticle; mostly 5 to 10 mm long. Ocelli absent, antennae with 4 segments, usually thick, cylindrical; tarsi with 2 segments. Aradids feed on wood-rotting fungi, about which large numbers of all instars often congregate, either on the outer surface of trees or beneath loose bark. The feeding stylets are extremely elongate, frequently longer than the entire body, apparently as an adaptation for penetrating fungal mycelia. The elongate stylets also allow the feeding insects to remain camouflaged on bark adjoining fungal masses. Abbreviation or absence of wings is common in aradids, either in both sexes or in females only.

Tingidae (lace bugs). Small, dorsally flattened species with raised, reticulate venation (Fig. 34.4*d*); usually pale-colored. Ocelli absent, antennae with four segments, tarsi with two segments; pronotum usually covering scutellum, often crested or hood-shaped. Tingids are gregarious, phytophagous bugs which usually feed on undersides of leaves. Honeydew is secreted, and some species are tended by ants; a few form galls. Nymphs of tingids are often strikingly different from adults, with spines or other outgrowths from the dorsum. Over 750 species, widely distributed.

Reduviidae (assassin bugs; Fig. 34.5*b*). Small to large; ocelli present (rarely absent); antennae with four or six to eight segments; rostrum with three segments, fitting into intercoxal stridulatory groove on prosternum; tarsi usually with three segments. Reduviidae is one of the largest families of Heteroptera, with over 3000 described species. These show an enormous variation in body form, size, color, and cuticular sculpturing, ranging from the gnatlike Emesinae to squat, robust types such as *Apiomerus.* Nearly all reduviids are predatory, but *Triatoma* (Fig. 15.1*d*), *Rhodnius* (Subfamily Triatominae), and a few others are ectoparasites of mammals and birds and vectors of *Trypanosoma cruzi*, which causes Chagas' disease in humans. The forelegs are raptorial in all predaceous reduviids, and may be greatly enlarged and specialized, as in *Phymata* (Fig. 34.2*b*), which ambushes prey from hiding places on flowers.

Enicocephalidae. Small, reduvioid bugs with head enlarged, globose behind eyes, antennae and rostrum with four segments, tarsi with one to two segments; forewings entirely membranous. Most species frequent damp leaf litter or other debris and are predaceous. Widespread but uncommon.

Miridae (plant bugs; Figs. 34.1, 34.5*c*). Delicate, oval or elongate bugs, mostly 2 to 10 mm long. Ocelli absent, antennae and rostrum with four segments; tarsi with three segments; forewings with distinct corium, clavus, and (usually) a large cuneus (Fig. 34.4*a*). Mostly phytophagous, feeding on nearly all higher plants, with a wide range in host specificity. *Lygus* is among a number of economically important species. Over 5000 species, distributed worldwide.

Anthocoridae (minute pirate bugs). Compact, short-legged, 1 to 5 mm long. Ocelli present, antennae with four segments, rostrum with three segments; tarsi with three segments, forewings with incomplete cuneus (Fig. 34.4*b*). Most species are predators of foliage-inhabiting insects, but others occur on bark, in leaf litter, on fungus, and even in stored products, where they attack mites and small insects or insect eggs. Over 300 species, widespread.

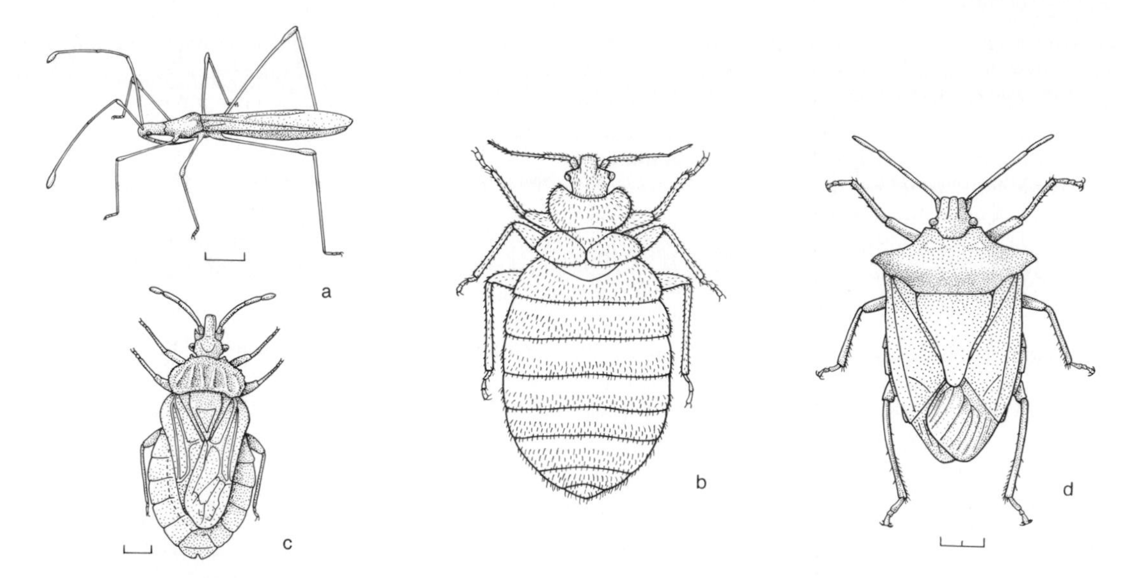

FIGURE 34.6 Representative Heteroptera: *a*, Berytidae (*Metapterus banksi;* scale equals 2 mm); *b*, Cimicidae (*Cimex lectularius*); *c*, Aradidae (*Aradus proboscideus;* scale equals 1 mm); *d*, Pentatomidae (*Euschistus conspersus;* scale equals 2 mm).

Microphysidae. Similar in appearance to Anthocoridae, but tarsi with 2 segments, rostrum with 4 segments. Most species, including the single North American representative, inhabit moss and lichen on or about dead trees, and are probably predators. *Embiophila* apparently is a predator of Embioptera, whose tunnels it infests, and related species probably attack spiders in their webs. Predominantly an Old World family.

Nabidae (damsel bugs). Superficially similar to Reduviidae, but without prosternal groove for rostrum. Ocelli present or absent, antennae with four or five segments, rostrum with four segments. These predaceous bugs commonly inhabit vegetation, and may be of some importance in controlling crop pests. Fertilization in some genera is directly into the hemocoel, as in Cimicidae. About 300 to 400 species, widely distributed.

Cimicidae (bedbugs; Fig. 34.6*b*). Oval, flattened ectoparasites, with forewings always reduced to short scalelike flaps. Ocelli absent, rostrum retractable into ventral groove, tarsi with three segments. Unlike lice, bedbugs contact their hosts only for periodic feeding, spending most of their time in nest material or adjoining cracks or crevices. Most are parasites of birds, some attack cave-dwelling bats, and one species is associated with humans. Fertilization is by a spermatophore, forcibly injected by the male, through a modified area of the female abdominal body wall (Caryon, 1964), a process termed *traumatic insemination.*

Polyctenidae. Highly specialized parasites of bats, mostly in tropical and subtropical regions. These bugs, which infest the bodies of their hosts, are highly modified for moving through fur (Fig. 15.1*c*). All species are viviparous, with only two nymphal instars. Two species occur in Southwestern United States.

Thaumastocoridae. Small flattened bugs with broad head, eyes frequently on short, lateral stalks; tibiae usually with membranous apical appendage. Predominantly a tropical family. New World species appear to be restricted to palms, feeding on the unfolding leaves. The single United States species occurs in Florida.

Dipsocoridae and Schizopteridae. Oval, predaceous bugs, 1 to 1.5 mm long; recognized by the enlarged basal antennal segments. Most species occur in leaf litter, moss, and lichen. Widely distributed but uncommon.

Saldidae. Oval, with broad head and large, prominent eyes. Ocelli present; rostrum with three segments, projecting below head, even in repose; tarsi with two segments; forewing with four or five long, parallel-sided cells in membrane (Fig. 34.4*c*). Active predatory bugs capable of rapid running, jumping, or flying. Saldids are widely distributed in wet habitats such as marshes and stream margins, including saline situations; *Aepophilus* (Europe) and *Omania* (tropical) are intertidal.

Leptopodidae. These insects are confined to tropical and subtropical regions of the Eastern Hemisphere, where they mostly inhabit dry areas along streams. One species, introduced to California, occurs under debris in arid regions.

Gerridae (water striders). Slender, elongate insects specialized for skating on the surface film of still or moving water, using the greatly elongate middle and hind legs; rostrum with four segments. Gerrids are most abundant on still or slowly moving water, where they scavenge and attack small arthropods. A number of species occur in marine situations, including the open seas at great distances from land.

Veliidae. Similar to Gerridae but with three-segmented rostrum and shorter middle and hind legs. *Velia* commonly frequents margins of quiet pools in streams; *Rhagovelia* (Fig. 34.7) inhabits riffles in fast-flowing streams, using the fanlike hemicircle of setae on the hind tarsi to swim against the current.

Mesoveliidae. Similar to Veliidae but with claws inserted apically. Mesoveliids frequent seeps, margins of quiet pools along streams, or ponds. In this family, as well as the Veliidae and Gerridae, wing polymorphism is common, and wingless, short-winged, or fully winged individuals may be encountered in the same population.

Hebridae. Small, stout-bodied bugs; antennae with four or five segments, ocelli present, rostrum with three visible segments, retractable into groove between coxae. Widespread in marshes, stream margins, etc., but uncommon.

Hydrometridae. Extremely elongate, linear bugs with threadlike legs and antennae. Hydrometrids are sluggish, slow-moving water striders, inhabiting quiet pools, marshes, and swamps, mostly in tropical regions.

Notonectidae (back swimmers). Stout, wedge-shaped bugs which swim on their backs, using the oarlike hind legs. Ocelli absent, antennae with four segments. Notonectids are free-swimming predators in open water or small pools or along stream and lake margins. Their bite is painful.

Pleidae. These insects are distinguished from notonectids by having three antennal segments, and by the partial fusion of the head and thorax. Both families are widely distributed, but pleids are much less commonly encountered.

Corixidae (water boatmen; Fig. 34.8*c*). Flattened, streamlined swimming bugs with fore- and hind legs oarlike, without tarsal claws. Rostrum short, concealed; antennae with three or four segments. Corixidae occur in almost all still water, including salt ponds and brackish pools; they are less abundant in

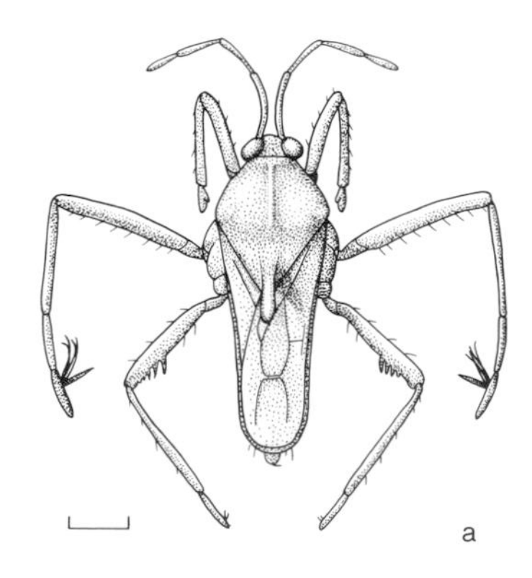

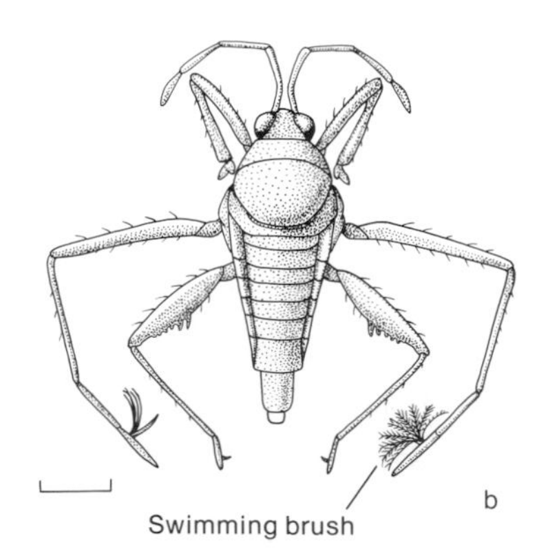

FIGURE 34.7 Veliidae (*Rhagovelia distincta*): *a*, winged form; *b*, apterous form. Scale equals 1 mm.

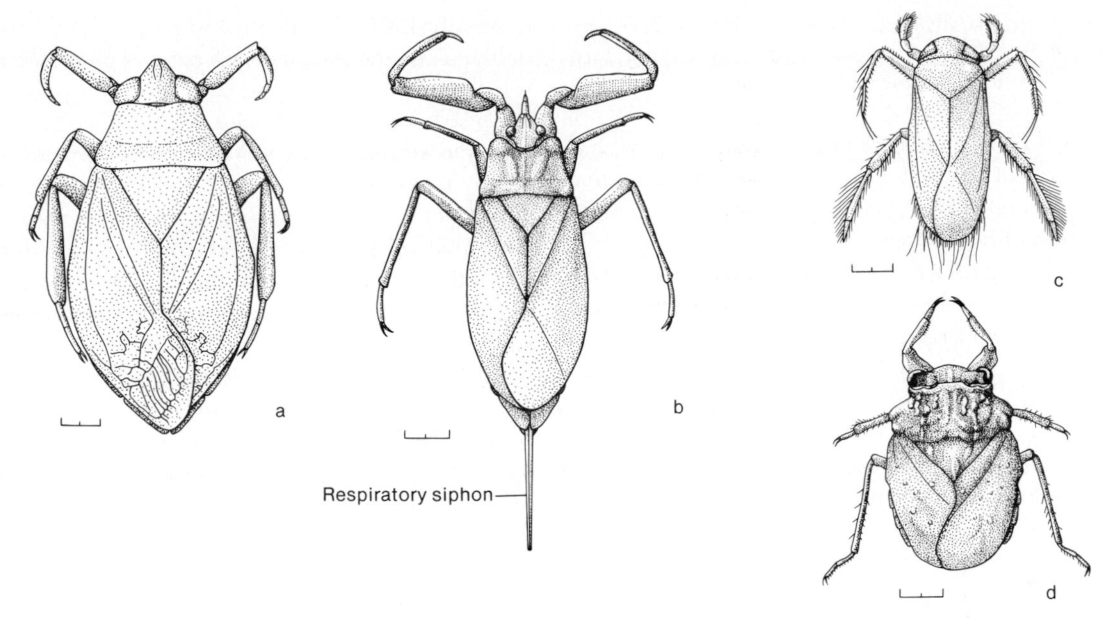

FIGURE 34.8 Representative Heteroptera: *a*, Belostomatidae (*Belostoma flumineum*); *b*, Nepidae (*Nepa appiculata*); *c*, Corixidae (*Hesperocorixa laevigata*); *d*, Gelastocoridae (*Gelastocoris oculatus*). Scales equal 2 mm.

streams. Corixids include a few predators, but most use the paddlelike forelegs to scoop up bottom detritus containing algae, diatoms, and other microorganisms, from which they suck the protoplasm.

Nepidae (water scorpions; Fig. 34.8*b*). Elongate, cylindrical or flattened with long, slender legs; anterior legs raptorial; posterior pairs adapted for walking. Antennae with three segments, tarsi with one segment, without claws on anterior legs. Nepids are sluggish predators which ambush prey from camouflaged positions on aquatic vegetation or debris. They are widespread, but more abundant in tropical regions.

Belostomatidae (giant water bugs; Fig. 34.8*a*). Broad, flattened, oval bugs with the middle and hind legs flattened, fringed for swimming, the anterior pair raptorial; antennae with four segments. Belostomatids are moderate to large (to about 100 mm long), free-swimming predators which kill small vertebrates as well as other arthropods, and bite fiercely if handled freely. Widely distributed.

Naucoridae. Similar to belostomatids in general appearance, but more rounded and streamlined in dorsal aspect; anterior femora extremely thick; middle and posterior legs cylindrical, spiney. Naucorids frequent vegetation, bottom debris, undersides of stones, and similar concealed situations, in both running and still water. Respiration is through a plastron, which occurs in no other Hemiptera. Mostly tropical in distribution. North American species are mostly in the southwest.

Gelastocoridae (Fig. 34.8*d*). Squat, stout-bodied, with prominent, bulging eyes and raptorial forelegs. Gelastocorids are commonly associated with sandy or muddy banks. They actively search for prey, which they capture by sudden, short hops. As in damselflies, the widely spaced, hemispherical eyes allow vision in all directions, and probably provide binocular depth perception.

Ochteridae. Similar to Gelastocoridae, but with longer antennae, long, four-segmented rostrum, and anterior legs similar to posterior pairs. Ochterids are widespread but infrequently encountered. They occupy the same situations as saldids and gelastocorids.

Suborder Homoptera (Leafhoppers, Planthoppers, etc.)

Head deflexed, with rostrum closely appressed to prosternal region, its base extending between anterior coxae; antennae usually with 2 to 10 segments;

prothorax usually small (exception, large in Membracidae, Peloridiidae), transverse or quadrangular; tarsi with 3 or fewer segments.

All Homoptera are phytophagous, but within the role of terrestrial herbivore they have become exceedingly diversified, both in body form and in details of life history.

Most species can be grouped into one of two broad categories on the basis of morphological and biological features. The Superfamilies Cicadoidea, Fulgoroidea, Cercopoidea, and Cicadelloidea comprise the Series[1] **Auchenorrhyncha.** In this group the antennae are very short and bristlelike, the rostrum is clearly articulated with the head, and the ovipositor is bladelike, for inserting the eggs into plant tissue. The Auchenorrhyncha are mostly active insects which live exposed on plant surfaces and jump to escape predators. Exceptions include cicadas, which cannot jump, but readily fly as adults, and nymphal cercopoids, which live immersed in a self-produced coating of froth as immatures, and have practically lost the ability to jump. The Auchenorrhyncha have simple life cycles. Immatures generally resemble adults, except for the absence of wings, although nymphs of cicadas are specialized for subterranean life and are superficially quite different from adults. Nearly all Auchenorrhyncha are bisexual, and they never have alternating generations, as in the Sternorrhyncha, discussed below. Males of many species produce sound from a pair of tymbals in the base of the abdomen. In cicadas the tymbals are associated with a resonating chamber. The chorusing noise produced by large aggregations of individuals may be almost deafening, and probably functions in confusing birds and other predators, as well as in courtship. Sounds produced by cicadellids are too feeble to be heard by humans without amplification; their function in most species is apparently in courtship.

The Superfamilies Psylloidea, Aphidoidea, Aleyrodoidea, and Coccoidea constitute the Series **Sternorrhyncha.** These Homoptera have relatively long, filamentous antennae, not bristlelike as in the Auchenorrhyncha. The rostrum is inserted on the extreme posteroventral part of the head, and is usually closely appressed to the thoracic sterna, from which it may appear to arise. Tarsi are one- or two-segmented. The ovipositor is usually undeveloped, and eggs are deposited on plant surfaces. Adult psyllids and aleyrodids are active jumping or flying insects, but the Sternorrhyncha are predominantly nonvagile, specialized for rapid feeding and reproduction. Parthenogenesis is widespread, and in Aphidoidea may be alternated with sexual reproduction at certain times of year. The development of complex life cycles in these groups is discussed in greater detail further on. Many Sternorrhyncha are notable for the tendency of the immature and adult stages to diverge morphologically. In aleyrodids and male coccoids the degree of differentiation is equivalent to that in endopterygote insects. The nonmotile "larvae," or neanides, with greatly reduced appendages and sense organs, are essentially a feeding stage. The motile adult is preceded by a nonfeeding "pupal" stage, and physiologically these insects are holometabolous. Many Sternorrhyncha form commensal relationships with ants, which are attracted to their copious exudates of honeydew, and which to some extent protect the homopterans from predators and move them about their host plants. Such relationships are most specialized with some aphids. For example, the corn root aphid, *Anuraphis maidiradicis,* is tended by ants (chiefly *Lasius*), in all instars. The eggs are collected in the fall and overwintered in the ant nests. The first-instar aphids are carried to the spring host (smartweed). Later generations are transferred to the summer host (corn). In many sternorrhynchous Homoptera dermal glands secrete a protective cover of wax, either in the form of characteristic filaments or plates, or as a powdery deposit. In some Coccoidea the eggs are enmeshed in a flocculent envelope of wax filaments, and wax is a constituent of the scale of diaspidids and other armored scales.

The Series **Coleorrhyncha** contains only the Family Peloridiidae, with about 20 species dis-

[1]A series is a taxonomic category used in the Hemiptera for classifying groups between a suborder and a superfamily.

tributed in Australia, New Zealand, southern South America, and a few intervening islands. Superficially the adults resemble tingids, with the reticulate wings held flat over the abdomen. The alimentary canal lacks a filter chamber, as in the Heteroptera, but the rostrum is inserted posteriorly on the head, as in Homoptera. The lateral prothoracic region and wing pads of nymphs are demarked by sutures (a characteristic of some fossil Homoptera) and superficially resemble the paranotal lobes of Palaeodictyoptera. All instars live and feed on mosses and liverworts. The morphological characteristics, as well as the discontinuous circum-antarctic distribution and restriction to primitive host plants, seem to indicate that the Peloridiidae are a relictual family, possibly the most primitive extant Hemiptera.

KEY TO SUPERFAMILIES AND MAJOR DIVISIONS OF THE SUBORDER Homoptera (Adults)

1. Tarsi with 1 or 2 segments, or legs absent 2
   Tarsi with 3 segments (series **Auchenorrhyncha**) 6

2(1). Pronotum with differentiated lateral lobes; wings held flat over abdomen (series **Coleorrhyncha**) **Peloridioidea**
   Pronotum without differentiated lateral lobes; wings (if present) held rooflike (series **Sternorrhyncha**) 3

3(2). Tarsi with 1 segment and 1 claw, or legs vestigial or absent; females larviform or gall-like or scalelike (Fig. 34.12*c* to *e*), males with mesothoracic wings and atrophied mouthparts (Fig. 34.12*b*) **Coccoidea** (p. 365)
   Tarsi with 2 segments and 2 claws; body insectlike, with recognizable legs, antennae, and mouthparts; wings 4 in number, if present 4

4(3). Tarsal segments subequal in size and shape; antennae with 7 to 10 segments 5
   Tarsi with first segment small, triangular, or absent; antennae with 1 to 6, usually with 5 or 6 segments **Aphidoidea** (p. 364)

5(4). Antennae with 7 segments; wings membranous, without sclerotized leading edge (Fig. 34.10*e*); body powdered with white wax **Aleyrodoidea** (**Aleyrodidae**; p. 364)
   Antennae usually with 10 segments; wings with sclerotized leading edge; body not powdered with wax **Psylloidea** (**Psyllidae**; p. 363)

6(1). Antennae inserted on frons, between eyes (Fig. 34.9*a,b*); pedicel subequal to scape in diameter; mesocoxae short, contiguous or close together 7
   Antennae inserted laterally, beneath eyes (Fig. 34.9*c,d*); pedicel enlarged, often globular; mesocoxae elongate, distant **Fulgoroidea** (p. 362)

7(6). Head with 3 ocelli; anterior femora thickened; body more than 15 mm long **Cicadoidea** (**Cicadidae**, p. 361)
   Head with 2 ocelli, or ocelli absent; anterior femora slender; body rarely more than 10 mm long 8

8(7). Metatibiae with 1 or 2 large spurs on shaft; metacoxae short, conical **Cercopoidea** (**Cercopidae**, p. 362)
   Metatibiae with numerous small spines on shaft; metacoxae transverse, platelike **Cicadelloidea** (p. 362)

SUPERFAMILY CICADOIDEA (cicadas). A single family, Cicadidae, of moderate to very large insects with membranous, transparent wings, sometimes banded or spotted.

Adult cicadas are familiar because of the high-volume sounds produced by the males (see Chap. 6; Fig. 6.7), but because of the ventriloqual quality of their calls they are never conspicuous. Most species are arboreal, and they are often gregarious. Females slash twigs with their ovipositors to insert their eggs, a habit which sometimes becomes economically damaging. Newly eclosed nymphs drop to the soil, where they lead subterranean lives feeding on roots. Most species require several years to mature, and often all the members of a single generation appear simultaneously. Usually, a generation matures each year, but in the "periodical cicadas" (*Magicicada*) of eastern North America, generations are restricted to certain years at a specific location. These species require either 13 or 17 years to mature, and consequently may unexpectedly appear after several years' absence. At most localities the individuals emerging any given year consist of two or three morphologically similar species with synchronized life cycles. The ecological and evolutionary significance of these synchronized mass emergences restricted to certain years is not entirely understood, but they may be a means of alternately "swamping" and starving potential predators. The Cicadidae are distributed throughout the world but are especially abundant and diverse in the Tropics and Subtropics and in Australia.

SUPERFAMILY FULGOROIDEA (fulgorids; Fig. 34.9*c,d*). Clypeus separated from frons by distinct frontoclypeal ridge, not extending posteriorly between eyes; antennae usually with second segment enlarged, with conspicuous sensory organs; 2 ocelli (occasionally 3) on sides of head near eyes; tegulae present on mesothorax.

Fulgoroidea range from small to large in size, and occur in a diverse and sometimes bizarre array of body forms. Many species (Cixiidae, Delphacidae, Achilidae) superficially resemble Cicadellidae. Others have broad, mothlike wings (Flatidae) or have the wings reduced and strap-shaped (Dictyopharidae, Issidae). Exaggeration of the proportions of various parts of the body, including head, wings, and legs, occurs in various species. The functions of such modifications are obscure.

Details of the biology and life history are unknown except for a very few fulgoroids. Most are phloem feeders, but some (Achilidae, Derbidae) feed on fungus. Most species probably spend their entire lives on foliage, but most Cixiidae apparently live underground as nymphs, feeding on grass roots, or in ant nests. Fulgoroids are primarily a tropical group, and most of the relatively few economically important species occur on sugar cane, coffee, and similar crops.

Recognition of fulgoroid families is rendered difficult by the great structural variation within many families, together with a repeated tendency toward convergence in external appearance, especially in the numerous flightless species from different families. About 20 families, many obscure or principally tropical, are recognized by most authorities. About half of these occur in North America. The interested student is referred to Kramer (1960) for the most complete discussion of the superfamily.

SUPERFAMILY CERCOPOIDEA (spittlebugs). Adults superficially similar to Cicadellidae; best distinguished by the tibial characters given in the key.

The soft, whitish nymphs live within frothy masses of "spittle," which is produced by blowing air through a viscous mucosaccharide secreted by the Malpighian tubules and released through the anus. Air is expelled by contracting a cavity formed by the ventral extension of the abdominal tergites beneath the abdomen. The spittle serves both as a protective device and as a means of reducing evaporation (Marshall, 1966). Nymphs of some Australian species construct hardened tubes from the anal secretion, attaching these to twigs or branches. All instars are phloem feeders. The single family, CERCOPIDAE, is cosmopolitan, but most species are tropical.

SUPERFAMILY CICADELLOIDEA (leafhoppers and treehoppers; Fig. 34.9*a,b*). Distinguished by the tibial and coxal characters listed in the key. Cicadelloids also differ from other superfamilies in venational characters and in having the tentorium reduced.

Cicadelloids are all phloem feeders, almost always on higher plants. Many species are cryptically colored, and some, especially membracids, are structurally modified to resemble thorns or other plant parts. Membracid nymphs are often gregarious and may be tended by ants. The great majority of Cicadelloids have simple life cycles with many generations per year. Adults, especially of Cicadellidae, are highly mobile and may

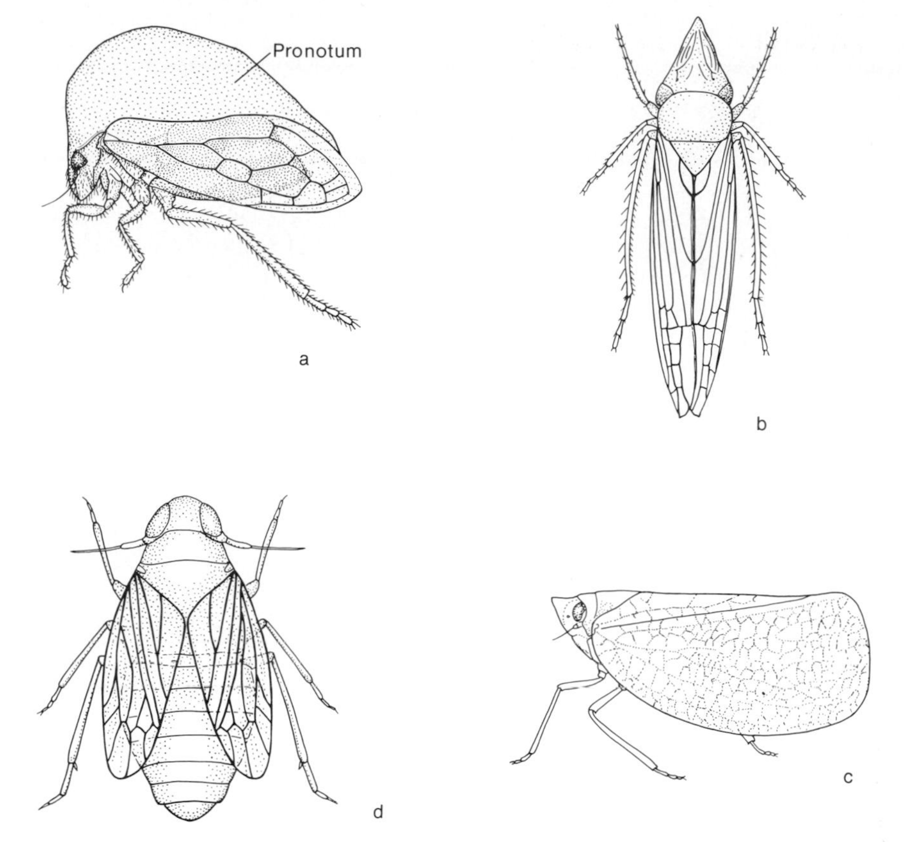

FIGURE 34.9 Representative Homoptera: *a,* Membracidae (*Stictocephala inermis*); *b,* Cicadellidae (*Draeculacephala mollipes*); *c, d,* Fulgoroidea; *c, Acanalonia bivittata* (Acanaloniidae); *d, Delphacodes campestris* (Delphacidae). (a *from Yothers,* 1934; b to d *from Osborn,* 1938.)

migrate long distances, particularly when local food sources (such as ephemeral native plants) wither. In large part because of their tendency to move among hosts, cicadellids are extremely important as transmitters of plant diseases, especially viruses.

Two families occur in North America. MEMBRACIDAE are characterized by the great enlargement of the pronotum, which extends backwards over the scutellum and abdomen (Fig. 34.9*a*). In CICADELLIDAE the pronotum is transverse or quadrangular, and the scutellum is clearly visible (Fig. 34.9*b*).

SUPERFAMILY PSYLLOIDEA (psyllids; Fig. 34.10*a,b*). Minute to small Homoptera with long, bristlelike antennae, 3 ocelli; forewings with venation reduced by fusion of R, M, and Cu veins; metacoxae enlarged, elongate.

Psyllids often bear a striking resemblance to minute cicadas. As adults they are highly active, readily jumping and flying if disturbed. Nymphs are sluggish and are morphologically rather distinct from adults, with squat, flattened bodies. They mature on foliage, usually of woody plants. A few species cause the growth of galls. The life cycle is simple, with one to several bisexual generations per year. Psyllids are of considerable importance to specific crops because of disease transmission or toxic effects of the saliva which is injected into the hosts. PSYLLIDAE, the single family, is cosmopolitan.

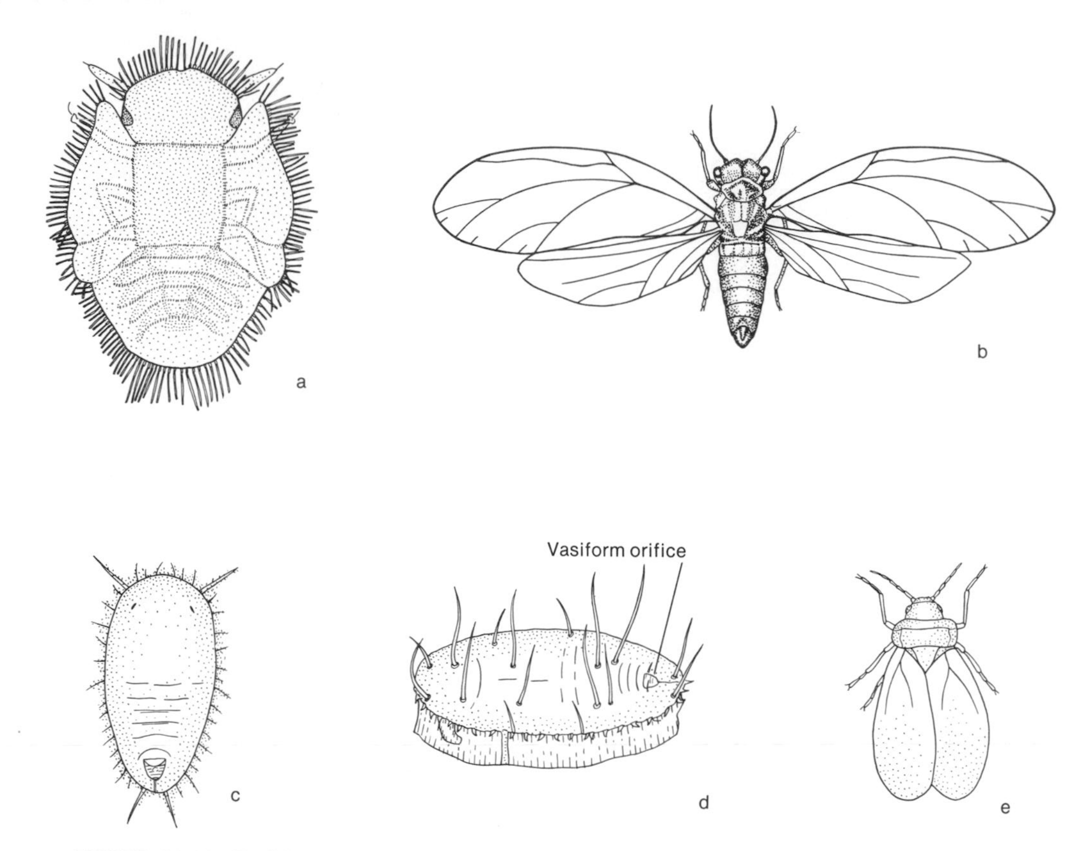

FIGURE 34.10 Psyllidae and Aleyrodidae: *a, b,* Psyllidae, nymph and adult, respectively (*Paratriazoa cockerelli*). *c* to *e,* Aleyrodidae, nymph, pupa, and adult, respectively (*Trialeurodes vaporariorum*). (a, b *from Knowlton and Janes,* 1931; c, d, e *from Lloyd,* 1922.)

SUPERFAMILY ALEYRODOIDEA (whiteflies; Fig. 34.10*c* to *e*). A single family, ALEYRODIDAE. Adults similar in general size and body form to Psyllidae; antennae long, bristlelike; wings opaque with dusting of white wax, venation greatly reduced, with 2 or 3 longitudinal veins, no clavus; all instars with characteristic *vasiform orifice* on last abdominal tergite (Fig. 34.10*d*).

These minute insects are phloem feeders, usually on the undersides of the leaves of angiosperms. A few are known from ferns, none from gymnosperms. Eggs, attached by short pedicels to the undersurfaces of leaves, are frequently arranged in arcs or circles. First instars are flattened, ovoid, with short legs. The second and third instars (larvae) are sessile, with atrophied legs. The fourth instar ("pupa") is similar to the larva but feeds only briefly, then becomes inactive. Before the final molt, the developing adult organs become visible through the pupal cuticle. The pupal stage is frequently ornamented by wax plates or filaments in highly individualized patterns, and the taxonomy of Aleyrodidae is based mainly on these structures. Reproduction may be either sexual or asexual; in some species unfertilized eggs produce males, but parthenogenetic races of only females are known. Aleyrodids are primarily tropical, and the most economically important species in North America attack citrus or greenhouse plants.

SUPERFAMILY APHIDOIDEA (aphids; plant lice; Fig. 34.11). Soft-bodied, globular or flattened insects with abdominal segmentation indistinct; either winged or apterous as adults, forewings with 3 to 5 longitudinal veins and distinct clavus; cornicles usually present but sometimes reduced, cone-shaped or ring-shaped.

The Aphidoidea are remarkable for the development of highly complex life cycles, which may involve cyclical parthenogenesis, cyclomorphosis (seasonal change in body form), and alternation of hosts. These are apparently adaptations to survive in seasonal climates, and in highly temperate areas all three phenomena may occur in a single species. For example, in Northern states, the woolly apple aphid (*Eriosoma lanigerum*, Fig. 34.11) overwinters as eggs on elm trees, where one or more spring generations of wingless, individuals mature. In early summer winged individuals appear and migrate to apple trees and related plants such as hawthorne, where an indefinite number of wingless generations follow. All these spring and summer generations are parthenogenetic and viviparous. In autumn, winged individuals migrate back to elm trees and parthenogenetically produce males and females, which reproduce sexually, each female laying a single egg. The annual cycle of *Rhopalosiphum padi* is illustrated in Fig. 4.8.

Many species with alternating generations occupy hosts which are ephemeral annuals or otherwise unsuitable during part of the year. Production of sexual forms is predominantly controlled by changes in temperature and photoperiod, and under subtropical or artificial conditions sexual forms may never appear. The influence of other factors involved in control of aphid polymorphism is discussed in Chap. 11 of this book, and in greater detail by Hille Ris Lambers (1966).

Taxonomy of Aphidoidea is unsettled, with one to nine families recognized by different specialists. The more important differences among families involve features of the life cycle, rather than morphological characters. APHIDIDAE is the dominant family, with over 3000 species, the great majority from north temperate areas and very few from the Tropics. Two other small families are commonly recognized: ADELGIDAE cause conelike galls on conifers, with at least one generation occurring on spruce (*Picea*); PHYLLOXERIDAE occur on oak (*Quercus*), grape (*Vitus*), and other plants, usually forming galls on at least one part of the plant. In the latter two families the cornicles are reduced or vestigial.

Aphidoidea are of exceptional economic importance. Besides the direct effects caused by sucking plant sap, many species damage their hosts by injecting toxic saliva or transmitting diseases, especially viruses. For example, *Myzus persicae,* the peach aphid, vectors at least 100 different viruses.

SUPERFAMILY COCCOIDEA (scale insects; mealybugs; Fig. 34.12). Minute to small, sexually dimorphic Homoptera; mature females with ovoid to globular body, frequently with segmentation indistinct, eyes small, arranged in groups of simple lenses; antennae 1-

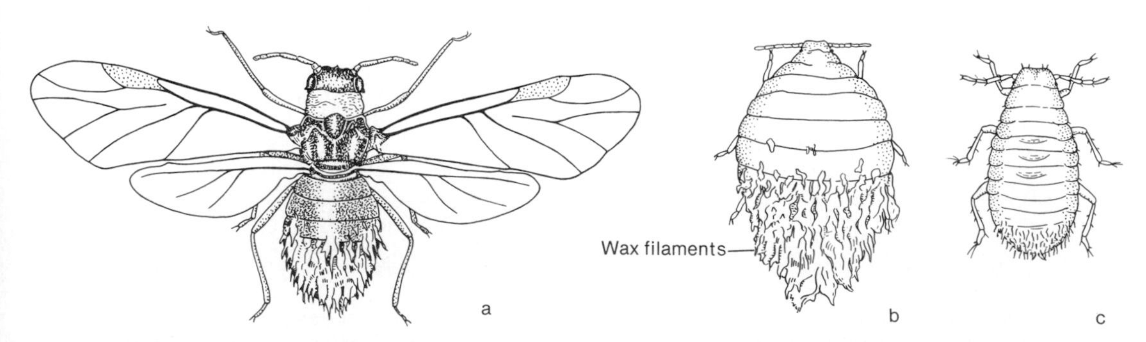

FIGURE 34.11 Aphidoidea, Aphididae (*Eriosoma lanigerum*, the woolly apple aphid). *a*, Winged female; *b*, apterous female; *c*, male (*From Baker*, 1915.)

to 9-segmented or vestigial; legs segmented, with single tarsal claw, or atrophied; rostrum 1- to 2-segmented, short; stylets extremely long, coiled in body; body frequently protected by waxy, cottony or hard, scalelike covering. Mature male gnatlike, with long, multiarticulate antennae, lateral eyes, enlarged mesothorax bearing 1 pair of wings with 1 or 2 longitudinal veins; hind wings reduced to halteres.

Morphologically Coccoidea are among the most highly modified insects, being specialized for a sessile or nearly sessile life attached to their host plants. The first instars, or crawlers (Fig. 34.12*a*), are always motile and function in dispersal. PSEUDOCOCCIDAE (Fig. 34.12*e*) (mealybugs) possess functional legs in all instars, but in other families of scale insects at least one instar is nonmotile. The crawlers can survive up to several days without feeding, during which period surprisingly large distances may be traversed. If a spot suitable for feeding is located, the first instar inserts its mouthparts and, depending on the family, becomes sessile or nearly so after the first molt, with atrophied legs, antennae, and eyes. At this time the insect begins secreting the protective covering, or scale, which, in different species, varies enormously in hardness, texture, and shape. Males pass through one more instar than females, which are apparently neotenic. The last two instars in males display external wing pads and other adult features, much as in endopterygote pupae, and the last instar does not feed. Reproduction is oviparous or viviparous. Oviparous forms usually cover the eggs with waxen oothecae or retain them beneath the mother's scale or body. Parthenogenesis occurs in various families, but its extent is uncertain.

Various scale insects have one or more instars specialized beyond the description given above. In MARGARODIDAE intermediate instars of females are legless "cysts," capable of withstanding periods of up to several years without food or water. Some members of the ERIOCOCCIDAE form galls, within which the insects mature. In some species the male and female galls are strongly differentiated, while in others the male remains within the parent gall. First instars of

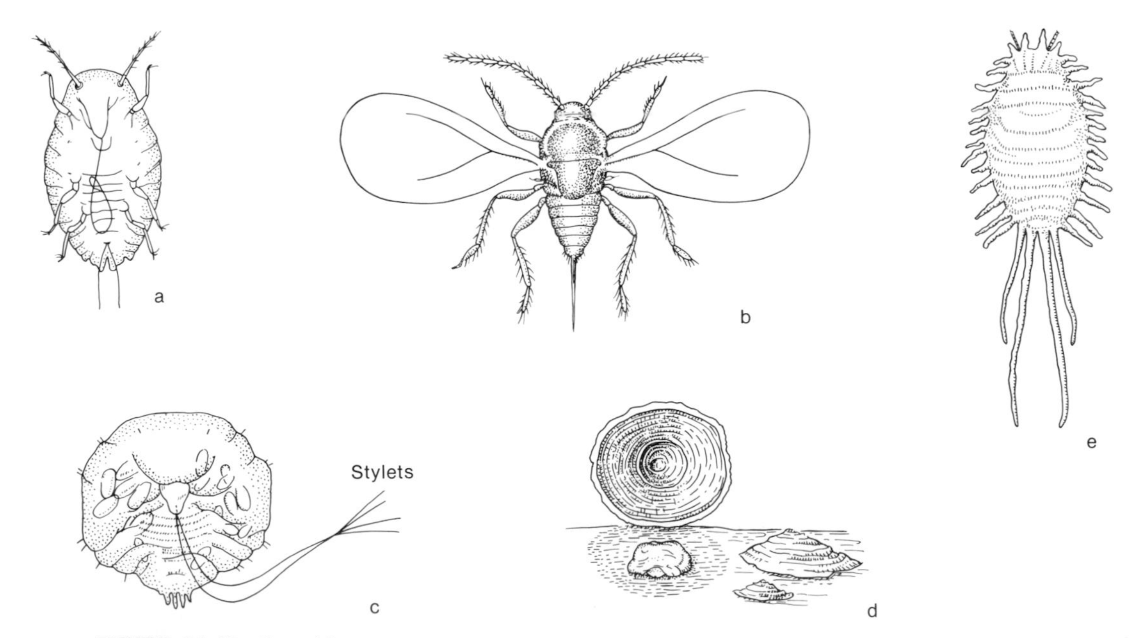

FIGURE 34.12 Coccoidea; *a* to *d*, various instars of the San José scale. *Quadraspidiotus perniciosus* (Diaspididae); *a*, first-instar nymph or crawler; *b*, adult male; *c*, adult female, ventral, extracted from scale revealing eggs inside body; *d*, scales; *e*, *Pseudococcus longispinus* (Pseudococcidae). (a to d *from U.S. Department of Agriculture;* e *from Green,* 1896–1922.)

some species, such as the coccid, *Walkeriana*, bear elongate filaments enabling them to travel on breezes, in the manner of a dandelion seed.

It should be pointed out that although many coccoids are serious economic pests, several useful products, such as coccineal, lac, and wax, have been derived from certain species, and *Dactylopius* (Eriococcidae) has been used in biological control of prickly-pear cactus (*Opuntia*).

Many species of armored scales (Diaspididae) may be recognized by the characteristic scales they produce (Fig. 34.12*d*), but family and species classification and identification of most Coccoidea are extremely difficult, requiring removal of the scale, clearing, staining, and mounting on slides. About a dozen families are recognized by most authorities. For identification of the common economic species, the reader is referred to Metcalf et al. (1962). The most complete taxonomic treatments of North American species are those of Ferris (1937–1953) and McKenzie (1967).

SELECTED REFERENCES

GENERAL

Blackman, 1974 (biology of aphids); Myers, 1929 (biology of cicadas); Poisson and Pesson, 1951; Southwood and Leston, 1959 (Heteroptera); Van Emden, 1972*a*.

MORPHOLOGY

Ekblom, 1926 (Heteroptera); Kramer, 1950 (Homoptera); Poisson and Pesson, 1951; Spooner, 1938 (head capsule); Williams and Kosztarab, 1972 (Coccoidea).

BIOLOGY AND ECOLOGY

Alexander and Moore, 1962 (periodical cicadas); Beardsley and Gonzalez, 1975 (armored scales); Blackman, 1974 (aphids); Delong, 1971 (Cicadellidae); Hille Ris Lambers, 1966 (aphid polymorphism); Kennedy and Stroyan, 1959 (aphids); Miller, 1956 (Heteroptera); Myers, 1929 (cicadas); Ossiannilsson, 1949 (sound production in Homoptera); Way, 1963 (mutualism between Homoptera and ants).

SYSTEMATICS AND EVOLUTION

Blatchley, 1926 (eastern North American species); W. E. Britton, 1923 (Connecticut species); A. R. Brooks and Kelton, 1967 (Canadian species); Cobben, 1968 (systematic and evolutionary importance of eggs); DeCoursey, 1971 (nymphs); J. W. Evans, 1963 (phylogeny of Homoptera); Ferris, 1937–1955 (scale insects of North America); Froeschner, 1960 (Cydnidae); Herring and Ashlock, 1971 (nymphs); Hungerford, 1959 (aquatic species); Kramer, 1950, 1960 (phylogeny of Homoptera); McKenzie, 1967 (Coccoidea of California); Stannard, 1956 (phylogenetic relationships); Usinger, 1956*b* (aquatic species of California); Williams and Kosztarab, 1972 (Coccoidea of Virginia).

35

ORDER THYSANOPTERA (Thrips)

Minute to small Exopterygota with slender, elongate body (Fig. 35.1*a*). Head elongate, hypognathous; mouthparts asymmetrical, with maxillae and left mandible modified as piercing stylets; eyes prominent, round or kidney-shaped, with large, round facets; antennae 4- to 9-segmented, inserted anteriorly. Thoracic segments subequal or prothorax largest; mesothorax and metathorax fused; wings narrow, straplike, fringed with long setae; legs short, adapted for walking; tarsi 1- or 2-segmented with eversible, glandular vesicle at apex. Abdomen 11-segmented, terminating in a tubular apex or with a large, serrate ovipositor.

These highly distinctive insects are remarkable for their ubiquity, inhabiting all sorts of vegetation, as well as leaf litter, fungus, and the subcortical region of decaying trees. Most foliage-inhabiting species imbibe plant juices released from cells pierced by the oral stylets. This method of feeding produces a characteristic mottling or silvering of the affected surfaces. Many phytophagous species preferentially infest the reproductive structures of plants, and the most serious economic damage involves buds, flowers, and young fruits. Various species transmit bacterial, fungal, or viral diseases of plants, the most serious being spotted wilt of tomatoes and other plants. Many species are predators of aphids or other small, soft-bodied arthropods, especially eggs of mites and Lepidoptera. Some feed on fungal spores, and a few cause formation of galls. Flower-inhabiting thrips commonly suck the liquid contents from pollen grains, and the larger Tubulifera can ingest small pollen grains and fungal spores. Thrips occasionally create a nuisance by biting humans.

The unique, asymmetrical mouthparts (Fig. 35.1*d*) of thrips share several details with the specialized styletiform mouthparts of Hemiptera. The single, awl-shaped mandible and the maxillary stylets (Fig. 35.1*b,c*) slide through a conical guide formed by the labrum, labium, and external maxillary plates. Internally the maxillary stylets are attached to the maxillary plates by a slender lever bearing the attachments of the protractor and retractor muscles, as in the Hemiptera. In feeding, the oral cone is appressed to the food and the stylets are punched a short distance into the tissues. The exuding fluids are sucked into the alimentary canal by a cibarial pump.

The fringed wings are a specialization which appears repeatedly in very small insects, frequently in conjunction with reduction of the area of the wing membrane. Winged thrips are competent fliers which may be disseminated long distances by favorable air movements; mass flights are recorded for several species. The tarsus bears an adhesive vesicle which is protruded during walking, enabling thrips to traverse very smooth surfaces. Such adhesive organs are apparently an adaptation for clinging to exposed plant surfaces, and analogous structures occur in a variety of phytophagous species in different orders. Most thrips are bisexual, but parthenogenesis is common, and males are rare or unknown in some species. Polymorphism is also of frequent occurrence, being most evident in the

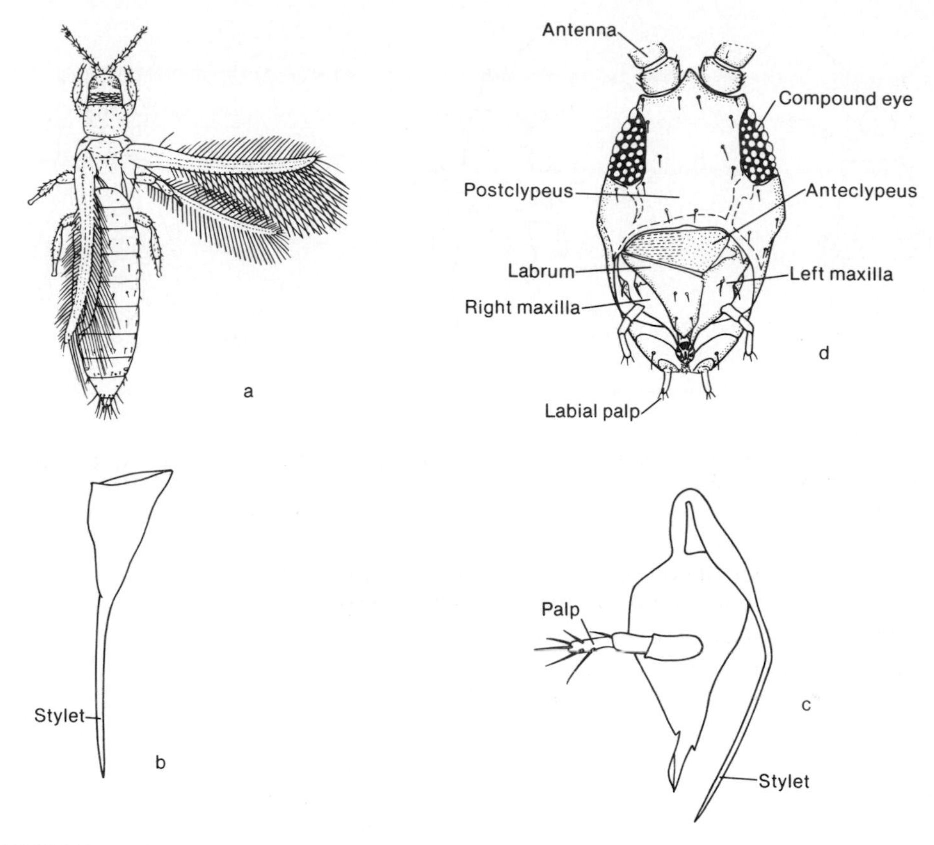

FIGURE 35.1 Thysanoptera: *a, Anaphothrips striata,* dorsal aspect; *b,* mandible, same; *c,* maxilla, same; *d,* frontal view of head, *Chirothrips hamatus.* (a, b, c *from Hinds,* 1900; d *from Jones,* 1954.)

degree of development of the wings, which may be present, reduced, or absent, especially in males.

Female thrips are diploid, while males are produced from unfertilized eggs and are haploid. This type of parthenogenetic reproduction, termed *arrhenotoky,* is universal in Hymenoptera (Chap. 46). Eggs are large in relation to body size. In different species, numbers of eggs vary from as few as two to several hundred. In the Suborder **Tubulifera**, in which the ovipositor is not developed, eggs are deposited in cracks, crevices, under bark, or exposed on foliage surfaces. In the Suborder **Terebrantia**, the serrate, bladelike ovipositor is used to insert the eggs into plant tissue. A few species are ovoviviparous. Eggs hatch after about 2 to 20 days, depending on temperature. The newly emerged nymphs resemble adults except in the absence of wings and in the reduced number of antennal segments. Two feeding instars are followed by a "prepupal" instar and one "pupal" instar in the Terebrantia, or by one prepupal and two pupal instars in the Tubulifera (Fig. 35.2). The pupal stage is usually passed in a cell prepared by the previous instar from soil or litter particles, but some species construct a silk cocoon or transform naked in cracks or crevices or on foliage. The prepupal and pupal stages are analogous to the pupae of endopterygote insects, and histolysis of various organs, especially head muscles, has been demonstrated in various species. Unlike metamorphosis in endopterygotes, external growth of wing pads in thrips occurs in more than one

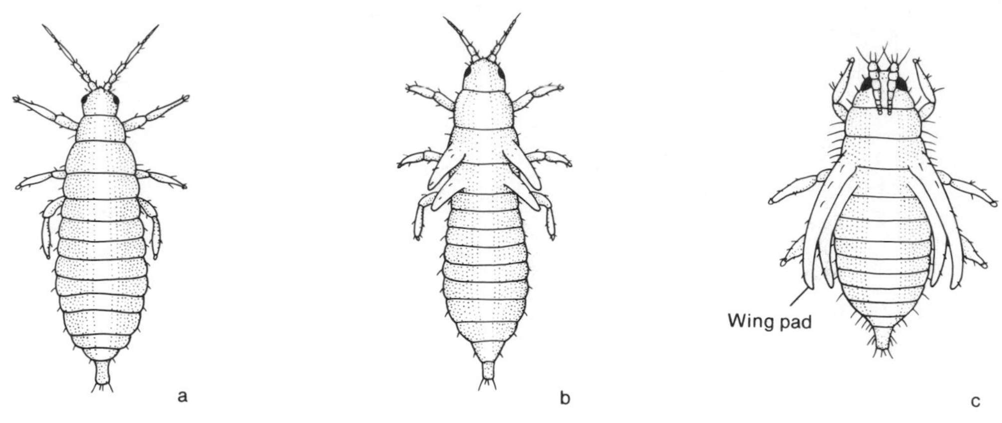

FIGURE 35.2 Developmental stages of Thysanoptera (*Hercothrips fasciatus*): *a*, last-instar larva; *b*, prepupa; *c*, pupa. (From Bailey, 1932.)

instar. Furthermore, their specialized mouthparts, reduced wing venation, and the organization of internal organs indicate that thrips represent a specialized derivative of hemipteroid insects, rather than primitive Endopterygota.

Many thrips pass through numerous generations annually, requiring as few as 10 days to develop from egg to adult. Others with a single generation spend most of the year in pupal diapause. Nymphs, pupae, or adults may overwinter in various species.

KEY TO THE FAMILIES OF
Thysanoptera

1. Terminal abdominal segment cylindrical, tubular (Fig. 35.3*a*); forewing (if present) with membrane smooth, glabrous (Suborder Tubulifera) **Phlaeothripidae**
   Terminal abdominal segment rounded or conical (Fig. 35.3*b*); forewing (if present) with membrane bearing minute hairs (microtrichia) (Suborder Terebrantia) 2

2(1). Antennae with longitudinal patches of sensilla

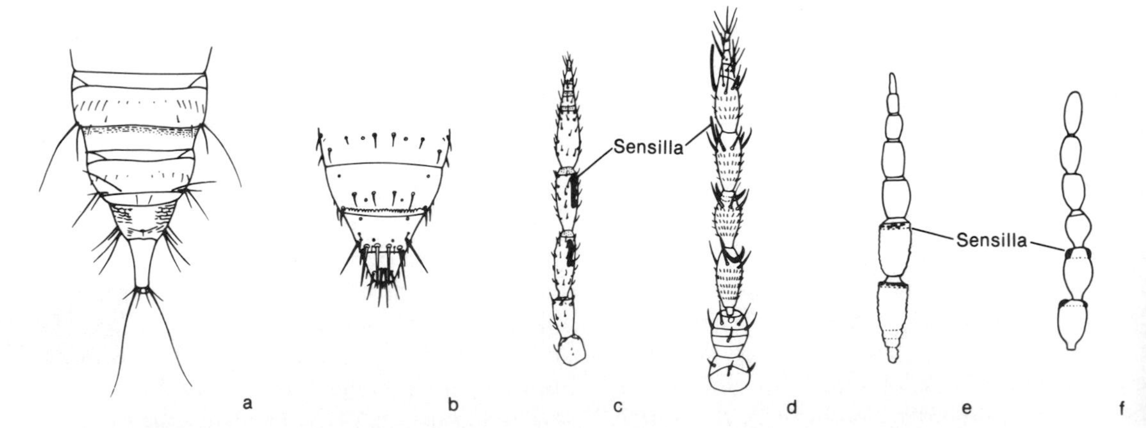

FIGURE 35.3 Taxonomic characters of Thysanoptera: *a*, Suborder Tubulifera, abdominal apex (*Hoplandothrips irretius*, Phlaeothripidae); *b*, Suborder Terebrantia, abdominal apex (*Anaphothrips striata*, Thripidae). *c* to *f*, Antennae, showing arrangement of sensilla; *c*, Aeolothripidae (*Aeolothrips scabiosatibia*); *d*, Thripidae (*Anaphothrips striata*); *e*, Heterothripidae (*Heterothrips salicis*); *f*, Merothripidae (*Merothrips morgani*).

on segments 3 to 4 (Fig. 35.3*c*); females with ovipositor curved upward **Aeolothripidae**
Antennae with round or transverse patches of sensilla on segments 3 to 4 (Fig. 35.3*d* to *f*) 3

3(2). Antennae with sensilla on segments 3 to 4 arranged as simple or forked cones (Fig. 35.3*d*) **Thripidae**
Antennae with sensilla on segments 3 to 4 as continuous or interrupted bands encircling apices of segments (Fig. 35.3*e,f*) 4

4(3). Antennae with 8 segments **Merothripidae**
Antennae with 9 segments **Heterothripidae**

Most common thrips, including those of economic significance, are in the Families Thripidae and Phlaeothripidae. Aeolothripidae are sometimes abundant on flowers or in litter. Most species have banded or maculated wings. Merothripidae and Heterothripidae are poorly represented in North America, a few species occurring in litter or on fungus and in flowers, respectively.

SELECTED REFERENCES

Bailey, 1957 (systematics, California species); Jones, 1954 (morphology); Lewis, 1973 (comprehensive general work); Sharov, 1975 (phylogenetic relationships); Stannard, 1957 (systematics of Tubulifera) and 1968 (Illinois species).

36 ORDER MEGALOPTERA (Dobsonflies)

Moderate-sized to large Endopterygota. *Adults* (Fig. 36.1*a*) with strong mandibles, maxillae, and labium, large compound eyes, ocelli present or absent. Thorax with 3 subequal segments; mesothorax and metathorax with large, subequal wings; metathoracic wings with pleated anal region folded over abdomen at rest. *Larva* (Fig. 36.2) with large head, large prognathous mandibles, large, sclerotized prothorax, smaller sclerotized mesothorax and metathorax; abdomen elongate, with segmented or unsegmented gills on segments 1 to 7 or 8; terminal segment bearing elongate medial appendage or paired prolegs. *Pupa* decticous, motile.

Megaloptera are morphologically similar to Neuroptera as adults, and in some classifications are included as a suborder of Neuroptera. The most reliable differentiating feature is the presence of an anal fold in the hind wing (Fig. 36.1*c*) (no folding in Neuroptera). In contrast, larvae of the two groups are well differentiated. Whereas nearly all neuropterous larvae have specialized suctorial mouthparts (Fig. 38.1*c*), in the Megaloptera mandibles and maxillae are unmodified and adapted for chewing (Fig. 36.2). The alimentary canal is complete in Megaloptera (end-

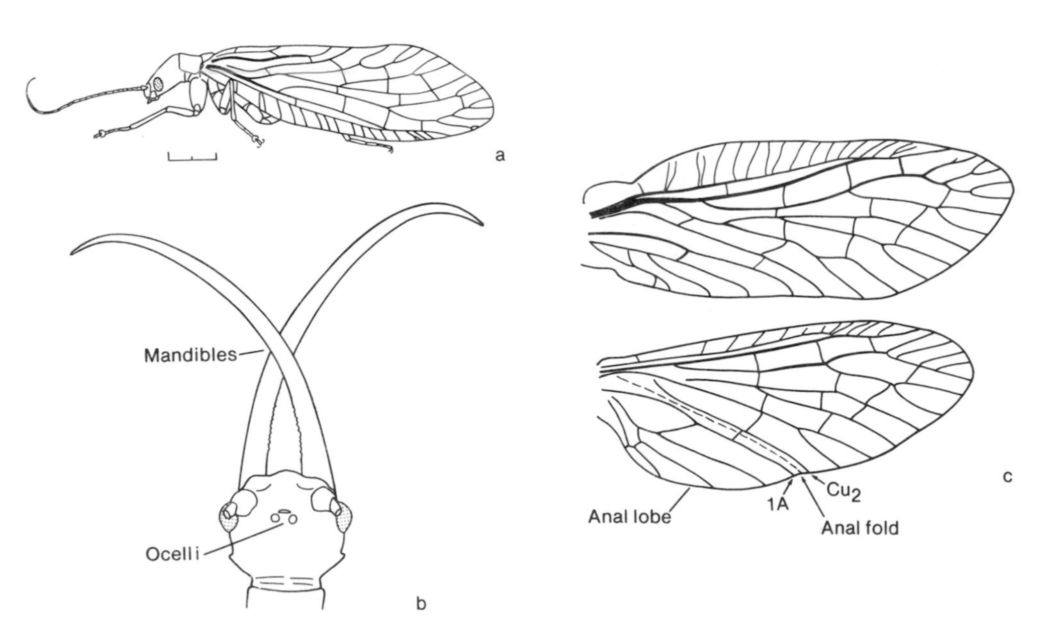

FIGURE 36.1 Adult Megaloptera: *a, Sialis californicus* (Sialidae; scale equals 2 mm); *b,* head of male *Corydalus,* antennae removed (Corydalidae); *c,* wings, *Sialis californicus.*

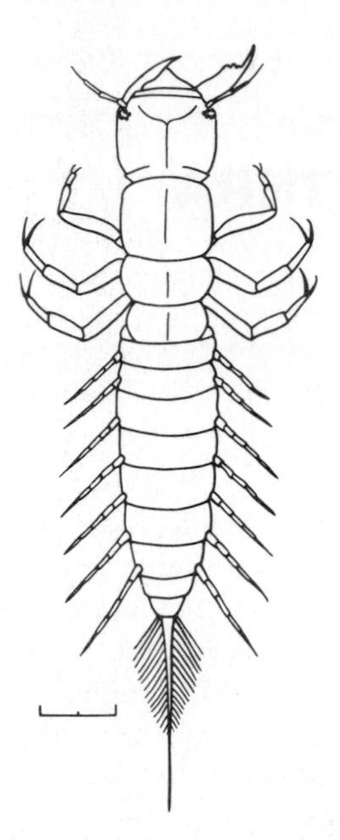

FIGURE 36.2 Larva of *Sialis* (Sialidae). Scale equals 2 mm.

ing in the blind midgut in Neuroptera), and pupation occurs in an earthen cell (within a silk cocoon in Neuroptera). In general structure (Fig. 36.2), larvae of Megaloptera are more similar to adephagous beetle larvae than to Neuroptera, and separation from Gyrinidae and Dytiscidae is sometimes difficult.

Megaloptera are familiar insects around permanent water, being especially prevalent about cool, well-oxygenated streams. Adults are weak fliers, seldom straying more than a few hundred meters from water. Apparently they do not feed, despite their well-developed mandibles, which are greatly enlarged in males of some Corydalidae (Fig. 36.1*b*). Mating occurs on streamside vegetation or on the ground. Single females may produce as many as several thousand eggs, and communal egg deposition sites on vegetation or other objects overhanging water may be used by many individuals, resulting in large egg masses.

Larvae shelter beneath stones, sunken vegetation, or other protected situations, where they attack any aquatic organisms small enough to subdue, and in turn are consumed by fish, frogs, and other predators. There are numerous larval instars, and full growth apparently requires a year in Sialidae and several years in Corydalidae. The mature larvae leave the water to excavate a pupal chamber in moist sand or soft earth beneath stones or driftwood. Pupae retain full mobility and can defend themselves with the mandibles.

KEY TO THE FAMILIES OF **Megaloptera**

1. Adults with 3 ocelli (Fig. 36.1*b*), 4th tarsal segment cylindrical, wing span 45 to 100 mm; larvae with 8 pairs of unsegmented abdominal gills, prolegs on last abdominal segment **Corydalidae**

   Adults without ocelli, 4th tarsal segment bilobed, wingspan 20 to 40 mm; larvae with 7 pairs of segmented abdominal gills, terminal filament on last abdominal segment (Fig. 36.2) **Sialidae**

Corydalidae. Widespread, with species throughout North America. Larvae occur predominantly in running water, especially fast-flowing streams. Adults are partly nocturnal, commonly appearing about lights.

Sialidae. Widespread in North America, most diverse in the west. Larvae occur in ponds and sluggish watercourses, as well as in swiftly flowing streams. Adults diurnal.

SELECTED REFERENCES

Chandler, 1956 (California species); Kelsey, 1954, 1957 (morphology); Merritt and Cummins, 1978 (general); Parfin, 1952 (Minnesota species); Ross, 1937*b* (systematics of *Sialis*).

37

ORDER RAPHIDIOPTERA (Snake Flies)

Medium-sized Endopterygota. *Adult* (Fig. 37.1*a*) with strong mandibles, maxillae, and labium, prominent compound eyes, ocelli present or absent. Prothorax slender, elongate, necklike; mesothorax and metathorax subequal; wings similar in size and venation, with large pterostigma; female with long, slender ovipositor. *Larva* (Fig. 37.1*b*) with large head, strong prognathous mandibles; prothorax sclerotized, slightly larger than membranous mesothorax and metathoras; abdomen fleshy, cylindrical. *Pupa* decticous, motile.

In most features snake flies resemble Megaloptera (Fig. 36.1); they are sometimes considered specialized terrestrial members of that order. Unlike the Megaloptera, which do not feed as adults, snake flies are voracious predators throughout life, and the elongate head capsule and prothorax are adaptations to increase mobility of the head, which is used to strike the prey, in a snakelike fashion. Fossils similar to modern Raphidioptera date from the Jurassic Period,

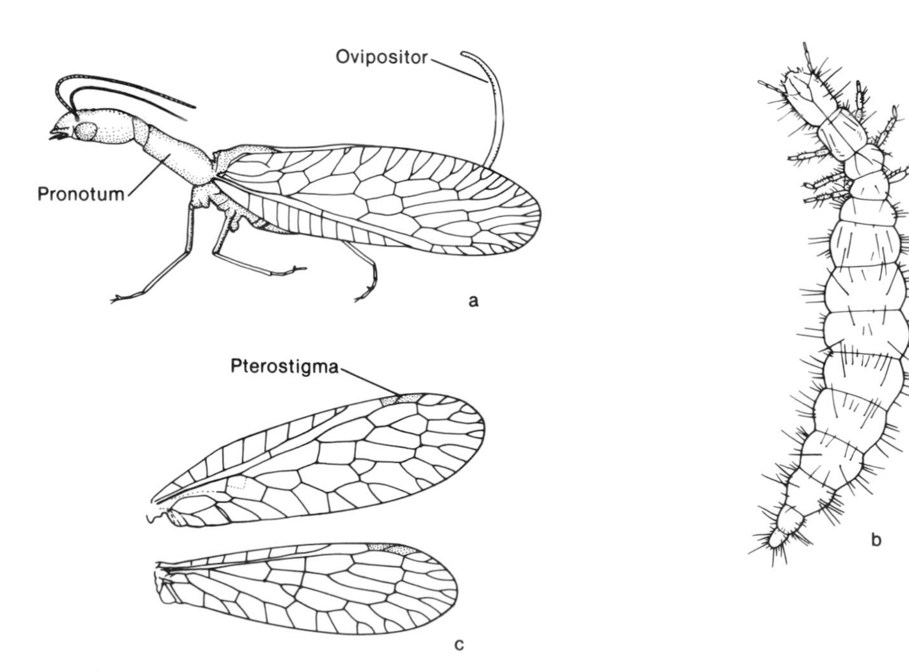

FIGURE 37.1 Raphidioptera, Raphidiidae: *a*, adult of *Agulla bracteata; b*, larva of *Agulla bracteata; c*, wings of *Agulla astuta*. (*From Woglum and McGregor*, 1958, 1959.)

suggesting a very early divergence from the Megaloptera.

Raphidioptera predominantly inhabit woodlands, where the adults are encountered on foliage, flowers, tree trunks, or similar places. Eggs are inserted into crevices in bark or rotting wood. Larvae commonly occur under loose bark, in porous rotten wood, leaf litter, and similar places. As in Megaloptera, pupation is in a rough cell constructed by the last-instar larva, without a cocoon. The pupa retains full mobility and frequently moves to a second location before adult emergence.

Snake flies occur throughout most of the world except Australia. Two families occur in North America, both restricted to the region west of the Rocky Mountains. In Raphidiidae (Fig. 37.1) ocelli are present, antennal segments are basally constricted, and antennae are usually less than half as long as the body. In Inocellidae ocelli are absent, antennal segments are cylindrical, and antennae are usually about as long as the body.

SELECTED REFERENCES

Achtelig and Kristensen, 1973 (phylogenetic relationships); Aspöck and Aspöck, 1975 (general); Carpenter, 1936 (systematics of North American species); Ferris and Pennebaker, 1939 (morphology); Woglum and McGregor, 1958, 1959 (life history).

38

ORDER NEUROPTERA (Lacewings, Ant Lions, etc.)

Small to large Endopterygota. *Adult* with chewing mouthparts, large lateral eyes, ocelli present or absent; antennae multiarticulate, usually filiform or moniliform. Mesothorax and metathorax similar in structure, with subequal wings, usually with similar venation. Abdomen without cerci. *Larvae* with clearly defined head capsule; mandibles and maxillae usually elongate, slender, modified for sucking (Fig. 38.1*c*); thoracic segments with walking legs with 1-segmented tarsus usually bearing 2 claws; abdomen frequently bearing suction disks on last 2 segments, without cerci. *Pupa* exarate, decticous, enclosed in silken cocoon.

The Neuroptera comprise a small but highly variable order of predominantly predatory insects which display a mixture of primitive and specialized features. Permian fossils include several families of Neuroptera, and by the mid-Mesozoic Era forms similar to most modern families had appeared. Neuroptera occur in all parts of the world, but many families show relict distributions. The fauna of Australia is especially rich, including several endemic families as well as a great variety of primitive forms from other families. Ithonidae are restricted to North America and Australia, and Polystoechotidae to North America and southern South America. Most North American Neuroptera are green or various shades of gray or brown, sometimes with dark wing maculations; they are mostly rather obscure insects. Many Australian and oriental species are strikingly colored or patterned.

Adult Neuroptera are soft-bodied insects of generalized body plan (Fig. 38.1*a*). The mouthparts are adapted for chewing, with strong mandibles and maxillae, and small labium. The thoracic structure and wing venation are especially generalized, compared with those of most other Endopterygota. In most species the thoracic segments are subequal, with the prothorax freely movable, the wings (Fig. 38.3) are similar in size, shape, and venation, except in Coniopterygidae, in which the hind wings may be much reduced. In Ithonidae, as well as several exotic families, characteristic sensory structures (*nygmata*) occur in the wing membrane. Wing coupling occurs in several families, especially those with strong flying members. In Hemerobiidae coupling is by a bristlelike frenulum on the base of the hind wing. Similar coupling occurs in some Chrysopidae and Mantispidae. The wings of Coniopterygidae are coupled by hamuli-like setae on the costal margin of the hind wings. Abdominal scent glands, usually situated on the first or fifth to seventh abdominal tergites, occur in several families, including Myrmeleontidae and Ascalaphidae. In Chrysopidae these glands occur on the prothorax and offer some protection against vertebrate and invertebrate predators.

Immatures of Neuroptera (Figs. 38.1*b*, 38.2) are all predatory, frequently specialized for particular types of prey. A peculiarity of all species is the blindly ending midgut. During pupation the alimentary canal becomes complete, and the accumulated fecal material is voided shortly after adult emergence.

In Polystoechotidae and Ithonidae larval mandibles and maxillae are short and blunt. The larva of *Polystoechotes* is somewhat similar in general morphology to larvae of Megaloptera

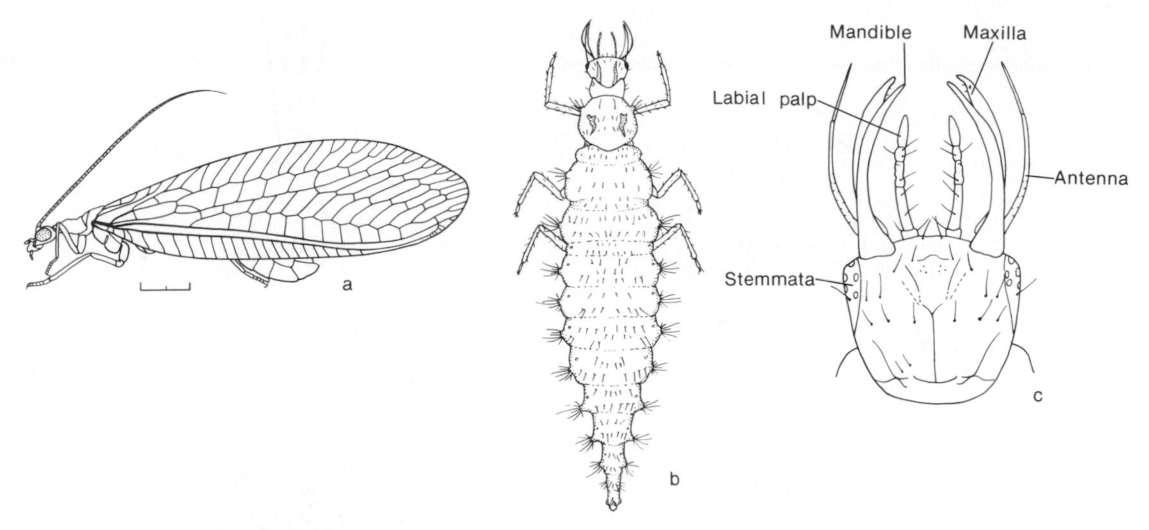

FIGURE 38.1 Neuroptera, Chrysopidae (*Chrysopa carnea*); *a*, adult (scale equals 2 mm); *b*, mature larva; *c*, head of larva. (b *from Tauber*, 1974; c *from Withycombe*, 1924.)

(Withycombe, 1925), and Polystoechotidae are probably among the most primitive extant Neuroptera. In most other larvae the mandibles are elongate, with a deep mesal groove which is covered by the maxilla to form an efficient piercing and sucking apparatus (Fig. 38.1*c*). Configuration of the larval body is related to mode of life. Chrysopidae (Fig. 38.1*b*), Hemerobiidae, and Coniopterygidae are freely roving predators of small, soft-bodied arthropods, especially Homoptera and mites. Their legs are relatively long with large empodia; the abdomen is slender, tapering, and prehensile, an adaptation for clinging to vegetation. In contrast, in the Myrmeleontidae (Fig. 38.2*a*) and Ascalaphidae, the abdomen and thorax are either globular or broadly flattened and are relatively inflexible. The legs are short and in Myrmeleontidae are adapted for backward movement and for digging. The head is broad and flat, with enormous, sickle-shaped jaws. These larvae are sedentary predators which ambush prey from concealed positions. Many Myrmeleontidae construct conical pitfalls in dry, friable soil, the larva buried at the bottom with open jaws. Ascalaphid larvae sit motionless in litter or on vegetation, snapping the jaws closed on any suitably sized prey. Larvae which frequent exposed situations on vegetation sometimes bear specialized hooked or spatulate setae which anchor a camouflaging layer of debris, including the dry husks of prey. This habit is shared with some chrysopid larvae.

Larvae of Sisyridae (Fig. 38.2*b*) are superficially similar to those of Chrysopidae and Hemerobiidae but are specialized predators of freshwater sponges. They are distinguished by their straight, needlelike mandibles and jointed abdominal gills. Mantispid larvae are specialized parasitoids of spider egg masses or of immature vespid wasps. The motile first instar, which is similar to the larvae of chrysopids, searches for a suitable host before molting to the grublike second instar, with short, ineffective legs and reduced head capsule. Larvae of Berothidae are predators of ants and termites, in whose nests they live.

Many adult Neuroptera are predators of a variety of insects; other species imbibe nectar, pollen, or the honeydew secreted by homopterous insects. Adults are relatively short-lived except for those which diapause. Most of them are weak, erratic fliers, but flight is strong and direct in many Ascalaphidae. Eggs are deposited in soil in ground-dwelling forms, or on foliage in

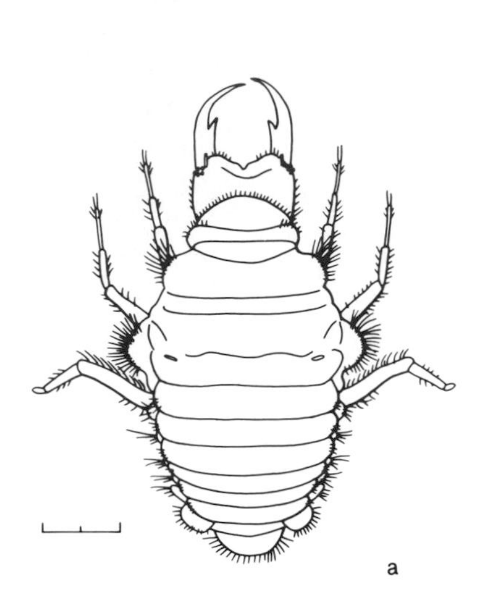

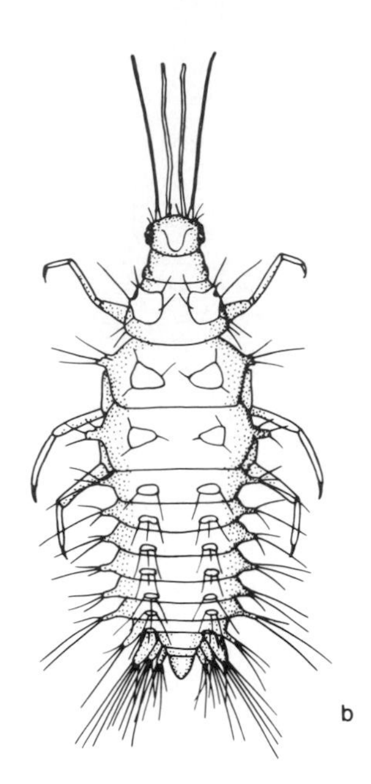

FIGURE 38.2 Larvae of Neuroptera: *a*, Myrmeleontidae (*Vella;* scale equals 2 mm); *b*, Sisyridae (*Climacia areolaris*). (b *from Parfin and Gurney,* 1956.)

arboreal types, being sessile (Hemerobiidae, Dilaridae, Coniopterygidae) or attached by a short or long stalk (Mantispidae, Chrysopidae, Berothidae). Three larval instars are typical of most families, with four or five instars reported in Ithonidae. The length of larval life is correlated with type of prey and prey availability. In those species which consume relatively abundant prey, such as Homoptera, full growth may be attained after only a few weeks. In Myrmeleontidae and Ascalaphidae, where hunting success is highly unpredictable, larvae are capable of ingesting very large meals, then fasting for many months if necessary. In such forms larval life may last several years.

All Neuroptera pupate within shelters spun from silk produced by the Malpighian tubules. In Hemerobiidae the cocoon is usually wispy, but Chrysopidae and soil-dwelling types may incorporate sand or debris to form a tough protective covering. In some families the cocoon is double-layered. The movable pupal mandibles are sharply toothed, serving to open the cocoon. It may be noted that pupal mandibles in Trichoptera and a few primitive Lepidoptera have the same function.

KEY TO NORTH AMERICAN FAMILIES OF

Neuroptera

1\. Wings with few veins and less than 10 closed cells (Fig. 38.3*a*); small to minute insects covered with whitish exudate **Coniopterygidae**

Wings with numerous veins; more than 10 closed cells (usually very numerous); body without whitish exudate 2

2(1). Forelegs raptorial (Fig. 38.4*b*) **Mantispidae**

Forelegs not raptorial 3

3(2). Antennae filiform, moniliform, or pectinate (see Fig. 2.7) 5

Antennae gradually or abruptly clubbed 4

4(3). Antennae nearly as long as body, with abrupt knob at apex **Ascalaphidae**

Antennae about as long as head and thorax, gradually thickened (Fig. 38.4*a*) **Myrmeleontidae**

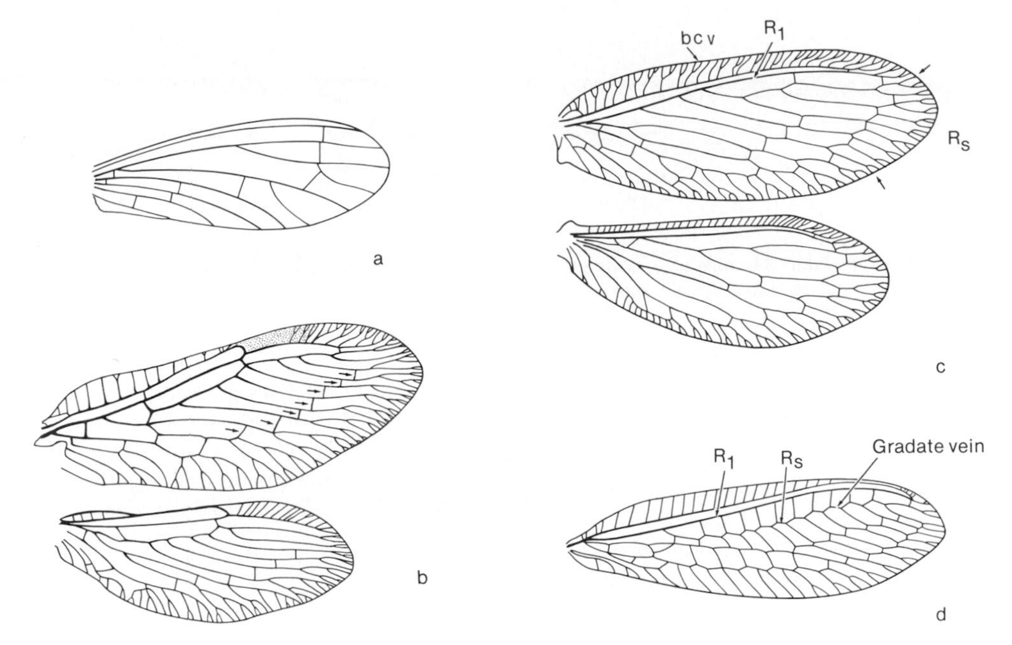

FIGURE 38.3 Wings of Neuroptera: *a,* forewing of Coniopterygidae (*Conwentzia hageni*); *b,* Mantispidae (*Plega signata*); *c,* Hemerobiidae (*Hemerobius ovalis*); *d,* forewing of Chrysopidae (*Chrysopa carnea*). *bcv,* bifurcate costal crossveins; extent of R_s marked by marginal arrows. (a *from Banks,* 1907.)

5(3). Wings with sensory spots (nygmata) in membrane between branches of radial vein; hind wing abruptly narrowed at base **Ithonidae**
Wings without sensory spots; hind wing gradually narrowed at base (Fig. 38.3*b* to *d*) 6

6(5). Antennae pectinate (see Fig. 2.7) (male), or with ovipositor exserted, as long as abdomen (female) **Dilaridae**
Antennae filiform or moniliform; ovipositor internal, not visible 7

7(6). Wings with numerous crossveins between veins R_1 and R_s (Fig. 38.3*d*) or R_s arising from 2 to 3 stems (Fig. 38.3*c*) 8
Wings with 1 to 4 crossveins between veins R_1 and R_S (as in Fig. 38.3*b*) 9

8(7). Forewing with some costal crossveins bifurcate (Fig. 38.3*c*); body and wings usually brown **Hemerobiidae**
Forewing without bifurcate costal crossveins (Fig. 38.3*d*); body and wings usually green **Chrysopidae**

9(7). Forewing with numerous parallel branches of radial vein, wingspan at least 40 mm **Polystoechotidae**
Forewing with 4 to 7 branches of radial vein; wingspan less than 30 mm 10

10(9). Forewing with some costal crossveins bifurcate (as in Fig. 38.3*b,c*); gradate veins present (as in Fig. 38.3*b*) **Berothidae**
Forewing without bifurcate crossveins (Fig. 38.3*d*); gradate veins absent **Sisyridae**

Coniopterygidae. Small or minute insects, superficially resembling whiteflies (Aleyrodidae) because of the reduced wing venation (Fig. 38.3*a*) and the whitish powder which covers the body and wings. Coniopterygids are common, widespread insects but usually escape notice because of their small size.

Ithonidae and Polystoechotidae. A single, rare ithonid occurs in southern California; the two North American polystoechotids are more widespread but also rare. Adults of both families are relatively large (30 to 70 mm) with broad wings and robust body.

Dilaridae. Small lacewings (wingspan, 6 to 12 mm) whose larvae hunt on tree trunks. Widely distributed, with 2 uncommon species in eastern North America.

Berothidae. Small lacewings, superficially similar to Hemerobiidae, usually distinguished by scalloped

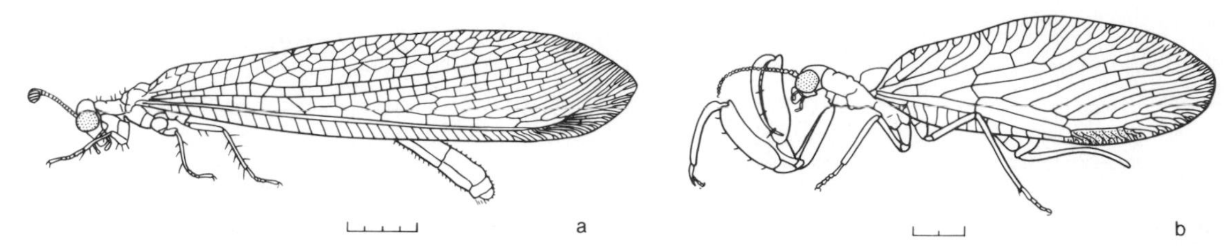

FIGURE 38.4 Adult Neuroptera: *a*, Myrmeleontidae (*Myrmeleon immaculatus;* scale equals 4 mm); *b*, Mantispidae (*Plega dactylota*, female; scale equals 2 mm).

outer wing margins. The North American fauna includes less than 15 uncommonly encountered species.

Sisyridae. Widespread but seldom encountered, being restricted by the occurrence of freshwater sponges, their only hosts. Eggs are deposited on foliage overhanging water, the newly emerged larvae dropping into the water. Pupation occurs on emergent vegetation or on other objects near the water. At least one of several species occurs in most regions of North America.

Hemerobiidae and Chrysopidae (lacewings; Fig. 38.1). The superficially similar adults are reliably distinguished by the presence or absence of bifurcate costal crossveins. Wings of Hemerobiidae are frequently hairy, those of Chrysopidae bare. These two families represent one of the dominant groups of Neuroptera, with many North American species.

Mantispidae (Fig. 38.4*b*). Distinguished from all other Neuroptera by the enlarged raptorial forelegs. Nocturnal species are gray or brown; diurnal species include bright green forms as well as others patterned in red, yellow, and black, apparently mimicking vespid wasps. Mantispidae are most abundant in tropical and subtropical regions.

Myrmeleontidae and Ascalaphidae. Differentiated from other Neuroptera by the elongate body form, similar to that of Odonata. Ascalaphidae are not familiar insects in North America, but Myrmeleontidae (ant lions, Figs. 38.2*a*, 38.4*a*) are among the commonest Neuroptera, especially in arid regions, where large numbers of the nocturnal adults often collect about lights.

SELECTED REFERENCES

MORPHOLOGY

Ferris, 1940; Mickoleit, 1973 (ovipositor); Withycombe, 1925.

BIOLOGY AND ECOLOGY

McKeown and Mincham 1948 (Mantispidae); New, 1975 (Chrysopidae); R. C. Smith, 1922 (Chrysopidae); Tauber, 1969 (Chrysopidae); Toschi, 1965 (Chrysopidae); Wheeler, 1930 (Myrmeleontidae); Withycombe, 1925.

SYSTEMATICS

Banks, 1927 (nearctic Myrmeleontidae); MacLeod and Adams, 1967 (Berothidae); Meinander, 1972 (Coniopterygidae); Carpenter, 1940 (general); Mickoleit, 1973 (phylogeny); Peterson, 1951 (larvae); Rehn, 1939 (North American Mantispidae).

ORDER COLEOPTERA (Beetles)

39

Adult. Minute to large Endopterygota. Compound eyes present (absent in many specialized forms), ocelli almost always absent; antennae almost always with 11 or fewer segments; mouthparts adapted for chewing, with recognizable mandibles, maxillae, and labium. Prothorax large, mobile, with ventral extensions of notum usually extending to vicinity of coxae; mesothorax small, fused with metathorax ventrally; mesonotum visible externally as scutellum; mesothoracic wings developed as sclerotized, rigid, elytra with specialized longitudinal venation; metathorax large (small in flightless forms), with strongly sclerotized sternum, weakly sclerotized tergum. Abdomen usually with sterna strongly sclerotized, terga membranous or weakly sclerotized with 2 or more terminal segments usually reduced; cerci absent.

Larva. Body form extremely variable but almost always with sclerotized head capsule, antennae with 2 to 4 segments, mouthparts mandibulate; thoracic segments with 4 or 5 segmented legs bearing 1 or 2 terminal claws; abdomen without legs, but frequently bearing segmented or unsegmented, sclerotized urogomphi on terminal segments. Wood-boring or other internally feeding larvae sometimes have legs, sensory organs, and head capsule reduced.

Pupa. Adecticous, exarate, similar to adult in general form.

The Coleoptera, with more than 250,000 described species, constitute the largest order of insects. Five families of beetles number over 20,000 species, weevils (Curculionidae) topping the list with at least 60,000 members, substantially more than most individual phyla of animals. As might be expected in such a large assemblage, Coleoptera are exceedingly variable ecologically and biologically. The majority of beetles are terrestrial herbivores, but several entire families and portions of others are predatory, frequently with highly specialized host ranges or life cycles. Either larvae or adults or both may be aquatic, with numerous freshwater and a few marine (intertidal) species known. In addition, beetles have exploited an extremely diverse array of narrow, specialized niches, such as seed predation; boring in leaves, stems, or other restricted plant parts; formation of galls or tumors on plants; life inside ant or termite colonies, and many others. Certain relatively specialized habitats that have been extensively utilized, often to the near exclusion of other insects, are (1) the subcortical region of woody plants; (2) leaf litter and the top layers of soil; and (3) fungi, including fungi in soil, subcortical habitats, dung, and carrion, as well as the fruiting bodies of mushrooms and shelf fungi. One major source of food that has been scarcely exploited by Coleoptera is nectar from flowering plants, probably because beetles are relatively inefficient fliers.

To a major extent, the evolutionary success of Coleoptera seems to stem from the protection from physical trauma that is provided by the high degree of sclerotization and body compaction. The most obviously affected structures are the forewings or elytra, which are modified as protective covers for the hind body (Fig. 39.1). Elytra are almost always present, even in wingless forms, but may be abbreviated, in which case the abdominal terga are sclerotized. In most beetles the elytra form hard sheaths, molded to the shape of the abdomen and interlocking with one another along the midline at rest. The movement of the elytra is limited to opening and

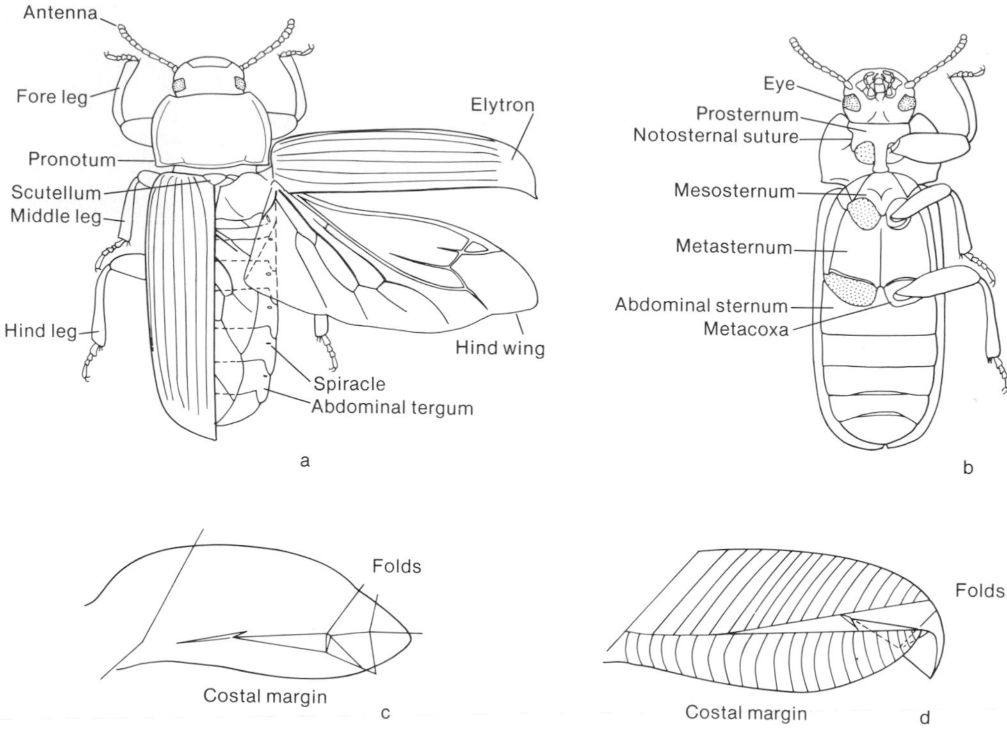

FIGURE 39.1 Anatomy of Coleoptera: *a*, dorsal aspect of *Tenebrio molitor* (Tenebrionidae), right elytron and hind wing extended; *b*, ventral aspect of same; *c*, diagram of folds in hind wing of same; *d*, folding of wing of same.

closing at times of flight. In flightless forms the elytra often interlock immovably along the midline and may dovetail with the abdominal sterna laterally.

Except during flight, the metathoracic wings are concealed beneath the elytra. This is accomplished by a complex series of foldings in the wing membrane (Figs. 39.1*c*, 39.3*b*), conforming the size and shape of the wing to the space available in the subelytral cavity. Wing venation is specialized to allow the necessary folding, with the few closed cells in the unfolded anal portion of the wing (Fig. 39.1*a*).

In various families body compaction may be increased in a variety of ways. Commonly the head is more or less deeply retracted into the thorax. In many families the coxae are enclosed in cavities in the sternal sclerites, reducing mobility but increasing the strength of the coxal articulations. Antennae and the distal joints of legs may also be withdrawn into concavities, and in strongly modified species the heavily armored body is practically without projections when the head and appendages are retracted.

The spiracles open into the subelytral cavity, with the exception of the mesothoracic pair, which is concealed between the prothorax and mesothorax. The enclosed position of the spiracles creates a moisture gradient, greatly enhancing water retention, and Coleoptera are usually a conspicuous element of arid habitats. Arid-adapted forms are often flightless, with elytra tightly interlocking with the abdominal sternites. Spiracles may be provided with special closing mechanisms, and Malpighian tubules may be cryptonephric, further increasing water-retention efficiency.

In general, body form in Coleoptera reflects the mode of life. Subcortical species are often dorsoventrally flattened, sometimes extremely so, allowing entry between tightly adherent bark and wood. In these species the legs are laterally directed and are usually short. Species which burrow through substrates tend to be very compact, either circular or oval in cross section. Swimming forms are streamlined.

Many Coleoptera, especially those which lead exposed lives, store defensive substances which are self-produced or obtained from plants. These may be contained in the hemolymph (Coccinellidae, Lycidae) or held in reservoirs and forcibly ejected when needed (Carabidae, Tenebrionidae). Compounds produced range from alkaloids to various organic acids, aldehydes, and quinones. Coleoptera include the most familiar bioluminescent organisms (fireflies, glowworms), in most of which luminescence is used in mate location and courtship when one sex is flightless.

LIFE HISTORY. Sexes are generally similar except for minor differences in size or development of sense organs or appendages. Striking exceptions include many Lampyridae (fireflies) and related families, in which females are wingless and frequently larviform. In some Scarabaeidae (June beetles) and Lucanidae (stag beetles), the males bear elaborate horns or greatly enlarged mandibles.

Copulation almost always proceeds with the male mounted dorsally on the female, and may be preceded by specialized courtship behavior, such as stridulation (Scolytidae) or palpation and antennation (Meloidae). Eggs are usually simple and ovoid, with relatively thin, unsculptured chorion, with oviposition on or close by the larval food. Eggs are laid singly or in small batches in many Coleoptera, but egg masses are deposited by some (e.g., Coccinellidae). Immatures are usually of limited mobility, but many predatory larvae are active hunters; in parasitoid Coleoptera the highly mobile first instars (triungulins) function in locating the host. Normally, all larval instars, which vary greatly in number, are similar, but hypermetamorphosis is the rule among parasitoid forms, with later instars grublike. Body form among larvae reflects mode of life and is at least as variable as in adults. However, most coleopterous larvae can be classified as (1) *campodeiform* (Fig. 39.4*b*): active, predatory, with prognathous head, long legs, and elongate urogomphi (Carabidae; Staphylinidae); (2) *eruciform* (Fig. 39.15*h*): caterpillarlike, with cylindrical body, short legs, hypognathous head (Coccinellidae; Erotylidae); (3) *scarabaeiform* (Fig. 39.7*d*): obese, C-shaped body with moderate legs; burrowing in soil or rotting wood (Scarabaeidae; Lucanidae); (4) *cucujiform* (Fig. 39.15*i*): flattened or subcylindrical body, prognathous head, moderate legs, usually with well-developed urogomphi of diverse configuration (Cucujidae, Nitidulidae, Colydiidae); (5) *apodous* (Fig. 39.17*e*): grublike, with legs, eyes, antennae, and urogomphi reduced or absent (Curculionidae, Bruchidae). It should be stressed that many other body forms are associated with specialized modes of life, and that the larval types described above correspond to general modes of adaptation, not evolutionary relationships.

Pupation usually occurs in a cell constructed by the larva in the feeding substrate or nearby. Cocoons are uncommon but are formed by some Curculionidae as well as a few other families from silk produced by the Malpighian tubules. Limited pupal mobility is enabled by abdominal movements. Various parts of the body frequently bear tubercles, knobs, or setae characteristic of the pupa. Adjacent abdominal segments sometimes have opposable "gin traps," pointed sclerotized prominences supposedly used in defense. Pupae which occur on exposed surfaces (e.g., Coccinellidae) often remain within the tough larval cuticle. The pupal stage usually lasts 2 to 3 weeks; pupal diapause occurs in forms which pupate in protected cells (e.g., Cerambycidae) but seems to be relatively uncommon in Coleoptera, which diapause more frequently as larvae or adults.

KEY TO THE SUBORDERS OF Coleoptera

1. Pronotum with notopleural sutures (Fig. 39.2) **2**
   Pronotum without notopleural sutures (Fig. 39.1*b*)[1] **Polyphaga** (p. 386)

[1] *Micromalthus*, with one uncommon species in eastern North America, lacks notopleural sutures. It will key to Staphylinidae.

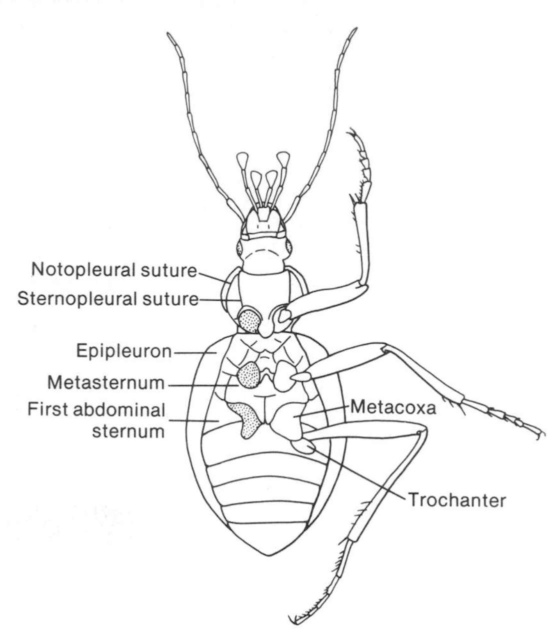

FIGURE 39.2 Ventral aspect of *Scaphinotus* (Carabidae); scale equals 2 mm.

2. Metacoxae dividing first visible abdominal sternite (Fig. 39.2) **Adephaga** (p. 385)
   Metacoxae not dividing first visible abdominal sternite (Fig. 39.1*b*) 3

3. Antennae clubbed (two small families in North America, Sphaeriidae and Hydroscaphidae) **Myxophaga** (p. 384)
   Antennae filiform (two small families, Cupesidae and Micromalthidae) **Archostemata** (p. 384)

Suborder Archostemata

Archostemata, including only the Families Cupesidae and Micromalthidae, are most notably distinguished from other beetles by having the wing membrane spirally rolled in repose, rather than folded (Fig. 39.3*b*). The presence of distinct pleural sclerites in the prothorax (Cupesidae) and closed cells in the cubitomedian area of the wing are primitive features shared with the Myxophaga and Adephaga.

CUPESIDAE are wood-inhabiting insects of moderate size (5 to 15 mm), which bore in decaying trunks of both conifers and angiosperms. They are seldom encountered but are locally abundant. Males of some species are attracted to chlorine bleaches, which apparently mimic a sex pheromone. Adults are characterized by the elongate, parallel-sided body, with reticulately sculptured elytra (Fig. 39.3*a*). Larvae are eruciform, with five-segmented legs with one or two claws. Only 25 species are known; five occur in North America. The earliest fossils definitely assignable to the Coleoptera, impressions of elytra from Permian beds in Asia, are very similar to elytra of extant Cupesidae. The genus *Tshekardocoleus,* described by Rohdendorf (1944), appears to have venation intermediate between that of Coleoptera and Megaloptera.

The MICROMALTHIDAE are minute beetles (1 to 2 mm) known from the single species *Micromalthus debilis*, which occurs in rotting wood in eastern North America and has been introduced to several other continents. The life history of *Micromalthus* is extremely complex, with at least five distinct larval forms, including one which is neotenic and another which is ectoparasitic upon the mother (Pringle, 1938). Females are diploid, males haploid.

Suborder Myxophaga

Four small families are included in the Myxophaga. The wings are folded basally, but rolled apically, and contain a closed cell (oblongum) in the cubitomedian area of the wing; the prothoracic notopleural sutures are distinct, and the mandible has a distinct molar lobe. These features suggest relationships to both Archostemata and Polyphaga. Larvae are ovate with broad, deflexed head, 5-segmented legs with single claws, and tracheal gills on the abdomen. All species which have been studied are aquatic, apparently feeding on algae.

Two families occur in North America. HYDROSCAPHIDAE inhabit margins of small sluggish streams and springs, including hot springs, in western North America. Adults are 1 to 2 mm long, broad and flattened, with truncate elytra exposing two or three abdominal tergites (Fig. 39.3*c*). Five species are known, with a single North American representative. SPHAERIIDAE are minute (0.5 to 1 mm long), highly convex beetles

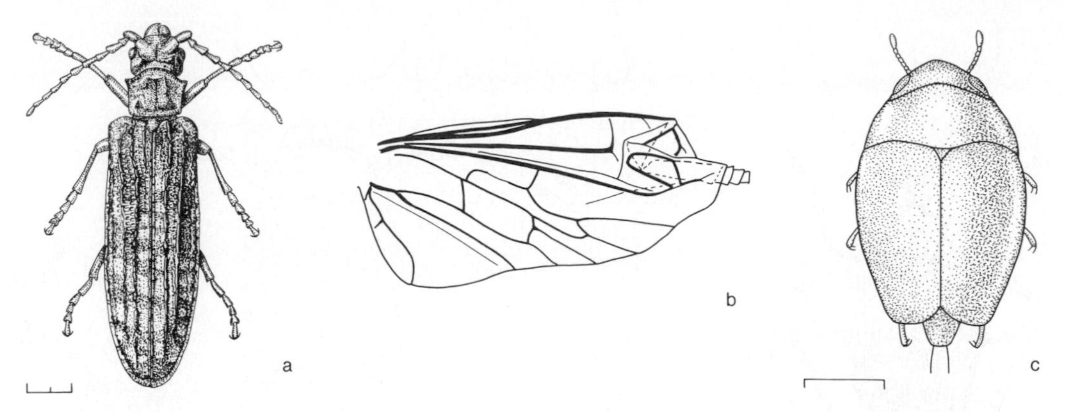

FIGURE 39.3 Coleoptera, Archostemata and Myxophaga; *a*, Archostemata, Cupesidae (*Priacma serrata;* scale equals 2 mm); *b*, rolled wing of *P. serrata; c*, Myxophaga, Hydroscaphidae (*Hydroscapha natans;* scale equals 0.25 mm).

which occur in wet gravel and debris along stream margins. The elytra completely cover the abdomen, differentiating sphaeriids from hydroscaphids. There are two North American species, 11 worldwide.

Suborder Adephaga

These predominantly predatory beetles represent one of the two major evolutionary lines of Coleoptera, with two large, diverse families and several smaller, more specialized ones. Adephaga have distinct pleural sclerites in the prothorax (Fig. 39.2), and usually have a distinct oblongum in the wing, but fold the wing membrane in repose. They are distinguished from all other Coleoptera by the immovable metacoxae, which are fused to the first abdominal sternite, and by the enlarged trochanters of the hind legs (Fig. 39.2). In all Adephaga the molar lobe of the mandible is reduced or absent, a common modification in predatory species of Polyphaga as well. Other useful distinguishing features of Adephaga include large, round (not emarginate) eyes; long, filiform or moniliform antennae; abdomen with six visible sternites; tarsal segmentation 5-5-5. Larvae are campodeiform, with five-segmented legs with two claws.

The dominant family is the CARABIDAE, (ground beetles; Fig. 39.4*a*), with about 20,000 species distributed throughout the world. Carabidae are most abundant in moist situations and tend to be nocturnal, but there are many exceptions. Size ranges from about 1 to 50 mm. Adults and larvae of most species are probably nonspecific predators, but diets are highly restricted in others (e.g., *Brachinus,* which attacks hydrophilid prepupae or pupae). About 2500 North American species are known.

Cicindelidae and Rhysodidae are closely related to Carabidae, although Rhysodidae superficially appear quite different. CICINDELIDAE (tiger beetles) are fast-running, diurnal predators which are among the most efficient flying beetles. Larvae are sessile predators which construct deep burrows in friable soil, usually near water. RHYSODIDAE are sluggish, heavily sclerotized beetles which inhabit moist, decaying wood.

HALIPLIDAE inhabit slowly moving or standing water, especially the shallows along streams or pond margins. Larvae feed exclusively on algae. Adults mingle this diet with occasional small arthropod prey.

The Dytiscidae, Gyrinidae, and Amphizoidae are aquatic predators. DYTISCIDAE (Fig. 39.4*b,c*) is the dominant family, with about 4000 species (330 in North America) occupying all freshwater habitats. Adults and larvae must periodically surface to renew their air supply, somewhat

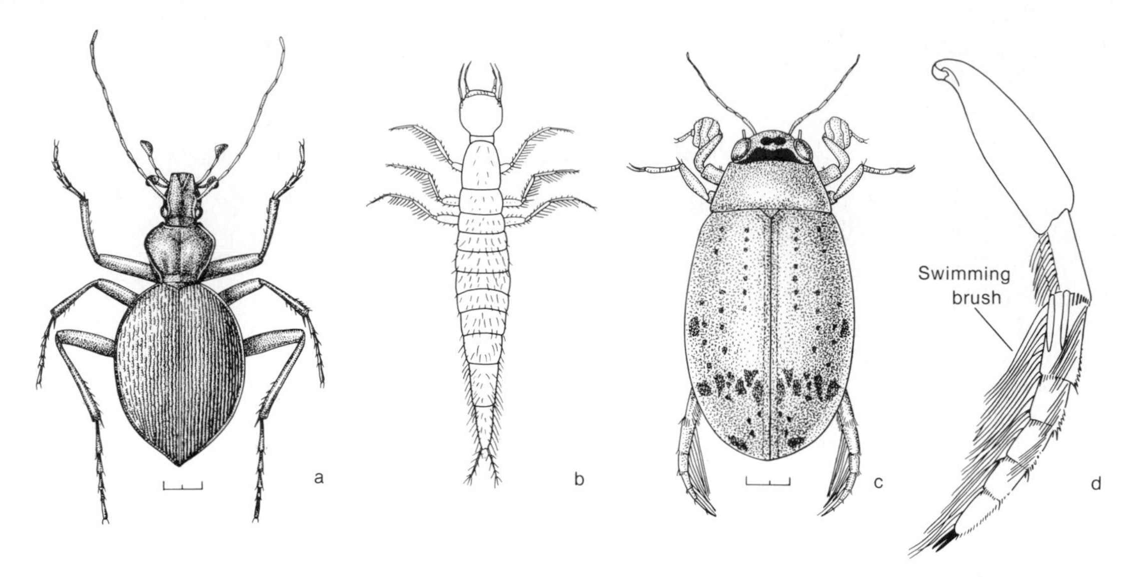

FIGURE 39.4 Representative Adephaga: *a*, Carabidae, adult (*Scaphinotus*); *b*, Dytiscidae, larva (*Hydaticus*); *c*, Dytiscidae (*Eretes sticticus*); *d*, hind leg of *E. sticticus* (Dytiscidae). (b *from Böving and Craighead,* 1931.) Scales equal 2 mm.

restricting them to shallow water. Size ranges from about 1 mm to more than 25 mm.

GYRINIDAE (whirligigs) are specialized for surface swimming, with the eyes divided, allowing simultaneous submarine and aerial vision. Adults tend to be highly gregarious, often congregating in dense swarms in eddies or quiet pools, where they gyrate rapidly and erratically about on the water surface. Their antennae are sensitive to surface film disturbances, enabling the beetles to avoid one another and to locate the floating insects on which they prey. Larvae frequent aquatic vegetation.

AMPHIZOIDAE include only five species, all but one from North America. Adults and larvae inhabit floating debris trapped in eddies or entangled in shoreline vegetation, mostly along cold, fast-flowing streams.

KEY TO THE FAMILIES OF THE SUBORDER
Adephaga

1. Metacoxae covering no more than first visible abdominal sternite 2
   Metacoxae expanded as plates covering at least first 3 sternites **Haliplidae**

2(1). Hind coxae extending laterally to meet elytra 3
   Hind coxae not extending laterally as far as elytra (Fig. 39.2) 5

3(2). Legs with long hairs fringing tibiae and tarsi (Fig. 39.4*d*) 4
   Legs without long fringing hairs **Amphizoidae**

4(3). Eyes divided into dorsal and ventral lobes **Gyrinidae**
   Eyes rounded or emarginate, not divided **Dytiscidae** (including **Noteridae**)

5(2). Metasternum without transverse suture in front of metacoxae; antennae filiform or serrate 6
   Metasternum with transverse suture just before metacoxae; antennae thick, moniliform **Rhysodidae**

6(5). Clypeus extending laterally in front of antennal insertions **Cicindelidae**
   Clypeus not extending laterally as far as antennal insertions **Carabidae**

Suborder Polyphaga

This is an exceedingly diverse assemblage, including more than 90 percent of the Coleoptera.

The wings lack a closed cell (oblongum) in the cubitomedian field, and are always folded in repose. The pronotum extends ventrally to the vicinity of the lateral coxal articulation, with the reduced pleural sclerites internal. Larvae are variable but have legs with four segments and a single claw (or legs vestigial or absent).

The Polyphaga are divided into numerous superfamilies. Most of these are relatively homogeneous, in terms of both morphological and ecological characteristics, the major exception being the Cucujoidea. The following key is modified from that of Britton (1970); the taxonomic arrangement is essentially that of Crowson (1960). The most important taxonomic characters include the configuration of the antennae (Fig. 2.7), the number of abdominal segments, and tarsal structure. The numbers of tarsal segments on the fore- middle, and hind legs, respectively, are indicated by three-digit formulas in the key (e.g., tarsal segmentation 5-5-4). The procoxal cavities are considered *open* if not enclosed posteriorly by the prothorax (Fig. 39.14*g*); they are *closed* if a sclerotized portion of the notum extends medially behind them to the procoxal process (Fig. 39.14*c*). It should be stressed that keying members of the Polyphaga requires great care because of the many exceptional genera. For a more detailed treatment of North American Coleoptera, Arnett (1960) is the most complete reference.

KEY TO THE SUPERFAMILIES OF THE SUBORDER
Polyphaga

1. Antennae with terminal 3 to 7 segments enlarged and flattened as a discrete, lamellate, one-sided club (Fig. 39.7*b*) **Scarabaeoidea** (p. 391)
   Antennae variable, but not lamellate, *or* if rarely lamellate, more than 7 segments involved — 2

2(1). Antennae with segment 6 (occasionally segment 4 or 5) cup-shaped, transverse (Fig. 39.5*b*); maxillary palpi frequently as long as antennae or longer (Fig. 39.5*b*) **Hydrophiloidea** (p. 389)
   Antennae with segment 6 unmodified; maxillary palpi rarely elongate — 3

3(2). Elytra truncate, exposing 2 or more abdominal tergites, which are sclerotized (Fig. 39.6*a* to *d*) — 4
   Elytra usually covering entire abdomen, or rarely exposing 1 or 2 tergites — 10

4(3). Wings at rest extending beyond elytra, without transverse folds — 5
   Wings at rest folded transversely beneath elytra or absent — 6

5(4). Maxillary palpi flabellate; eyes nearly contiguous dorsally — 15
   Maxillary palpi filiform; eyes widely separated — 27

6(4). Claws dentate (Figs. 39.13*c*, 39.14*d*) or appendiculate (Fig. 39.14*e*, *i*) — 10
   Claws unmodified — 7

7(6). Antennae elbowed with last 3 segments forming capitate club **Histeroidea** (p. 389)
   Antennae not elbowed — 8

8(7). Abdomen with 6 or 7 visible sternites; antennae usually clavate or filiform **Staphylinoidea** (p. 390)
   Abdomen with 5 visible sternites; antennae with capitate club — 9

9(8). Abdomen with 3 or more tergites exposed **Staphylinoidea** (p. 390)
   Abdomen with 2 or fewer tergites exposed — 10

10(3,6,9). Abdomen with 6 or 7 visible sternites — 11
   Abdomen with 3 to 5 visible sternites — 16

11(10). Tarsal segmentation 3-3-3 **Cucujoidea (Coccinellidae,** p. 398)
   Tarsal segmentation 5-5-5 — 12

12(11). Antennae with distinct, usually capitate club — 13
   Antennae filiform, serrate, or pectinate — 15

13(12). Tarsi with segment 4 deeply bilobed; claws frequently toothed (Fig. 39.13*c*) **Cleroidea** (p. 398)
   Tarsi with segment 4 not bilobed; claws simple — 14

14(13). Antennal club with 3 segments **Staphylinoidea** (p. 390)
Antennal club with 5 segments **Hydrophiloidea** (p. 389)

15(5,12). Tarsi filiform (Fig. 39.14*f*), slender, at least as long as tibiae **Lymexyloidea** (p. 398)
Tarsi stout, not filiform, shorter than tibia **Cantharoidea** (p. 395)

16(10). Metacoxae with posterior face vertical, concavely excavated for reception of femora (Fig. 39.9*b*) **17**
Metacoxae flat or convex, without excavated posterior face (Fig. 39.12*a*) **23**

17(16). Procoxae globular, round or oval (Fig. 39.14*g*), not projecting from coxal cavities **18**
Procoxae transversely or dorsoventrally elongate (Fig. 39.14*c*), sometimes projecting from coxal cavities (Fig. 39.6*d*) **19**

18(17). Abdomen with first 2 sternites fused, sutures obscured **Buprestoidea** (p. 394)
Abdomen with all sternites free, separated by distinct sutures **Elateroidea** (p. 394)

19(17). Antennae with last 3 to 5 segments differentiated as strong, capitate club, or greatly elongate **20**
Antennae filiform, serrate, or clavate, but not with last 3 segments strongly differentiated **21**

20(19). Antennae with last 3 segments greatly elongate **Bostrichoidea** (p. 397)
Antennae with last 3 to 5 segments forming short, capitate club (Fig. 39.12*c,d*) **Dermestoidea** (p. 396)

21(19). Tarsi with next to last segment deeply bilobed (Fig. 39.8*e*) **Dascilloidea** (p. 392)
Tarsi with next to last segment not bilobed; tarsi usually filiform (Fig. 39.14*f*) **22**

22(21). Head with clypeus distinctly separated from frons by fine suture **Dryopoidea** (p. 393)
Head with clypeus fused with frons **Byrrhoidea** (p. 392)

23(16). Tarsal segmentation 5-5-5 **24**
Tarsal segmentation 5-5-4, 4-4-4, or 4-4-3 **27**

24(23). Prothorax usually hoodlike, produced forward over head (Fig. 39.12*a*); legs usually with trochanters elongate; antennal insertions usually very close **Bostrichoidea** (p. 397)
Prothorax not hoodlike; trochanters small; antennal insertions distant **25**

25(24). Tarsi with large, bisetose empodium (Fig. 39.13*f*); procoxae conical, projecting from cavities; body usually with numerous, erect bristles **Cleroidea** (p. 398)
Tarsi with empodium absent or small, inconspicuous (Fig. 39.13*c*); procoxae not projecting; body usually without erect bristles **26**

26(25). Tarsi with fourth segment minute, concealed in groove of bilobed third segment (Fig. 39.16*b*) **30**
Tarsi with fourth segment subequal to third; if small, not concealed in deep groove of third segment **Cucujoidea** (p. 398)

27(5,23). Tarsal segmentation 5-5-4 **Cucujoidea** (p. 398)
Tarsal segmentation 4-4-4 or 3-3-3 **28**

28(27). Tarsal segments not lobed beneath **Cucujoidea** (p. 398)
Tarsi with 1 to 3 basal segments lobed beneath (Figs. 39.8*e*, 39.13*c*) **29**

29(28). Metasternum without transverse suture **Cleroidea** (p. 398)
Metasternum with transverse suture just before metacoxae **30**

30(26,29). Head prolonged as a beak on which the antennae are inserted (Fig. 39.17*a,b*) **Curculionoidea** (p. 404)
Head not prolonged as a beak; antennae inserted on frons near eyes (Fig. 39.17*d*) **31**

31(30). Antennae filiform, serrate, or occasionally clavate **Chrysomeloidea** (p. 402)
Antennae strongly capitate 32

32(31). Antennae elbowed **Curculionoidea** (p. 404)
Antennae straight 33

33(32). Body densely covered by scales or hairs **Curculionoidea (Anthribidae** p. 404)
Body glabrous or sparsely pubescent **Cucujoidea (Erotylidae, Languriidae)** (p. 398)

SUPERFAMILY HYDROPHILOIDEA (Fig. 39.5). Predominantly aquatic beetles, characterized by short antennae, with elongate scape and compact club with 3 to 5 segments. In many species the maxillary palpi are elongate, frequently longer than the antennae (Fig. 39.5*b*).

Hydrophilidae (Fig. 39.5*a,b*) occupy all aquatic situations. The most familiar species are relatively large, free-swimming forms. Others burrow in bottom debris or hide in aquatic vegetation. Many are littoral forms, especially along gravelly stream margins. The numerous members of the Subfamily Sphaeridiinae inhabit dung or moist humus in terrestrial habitats. Hydrophilids are mostly scavengers as adults, predatory as larvae.

Limnebiidae (Fig. 39.5*c*) are small beetles (1 to 2 mm) which occur in a variety of littoral habitats, including seeps, wet gravel, or coarse sand banks, splash zones, and marine rock pools. Georyssidae creep about mud or sand bars along streams. Adults coat the elytra with a camouflaging of mud. Both families are widespread within their restricted habitats.

KEY TO THE FAMILIES OF
Hydrophiloidea

1. Abdomen with 5 visible sternites; antennae with club 3-segmented 2
   Abdomen with 6 to 7 visible sternites; antennae with club 5-segmented **Limnebiidae**

2. Tarsal segmentation 4-4-4 **Georyssidae**
   Tarsal segmentation 5-5-5 **Hydrophilidae**

SUPERFAMILY HISTEROIDEA (Fig. 39.6*c*). Adults distinguished by compact body, frequently with retractile appendages; antennae elbowed with abrupt, capitate club. Appendages frequently bear enlarged, spinose setae in characteristic patterns, and the cuticle is typically very hard, black, and polished.

Histeridae are cosmopolitan predators, abundant about carrion and dung, where they are most frequently encountered. Numerous spe-

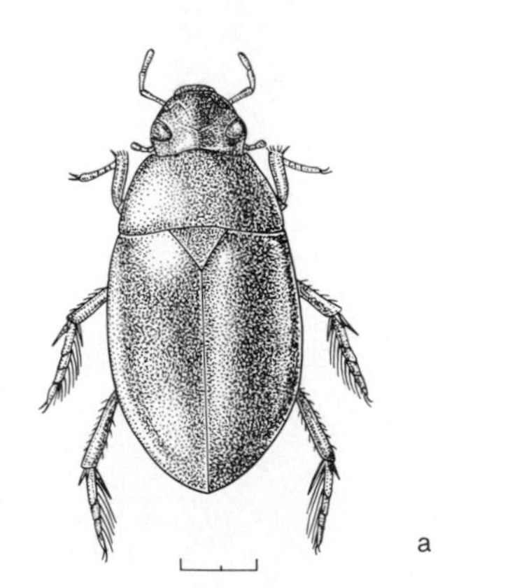

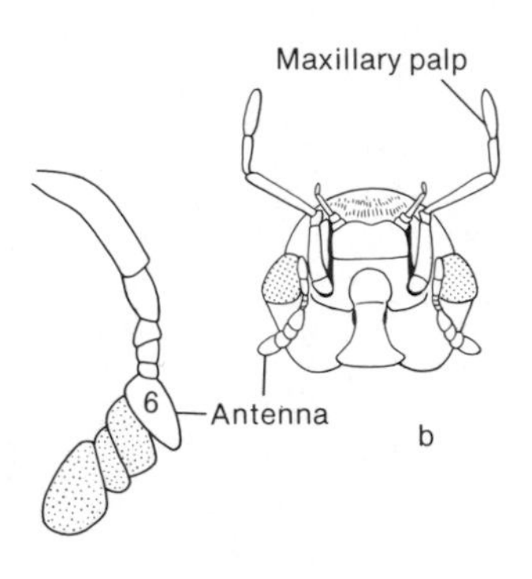

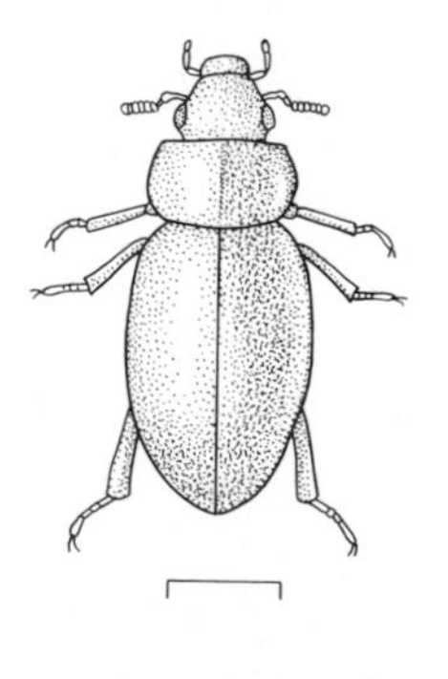

FIGURE 39.5 Hydrophiloidea: *a*, Hydrophilidae (*Tropisternus ellipticus;* scale equals 2 mm); *b*, ventral aspect of head of *Tropisternus* with cup-shaped sixth antennal segment marked; *c*, Limnebiidae (*Ochthebius rectus;* scale equals 0.5 mm).

cies burrow in sand, and others are restricted to mammal burrows or ant or termite nests. Many of the free-living species are predators of fly larvae or pupae. SPHAERITIDAE, with three known species, are occasionally found in association with decaying plant and animal products. Their biology is unknown. Sphaeritids are distinguished from histerids in having the antennae straight (not elbowed) and the front tibiae simple (not dentate).

SUPERFAMILY STAPHYLINOIDEA (Fig. 39.6*a,b,d*). Adults with wing venation reduced by loss of the mediocubital loop and with only 4 Malpighian tubules. Larvae usually with articulated urogomphi and 4-segmented legs with single claws. In external features the Staphylinoidea are extremely variable and difficult to recognize. Most have the elytra abbreviated, exposing 3 or more abdominal tergites, which are sclerotized, but there are many exceptions.

Staphylinoid beetles are found mainly in moist habitats, frequently in association with fungi or in leaf litter. PTILIIDAE, PSELAPHIDAE, SCYDMAENIDAE, SCAPHIDIIDAE, and LEIODIDAE are largely restricted to fungus–leaf mold environments, but in all these families a few species are known to be commensals in ant or termite nests.

SILPHIDAE, as well as some Leiodidae and many Staphylinidae, are associated with carrion, either as scavengers or as predators, especially of Diptera. *Nicrophorus* (Fig. 39.6*b*) (Silphidae) is well known for burying corpses of small vertebrates, which are then prepared as food for the grublike, sedentary larvae. Many Leiodidae are associated with mammal nests, and the LEPTINIDAE are all specialized as nest inhabitants or ectoparasites of mammals. Adult leptinids have reduced eyes and antennae, flattened body with leathery cuticle, and mouthparts adapted for feeding on cutaneous debris.

The STAPHYLINIDAE is a diverse group, encompassing all the modes of life described above. Perhaps the largest number of species are associated with fungi, including molds and rusts.

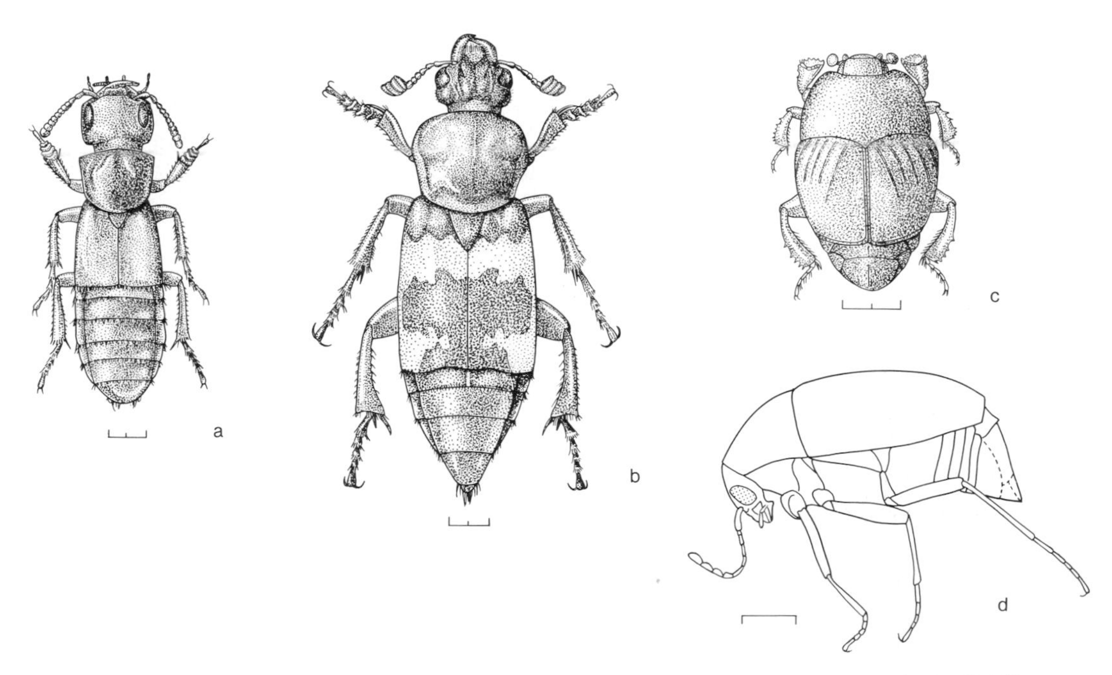

FIGURE 39.6 Staphylinoidea and Histeroidea; *a*, Staphylinidae (*Platycratus maxillosus*); *b*, Silphidae (*Nicrophorus marginatus*); *c*, Histeridae (*Saprinus lugens*); *d*, Scaphidiidae (*Scaphisoma quadriguttatum*). Scales equal 2 mm, except that of *d* equals 0.5 mm.

Many are found near water, and staphylinids are the commonest marine Coleoptera. Very few staphylinoid beetles are phytophagous; although none is presently of economic significance, many Staphylinidae are predaceous, especially on Diptera, and may become important as biological control agents.

KEY TO THE FAMILIES OF
Staphylinoidea

1. Elytra truncate, exposing at least 3 abdominal tergites — 2
   Elytra exposing no more than 2 tergites, frequently covering entire abdomen — 3

2(1). Abdomen swollen, at least twice width of pronotum; tarsi with 1 claw or with claws unequal — **Pselaphidae**
   Abdomen slender, less than 1.5 times width of pronotum; tarsi with claws of equal size — **Staphylinidae** (Fig. 39.6*a*)

3(1). Last abdominal tergite conical, as long as 3 preceding segments (Fig. 39.6*d*); first abdominal sternite as long as 2 to 4 combined — **Scaphidiidae**
   Last abdominal tergite about equal to preceding segment; first sternite about equal to second — 4

4(3). Metacoxae with posterior surface concave, excavated for reception of femora; length less than 1.5 mm — **Ptiliidae** (incl. **Limulodidae**)
   Metacoxae with posterior surface flat or convex; length greater than 1.5 mm — 5

5(4). Procoxae conical, prominent (Fig. 39.6*d*) — 6
   Procoxae globular, not projecting from cavities — **Leptinidae**

6(5). Metacoxae contiguous or approximate — 7
   Metacoxae widely separated — **Scydmaenidae**

7(6). Tibial spurs small, indistinct — 8
   Tibial spurs large, conspicuous; body frequently convex, capable of being rolled into a ball — **Leiodidae** (incl. **Leptodiridae**)

8(7). Antennae clavate with segment 8 smaller than 7 or 9 — **Leiodidae**
   Antennae clavate or capitate; segments 7 to 9 subequal — **Silphidae** (Fig. 39.6*b*)

SUPERFAMILY SCARABAEOIDEA (Fig. 39.7). Distinguished from nearly all other Coleoptera by short antennae with asymmetrical, lamellate club (Fig. 39.7*b*). Most scarabaeoid beetles are stout-bodied, with the head sunk deeply into the prothorax and anterior tibiae expanded and serrate for digging. Larvae are stout, with large, hypognathous head, well-developed legs, and large, soft abdomen, usually curled ventrally beneath the forebody (Fig. 39.7*d*).

LUCANIDAE (stag beetles) and PASSALIDAE inhabit decaying wood, where larvae and adults are usually associated in common galleries. The mandibles of most male Lucanidae vary allometrically, being similar to those of the female in small individuals, and disproportionately enlarged in large individuals. They are apparently used in agonistic encounters between males in some species. Passalids are flattened and heavily sclerotized, with striate elytra. Larvae, which always occur with adults, stridulate by rubbing the tiny, digitate hind leg over the microscopically ridged mesocoxa. Adults apparently macerate food for the larvae.

SCARABAEIDAE (chafers, June beetles, etc.) is the dominant family. Some scarabaeid larvae feed on rotting wood, but mammal dung, humus, carrion, and roots of living vegetation are consumed by others. Adults of many species, such as the Japanese beetle (*Popillia japonica*), attack living foliage, especially of the Rosaceae. A few species are specialized as predators or commensals in ant or termite nests. Scarabs range in size from burrowing species less than 2 mm long to giants over 10 cm in length, including the bulkiest extant insects.

KEY TO THE FAMILIES OF
Scarabaeoidea

1. Lamellae of antennal club thick, rounded, not capable of close apposition — 2
   Lamellae of antennal club flattened, apposable (Fig. 39.7*b*) — **Scarabaeidae**

2(1). Antennae elbowed, first segment longer than next 5 segments combined — **Lucanidae**
   Antennae straight, first segment shorter than next 4 combined — **Passalidae**

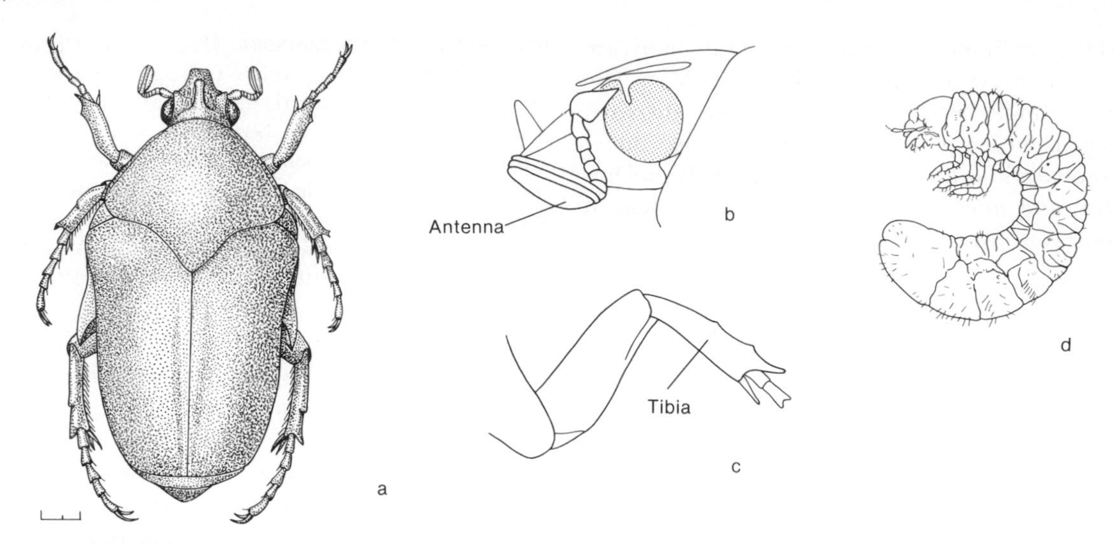

FIGURE 39.7 Scarabaeoidea, Scarabaeidae: *a,* adult *Cotinis nitida* (scale equals 2 mm); *b,* antenna and *c,* foreleg of *Cotinis nitida; d,* larva of *Popillia japonica.* (d *from Böving and Craighead,* 1931.)

SUPERFAMILY DASCILLOIDEA (Fig. 39.8). Antennae filiform or serrate, procoxae conical, projecting from cavities; procoxal cavities open; intercoxal process narrow; metacoxae posteriorly concave for reception of femora; tarsal segmentation 5-5-5; five visible abdominal sternites. The CLAMBIDAE, included in the Dascilloidea because of similarities of the larvae to that of *Eucinetus* (Crowson, 1960), have 4-4-4 tarsal segmentation and clubbed antennae. Adults are capable of rolling into a ball.

The Dascilloidea contains a diverse assemblage of beetles from a few small families. Adults of most tend to occur in moist situations, frequently around water, but only the HELODIDAE (larvae) are aquatic. Their multiarticulate antennae are unique among immatures of endopterygote insects. Larvae of DASCILLIDAE are scarabaeiform and occur in soil (*Dascillus*).

RHIPICERIDAE adults (Fig. 39.10*c*) are robust, elongate, brown or black beetles 16 to 24 mm long. Their scarabaeiform larvae are internal parasitoids of cicada nymphs. EUCINETIDAE occur under bark in association with fungi or slime molds. Adults are flattened, ovoid beetles with the head deflexed. The life histories and biology of clambids are almost unknown.

KEY TO THE FAMILIES OF

Dascilloidea

1. Length less than 1 mm; antennae clubbed **Clambidae**
   Length greater than 3 mm; antennae filiform, serrate (Fig. 39.8*a*), or lamellate (Fig. 39.10*c*) 2

2(1). Antennae filiform or serrate 3
   Antennae lamellate **Rhipiceridae**

3(2). Metacoxae broad, platelike, oblique **Eucinetidae**
   Metacoxae narrow, transverse 4

4(3). Tarsi with segments 2 to 4 lobed ventrally (Fig. 39.8*e*) **Dascillidae**
   Tarsi with only segment 4 lobed **Helodidae**

SUPERFAMILY BYRRHOIDEA. Strongly convex beetles with the head strongly deflexed, the clypeus fused with the frons, antennae filiform; procoxae transverse, intercoxal process broad, procoxal cavities open; tarsal segmentation 5-5-5; 5 visible abdominal sternites.

The single family, BYRRHIDAE, includes about 40 North American species. They are small to moderate in size. Adults have retractile appen-

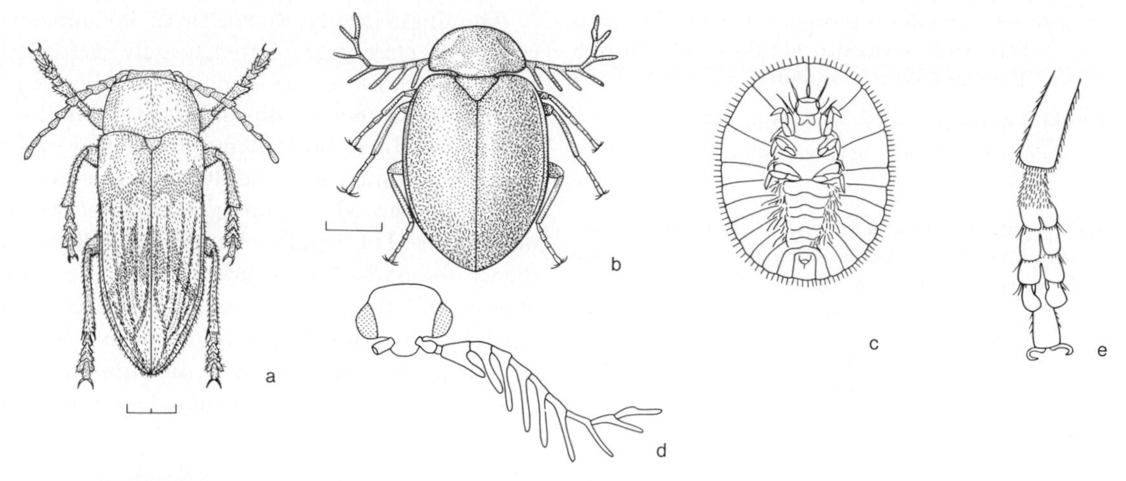

FIGURE 39.8 Dascilloidea and Dryopoidea: *a,* Dascillidae (*Dascillus davidsoni;* scale equals 2 mm); *b,* Psephenidae, adult (*Eubrianax edwardsi;* scale equals 1 mm); *c,* Psephenidae, larva (*Eubrianax*); *d,* antenna of adult *Eubrianax; e,* tarsus of *Dascillus.*

dagcs, accounting for the common name, "pill bugs." Adults and larvae are herbivorous. Larvae are scarabaeiform, occurring in turf, humus, moss, or sand.

SUPERFAMILY DRYOPOIDEA (Fig. 39.8). Procoxal cavities open posteriorly, prosternal process fitting into concavity in mesosternum; metacoxae concavely excavate posteriorly for reception of femora, tarsal claws very elongate; abdomen with 5 visible sternites. Larvae eruciform, with retractable anal gills, or highly modified [Psephenidae. (Fig. 39.8*c*)]

Dryopoidea are aquatic or subaquatic in some stage of their lives, with the exception of Callirhipidae and Eurypogonidae. Adult Eurypogonidae are encountered on foliage, and the larvae live in moss. Adults and larvae of Callirhipidae occur under the bark or in the wood of dead logs. The single North American species is in Florida. Heteroceridae and Limnichidae occur along stream margins. Adults and larvae of heterocerids burrow in mud or sand. Limnichids are most frequently encountered on mud or sand banks or on low vegetation near water.

Dryopidae are unique among insects in having terrestrial larvae and aquatic adults. The larvae burrow in damp sand or soil. Newly emerged adults fly to streams, where they cling to the undersides of submerged stones or logs, often in association with elmid adults and psephenid larvae (water pennies). In contrast to the Hydrophiloidea, aquatic dryopoids respire with tracheal gills (larvae) or cuticular plastrons (adults), and may remain submerged indefinitely without replenishing their air supply. Most of thesc beetles are infrequently collected, but many are common stream inhabitants.

KEY TO THE FAMILIES OF

Dryopoidea

1. Procoxae conical, projecting from cavities — 2
   Procoxae transverse or rounded, not projecting — 6

2(1). Maxillary palpi with second segment as long as next 2 segments combined; abdomen usually with 6 to 7 visible sternites — **Psephenidae**
   Maxillary palpi with second segment subequal to third; abdomen with 5 visible sternites — 3

3(2). Antennae flabellate (Fig. 2.7), inserted on a strong protuberance on the frons — **Callirhipidae**

Antennae filiform, serrate, or clavate (Fig. 2.7), frequently concealed within prosternal cavity; frons without antennal protuberance **4**

4(3). Posterior margin of pronotum wrinkled **5**
Posterior margin of pronotum not wrinkled **Eurypogonidae**

5(4). Ventral surface of body with grooves for retraction of legs and antennae **Chelonaridae**
Ventral surface of body without grooves for retraction of appendages **Ptilodactylidae**

6(1). Tarsi with 4 segments **Heteroceridae**
Tarsi with 5 segments **7**

7(6). Tarsi with last segment longer than preceding segments combined **8**
Tarsi with last segment shorter than preceding segments combined **Limnichidae**

8(7). Antennae filiform, elongate **Elmidae**
Antennae short, with stout, pectinate club **Dryopidae**

SUPERFAMILY BUPRESTOIDEA (Fig. 39.9). Fore coxae globular, procoxal cavities open posteriorly, prosternal process articulated with mesosternum; metasternum with transverse suture; metacoxae grooved for reception of femora; tarsal segmentation 5-5-5; abdomen with five visible sternites, the first two rigidly fused.

The single family, BUPRESTIDAE, is characterized by the closely articulated, heavily sclerotized body, with short, serrate antennae and short legs. Buprestids are active, diurnal insects, found either about their host plants or on flowers or foliage, on which most adults feed. Larvae are narrowly adapted for boring in tissues of perennial plants. The head and thorax are flattened (hence the name "flat-headed borers"), legs are absent, and the abdomen is long, slender, and desclerotized. Most buprestids bore in wood of trees or shrubs, but some mine the pithy stems of rosaceous plants, and a few mine leaves or form galls.

SUPERFAMILY ELATEROIDEA (Fig. 39.10). Fore coxae globular or rounded, cavities open posteriorly, prosternal process elongate, articulated with mesosternum; metasternum without transverse suture; metacoxae grooved for reception of femora; tarsal segmentation 5-5-5; 5 visible abdominal sternites, all freely articulated. Larvae elongate, cylindrical, sclerotized; prognathous with well-developed legs; urogomphi frequently present.

This superfamily includes one large, familiar family, the ELATERIDAE (click beetles), and several small, obscure families. Most elateroid bee-

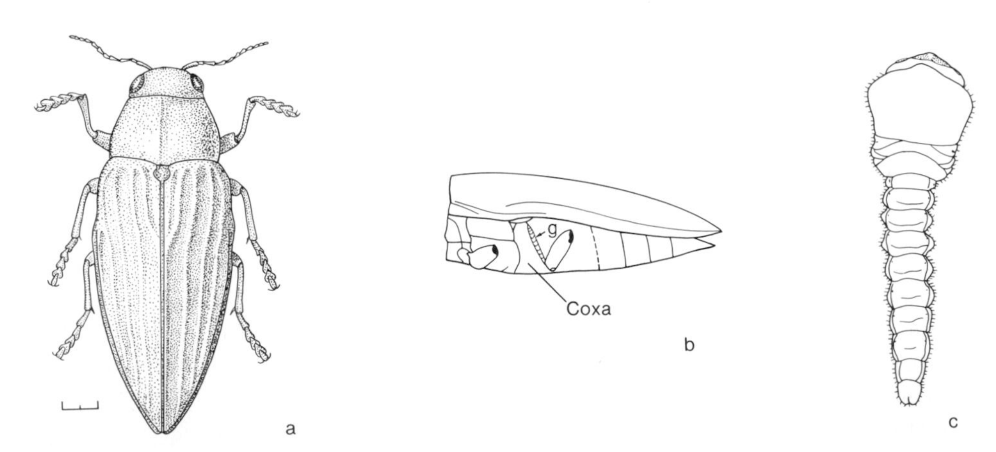

FIGURE 39.9 Buprestoidea, Buprestidae: *a, Buprestis aurulenta; b,* metacoxal region of *Buprestis; c,* dorsal aspect of larva of *Chrysobothris.* (c *from Böving and Craighead,* 1931.) *g,* groove in coxa for reception of femur.

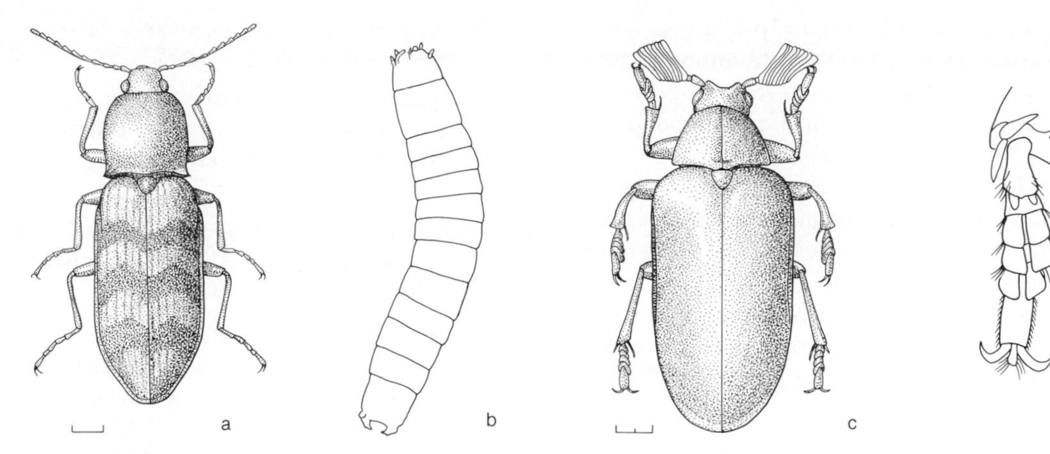

FIGURE 39.10 Elateroidea and Dascilloidea *a*, Elateridae (*Ctenicera tigrina;* scale equals 1 mm); *b*, Elateridae, larva (*Cryptohypnus*); *c*, Rhipiceridae (*Sandalus niger*, male; scale equals 2 mm); *d*, tarsus of *Sandalus* (Rhipiceridae). (b *from Böving and Craighead*, 1931.)

tles are associated with wood, at least as larvae, but the Family Elateridae has many soil-dwelling forms. Many of these wireworms were important pests of cereal and root crops before the advent of soil insecticides. Other elaterid larvae are predatory or omnivorous. Adult elaterids are distinguished from all other beetles by the loosely articulated prothorax and the modification of the prosternal process and mesosternum into a jumping or clicking mechanism (suggesting the name "click beetles").

KEY TO THE FAMILIES OF
Elateroidea[1]

1. Head with labrum freely articulated; antennae inserted close to eyes **Eucnemidae**
   Head with labrum fused to clypeus, indistinguishable; antennae closer to mandibles than eyes 2

2(1). Mandibles large, prominent, sickle-shaped **Cebrionidae**
   Mandibles small, not noticeably protruding from mouth. 3

3(2). Prothorax loosely joined to mesothorax **Elateridae**
   Prothorax tightly joined to mesothorax; prosternal process fused with mesosternum **Throscidae**

[1]Perothopidae and Cerophytidae, each represented by a single rare genus in North America, are not included in the key.

SUPERFAMILY CANTHAROIDEA (Fig. 39.11). Procoxae prominent, conical, projecting; procoxal cavities open posteriorly, intercoxal process reduced, seldom reaching mesosternum; tarsal segmentation 5-5-5; abdomen almost always with 6 to 7 visible sternites. Larvae eruciform (Fig. 39.11*c*) or highly modified, with suctorial mandibles.

The cantharoid beetles are characterized by their leathery, flexible integument and loosely articulated, flattened bodies. Antennae are usually serrate, more uncommonly pectinate or flabellate. The elytra are approximated in repose, but not interlocked as in many other beetles. Larvae are eruciform, with small head and soft abdomen, or have the body segments platelike and usually flattened dorsally. Some of the abdominal segments may be strongly differentiated from the thoracic region, as in Lycidae and Brachypsectridae, resulting in distinctive "trilobite larvae." Nearly all cantharoid beetles are predators in all instars, and the larvae have the mandibles medially channeled for uptake of liquid food. CANTHARIDAE

(Fig. 39.11*a*) are nonspecific predators, while Lampyridae (Fig. 39.11*b*) feed almost exclusively on terrestrial snails, and some phengodids attack only millipedes, which are immobilized by toxin produced by oral glands.

The hemolymph of cantharids and lycids commonly contains the irritant cantharidin, and the bright, conspicuous coloration of these diurnal insects is presumably aposematic. Many members of both families apparently serve as models for mimicking insects ranging from Hemiptera to Lepidoptera as well as other Coleoptera, especially Cerambycidae. Many Lampyridae luminesce from abdominal organs and use flash patterns in communication between the sexes, which are strongly dimorphic, with wingless, larviform females. Some larval and adult female phengodids also luminesce, but do not communicate by flash patterns.

KEY TO THE FAMILIES OF Cantharoidea

1. Abdomen with 5 sternites visible **Brachypsectridae**
   Abdomen with 6 to 7 sternites visible 2

2(1). Elytra with reticulate sculpture; mesocoxae distant **Lycidae**
   Elytra striate, punctate, or smooth; mesocoxae usually contiguous 3

3(2). Antennae inserted laterally, in front of eyes; maxillary palpi greatly elongate **Telegusidae**
   Antennae inserted on frons between eyes; maxillary palpi normal 4

4(3). Head retracted beneath pronotum, concealed in dorsal aspect (Fig. 39.11*b*) **Lampyridae**
   Head fully visible in dorsal aspect 5

5(4). Antennae with 12 segments, usually flabellate **Phengodidae**
   Antennae with 11 segments, usually filiform or serrate **Cantharidae**

SUPERFAMILY DERMESTOIDEA (Fig. 39.12*c,d*). Antennae abruptly capitate, procoxae prominent, projecting, or transverse; metacoxae excavated for reception of femora; tarsal segmentation 5-5-5, with simple tarsomeres; abdomen with 5 visible sternites. Larvae eruciform, usually with long spinose setae on the tergites.

These small to moderate-sized beetles are compact, oval, and usually densely covered with setae or appressed scales. Many species bear a single median ocellus. Typically the posterior margin of the pronotum is medially produced,

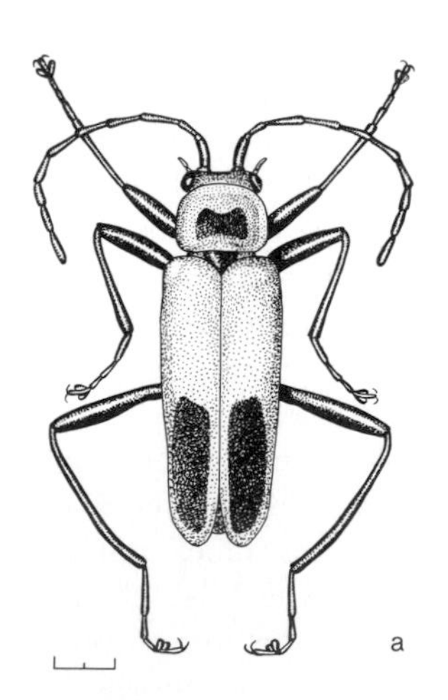

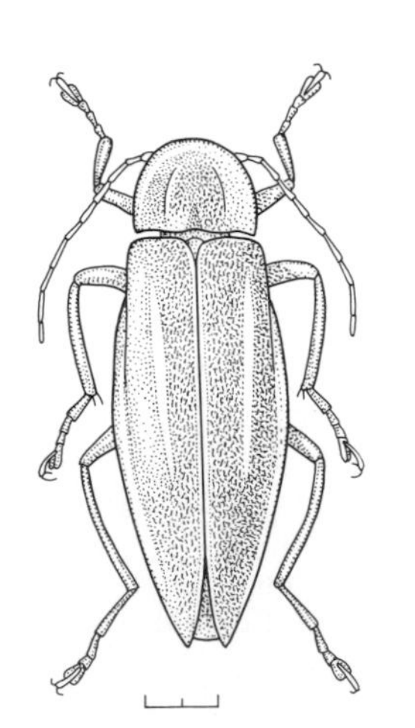

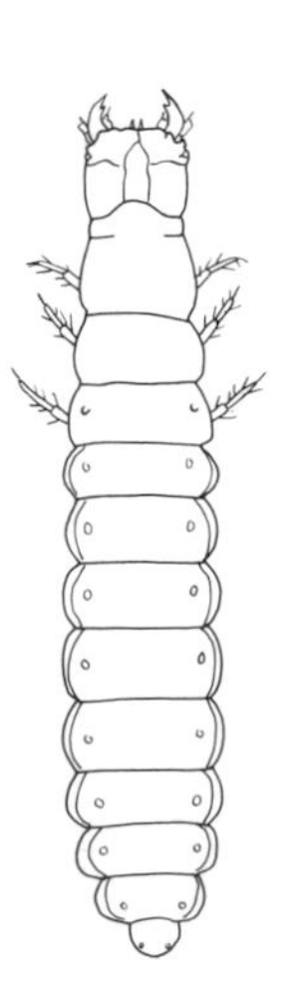

FIGURE 39.11 Cantharoidea: *a*, Cantharidae (*Chauliognathus pennsylvanicus*); *b*, Lampyridae (*Photurus pennsylvanicus*); *c*, larva of *Chauliognathus*. (c *from Böving and Craighead*, 1931.) Scales equal 2 mm.

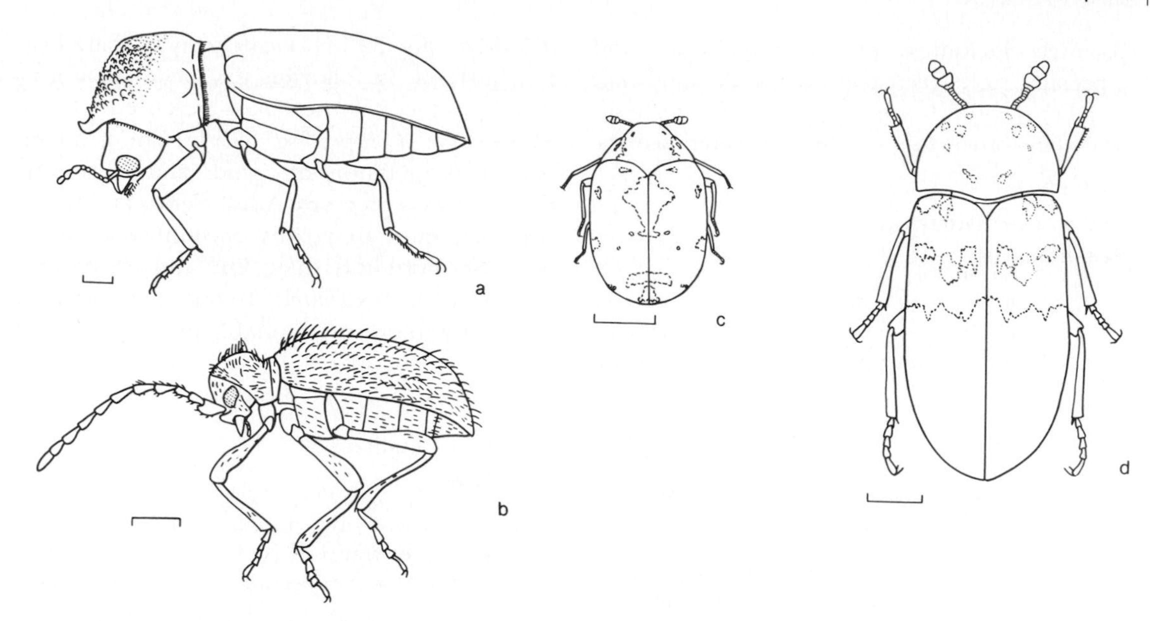

FIGURE 39.12 Bostrychoidea: *a,* Bostrychidae (*Apatides fortis*); *b,* Ptinidae (*Ptinus clavipes*); *c,* Dermestidae (*Anthrenus lepidus*); *d,* Dermestidae (*Dermestes lardarius*). Scales equal 1 mm, except that of *b* equals 0.5 mm.

concealing the scutellum. Three families are included—Dermestidae, Derodontidae, and Nosodendridae.

As larvae, DERMESTIDAE (Fig. 39.12*c,d*) are scavengers of both plant and animal material but are especially abundant on drying skin, feathers, fur, and other proteinaceous substances. The smaller species frequently occur in old nests of Hymenoptera, feeding on cast skins, pollen, and dead insects. Adults of many dermestids are common on flowers.

NOSODENDRIDAE mainly occupy sap flows from trees, both conifers and angiosperms. For example, *Nosodendron californicum* is common in sap flows on white fir, *Abies concolor.*

DERODONTIDAE are rarely collected beetles which inhabit the fruiting bodies of slime molds and wood-rotting fungi, frequently under the bark of dead trees.

KEY TO THE FAMILIES OF
Dermestoidea

1. Head with pair of ocelli near inner margins of compound eyes **Derodontidae**
   Head with single median ocellus or without ocelli 2

2(1). Forelegs with tibiae flat, serrate; median ocellus absent **Nosodendridae**
   Forelegs with tibiae cylindrical or oval in cross section, with straight margins; median ocellus frequently present **Dermestidae**

SUPERFAMILY BOSTRICHOIDEA (Fig. 39.12*a,b*). Antennae filiform, frequently with last three segments enlarged or elongate; head usually deflexed, frequently concealed dorsally by hoodlike prothorax, tarsal segmentation 5-5-5; abdomen with 5 visible sternites. Larvae weakly sclerotized, glabrous, with reduced head capsule and legs; urogomphi absent; body stout, C-shaped.

The beetles of this superfamily mostly feed on dead plant material. BOSTRICHIDAE, LYCTIDAE, and some ANOBIIDAE bore in dead wood. Bostrychids mostly attack recently felled trees, while anobiids and lyctids may enter seasoned timbers. Ptinids and some anobiids are more polyphagous, consuming a variety of plant and animal products, including spices, certain drugs, and especially stored cereal products. A significant number of ptinids are commensals in ant nests, and many others frequent bird or mammal nests.

KEY TO THE FAMILIES OF
Bostrichoidea

1. Metacoxae with posterior face excavated, concave, for reception of femora **Anobiidae**
   Metacoxae with posterior face flat or slightly convex 2

2(1). Antennae filiform, inserted in close proximity between eyes (Fig. 39.12*b*) **Ptinidae**
   Antennae clubbed or with 2 to 3 terminal segments longer than preceding 7 segments 3

3(2). Prothorax hoodlike, concealing head in dorsal aspect (Fig. 39.12*a*), or, if not hoodlike, first abdominal sternite subequal to second **Bostrichidae**
   Prothorax not hoodlike; first abdominal sternite nearly as long as second and third combined **Lyctidae**

SUPERFAMILY CLEROIDEA (Fig. 39.13). Head usually prognathous, with antennae clubbed or filiform; procoxae transverse, metacoxae usually extending laterally beyond metasternum, prominent or flat, but not posteriorly concave for reception of femora; tarsal segmentation 5-5-5, with large bisetose empodium frequently present (Fig. 39.13*f*); abdomen with 5 (rarely 6) visible sternites. Larva eruciform, with prognathous head, sclerotized prothorax, well-developed legs, urogomphi arising from sclerotized tergal plate.

Cleroids are nearly all predatory, at least as larvae. Exceptions include the Corynetinae (CLERIDAE) and several TROGOSITIDAE (*Ostoma, Calitys*). The Corynetinae feed on carrion, including preserved meat; hence the name "ham beetles." *Ostoma* (Fig. 39.13*e*) and *Calitys* are fungus feeders. Many clerids and trogositids, such as *Temnochila* (Fig. 39.13*d*), are broad-spectrum predators, while others specialize on one or a few prey organisms. Some species of *Enoclerus* (Cleridae; Fig. 39.13*a*) locate scolytid (bark beetle) burrows by following sex pheromone gradients produced by the female bark beetles. Some MELYRIDAE (*Collops;* Fig. 39.13*b*) are of minor benefit as predators of aphids and other soft-bodied insects on crops. Adult clerids and trogositids are most frequently encountered around dead or moribund trees, but also on flowers. Melyridae are extremely abundant on flowers, and many species are probably phytophagous as adults.

KEY TO THE FAMILIES OF
Cleroidea

1. Procoxae prominent, conical, projecting 2
   Procoxae transversely elongate (as in Fig. 39.14*c*), not projecting **Trogositidae**

2(1). Antennae usually clubbed; tarsi with segments 1 to 4 ventrally lobed (Fig. 39.13*c*) **Cleridae**
   Antennae usually filiform, rarely with loose 2- to 3-segmented club; tarsi simple, filiform, or with fourth segment lobed **Melyridae**

SUPERFAMILY LYMEXYLOIDEA. Elongate, with 5 to 7 visible abdominal sternites; procoxal cavities confluent, mesocoxal cavities contiguous; antennae very short; tarsi at least as long as tibiae, with 5-5-5 segmentation. A single family, LYMEXYLIDAE.

The extremely slender, elongate larvae tunnel through hard wood of moribund or dead trees or sawn lumber, apparently feeding on fungus. Adult males are remarkable for the elaborate, flabellate maxillary palpi. The elytra of *Atractocerus* are extremely short, almost completely exposing the longitudinally folded wings.

SUPERFAMILY CUCUJOIDEA (Figs. 39.14, 39.15). Antennae filiform, moniliform, or clubbed, very rarely serrate; metacoxae flat or convex, lacking concave posterior face for reception of femora; tarsal segmentation usually 5-5-5 or 5-5-4, occasionally 4-4-4 or 3-3-3; abdomen almost always with 5 sternites, rarely with 6 or 7. Larvae eruciform or highly modified; head

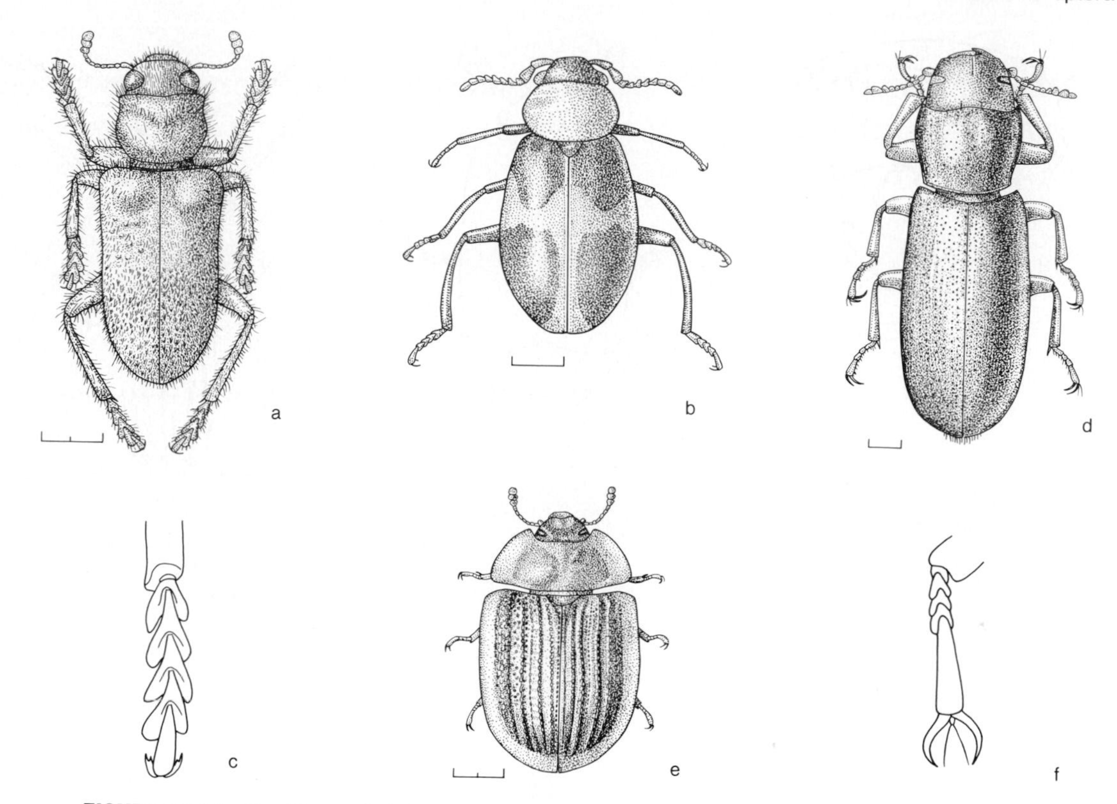

FIGURE 39.13 Cleroidea; *a*, Cleridae (*Enoclerus moestus;* scale equals 2 mm); *b*, Melyridae (*Collops histrio;* scale equals 1 mm); *c*, tarsus of *Enoclerus* (Cleridae); *d*, Trogositidae (*Temnochila virescens;* scale equals 1 mm); *e*, Trogositidae (*Ostoma pippingskoeldi;* scale equals 2 mm); *f*, tarsus of *Ostoma* (Trogositidae).

capsule sclerotized, thoracic legs well developed; abdomen usually with urogomphi on one or more terminal segments.

The Cucujoidea, containing about 50 families, is morphologically and ecologically the most diverse superfamily of beetles, yet the number of species is relatively low, including only about one-tenth of all the Coleoptera. Only a single family, TENEBRIONIDAE (darkling ground beetles), is large (about 15,000 species). COCCINELLIDAE, MELOIDAE, and NITIDULIDAE each contain about 2000 to 5000 species; the other families are mostly small, and frequently quite restricted biologically, to narrow, specialized niches.

Cucujoidea are preeminently wood-inhabiting insects, frequenting the subcortical region of dead trees and shrubs. This habitat is extremely rich in fungi, molds, and other saprophytic organisms, and many Cucujoidea are fungivorous. For example, Ciidae feed exclusively on wood-rotting polypores, and Erotylidae and Endomychidae are strictly dependent on fungi. Many other families are less obviously associated with fungi but probably feed on molds, yeasts, or other saprophytes, at least as larvae (e.g. RHIZOPHAGIDAE, MYCETOPHAGIDAE, CRYPTOPHAGIDAE). Several families, such as COLYDIIDAE and CUCUJIDAE, include predatory forms. The biology of many families, though clearly associated with the subcortical habitat, is almost unknown.

The most successful cucujoid beetles (by number of species) are those which are not primarily associated with wood. Tenebrionidae, by far the largest family, are mostly ground-dwelling, phytophagous organisms, although a

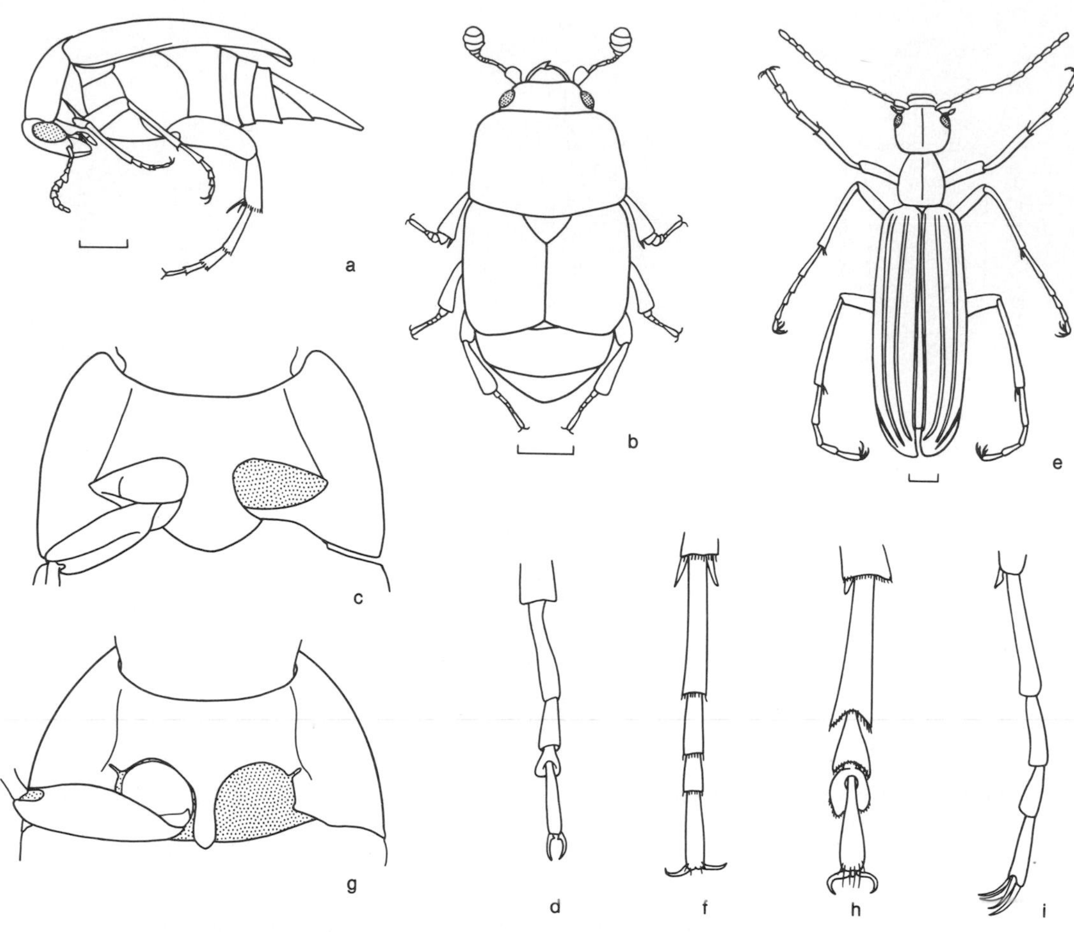

FIGURE 39.14 Cucujoidea: *a*, Mordellidae (*Mordella albosuturalis;* scale equals 1 mm); *b*, Nitidulidae (*Carpophilus hemipterus;* scale equals 0.5 mm); *c*, ventral aspect of prothorax of *Carpophilus* (Nitidulidae); *d*, tarsus of *Ichthidion* (Pedilidae); *e*, Meloidae (*Pleuropompha tricostata;* scale equals 2 mm); *f*, tarsus of Tenebrionidae (*Telacis opaca*); *g*, ventral aspect of prothorax of *Synchroa* (Melandryidae); *h*, tarsus of *Ditylus quadricollis* (Oedemeridae); *i*, tarsus of *Pleuropompha tricostata* (Meloidae).

sizable minority of the species live in wood or fungi. Coccinellidae (Fig. 39.15*a,h*) are predominantly predators of Homoptera, and Meloidae (Fig. 39.14*e*) are specialized predators of immature Hymenoptera and Orthoptera. The other major cucujoid families (NITIDULIDAE, ANTHICIDAE) are biologically diverse, including phytophagous, predatory, and saprophagous species, but with relatively few occurring under bark.

Classification of the Cucujoidea is extremely complex, and identification is difficult without a reference collection. Many families are only rarely encountered or are poorly differentiated. The following key includes only the more familiar groups, and should be used with care. For identification of the smaller, more obscure families, Arnett (1960) and Crowson (1960) are the most useful references.

KEY TO THE COMMON FAMILIES OF **Cucujoidea**

1. Tarsal segmentation 5-5-4 — 2
   Tarsal segmentation 5-5-5, 4-4-4, or 3-3-3 — 12

2(1). Procoxal cavities closed posteriorly by a process of the hypomeron (Fig. 39.14*c*) **3**
Procoxal cavities open posteriorly (Fig. 39.14*g*) **5**

3(2). Antennae capitate, club with 2 to 3 segments **4**
Antennae filiform, serrate, or clavate **Tenebrionidae** (incl. **Alleculidae, Lagriidae**) (Fig. 39.1)

4(3). Antennae with 10 segments **Rhizophagidae**
Antennae with 11 segments **Nitidulidae** (Fig. 39.14*b*)

5(2). Head abruptly constricted behind eyes, exserted **9**
Head gradually narrowed behind eyes, usually retracted into prothorax **6**

6(5). Tarsi with fourth segment dilated, ventrally pubescent (Fig. 39.14*h*) **Oedemeridae**
Tarsi with no dilated segments **7**

7(6). Mesocoxal cavities closed laterally by mesosternum (as in Fig. 39.12*a,b*) **Cryptophagidae**
Mesocoxal cavities closed laterally by mesepimeron (Fig. 39.1*b*) **8**

8(7). Body very strongly flattened **Cucujidae** (Fig. 39.15*g*)
Body rounded in transverse section **Melandryidae**

9(5). Prothorax with sharply defined lateral margins; last abdominal tergite conical, projecting beyond elytral apex **Mordellidae** (Fig. 39.14*a*)
Prothorax rounded, without distinct lateral margins **10**

10(9). Base of pronotum narrower than elytra **11**
Base of pronotum at least as wide as elytra. **Rhipiphoridae**

11(10). Tarsal claws toothed or serrate with ventral appendages (Fig. 39.14*i*) **Meloidae** (Fig. 39.14*e*)
Tarsal claws simple, not toothed or serrate **Anthicidae** and **Pedilidae** (Fig. 39.15*b*)

12(1). Tarsal segmentation 5-5-5 **20**
Tarsal segmentation 4-4-4 or less **13**

13(12). All tarsi with 4 segments **15**
Tarsi with 3 segments or fewer **14**

14(13). Elytra truncate, exposing terminal abdominal segment (Fig. 39.14*b*) **15**
Elytra covering entire abdomen **16**

15(13,14). Procoxae transversely elongate (Fig. 39.14*c*) **Nitidulidae** (Fig. 39.14*b*)
Procoxae globular, round (Fig. 39.14*g*) **Rhizophagidae**

16(14). Tarsi with second segment slender **18**
Tarsi with second segment dilated (Fig. 39.15*c*) **17**

17(16). First abdominal sternite with impressed lines parallel to coxal cavities **Coccinellidae** (Fig. 39.15*a*)
First abdominal sternite without impressed lines **Endomychidae**

18(16). Abdominal sternites flexibly joined, sutures distinct **19**
Abdominal sternites fused, sutures usually obscured **Colydiidae** (Fig. 39.15*e*)

19(18). Antennae filiform; body strongly flattened **Cucujidae** (Fig. 39.15*g*)
Antennae capitate, with basal segment usually large, globular (Fig. 39.15*f*); minute to small in size **Lathridiidae**

20(12). Antennae filiform or moniliform; body usually strongly flattened **Cucujidae** (Fig. 39.15*g*)
Antennae clubbed; body usually convex **21**

21(20). Epipleura continuing to elytral apices **23**
Epipleura terminating before elytral apices **22**

22(21). Antennae with 10 segments (10 and 11 fused) **Rhizophagidae**
Antennae with 11 segments **Nitidulidae** (Fig. 39.14*b*)

23(21). Body convex, short, hemispherical; claws appendiculate or toothed (Figs. 39.14*i*, 39.15*c*) **Phalacridae**
Body elongate, ovoid; claws simple **24**

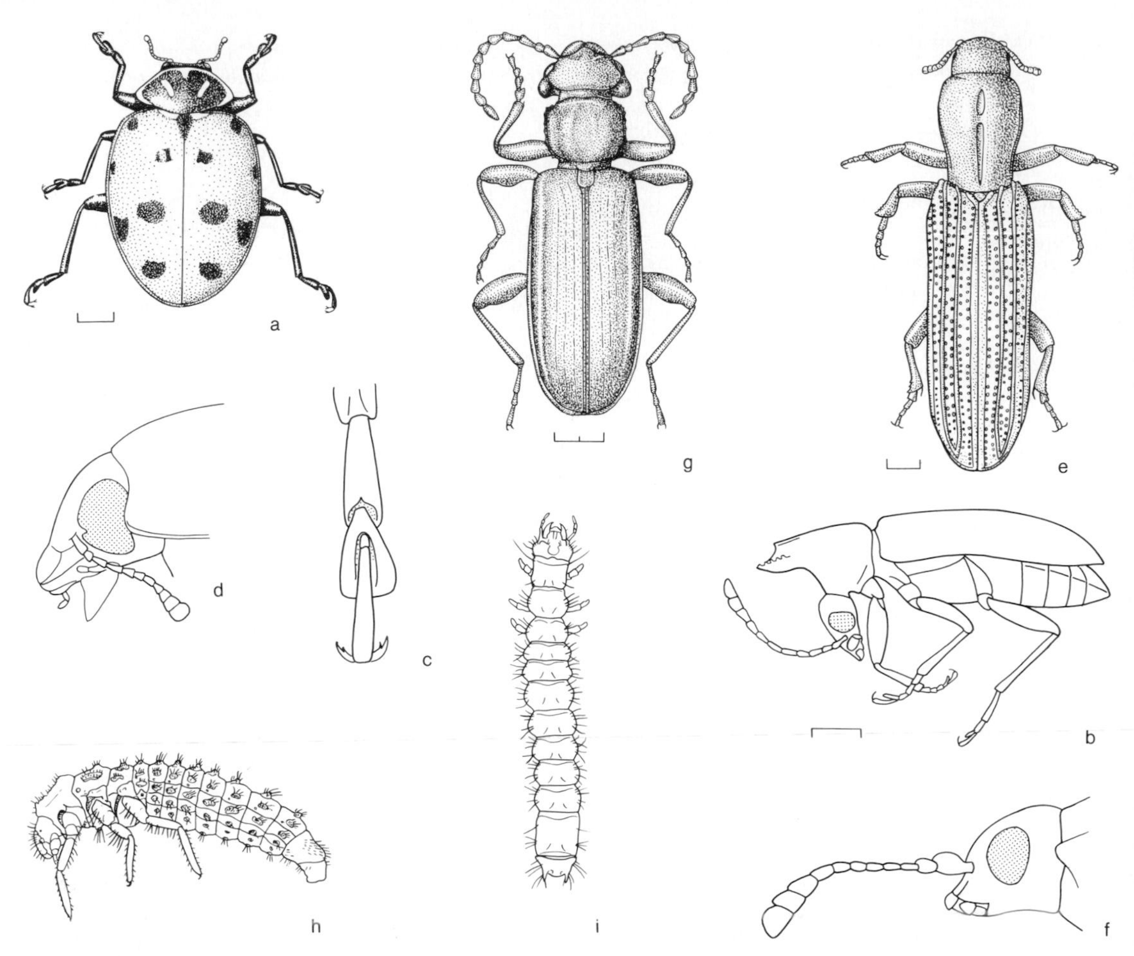

FIGURE 39.15 Cucujoidea: *a,* Coccinellidae (*Hippodamia convergens;* scale equals 1 mm); *b,* Anthicidae (*Notoxus monodon;* scale equals 0.5 mm); *c, d,* tarsus and antenna of *Hippodamia* (Coccinellidae); *e,* Colydiidae (*Deretaphrus oregonensis;* scale equals 1 mm); *f,* antenna of *Enicmus* (Lathridiidae); *g,* Cucujidae (*Cucujus clavipes;* scale equals 2 mm); *h,* larva of Coccinellidae (*Coccinella*); *i,* larva of Cucujidae (*Cucujus*). (h, i. *after Böving and Craighead,* 1931.)

24(23). Body with upper surface smooth, without setae **25**
Body with upper surface pubescent **Cryptophagidae**

25(24). Procoxal cavities closed posteriorly (Fig. 39.14*c*) **Erotylidae**
Procoxal cavities open posteriorly (Fig. 39.14*g*) **Languriidae**

SUPERFAMILY CHRYSOMELOIDEA (Fig. 39.16). Antennae filiform, serrate, or clavate, rarely capitate; metacoxae flat or convex, without posterior concavity for retraction of femora; tarsal segmentation 5-5-5, but with fourth segment extremely small, concealed in bilobed third segment (Fig. 39.16*b*); abdomen with 5 visible sternites. Larvae eruciform, with 3 segmented antennae and well-developed legs (externally feeding forms) or with antennae and legs reduced (borers, internally feeding forms).

Nearly all the 40,000 members of this superfamily are phytophagous. As a group they feed on all parts of higher plants, from the roots on the one hand to the flowers and seeds on the other.

Most species of higher plants are probably utilized by at least one of these beetles, and the economic importance of chrysomeloid beetles is almost entirely due to their consumption of plant parts. CHRYSOMELIDAE such as *Diabrotica* (cucumber beetles, corn root worms; Fig. 39.16*a*), *Haltica* and *Epitrix* (flea beetles), and *Leptinotarsa* (Colorado potato beetle) are significant pests on many crops. A few chrysomelids, such as *Diabrotica,* have been implicated in transmission of plant viruses. CERAMBYCIDAE (Fig. 39.16*e*) include many minor forest pests. The apple tree borer, *Saperda candida,* infests the living trunks and large branches of several rosaceous trees.

Chrysomelidae, popularly called "leaf beetles," actually attack most parts of plants. The majority of species are external feeders on leaves, flowers, buds, or stems, but many occur in the soil as larvae, feeding on roots or underground stems. Chrysomelids include a large number of leaf miners (mostly tropical), a number of species which form galls, species with aquatic larvae, and commensals in ant nests.

Cerambycidae as larvae are primarily borers in woody plants, either living or dead. They attack trunks, branches, stems, and roots. Some feed on stems or roots of herbs, a few mine in leaf petioles; there appear to be no true leaf miners. BRUCHIDAE (bean weevils; Fig. 39.16*c*) are borers in seeds, mostly of Leguminosae and Palmaceae, but many other plant families are attacked. The great majority oviposit in immature seed pods or capsules, but *Acanthoscelides obtectus* attacks dried peas and beans and is an important pest of stored products.

KEY TO THE FAMILIES OF **Chrysomeloidea**

1. Antennae usually inserted on prominent tubercles, capable of being reflexed backward over

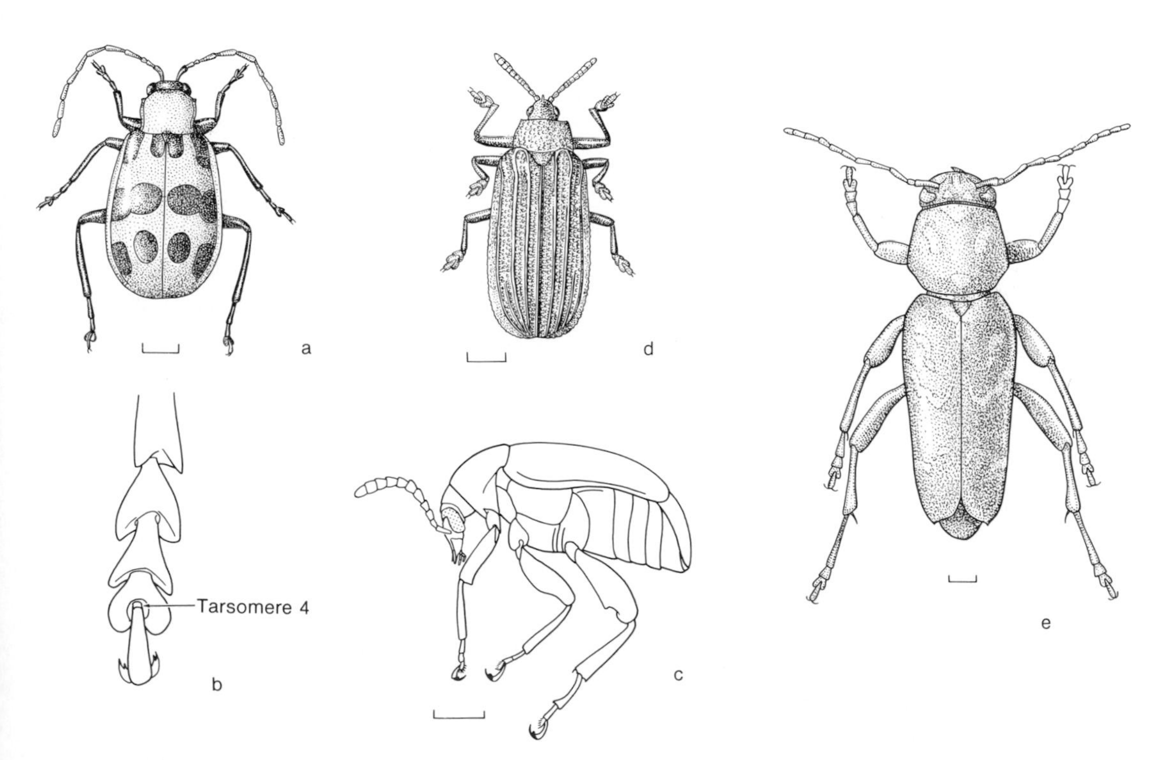

FIGURE 39.16 Chrysomeloidea: *a,* Chrysomelidae (*Diabrotica undecimpunctata*); *b,* tarsus of *Chrysochus* (Chrysomelidae); *c,* Bruchidae (*Bruchus rufimanus*); *d,* Chrysomelidae (*Odontata dorsalis*); *e,* Cerambycidae (*Xylotrechus nauticus*). Scales equal 1 mm.

body, and frequently very elongate **Cerambycidae**

Antennae not usually inserted on tubercles, usually less than half length of body, and not reflexible over body 2

2(1). Head usually deeply retracted into prothorax; fore coxae usually distinctly separated **Chrysomelidae**

Head prominent, not deeply set in prothorax; fore coxae usually contiguous **Bruchidae**

SUPERFAMILY CURCULIONOIDEA (Fig. 39.17). Antennae elbowed and capitate, or occasionally filiform, serrate, or clavate; oral region more or less prolonged as a distinct rostrum; prothorax without distinct lateral margins; metacoxae flat or convex, without posterior concavity for reception of femora; tarsal segmentation 5-5-5, with fourth segment minute, hidden in bilobed third; abdomen with 5 visible sternites. Larvae apodous (with reduced legs in a few primitive genera), with papilliform 1- or 2-segmented antennae; no urogomphi.

The Curculionoidea, comprising over 65,000 species, is easily the largest superfamily group of Coleoptera, and the most important from an economic standpoint. The evolutionary success of these beetles apparently stems at least in part from the elongation of the stomal region of the head into a rostrum, which is used in preparing oviposition holes as well as in feeding. The rostrum is most specialized in the CURCULION-

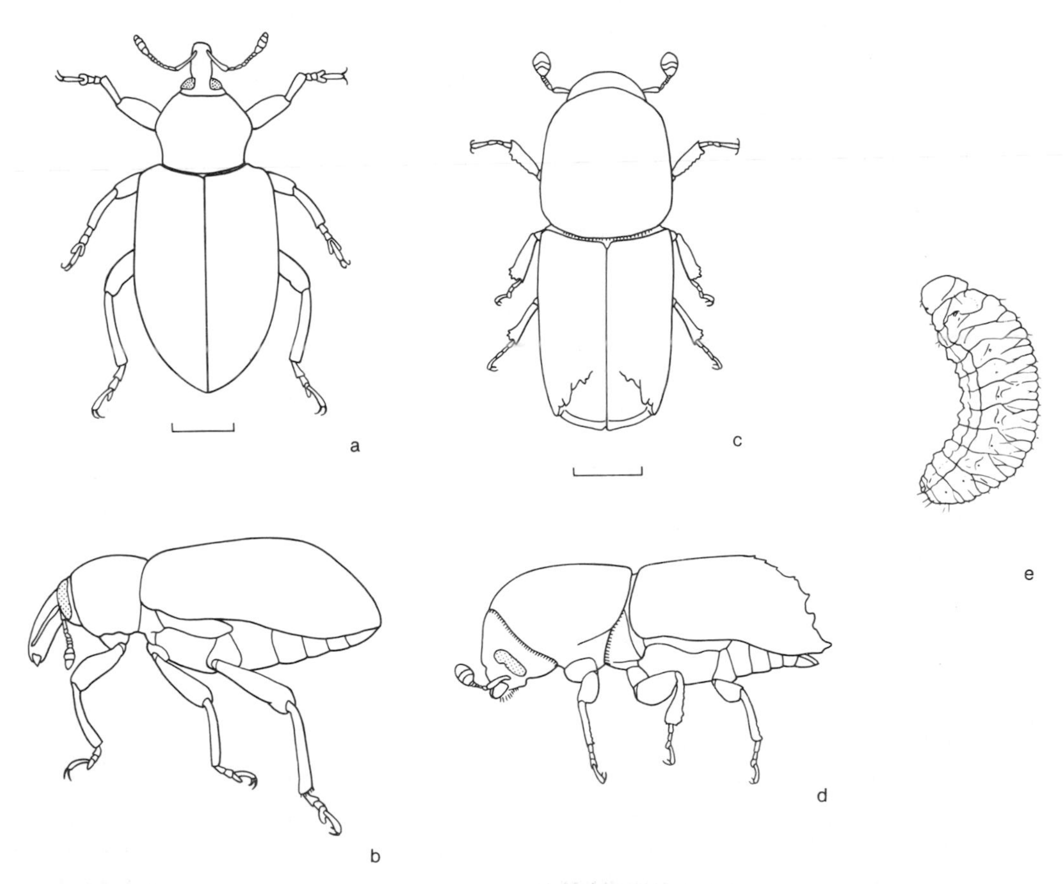

FIGURE 39.17 Curculionoidea: *a, b,* Curculionidae (*Hypera postica*); *c* to *d.* Scolytidae (*Ips*); *a, c,* adults, dorsal; *b, d,* adult, lateral; *e,* larva of Curculionidae (*Naupactus*). Scales equal 1 mm. (e *from Böving and Craighead,* 1931.)

IDAE, where the labrum is absent and maxillae and labium are reduced and partly concealed (Endpaper 1*d*).

The Cuculionidae (weevils), with an estimated 60,000 described species, are probably the most highly evolved family of Coleoptera. Weevils are typically found on foliage or flowers as adults, but some species are ground dwellers, some burrow in sand dunes, and a few are aquatic or marine. Larvae are mostly internal feeders or subterranean, and are rarely encountered without searching, but some (e.g., *Hypera*, alfalfa and clover weevils) feed externally on foliage.

Weevils include many serious economic pests, such as the alfalfa and clover weevils (*Hypera*), the cotton-boll weevil (*Anthonomus grandis*), Fuller's rose weevil (*Pantomorus godmani;* infests many ornamentals), strawberry weevils (*Brachyrhinus*; infests roots of ornamentals and row crops), and the plum curculio (*Conotrachelus nenuphar*; infests fruit of many orchard crops). Species of *Sitophilus* are important pests of stored whole grains.

BRENTIDAE and ANTHRIBIDAE are relatively small families, closely related to Curculionidae. Brentidae feed in decaying wood as larvae, and the adults are usually found under bark. Anthribidae feed in fungi, decaying wood, or, occasionally, seeds or fruit. Both families are primarily tropical in distribution.

The SCOLYTIDAE (bark beetles) attack woody plants by boring into the cambial tissue, where adults feed and prepare oviposition galleries. Most species infest necrotic or freshly killed plants, but some invade apparently healthy trees and are the world's major forest pests. In North America species of *Ips* and *Dendroctonus* cause severe damage on conifers. *Scolytus multistriatus* is responsible for transmitting a fungal disease which has largely exterminated American elm trees. Morphologically, bark beetles are virtually indistinguishable from some weevils, and it may be pointed out that Cossoninae (Curculionidae) are biologically very similar to bark beetles. The two groups are treated separately here because of the massive amount of applied literature keyed to traditional classifications.

KEY TO THE COMMON FAMILIES OF
Curculionoidea[1]

1. Maxillary palpi segmented; labrum a distinct sclerite; rostrum broad, flattened, often short — **Anthribidae**
   Maxillary palpi rigid, inflexible, frequently concealed; labrum fused with rostrum — 2

2. Antennae capitate, elbowed, with basal segment longer than next 3 segments combined (Fig. 39.17*a*) — 3
   Antennae not elbowed; filiform, clavate, or capitate — 4

3. Rostrum usually elongate; antennal insertions distant from eyes (Fig. 39.17*a*) — **Curculionidae**
   Rostrum usually undeveloped, short; antennal insertions contiguous to eyes (Fig. 39.17*d*) — **Scolytidae** (incl. **Platypodidae**)

4. Antennae capitate — **Curculionidae (Apioninae)**
   Antennae filiform or clavate — **Brentidae**

[1]Belidae, Nemonychidae, Oxycorynidae, Apionidae, and several other small families recognized by some authors are not distinguished here.

SELECTED REFERENCES

GENERAL

Arnett, 1960 (biology and systematics of North American species, with excellent bibliographies); Britton, 1970 (excellent general discussion); Coleopterists Bulletin, 1947 to date; G. Evans, 1975 (biology); Hinton, 1945 (species associated with stored products).

MORPHOLOGY

Doyen, 1966 (*Tenebrio*); Duporte, 1960 (cranial structures); Evans, 1961 (musculature, reproductive systems); Larsen, 1966 (locomotor organs).

BIOLOGY AND ECOLOGY

Alexander et al., 1963 (stridulation); Balduf, 1935 (bionomics of predaceous species); G. Evans, 1975 (general); Hinton, 1955*b*, 1969*a* (respiration); Hodek, 1973 (Coccinellidae); Linsley, 1959*a* (mimicry) and 1959*b* (Cerambycidae); Lloyd, 1971 (bioluminescence, communication); Ritcher, 1958

(Scarabaeidae); Rudinsky, 1962 (Scolytidae); Tschinkel, 1975*a,b* (chemical defenses).

SYSTEMATICS AND EVOLUTION

Arnett, 1960 (North American genera); Böving and Craighead, 1930 (larvae); Britton, 1974 (larvae); Crowson, 1955, 1960 (higher classification); Dillon and Dillon, 1961 (Eastern United States species); Hatch, 1957–1973 (Pacific Northwest species); Hinton, 1945 (species associated with stored products); Leech and Chandler, 1956 (aquatic families); Leech and Sanderson, 1959 (aquatic families); Merritt and Cummins, 1978 (aquatic families); Peterson, 1951 (larvae).

ORDER STREPSIPTERA

40

Minute to small Endopterygota specialized for endoparasitism. Males (Fig. 40.1*a*) highly distinctive, with large, transverse head; antennae with 4 to 7 pectinate segments; bulging eyes with large facets; ocelli absent. Prothorax and mesothorax very small, mesothorax with reduced, elytralike wings without veins; metathorax very large, bearing broad, fan-shaped wings with a few thick, radiating veins; legs without trochanters, usually without tarsal claws. Abdomen cylindrical, tapering, concealed basally by large postscutellum. Females wingless, coccidlike (Mengeidae) (Fig. 40.2*c*) or larviform, grublike. First-instar larva (Fig. 40.1*b,c*) (triungulin) spindle-shaped, lacking antennae, mandibles; 3 pairs thoracic legs without trochanters; abdomen with bristlelike caudal styles. Later instars grublike (Fig. 40.1*d*), without appendages or distinct mouthparts. Pupa exarate, adecticous, pharate in puparium formed by cuticle of last larval instar.

Strepsiptera are among the very few insects which internally parasitize other arthropods. They infest or stylopize a variety of insects, including Thysanura, Blattodea, Mantodea, Orthoptera, Hemiptera, Diptera, and Hymenoptera. Of these, Hemiptera and Hymenoptera are the most common hosts. Unlike parasitoid Hymenoptera and Diptera, which eventually kill their host and usually consume most of its internal organs, Strepsiptera mature within active, living insects, which the females of the great majority of species never leave.

Morphologically Strepsiptera are perhaps more similar to Coleoptera than to any other order of insects. The antennae and elytriform forewings resemble the same structures in parasitoid beetles (Rhipiphoridae), and the first-instar larvae are similar to triungulins of Meloidae and Rhipiphoridae. Furthermore, subsequent larval instars of parasitoid beetles are grublike, as in Strepsiptera. However, the reduced prothorax and enlarged postscutellum are not features which occur in Coleoptera. Strepsiptera are unique in lacking trochanters in all instars.

The life cycle is strongly modified for endoparasitism. Males spend their life of 1 or 2 days seeking females, which in typical species remain pharate within the larval cuticle, with only the cephalothorax protruding outside the host. Virgin females secrete a sex pheromone which provides an efficient means of collecting males, which will be attracted for many days unless copulation is allowed, whereupon pheromone production ceases. Insemination is through an opening (brood passage) in the female cephalothorax (Fig. 40.2*a,b*). Numerous eggs are produced (more than 1000 have been recorded), being retained within the female until hatching. The minute triungulins emerge through the brood passage to seek appropriate hosts, which are usually immature instars of the proper species. Movement is by crawling or by springing, which is effected by curving the abdomen ventrally, then suddenly snapping it backward. If an appropriate host is encountered, the triungulin immediately enters its body, aided by an oral secretion which softens the host cuticle. The triungulin soon molts, producing the legless, grublike second instar; subsequent growth and development occur within the host body cavity. The fully grown larva extrudes its anterior end

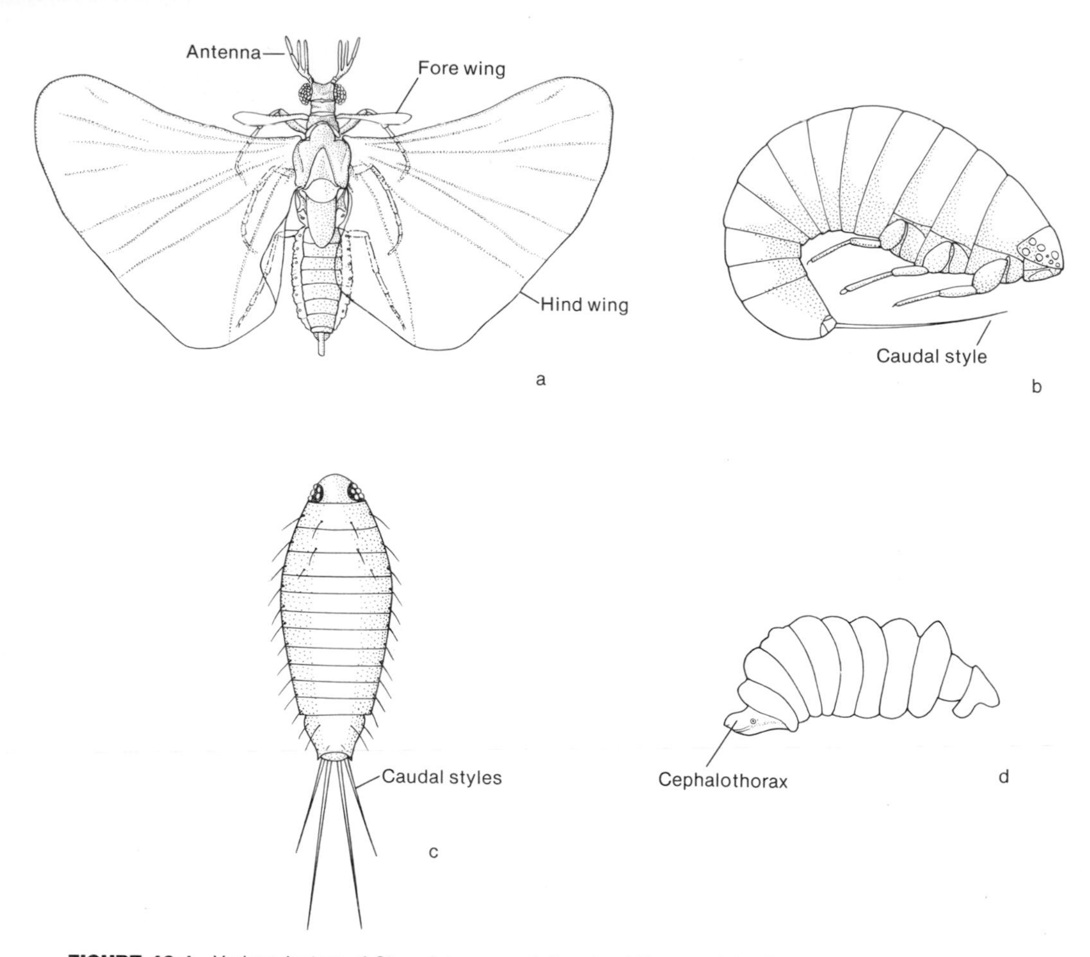

FIGURE 40.1 Various instars of Strepsiptera: *a,* adult male of *Eoxenos laboulbenei* (Mengeidae); *b,* lateral aspect of first-instar larva or triungulin of *Halictophagus tettigometrae* (Halictophagidae); *c,* dorsal aspect of first instar of *Eoxenos* (Mengeidae); *d,* lateral aspect of third-instar larva of *Halictophagus* (a, c *from Parker and Smith,* 1934; b, d *from Silvestri,* 1940.)

through the intersegmental membrane of the host abdomen (Fig. 40.2*d*), and pupation occurs in this position. The adult male later emerges by pushing off the cap of the protruding pupal cephalothorax.

The typical life cycle described above differs in certain species. In Mengeidae the females, though wingless, have well-developed legs and live free of their hosts, which are Lepismatidae (Thysanura). The Mengeidae undoubtedly represent the most primitive Strepsiptera, and the only fossil record for the order consists of Eocene mengeids.

In Stylopidae, which parasitize bees and wasps, the triungulins are deposited on flowers. When adults of suitable host species visit the flowers, the triungulins usually cling to the bee or wasp, or in *Stylops pacifica* (*see* Chap. 14, Fig. 14.4), the triungulins are ingested with nectar. Once in the host's nest, the triungulins disembark (or are regurgitated) and parasitize the host's eggs or larvae. It may be noted that many meloid and rhipiphorid triungulins are likewise phoretic on adults of their hosts.

Stylopized hosts frequently differ from unparasitized members of the same species in mor-

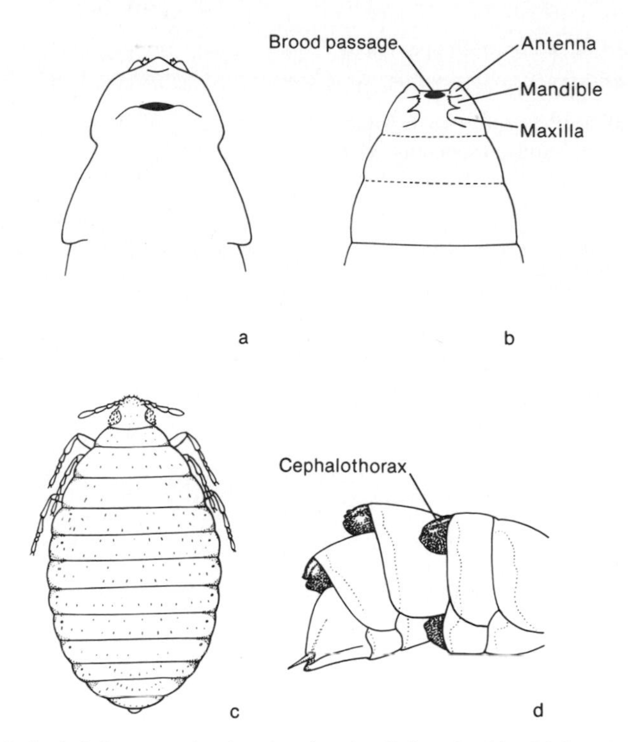

FIGURE 40.2 *a, b,* Cephalothoraces of mature females of: *a,* Halictophagidae (*Halictophagus tettigometrae*); *b,* Corioxenidae (*Triozocera mexicana*); *c,* mature female, Mengeidae (*Eoxenos laboulbenei*); *d,* puparia of Stylopidae (*Xenos vesparea*) in the abdomen of a vespid wasp (*Polistes nympha*). (b *from Silvestri,* 1940; c *from Parker and Smith,* 1940; d *from Szekessy,* 1960.)

phological details. Especially affected are secondary sexual features, which may be modified to resemble homologous structures of the opposite sex. Internal organs, especially the gonads, are stunted, suggesting the term "parasitic castration." However, at least male hosts appear to be capable of reproduction. It is known that host life expectancy is sometimes increased by stylopization, but it is not clear whether differential survival has a physiological basis or is mediated by behavioral changes which decrease predation.

KEY TO NORTH AMERICAN FAMILIES OF
Strepsiptera

1. Males with tarsi 5-segmented, the terminal segment narrow with minute claws; females with brood passage opening terminally (Fig. 40.2*b*) **Corioxenidae**[1]

   Males with tarsi 2- to 4-segmented, without claws; females with brood passage opening preterminally (Fig. 40.2*a*) 2

2(1). Males with tarsi 4-segmented; parasites of Hymenoptera **Stylopidae**

   Males with tarsi 2- or 3-segmented; parasites of Hemiptera or Orthoptera 3

3(2). Males with tarsi 2-segmented; parasites of Fulgoroidea **Elenchidae**

   Males with tarsi 3-segmented; parasites of Homoptera, Orthoptera **Halictophagidae**

[1]Included in Mengeidae in some classifications.

Corioxenidae. In the United States occurring uncommonly in the Southern states. Hosts are Heteroptera.

Stylopidae. Common parasites of bees (Andrenidae, Halictidae, Colletidae) and wasps (Sphecidae, Vespidae). Over 150 species occur in North America.

Halictophagidae. In North America locally common parasites of Homoptera; pygmy grasshoppers (Tetrigidae) are also infested, and exotic species parasitize Tridactylidae and Blattodea.

Elenchidae. Similar to Halictophagidae. North American species parasitize Fulgoroidea, especially Delphacidae; uncommon.

SELECTED REFERENCES

Bohart, 1941 (North American species); Jeannel, 1945 (phylogenetic relationships); Kinzelbach, 1971*a, b* (general); Linsley and MacSwain, 1957 (biology of *Stylops*); Pierce, 1918 (morphology); Ulrich, 1966 (classification).

ORDER MECOPTERA (Scorpion Flies)

41

Small to medium-sized Endopterygota. *Adult* (Fig. 41.1*a*) with slender, elongate body; elongate, hypognathous head capsule; mandibles and maxillae slender, elongate, apically serrate; labium elongate, with fleshy 1- to 3-segmented palps. Thoracic segments subequal or prothorax reduced, mesothorax and metathorax fused; wings narrow, elongate, subequal, crossveins numerous. Abdomen cylindrical, elongatc, with 1- or 2-segmented cerci. *Larvae* (Fig. 41.1*c*) eruciform, or scarabaeiform with sclerotized head capsule, mandibulate mouthparts, lateral compound eyes; thoracic segments subequal, prothorax with distinct notal sclerite; thoracic legs short with fused tibia and tarsus and single claw; abdomen usually with prolegs on segments 1 to 8, segment 10 modified as a suction disk or with a pair of hooks *(Nannochorista)*. *Pupa* decticous, exarate, nonmotile.

In their simple mandibulate mouthparts and membranous wings of similar shape and venation, scorpion flies are similar to the primitive neuropteroid Orders Neuroptera and Megaloptera. As in those orders, mecopteran fossils first appear in the Permian Period. Before the end of the Paleozoic Era a diverse fauna, including three suborders and numerous families, had developed. Most of these taxa, including a fourth suborder which appeared in the Mesozoic Era, are now extinct. Mecoptera are presently widespread, especially in humid temperate and subtropical regions, but a high degree of endemism characterizes local regions. The Australian fauna is highly distinct, with many primitive species and one endemic family, Choristidae. Four families occur in North America, including Meropeidae, otherwise known only from Australia.

Despite the superficial similarity of Mecoptera to Neuroptera, the structural details of the mouthparts of adult scorpion flies indicate a relationship with Diptera. The slender mandibles and maxillae and elongate face of bittacids (Fig. 41.1*b*) are not greatly different from the homologous structures in primitive Diptera (compare with Fig. 42.3*a,b*), and the labial palps of *Chorista* (Choristidae) are enlarged and fleshy, suggesting a primitive labellum. In addition, the fossils *Permotanyderus* and *Choristotanyderus* are essentially intermediate between Mecoptera and Diptera. In contrast, the larvae of Mecoptera are similar to those of Lepidoptera and Trichoptera, a relationship which is also indicated by certain fossils.

Most Mecoptera prefer moist conditions, where the adults are to be found on rank vegetation or flying about. In species which inhabit arid or hot regions, such as *Apterobittacus* in California, adult activity is usually restricted to the cool, wet part of the year. Most scorpion flies are winged, but various genera are brachypterous [*Boreus;* Boreidae (Fig. 41.2*a*)] or completely wingless [*Apterobittacus;* Bittacidae (Fig. 41.2*b*)]. *Boreus* is a winter form, most commonly found about moss, which is used as adult and larval food. *Apterobittacus* frequents low vegetation, clambering nimbly about with the long legs. In *Brachypanorpa* (Panorpidae) females are brachypterous, males fully winged.

Bittacidae are predatory as adults, but most Mecoptera, such as *Boreus,* mentioned above, are apparently herbivores or scavengers. In Bittacidae the single large tarsal claws fold bladelike

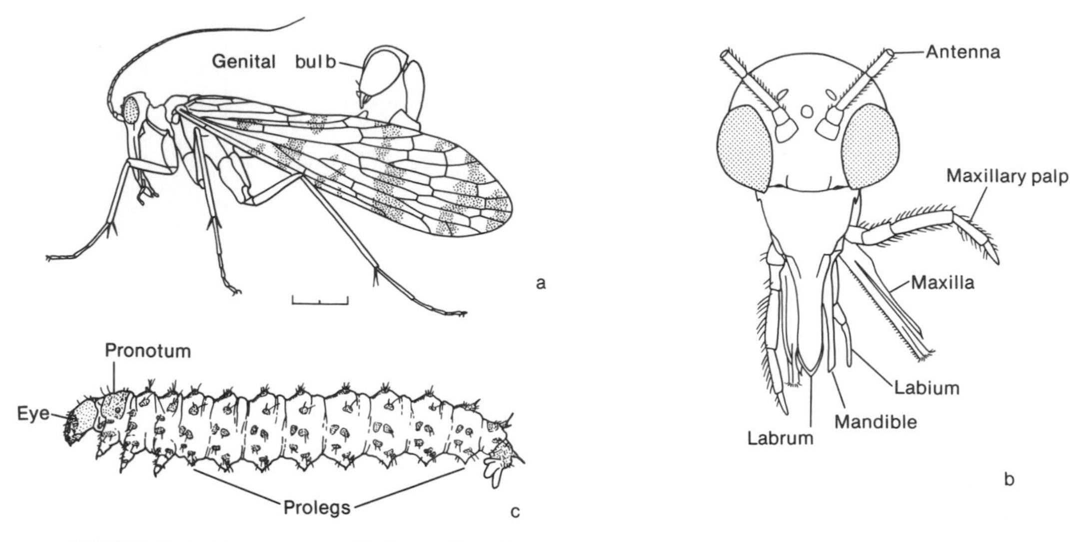

FIGURE 41.1 Mecoptera: *a*, adult Panorpidae (*Panorpa anomala*, male; scale equals 2 mm); *b*, facial aspect of head and mouthparts of *Bittacus chlorostigma*, antennae partly removed (Bittacidae); *c*, larva of *Panorpa nuptialis* (Panorpidae). (c *from Byers*, 1953.)

against the terminal tarsomeres (Fig. 41.2*c*), and are used to snare small, soft-bodied insects. Panorpidae occasionally visit flowers, probably for nectar, but feed chiefly on dead insects and snails.

Life histories have been described for only a few species. Copulation takes place on vegetation or the soil. In *Bittacus* mating is accompanied by complex courtship behavior involving a sex attractant secreted by abdominal glands of males. During copulation males of some species present the female with a previously captured prey item, which she partially consumes. Egg deposition is variable. Panorpidae and Choristidae oviposit small or large batches of delicate, smooth eggs in crevices, while Bittacidae scatter their tough, leathery, polygonal or spherical eggs on the soil. Larvae emerge after a brief period, or as in *Harpobittacus* and *Apterobittacus*, after a diapause of several months. Habits of immatures are poorly known; most larvae apparently occur in moist litter, consuming dead insects or plant material. The larva of *Nannochorista* (Australia and Chile) is aquatic. Where known, pupation occurs in a cell in soil, decaying wood, or litter. Prepupal diapause of several months is recorded in *Harpobittacus* and in *Panorpa*.

KEY TO NORTH AMERICAN FAMILIES OF
Mecoptera

1. Tarsi with single large claw about as long as apical tarsomere (Fig. 41.2*b,c*) **Bittacidae**
   Tarsi with claws paired, usually much shorter than apical tarsomere 2

2(1). Ocelli absent or very small; wings with reticulate crossveining in costal space, or rudimentary, without veins 3
   Ocelli present; wings with a few simple crossveins in costal space **Panorpidae**

3(2). Wings large, broad with complex venation; body dorsoventrally flattened **Meropeidae**
   Wings scalelike or bristlelike; body cylindrical or laterally flattened (Fig. 41.2*a*) **Boreidae**

Bittacidae (Fig. 41.2*b*). Slender-bodied, long-legged Mecoptera resembling craneflies. Widespread and locally common in woodlands throughout the world, with 9 species in North America.

Panorpidae (Fig. 41.1*a,c*). The most abundant and diverse family in the Northern hemisphere, with about 40 species of *Panorpa* in Eastern North America. The common name, "scorpion fly," refers to the enlarged genital bulb of males. *Brachypanorpa* occurs in the Southeast and Northwest.

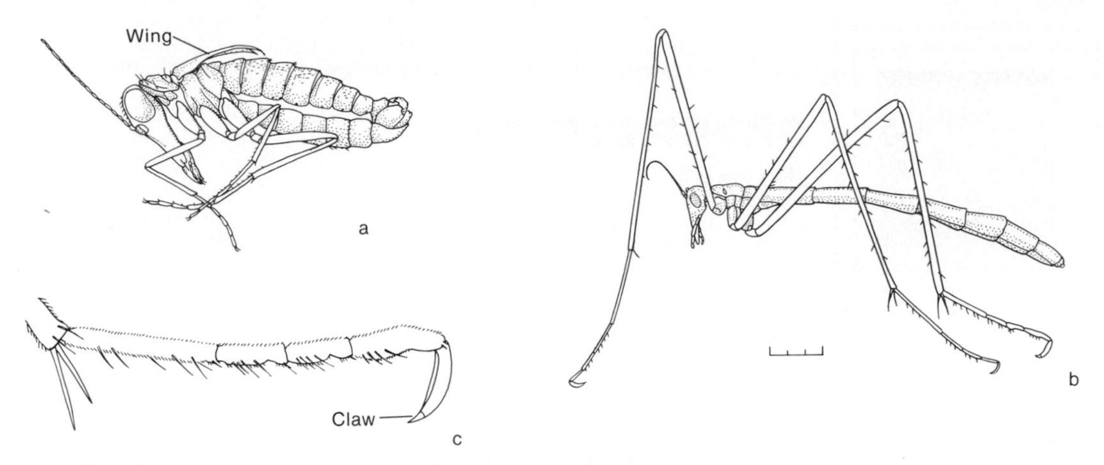

FIGURE 41.2 Mecoptera: *a*, Boreidae (*Boreus brevicaudus*); *b*, tarsus of *Apterobittacus; c*, Bittacidae (*Apterobittacus apterus;* scale equals 3 mm). (a *from Byers*, 1961.)

Meropeidae. A single species, Eastern states, and one species in Australia. Body relatively robust, flattened.

Boreidae (Fig. 41.2*a*). Restricted to the Northern Hemisphere, with about a dozen North American species, mostly in western mountain ranges. The dark colored adults, about 4 mm long, sometimes crawl about on the surface of snow, hence the name "snow fleas."

SELECTED REFERENCES

Byers, 1963 (life history), and 1965, 1971 (higher classification); Carpenter, 1931 (systematics of North American species); Cooper, 1972 (morphology) and 1974 (life history); Hepburn, 1969 (morphology); Hinton, 1958 (phylogenetic relationships); Penny, 1975 (evolution and phylogeny); Setty, 1940 (morphology, biology).

42

ORDER DIPTERA (Flies)

Adult. Minute to moderately large Endopterygota, usually with functional wings on the mesothorax. Head free, mobile, usually with large compound eyes, 3 ocelli; antennae multiarticulate, filiform or moniliform, or reduced, with terminal segment bristle-shaped; mouthparts diversely modified, usually as a proboscis for imbibing liquid food (Fig. 42.3). Prothorax reduced to a small, collarlike region, usually immovably fused to the mesothorax; mesonotum large, convex, usually divided into prescutum, scutum, and scutellum by transverse sutures (Fig. 42.2); metathorax reduced to a narrow, transverse band or lateral plates bearing halteres and metathoracic spiracles; wings with few or numerous longitudinal veins, few closed cells; legs usually with pulvilli, empodia, or other pretarsal structures.

Larva (Fig. 42.5). Usually maggotlike or vermiform, without legs, but occasionally with unsegmented pseudopods on various segments, especially in Nematocera; head capsule sclerotized as discrete tagma or undifferentiated; eyes and antennae reduced or absent. Thorax usually undifferentiated from abdominal region; abdomen almost always without cerci or urogomphi.

Pupa. Adecticous; obtect (Nematocera, Orthorrhapha) or exarate, enclosed in sclerotized *puparium* (Cyclorrhapha).

The Diptera comprise approximately 85,000 species, considerably fewer than in the Coleoptera, Hymenoptera, or Lepidoptera. Yet the flies must be considered one of the dominant groups of insects because of their extreme abundance over a great variety of ecological situations. Moreover, because of the small size and obscure habits of many flies, it seems likely that the eventual number will exceed 150,000 species.

Diptera include predators, parasites, and parasitoids, but the majority of species are saprophytic, and Diptera are usually the dominant invertebrate consumers of decaying vegetation or decomposing animal products. Flies are common visitors to flowers, which provide the only food for adults of some families (e.g., Bombyliidae, Conopidae, Acroceridae, Apioceridae). However, few Diptera have engaged in the intimate type of symbiotic relationship displayed by the angiosperm plants and Hymenoptera, and flies appear to be of sporadic importance in pollination except in arctic and high montane regions. Almost no adult Diptera directly damage living plants, except by oviposition.

Larvae occur predominantly in moist or subaquatic situations, less frequently in either dry or strictly aquatic habitats. Many species of arid or

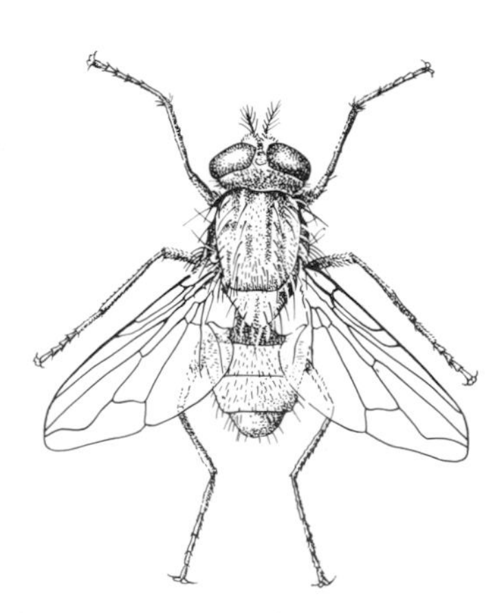

FIGURE 42.1 *Musca domestica,* the housefly (Muscidae). (From U.S. Department of Agriculture.)

subarid regions exist in moist microhabitats, such as animal burrows, rotting succulent vegetation, or feces, and many are specialized for quick exploitation of ephemeral sources of food and water. In such species the larval growth phase may span 1 to 2 weeks or less. Immatures of several families are internal parasites of vertebrates, the only significant group of insects to acquire this habit. A relatively small number of fly larvae attack living plants, as miners or borers in various plant parts, or by causing galls.

The chief economic importance of Diptera stems from their transmission of virulent diseases of humans and domestic animals. Malaria, yellow fever, and encephalitis, all transmitted exclusively by mosquitoes, have been three of the most persistent and debilitating human diseases, particularly in tropical climates. Houseflies (Fig. 42.1), as well as other species, mechanically vector numerous microorganisms, especially those causing dysenteries and other enteric ailments. With exceptions such as the Hessian fly [*Mayetiola destructor* (Cecidomyiidae)], root maggots (Psilidae), and fruit flies (Tephritidae, Otitidae), flies cause relatively minor damage to crops, as would be expected from their predominance as decomposers rather than herbivores. In addition, many phytophagous Diptera, such as fruit flies (Tephritidae), tend to be highly host-specific and susceptible to control by cultural practices such as staggering crops, isolating plantings, and removing crop debris.

MORPHOLOGICAL ADAPTATIONS. The most conspicuous characteristic of the body plan of adult Diptera is the high degree of adaptation for rapid, efficient flight. For most flies, flight is required for location of food or other resources, escape from predators, and successful reproduction. Diptera, along with Hymenoptera and Lepidoptera, are notable for the strong partitioning of resources between immatures and adults. For most species, highly developed flying ability has facilitated such partitioning.

Obvious adaptations to aerial life include foreshortening and streamlining of the body and shortening of the antennae, as in most Cyclorrhapha. More profound modifications involve structural changes in the thorax. In all Diptera, the wing-bearing mesothorax has become enormously enlarged, while the prothorax and metathorax are reduced to narrow, collarlike regions (Fig. 42.2). The prothorax, marked by the attachments of the forelegs, is largest in Nematocera; in most Brachycera it is represented by narrow pleural sclerites and dorsolateral prominences, or *humeri*. The metathorax is represented in most species by small lateral sclerites supporting the halteres. The metapleura and sterna are closely associated with the mesothorax and difficult to distinguish.

The architecture of the mesothorax is complicated by the development of secondary sutures, as well as by distortion caused by enlargement, especially of the notal region. Both the notum and postnotum are usually distinct (Fig. 42.2*a,b*). The notum is divided into *prescutum, scutum,* and *scutellum* by transverse sutures, which are represented internally by ridges, providing braces against the stress of the highly developed flight muscles. In Nematocera the postnotum is a relatively large, dorsal sclerite (Fig. 42.2*b*). In Brachycera the scutellum is prominent, becoming especially enlarged in Cyclorrhapha, where it may be divided into distinct scutellum and subscutellum (Figs. 42.2*c*, 42.18). In these flies the postnotum is represented by small, lateral plates. Enlargement of the scutellum probably provides additional area for muscle attachment, which is further increased in most Diptera by large, laminate *phragmata* which project ventrally into the thoracic cavity (Fig. 42.2*d*). The *prephragma* is often small or ridge-shaped, but the *postphragma* (postnotum) is almost always a prominent, convex plate, sometimes nearly separating the internal thoracic and abdominal spaces.

The pleural region of the mesothorax is exceedingly variable and complex, frequently with secondary sutures, incorporation of portions of the coxae, and radical shifts in size, shape, or position of certain sclerites. The principal landmarks are (1) the pleural wing suture, straight in Nematocera but sinuous in most Brachycera; and (2) the metathoracic spiracle, located at the

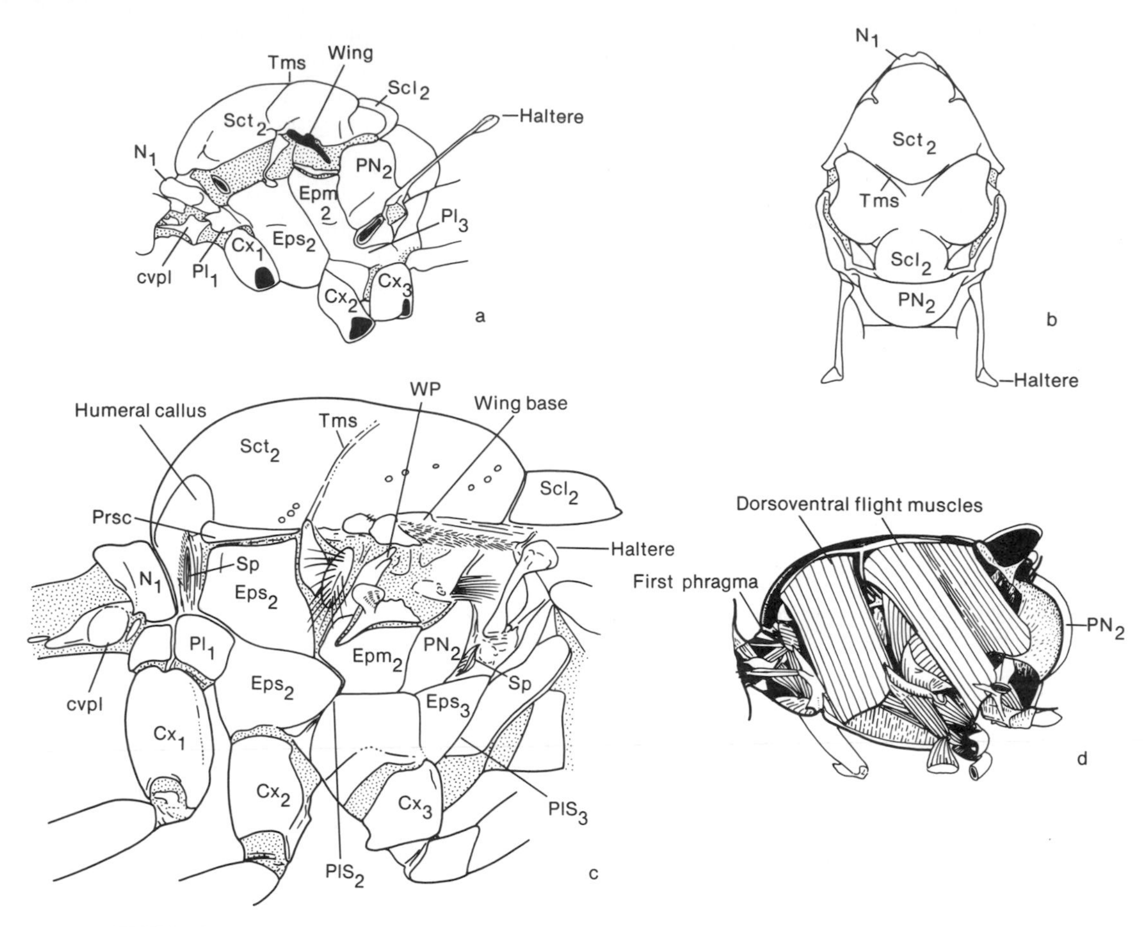

FIGURE 42.2 Structural features of Diptera: *a*, lateral aspect of pterothorax of Nematocera (Tipulidae, *Holorusia*); *b*, dorsal aspect of same; *c*, lateral aspect of pterothorax of Brachycera (Asilidae, *Mallophora*); *d*, longitudinal section through pterothorax of *Tabanus* (Tabanidae). *cvpl*, cervical plates; *Cx*, coxa; *Epm*, epimeron; *Eps*, episternum; *N*, notum; *Pl*, pleuron; *PlS*, pleural suture; *PN*, postnotum; *Scl*, scutellum; *Sct*, scutum; *Sp*, spiracle; *Tms*, transmesonotal suture; *WP*, pleural wing process. (c *from Bonhag*, 1949; d *modified from Cole*, 1969.)

anterior dorsal margin of the metapleuron, usually just below the haltere. The homology of certain structures and their functional significance are obscure, but the topographic features illustrated in Fig. 42.2 are important for identification of many taxa.

The thoracic specializations of Diptera are correlated, of course, with two-winged flight. Whereas the wings of most Hymenoptera and Lepidoptera are coupled during flight, functioning as a single airfoil, the hind wings of Diptera are transformed into *halteres* (Fig. 42.2) which are vibrated rapidly during flight, apparently functioning as gyroscopic balancing organs. In some Brachycera (the Calyptratae) the halteres are enclosed in cavities covered by *calypters,* foldings of the trailing edge of the wing base (Fig. 42.18*d*). Venation is highly variable. Some Nematocera have extensive, complex venation, not greatly different from that in Mecoptera (Figs. 42.6*a*, 42.7*a*). In most families, some reduction in number of veins and closed cells is apparent (Figs. 42.10, 42.14), often accompanied by a crowding of veins (usually the costal, subcostal, and first radial) into the leading edge of the wings, a strengthening configuration which occurs repeatedly in insects with a rapid wingbeat. Also correlated with the development of more efficient flight is an increasing consolidation of the nervous system. In Nematocera several abdominal and three thoracic ganglia are usually present. In Brachycera these numbers are variously reduced, and in the most specialized Cyclorrhapha all are incorporated into a single, large ganglionic mass in the thorax.

Diptera all imbibe liquid food, but unlike the Hemiptera, which are narrowly specialized for piercing and sucking, flies have evolved a variety of modes of feeding, and some Syrphidae and Bombyliidae ingest finely particulate material such as pollen along with liquids. In certain Nematocera mouthparts (Fig. 42.3*a*) are similar to those of Mecoptera, with bladelike mandibles, maxillae with lacinia and multiarticulate palpi, and elongate labium with one- or two-segmented fleshy palpi or *labellar* lobes. In Tabanidae and some bloodsucking Nematocera, the bladelike mandibles, maxillae (laciniae), labrum, and hypopharynx are enclosed in the stout U-shaped labium, which acts as a guide for the cutting mouthparts. The labellum bears numerous minute surface channels, or *pseudotracheae,* for sponging up oozing blood by capillarity. In Culicidae the mandibles and laciniae are fine, styliform blades, and the labrum and hypopharynx are delicate tubes for uptake of food and delivery of saliva (Fig. 42.3*b*). All are ensheathed by the flexible labium and form a piercing structure analogous to that of Hemiptera (see Fig. 34.2*e* to *g*). In predatory Brachycera (Asilidae, Empididae), the labial sheath is rigidly sclerotized, and the laciniae, labrum, and hypopharynx are stout blades used to pierce the integument of prey. Mandibles are absent, and the labellum vestigial.

In many flies, including nearly all Cyclorrhapha, the mouthparts consist of the labrum, labium, and hypopharnyx combined to form a thick, fleshy *rostrum,* or proboscis, terminated by a greatly enlarged labellum (Fig. 42.3*c*). Soluble constituents of solid foods are dissolved in saliva or regurgitated gut contents before being consumed. Predation and parasitism in these specialized forms, which have lost the laciniae, are accomplished by the secondary development of sharp, sclerotized projections of the pseudotracheae or preoral region of the labellum, which are used to pierce the host integument. In the highly specialized tsetse fly (*Glossina*), the rostrum has secondarily become slender and elongate,

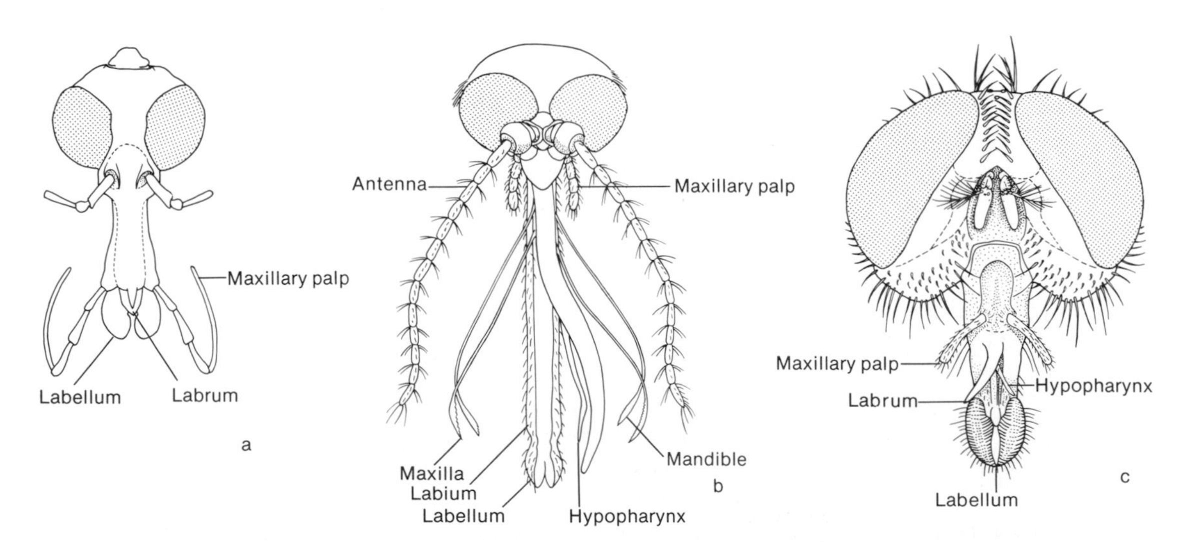

FIGURE 42.3 Mouthparts of adult Diptera: *a,* primitive nematoceran type (Tipulidae, *Ctenacroscelis*); *b,* piercing–sucking type (Culicidae, *Culex*); *c,* sponging type (Muscidae, *Musca*). (*From Crampton and Snodgrass.*)

with highly reduced labella. The stout, elongate maxillary palpi are closely applied to the rostrum, providing additional rigidity.

Most Diptera have one or more diverticula of the foregut, which function as a crop. Liquid meals, large in volume, may be temporarily held in the diverticula, to be gradually released into the intestine.

A peculiarity of the order is the rotation of the male genitalia 90 to 360° from their original position. This condition, termed *torsion,* occurs during pupal development and entails a permanent twisting of the sperm duct around the hindgut. The function of torsion is not understood, but the degree of twisting is of taxonomic significance, as explained further on.

Diptera are remarkable for the profound morphological difference between adults and immatures. All dipterous larvae are without true legs, and the great majority lack trunk appendages entirely. Leglessness among insect larvae is mostly restricted to borers in plant tissues (Coleoptera, Lepidoptera, Hymenoptera), parasitoids (Hymenoptera, Coleoptera), and those inhabiting other protected microenvironments with assured food supply (nest-making Hymenoptera). Reduction of locomotory and cranial structures in fly larvae is probably related to their widespread occurrence in rich nutrient sources, where it is advantageous to feed and grow as rapidly as possible. Under such circumstances extreme divergence between adults and larvae is possible, the larvae being specialized as a feeding stage, the adults as a reproductive and dispersal stage. Divergence not only allows specialization for rapid exploitation of temporary food sources, but also decreases competition between larvae and adults.

Larvae of the major taxonomic groups of Diptera differ in important morphological features. In Nematocera the *eucephalic* head capsule is sclerotized, with functional mandibles, maxillae, and stemmata and antennae (Fig. 42.4*a*). Mandibles move in a transverse plane, as in most mandibulate adult insects. In Orthorrhapha the head capsule is unsclerotized posteriorly *(hemicephalic)* (Fig. 42.4*b*) and usually retracted within the thorax, which is tough and leathery. Mandibles move in a more nearly vertical plane. Cyclorrhapha are *acephalic,* without a differentiated head. Mouthparts consist of a pair of protrusible, curved "mouthhooks" and associated internal sclerites (Fig. 42.4*c*). To a certain extent larval morphology corresponds to habitat. For example, free-swimming nematoceran larvae frequently have pseudopodial appendages, sometimes jointed and bearing terminal bristles or hairs (Fig. 42.5*a,b*). Internal parasitoids and larvae inhabiting rich food sources, such as rotting flesh or dung, usually lack appendages (Fig. 42.5*c*).

LIFE CYCLE. Diptera are predominantly bisexual, with most cases of parthenogenesis recorded

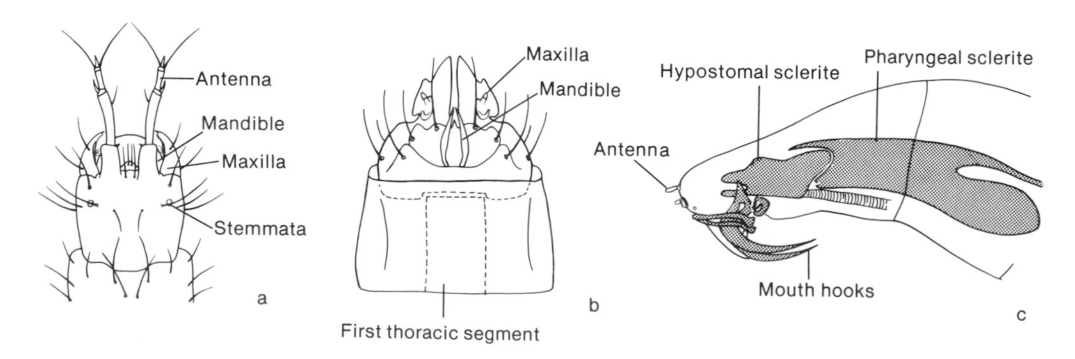

FIGURE 42.4 Cephalic structure of larval Diptera: *a,* Nematocera (Chironomidae, *Tanytarsus fatigans*); *b,* Brachycera, Orthorrhapha (Asilidae, *Dasyllis*), *c,* Brachycera, Cyclorrhapha (Anthomyiidae, *Spilogona riparia*). (a *from Branch,* 1923; b *from Malloch,* 1917; c *from Keilin,* 1917.)

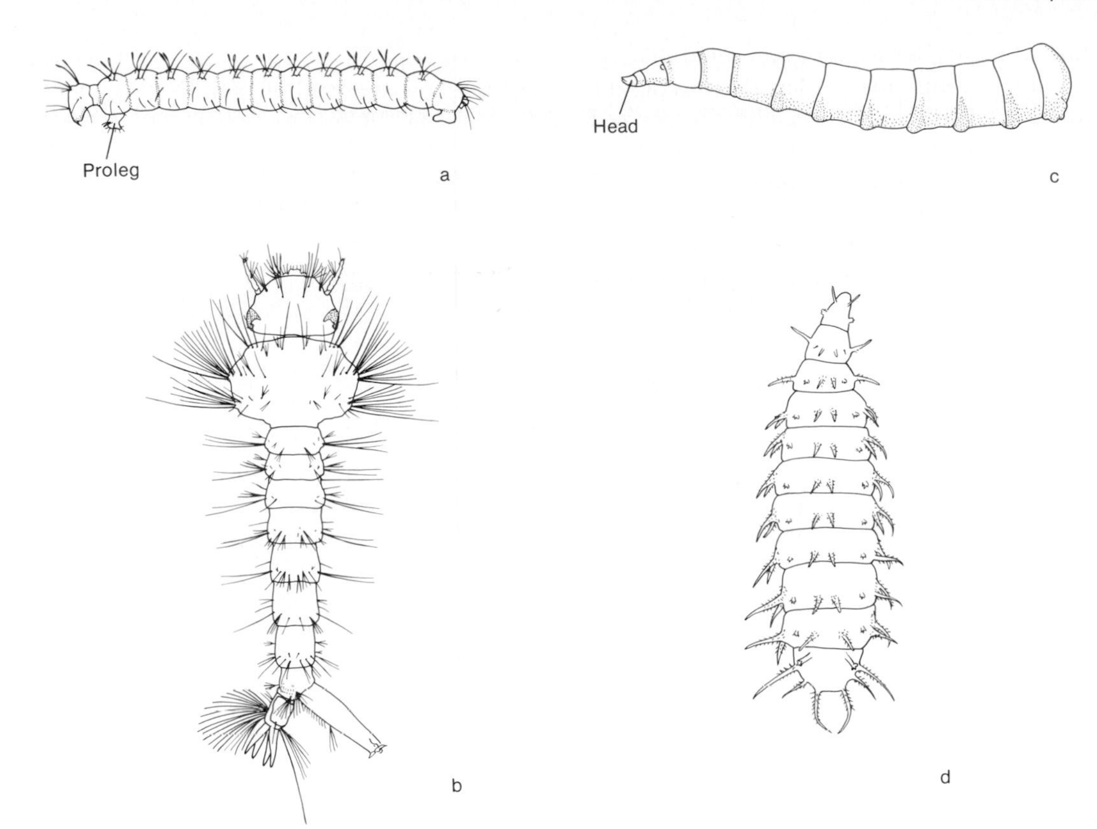

FIGURE 42.5 Representative Diptera larvae: *a,* Nematocera (Ceratopogonidae, *Forcipomyia specularis*); *b,* Nematocera (Culicidae, *Aëdes stimulans*); *c,* Brachycera, Cyclorrhapha (Muscidae, *Musca domestica*); *d,* Brachycera, Cyclorrhapha (Anthomyiidae, *Fannia canicularis*). (a *from Malloch,* 1917; b *from Matheson,* 1945; c *from Patton,* 1931; d *from Detweiler,* 1929.)

from the Nematocera. Eggs are usually small and ovoid, with thin, unsculptured chorion, and are typically deposited singly or in small masses on or near the larval food. Eggs usually hatch quickly, and egg diapause is infrequent. Cyclorrhapha include several ovoviviparous forms, as well as others which oviposit eggs that hatch immediately after deposition. Larval development in many families is extremely rapid, with many generations per year, but many aquatic species and free-living predators, such as Therevidae and Asilidae, have a single generation each year. Four larval instars are typical, more being recorded for some species and only three in Cyclorrhapha. Pupation occurs either in the larval substrate or a short distance away where drier conditions prevail. Obtect pupae frequently bear tubercles, horns, or other prominences on various parts of the body. In most cases these are probably used by the pharate adult to work its way through loose substrates before emerging from the pupal skin. In forms with coarctate pupae the puparium is burst along a transverse line of weakness by expansion of an eversible elastic sac, or *ptilinum,* which is extruded through the *ptilinal suture,* just above the antennal bases. In the mature adult this suture is permanently sealed.

Adult Diptera characteristically have rather short life-spans of a few days to a few weeks, and some ephemeral forms do not feed. Courtship and mating commonly take place in the air, at least in part. Aerial swarming behavior, in which males form dense, dancing masses, often near characteristic landmarks or over water, precedes mating in many Nematocera and some Brachycera, notably Empididae. A female entering the swarm pairs with a male; usually the coupled flies then leave the swarm. Elaborate courtship behavior typifies various families. Empidid males frequently present females with corpses of prey. Many drosophilids and tephritids show complex premating behavioral displays involving posturing or visual presentation of the pictured wings by the males. In both families courtship differences are important isolating mechanisms among species.

CLASSIFICATION. There has been continuing proliferation of family names in the Diptera, often as designations for one or a few small genera which did not entirely conform to larger, more well-known and accepted families. There have also been recent proposals for revolutionary reclassifications of all or part of the order (Griffiths, 1972). The classification presented here is conservative on both counts. Most of the family names omitted apply to very uncommonly encountered insects, often with extremely specialized habits. Exhaustive treatments of North American Diptera include Curran's (1934) and Cole's (1969).

KEY TO THE SUBORDERS OF
Diptera

1. Antennae with at least 6 segments, usually filiform or moniliform and frequently longer than thorax; maxillary palpi with 3 to 5 segments; pleural suture straight (Fig. 42.2*a*) **Nematocera** (p. 420)
   Antennae with fewer than 6 segments, terminal segment usually either elongate or with bristle-like style or arista; maxillary palpi with 1 to 2 segments; pleural suture with two right-angle bends (Fig. 42.2*c*) **Brachycera** (p. 425)

Suborder Nematocera

Adults. Antennae with 6 to 14 or more segments, usually filiform or moniliform, frequently longer than head and thorax, never with style or arista; maxillary palpi usually elongate, drooping, with 3 to 5 segments; mesothorax with pleural suture straight or slightly curved (Fig. 42.2*a*); wings usually without discal cell, and with veins Cu_2 and 2A never intersecting, usually diverging (Fig. 42.7*a*). *Larvae.* Head capsule sclerotized, with labium, maxillae, and horizontally biting, toothed mandibles. *Pupa.* Obtect, free (not enclosed in larval cuticle).

The flies comprising the Suborder Nematocera are mostly frail-bodied, gnatlike insects with long, slender legs and thin, delicate cuticle. Body size is commonly about 1 cm or less, with the notable exception of some Tipulidae, which have wingspans greater than 100 mm. The great majority of Nematocera are dull yellowish, brown, or black, usually nocturnal insects.

Structurally, Nematocera include the most primitive Diptera. In adult Blephariceridae, which are predatory, the mandibles are stout, bladelike cutting organs. The maxillary palpi are long, with four segments, and the labium is elongate, not distally enlarged or specially textured as a labellum. Such mouthparts, which superficially are not greatly different from those of Bittacidae (Mecoptera) (see Fig. 41.1*b*), except in the development of the maxillae, may represent the primitive condition in Diptera. In Simuliidae, Culicidae, Ceratopogonidae, and Psychodidae, the mouthparts are only slightly modified from this condition, and are used to puncture the integument, usually of vertebrate hosts. Specialization for ectoparasitism on vertebrates probably represents the initial adaptive radiation of the Diptera. In nonbiting Nematocera, the mouthparts are either vestigial (Deuterophlebiidae, Cecidomyiidae, Chironomidae) or adapted for imbibing nectar or other free liquids (Tipulidae, Scatopsidae, Bibionidae, Anisopodidae), with reduced mandibles, maxillae, and labellum, as in the Suborder Brachycera.

Relatively large, articulated prothoracic sclerites are present in Tipulidae, Tanyderidae, Bibionidae, and Scatopsidae; a distinct metanotal sclerite is present in Psychodidae. Wing venation is highly variable. Four radial branches are characteristic of several families, and five radials are present in Tanyderidae, together with four branches of the media (Fig. 42.6*a*). This is prob-

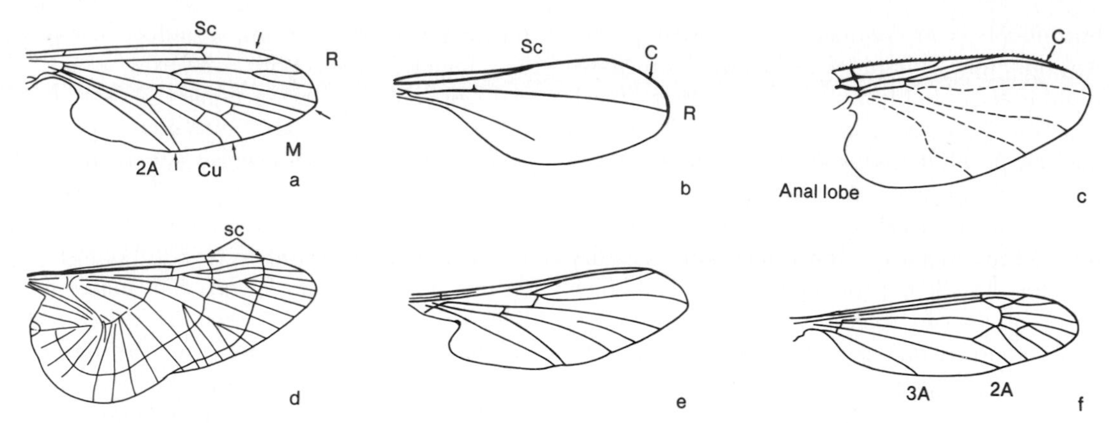

FIGURE 42.6 Wings of Nematocera: *a,* Tanyderidae (*Protoplasa*); *b,* Cecidomyiidae (*Contarinia sorghicola*); *c,* Simuliidae (*Simulium argus*); *d,* Deuterophlebiidae (*Deuterophlebia*); *e,* Blephariceridae (*Agathon comstocki*); *f,* Tipulidae (*Nephrotoma wulpiana*). *sc,* secondary creases; arrows at margin of wings delimit branches of major veins. (a *from Alexander;* d *modified from Cole,* 1969.)

ably the most generalized venation in the order; most families show fusion or loss of certain veins (Figs. 42.6*b*, 42.11, 42.14), and in Cecidomyiidae and Simuliidae all but two or three anterior veins are extremely weak or absent. Foldings or creases in the wing membrane form a faint network between the normal veins in Blephariceridae and Deuterophlebiidae (Fig. 42.6*d*). This peculiarity is apparently an adaptation to allow the fully hardened, folded wings to be used immediately after the adults emerge from the torrents and cataracts where pupation occurs.

Adult Nematocera are predominantly nocturnal, or if diurnally active, usually restrict their activity to deep shade or to periods of overcast skies. The major exception is the Family Bibionidae, which consists of diurnal, flower-visiting members. Food habits of most adults are obscure,

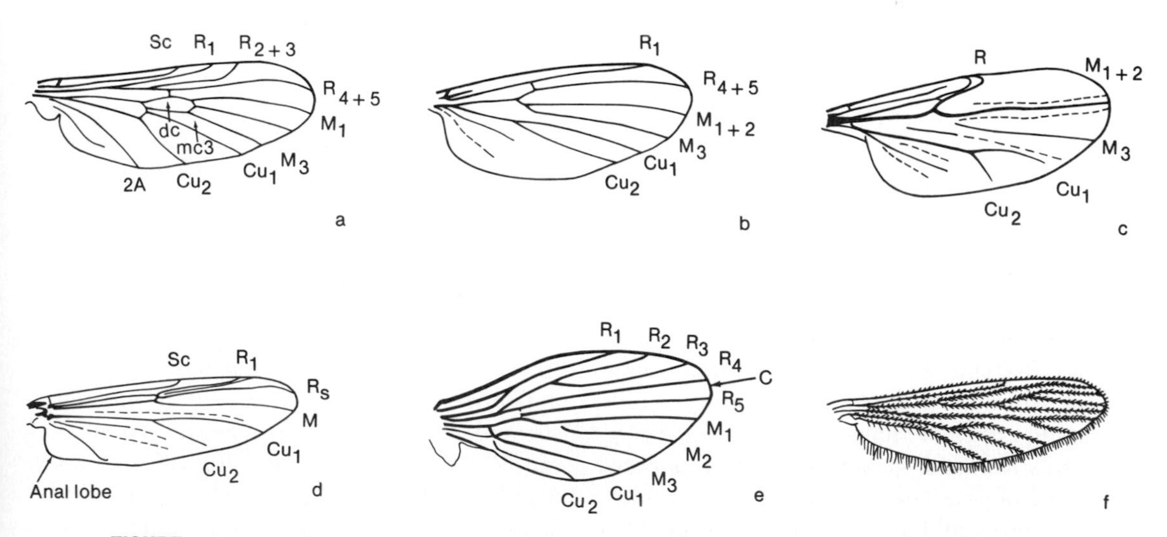

FIGURE 42.7 Wings of Nematocera: *a,* Anisopodidae (*Sylvicola fenestralis*); *b,* Mycetophilidae (*Mycetophila fungorum*); *c,* Ceratopogonidae (*Culicoides variipennis*); *d,* Chironomidae; *e,* Psychodidae (*Pericoma truncata*); *f,* Culicidae (*Culex tarsalis*). *dc,* discal cell or cell M_2; *mc*3, cell M_3.

but members of families such as Scatopsidae, Anisopodidae, Tipulidae, Sciaridae, and Mycetophilidae probably imbibe liquid decay products from plant or animal material. Chironomidae, Deuterophlebiidae, Cecidomyiidae, as well as some members of other families do not feed as adults. Blephariceridae prey on soft-bodied flying insects, including other flies and mayflies. Ceratopogonidae, commonly known for their habit of sucking blood from vertebrates, are predominantly parasites of other insects, usually much larger than themselves. Female Simuliidae (blackflies) and Culicidae (mosquitoes) use vertebrate blood as food, attacking reptiles, birds, and mammals. In many species a blood meal is required for production of fertile eggs. Members of both families transmit serious diseases of humans (see Chap. 16), constituting the chief economic importance of nematocerous flies.

Swarming behavior occurs in many families, being especially pronounced in Chironomidae, Bibionidae, and Mycetophilidae. Eggs are deposited on or near the larval substrate, usually singly, occasionally in adhering masses. Egg rafts, formed by one to many females, are characteristic of many mosquitoes. The great majority of Nematocera reproduce bisexually, but paedogenetic parthenogenesis occurs in *Miastor* (Cecidomyiidae), in which larviform females reproduce ovoviviparously. *Miastor* inhabits the fermenting phloem beneath bark of dead trees. Parthenogenesis is apparently a mechanism for rapidly exploiting this rich food source. When the tree begins to dry out, winged, bisexual adults appear.

Larvae of Nematocera include aquatic, subaquatic, and terrestrial forms. Blephariceridae and Deuterophlebiidae occupy torrents and cataracts, clinging to stones with suckers and grazing on diatoms and algae. Simuliidae occur primarily in moving water, including rapids, attaching either to stones or larger aquatic organisms with the single sucker on the abdominal apex. All pupate near the site of larval attachment. Chironomidae and Culicidae occur in a diverse array of aquatic habitats. Chironomid larvae are usually bottom dwellers, frequently forming tubes in mud, silt, or under stones in still or moving water. Both well-oxygenated and stagnant waters are occupied. *Clunio* larvae (Chironomidae) inhabit intertidal cracks and crevices throughout the world. Emergence of the terrestrial adults is usually synchronized with full moons, ensuring favorable low tides. *Pontomyia,* inhabiting coral reefs in the Indo-Pacific region, is marine in all instars. The highly modified males superficially resemble Veliidae (Hemiptera). They skate about on the surface of tide pools, carrying the vermiform females.

Culicid larvae, or mosquito wrigglers, are primarily inhabitants of standing water, but backwaters or eddies in slowly moving streams, rivers, or ditches offer suitable habitats as well. Many species are specialized for utilizing very small, enclosed bodies of water, such as rot holes in trees, potholes in rocks, crab burrows, or the water entrapped by bromeliads, pitcher plants, or certain other plants. Rainwater held by discarded cans and bottles provides the preferred larval habitat for some species, which have, in effect, become domesticated. Mosquito larvae are mostly filter feeders on plankton; a few species, including the entire Subfamily Chaoborinae (phantom midges), are predatory.

Most dipterous larvae, aquatic or not, require atmospheric oxygen for respiration. The major exceptions are the Blephariceridae, Deuterophlebiidae, and Chironomidae, which respire either through tracheal gills or, occasionally, with plastrons. Some mosquito larvae can respire slowly under water and may overwinter in this way, but most other "aquatic" Nematocera will drown if long separated from the air. Most of these *subaquatic* larvae inhabit littoral regions of streams or ponds, semiliquid decaying vegetation or sewage, or other wet situations. A few, such as Ptychopteridae, extend a flexible breathing tube to the surface for air, an adaptation encountered again in some syrphid larvae, but most have only short breathing tubes or none. Many Tipulidae and Ceratopogonidae, as well as the Psychodidae, Anisopodidae, Thaumaleidae, Tanyderidae, and Trichoceridae can properly be classified as subaquatic. Within their

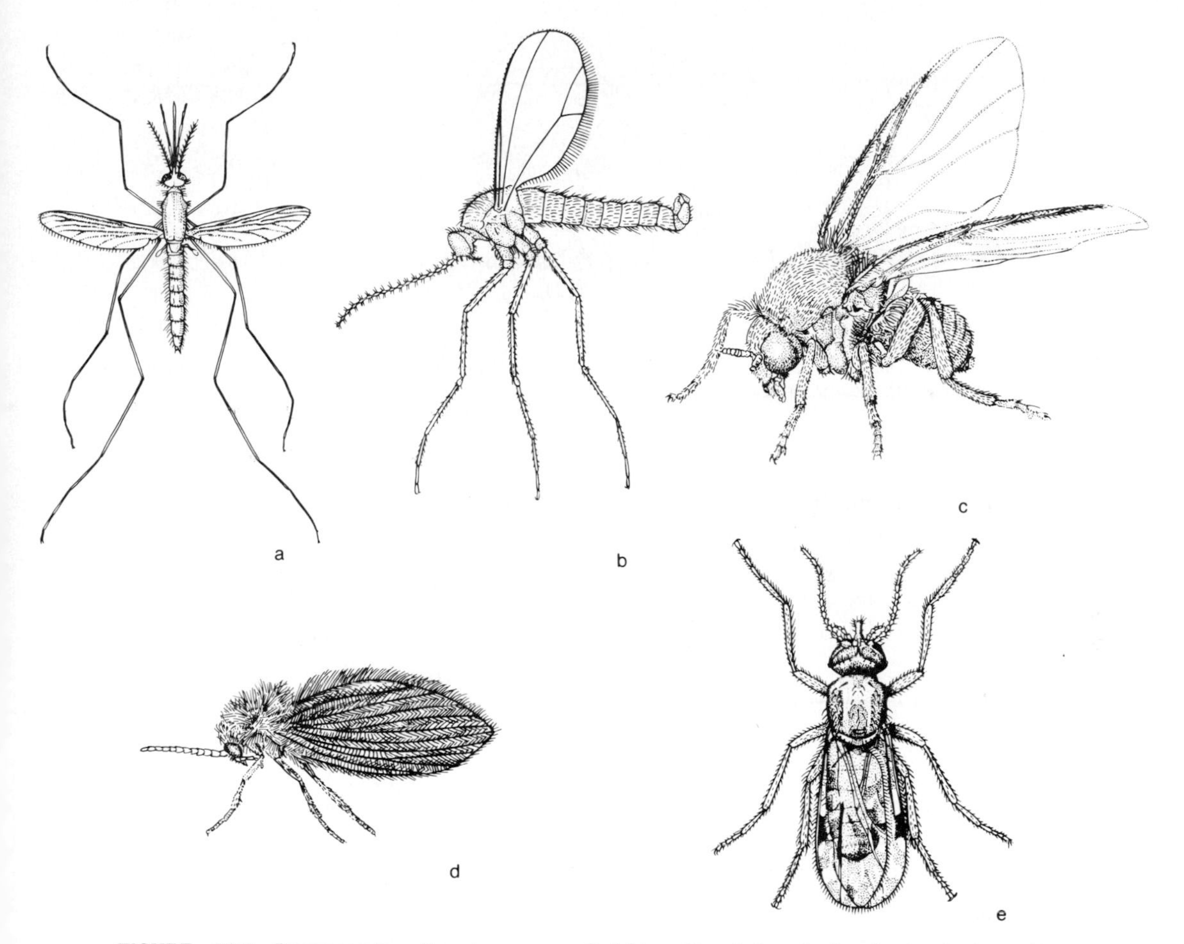

FIGURE 42.8 Representative Nematocera: *a,* Culicidae (*Anopheles freeborni*); *b,* Cecidomyiidae (*Mayetiola destructor*); *c,* Simuliidae (*Prosimulium mixtum*); *d,* Psychodidae (*Psychoda*); *e,* Ceratopogonidae (*Culicoides variipennis*). (a *from Wilson,* 1904; c *from Peterson,* 1970; b, d *from Cole,* 1969; e *from Hope,* 1932.)

semiliquid habitats these larvae function mostly as decomposers, less frequently as herbivores or predators.

The Families Scatopsidae, Bibionidae, Sciaridae, and Mycetophilidae, as well as some Ceratopogonidae and Tipulidae, are best considered terrestrial, though the larval medium is usually moist. Larvae in this category are predominantly decomposers, differing chiefly in the type of material consumed. Mycetophilidae and Sciaridae are commonly associated with fungi, ranging from molds to mushrooms and bracket fungi; Scatopsidae frequent dung, and Bibionidae, moist humus. In all these families exceptional members are predatory.

The Cecidomyiidae cannot be accommodated in a simple classification by larval habitat. These minute, fragile flies are best known as the causative agents of galls, which the larvae produce on a wide variety of herbaceous plants. Other members mine leaves or stems, and include agriculturally important species, such as the Hessian fly (*Mayetiola*). A few are predators of aphids and other soft-bodied insects, and some live in forest litter and soil, probably feeding on fungi.

In Simuliidae and Mycetophilidae pupation occurs within silken cocoons, but the obtect pupae of most families are naked. Blephariceridae, Deuterophlebiidae, and Simuliidae, which often pupate in waterfalls or rapids,

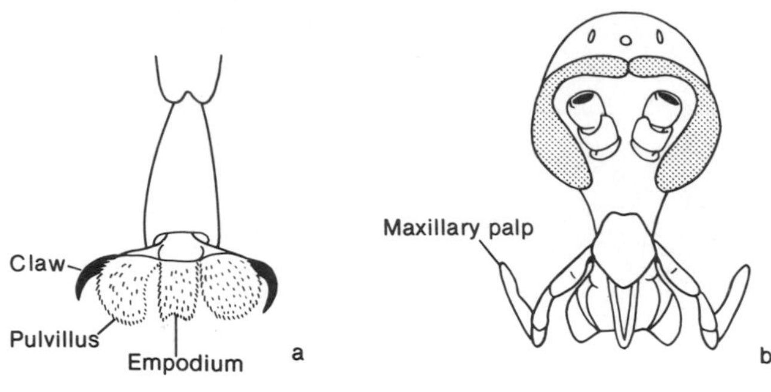

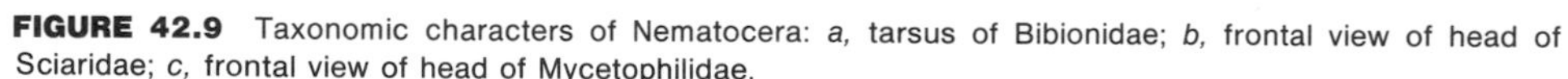
FIGURE 42.9 Taxonomic characters of Nematocera: *a*, tarsus of Bibionidae; *b*, frontal view of head of Sciaridae; *c*, frontal view of head of Mycetophilidae.

emerge as functional adults, bobbing to the surface and immediately taking flight without an initial period of wing expansion and hardening of the cuticle. Culicids are exceptional in retaining full mobility throughout the pupal period.

KEY TO THE FAMILIES OF THE SUBORDER
Nematocera

1. Wings with secondary network of fine creases or folds in addition to venation (Fig. 42.6*d*)[1]; mesonotum with transverse suture straight
   Deuterophlebiidae and **Blephariceridae**
   Wings without secondary network of creases or folds; mesonotum with transverse suture straight or V-shaped — 2

2(1). Mesonotum with transverse suture V-shaped, extending to vicinity of scutellum (Fig. 42.2*b*) — 3
   Mesonotum with transverse suture straight or weakly curved — 5

3(2). Head with 2 to 3 ocelli — **Trichoceridae**
   Head without ocelli — 4

4(3). Wing with 2 anal veins reaching margin (Fig. 42.6*f*); antennae usually with 12 to 13 segments — **Tipulidae**
   Wing with 1 anal vein reaching margin (Fig. 42.6*a*); antennae with 15 to 25 segments
   Ptychopteridae and **Tanyderidae**

[1]Apterous species are known from several families.

5(2). Ocelli present — 6
   Ocelli absent — 12

6(5). Costal vein ending at or near wing tip (Fig. 42.6*c* to *f*) — 7
   Costal vein weakened behind wing tip but continuing around posterior margin of wing (Fig. 42.6*b*); wings with 3 to 6 longitudinal veins, usually without crossveins; minute flies with long, many-segmented antennae — **Cecidomyiidae** (Fig. 42.8*b*)

7(6). Wing with discal cell (M_2) present (Fig. 42.7*a*) — 8
   Wing without discal cell (Fig. 42.7*b*) — 9

8(7). Wing with cell M_3 open externally (Fig. 42.7*a*)
   Anisopodidae (incl. **Pachyneuridae**)
   Wing with cell M_3 closed (Fig. 42.11*a*)
   Xylophagidae

9(7). Tibiae without apical spurs (tibial apex may be pointed, spurlike); wings with anterior veins much thicker than posterior veins
   Scatopsidae (incl. **Hyperoscelididae**)
   Tibia with articulated spurs near apex, wings usually with all veins of about same thickness — 10

10(9). Tarsi with broad, fleshy pads (pulvilli) beneath claws (Fig. 42.9*a*) — **Bibionidae**
   Tarsi without pulvilli — 11

11(10). Eyes meeting above antennal bases (Fig. 42.9*b*); coxae moderate in length (Fig. 42.8*b*); occasionally elongate — **Sciaridae**

Eyes separated dorsally (Fig. 42.9*c*); coxae nearly always greatly elongate **Mycetophilidae**[1]

12(5). Costal vein ending at or near wing tip (Fig. 42.7*c,d*) **13**

Costal vein weakened behind wing tip, but continuing around posterior margin (Fig. 42.7*e*) **15**

13(12). Wings broad, oval, with large anal lobe; anterior veins always much stronger than faint posterior veins (Fig. 42.6*c*) **Simuliidae** (Fig. 42.8*c*)

Wings usually lanceolate, elongate, with anal lobe small or absent (Fig. 42.7*c, d*); venation variable **14**

14(13). Wing with radial veins intersecting costal margin near midpoint (Fig. 42.7*c*); radial veins usually stronger than posterior veins; mouthparts of piercing type **Ceratopogonidae** (Fig. 42.8*e*)

Wing with radial veins extending nearly to apex (Fig. 42.7*d*); radial and posterior veins similar; mouthparts without piercing mandibles **Chironomidae**

15(12). Wings broad, with pointed tip; membrane usually densely hairy; radial vein usually with 5 branches (Fig. 42.7*e*) **Psychodidae** (Fig. 42.8*d*)

Wings usually narrow, lanceolate, and bare (veins may bear short hairs); radial vein usually with 3 to 4 or fewer branches **16**

16(15). Wings with 7 or fewer longitudinal veins reaching margin (Fig. 42.6*b*); minute, frail flies some **Cecidomyiidae** and **Thaumaleidae**

Wings with at least 9 longitudinal veins reaching margin (Fig. 42.7*f*) **Culicidae** (Fig. 42.8*a*)

Suborder Brachycera

Adults. Mostly stouter-bodied than Nematocera, with antennae short, with terminal style or arista; maxillary palpi usually short, with 1 to 2 porrect segments; pleural suture bent around sternopleural sclerite (Fig. 42.2*c*); wings with discal cell present (Orthorrhapha) or absent (Cyclorrhapha), anal cell narrowed before wing margin (Figs. 42.10, 42.11, 42.14). *Larvae.* Head capsule incomplete or vestigial, usually retractable into thorax; mandibles biting in vertical plane (Fig. 42.4*b*), or mouthparts represented by specialized "mouthhooks" (Fig. 42.4*c*).

[1]A few Anisopodidae and Pachyneuridae will key out here.

Brachycera comprise the great majority of flies, including many familiar groups such as horseflies, houseflies, and blowflies. The suborder consists of two groups of families, termed *divisions,*[2] which are difficult to separate by adult features but are well differentiated by larval and pupal characteristics. The Division **Orthorrhapha** is characterized by hemicephalic larvae with vertically biting mandibles. The pupa is obtect (exception: exarate, coarctate in Stratiomyidae). In the Division **Cyclorrhapha,** larvae are acephalic, without true mandibles. The pupa is coarctate, enclosed in a hardened puparium formed from the cuticle of the last larval instar.

Cyclorrhapha are further subdivided as the Series **Schizophora,** characterized by the presence of the ptilinal suture just above the antennae. In the Series **Aschiza** the ptilinum and ptilinal suture are absent, although pupation is within a puparium. In the following key the Aschiza segregate with the Orthorrhapha, with which they share several adult features.

KEY TO THE MAJOR DIVISIONS OF THE SUBORDER Brachycera

1. Head with U-shaped ptilinal suture just above antennae (Fig. 42.15*a*); antennae usually with 3 segments, terminal segment with arista or style attached dorsally before apex (Fig. 42.15*a* to *h*) **Cyclorrhapha, Series Schizophora** (p. 431)

Head without U-shaped ptilinal suture; antennae 3- to 5-segmented, terminal segments elongate or with style or arista attached terminally (Fig. 42.13*c*) **Orthorrhapha** and **Cyclorrhapha, Series Aschiza** (p. 426)

[2]The taxonomic divisions of Diptera are groups of families and should not be confused with the Divisions Exopterygota and Endopterygota.

Division Orthorrhapha

Differentiated from Cyclorrhapha primarily by the obtect pupae (pupae exarate, enclosed in puparium in Cyclorrhapha). Exception: Stratiomyidae, with pupa obtect, enclosed in unmodified larval cuticle.

The Orthorrhapha comprise an extremely diverse assemblage of flies, including several of the largest families. Body form varies from rather frail, slender types similar to Nematocera (Rhagionidae) to those with stout, compact shape resembling that of Cyclorrhapha (Tabanidae, Dolichopodidae). Many Empididae and Dolichopodidae are minute, but body size in Orthorrhapha averages larger than in Nematocera or Cyclorrhapha. Members of the Asilidae (robber flies), Mydaidae, and Apioceridae include the largest flies, with body lengths up to 60 mm. Most Orthorrhapha are active diurnal insects, often conspicuous on flowers or foliage.

Most of their morphological features suggest that the Orthorrhapha are intermediate between the Nematocera and Cyclorrhapha, from which some Orthorrhapha cannot be distinguished on the basis of adult characteristics. For example, antennae in some Coenomyiidae have up to ten segments, and in many Rhagionidae have eight segments, recalling the multiarticulate antennae of Nematocera. In contrast, Dolichopodidae and many Empididae, as well as some members of several other families, have short antennae with three to four segments with a terminal arista, as in some Cyclorrhapha. This intermediacy involves all the characteristics used to separate the Cyclorrhapha and Nematocera, including mouthpart structure, thoracic structure and wing venation, genitalic features, and internal organ systems.

The Rhagionidae seem to represent the least specialized Orthorrhapha. Many of these usually soft-bodied flies have piercing-sucking mouthparts with well-developed, bladelike mandibles and maxillae. Similar mouthparts occur in Tabanidae (horseflies). In the predatory Asilidae and Empididae, at least the mandibles are absent, and usually only the hypopharynx is used to pierce the prey. Many Orthorrhapha [Bombyliidae (bee flies), Mydaidae, Apioceridae, Acroceridae, Nemestrinidae] have mouthparts adapted for sucking up nectar through an elongate proboscis formed by the labium. In a few Empididae and in Dolichopodidae the labium is produced as a short, flexible proboscis tipped by a fleshy labellum. These flies are predatory, the prey being pierced by cuticular pseudotracheal teeth along the medial groove of the labellum.

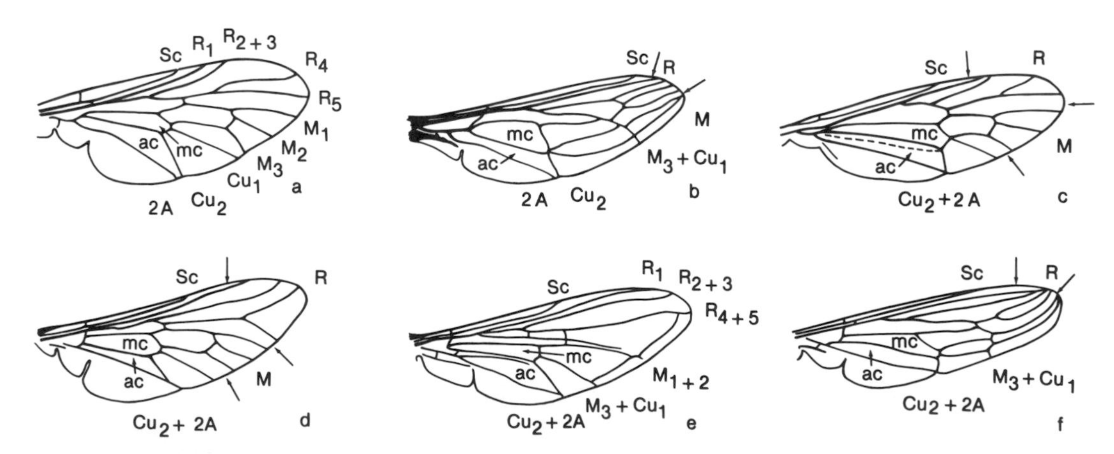

FIGURE 42.10 Wings of Brachycera: *a*, Rhagionidae (*Symphoromyia*); *b*, Nemestrinidae (*Neorhyncocephalus sackeni*); *c*, Xylophagidae (*Xylophagus decorus*); *d*, Tabanidae (*Tabanus laticeps*); *e*, Syrphidae (*Syrphus opinator*); *f*, Apioceridae (*Rhaphiomidas maehleri*). *ac*, anal cell; *mc*, median cell; marginal arrows delimit branches of major veins.

The wing venation is very generalized in several families, notably the Rhagionidae (Fig. 42.10*a*), with the radial vein four-branched, and the medial vein four-branched. In a number of families one or more branches of the radial or medial veins are looped forward to intersect the wing margin before its apex (Fig. 42.10*b,f*). In several of these families (e.g., Bombyliidae, Asilidae, Apioceridae, Mydaidae), venation is highly distinctive, and useful in identification. In many Empididae and Dolichopodidae, the venation is essentially the same as in the Cyclorrhapha, with the radius three-branched, the median two-branched, and the subcosta and R_1 extremely short (Fig. 42.11*g*). Calypters are typically small or absent in Orthorrhapha, but are large and conspicuous in Tabanidae and Acroceridae.

Two features in which Orthorrhapha are clearly intermediate between the more generalized Nematocera and the Cyclorrhapha are (1) torsion of the external genitalia in males and (2) the degree of consolidation of the central nervous system. In most Nematocera rotation is not evident; but torsion of 90 to 180°, frequently accompanied by asymmetry of genitalic structures, is typical of most Orthorrhapha. In Dolichopodidae and Cyclorrhapha, torsion has proceeded through 360°, so that the original dorsoventral relationships are restored, but marked asymmetry persists. The biological significance of torsion, which occurs in the pupal stage, is obscure; but the condition is almost always associated with a ventral folding of the genitalia beneath the abdomen.

The consolidation of the central nervous system is probably related to flying ability. Nematocera, which are often weak, periodic fliers, have three thoracic and six to seven abdominal ganglia. In Orthorrhapha a variable amount of condensation has occurred, and in Cyclorrhapha a single large thoracic ganglion or one thoracic and one abdominal ganglion remain.

BIOLOGY. Parasitism, of common occurrence in adult Nematocera, reappears in many Tabanidae and in a few Rhagionidae (*Symphoromyia*). In these flies, as in Nematocera, only females are bloodsuckers, the males imbibing nectar. Predation, encountered in such Nematocerous families as Ceratopogonidae and Blephariceridae, is universal among Asilidae and widespread among Empididae and Dolichopodidae. Asilidae are aggressive, aerial predators, attacking any insects small enough to subdue, including wasps and bees. The legs are long and robust, frequently bearing strong bristles useful in holding struggling prey. The posterior pairs are successively longer, and together the legs form a basket much as in the Odonata. The strong bristles about the eyes and face probably prevent thrashing prey from injuring the fly's head. Empididae and Dolichopodidae never achieve the large size of Asilidae, and tend to inhabit shadier or moister situations, frequently about seeps or running or standing water. Most members of both families are nimble runners, and many run down their prey, rather than pursuing it in the air.

Adults of Acroceridae, Bombyliidae, Mydaidae, Nemestrinidae, Apioceridae, and occasional species of several other families of Orthorrhapha feed on nectar, and some Bombyliidae eat pollen. Nectar feeding is sporadic in Nematocera (typical of Bibionidae), but recurs in many Cyclorrhapha. Some of these nectar-feeding species are equipped with elongate proboscises and hover above flowers while feeding. Other species have only a short proboscis or labellum, and alight on the flower. In several families, notably Therevidae, Scenopenidae, Coenomyiidae, and Xylophagidae, adult food is unknown and some of these may not feed as adults. Saprophagy, extremely widespread among adult Cyclorrhapha, is uncommon among Orthorrhapha.

Most Orthorrhapha are solitary, but male swarming is highly developed in Empididae and occurs in some Rhagionidae. During courtship, males of many empidids present females with prey, usually smaller Diptera. In some species courtship behavior has become ritualized to the extent that males (1) present the female with the prey plastered to the side of a silken or frothy balloon; (2) present the balloon with a non-prey

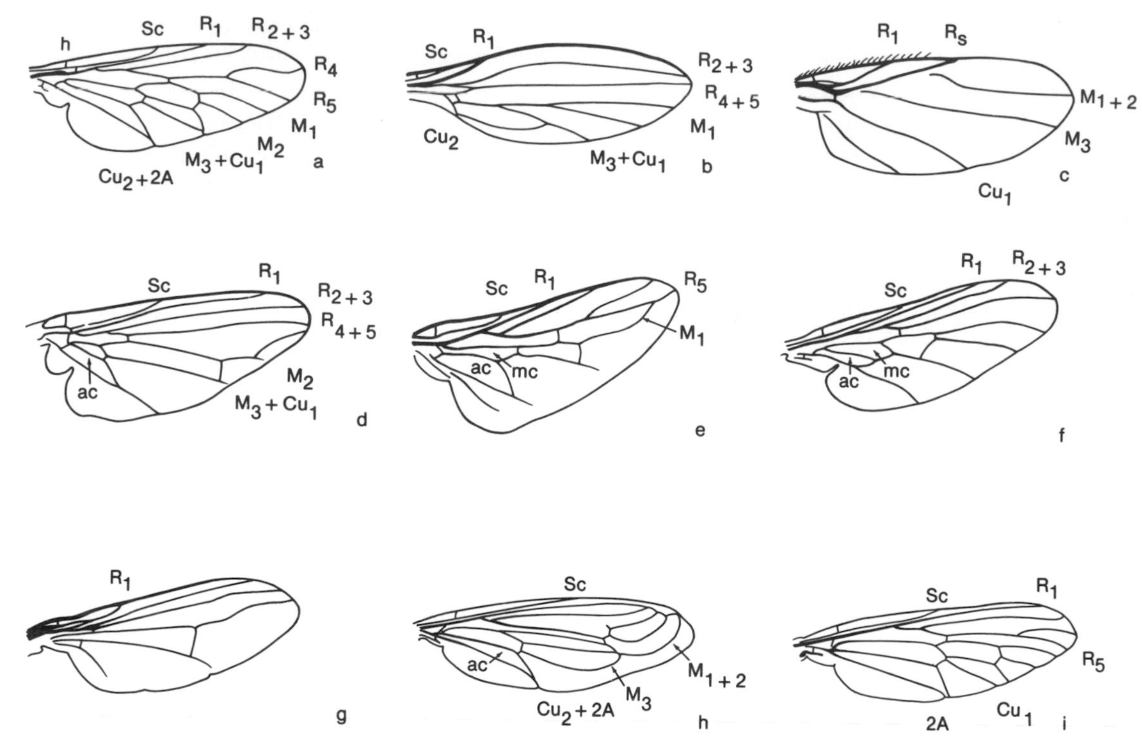

FIGURE 42.11 Wings of Brachycera: *a*, Therevidae (*Thereva vialis*); *b*, Lonchopteridae (*Lonchoptera*); *c*, Phoridae (*Chaetoneurophora variabilis*); *d*, Platypezidae (*Calotarsa insignis*); *e*, Scenopinidae (*Metatrichia bulbosa*); *f*, Empididae (*Rhamphomyia*); *g*, Dolichopodidae (*Tachytrechus angustipennis*); *h*, Mydaidae (*Nemomydas pantherinus*); *i*, Asilidae (*Cyrtopogon montanus*). *ac*, anal cell; *mc*, median cell.

item (small stick, sand grain); (3) present the balloon alone or present a non-prey item alone. Courtship in Dolichopodidae often involves signaling by the males, using the brushes and flanges on the tibiae and tarsi.

Most Orthorrhapha oviposit moderate numbers of eggs on or near the larval substrate. Bombyliidae, Acroceridae, and Nemestrinidae, whose larvae are internal parasitoids of other insects, oviposit extremely large numbers of eggs (up to 900 recorded in Acroceridae) in habitats where the active first-instar larvae are likely to encounter hosts. Most Orthorrhapha apparently deposit the eggs singly or in small batches, but horseflies form tiered egg masses on emergent aquatic vegetation or other objects near aquatic habitats. *Atherix* (Rhagionidae) forms communal egg masses on twigs, small branches, culverts, or other objects overhanging water. The flies die in place after oviposition, their bodies combining with the eggs to form incrustations that were formerly used as food by some Indians.

Larvae of Orthorrhapha occupy a broad array of habitats, ranging from aquatic to terrestrial. Aquatic larvae (some Rhagionidae, Stratiomyidae, Tabanidae, Empididae, and Dolichopodidae) require atmospheric oxygen, restricting them to shallow water. Most are inhabitants of seeps, marshes, or the littoral regions of lakes or streams, rather than of open-water situations. Empididae, Dolichopodidae, Asilidae, Coenomyiidae, and Xylophagidae usually inhabit moist soil, leaf litter, or rotten wood. Therevidae, Apioceridae, and Mydaidae frequent rotten wood, soil, or sand, frequently in quite arid circumstances. Scenopenid larvae are

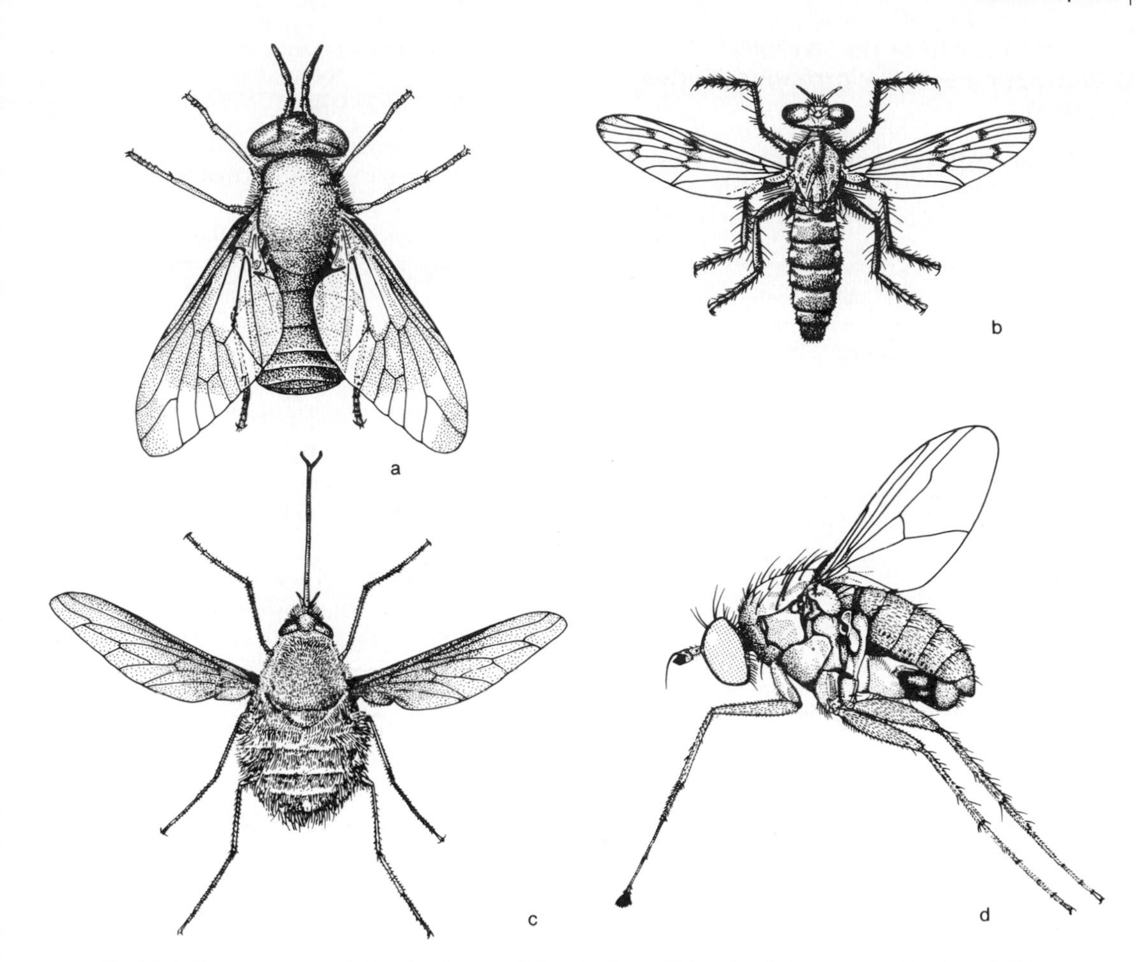

FIGURE 42.12 Representative Brachycera, Orthorrhapha: *a*, Tabanidae (*Chrysops proclivus*); *b*, Asilidae (*Metapogon pictus*); *c*, Bombyliidae (*Bombylius lancifer*); *d*, Dolichopodidae (male *Dolichopus*). (c *from Cole*, 1969.)

most frequently encountered in carpets, where they feed on moth larvae and other insects.

Larvae of Orthorrhapha are commonly predaceous, but several families include both predatory and herbivorous species (Rhagionidae, Stratiomyidae, Tabanidae, Asilidae), or scavengers (Empididae, Dolichopodidae, Asilidae). Larvae of Acroceridae, Nemestrinidae, and Bombyliidae are all internal parasitoids of other arthropods. Acroceridae utilize spiders exclusively; Nemestrinidae and Bombyliidae attack a variety of insect hosts.

Pupation occurs in or near the larval substrate. The conspicuous, posteriorly directed spines of many orthorrhaphous pupae aid the pharate adult in working its way to the surface of loose sand or soil, from which the cast pupal skins may frequently be seen protruding. Stratiomyidae are unique in having the exarate pupa enclosed in the cuticle of the last larval instar, which is not modified as a puparium as in the Cyclorrhapha. This arrangement probably functions in floating the pupa in the stagnant waters where many Stratiomyidae develop, and is apparently not homologous to the puparium of Cyclorrhapha.

KEY TO THE FAMILIES OF THE DIVISIONS

Orthorrhapha and Cyclorrhapha (Series Aschiza)

1. Tarsi with 3 subequal pads below claws (Fig. 42.9*a*) 2
   Tarsi with 2 or fewer pads below claws (Fig. 42.13*a*) 8

2(1). Head very small, thorax and abdomen inflated; calypters very large **Acroceridae**
   Body not so shaped; calypters small or vestigial 3

3(2). Antennae with third segment annulate (Fig. 42.13*b*) 5
   Antennae with third segment not annulate; usually elongate or with terminal bristle (Fig. 42.13*c*) 4

4(3). Wings with branches of median vein parallel to posterior margin (Fig. 42.10*b*) **Nemestrinidae**
   Wings with branches of median vein intersecting posterior margin (Fig. 42.10*a*) **Rhagionidae**

5(3). Calypters large, conspicuous 6
   Calypters small or vestigial 7

6(5). Anal cell open (as in Fig. 42.10*a*); eyes densely hairy **Coenomyiidae**
   Anal cell closed at or before wing margin (Fig. 42.10*c,d*); eyes not usually hairy **Tabanidae** (Fig. 42.12*a*)

7(5). Tibial spurs present, at least on middle legs **Xylomyidae** and **Xylophagidae**
   Tibial spurs absent **Stratiomyidae**

8(1). Head hemispherical, larger than thorax, with antennae usually inserted below midline of long, narrow face **Pipunculidae**
   Head much smaller than thorax, usually with broad face; antennae inserted above midline of head 9

9(8). Cu_2 long, reaching wing margin or joining 2A near margin; anal cell much longer than median cell (Fig. 42.10*e*) 10
   Cu_2 absent (Fig. 42.11*c*), or, if present, joining 2A more than a quarter of its length in from margin (Fig. 42.11*d,e,f*); anal cell subequal to median cell (Fig. 42.11*d,e,f*) 15

10(9). Medial veins looped forward, with M_{1+2} joining R_{4+5} before wing margin (Fig. 42.16*a*); antennae usually with dorsal bristle **Syrphidae** (Fig. 42.16*a*)
   Medial veins not joining R_{4+5} (Fig. 42.10*f*) (but radial veins may unite before wing margin); antennae usually with terminal style 11

11(10). Vertex concave between eyes (Fig. 42.13*d*) 12
   Vertex flat or convex between eyes (Fig. 42.15*a* to *c*) 13

12(11). Head with 3 ocelli; antennae usually with 3 segments **Asilidae** (Fig. 42.12*b*)

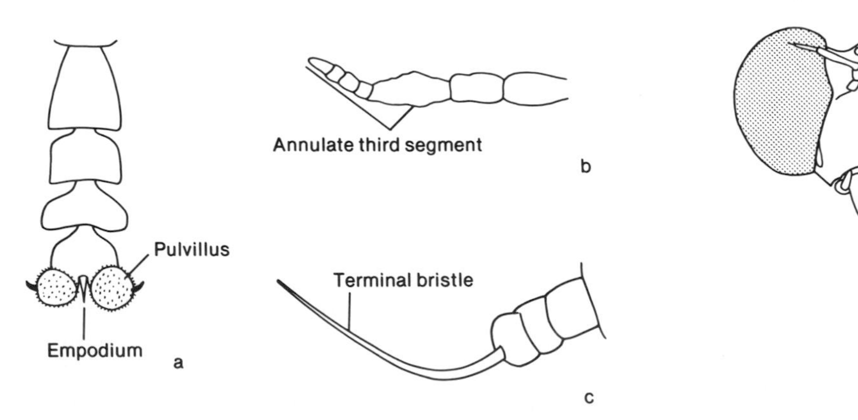

FIGURE 42.13 Taxonomic characters of Brachycera, Orthorrhapha: *a*, tarsus of Syrphidae; *b*, antenna of Tabanidae; *c*, antenna of Rhagionidae; *d*, head of Asilidae. (b *from Brennan*, 1935.)

Head with 1 or no ocelli; antennae with 4 segments **Mydaidae**

13(11). Veins R_5 and M_1 looped forward, terminating anterior to wing apex (Fig. 42.10*f*) **Apioceridae**

Veins R_5 and M_1 intersecting wing margin posterior to apex (Fig. 42.11*a*) 14

14(13). Wings with 3 median and cubital cells (Fig. 42.12*c*); usually scaley or hairy flies with long proboscis **Bombyliidae** (Fig. 42.12*c*)

Wings with 4 median and cubital cells (Fig. 42.11*a*); body bare or pubescent; proboscis short **Therevidae**

15(9). Wings with pointed apex; crossveins restricted to extreme basal portion of wing (Fig. 42.11*b*) **Lonchopteridae**

Wings with rounded apex; distal crossveins present or absent 16

16(15). Radial veins much thicker than medial and cubital veins, which are without crossveins (Fig. 42.11*c*); antennae appearing 1-segmented **Phoridae**

Veins all of approximately equal thickness; mediocubital region of wing with at least 1 crossvein; antennae with 2 or more segments 17

17(16). Hind tibiae and tarsi dilated, flattened; wings with anal cell pointed distally (Fig. 42.11*d*). **Platypezidae**

Hind tibiae and tarsi cylindrical or rounded; if dilated, anal cell not pointed 18

18(17). Vein M_1 curved anteriorly, intersecting wing margin before apex (Fig. 42.11*e*); antennae without arista or style **Scenopinidae**

Vein M_1 straight or curved posteriorly, intersecting wing margin behind apex (Fig. 42.11*f*); arista or style frequently present 19

19(18). Wings with r-m crossvein very close to base; R_1 usually joining C in basal half of wing (Fig. 42.11*g*); calypters small; proboscis usually a pointed beak **Dolichopodidae** (Fig. 42.12*d*)

Wings with r-m crossvein located more than one-fourth distance from base to apex; R_1 usually joining C in distal half of wing (Fig. 42.11*f*); calypters usually conspicuous, fringed; proboscis usually terminating as a fleshy labellum **Empididae**

Division Cyclorrhapha

Differing from Orthorrhapha chiefly in having the pupae coarctate and enclosed in a puparium formed by the modified cuticle of the last-instar larva.

The Cyclorrhapha contain a vast diversity of mostly small, compactly built flies. The great majority of species emerge by pushing off the end of the puparium with the ptilinum. The ptilinal opening is represented by a faint crescentic scar just above the antennal sockets in the fully hardened flies, hence the series name Schizophora. A few families, comprising the Series Aschiza, lack the ptilinum.

The Series Schizophora may be further divided into sections (groups of families). The Section Calyptratae includes flies with relatively large calypters (muscids, tachinids, etc; Fig. 42.18*d*). The Section Acalyptratae contains flies with the calypters small or absent. Included here are numerous families of small flies that are usually difficult to identify.

Mouthparts are much reduced in all Cyclorrhapha, with mandibles absent and maxillae represented by palpi. The labium is modified as a labellum which bears sharp pseudotracheal or prestomal teeth in predatory or parasitic forms. The head typically bears bristles in characteristic positions which are of great taxonomic utility.

Wing venation is reduced, leaving no more than three radial and three medial branches. The costa is often broken at one or more places; in several families one or more radial or medial veins are looped anteriorly, sometimes joining the next anterior vein. Such anomalies produce highly characteristic venational patterns, which are very useful in identifying many families.

The abdomen of Cyclorrhapha is characteristically divided into an anterior *preabdomen* and the *postabdomen,* which is flexed anteroventrally beneath the preabdomen in males. Male genitalia are rotated 360°. This extreme degree of torsion, termed *circumversion,* differentiates Cyclorrhapha from most Orthorrhapha.

BIOLOGY. The most common feeding mode of cyclorrhaphous larvae is decomposition; these larvae are the primary insect consumers of decaying plant and animal material. High vagility of adults allows prompt location and exploitation of temporary food sources. High mobility may also be used to exclude competitors. An extreme example is the horn fly *(Haematobia irritans),* which resides on cattle, immediately ovipositing in freshly dropped dung before other insects are attracted.

The variety of feeding substrates of saprophagous Cyclorrhapha is very great. Some species are quite catholic, but most restrict themselves to either plant or animal material, and many are highly specific. For example, Neriidae (in North America) are limited to rotting cacti, Coelopidae to decaying seaweed, and many species of other families have narrow substrate ranges, including such items as fermenting tree sap, frass in burrows of wood-boring beetles, or the debris which collects in bird nests or animal burrows. Many of these saprophagous species are probably feeding largely on fungi or bacteria, and some will mature on agar bacterial cultures. It may be noted that *Drosophila melanogaster,* an extremely important experimental animal for research in genetics and nutrition, is a decomposer of decaying fruits in nature, accounting for its easy maintenance in confined spaces.

The saprophytic Cyclorrhapha are of considerable economic significance because of their frequent close associations with humans or domesticated animals. Muscidae and Anthomyiidae (as well as several less significant families) commonly breed in animal dung, often reaching epidemic numbers in stock-raising areas. Many of these species are mechanical vectors of communicable diseases, ranging from dysenteries to poliomyelitis. A few species such as the horn fly, *Haematobia irritans,* the stable fly *(Stomoxys calcitrans),* and the tsetse flies *(Glossina)* have secondarily acquired bloodsucking habits. *Glossina* are intermediate hosts for several trypanosomiases, including sleeping sickness of humans.

The Calliphoridae and Sarcophagidae are known as blowflies and flesh flies, from their habit of ovipositing on carrion. Some species, especially Calliphoridae, may oviposit in wounds of living animals, including humans. Some of these, such as *Cochliomyia hominovorax* and *Lucilia cuprina,* cause serious livestock losses, and the former has been the subject of intense control efforts using sterile-male techniques (see Chaps. 9, 15).

Oestridae, Cuterebridae, and Gasterophilidae, the botflies and warbles, are almost unique among insects in parasitizing mammals internally. Gasterophilid larvae attach to the stomach lining of horses, donkeys, rhinoceroses, and elephants. Most oestrid and cuterebrid larvae inhabit the nasal or respiratory passages of rodents and ungulates. Other species occupy self-formed cavities in subcutaneous tissue, breathing through a small hole maintained in the skin. The tropical *Dermatobia hominis,* attacking humans as well as livestock, is unusual in that the adult females oviposit on mosquitoes or other flies. When the mosquito feeds on a warm-blooded animal, the host's body heat stimulates the egg to hatch. The larvae then drop to the host and penetrate the skin. Other species oviposit directly on the host's body, usually cementing the eggs to hairs. It seems likely that these parasitic species evolved from flesh flies, and some authorities classify *Dermatobia* as a calliphorid.

Adults of three specialized families of Cyclorrhapha are ectoparasitic. Hippoboscidae are widespread on birds, with a lesser number infesting mammals, especially ungulates, but also dogs. Nycteribiidae and Streblidae are restricted to bats. All these flies are superficially similar to lice, and many species dehisce their wings after reaching a host.

Braulidae live in the hives of honey bees. Adults are apterous and are usually found on the queen or worker bees. The larvae tunnel in the wax combs.

Parasitoid Cyclorrhapha attack a variety of insect orders, as well as spiders; Sciomyzidae are unusual in attacking freshwater and terrestrial snails. Parasitoid species oviposit directly on or in the host (Chloropidae, Pyrgotidae, Tachinidae) or in locations where the first-instar larvae will encounter the host (Tachinidae, Sarcophagidae). Internal parasitoids frequently tap the host tracheal system, emphasizing the dependence of most flies on atmospheric respiration.

Relatively few Cyclorrhapha are predatory. Carnivorous species include Chamaemyiidae and many Syrphidae (larvae predatory on Homoptera), some Sciomyzidae (larvae predatory on snails), as well as occasional members of many other families. Predatory adults (some Anthomyiidae) usually attack organisms much smaller than themselves.

Eggs of most Cyclorrhapha are deposited directly into appropriate food sources, hatching after a brief time. Most Sarcophagidae, many Calliphoridae, as well as occasional other species, are viviparous or ovoviviparous, depositing larvae onto decomposing organic material. Larvae of *Glossina* (Muscidae), Hippoboscidae, Nycteribiidae, and Streblidae are retained internally by the female until fully mature. Larval nourishment is by means of a placentalike organ in the common oviduct. The newly deposited larvae pupate immediately.

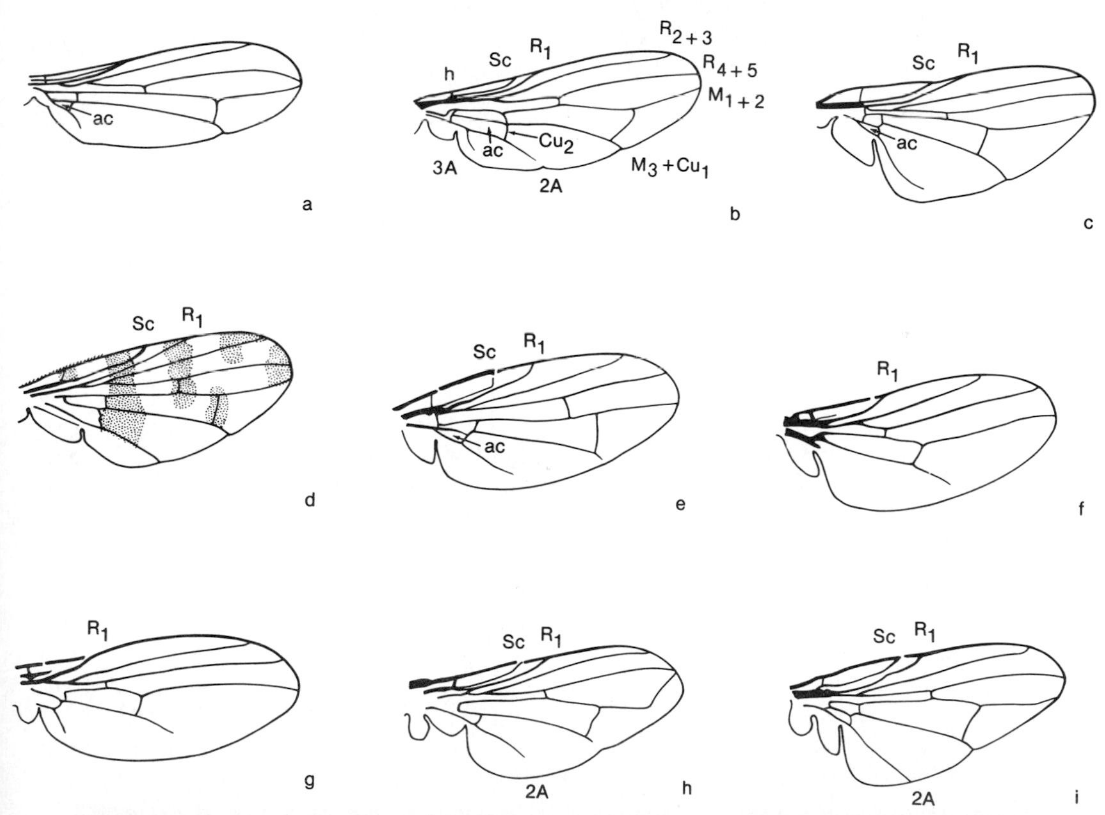

FIGURE 42.14 Wings of Brachycera, Cyclorrhapha: *a,* Micropezidae (*Compsobata mima*); *b,* Sciomyzidae (*Sepedon pacifica*); *c,* Chamaemyiidae (*Leucopis bivitta*); *d,* Otitidae (*Ceroxys latiusculus*); *e,* Tephritidae (*Trupanea nigricornis*); *f,* Chloropidae (*Thaumatomyia pulla*); *g,* Agromyzidae (*Cerodontha dorsalis*); *h,* Muscidae (*Musca domestica*); *i,* Anthomyiidae. *ac,* anal cell.

Cyclorrhapha pass through three functional instars, the fourth instar being suppressed in conjunction with the development of the puparium. Many species pupate in the larval food, protected by the hard, rigid puparial shell, which deters cannibalism, as well as offering protection against predation, parasitism, and physical factors. In contrast to most cocoons and other pupation shelters, the puparium requires no external support or material in its construction; it is probably responsible in large part for the success of the Cyclorrhapha.

KEY TO THE COMMON FAMILIES OF THE DIVISION

Cyclorrhapha (Series Schizophora)

1. Body flattened, louselike, with hind coxae widely separated or mesonotum reduced, similar to abdominal segments **Hippoboscidae, Braulidae, Nycteribiidae, Streblidae**

   Body not louselike; middle and hind coxae closely approximated 2

2(1). Mouthparts vestigial or lacking; mouth cavity small (Fig. 42.18*a*) pubescent flies at

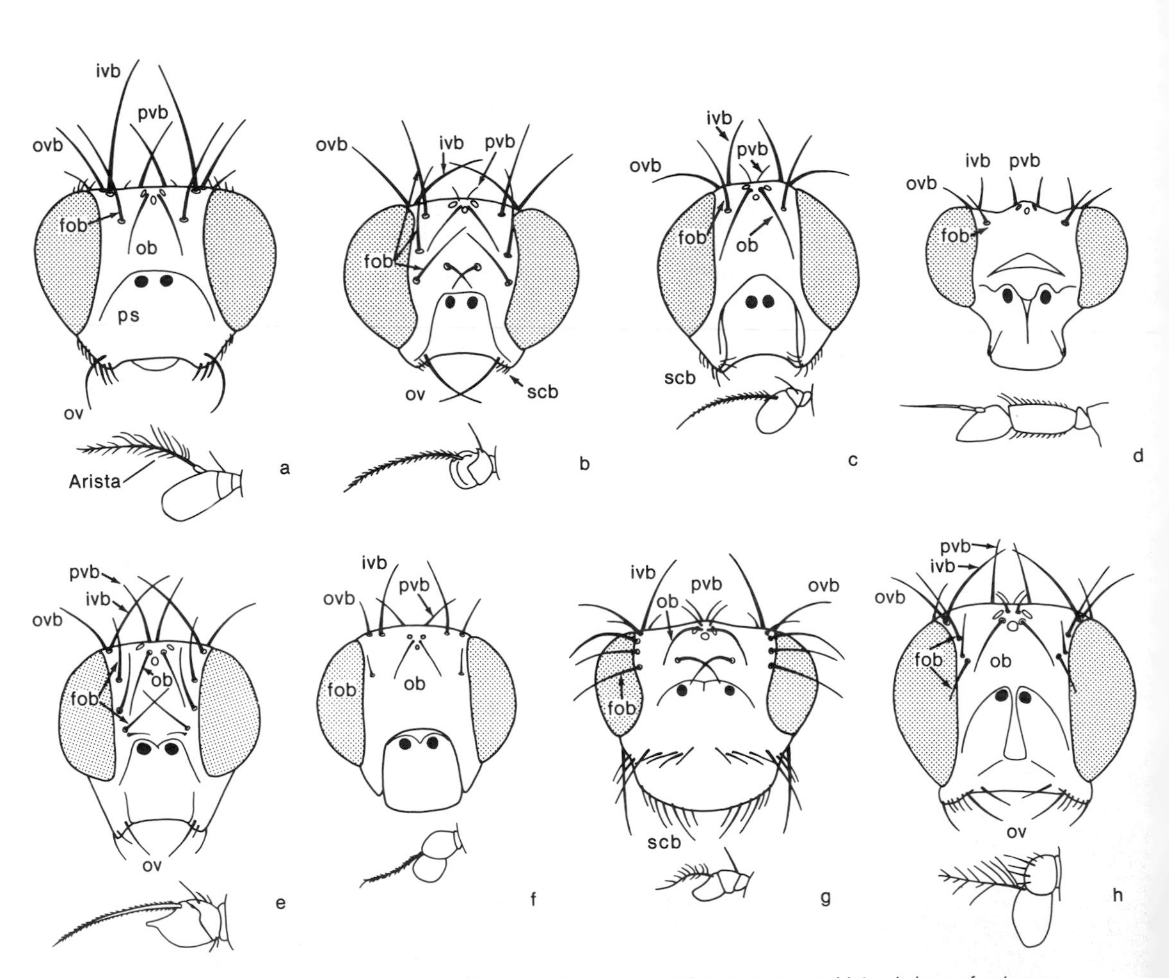

FIGURE 42.15 Heads of Brachycera, Cyclorrhapha, showing bristle patterns and lateral views of antennae; *a*, Heleomyzidae (*Suillia limbata*); *b*, Clusiidae (*Clusia occidentalis*); *c*, Lonchaeidae (*Earomyia brevistylata*); *d*, Sciomyzidae (*Sepedon pacifica*); *e*, Agromyzidae (*Cerodontha dorsalis*); *f*, Psilidae (*Psila microcera*); *g*, Ephydridae (*Paracoenia bisetosa*); *h*, Drosophilidae (*Drosophila simulans*). *fob*, fronto-orbital bristles; *ivb*, inner vertical bristles; *ob*, ocellar bristles; *ov*, oral vibrissae, *ovb*, outer vertical bristles; *ps*, ptilinal suture; *pvb*, postvertical bristles; *scb*, subcranial bristles.

least 15 mm long
Gasterophilidae, Oestridae, Cuterebridae
Mouthparts functional, usually with large labellum, and labial palps protruding from broad mouth cavity (Fig. 42.3*c*); usually less than 15 mm long 3

3(2). Mesothorax with complete transverse suture (sometimes faint) anterior to wing attachments (Figs. 42.18*b,c*; 42.17*d*); lower calypter usually large (Section Calyptratae) **29**
Mesothorax with transverse suture present at lateral margins or absent (Fig. 42.17*a*); lower calypter small or vestigial (Section Acalyptratae) 4

4(3). Vein Sc reaching costal margin of wing (Fig. 42.14*a* to *d*), separate from R_1 except at extreme base; anal cell present (Fig. 42.14*a* to *d*) 5
Vein Sc not reaching costal margin, frequently fused with R_1 for most of its length (Fig. 42.14*f,g*); anal cell absent (Fig. 42.14*f*) 21

5(4). Mesonotum strongly flattened; head closely appressed to thorax, legs bristly **Coelopidae**
Mesonotum convex, if flattened, head loosely attached to thorax, legs not conspicuously bristly 6

6(5). Head with at least a pair of stout bristles (oral vibrissae) just in front of mouth cavity (Fig. 42.15*a,b*) 7
Head without oral vibrissae (as in Fig. 42.15*g*) 11

7(6). Palpi vestigial; metathoracic spiracle with at least 1 peripheral bristle; abdomen usually narrowed at base **Sepsidae** (Fig. 42.17*a*)
Palpi functional, carried lateral to rostrum 8

8(7). Head with postvertical bristles parallel or divergent (sometimes absent) (Fig. 42.15*b* to *d*) 9
Head with postvertical bristles converging distally; ocellar bristles between median and lateral ocelli (Fig. 42.15*a*); tibiae with dorsal spines near apex; costal vein spinose **Heleomyzidae**

9(8). Antennae with subapical arista on third segment; eyes round; head with 2 to 4 pairs of fronto-orbital bristles (Fig. 42.15*b*) **Clusiidae**
Antennae with arista attached near base of third segment; eyes elongate, oval; head with 0 to 2 pairs of fronto-orbital bristles **10**

10(9). Antennae with third segment at least 2 to 3 times as long as broad (Fig. 42.15*c*); vein 2A slightly sinuous, reaching wing margin; tibiae usually without preapical bristles; usually stout black or blue-black flies less than 5 mm long **Lonchaeidae**
Antennae with third segment no more than twice as long as broad (as in Fig. 42.15*b*); vein 2A straight or curved, not reaching

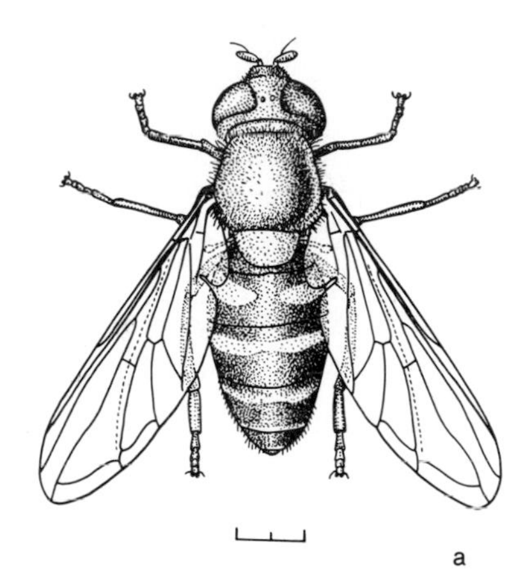

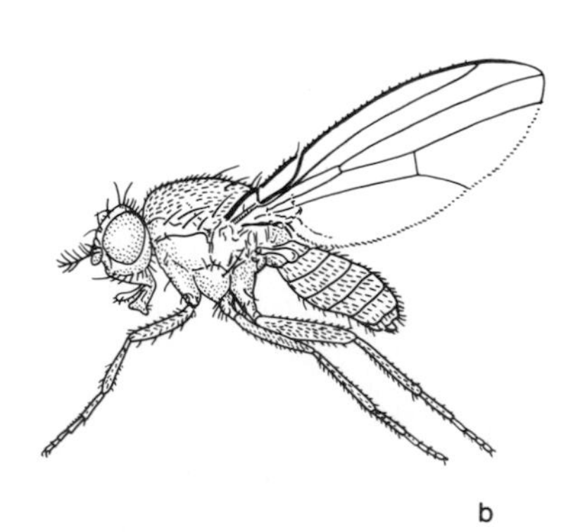

FIGURE 42.16 Representative Brachycera, Cyclorrhapha: *a*, Syrphidae (*Syrphus opinator*, scale equals 2 mm); *b*, Drosophilidae (*Drosophila busckii*).

wing margin; costal vein never spinose **Piophilidae**

11(6). Wings with vein M_{1+2} curving anteriorly, approaching or joining R_{4+5} (Fig. 42.14*a*); legs often long, stiltlike 12

Wings with vein M_{1+2} not closely approaching R_{4+5} (Fig. 42.14*b*), or legs not long, stiltlike 13

12(11). Proboscis long, slender, jointed, and bent anteriorly (Fig. 42.17*b*); body stout or slender **Conopidae** (Fig. 42.17*b*)

Proboscis short, thick, not jointed as above; body slender, legs long, slender **Neriidae, Micropezidae, Tanypezidae**

13(11). Some or all tibiae with dorsal bristles near apex; ovipositor short, membranous, retractile 14

Tibia usually without preapical bristles; if present, ovipositor long, sclerotized 16

14(13). Head with postvertical bristles parallel or diverging (rarely absent) (Fig. 42.15*d*); antennae without dorsal bristle on second segment 15

Head with postvertical bristles converging (Fig. 42.15*a*); antennae with dorsal bristle on second segment. **Lauxaniidae**

15(14). Femora without bristles; vein R_1 intersecting costa beyond middle of wing **Dryomyzidae** and **Helcomyzidae**

Femora with fine setae and stout bristles, usually with bristle near middle of anterior surface of mid-femur; vein R_1 intersecting costa near middle of wing (Fig. 42.14*b*) **Sciomyzidae**

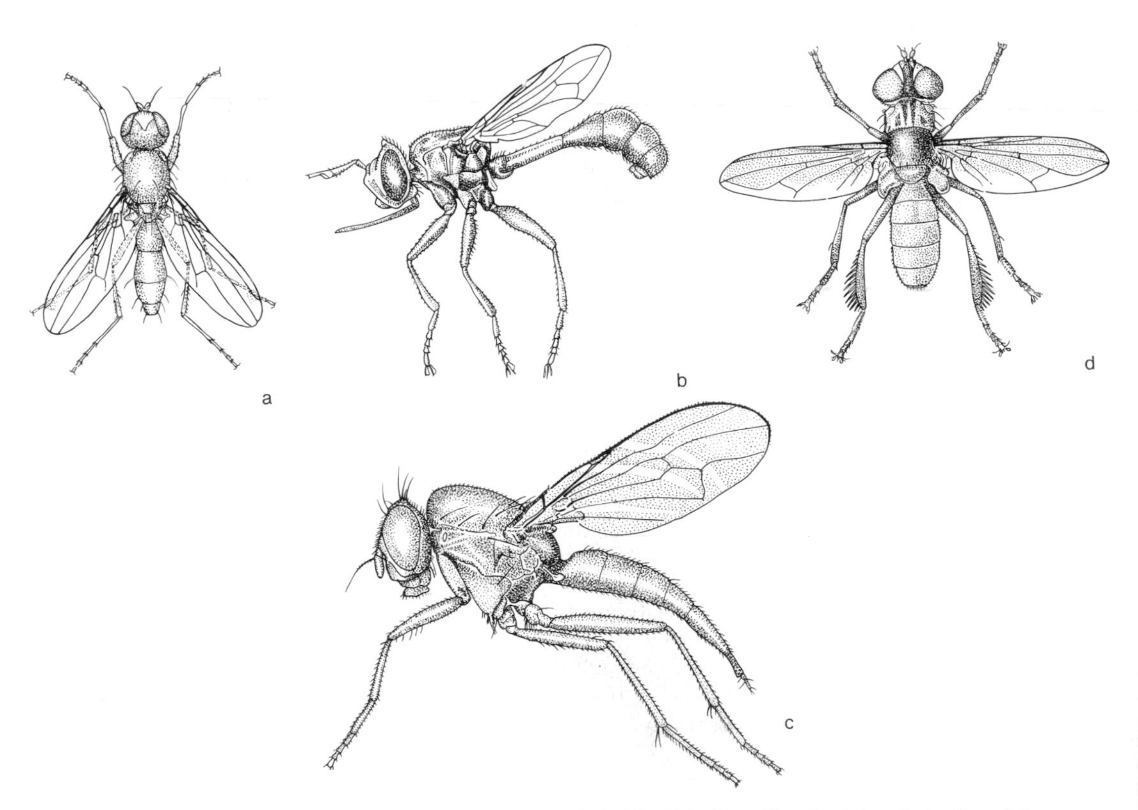

FIGURE 42.17 Representative Brachycera, Cyclorrhapha: *a*, Sepsidae (*Sepsis violacea*); *b*, Conopidae (*Physocephala texana*); *c*, Otitidae (*Tritoxa flexa*); *d*, Tachinidae (*Trichopoda plumipes*).

16(13). Costa with break near intersection of Sc (Fig. 42.14*e*) **19**

Costa not broken near intersection of Sc (Fig. 42.14*c,d*) **17**

17(16). Vein Cu_2 straight, meeting 2A at about a right angle (Fig. 42.14*c*); vein R_1 bare; ovipositor short, membranous; small gray flies **Chamaemyiidae**

Vein Cu_2 angulate, meeting 2A at an acute angle (Fig. 42.14*d*); vein R_1 usually bearing setae; ovipositor sclerotized, usually projecting **18**

18(17). Ocelli absent; base of ovipositor conical **Pyrgotidae**

Ocelli present; ovipositor flattened; wings usually banded or spotted **Otitidae** (Fig. 42.17*c*)

19(16). Vein Sc bent apically toward costa at nearly a right angle, usually not reaching costa (Fig. 42.14*e*); wings frequently patterned **Tephritidae**

Vein Sc bent toward costa at much less than a right angle, usually meeting costa at about 45° angle (Fig. 42.14*d*) **20**

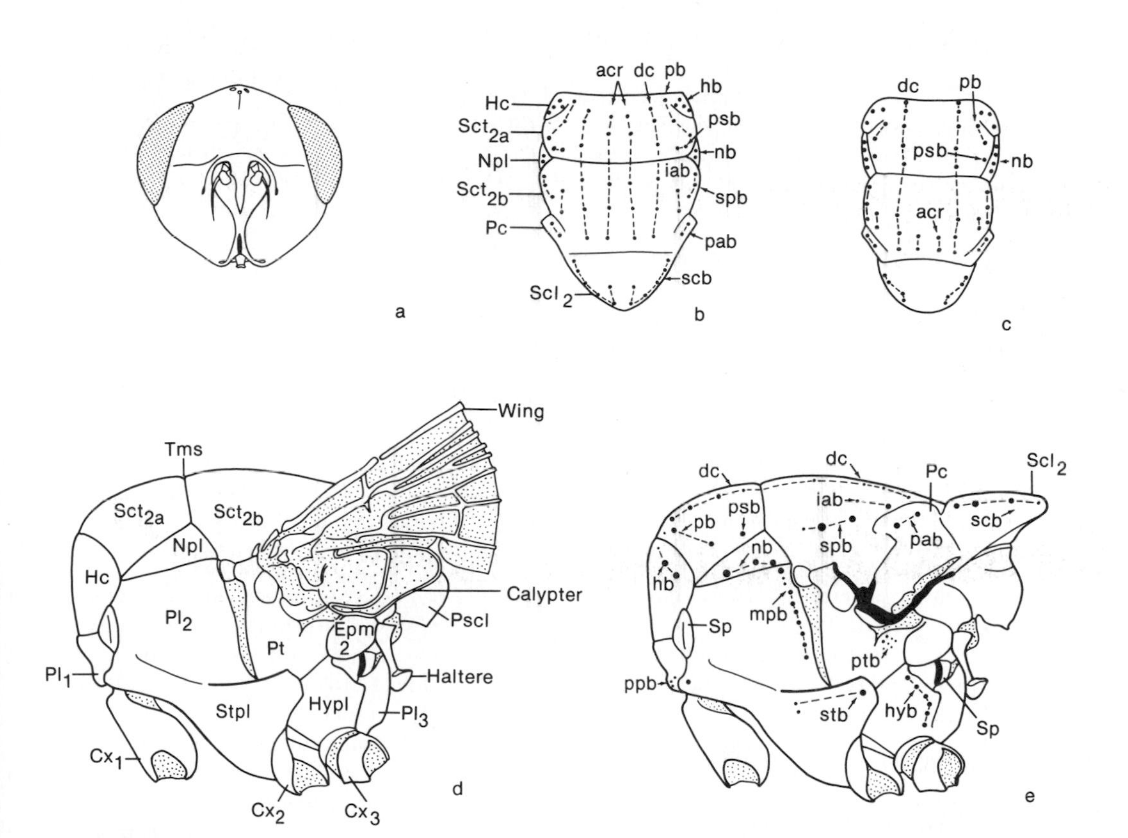

FIGURE 42.18 Taxonomic characters of Brachycera, Cyclorrhapha: *a*, head of Cuterebridae (*Cuterebra*); *b*, dorsal view of pterothorax of Calliphoridae; *c*, dorsal view of pterothorax of Sarcophagidae; *d, e*, lateral view of thorax of Calliphoridae (*Phormia regina*). *acr*, acrostichal bristles; *Cx*, coxa; *dc*, dorsocentral bristles; *Epm*, epimeron; *hb*, humeral bristles; *Hc*, humeral callus; *hyb*, hypopleural bristles; *Hypl*, hypopleuron; *iab*, intra-alar bristles; *mpb*, mesopleural bristles; *nb*, notopleural bristles; *Npl*, notopleuron; *N*, notum; *pab*, postalar bristles; *Pc*, postalar callus; *pb*, posthumeral bristles; *Pl*, pleuron; *psb*, prescutal bristles; *Pscl*, postscutellum; *ppb*, propleural bristle; *ptb*, pteropleural bristles; *Pt*, pteropleuron, *scb*, scutellar bristles; *scl*, scutellum; *Sp*, spiracle; *stpl*, sternopleural bristles; *Stpl*, sternopleuron; *spb*, supra-alar bristles; *Tms*, transmesonotal suture.

20(19). Femora thickened, spiny; abdomen with bristles on second segment **Otitidae** (Fig. 42.17*c*)

Femora slender, not spiny; abdomen without bristles on second segment **Lonchaeidae**

21(4). Vein Sc bent apically toward costa at nearly a right angle, usually not reaching costa (Fig. 42.14*e*); Cu_2 angulate, meeting 2A at an acute angle **Tephritidae**

Vein Sc bent toward costa at much less than a right angle, usually meeting costa at about 45° (Fig. 42.14*f*); Cu_2 straight, meeting 2A at about a right angle 22

22(21). Hind legs with 2 basal tarsal segments thickened, larger than succeeding segments; first segment shorter than second **Sphaeroceridae**

Hind legs with 2 basal tarsal segments not markedly thicker than succeeding segments; first segment usually longer than second 23

23(22). Costal vein with single break near Sc or R_1 (Fig. 42.14*f*) 27

Costal vein with breaks near Sc and near humeral crossvein (Fig. 42.14*e*) 24

24(23). Wings with anal cell closed or nearly closed Fig. 42.14*d,e*); antenna with arista covered with short hairs 25

Anal cell absent (Fig. 42.14*f*); arista variable; ocellar triangle large, conspicuous **Chloropidae**

25(24). Head with postvertical bristles converging apically (Fig. 42.15*a*) **Tethinidae, Anthomyzidae**

Head with postvertical bristles diverging (Fig. 42.15*c*) or absent 26

26(25). Head with pair of stout bristles (oral vibrissae) just in front of mouth cavity (Fig. 42.15*e*); postvertical bristles diverging (Fig. 42.15*e*); small, gray or black flies with ovipositor sclerotized, not retractable **Agromyzidae**

Head without oral vibrissae; postvertical bristles frequently absent (Fig. 42.15*f*); small brownish or yellowish flies with very strong break at basal third of costa **Psilidae**

27(23). Head with postvertical bristles diverging (Fig. 42.15*g*); pair of stout bristles (oral vibrissae) usually absent in front of mouth cavity (Fig. 42.15*g*); anal cell absent **Ephydridae**

Head with postvertical bristles parallel or converging (Fig. 42.15*h*); (occasionally absent); oral vibrissae usually present; anal cell present or absent 28

28(27). At least 1 pair of fronto-orbital bristles converging toward midline of head (Fig. 42.15*b,e*); antennae with arista usually bare **Milichiidae**

Fronto-orbital bristles not directed medially (Fig. 42.15*g,h*); arista plumose **Drosophilidae** (Fig. 42.16*b*) (incl. **Camillidae, Curtonotidae, Diastatidae**)

29(3). Meropleuron with row or cluster of large hypopleural bristles below metathoracic spiracle (Fig. 42.18*e*); wing with vein M_1 almost always abruptly curving forward (Figs. 42.14*h*, 42.17*d*) 30

Meropleuron usually without bristles; if bristled, the pteropleuron lacks bristles and vein M_1 is straight or gently curved (Fig. 42.14*i*) 32

30(29). Thorax with postscutellum large, convex, transverse (Fig. 42.18*d,e*); abdominal terga usually bearing strong bristles medially **Tachinidae** (Fig. 42.17*d*)

Thorax with postscutellum small; abdominal terga rarely with medial bristles 31

31(30). Body usually metallic green or blue; arista almost always plumose along entire length; mesothorax with 2 or 3 notopleural bristles (Fig. 42.18*b*) **Calliphoridae**

Body not metallic, usually black or gray with dark longitudinal stripes on thorax; arista usually plumose in basal half; mesothorax with 4 notopleural bristles (Fig. 42.18*c*) **Sarcophagidae**

32(29). Vein $Cu_2 + 2A$ not reaching wing margin (Fig. 42.14*h*); lower calypter longer than upper calypter **Muscidae**

Vein $Cu_2 + 2A$ reaching wing margin (Fig. 42.14*i*); lower calypter subequal to or smaller than upper calypter **Anthomyiidae**

SELECTED REFERENCES

GENERAL

Askew, 1971 (parasitic families); Felt, 1940 (gall-forming families); Gillett, 1971, 1972 (mosquitoes); Oldroyd, 1964 (comprehensive discussions of biology and ecology); Séguy, 1950, 1951.

MORPHOLOGY

Bonhag, 1949 (horsefly); Crampton, 1942; Séguy, 1951; Snodgrass, 1944 (mouthparts).

BIOLOGY AND ECOLOGY

Bateman, 1972 (fruit flies); Demerec, 1950 *(Drosophila);* Downes, 1969 (swarming behavior) and 1971 (bloodsucking); Gillett, 1972 (mosquitoes); Hering, 1951 (leaf miners); Horsfall et al., 1973 (Bionomics of *Aëdes*); James, 1947 (myiasis); Norris, 1965 (blowflies); Oldroyd, 1964 (general); Oliver, 1971 (Chironomidae); Sacca, 1964 (houseflies); Séguy, 1950, 1951 (general); West, 1951 (housefly).

SYSTEMATICS AND EVOLUTION

Cole, 1969 (Western United States species); Curran, 1934 (North American genera); Diptera of Connecticut, 1942–1964; Griffiths, 1972 (higher classification); Hall, 1948 (Calliphoridae of North America); Hennig, 1948–1952 (larvae); Hinton, 1958 (phylogenetic relationships); Hull, 1962 (Asilidae); Merritt and Cummins, 1978 (aquatic families); Peterson, 1951 (larvae); Steffan, 1966 (Sciaridae of North America); Wirth and Stone, 1956 (aquatic families).

43 ORDER SIPHONAPTERA (Fleas)

Minute to small, wingless Endopterygota with laterally flattened body. *Adults* (Fig. 43.1*a*) with hypognathous head capsule, 3-segmented antennae, frequently with small lateral ocelli (no compound eyes or dorsal ocelli); labrum and laciniae slender, bladelike (Fig. 43.1*b*); maxillary and labial palps usually 4- to 5-segmented, mandibles absent. Thoracic segments subequal or metanotum slightly enlarged; legs stout with large coxae; hind legs enlarged, modified for jumping. Abdomen 10-segmented, tenth tergite bearing a complex sensilium and associated guard hairs (Fig. 43.1*c*). *Larva* apodous, vermiform, usually with distinct head capsule, 1-segmented antennae, mandibulate mouthparts (Fig. 43.1*d*). *Pupa* adecticous exarate, in wispy cocoon.

Fleas occur throughout the world, parasitizing mostly mammals (94 percent), but about 100 species (6 percent) infest various birds, including domestic fowls. Unlike lice, which depend on the host for shelter, warmth, and oviposition sites as well as food, fleas are intermittently parasitic. Larvae are mostly nest inhabitants rather than parasites, but larvae of a few species are ectoparasites, and one is an endoparasite. Adults frequently leave the host for various periods of time. Host specificity varies greatly. Species with broad host ranges readily change host species, and many fleas with restricted host ranges will temporarily infest unsuitable hosts, particularly if hungry. Transfers among individuals of the same species may be frequent, especially among colonial or gregarious hosts. Dead animals are quickly deserted, and in general, high vagility and infidelity to a single host are responsible for the importance of fleas in transmitting bubonic plague from rodents to humans (see Chap. 16).

Adult fleas are so highly modified that their relationship to other orders of insects is uncertain. However, the absence of mandibles and presence of suctorial mouthparts suggest a possible origin from Diptera or Mecoptera, and flea larvae strongly resemble immatures of the dipterous Suborder Nematocera.

Adult morphology is strongly correlated with the ectoparasitic life. The laterally compressed body and short, retractable antennae allow easy passage through the host's pelage or plumage, and the wedge-shaped coxae appear to be adapted for pushing through hair. Various parts of the body, especially the pronotum, bear combs of stout, backwardly directed, modified setae, or *ctenidia,* which further aid forward progress. Smaller bristles on various parts of the body serve the same function. In general, fleas avoid the preening or scratching of the host by rapidly running over the host's body. As in most other insect ectoparasites of vertebrates, the cuticle is tough and leathery, protecting the internal organs by a hydrostatic cushion provided by the enclosed hemolymph. The jumping hind legs are used to escape or to mount a potential host.

Adult life-spans as long as a year have been recorded, and *Pulex irritans,* the human flea, has survived over 4 months without feeding. Mating and oviposition take place on the host or within its nest or burrow. In species associated with solitary animals, reproduction may be synchron-

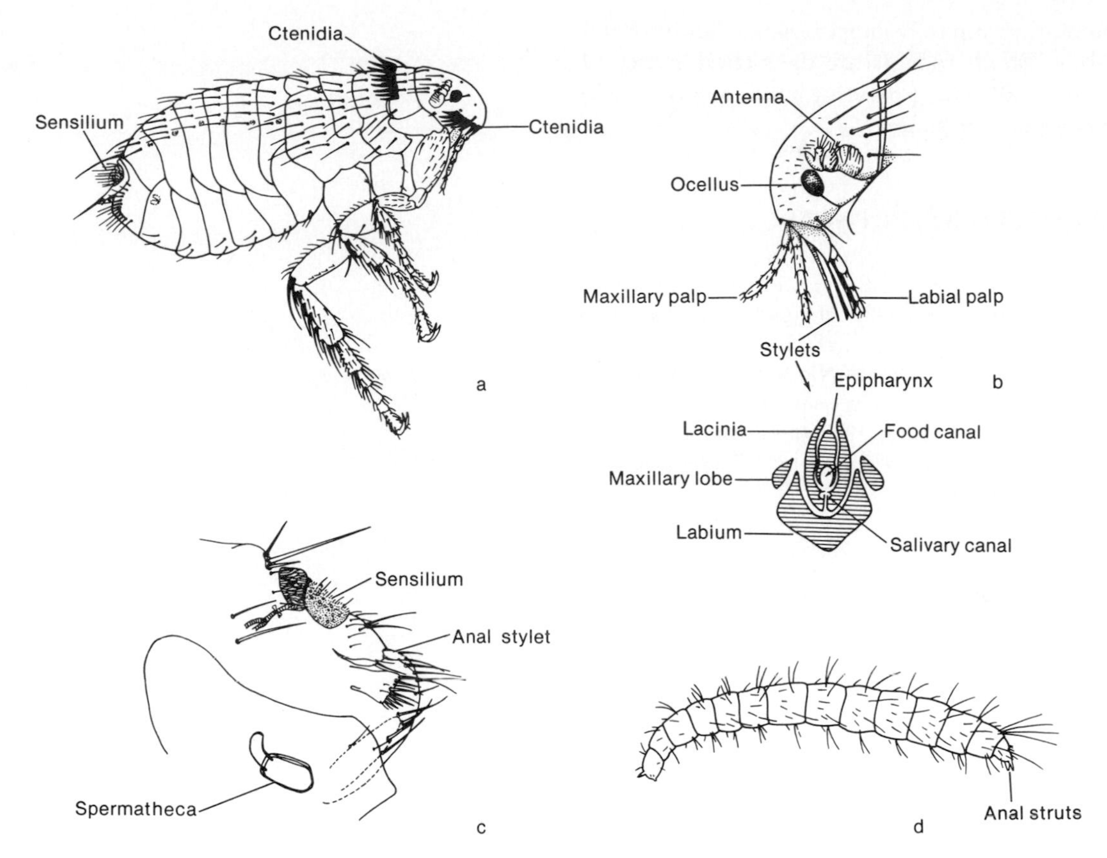

FIGURE 43.1 Siphonaptera: *a,* adult of *Ctenocephalides felis,* the cat flea; *b,* mouthparts of *Xenopsylla cheopis; c,* abdominal sensilium of *Megabothris; d,* larva of *Xenopsylla cheopis.* (a; *courtesy of D. P. Furman;* b, c *from Furman and Catts,* 1961; d *from Patton,* 1931.)

ized with hormonal changes in the host, so that a fresh generation of fleas is available to infest the young animals leaving the nest. Blood meals are probably necessary for egg maturation in all species.

Pulex irritans may produce over 400 eggs per individual, and cat fleas produce up to 1000 eggs. Eclosion ensues after a few days or weeks, and larval development is complete after about 1 to 3 weeks under favorable conditions. Under harsher conditions, especially low temperatures, larval growth may require up to 6 months. Larvae are scavengers and predators, feeding on dried flea excrement, small arthropods, and other debris which collects in animal nests. In houses the cracks and crevices in floors covered by carpets are favorable developmental sites. The third-instar larva spins a thin cocoon, to which dust and debris adhere. Pupation requires a few weeks in most species. Newly emerged adults frequently remain in the cocoon for some time, especially if no host is available. In some species massive emergence of such quiescent adults is apparently stimulated by mechanical disturbances such as movements of a potential host—e.g., swallows returning to their nest in spring.

Positive identification of fleas is difficult, requiring slides of cleared specimens. The North American fauna includes 9 of 17 families. Nearly all the familiar species which attack humans and domestic animals belong to a single family, PULICIDAE. Small rodents are the richest source of additional taxa, providing hosts to about 200 species in 7 families.

SELECTED REFERENCES

Askew, 1971 (biology); Ewing and Fox, 1943 (North American species); G. P. Holland, 1949 (Canadian species), and 1964 (evolution and classification); Hubbard, 1947 (western North American species); Layne, 1971 (Florida species); Riek, 1970*b* (fossil record); Rothschild, 1965 (hormonal interaction with hosts), and 1975 (general); Snodgrass, 1946 (morphology).

ORDER LEPIDOPTERA (Butterflies and Moths)

44

Adult (Figs. 44.1 to 44.3). Minute to large Endopterygota with body and wings covered with scales or hairs, and mouthparts almost always modified as a proboscis. Head hypognathous with large lateral eyes, frequently with ocelli and/or *chaetosemata;* antennae with numerous segments; mouthparts (Fig. 44.2*c*) consisting of elongate galeae, maxillary palpi (frequently very small), and large labial palpi. Prothorax usually small, collarlike, with *patagia* dorsolaterally; mesothorax large with clearly defined scutum and scutellum, and with large lateral *tegulae* covering wing bases; metathorax usually much reduced, with lateral scutal sclerites. Wings covered with scales (flattened *macrotrichia*); hind wing coupled to forewing by *frenulum, jugum* (Fig. 44.3), or basal overlap; venation mostly of longitudinal veins with few crossveins, usually with large *discal cell* in basal half of wing (Fig. 44.10*c*). Legs adapted for walking, usually with 5 tarsomeres; prothoracic legs occasionally much reduced. Abdomen with 10 segments, first segment reduced, with vestigial sternum; segments 9 and 10 strongly modified in relation to genitalia.

Larva (Fig. 44.5). Eruciform with sclerotized head capsule; hypognathous, mandibulate mouthparts, short, 3-segmented antennae and usually 6 lateral ocelli; thorax with short, 5-segmented legs with single tarsal claws; abdomen 10-segmented, with short fleshy prolegs on some or all of segments 3 to 6 and 10.

Pupa (Fig. 44.5). Rarely decticous, exarate; typically adecticous, obtect, with body parts fused, except for 2 or more free abdominal segments, which remain movable; pupa usually enclosed in silken cocoon or shelter.

Members of the Order Lepidoptera have received more attention from naturalists and entomologists than any other group of insects. Most of this effort has been directed toward the butterflies and showier moths, and knowledge of the smaller, more obscure species which constitute the great majority of the order is largely lacking.

Biologically, as well as morphologically, the Lepidoptera are perhaps the most uniform of the large holometabolous orders. Adults, with few exceptions, feed on nectar, honeydew, fermenting sap, or other similar products, or have the mouthparts atrophied. Larvae of nearly all species feed on angiosperms and gymnosperms; it is larval feeding which is almost entirely responsible for the great economic importance of the order. It seems likely that nearly all higher plants are host to at least one species of Lepidoptera. In general, Lepidoptera are the dominant insect herbivores in most habitats, having extensively invaded all available niches except that of sucking plant sap. In North America larvae of about 175 species feed on fungi or detritus.

Certain Lepidoptera have been important in the development of ideas in evolutionary biology and ecology. The butterfly *Papilio dardanus,* whose females mimic danaine nymphalids of several species, is one of a few organisms for which the genetic basis of polymorphism is understood. Sex-limited genes were first described in the genus *Colias.* The evolution of concealing coloration has been intensively studied by Kettlewell and others in the geometrid moth *Biston betularia* (see Chap. 15), and continuing investigations of the population biology and adaptive ecology of *Euphydryas editha* (Nymphalidae) are important in explanations of how populations differentiate (Part II, see Chap. 10). These are just a few

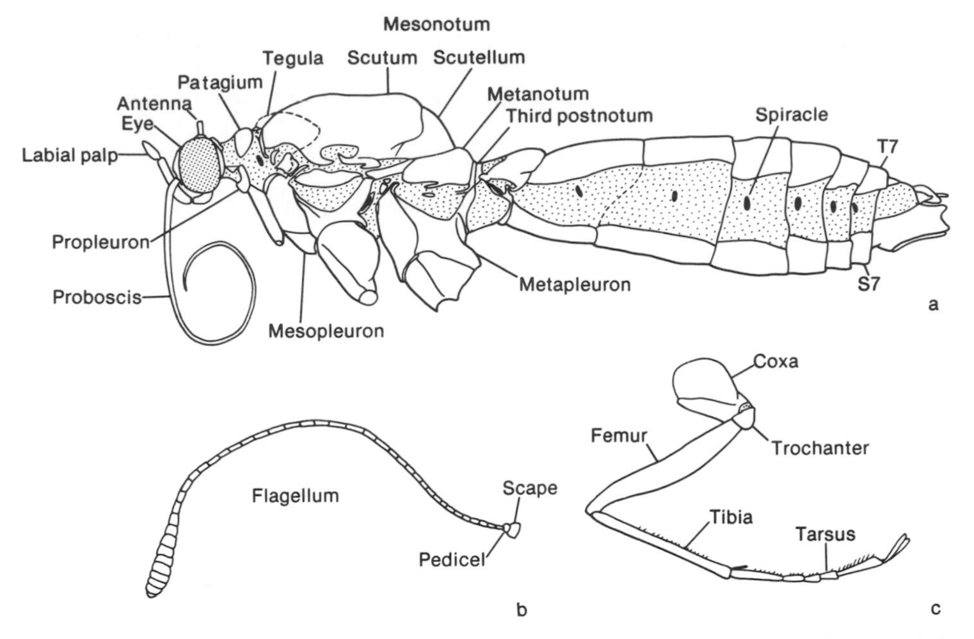

FIGURE 44.1 Anatomy of Lepidoptera (*Danaus plexippus*, Nymphalidae, from Ehrlich, 1958): *a*, lateral view of body; *b*, antenna; *c*, mesothoracic leg; *T*, tergum; *S*, sternum.

examples. In general, butterflies have been favorite subjects for much recent experimental and analytical work, especially regarding plant-herbivore coevolution, geographic variation, and population structure.

A distinctive morphological feature of most Lepidoptera is the body covering of scalelike macrotrichia. In most species these are of dull brown or gray hues in mottled patterns which help to conceal the resting insects. In butterflies the scales produce bright contrasting patterns which may function aposematically or may provide characteristic patterns allowing easy recognition between the sexes. Scales are easily detached, and probably enable these insects to slip from the grasp of many potential predators. In some of the larger moths scales may provide an insulation which allows the internal body to more quickly reach the elevated temperatures at which flight occurs.

The majority of Lepidoptera have the mouthparts adapted as a long coiled proboscis through which liquid food is ingested (Fig. 44.2*c*). The proboscis consists of the greatly elongate maxillary galeae, structured as a series of sclerotized rings connected by membranes and two sets of muscles, which provide the flexibility required for coiling and uncoiling. The two galeae are tightly interlocked by hooks and spines, forming a double-walled tube. Small maxillary palpi are usually present at the base of the proboscis, and large 3- to 4-segmented labial palpi are present on all but some nonfeeding forms. Other oral structures are absent in typical Lepidoptera except as vestiges, but functional mandibles, accompanied by maxillae with both laciniae and unmodified galeae, occur in Micropterygidae (Fig. 44.2*a*,*b*). Reduced, toothless mandibles also occur in Eriocraniidae, in combination with a short proboscis whose galeae are held together only while feeding.

As in most four-winged insects adapted to an aerial existence, Lepidoptera have evolved mechanisms to synchronize their wingbeats. The simplest type of wing coupling consists of a *jugum*, a lobe projecting from the posterior base of the forewing and overlying the hind wing (Fig. 44.3*b*). In most species coupling is achieved by

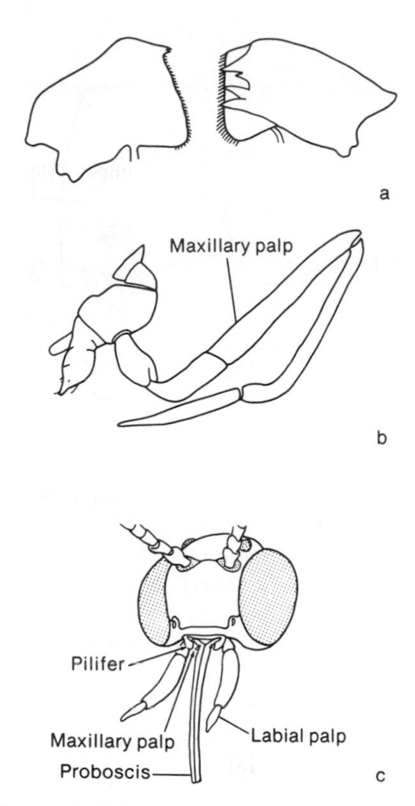

FIGURE 44.2 Mouthparts of Lepidoptera: *a,* mandibles of Zeugloptera (Micropterygidae, *Palaeomicroides*); *b,* maxilla of same; *c,* head of moth, *Synanthedon.* (a, b *from Issiki,* 1931; c *from Snodgrass,* 1935.)

one or more stout bristles (the *frenulum*) which arise from the base of the hind wing and interlock with a cuticular flap or row of setae (the *retinaculum*) on the underside of the forewing (Fig. 44.3*a*). In Papilionoidea and a few other Ditrysia, coupling is *amplexiform.* In these species the forewing, especially in the basal region, broadly overlaps the hind wing, which may bear special humeral veins which stiffen the region of overlap (Fig. 44.10*e*).

Female Lepidoptera are remarkable for the complexity of the internal reproductive system. Distinct organizational plans typify the Suborders Monotrysia and Ditrysia. In all Lepidoptera, sperm are transmitted as a spermatophore which is stored in the *bursa copulatrix.* In Monotrysia (Fig. 44.4*a*) the bursa and common oviduct are confluent and may join the alimentary canal in a cloaca. Eggs are fertilized from the bursa copulatrix as they pass down the common oviduct. In Ditrysia (Fig. 44.4*b*) the bursa, common oviduct, and rectum open independently. The bursal opening is situated medially on sternite 8, which may be highly modified, along with segments 9 and 10. As in the Monotrysia, spermatophores are stored in the bursa, but the spermatozoa migrate via the *ductus seminalis* to the spermatheca, from whence the eggs are fertilized. In Hepialoidea the openings of the bursa and common oviduct are situated close together on segments 9 and 10, and connected by a groove along which the sperm migrate to the spermatheca. Ontogeny of the ditrysian structures during metamorphosis suggests that a single opening on the eighth sternite represents the primitive configuration, the common oviduct secondarily extending posterad to the region of the anus (Dodson, 1937).

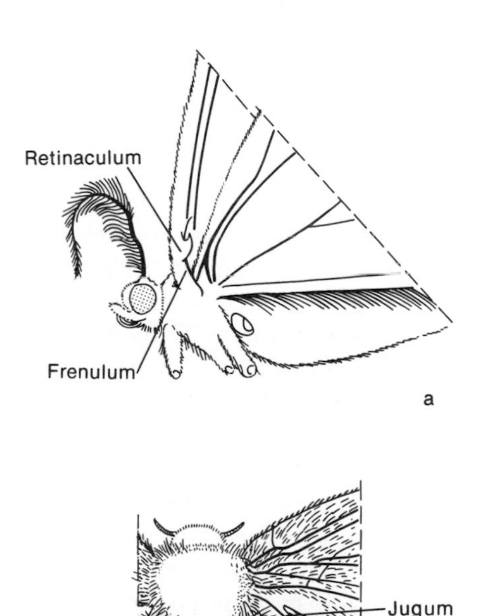

FIGURE 44.3 Wing-coupling mechanisms of Lepidoptera: *a,* frenulum (Geometridae, *Neoalcis californica*); *b,* jugum (Hepialidae, *Hepialus*).

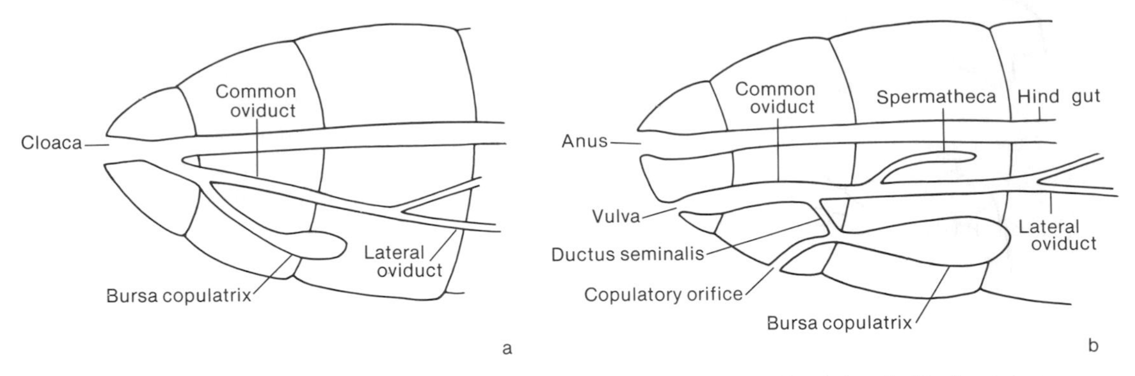

FIGURE 44.4 Reproductive systems of Lepidoptera (diagramatic): *a*, monotrysian system; *b*, ditrysian system.

The body plan shared by nearly all Lepidopterous larvae comprises an elongate trunk with large, sclerotized head capsule, three pairs of thoracic legs, and up to five pairs of unjointed abdominal appendages or prolegs (Fig. 44.5). The prolegs, which typically bear rows or circles of short hooked spines, or *crochets* (Fig. 17.5*b*), assist the larvae in clinging to the exposed surfaces of vegetation. Special tarsal modifications occur in a variety of other insects which frequent smooth surfaces of leaves. Abdominal segments 3 to 6 and 10 typically bear prolegs, but in many families, such as Geometridae, the number is reduced. Crochets are lacking only in a few burrowing forms and in highly specialized leaf-mining families such as Nepticulidae, Heliozelidae, and Eriocraniidae (Fig. 44.5*c*,*d*), in which the flattened body may lack legs entirely.

Silk production is universal among lepidopterous larvae, being utilized in construction of a wide variety of larval shelters or galleries, as well as in formation of the pupal shelter or cocoon. Silk is produced by large paired glands, usually extending beside the gut posteriorly into the abdomen and opening through a median spinneret on the labium. The size and configuration of the glands are extremely variable, but the evolutionary significance of this variation is not known.

Pupae are exarate in many primitive forms, including Zeugloptera and Dacnonypha. Obtect pupae (Fig. 44.5*e*), with the appendages variably fused to the body, are characteristic of the higher Lepidoptera. In specialized families, appendages may be received in grooves in the body wall, and the body regions may be ill defined because of loss of external sutures. Rudiments of mandibles, maxillae, or other mouthparts are present in many families; functional mandibles, occurring in the Suborders Zeugloptera and Dacnonypha, are used to open the cocoon just before adult emergence. At least one abdominal joint remains flexible, and wriggling of the abdomen is the only means of motility in most pupae. Segment 10 may be modified as a *cremaster,* a series of hooked setae used to anchor pupae of some specialized families to the cocoon or substrate.

BIOLOGY. Adult Lepidoptera are mostly rather short-lived. Many species with atrophied mouthparts do not feed. Others imbibe nectar, honeydew, or exudates from fermenting fruit or sap, while a very few are known to suck blood or pierce fruit, using the proboscis as a piercing organ. Both sexes of nearly all species possess functional wings, but many female Psychidae and Lymantriidae, as well as occasional species of other families are flightless, especially those that are active in winter or at high elevations. Brachyptery in males is limited to a very few species, mostly on oceanic islands. Activity is predominantly nocturnal or at dusk and dawn,

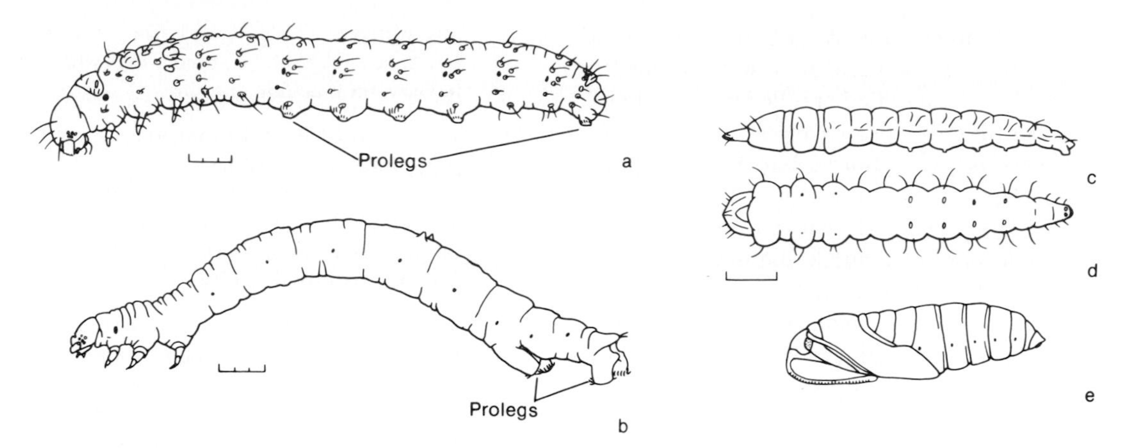

FIGURE 44.5 Larvae and pupae of Lepidoptera: *a*, larva of Hepialidae (*Hepialus;* scale equals 3 mm); *b*, larva of Geometridae (*Synaxis pullata;* scale equals 3 mm); *c, d*, lateral and ventral views, respectively, of a leaf-mining larva (Tischeriidae, *Tischeria;* scale equals 1 mm); *e*, pupa of Sphingidae (*Manduca*).

but nearly all members of the Papilionoidea (butterflies) are sunshine-loving, and certain families or genera of moths are diurnal. In general, nocturnal species tend toward brown or gray and white mottled color patterns, which are important in concealing the insects on their daytime perches. Tree trunks or branches, lichen, and other irregular surfaces are favorite spots for daytime sheltering, and the scale or hair tufts on many moths enhance their camouflage. At least some moths are able to select backgrounds matching their own ground color (Kettlewell, 1973). The brightly colored hind wings of some nocturnal species are covered by the forewings at rest, being suddenly revealed if the moth is disturbed. Such hind-wing patterns sometimes strikingly resemble eyes of vertebrates, and are apparently used to startle birds or other potential predators (Fig. 15.3).

Lepidoptera are almost exclusively bisexual. Female sex pheromones serving to attract males have been isolated from a great number of species, especially moths, and are probably present in most species. Pheromones may be effective over relatively short distances, as in many of the smaller moths, or may attract males from distances of hundreds of meters, as in Saturniidae. Many butterflies apparently rely on visual cues for initial recognition, frequently followed by behavioral sequences or release of short-range pheromones by males. Copulation takes place on the ground or other perches.

Eggs are characteristically deposited on or near the host plant, occasionally inserted into plant tissue, the number per individual varying from about a dozen to many thousands. Eggs are extremely variable, ranging from cylindrical or barrel-shaped to flat wafers. The chorion may be smooth and featureless or regularly or irregularly sculptured with ribs, tubercles, or pits. Eggs may be scattered singly or in small groups, or deposited in large masses. Egg diapause is a common feature of lepidopteran life cycles.

Larvae are predominantly herbivores, but many scavengers (many Tineidae, Oecophoridae) and a few predators (a few Lycaenidae, Cosmopterygidae, Noctuidae) are known. Epipyropidae are unique in living externally on the bodies of leafhoppers, especially Fulgoroidea, probably as ectoparasitoids. Roots, stalks and stems, leaves, flowers, fruits, and seeds are consumed by phytophagous forms. In general, larvae of specialized families of the Suborder Ditrysia live in exposed situations and are frequently colored or patterned to resemble their host plants. Larvae of primitive families of Ditrysia as well as those of the Suborders Zeugloptera, Dacnonypha, and Monotrysia frequently feed in leaf mines, tunnels, galls, self-constructed shelters, or other concealed situations. Shelters commonly consist of tied or rolled leaves or twigs, parchmentlike shields or covers, or silken webs. Most species are solitary, but Lasiocampidae (tent caterpillars), among others, are highly gregarious. Host specificity varies from acceptance of a single part of a single plant species, as in many leaf miners, to extreme polyphagy, as in many Noctuidae and Geometridae. Lepidoptera are heavily parasitized, probably because of the exposed position of many of the larvae.

Pupation frequently occurs on the host plant, usually in a cocoon formed by the prepupal larva. Leaf miners often pupate within the larval mines, and borers in the larval tunnels, with or without a silken shelter. Noctuidae, Sphingidae, and some others pupate naked in soil or litter, and most Papilionoidea (butterflies) and occasional members of other groups hang exposed on plants or other erect objects.

CLASSIFICATION. A confusing variety of classifications is currently available. Older arrangements recognize the Frenatae and Jugatae as suborders, based on the type of wing coupling. The terms macrolepidoptera and microlepidoptera refer primarily to body size, which is without phylogenetic significance, and the Suborders Rhopalocera and Heterocera refer to antennal shape (clubbed or not clubbed). The classification presented here emphasizes the important structural features of pupal and adult mouthparts and of the female reproductive system. The number of families recognized is conservative, many of the small, obscure families being combined with more generally accepted taxa.

KEY TO THE NORTH AMERICAN SUBORDERS OF Lepidoptera

1. Forewings and hind wings ovate, similar in venation, size, and shape (Fig. 44.6*a*); hind wings with at least 10 veins reaching margin 2

   Forewings and hind wings conspicuously different in venation (Figs. 44.7 to 44.11); usually different in shape and size, or narrow, lanceolate (Fig. 44.8*f*); hind wings with 9 or fewer veins reaching margin 4

2(1). Large moths at least 10 mm long; maxillary palpi minute or vestigial **Monotrysia,** in part (p. 449)

   Small moths less than 6 mm long; maxillary palpi 5-segmented, conspicuous (Fig. 44.2*b*); iridescent, diurnal moths 3

3(2). Middle tibiae without spurs (apical tufts of hairs may be present); proboscis absent **Zeugloptera** (p. 448)

   Middle tibiae with single spur; proboscis short, noncoiled **Dacnonypha** (p. 449)

4(1). Wings narrow, lanceolate (Fig. 44.8*f*); hind wings with fringe of setae broader than wing membrane; small moths less than 6 mm long 5

   Wings broader, ovate or truncate; hind wings with setal fringe narrower than membrane; small to very large lepidopterans 6

5(4). Proboscis absent or vestigial; maxillary palpi 5-segmented, conspicuous; female with sclerotized, large external ovipositor **Monotrysia,** in part (p. 449)

   Proboscis present; maxillary palpi usually very small, 1- to 4-segmented; ovipositor membranous, inconspicuous or absent **Ditrysia,** in part (p. 450)

6(4). Maxillary palpi large, folded, with 5 segments (as in Fig. 44.7*c*); if palpi inconspicuous, antennae much longer than body; ovipositor large, sclerotized, conspicuous; small moths 5 to 10 mm long **Monotrysia,** in part (p. 449)

   Maxillary palpi usually inconspicuous, with 1 to 4 segments (Figs. 44.7*f*, 44.8*b*); if palpi 5-segmented, then antennae shorter than body and ovipositor membranous, inconspicuous or absent; small to very large lepidopterans 3 to 50 mm long **Ditrysia,** in part (p. 450)

Suborder Zeugloptera

Small moths without proboscis but with functional mandibles in pupa and adult (Fig. 44.2*a*); adult with ocelli, 5-segmented maxillary palp, 4-segmented labial palp; subcosta branching near middle, jugum with humeral vein. Larva sluglike with prolegs on first 8 abdominal segments, large suckers on ninth and tenth segments. Pupa exarate.

The single family, MICROPTERYGIDAE, contains diurnal, frequently iridescent moths which occur sporadically throughout the world. Adults feed on pollen of various flowers; larvae feed on moss, liverworts, and possibly detritus. Pupation is in a tough, parchmentlike cocoon. The North American fauna includes three or four species of *Epimartyria,* which are mostly encountered in damp woodlands and bogs. Thirty-five species worldwide.

Suborder Dacnonypha

Small moths with or without proboscis, mandibles present or absent; maxillary palpi 5-segmented, labial palpi 3- to 4-segmented; forewing with fingerlike jugum. Larva apodous; pupa exarate with large functional mandibles.

Dacnonypha are small or very small, mostly diurnal moths, often with iridescent vestiture. The suborder contains an assemblage of primitive moths with many generalized features. *Eriocraniella* and *Dyseriocrania,* with species throughout North America, have vestigial adult mandibles and maxillae lacking the laciniae. The galea is modified as a short proboscis. Larvae initiate a linear mine in leaves of oak, chestnut, or other woody plants, later expanding the mine into a large blotch, then leaving the mine to pupate in the soil. The Australian *Agathiphaga* lacks a proboscis but possesses functional mandibles and laciniae. Larvae mine in the seeds of *Agathis* (Auricareaceae), an archaic group of conifers. Adult food is unknown but probably includes pollen. The world fauna of Dacnonypha numbers only a few hundred species. North American species belong in the Family ERIOCRANIIDAE.

Suborder Monotrysia

Very small to large moths with mandibles and laciniae absent, galeae as a short proboscis or absent; maxillary palpi 1- to 5-segmented, frequently minute; labial palpi with 2 to 3 segments. Larva caterpillarlike (Fig. 44.5*a*) or flattened, apodous (Fig. 44.5*c*,*d*). Pupa adecticous, obtect.

The Monotrysia includes extremely diverse taxa and is probably polyphyletic. HEPIALIDAE are moderate or large moths with hairy, elongate bodies and long narrow wings (Fig. 44.9*a*). The proboscis is vestigial or absent, and adults do not feed. Venation is very similar in the fore- and hind wings (Fig. 44.6*a*); wing coupling is with a jugum (Fig. 44.3*b*). The family is cosmopolitan but especially well represented in Australia, where larval feeding habits include external foliage feeding as well as boring in stems or trunks of woody plants. North American species are borers in various shrubs and trees. Hepialids are unusual in broadcasting the numerous eggs in the general vicinity of the larval food, rather than ovipositing on the host plant.

NEPTICULIDAE and OPOSTEGIDAE are small to exceedingly small moths with linear wings with reduced venation, characteristics which are common to various unrelated small lepidopterous leaf miners. Adults have rudimentary mouthparts, and many probably do not feed. Larvae of a few species cause twig or petiole galls or mine in bark, but the great majority excavate serpentine mines in the leaves of a broad array of woody plants. Larvae leave the mine to pupate in a tough cocoon in soil or leaf litter.

Habits of the Superfamily Incurvarioidea are heterogeneous. HELIOZELIDAE, TISCHERIIDAE, and some INCURVARIIDAE are leaf miners as early instar larvae, later leaving the mine and feeding externally, usually cutting out oval shelters from the mine. Larvae of *Adela* (Incurvariidae) (Fig. 44.9*b*) first feed in developing ovules, later inhabiting cases constructed from leaf or flower parts, and feeding externally, often as detritus feeders. Other Incurvariidae (*Tegeticula* and *Prodoxus*) bore the seed pods or stems of *Yucca,* the adult moths ovipositing in the immature flower ovaries or stalks. *Tegeticula* females have specialized, elongate maxillary palpi which are used to collect the glutinous pollen of *Yucca,* which is pollinated only by these moths (see Fig. 13.4). The Suborder Monotrysia probably includes less than 1000 species worldwide.

KEY TO THE NORTH AMERICAN FAMILIES OF THE SUBORDER
Monotrysia

1. Mouthparts atrophied or absent; wings with fingerlike jugum (Fig. 44.3*b*); head small, attached anteroventrally **Hepialidae**
   Mouthparts with haustellum and maxillary and labial palps; wings with frenulum; head large, attached anteriorly. 2

2(1). Antenna with scape expanded into broad eye cap (Fig. 44.7*b*); forewing lacking discal cell (Superfamily Nepticuloidea) 3
   Antennae with scape not expanded; forewing with discal cell (Fig. 44.7*e*) (Superfamily Incurvarioidea) 4

3(2). Forewing without branched veins **Opostegidae**
   Forewing with some veins branched **Nepticulidae**

4(2). Cranium with erect, irregular hairs **Incurvariidae** and **Tischeriidae**
   Cranium with smooth, flattened scales **Heliozelidae**

Suborder Ditrysia

An exceedingly variable assemblage characterized by the ditrysian organization of the female reproductive organs (Fig. 44.4). Ditrysia have the galea produced as a proboscis (occasionally atrophied). The wings are scaled (rarely with a few aculeae), and wing coupling is either frenulate or amplexiform (never jugate). Wing venation is usually strikingly different in the fore- and hind wings (*heteroneurous*), in contrast to the similar, *homoneurous* venation of the more primitive suborders.

The Ditrysia comprise the great majority of Lepidoptera, including nearly all the familiar or economically important species. Venation, an important differentiating feature for many families, is frequently invisible without descaling the wings. The following key, based as far as possible on characteristics other than venation, is modified in part from Common (1970). The assistance of J. A. Powell is gratefully acknowledged.

KEY TO THE COMMON NORTH AMERICAN SUPERFAMILIES OF THE SUBORDER
Ditrysia

1. Wings absent or vestigial 24
   Wings large, suitable for flying 2

2(1). Antennae threadlike, with abruptly swollen knob at tip, or with tip abruptly hooked (Fig. 44.6*b,c*); frenulum and ocelli absent 3
   Antennae filiform, feathery, or gradually enlarged, very rarely knobbed; if thickened apically, frenulum present; ocelli present or absent 4

3(2). Distance between bases of antennae greater than diameter of eye (Fig. 44.6*c*); hind tibia usually with intermediate spur; all veins arising individually from discal cell in both wings (Fig. 44.10*a*) **Hesperioidea** (p. 460)
   Distance between bases of antennae less than diameter of eye; hind tibiae without intermediate spur; forewings with some peripheral veins branched (Fig. 44.10*b*) **Papilionoidea** (p. 460)

4(2). Antennae spindle-shaped, thickest near middle or in apical third (Fig. 44.6*d*); body long, stout, pointed at both ends; ocelli absent; heavy-bodied moths more than 20 mm long **Sphingoidea** (p. 465)
   Antennae threadlike, featherlike, or comblike, never spindle-shaped; body rarely spindle-shaped; ocelli present or absent 5

5(4). Wings deeply cleft (Fig. 44.9*k*) with hind wing divided into 3 or more lobes 6
   Wings not deeply cleft; hind wings consisting of single undivided lobe 7

6(5). Hind wing divided into 3 lobes (Fig. 44.9*k*) **Pterophoroidea** (p. 458)
   Hind wing divided into 6 to 7 lobes **Copromorphoidea** (**Alucitidae**, p. 456)

7(5). Hind wing narrow or linear (Fig. 44.8*f*), with marginal fringe of hairs or scales subequal

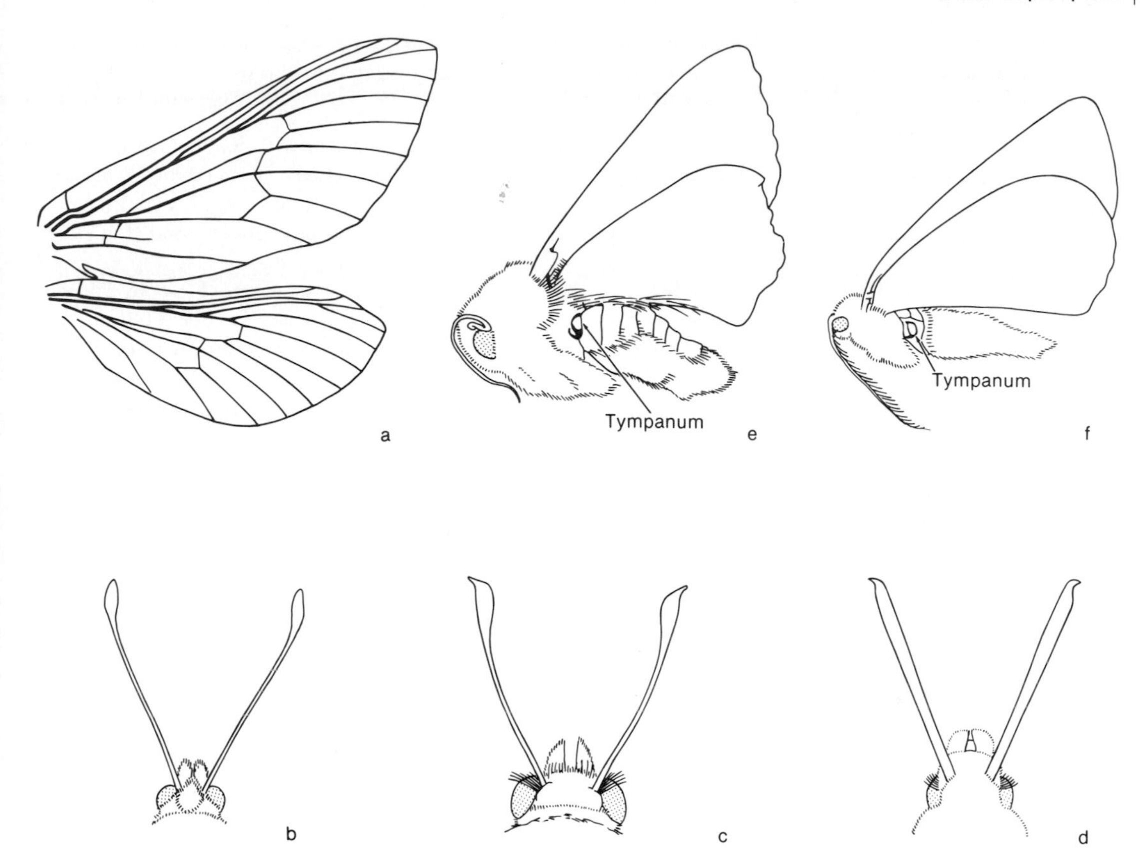

FIGURE 44.6 Taxonomic features of Lepidoptera: *a*, wings of Hepialidae (*Hepialus sequoiolus*); *b*, antenna of Papilionidae (*Papilio*); *c*, antenna of Hesperiidae (*Hesperia*); *d*, antenna of Sphingidae (*Celerio*); *e*, tympanal organ of Noctuidae (*Orthosia pacifica*); *f*, tympanal organ of Geometridae (*Drepanulatrix*).

to or broader than wing membrane; venation frequently reduced; small to minute moths with wingspan almost always less than 15 mm **8**

Hind wing oval or triangular, with marginal fringe of scales much shorter than wing breadth; venation usually complete; wingspan usually greater than 10 mm **12**

8(7). Hind wing with vein Sc + R_1 fused with R_s beyond discal cell, then diverging (Fig. 44.8*d*); abdomen with paired tympana on first sternite (Fig. 44.6*f*) **Pyraloidea** (p. 458)

Hind wing with veins Sc + R_1 and R_s separate beyond discal cell (Fig. 44.8*e,g*) or Sc occasionally joined to discal cell by short R_1 (Fig. 44.8*h*); abdomen without tympana **9**

9(8). Proboscis not covered by scales; maxillary palpi large or small **10**

Proboscis densely covered by overlapping scales, at least basally (Fig. 44.8*b,c*); maxillary palpi short, 4-segmented, folded about base of proboscis (Fig. 44.8*c*) **Gelechioidea,** in part (p. 455)

10(9). Antenna with scape enlarged, cup-shaped on under side, forming cover over eye (Fig. 44.7*b*) **Tineoidea,** in part (p. 454)

Antennae with scape not expanded, cup-shaped **11**

11(10). Maxillary palpi usually long, 5-segmented, folded in repose (Fig. 44.7*c*); hind tibiae with long, stiff bristles only on dorsal surface, or smooth-scaled **Tineoidea,** in part (p. 454)

Maxillary palpi with 3 to 4 segments, not folded in repose (Fig. 44.8*a*); hind tibiae frequently with whorls of long stiff bristles **Yponomeutoidea,** in part (p. 455)

12(7). Tympanal organs present in thorax or abdomen (Fig. 44.6*e,f*) 13
Tympanal organs absent 15

13(12). Tympanal organs in thorax (Fig. 44.6*e*) **Noctuoidea** (p. 465)
Tympanal organs in abdomen (Fig. 44.6*f*) 14

14(13). Maxillary palpi minute or vestigial; proboscis bare, not covered with scales **Geometroidea** (p. 463)
Maxillary palpi with 3 or 4 segments, usually about as long as head and held horizontally in front of head; proboscis densely covered with scales at base (Fig. 44.8*b*) **Pyraloidea** (p. 458)

15(12). Vein M present in discal cell of one or both wings (Fig. 44.7*d*) 16
Vein M absent from discal cell of both wings (Fig. 44.7*e*) 20

16(15). Vein M branched in discal cell of both wings; R with 5 branches in forewing (Fig. 44.7*d*); proboscis vestigial or absent; large, heavy-bodied moths at least 20 mm long **Cossoidea** (p. 452)
Vein M not branched in discal cell of both wings; R usually with 4 branches in forewing (Fig. 44.7*a*); mostly small, frail moths less than 15 mm long 17

17(16). Hind wing with marginal hairs or scales much longer near wing base than at apex 18
Hind wing with marginal hairs or scales near wing base subequal in length to those at apex 19

18(17). Labial palpi held horizontally in front of head (Fig. 44.7*f*); frequently longer than head **Tortricoidea,** in part (p. 453)
Labial palpi drooping or upright, not held horizontally (Fig. 44.7*c*); usually short **Tineoidea,** in part (p. 454)

19(17). Forewing with vein Cu_2 intersecting 1A before margin **Tineoidea,** in part (p. 454)
Forewing with vein Cu_2 not intersecting 1A, separate to margin (as in Fig. 44.8*g* to *i*) **Zygaenoidea** (p. 457)

20(15). Proboscis bare or with few, sparse scales near base (Fig. 44.8*a*); if scaled, ocelli are large and prominent and maxillary palpi are not folded about base of proboscis (Fig. 44.8*b*) 21
Proboscis densely covered by overlapping scales, at least basally (Fig. 44.8*b,c*); ocelli always small or absent; maxillary palpi short, folded around base of proboscis (Fig. 44.8*c*) **Gelechioidea** (p. 455)

21(20). Hind wing with marginal hairs or scales much longer near wing base than at apex 22
Hind wing with marginal hairs or scales near wing base subequal in length to those at apex 23

22(21). Labial palpi held horizontally in front of head, frequently as long as head or longer (Fig. 44.7*f*) **Tortricoidea**[1] (p. 453)
Labial palpi drooping or upright, not held horizontally (Fig. 44.8*a*), usually shorter than head **Yponomeutoidea** (p. 455)

23(21). Forewing at least 4 times as long as broad; wings held together by recurved bristles along margins and by frenulum (as in Fig. 44.3*a*) **Yponomeutoidea (Sesiidae)** (p. 455)
Forewing no more than twice as long as broad; wings lacking both marginal bristles and frenulum, or with vestigial frenulum **Bombycoidea** (p. 464)

24(1). Legs greatly reduced, nonfunctional; moth developing in case constructed by larva **Tineoidea** (p. 454)
Legs well developed, used in walking; moth not developing in larval case 25

25(24). Proboscis vestigial or absent **Noctuoidea** (p. 465)
Proboscis present, longer than head **Geometroidea** (p. 463)

SUPERFAMILY COSSOIDEA. Small to large moths, North American species moderate or large (Fig. 44.9*c*). Body

[1]Carposcinidae (Copromorphoidea) with seven North American species will also key here.

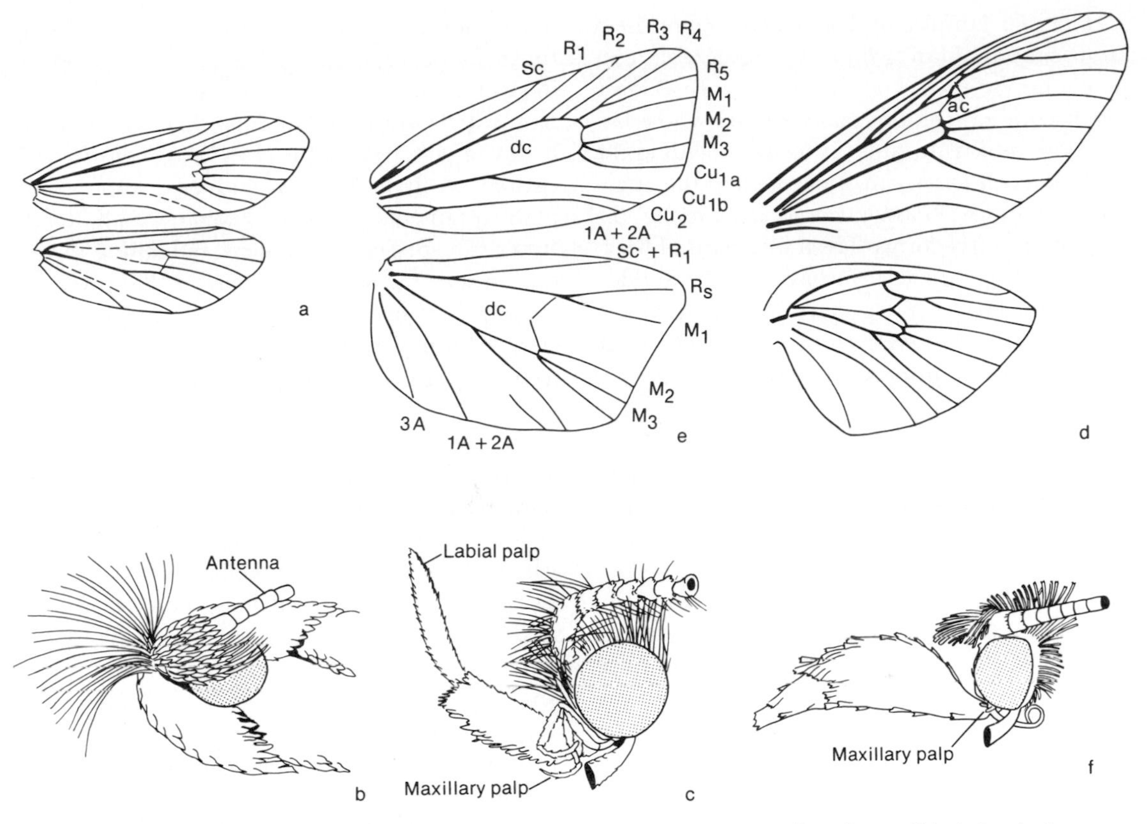

FIGURE 44.7 Taxonomic characters of Lepidoptera: *a,* wings of Tineidae (*Scardia gracilis*); *b,* head of Lyonetiidae (*Bucculatrix athaeae*); *c,* head of Tineidae (*Scardia gracilis*); *d,* wings of Cossidae (*Prionoxystus robinae*); *e,* wings of Tortricidae (*Sparganothis*); *f,* head of Tortricidae (*Sparganothis senecionana*). Left labial palp (near side) removed on heads. *ac,* accessory cell; *dc,* discal cell.

stout, long, usually exceeding hind wings; head with erect hairs or scales, ocelli absent, proboscis very short or absent, antennae bipectinate in male; forewing with large accessory cell, discal cell divided by forked M vein (Fig. 44.7*d*). Larva stout, cylindrical, with prognathous mandibles; thorax with large, sclerotized dorsal shield.

The Family Cossidae occurs in North America. These moths show many primitive features, particularly the extensive wing venation with poorly defined discal cell, and the chaetotaxy of the larva. At the same time they are specialized in the reduction of the mouthparts and loss of ocelli. Adult cossids are short-lived. They produce large numbers of eggs (up to 18,000 have been recorded from a single individual). Larvae bore in heartwood or beneath the bark of various trees or shrubs, or invade the crowns of woody herbs, requiring up to several years to mature. Pupation is within the larval gallery. The family is particularly diversified in Australia. About 40 species are known in North America.

SUPERFAMILY TORTRICOIDEA. Small moths, mostly with broad, truncate, gray, brown, or tan mottled wings (Fig. 44.9*e*). Head with rough scales on vertex, ocelli and chaetosemata usually present, proboscis short or vestigial (Fig. 44.7*f*); maxillary palpi minute, labial palpi 3-segmented, porrect. Forewings with accessory cell present (Fig. 44.7*d*), discal cell sometimes divided by unbranched M.

The two families of Tortricoidea comprise a large cosmopolitan group of considerable economic importance. The TORTRICIDAE (including Olethreutidae) contain many species whose larvae bore into fruit, nuts, or seeds, including the codling moth *Laspeyresia pomonella* and the oriental fruit moth, *Grapholitha molesta,* among the more destructive forms. The movements of Mexican jumping beans are caused by the sudden jerks of the larvae of a species of *Laspeyresia* which lives in the seeds. Many other tortricid larvae feed externally, usually sheltering in rolled or tied leaves. These external feeders also include many economically important species, such as the spruce budworm, *Choristoneura fumiferana,* and the fruit tree leaf roller, *Archips argyrospilus.* Pupation usually occurs within the larval food or shelter.

In the PHALONIIDAE the forewing is usually buckled and bent downward at the end of the discal cell, distinguishing this family from most Tortricidae, which have flat wings. The larvae hollow seeds, root crowns, or foliage terminals on a variety of herbs, shrubs, and trees.

SUPERFAMILY TINEOIDEA. Very small to medium-sized moths, usually somber gray or brown with elongate rounded or pointed wings (Fig. 44.9*d*). Head rough-scaled (most families, Fig. 44.7*c*) or smooth-scaled (Gracilariidae); ocelli present or absent, chaetosemata absent; proboscis usually present, without covering of scales; maxillary palpi large, usually 5-segmented, labial palpi short, drooping or ascending (not porrect) (Fig. 44.7*c*). Forewings with extensive venation (Fig. 44.7*a*), frequently including branched or unbranched M vein in discal cell; accessory cell absent.

Both generalized and highly specialized taxa are included in this superfamily. As in most other primitive groups of Lepidoptera, the larvae tend to occur in protected situations. For example, TINEIDAE includes species which tunnel in fungi, others which scavenge beneath loose bark, in bat caves and animal burrows, or in human habitations. Many members of this family consume material of animal origin, especially fur and feathers, including articles of apparel; hence their common name, "clothes moths." Many tineid larvae construct tubes over their food or inhabit a portable protective case. In the PSYCHIDAE, which are plant feeders, this case-making behavior is universal, the cases being constructed entirely of silk in some psychids or incorporating plant fragments in others. Females of all North American species are wingless, remaining in the case, where the eggs are deposited. However, in primitive forms, including many Australian species, the females are winged. Psychidae do not feed as adults, and the proboscis is vestigial or absent.

Larvae of LYONETIIDAE and GRACILARIIDAE are phytophagous. Most species mine leaves; they may be highly modified with legs reduced or absent. Some species form serpentine mines, while others later enlarge the mine into a blotch. *Bucculatrix,* probably the most familiar genus, initiates a serpentine mine, then abandons the mine to tie adjacent leaves together and skeletonize them. Adult lyonetiids, like many other very small moths, have narrow, lanceolate wings with reduced venation. The wing tip is suddenly attenuate in many species. Many gracilariid larvae show striking hypermetamorphosis. The early leaf-mining instars often lack both prolegs and thoracic legs, and have reduced head capsule and mouthparts. The full-grown larva has normal head and mouthparts, thoracic legs with claws, and prolegs on abdominal segments 3 to 5 and 10, and they skeletonize rolled leaves.

KEY TO THE NORTH AMERICAN FAMILIES OF
Tineoidea

1. Wings about as long as body or longer — 2
   Wings absent or shorter than body — **Psychidae** (♀♀ only)

2(1). Antenna with scape enlarged, ventrally concave, forming eyecap (Fig. 44.7*b*); very small moths(<4 mm long) with lanceolate wings — **Lyonetiidae**
   Antenna with scape slender, cylindrical or slightly flattened, not forming cover for eye; size variable — 3

3(2). Head with erect, bushy scales covering vertex and face (Fig. 44.7*c*) 4
Head with smooth, appressed scales at least on face (as in Fig. 44.7*b*); vertex smooth-scaled or with posterior fringe of erect scales; very small moths (<4 mm long) with lanceolate wings **Gracilariidae**

4(3). Proboscis and maxillary palps vestigial; antennal scape without pecten **Psychidae** (♂♂)
Proboscis and maxillary palps almost always large, functional (Fig. 44.7*c*); if vestigial, pecten present on antennal scape **Tineidae**

SUPERFAMILY YPONOMEUTOIDEA (Fig. 44.9*f,h*). Small or occasionally moderate in size; head with smooth scales; ocelli and chaetosemata present or absent; proboscis moderate, without covering of scales; labial palpi short, either drooping or ascending (Fig. 44.8*a*). Forewing usually elongate with rounded or angulate apex; discal cell rarely divided by M vein.

The moths of this superfamily are mostly uncommon or obscure, with some notable exceptions mentioned below. Many species, including all Sesiidae and most Glyphipterygidae and Heliodinidae, are brightly colored, diurnal insects. In contrast to the morphologically similar Tineoidea, which contains many scavengers, nearly all Yponomeutoidea are phytophagous. *Euclemensia* (Heliodinidae) is unusual in being an internal parasitoid of *Kermes* (Coccidae).

Douglasiidae mine leaves or stems as larvae, recalling the feeding mode of many Tineoidea and other primitive moths, and sesiids are borers, but most Yponomeutoidea are external feeders on foliage, usually spinning a more or less extensive web over the food. Although many are widely distributed, as the cosmopolitan *Plutella xylostella* (Yponomeutidae) which feeds on most Cruciferae, few are of economic significance.

In contrast, the Sesiidae (Fig. 44.9*h*) include several species destructive to agricultural crops. The moths of this family are mostly mimics of stinging Hymenoptera, and have the wing membrane free of scales except at the margins. Larvae bore in trunks, stems or roots of herbs, shrubs, or trees, causing significant damage to peach, currant, gooseberry, blackberry, and squash and other cucurbits.

KEY TO THE NORTH AMERICAN FAMILIES OF Yponomeutoidea

1. Wings with scales only on margins, central region transparent **Sesiidae**
Wings entirely covered with scales, opaque 2

2(1). Tibia and tarsus of hind legs bearing stiff, erect bristles **Heliodinidae** and **Epermeniidae**
Tibia and tarsus of hind legs with smooth scales 3

3(2). Ocelli large, conspicuous; maxillary palpi minute (as in Fig. 44.7*f*) 4
Ocelli absent; if ocelli present, maxillary palpi large (Fig. 44.8*a*) **Yponomeutidae**

4(3). Body length less than 3 mm; hind wings narrow, lanceolate, with fringe of setae broader than membrane **Douglasiidae**
Body length greater than 3 mm; hind wings with membrane broader than fringe of setae **Glyphipterygidae**

SUPERFAMILY GELECHIOIDEA (Fig. 44.9*g,l*). Small or very small moths; ocelli present or absent, chaetosemata absent. Proboscis clothed with scales, at least basally, and embraced by small, 3- to 4-segmented maxillary palpi; labial palpi recurved over head, usually extending to vertex. (Fig. 44.8*c*). Forewing with discal cell undivided by M vein.

The Gelechioidea represents the largest superfamily of "microlepidoptera," the Family Gelechiidae alone containing at least 4000 species. North American forms are predominantly small, somber-colored, rather obscure moths, but some Old World species reach 75 mm in wingspan, with bright, conspicuous markings. In terms of larval feeding the Gelechioidea are perhaps more diverse than any other superfamily. Included are leaf miners, tiers, and rollers; borers in stems, seeds, nuts, tubers, and fruits; gall makers, scavengers, and predators of scale insects. In addition many species feed exposed or protected by a silken web or case. Most gelechioid families, especially those with large numbers of species, are variable both in feeding mode and host-plant preferences. Several of the smaller, more specialized families are more restricted. For example, COLEOPHORIDAE are exclusively case bearers, at least in later instars. ELACHISTIDAE are predominantly miners in grasses and other monocotyledonous plants. Pupation occurs in the larval shelter in many species, or exposed in others, sometimes with a silk girdle attaching the pupa to the substrate.

Several families, notably Gelechiidae, and to a less extent OECOPHORIDAE and COSMOPTERYGIDAE, show a marked tendency to attack the reproductive structures of their hosts—buds, flowers, tubers, etc. Most economically important species display this preference, including the pink bollworm, *Pectinophora gossypiella* (Gelechiidae), possibly the most important insect pest on cotton. Gelechiids of lesser importance include the potato tuber worm, *Phthorimaea operculella* (Fig. 44.9*g*), which mines leaves as well as boring into the tubers, and the Angoumois grain moth, *Sitotroga cerealella,* which infests various whole grains, either fresh or dried. A few Oecophoridae and Cosmopterygidae occasionally cause minor damage on a few crops or ornamentals.

KEY TO THE COMMON NORTH AMERICAN FAMILIES OF
Gelechioidea

1. Hind wings narrow, linear (Fig. 44.8*e,f*), with posterior fringe of setae much wider than membrane — 2
   Hind wing broader, oval, with setal fringe narrower than membrane — 7

2(1). Forewing with thickening (pterostigma) on costal margin near apex (Fig. 44.8*e*); vein R_2 arising from near apex of discal cell; antennae with scape usually expanded, sometimes concave beneath (Fig. 44.8*c*) — **Blastobasidae**
   Forewing without pterostigma; antenna with scape slender, never concave beneath — 3

3(2). Maxillary palpi with two segments, exceedingly small, apparently absent — 4
   Maxillary palp with 3 to 4 segments; folded around base of proboscis (as in Fig. 44.8*c*) — 5

4(3). Forewings with vein 2A forked at base (Fig. 44.8*f*) — **Coleophoridae**
   Forewings with vein 2A undivided at base — **Elachistidae**

5(3). Forewing with vein R_2 arising well before apex of discal cell (Fig. 44.8*g*) — 6
   Forewing with vein R_2 arising very near apex of discal cell (as in Fig. 44.8*e*); mostly small, dark moths; hind wing with closed discal cell — **Scythridae**

6(5). Hind wing with veins R_s and M_1 arising separately on discal cell (Fig. 44.8*g*); abdomen often with dorsal spines on segments 1 to 7 — **Oecophoridae**
   Hind wing with veins R_s and M_1 arising conjointly from discal cell (as in Fig. 44.8*h*); abdomen without dorsal spines — **Cosmopterygidae (from Momphidae, Walshiidae)**

7(1). Both wings with vein 1A present, at least in distal half of wing; hind wing not terminating as a narrow appendage (Fig. 44.8*g*) — 8
   Both wings with vein 1A absent; hind wing often narrowed, apically appendiculate (Fig. 44.8*h*) — **Gelechiidae**

8(7). Hind wing with veins R_s and M_1 arising separately from discal cell (as in Fig. 44.8*g*) — **Oecophoridae and Ethmiidae**
   Hind wing with veins R and M_1 arising contiguously from discal cell, or basally fused (as in Fig. 44.8*h*) — **Stenomidae**

SUPERFAMILY COPROMORPHOIDEA. Small to moderate in size. Ocelli present or absent, chaetosemata absent, head with smooth, appressed scales; proboscis not clothed with scales; maxillary palpi minute, labial palpi usually recurved.

Two small families represent this group of rarely encountered moths in North America. ALUCITIDAE are readily distinguished from all other Lepidoptera in having the hind wing divided into at least six plumes. CARPOSINIDAE are superficially similar to Tortricidae, but lack vein M_1 in the hind wing. Known larvae of both families are internal feeders in fruits, buds, galls,

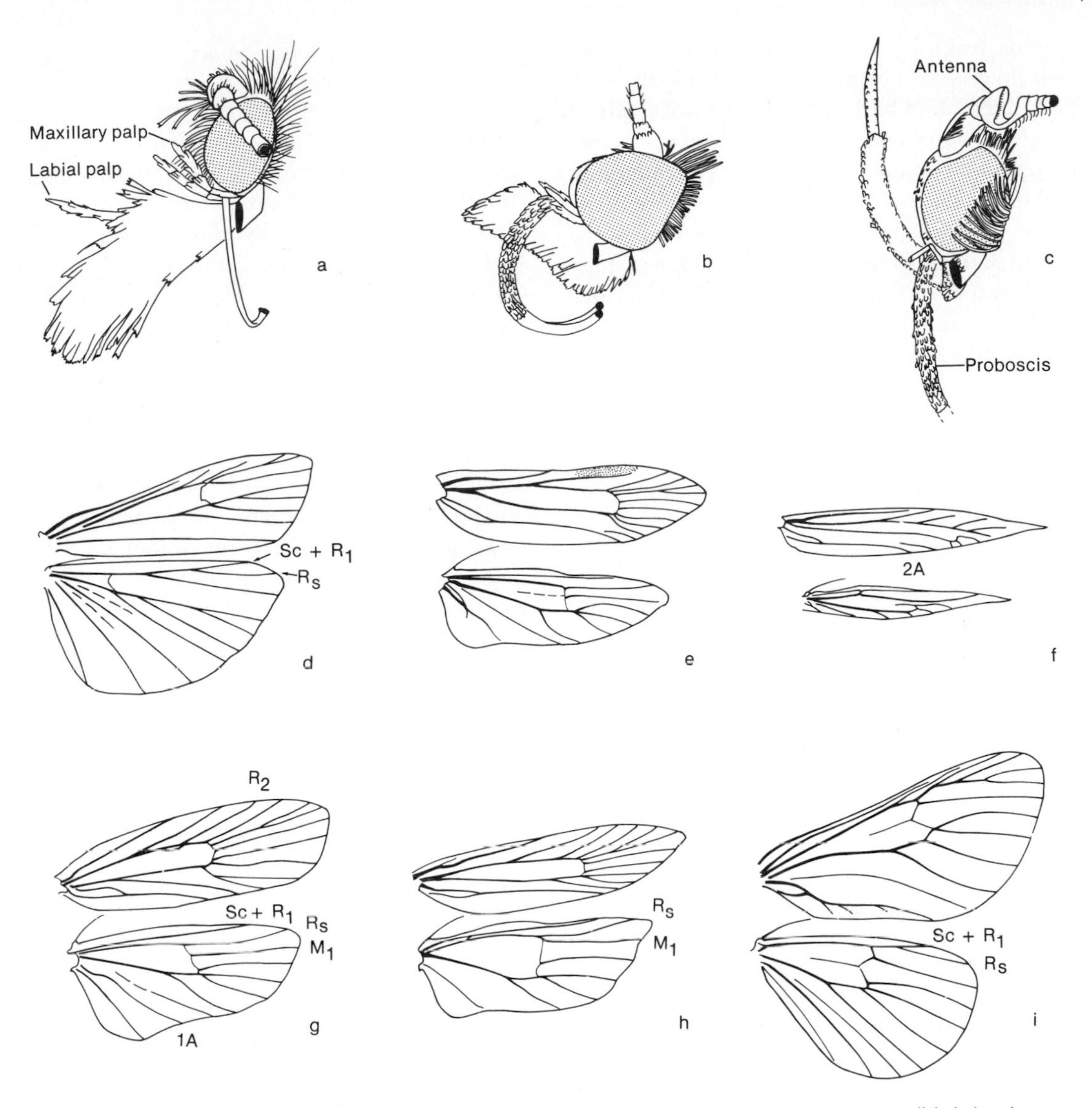

FIGURE 44.8 Taxonomic characters of Lepidoptera; *a*, head of Yponomeutidae (*Abebaea cervella*); *b*, head of Pyralidae (*Nomophila noctuella*); *c*, head of Blastobasidae (*Holcocera gigantella*); *d*, wings of Pyralidae (*Dioryctria*); *e*, wings of Blastobasidae (*Holocera*); *f*, wings of Coleophoridae (*Coleophora*); *g*, wings of Oecophoridae (*Agonopteryx fuscitermidella*); *h*, wings of Gelechilidae (*Filatima demissae*); *i*, wings of Megalopygidae (*Norape tener*). Left labial palp (near side) removed on heads.

or necrotic bark. Fewer than 10 species are known from North America.

SUPERFAMILY ZYGAENOIDEA. Small to medium-sized moths with bipectinate antennae, at least in males. Ocelli, chaetosemata present or absent; proboscis large (Pyromorphidae) or absent; maxillary palpi small or absent, labial palpi small or minute. Fore- and hind wings with discal cells divided by M vein (Fig. 44.8*i*).

Adults of this small superfamily are mostly densely hairy, heavy bodied moths, frequently brownish or tan in color. ZYGAENIDAE, however, include brightly colored, diurnal species which visit flowers for nectar. Larvae are highly modified, often sluglike in appearance with tufts or bands of hairs which may produce severe skin irritation on contact. Most zygaenid caterpillars are leaf skeletonizers. *Harrisina americana,* the grape leaf skeletonizer, sometimes causes economic damage. The larvae of EPIPYROPIDAE are ectoparasites of Fulgoroidea, the grublike caterpillar residing beneath one wing of the host. About 100 species of Zygaenoidea occur in North America, mostly in the family LIMACODIDAE.

KEY TO THE NORTH AMERICAN FAMILIES OF **Zygaenoidea**

1. Proboscis and maxillary palpi vestigial or absent; vertex without chaetosemata — 2
   Proboscis and maxillary palpi large, functional; chaetosemata large — **Pyromorphidae**

2(1). Hind wing with veins $Sc + R_1$ and R_S fused along most of the discal cell (Fig. 44.8*i*) — **Megalopygidae**
   Hind wing with veins $Sc + R_1$ and R_S separate or fused for a very short distance near base (as in Fig. 44.8*g*) — 3

3(2). Tibial spurs absent; all mouthparts absent or vestigial; forewing with all veins arising separately from discal cell — **Epipyropidae**
   Tibial spurs almost always present on middle and hind legs; maxillary and labial palpi usually apparent; forewing with at least one radial vein branched after leaving discal cell (as in Fig. 44.8*i*) — **Limacodidae**

SUPERFAMILY PYRALOIDEA. Small to large moths with ocelli and chaetosemata present or absent; maxillary palpi usually moderate, 4-segmented; labial palpi usually beaklike, held porrect in front of head (Fig. 44.8*b*); legs usually long, slender; wings without M vein in discal cell (Fig. 44.8*d*).

Two families occur in North America. THYRIDIDAE, including small, dusky moths, usually with transparent patches in the wings, are distinguished from Pyralidae in lacking vein CuP in the hind wing. Fewer than 10 species occur in North America. In contrast, the PYRALIDAE constitute an enormous grouping, with over 1100 species recorded north of Mexico. Adults are mostly small, dull-colored insects with elongate, triangular wings (Fig. 44.9*i,j*) but large, bright-colored or metallic forms occur, especially in the Tropics. Pyralidae are unique among "microlepidoptera" in possessing tympanal auditory organs on the abdomen. Analogous structures occur in Geometroidea and Noctuoidea.

Larval biology is highly variable, including aquatic as well as terrestrial forms, external feeders on webbed foliage, borers in stems, seeds, or fruit, gall-inhabiting forms, and scavengers. Numerous economically important species attack stored products, in addition to a wide array of agricultural crops. More destructive species in North America include *Ostrinia nubilalis,* the European corn borer, and *Desmia funeralis,* the grape leaf folder, as well as several cosmopolitan granary pests (*Pyralis farinalis, Anagasta kuehniella,* and *Plodia interpunctella*). *Galleria mellonella* and *Achroia grisella* infest beehives and are capable of digesting wax with the aid of intestinal bacteria. A few species are beneficial, including *Cactoblastis cactorum,* which has been utilized to control *Opuntia* cactus in Australia.

SUPERFAMILY PTEROPHOROIDEA. Small to moderate, ocelli and chaetosemata absent; maxillary palpi

FIGURE 44.9 Representative Lepidoptera: *a,* Hepialidae (*Hepialus sequoiolus*); *b,* Incurvariidae (*Adela trigrapha*); *c,* Cossidae (*Prionoxystus robinae*); *d,* Tineidae (*Tinea pellionella*); *e,* Tortricidae (*Archips argyrospilus*); *f,* Yponomeutidae (*Atteva punctella*); *g,* Gelechiidae (*Phthorimaea operculella*); *h,* Sesiidae (*Ramosia resplendens*); *i,* Pyralidae (*Udea profundalis*); *j,* Pyralidae (*Desmia funeralis*); *k,* Pterophoridae

a
b
c
d
e
f
g
h
i
j
k
l

minute, labial palpi long, held horizontally in front of head; fore- and hind wings each divided into 2 to 4 lobes.

The only family, Pterophoridae, includes slender-bodied moths with characteristic, deeply cleft wings which are held at right angles to the body at rest (Fig. 44.9*k*). Larvae are sometimes leaf miners initially, later folding or rolling leaves or boring in stems. Fewer than 100 species in North America.

SUPERFAMILY HESPERIOIDEA (Skippers). Medium-sized, with ocelli absent, chaetosemata present; antennae scaled, gradually clubbed, usually with hooked, bare tip (Fig. 44.6*c*); maxillary palpi absent, labial palpi large, erect; forewing with 5 radial veins arising separately from discal cell (Fig. 44.10*a*).

A single family, Hesperiidae (Fig. 44.11*a,b*). Over 3000 species of these stout-bodied butterflies are known, with about 200 in the United States. North American species are diurnal, but some tropical species fly at dusk or night. The wingbeat is rapid and flight direct, rather than fluttering, as in most butterflies (Papilionoidea). Adults sip nectar from flowers; the larvae mostly feed from webbed foliage. The more primitive species utilize a variety of dicotyledonous plants, but the more specialized ones, comprising about 60 percent of the superfamily, eat monocotyledonous plants, especially grasses. Larvae of the Subfamily Megathyminae (sometimes treated as a family) bore in stalks and roots of *Yucca* and *Agave.* Pupation is in a cocoon of webbed foliage or in the larval gallery.

SUPERFAMILY PAPILIONOIDEA (Butterflies) (Fig. 44.11*c* to *i*). Medium to large, occasionally small lepidopterans; ocelli absent, chaetosemata present; antennae gradually clubbed (Fig. 44.6*b*), rarely with hooked tip; maxillary palpi small, 1-segmented or absent; labial palpi short to much longer than head; forewing with 2 or more radial veins arising conjointly from discal cell (Fig. 44.10*b*).

The butterflies, because of the aesthetic appeal of their brilliant colors or striking patterns, have received more attention than any other group of insects. Accurate species lists exist for most temperate regions, and comprehensive biological information is available for many species. The Papilionoidea are among the most specialized Lepidoptera in terms of behavior and ecology. Perhaps more than any other insects, butterflies display aposematic or warning coloration and engage in mimicry relationships . In North America the most familiar examples involve (1) the monarch, *Danaus plexippus* (model), and the viceroy, *Limenitis archippus* (mimic; both Nymphalidae), and (2) *Battus philenor* (model, Papilionidae) and *Limenitis astyanax* (mimic, Nymphalidae). In the first relationship the model is known to store plant cardiac glycosides ingested by the larvae, which feed on milkweeds (Asclepiadaceae) or other toxic plants (see Chap. 15). Mimicry is especially common in tropical regions. At single localities in Central and South America up to 20 mimetic species of Ithomiinae and Heliconiinae (Nymphalidae) are commonly encountered, along with occasional members of other families.

Adults of several species are migratory. Most notable is the monarch, which has spread over the globe, following introductions of its host plants. In North America two-way migrations between Mexico and southern Canada annually cover at least 2000 km. In North America *Cynthia, Libythea,* and *Colias* migrate over shorter distances. Local movements include aggregation on hilltops and patrolling behavior in males, both involved with mate location. Males of many species have a home range in which they occupy centralized perches from which exploratory flights are made to locate females.

Larvae of butterflies are predominantly external foliage feeders. Most are cryptically colored, but species feeding on distasteful plants may be brightly ringed or banded. Papilionidae erect an odoriferous *osmeterium* when disturbed, while larvae of many Nymphalidae are protected by hairs or spines. Certain lycaenid larvae are myrme-

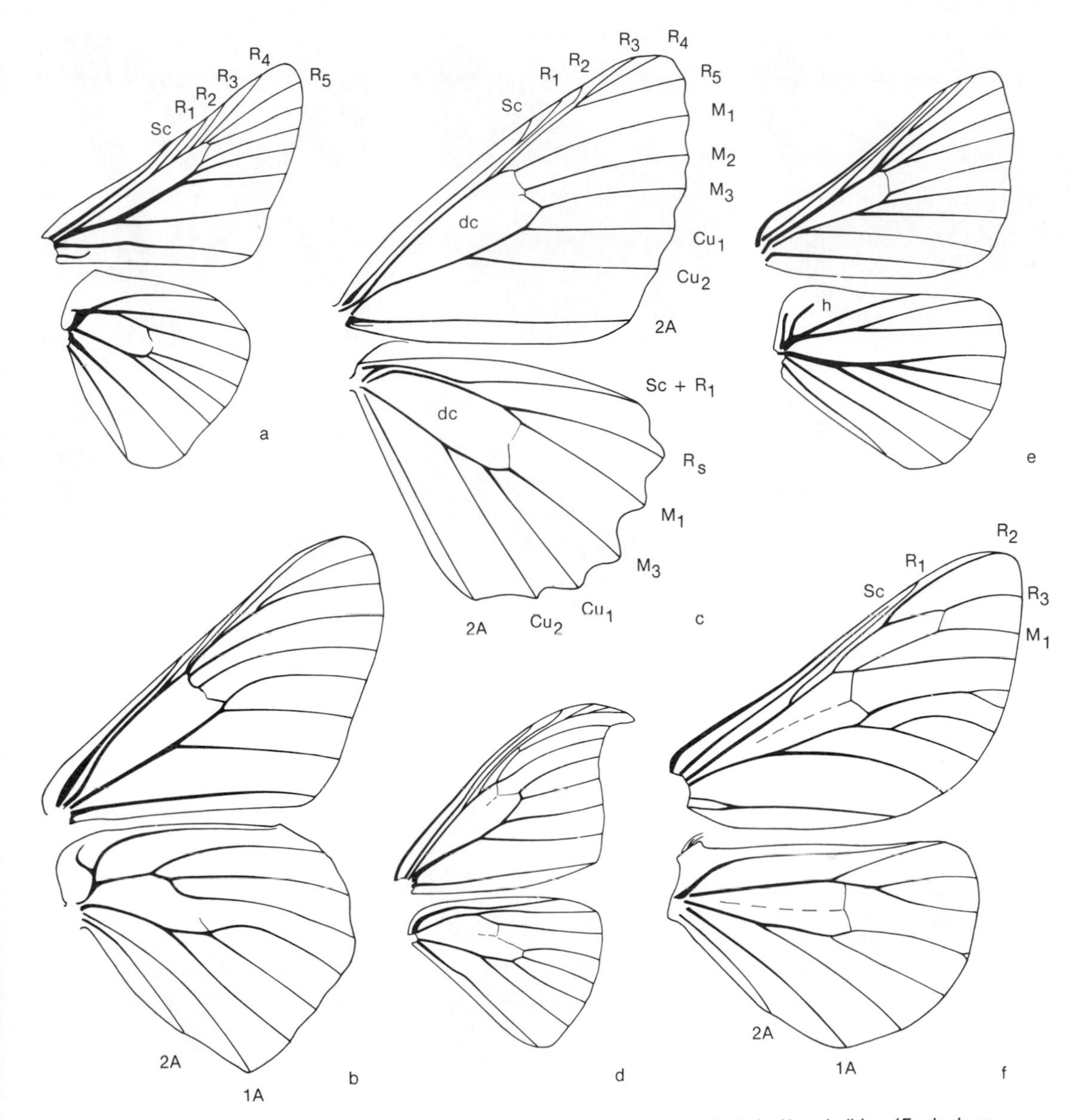

FIGURE 44.10 Wings of Lepidoptera: *a*, Hesperiidae (*Hesperia harpalus*); *b*, Nymphalidae (*Euphydryas chalcedona*); *c*, Geometridae (*Neoalcis californica*); *d*, Drepanidae (*Drepana*); *e*, Lasiocampidae (*Malacosoma californica*); *f*, Saturniidae (*Hemileuca nevadensis*). *dc*, discal cell; *h*, humeral veins.

cophilous, and a few species are predatory on aphids or coccids. Naked pupae or chrysalises, attached to the substrate by the cremaster, are typical of Papilionoidea. Some chrysalises are also supported by a silken girdle about the midregion.

The butterflies are frequently classified as the Suborder Rhopalocera, a special recognition which is contradicted by their overall similarity to certain families of moths. Some authorities recognize considerably more families than treated here, but recent comparative studies of

FIGURE 44.11 Representative Lepidoptera, Papilionoidea and Hesperioidea: *a,* Hesperiidae (*Hesperia harpalus*); *b,* Hesperiidae (*Pyrgus scriptura*); *c,* Papilionidae (*Papilio rutulus*); *d,* Nymphalidae (*Cynthia virginiensis*); *e,* Nymphalidae (*Euphydryas editha*); *f,* Pieridae (*Colias philodice*); *g,* Lycaenidae (*Strymon melinus*); *h,* Lycaenidae (*Apodemia mormo*); *i,* Satyridae (*Cercyonis peglerae*).

skeletal anatomy show that most of the obvious variation involves superficial characteristics.

KEY TO THE FAMILIES OF
Papilionoidea

1. Hind wing with 2 anal veins (Fig. 44.10*b*); fore tibiae without epiphysis (articulated spur) 2
   Hind wing with 1 anal vein; fore tibiae with epiphysis **Papilionidae**

2(1). Tarsal claws simple, not divided 3
   Tarsal claws bifid **Pieridae**

3(2). Labial palpi shorter than thorax; usually erect 4
   Labial palpi longer than thorax, held horizontally in front of head **Libytheidae**

4(3). Antennal base and eye contiguous; eye usually notched around antennal base **Lycaenidae**
   Antennal base and eye not contiguous; eye not notched around antenna **Nymphalidae**

SUPERFAMILY GEOMETROIDEA. Mostly small to medium moths with broad wings and slender body; ocelli usually absent; chaetosemata present, except in Drepanidae; maxillary palpi minute or absent; labial palpi usually short, ascending; abdomen with tympanal auditory organs located ventrolaterally in first sternite (Fig. 44.6*f*).

In North America this superfamily is represented predominantly by Geometridae (Fig. 44.13*d,e*), one of the largest families of moths, with about 12,000 species, 1200 recorded north of Mexico. Accumulations of moths at lights normally contain a high proportion of Geometridae. Many species assume characteristic postures with the wings pressed against the substrate or held at various angles to the long axis of the body, and spend the daylight hours resting on bark, lichen, twigs, or other surfaces where posturing apparently disrupts the body configuration, helping to camouflage the moths. The abdominal tympanal organs probably serve the same function as those of the Noctuidae—avoidance of bats by detection of the high-pitched sounds the mammals produce to echo-locate their prey. Females of several species are brachypterous, living only long enough to mate and deposit their eggs near the pupation site.

Geometrid larvae are commonly known as inchworms or measuring worms, in reference to their looping gait, which is necessitated by the lack of abdominal prolegs in the middle of the body (Fig. 44.5*b*). Many bear a strong resemblance to twigs, which is enhanced by the rigid postures adopted at rest. Numerous species, such as the omnivorous looper, cause minor feeding damage on ornamentals; the cankerworms *Paleacrita vernata* and *Alsophila pometaria* become serious pests. Pupation is usually in a flimsy cocoon.

Whereas most members of the superfamily have cryptically patterned brown, gray, or green wings and bodies, the tropical, diurnal Uraniidae, frequently with brilliantly metallic wings, often resemble papilionid butterflies. Larvae of many species feed on Euphorbiaceae, which contain alkaloids toxic to vertebrates. Presumably these toxins are retained by the adults.

KEY TO THE NORTH AMERICAN FAMILIES OF
Geometroidea[1]

1. Wings about as long as body 2
   Wings less than half as long as body, not functional in flying **Gcomctridac,** in part

2(1). Hind wings with veins Sc + R_1 and R_s diverging or parallel throughout length; frenulum always present **Epiplemidae**
   Hind wings with veins Sc + R_1 recurved toward vein R_s at or beyond discal cell (Fig. 44.10*c,d*); frenulum present or absent 3

3(2). Veins Sc + R_1 and R_s contiguous or fused beyond discal cell (Fig. 44.10*d*); proboscis usually absent; forewings usually with hooked apex **Drepanidae**
   Veins Sc + R_1 and R_s contiguous or fused before apex of discal cell (Fig. 44.10*c*); proboscis almost always present; forewings not usually hooked **Geometridae,** in part

[1]Sematuridae, lacking abdominal tympanic organs, are classified as Geometroidea. A single species occurs in southern Arizona.

SUPERFAMILY BOMBYCOIDEA. Medium to large stout-bodied moths with broad wings and without ocelli or chaetosemata (Fig. 44.13*g,h*); proboscis and maxillary palps rudimentary or absent; antennae bipectinate, featherlike; frenulum reduced or absent; tympanal organs absent.

Adult Bombycoidea are specialized for reproduction, with atrophied mouthparts. The frenulum is almost always absent, wing coupling being accomplished by overlap, as in butterflies. Nevertheless, many of these moths are strong fliers, the males traversing distances of several miles to reach females. Males fly upwind and locate nearby females by following gradients of pheromones released by the moths. Despite their inability to feed, the moths may remain active for many days, subsisting on stored fats. Females emerge with the eggs at an advanced stage of development, and usually remain near the pupation site until mating occurs.

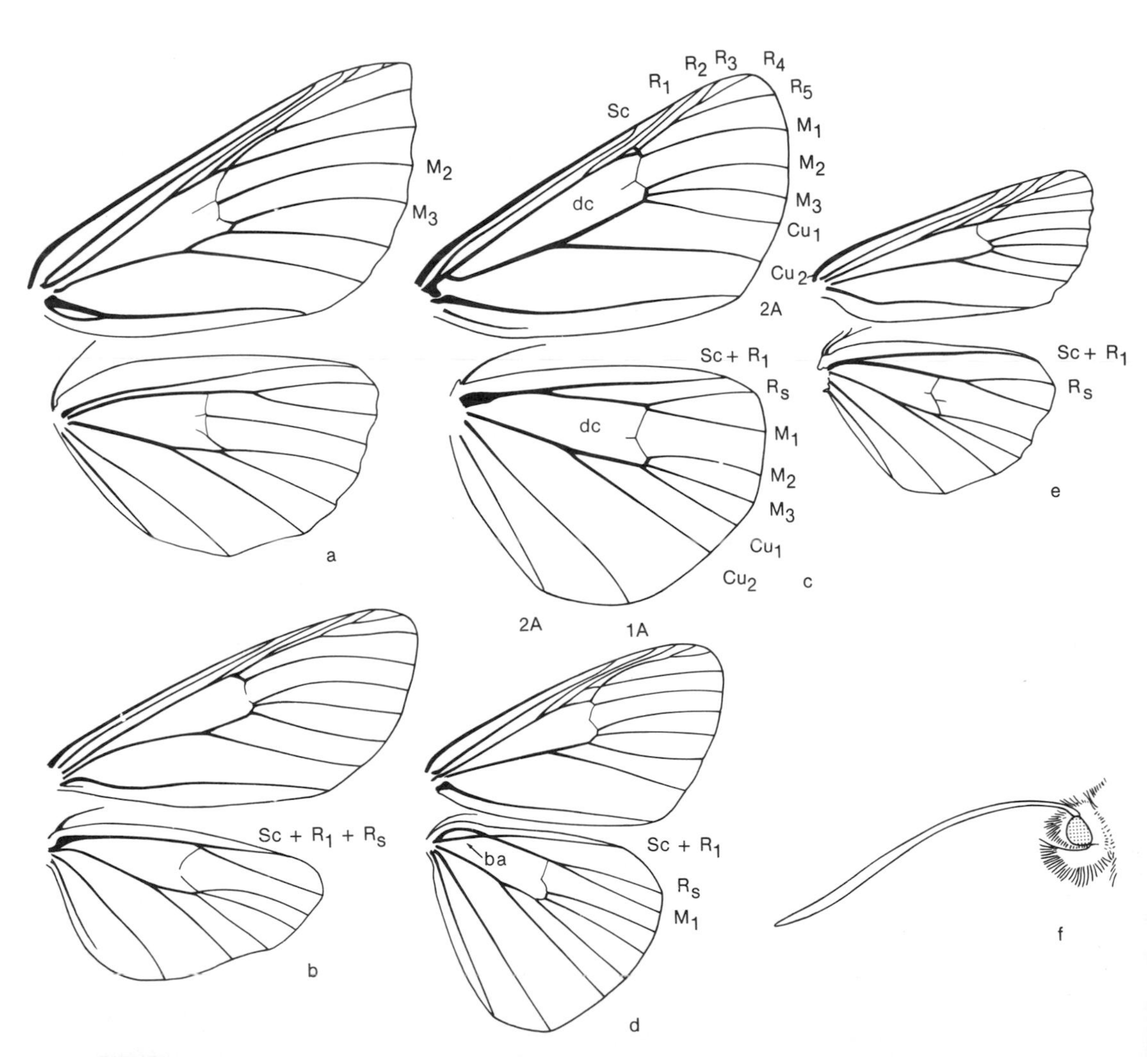

FIGURE 44.12 Taxonomic characters of Lepidoptera: *a* to *e*, wing venation; *a*, Notodontidae (*Nadata gibbosa*); *b*, Ctenuchidae (*Ctenucha rubroscapus*); *c*, Arctiidae (*Apantesis ornata*); *d*, Lymantriidae (*Orgyia vetusta*); *e*, Noctuidae (*Lacinipolia quadrilineata*); *f*, antenna of Agaristidae (*Alypia ridingsi*). *ba*, basal areole; *dc*, discal cell.

Larvae feed predominantly on foliage of shrubs and trees. Most species are solitary, but some Saturniidae feed gregariously, and Lasiocampidae, the tent caterpillars, construct large, communal webs or tents enveloping entire branches and frequently cause severe defoliation. They are the only significantly destructive members of the superfamily. In North America several species of *Malacosoma* (Fig. 44.13*h*) are important tent caterpillars.

The pupal period is passed in a tough cocoon, and *Bombyx mori,* the silkworm, is propagated commercially for its silk.

In North America SATURNIIDAE is the commonest family. Many of the species are elegant moths, subtly shaded in muted colors with contrasting eye spots in one or both wings (Fig. 44.13*g*).

KEY TO THE NORTH AMERICAN FAMILIES OF
Bombycoidea

1. Hind wing with costal margin greatly expanded basally and supported by 2 to 3 short, stout humeral veins (Fig. 44.10*e*) **Lasiocampidae**
   Hind wing with costal margin not broadened, without humeral veins (Fig. 44.10*f*) 2

2(1). Forewing with vein R with 3 branches (Fig. 44.10*f*); hind wing usually with 1 anal vein **Saturniidae (incl. Citheroniidae)**
   Forewing with vein R with 4 to 5 branches; hind wing with 2 anal veins **Lacosomidae, Eupterotidae,** and **Bombycidae**[1]

SUPERFAMILY SPHINGOIDEA. Medium to large, heavy-bodied moths with long, narrow forewings; ocelli and chaetosemata absent; proboscis usually very long (Fig. 13.1*a*); antennae usually thickened in middle (Fig. 44.6*d*); maxillary palpi 1-segmented, labial palpi stout, 2- or 3-segmented.

The single family, SPHINGIDAE, is most diverse in tropical regions, but the large, fast-flying moths are familiar insects throughout the world (Fig. 44.13*f*). The larger species exceed small hummingbirds in size and share the avian habit of extracting nectar while hovering above flowers. Certain plants, including some Onagraceae in temperate regions, are adapted for pollination by sphingid moths, with deep-throated flowers which open at dusk as the moths become active. Other sphingids are diurnal, some closely resembling bumblebees or other stinging Hymenoptera.

The stout-bodied larvae often bear a tapering caudal projection; hence the common name "horn worms." They consume prodigious quantities of foliage, and *Manduca quinquemaculata* and *M. sexta* are pests on solanaceous plants such as tobacco, tomato, and potato. Pupation occurs free in a cell in soil or in a wispy cocoon.

[1]*Bombyx mori,* the silkworm (Bombycidae) is sporadically reared in North America.

SUPERFAMILY NOCTUOIDEA. Small to large, mostly stout-bodied moths without chaetosemata, ocelli absent; proboscis long or reduced or absent, maxillary palpi minute or absent; tympanal organs present in metathorax (Fig. 44.6*e*).

The Noctuoidea constitute the largest group of Lepidoptera, with more than 20,000 species recognized. Included are many extremely familiar insects, such as the gypsy moth (*Porthetria dispar:* Lymantriidae, Fig. 44.13*i*) and various species of cutworms (Noctuidae, Fig. 44.13*a,b*). In terms of economic damage to crops, the Noctuoidea are easily the most important group of Lepidoptera in North America.

Adults are typically strong-flying, nocturnal insects which almost always constitute a major proportion of the insects attracted to lights. Certain species of NOCTUIDAE annually migrate distances of many hundreds of miles north from mild winter regions in Southeastern United States. Tympanal organs (Fig. 44.6*e*) are present in all but a few exceptional species. In Noctuidae they have been shown to function in detecting the high-pitched sounds emitted by bats to locate potential prey. The moths respond by evasive flight patterns, including sudden diving or erratic looping. Females of some species, including many Lymantriidae (tussock moths), are brachypterous.

Most Noctuoidea are mottled brown or black, very often with prominent tufts or mounds of hair or scales on the thorax or wings. As in other moths which rest on irregular surfaces, such patterning is important in camouflaging the body and disrupting its outline and contours. Many ARCTIIDAE (Fig. 44.13*c*) and most CTENUCHIDAE are contrastingly patterned in white, yellow or red and black. Many species, especially of CTENUCHIDAE, are diurnal, sometimes closely resembling stinging Hymenoptera.

Larvae of Noctuoidea are herbivores on a great variety of plants, including herbaceous and woody angiosperms and gymnosperms. Many species are highly polyphagous, greatly augmenting their destructive potential. For example, the noctuid *Heliothis zea* is known as the corn earworm, the tomato fruitworm, and the cotton bollworm in different parts of the country or by different growers. Several species of LYMANTRIIDAE, most notably the gypsy moth, are important defoliators of shade trees in eastern North America. Noctuid and agaristid larvae are typically smooth-skinned. Noctuids frequently hide under debris or litter, emerging at night to feed on or near the surface, where they often chew through stems—hence the name "cutworms." Arctiid, ctenuchid, and lymantriid larvae are closely covered with bristly or silky hairs, which are brightly colored and urticating in some species.

Pupation occurs naked in a cell in the ground or bark (AGARISTIDAE, some Noctuidae), or in a cocoon. Larval hair is usually incorporated into the cocoon, which may then contain little or no silk. The urticating hairs of some Lymantriidae are transferred by the adult female from the cocoon to her freshly deposited egg mass.

KEY TO COMMON NORTH AMERICAN FAMILIES OF Noctuoidea

1. Wings absent or much less than half length of body **Lymantriidae** (♀♀ only)
   Wings about as long as body or longer 2

2(1). Forewing with veins M_2 and M_3 parallel (Fig. 44.12*a*); tympanal organ directed ventrally **Notodontidae**[1]
   Forewing with veins M_2 and M_3 converging near discal cell, frequently adjacent (Fig. 44.12*b* to *e*); tympanal organ directed posteriorly 3

3(2). Hind wing with veins $Sc + R_1$ and R_s confluent (Fig. 44.12*b*); brightly colored, frequently metallic, or wasplike diurnal species **Ctenuchidae**
   Hind wing with veins $Sc + R_1$ and R_s separate beyond base (Fig. 44.12*c* to *e*); usually gray or brown, rarely metallic or wasplike 4

4(3). Hind wing with vein M_2 arising closer to M_3 than M_1; M_2 usually arched basally (Fig. 44.12*c* to *e*) 5
   Hind wing with vein M_2 arising midway between M_1 and M_3, or closer to M_1; M_2 not usually arched (as in Fig. 44.12*a*) **Arctiidae,** in part

5(4). Antennae gradually enlarged apically (Fig. 44.12*f*); veins $Sc + R_1$ and R_s confluent in basal fourth of discal cell in hind wing; black moths with large white or yellow wing spots **Agaristidae**
   Antennae threadlike or feathery, not gradually enlarged; veins $Sc + R_1$ and R_s variable 6

6(5). Ocelli present 7
   Ocelli absent 9

7(6). Hind wing with vein $Sc + R_1$ usually swollen, bulbous at base (Fig. 44.12*c*), usually fused with R_s to about middle of discal cell **Arctiidae,** in part
   Hind wing with vein $Sc + R_1$ not noticeably swollen at base; separating before middle of discal cell 8

8(7). Black moths with several large white dots on wings; tympanal hoods very large, prominent, dorsolateral on first abdominal segment **Pericopidae**
   Dark brown or gray moths with mottled color pattern or contrastingly banded or pat-

[1]Dioptidae key here. The single North American species, *Phryganidia californica,* occurs in California. The slender-bodied, diurnal adults are common about the larval food plants—various species of oaks.

FIGURE 44.13 Representative Lepidoptera: *a*, Noctuidae (*Trichoplusia brassicae*); *b*, Noctuidae (*Catocala andromache*); *c*, Arctiidae (*Apantesis ornata*); *d*, Geometridae (*Sabulodes caberata*); *e*, Geometridae (*Dichorda iridaria*); *f*, Sphingidae (*Celerio lineata*); *g*, Saturniidae (*Hemileuca electra*); *h*, Lasiocampidae (*Malacosoma californica*); *i*, Lymantriidae (*Porthetria dispar*).

terned, but not with large, white dots; tympanal hoods lateral **Noctuidae**

9(6). Forewings with smooth scaling 10
Forewings with tufts or lines of raised scales **Nolidae**

10(9). Proboscis vestigial or absent; hind wings with $Sc + R_1$ and R_s forming basal areole (Fig. 44.12*d*) **Lymantriidae**
Proboscis large, functional, coiled; veins $Sc + R_1$ and R_s not separated at base (Fig. 44.12*e*) **Noctuidae**

SELECTED REFERENCES

GENERAL

Common, 1970; Hering, 1951 (leaf miners); Journal of the Lepidopterists Society, 1947 to date; Journal of Research on the Lepidoptera, 1962 to date; Urquhart, 1960 (monarch butterfly).

MORPHOLOGY

Dugdale, 1974 (female genitalia); Ehrlich, 1958 (butterflies); Hinton, 1955*a* (structure and distribution of larval legs); Kiriakoff, 1963 (tympana); Kristensen, 1968 (mouthparts); Mosher, 1916 (pupae).

BIOLOGY AND ECOLOGY

Brower et al., 1971 (mimicry); Brower and Brower, 1964 (mimicry); Ehrlich and Raven, 1967 (coevolution of butterflies and plants); Gilbert and Singer, 1975 (butterfly ecology); Hering, 1951 (leaf miners); Kettlewell, 1959, 1961 (industrial melanism); Powell, 1964 (life histories of many Tortricidae); Powell and Mackie, 1966 (biology of yucca moths); Roeder, 1965, 1966 (bat evasion); Sheppard, 1962 (mimicry and polymorphism); Tietz, 1973 (bibliography); Urquhart, 1960 (monarch butterfly).

SYSTEMATICS AND EVOLUTION

Common, 1975 (higher classification); Dugdale, 1974 (higher classification); Ehrlich, 1958 (higher classification); Ehrlich and Ehrlich, 1961 (butterflies); Forbes, 1923–1960 (outstanding survey of comparative morphology of order); Fracker, 1930 (larvae); Grote, 1971 (noctuidae of North America); Hinton, 1946, 1955*a*, 1958 (phylogenetic relationships); Howe, 1975 (butterflies of North America); Holland, 1968 (large moths of North America); Klots, 1951 (butterflies); Kristensen, 1968, 1971 (higher classification); Mosher, 1916 (pupae); Moths of America North of Mexico, 1972 to date; Petersen, 1948 (larvae).

ORDER TRICHOPTERA (Caddis Flies)

45

Adult. Small to moderate-sized Endopterygota with 2 pairs of wings. Head free, mobile, with large compound eyes, 2 to 3 ocelli present or absent, antennae filiform, elongate, with numerous segments; mandibles vestigial; maxillae with single small lobe or mala, maxillary palpi 3- to 5-segmented; labium with large mentum and 3-segmented palps. Thorax with first segment reduced, collarlike; mesothorax largest, metathorax large to small (Fig. 45.1*a*); wings with macrotrichia or hairs densely covering both veins and membrane; venation mostly of longitudinal veins with very few crossveins (Fig. 45.1*b*); legs long, slender, with long coxae (Fig. 45.1*a*); tibiae with both apical and preapical spurs; tarsi with 5 segments; abdomen with 9 segments in males, 10 segments in females.

Larvae (Fig. 45.2). Eruciform or campodeiform, frequently occurring in protective case or covering; head capsule sclerotized, with strongly developed mandibles, maxillae, and labium, very short antennae and lateral ocelli. Three distinct, short thoracic segments of variable sclerotization; legs long, slender, usually with some segments subdivided; abdomen soft, weakly sclerotized, with hook-shaped prolegs on terminal segment.

Pupa. Decticous, exarate, occurring in larval case or in specially constructed pupal case or cocoon.

With an estimated 5000 species worldwide, over 1000 in North America, the Trichoptera are one of the smaller orders of Endopterygota. The adults are slender, mostly somber-colored insects which somewhat resemble moths but never have the maxillae modified to form a proboscis, and have hair rather than scales on the body and wings. The basal portions of the radial and median veins are not obliterated to form a discal cell, as in most Lepidoptera, and in Trichoptera veins 2A and 3A in the forewing are usually looped forward to vein 1A (Fig. 45.1*b*). A useful diagnostic feature is the length of the antennae, which exceeds that of the forewing in most Trichoptera, being shorter in most Lepidoptera.

By day adult caddis flies rest concealed on vegetation, under bridges or in culverts, and are seldom seen without searching. In contrast, the larvae, which often inhabit portable cases, are conspicuous in many aquatic habitats, where they are of considerable importance as food for fish and other predators.

Adult Trichoptera are relatively homogeneous in external morphological features, seemingly adapted for a relatively short life. Although the mouthparts appear to be adapted for lapping up liquids, few Trichoptera have been observed to feed, and most species probably reproduce without taking any sustenance. Most caddis flies are weak, fluttery fliers, but others may fly great distances. In general the weaker fliers have relatively broad forewings, broadly overlapping the hind wings or with the jugal lobe enlarged, as in primitive Lepidoptera. In stronger-flying species the macrotrichia of the hind-wing base are developed to form a frenulum, which interlocks with the jugal lobe of the forewing. In several families a row of macrotrichia along the costal margin of the hind wing are enlarged and upcurved, hooking over a ventral, longitudinal ridge on the anal region of the forewing. In

TABLE 45.1 Characteristics of Larval Trichoptera of North America

Family	Case structure	Case material	Current	Temperature	Substrate
Philopotamidae	Long tunnels or nets	Silk—no extraneous material	Rapids, riffles	Cold or cool	Stony
Psychomyiidae	Tunnels or nets	Silk	Streams, ponds, lakes	Variable	Sandy bottoms, stones, aquatic vegetation
Hydropsychidae	Nets	Silk	Rivers, streams moderate current	Variable	Stones, logs, debris
Rhyacophilidae					
Rhyacophilinae	Free-living		Riffles, moving water	Cold	Irregular bottoms, stones, debris
Glossosomatinae	Tortoise shell–shaped; open both ends	Small stones, pebbles	Moving water	Cold	Stony, irregular bottoms
Hydroptilidae	Barrel-shaped or purse-shaped, open both ends	Variable	Variable	Variable	Variable
Phryganeidae	Stout, tubular case	Grass stem segments, spirally arranged	Marshes, stream eddies	Cold	Plant debris, silt
Brachycentridae	Cylindrical or square cases	Mostly silk (cylindrical cases) or rocks or woody material (square cases)	Rapids, riffles in streams, rivers	Cold	Variable
Limnephilidae	Tubular; variable in proportions	Extremely variable	Variable	Variable	Variable
Lepidostomatidae	Four-sided cases or slender tubes, or irregular cases	Sticks, twigs, sand, or varied materials	Springs, streams, rivers, lake shores	Cold	Variable
Calamoceratidae	Tubular, often triangular in cross section	Sticks, leaf fragments	Springs, slow streams	Cool or cold	Variable
Helicopsychidae	Helical spiral, snail shell–shaped cases	Stone	Springs, rapid streams, lake margins	Variable	Variable
Odontoceridae	Elongate tubular cases	Minute stones, compactly webbed together	Rapid streams	Cold	Sand, gravel, stones
Sericostomatidae	Stout, tubular cases	Minute stones	Rapid streams, springs	Cold	Sand, gravel, stones
Leptoceridae	Variable	Variable	Lakes, streams	Variable, often warm	Variable
Molannidae	Flattened, oblong cases	Minute stones, sand	Lakes, streams	Cool to cold	Stony
Beraeidae	Curved, tapering cases	Sand grains	Springs, seeps, small streams	Cool	Plant debris

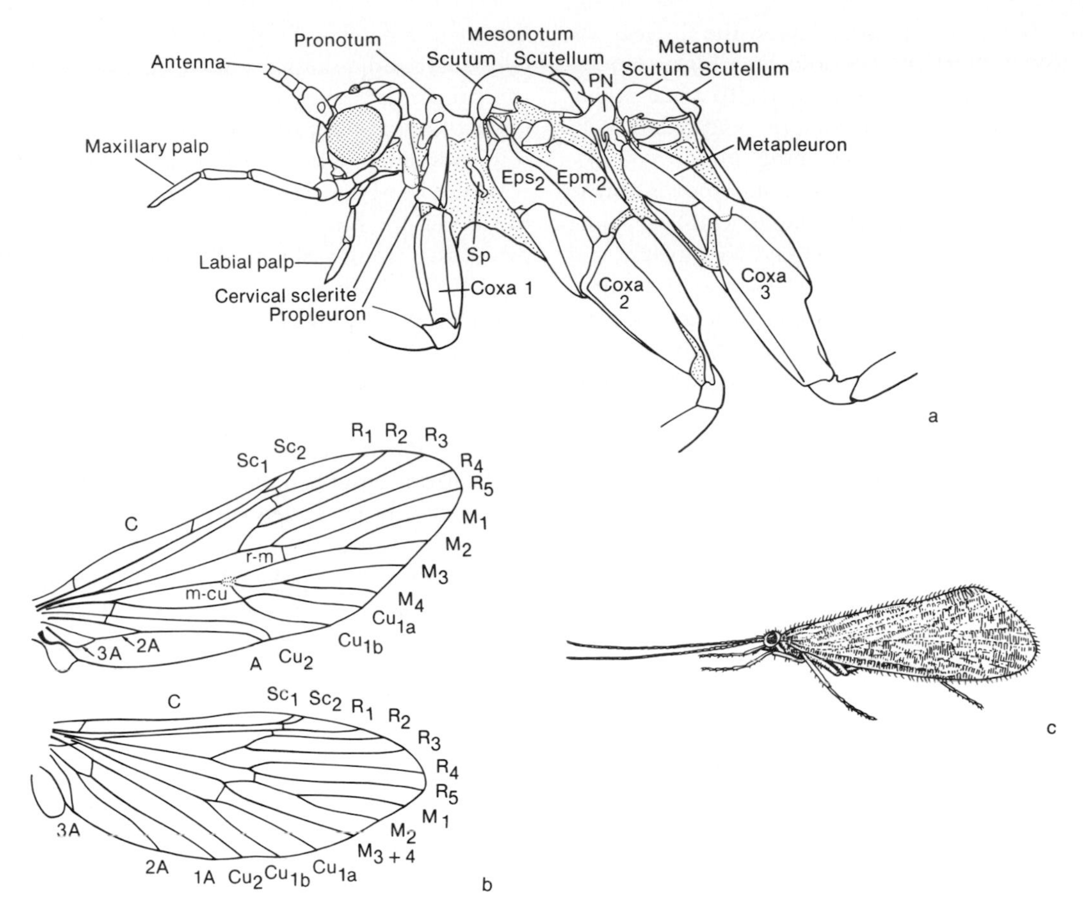

FIGURE 45.1 Trichoptera: *a*, head and thorax of *Rhyacophila torrentium* (Rhyacophilidae); *b*, wings of *Rhyacophila torrentium; c, Frenesia missa* (Limnephilidae). (a, b *from Schmid*, 1970; c *from Ross*, 1944). *Eps*, epsiternum; *Epm*, epimeron; *PN*, postnotum, *Sp*, spiracle.

Leptoceridae, which include many active, strong-flying diurnal species, and in Helicopsychidae, the modified hind-wing macrotrichia are developed as hamuli (Fig. 45.4*c*), much as in Hymenoptera, and the antennae are as much as three times the length of the body.

Larvae are of two general types. Free-living and net-spinning larvae are campodeiform, while those which inhabit cases tend to be eruciform. However, many intermediate forms are known. In Hydropsychidae, which appear to be most primitive, all the thoracic terga, as well as the prosternum, are sclerotized. The legs are robust and widely spaced, suited for walking, and the anal prolegs are stout with simple claws (Fig. 45.2*a*). Many of these non-case-making larvae are predatory and have slender elongate mandibles, superficially similar to those of Carabidae and other predatory Coleoptera.

Several modifications are apparent in larvae which construct cases. The terga of the meso- and metathorax and the prosternum are membranous. The thoracic segments are short, so that the legs are attached close together near the anterior end of the body, and can be easily protruded from the case (Fig. 45.2*b*). The middle and especially the hind legs are relatively long, further facilitating movement from within the case. In most caddis-fly larvae the forelegs are relatively short and are used more for manipulation of food than for walking. In some Rhyacophilidae the fore femora and tibiae are strongly raptorial, for grasping prey.

In case-making forms the abdomen is always extremely soft and vulnerable. Gills are more frequently present than in free-living species, and in some case makers a lateral fringe of abdominal hairs is used to circulate water through the case. Three tubercles, which position the larva within the case, are frequently present on the first abdominal segment (Fig. 45.2*b*). The terminal prolegs are fused basally, with the appendagelike distal portions ending in complexly pointed claws adapted for holding the larva inside the case.

Case construction and configuration are quite variable and probably arose independently several times. Most Rhyacophilidae are free-living, but members of its Subfamily Glossosomatinae construct saddle-shaped cases open at both ends, the protruding abdominal apex being used in locomotion. Hydroptilidae build somewhat similar cases, purse-shaped in general configuration, with the head and legs projecting from a slit in the anterior end, and the abdominal prolegs projecting from a slit in the posterior margin (Fig. 45.3*a*). Cases of other families are generally in the form of elongate tubes completely enclosing their inhabitants (Fig. 45.3*c* to *e*). Construction always involves binding particles from the substrate with silk produced from labial glands and extruded through a spinneret. Configuration varies from stout, straight tubes (many Limnephilidae) to long, slender, curved tubes (Leptoceridae). Cases are usually circular in cross section but may be polygonal [Brachycentridae, Leptoceridae (Fig. 45.3*e*)]. Members of the Helicopsychidae use sand grains or small pebbles to

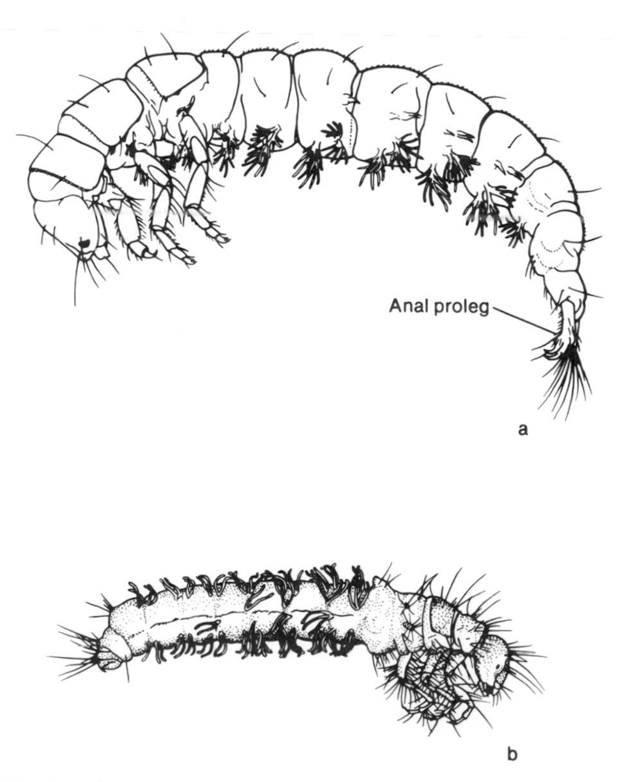

FIGURE 45.2 Trichoptera larvae: *a*, Rhyacophilidae (*Smicridea fusciatella*); *b*, Limnephilidae (*Philarctus quaeris*). (a *from Flint,* 1974; b *from Wiggins,* 1963.)

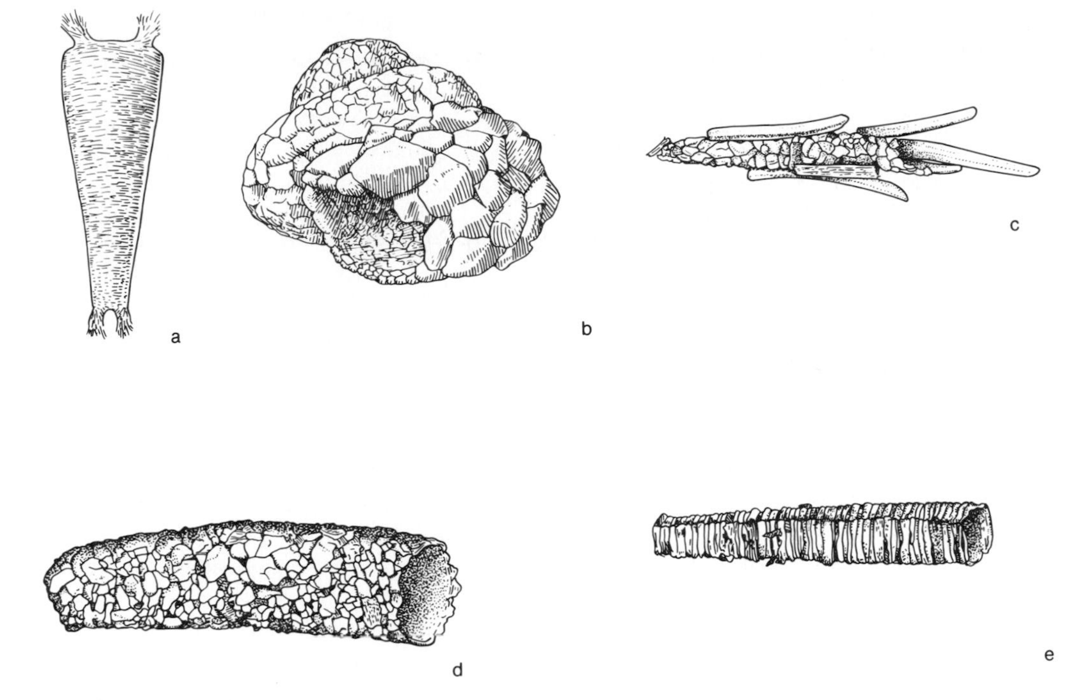

FIGURE 45.3 Cases of Trichoptera larvae: *a,* Hydroptilidae (*Oxyethira serrata*); *b,* Helicopsychidae (*Helicopsyche borealis*); *c,* Leptoceridae (*Mystacides sepulchralis*); *d,* Limnephilidae (*Pseudostenophylax edwardsi*); *e,* Brachycentridae (*Brachycentrus*). (a, b *from Ross,* 1944; c *from Yamamoto and Wiggins,* 1964; d *from Wiggins and Anderson,* 1968; e *from Wiggins,* 1965.)

build spiral tubes (Fig. 45.3*b*) which resemble snail shells, and Calamoceratidae build flat shelters from two pieces of leaf. In general, species inhabiting riffles and rapids build the most solidly constructed cases, those occupying lakes or ponds the loosest ones. Choice of building materials depends in part upon local availability, so that significant variation exists within species. Yet most Trichoptera choose particles within a certain size range and with a certain composition, and cases are often somewhat characteristic at the species level.

Caddis fly adults are encountered on vegetation and on stone overhangs and similar places near water. Most are nocturnal or fly at dusk or dawn, and many are attracted to lights. Copulation takes place in the air, at least in some species, and swarming behavior occurs in some families, including Leptoceridae. A few forms, such as *Dolophilodes* (Philopotamidae), are brachypterous or apterous. About 100 to over 1000 eggs, depending on the species, are laid in compact masses or long strings, directly into the water by most species, and females may oviposit while submerged. Limnephilidae may deposit egg masses on objects above the water, even on twigs high in trees. Apparently rain dissolves the matrix holding the eggs, which then drop into the water below.

Larvae pass through five to seven instars. Non-case-making forms may shelter beneath submerged stones or in debris (Rhyacophilidae) or may spin nets used for snaring food (Psychomyiidae, Hydropsychidae). Case-bearing species may occur exposed on sand or silt bottoms, shelter beneath stones or logs, or clamber over vegetation, depending on the species. Some Leptoceridae are capable of active swimming by using the fringed hind legs. Free-living and net-spinning forms tend to be predatory, while most case bearers are probably scavengers. Larval instars are similar in most species, but in the Hydroptilidae, the first four instars are free-living with short abdomen and long, slender abdominal prolegs. The fifth and sixth instars, which have a bloated, soft abdomen with short prolegs, inhabit cases. Early instars of some Lepidostomatidae build cases of sand grains, then switch to cases of leaf fragments as they grow.

Distribution of larval caddis flies is strongly influenced by the qualities of the substrate, as well as by current, presence and type of vegetation, and water temperature. Different portions of one stream or closely adjacent bodies of water which differ in one or more of these characteristics may support extremely different faunas.

Pupation occurs in water within a shelter—either the larval case, which may be shortened and closed, or a special pupal case. The pupal mandibles are used to open the case just before adult emergence. The pharate adult swims to the surface, using setal fringes on the middle legs, and immediately emerges and takes flight, or first climbs onto emergent vegetation or other objects.

Identification of most Trichoptera depends upon structures which shrivel in dried specimens; both larvae and adults should be collected into liquid preservatives. Pupae can frequently be identified by the features of the pharate adult. They are a valuable means of associating adults and larvae, since the cast skin of the last larval instar frequently remains in the case.

The following key is based on those of Ross (1944, 1959) and Wiggins (1977).

KEY TO THE NORTH AMERICAN FAMILIES OF **Trichoptera**

ADULTS

1. Wings without clubbed hairs; antennae about as long as wings or much longer; body length 5 to 40 mm — 2
   Wings with numerous erect clubbed hairs; antennae shorter than wings; densely hairy caddis flies less than 6 mm long — **Hydroptilidae**

2(1). Ocelli present — 3
   Ocelli absent — 8

3(2). Maxillary palpi with 3 to 4 segments — 4
   Maxillary palpi with 5 segments — 5

4(3). Maxillary palpi with 3 segments — **Limnephilidae** (♂♂)
   Maxillary palpi with 4 segments — **Phryganeidae** (♂♂)

5(3). Maxillary palpi with fifth (apical) segment subequal to fourth segment — 6
   Maxillary palpi with fifth segment 2 to 3 times longer than fourth segment — **Philopotamidae**

6(5). Maxillary palpi with first and second segments short, subquadrate, approximately equal in length — **Rhyacophilidae** (incl. **Glossosomatidae**)
   Maxillary palpi with second segment about twice length of first segment, distinctly longer than wide — 7

7(6). Fore tibiae with at least 2 spurs; middle tibiae with 4 spurs — **Phryganeidae** (♀♀)
   Fore tibiae with 1 spur or without spurs; middle tibiae with 2 to 3 spurs — **Limnephilidae** (♀♀)

8(2). Maxillary palpi with at least 5 segments — 9
   Maxillary palpi with 3 to 4 segments — 11

9(8). Maxillary palpi with fifth (apical) segment at least twice as long as fourth segment; fifth segment annulate, appearing multisegmented — 10
   Maxillary palpi with fifth segment subequal to fourth segment, not annulate — 11

10(9). Hind wings at least as broad as forewings, with anal region occupying one-third to one-half of wing; mesoscutum without raised, wartlike areas — **Hydropsychidae**
   Hind wings narrower than or occasionally subequal to forewings, with anal region usually occupying less than one-fourth of wing area; mesoscutum with a pair of raised, wartlike tubercles (as in Fig. 45.4*b*) — **Psychomyiidae** (incl. **Polycentropodidae**)

11(8,9). Middle tibiae without preapical spurs 12
Middle tibiae with preapical spurs 15

12(11). Antennae 1.5 to 3 times length of forewing; mesoscutum with irregular raised spots bearing setae (Fig. 45.4*a*) **Leptoceridae**
Antennae about as long as forewing; mesoscutum with small paired tubercles (Fig. 45.4*b*) or without tubercles 13

13(12). Hind wings with anterior margin straight or evenly curved 14
Hind wings with anterior margin straight basally, with row of stout curved setae, then abrubtly curved (Fig. 45.4*c*) **Helicopsychidae**

14(13). Middle tibiae with apical spurs about one-half length of basal tarsomere; distal segments of middle and hind tarsi with spines in apical crown on each segment **Beraeidae**
Middle tibiae with apical spurs about one-fourth length of basal tarsomere; distal segments of middle and hind tarsi with scattered spines **Sericostomatidae** and **Brachycentridae**

15(11). Middle femora with a row of 6 to 10 dark spines on apex of anterior face **Molannidae**
Middle femora with 2 or fewer dark spines on anterior face 16

16(15). Mesoscutellum small, arcuate (Fig. 45.4*d*); mesoscutum with a pair of longitudinal rows of setate spots **Calamoceratidae**
Mesoscutellum large, triangular or acutely rounded, pointed apically; mesoscutum with tubercles round or oval (Fig. 45.4*e*,*f*) 17

17(16). Mesoscutellum with a single large, raised area medially (Fig. 45.4*e*,*f*) 18
Mesoscutellum with paired, lateral raised areas 19

18(17). Tibial spurs hairy; mesoscutellum triangular, with medial raised area (Fig. 45.4*e*) some **Limnephilidae**

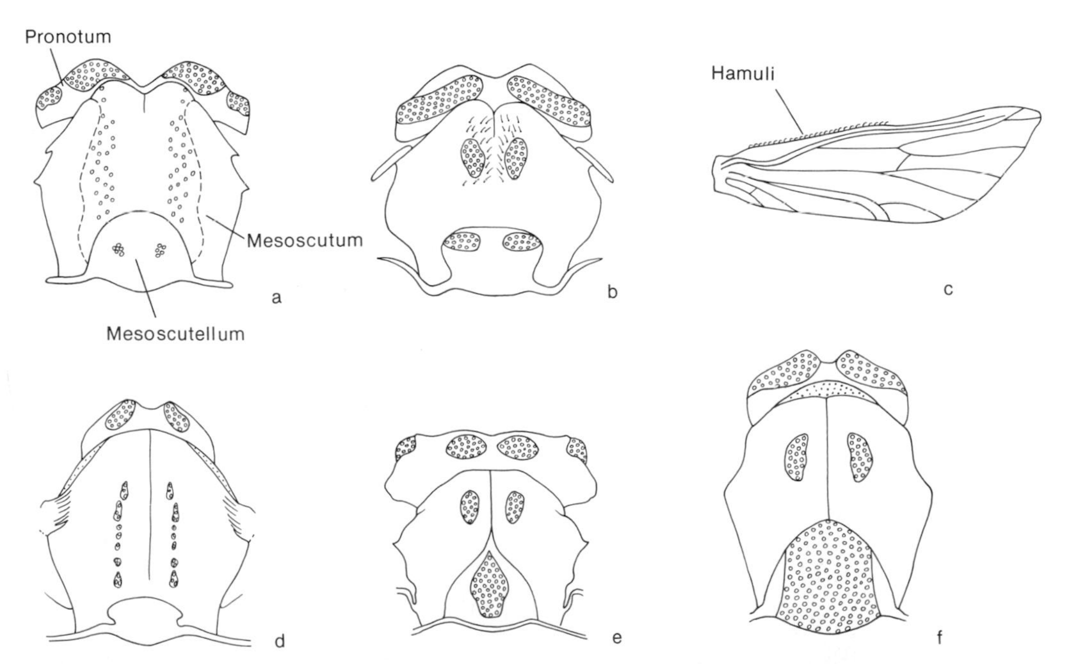

FIGURE 45.4 Taxonomic characters of Trichoptera, pronotum and mesonotum, and wings: *a*, Leptoceridae (*Athripsodes tarsipunctatus*); *b*, Helicopsychidae (*Helicopsyche borealis*); *c*, Helicopsychidae (*Helicopsyche*); *d*, Calamoceratidae (*Ganonema americanum*); *e*, Limnephilidae (*Goera calcarata*); *f*, Odontoceridae (*Psilotreta frontalis*). (*From Ross*, 1944.)

Tibial spurs without hairs; mesoscutellum acutely rounded with raised wart occupying most of area (Fig. 45.4*f*) **Odontoceridae**

19(17). Middle tibiae with irregular row of spines; middle tarsi with double row of spines **Brachycentridae**

Middle tibiae without spines; middle tarsi with sparse, scattered spines **Lepidostomatidae**

LARVAE

1. Metathorax with single tergal plate 2
Metathorax entirely membranous or with dorsum divided into at least 2 sclerites 3

2(1). Abdomen without gills; abdominal leg with 2 to 3 bristles at base of claw **Hydroptilidae**
Abdomen with numerous branched gills ventrally; abdominal leg with broad fan of setae at base of claw **Hydropsychidae**

3(1). Abdomen with sclerotized plate on ninth segment 5
Abdomen entirely membranous 4

4(3). Labrum membranous, narrowed basally **Philopotamidae**
Labrum sclerotized, broadest basally **Psychomyiidae** (incl. **Polycentropodidae**)

5(3). Abdominal prolegs fused basally, appearing as tenth abdominal segment; claws directed laterally, much shorter than basal portion of proleg **Rhyacophilidae** (incl. **Glossosomatidae**)
Abdominal prolegs free; claws directed ventrally, about as long as basal portion of proleg 6

6(5). Claws on all legs similar in structure 7
Claws of hind legs much smaller than those of front and middle legs, modified as short, setose clubs **Molannidae**

7(6). Antennae 1 to 4 times as long as wide 8
Antennae at least 8 times as long as wide **Leptoceridae**

8(7). Mesonotum with conspicuous sclerotized plates 9
Mesonotum membranous or with minute sclerites **Phryganeidae**

9(8). Mesonotum with a single pair of narrow, arcuate sclerites oriented longitudinally near midline (Fig. 45.5*a*) **Leptoceridae**
Mesonotum with 1 to several subquadrate, transverse, or oval sclerites 10

10(9). Labrum with about 16 long, stout setae in a closely set, transverse row across middle (Fig. 45.5*b*) **Calamoceratidae**
Labrum with 6 to 8 large setae arranged in a loose, transverse arc or irregularly scattered (Fig. 45.5*c*) 11

11(10). Larva inhabiting a spirally twisted case (Fig. 45.3*b*); claws of abdominal legs with a comb of subequal teeth **Heliocopsychidae**
Larval case not spirally twisted; claws of abdominal legs with teeth not arranged as a comb 12

12(11). Metanotum with broad anterior sclerite, longitudinal lateral sclerites, and narrow,

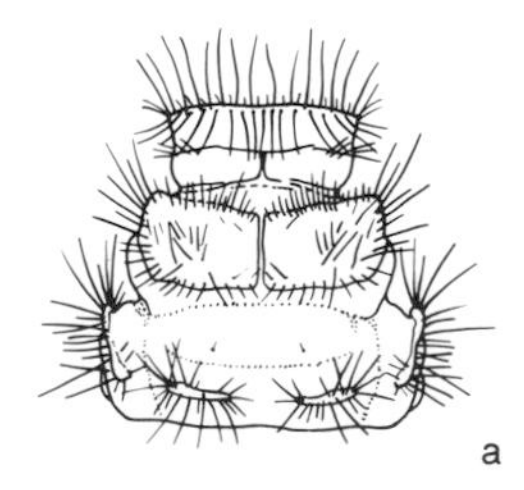

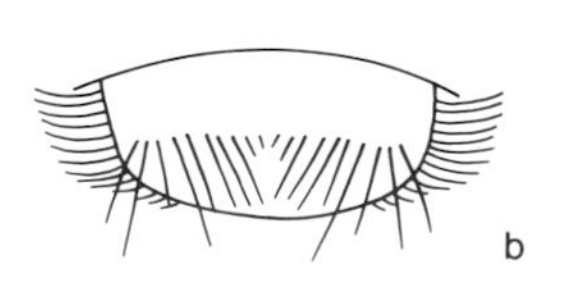

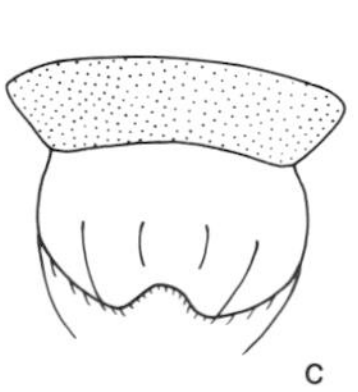

FIGURE 45.5 Taxonomic characters of Trichoptera larvae: *a*, thorax of Leptoceridae (*Athripsodes*); *b*, labrum of Calamoceratidae (*Ganonema*); *c*, labrum of Limnephilidae (*Limnephilus*). (*From Ross*, 1944.)

transverse posterior sclerite **Odontoceridae**

Metanotum usually with 1 to several small, rounded, poorly defined sclerites 13

13(12). Pronotal sclerite bearing strong, transverse furrow 14

Pronotal sclerite without transverse furrow 15

14(13). Hind leg with tarsal claw longer than tibia; claw without basal tooth **Beraeidae**

Hind leg with tarsal claw much shorter, usually about one-half length of tibia; claw with prominent basal tooth **Brachycentridae**

15(13). Antennae inserted adjacent to anterior margin of eye **Lepidostomatidae**

Antennae inserted midway between eyes and mandibles or closer to mandibles 16

16(15). Antennae inserted midway between eyes and mandibles; prosternum with distinct, medial hornlike projection **Limnephilidae** (including **Goeridae**)

Antennae inserted very close to base of mandibles; prosternum without hornlike projection **Sericostomatidae**

SELECTED REFERENCES

Denning, 1956 (California species); Lloyd, 1921 (biology of larvae); Merritt and Cummins, 1978 (North America); Ross, 1944 (Illinois species), 1956 (montane species), 1959 (North American species), 1964 (nets and cases of larvae), 1967 (biogeography); Wiggins, 1977 (larvae).

46

ORDER HYMENOPTERA (Bees, Wasps, Ants, etc.)

Adult. Minute to large Endopterygota, usually with 2 pairs of wings. Head free, mobile, with mandibles adapted for chewing; galeae or labium frequently elongate, suited for imbibing liquids; antennae multiarticulate, usually filiform or moniliform; compound eyes large, lateral; 3 ocelli usually present. Mesothorax enlarged, fused with small, collarlike pronotum and small metathorax (Fig. 46.1); wings membranous, stiff, hyaline, with relatively few veins, which usually delimit a few large, closed cells; forewings about 1.5 to 2 times as long as hind wings; wing coupling effected by *hamuli* (hooklike setae on leading edge of forewing). Abdomen with first segment inflexibly joined with metathorax, usually incorporated into thorax as *propodeum;* second abdominal segment usually constricted as narrow *petiole,* followed by swollen *gaster;* female with ovipositor modified as slicing or piercing organ for inserting eggs into tissue, or adapted as sting.

Larva. Eruciform, with distinct head capsule, chewing mandibles, 3 pairs thoracic legs and 6 to 8 pairs abdominal prolegs; or apodous, grublike, often with head capsule and mouthparts reduced or vestigial.

Pupa. Adecticous, usually exarate; frequently enclosed in silken cocoon secreted from labial glands.

Certain Hymenoptera, such as honeybees, ants, and some of the larger wasps, are among the most familiar insects. In contrast to most other orders, many Hymenoptera are beneficial to agriculture, through either pollination of crops or destruction of phytophagous insects. Parasitoid species have proved extremely useful in biological control, especially of introduced pests, and have been transported extensively about the earth. *Apis mellifera,* the honeybee, among the most thoroughly domesticated insects, is extremely important as a pollinator, and its products, honey and wax, are still commercially important, despite competition from alternative materials. Bee and wasp stings, annoyances from ants, and occasional damage by sawflies, gall wasps, or other phytophagous species are of only nuisance significance when compared with the positive attributes of this order. Hymenoptera comprise about 108,000 species, but this number is certain to be drastically increased as the numerous small, parasitoid species are discovered.

The most important adaptive features in Hymenoptera involve the mouthparts, thoracic and wing structure, and abdomen, including the ovipositor, all these structures apparently being modified in relation to predation. All Hymenoptera possess well-developed mandibles, which in many species are used for tasks other than feeding and may be modified for specific functions. For example, some megachilid bees and some ants cut foliage with the sharp scissorslike mandibles, while in the soldier castes of many ants the mandibles are designed for offense or defense. In most species, however, multiple functions are served. For instance, a worker ant may dig, transport food or soil particles, subdue or manipulate prey, defend the colony, or tend immatures—using the mandibles as a primary tool. Many Hymenoptera, especially the more advanced families of wasps and all bees, have evolved the ability to ingest liquid food, primarily nectar, through a proboscis formed by the glossae (part of the labium) and the maxillary

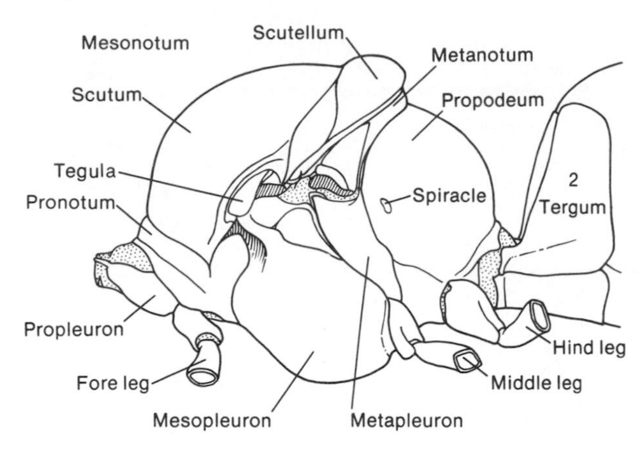

FIGURE 46.1 Lateral aspect of pterothorax of *Apis mellifera*. (*From Snodgrass*, 1942.)

palpi and galeae (Fig. 46.2*b*). In anthophorid and apid bees the proboscis may be longer than the body.

As in other aerial insects, the hymenopterous thorax is strongly modified for efficient flight. The mesothorax is much the largest segment, and in most species is solidly fused to the reduced pronotum and metathorax (Fig. 46.1). The head and forelegs are attached to the propleura, which are usually flexibly articulated to the pronotum and mesopleura. In general, the prothorax is best developed in ambulatory species such as ants, or in those which use the forelegs to dig (Pompilidae, Scoliidae). The metanotum is represented by a narrow sclerite situated between the mesopostnotum and the first abdominal tergite, which is inflexibly incorporated into the thorax as the *propodeum.* The propodeum may be recognized by the presence of the first abdominal spiracle (Fig. 46.1). The pleural sclerites are usually large, distinct plates, bearing the mesocoxal and metacoxal attachments. In general the extensive variation in size and shape of the various thoracic sclerites is distinctive of families and superfamilies and is of some use in classification.

The hind wings are always coupled to the much larger forewings by a series of hooked setae (*hamuli*) which extend forward from the costal margin of the hind wing and catch beneath a ventral fold of the anal margin of the forewing (Fig. 46.2*a*). The hamuli number from two or three in small species such as Chalcidoidea, to many in the larger wasps, where wing coupling may be very tenacious. Simplification of wing venation by reduction is typical of Hymenoptera, being broadly correlated with body size. Very small species, such as some Chalcidoidea, may retain only a single longitudinal vein in the forewing, with the hind wing veinless (Fig. 46.11*e*). In general, venation is highly variable and characteristic of various families or superfamilies, but aerodynamic significance of the different configurations is obscure.

The morphological specializations most responsible for the success of the Hymenoptera involve the abdomen. In the relatively primitive Suborder Symphyta the abdomen is broadly and relatively inflexibly joined to the thorax. The ovipositor is bladelike or sawlike in most symphytan families, adapted for slicing plant tissues. In the Suborder Apocrita the abdomen is strongly constricted at the second segment, greatly increasing its mobility (Fig. 46.1). The ovipositor is always developed as a piercing organ, which can be inserted deeply into host (usually animal) tissues. Morphologically the ovipositor consists of three pairs of gonapophyses or *valves* (Fig. 46.2*c* to *e*). The third valvulae form an external cover, while the first and second valvulae are driven into the host and convey the eggs and paralytic fluids in parasitic species. In

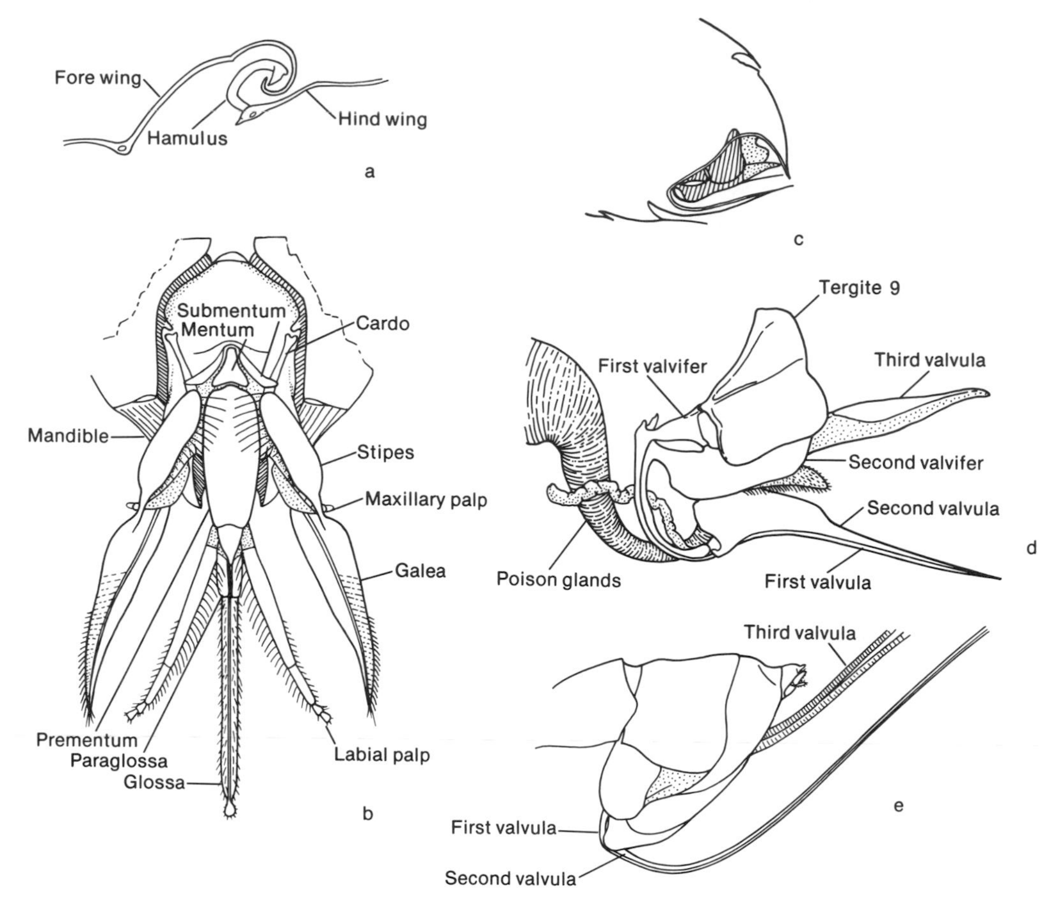

FIGURE 46.2 Structural features of Hymenoptera: *a*, diagramatic cross section of wings of *Apis mellifera* (Apidae), showing wing coupling; *b*, mouthparts of *Apis mellifera; c*, medial section through abdomen of *Apis mellifera*, showing sting in repose; *d*, dissected sting of *Apis mellifera; e*, end of abdomen and ovipositor of *Megarhyssa atrata* (Ichneumonidae). (*From Snodgrass*, 1933, 1942.)

wasps and bees the first and second pair of valves are further specialized as a stinger, discharging only venom. In most wasps the stinger is used to immobilize prey, but in Vespoidea (yellow jackets, hornets) and Apoidea (bees), its primary function is defense.

In general, larvae of Hymenoptera conform to two body plans. Larvae of many Symphyta, such as sawflies, are highly similar to lepidopterous caterpillars, and, like most caterpillars, are external foliage feeders (Fig. 46.3*a*). The large chewing mandibles are set in a strongly sclerotized head capsule. The thick, cylindrical trunk bears appendages on most abdominal segments, as well as on the thoracic segments. The abdominal legs, unlike those of lepidopterous larvae, lack grasping spines or crochets. Other useful differentiating characters include the presence of a single stemmata (six stemmata usually present in lepidopterous larvae) and the reduction of the antennae to papillae (antennae usually three-segmented in lepidopterous larvae).

The symphytan Families Siricidae and Cephidae, as well as certain Tenthredinidae are internal plant feeders which lack abdominal legs and have the thoracic legs reduced. All larvae of the Suborder Apocrita are apodous, with soft, maggotlike bodies and sometimes with the head

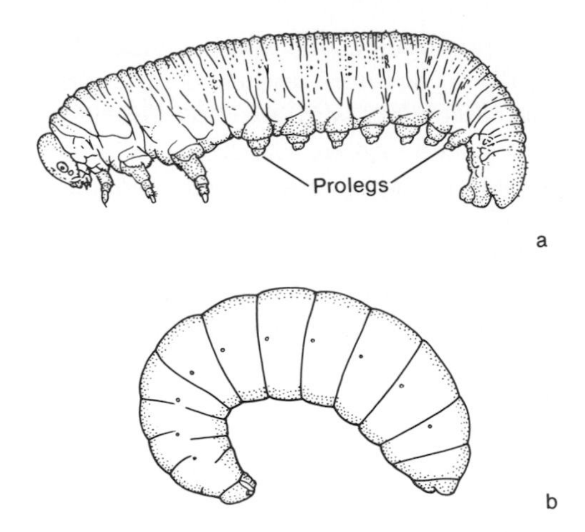

FIGURE 46.3 Larvae of Hymenoptera: *a*, eruciform larva of Symphyta (Diprionidae, *Neodiprion lecontei*); *b*, apodous larva of Apocrita (Apidae, *Xylocopa virginica*). (a *from Middleton*, 1921; b *from Stephen et al.*, 1969.)

capsule unsclerotized or retracted into the thorax (Fig. 46.3*b*). These larvae all inhabit protected environments (usually the body cavity of a host insect, or a cell prepared by the mother); hence the structural simplification. The midgut is not connected with the hindgut in most apocritan larvae. This is evidently an adaptation for avoiding fecal contamination of the host body or larval cell. At the last larval molt the alimentary canal becomes complete, and the accumulated fecal material, or *meconium,* is voided just before pupation.

Various parasitoid Hymenoptera are hypermetamorphic, the first instar, or occasionally more than one instar, being differentiated from the later maggotlike stages. For example, in some chalcidoid wasps which do not deposit their eggs in a host, the first instar is a *planidium* (Fig 46.4*a*). Planidia resemble the first-instar triungulin larvae of Coleoptera (Meloidae, Rhipiphoridae) and Strepsiptera, and perform the same function—host location. Locomotion is accomplished with the ventral extensions of the segmental sclerites, as legs are absent. *Caudate larvae,* found in some Ichneumonidae and Braconidae, and *nauplioid larvae* (Fig. 46.4*b*, *c*), in some Proctotrupoidea, hatch internally in the host. They may function in eliminating supernumerary parasites, as suggested by the large, sickle-shaped mandibles of the nauplioid type. A variety of other first instar forms—sometimes bizarrely spined or shaped—of uncertain function occur in different families of parasitoids.

Adult Hymenoptera, with the notable exception of some nocturnal ichneumonids and bees, are largely sunshine-loving creatures which frequent flowers, sunlit vegetation, or open soil. However, as parasitoids and predators, they have followed other insects into almost every niche imaginable. As adults, many Hymenoptera consume nectar and pollen, but many sawflies (Tenthredinidae) are predators of small, soft-bodied insects, and some parasitoid forms imbibe hemolymph which oozes from wounds made during oviposition.

Mating may occur in the air (as in honeybees, ants) or on some substrate. In Hymenoptera, males (if present) are produced parthenogenetically from unfertilized eggs, and are haploid, while the diploid females are produced from fertilized eggs. Sperm are stored in a spermatheca, where they may remain viable for long periods, their release being controlled by the female. Female control of fertilization allows unusual sex ratios, which may be extreme in

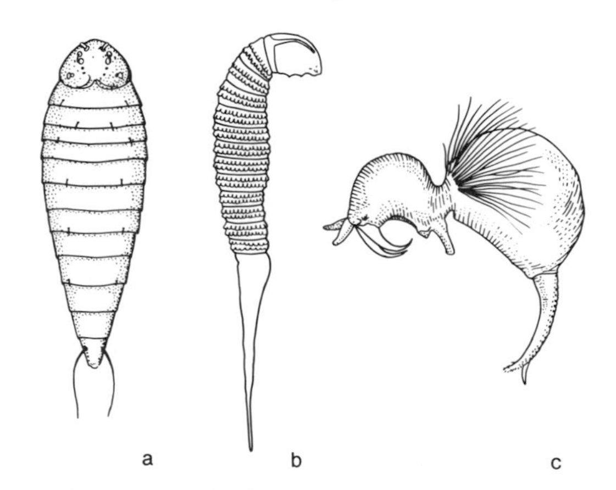

FIGURE 46.4 First-instar larvae of parasitoid Hymenoptera: *a*, planidium larva of Chalcidoidea (Perilampidae, *Perilampus hyalinus*); *b*, caudate larva of Ichneumonoidea (Ichneumonidae, *Cremastus flavoorbitalis*); *c*, eucoiliform larva of Proctotrupoidea (Scelionidae, *Scelio fulgidus*). (a *from Smith*, 1912; b *from Bradley*, 1934; c *from Noble*, 1938.)

social species, where relatively small numbers of males are produced only once or a few times a year. Various species of Hymenoptera are entirely parthenogenetic, being known only from diploid females. Cynipidae (gall wasps) are *cyclically parthenogenetic,* with generations of bisexual males and females alternating with generations of asexual females.

Several distinct types of life history occur within the order. As mentioned above, most or all members of the Suborder Symphyta are strictly herbivorous as larvae, and this mode of feeding is apparently primitive for the order. The vast majority of Apocrita are zoophagous as larvae. The most primitive species are parasitoids, living internally or sometimes externally on a host which is almost always selected by the mother. The host is always at least as large as the parasitoid, so that a single victim is sufficient to allow a single parasitoid to develop. It is uncertain how parasitoidism arose from phytophagous ancestors, but *inquilinism* could have been an intermediate stage. Inquilines inhabit galls or nests of other species, consuming the food intended for the original occupant, which is incidentally killed and eaten as well, usually as an egg or early-instar larva. Several primitive families, principally in the Superfamilies Proctotrupoidea and Chalcidoidea, include inquiline species.

Parasitoidism is practiced by about half the families of Hymenoptera, including some eminently successful taxa such as Ichneumonidae, Braconidae, and several families of Chalcidoidea. Most orders of insects as well as spiders and ticks are attacked. All instars, including eggs, are utilized as hosts. In typical life histories, larval development of the parasite begins soon after oviposition, and in some species eggs hatch during oviposition. Eggs of some species are deposited in immature hosts (eggs or early-instar larvae), where they may eclose, but larval development is postponed, sometimes as much as a year, until the host is nearly full-grown. Such *delayed parasitoidism* occurs in Ichneumonoidea and especially in the Proctotrupoidea. Many chalcidoids, as well as occasional members of other superfamilies, are *hyperparasitoids.* The host is another parasitoid rather than a primary victim. Secondary and even tertiary levels of hyperparasitoidism are known.

Most of the stinging wasps are predators that immobilize or kill one or more prey, which are installed in a specially constructed cell in which the larva matures. The Pompilidae (spider wasps), Bethylidae, Scoliidae, and Tiphiidae clearly bridge the difference between parasitoidism and typical predation. For example, some pompilids oviposit on a temporarily paralyzed host, which may partially regain motility before being consumed. Some species attack spiders in their burrows, where larval development occurs, while others prepare their own burrow, either before or after subduing the prey. In typical predation, most highly developed in the Superfamilies Vespoidea and Sphecoidea, nest preparation precedes prey capture, and in most species more than a single prey organism is required for larval growth.

Bees (Superfamily Apoidea) are secondarily phytophagous. Like the morphologically similar Sphecoidea, bees construct larval galleries, but they provision their larvae with pollen and nectar, rather than with paralyzed insects. The Families Megachilidae, Apidae, and Anthophoridae include some *cleptoparasitic* species which appropriate the nests and provisions of phytophagous species. Bees are notable, along with the ants and vespoid wasps, in having evolved complex societies. In all these taxa, sociality involves repression of reproduction in most female individuals, which function in obtaining food and maintaining and defending the colony. Small numbers of males are produced periodically, functioning only in reproduction. Sociality is discussed in greater detail in Chap. 7.

KEY TO THE SUBORDERS OF **Hymenoptera**

1. Abdomen with base broadly attached to thorax (Fig. 46.7); thoracic nota almost always with cenchri (Fig. 46.7*b*); hind wings almost always

with 3 basal cells (Fig. 46.6*b* to *d*)
Suborder Symphyta (p. 483)

Abdomen with base narrowly constricted at attachment to thorax (Fig. 46.9*c,d*); thoracic nota without cenchri; hind wings with 2 basal cells or fewer (Fig. 46.9*e*) **Suborder Apocrita** (p. 486)

Suborder Symphyta

Abdomen with base broadly attached to thorax, but with first abdominal tergite distinct, not incorporated as part of thorax; metanotum bearing raised, roughened, knoblike *cenchri* (Fig. 46.7*b*) (absent in Cephidae); hind wings usually with 3 basal cells (Fig. 46.6*c-d*) (except in Orussoidea, Fig. 46.8). Larvae eruciform, caterpillarlike, with large, round head capsule and 6 to 8 pairs of abdominal legs (external feeders, Fig. 46.3*a*), or abdominal legs absent (some internal feeders).

Symphyta include the most generalized Hymenoptera, both in terms of morphological characteristics and biological adaptations. Larvae are herbivorous, with the possible exception of the Orussoidea. In number of species, only the Superfamily Tenthredinoidea, whose members are largely adapted as external foliage feeders, is dominant today.

KEY TO THE NORTH AMERICAN SUPERFAMILIES OF **Symphyta**

1. Antennae inserted beneath a transverse ridge below the eyes; wings with veins strong, thick in basal half, faint apically (Fig. 46.8)
Orussoidea (p. 484)

Antennae inserted between the eyes; veins equally developed throughout wing 2

2(1). Anterior tibia with 1 apical spur 3

Anterior tibia with 2 apical spurs 4

3(2). Pronotum much wider than long (Fig. 46.5*a,b*), often with posterior margin strongly incurved; abdomen cylindrical
Siricoidea (p. 483)

Pronotum longer than wide or subquadrate (Fig. 46.7*c*); abdomen slightly compressed in lateral plane **Cephoidea** (p. 485)

4(2). Costal cell divided by intercostal vein, or if fused with subcostal vein, forewing with 3 marginal cells (Fig. 46.6*d*) **Xyeloidea** (p. 483)

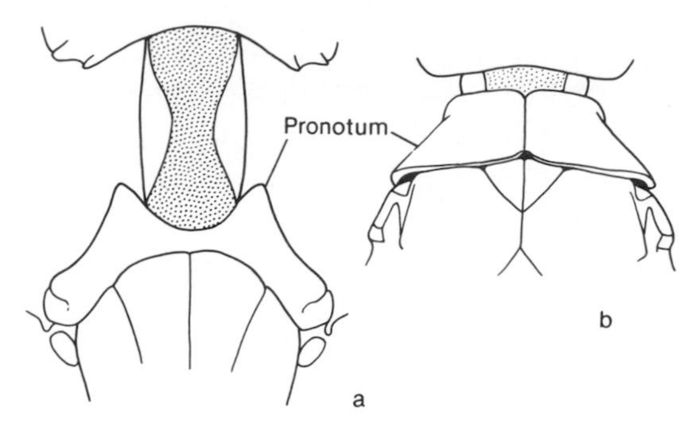

FIGURE 46.5 Taxonomic characters of Hymenoptera: *a, b,* dorsal aspects of pronota of Xiphydriidae and Syntexidae, respectively.

Costal cell without intercostal vein (Fig. 46.6*c*); forewing with 1 or 2 marginal cells
Tenthredinoidea (p. 485)

SUPERFAMILY XYELOIDEA. Small to moderate sawflies with the wing venation very generalized (Fig. 46.6*d*) and the ovipositor short or rudimentary.

The earliest known fossil hymenopteran, *Liadoxyela praecox,* from the Lower Triassic Period, is very similar to existing xyelids, which have the most generalized wing venation in the order. The superfamily contains two families in North America. Xyelidae are distinguished by the elongate third antennal segment. The larvae, which have prolegs on all abdominal segments, feed on staminate (male) cones of pines, in the case of *Xyela,* or attack needle buds of firs, in the case of *Macroxyela,* or foliage of other trees. Adults appear very early in the spring and may visit early flowers, especially those with massed blooms, such as chokecherry. Pamphiliidae have at least 13 antennal segments of approximately equal length. Larvae, which lack abdominal legs, tie or roll leaves of various trees, or feed beneath self-constructed webs.

SUPERFAMILY SIRICOIDEA (woodwasps, horntails). Moderate or large wasps with robust, cylindrical body; anterior tibia with single apical spur.

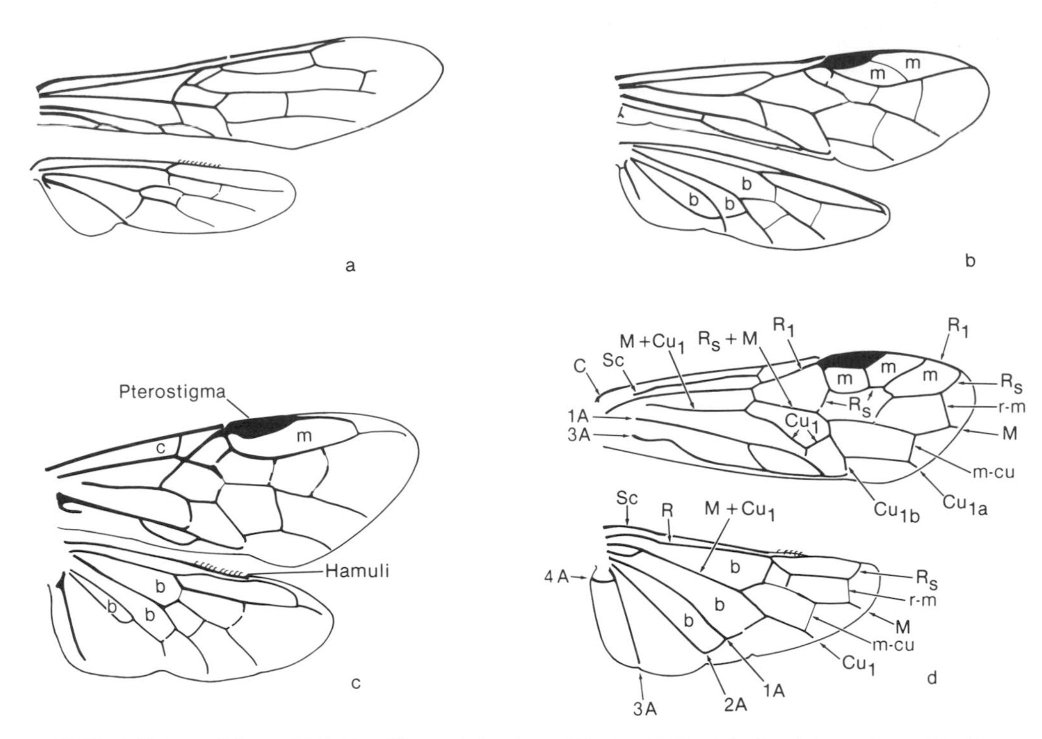

FIGURE 46.6 *a,* Wings of Siricidae (*Tremex*); *b,* wings of Tenthredinidae (*Tenthredo*); *c,* wings of Argidae (*Arge dulciana*); *d,* wings of Xyelidae (*Macroxyela ferruginea*). *b,* basal cell; *c,* costal cell; *m,* marginal cell; *Sc,* intercostal vein (subcosta).

The Siricoidea are associated with wood, where the short-legged larvae excavate tunnels. Females possess a cylindrical ovipositor, which is used to bore oviposition holes into living or felled trees. The most familiar insects in this superfamily are the Siricidae (Fig. 46.7*a*), whose larvae bore into the heartwood of various trees, occasionally causing minor damage. Syntexidae includes a single species whose larvae tunnel incense cedar (*Libocedrus*) in the Pacific Coast states. Larvae of Xiphydriidae bore in angiosperm trees, including birch, maple, oak, elm, and poplar. The wasps are fairly common in the Eastern states.

KEY TO THE FAMILIES OF

Siricoidea

1. Abdomen with last segment modified as short, upturned prong or spine (Fig. 46.7*a*); mesonotum bearing 2 diagonal grooves converging toward scutellum **Siricidae**

   Abdomen with last segment not modified as a prong or spine; mesonotum without grooves **2**

2(1). Pronotum with posterior margin strongly incurved (Fig. 46.5*a*) **Xiphydriidae**

   Pronotum with posterior margin nearly straight (Fig. 46.5*b*) **Syntexidae**

SUPERFAMILY ORUSSOIDEA (Fig. 46.8). A single family, Orussidae, of uncommon wasps whose biology and evolutionary relationships are unclear.

Adults are similar to other Symphyta in having a broad thoracic-abdominal articulation. Unlike Symphyta, orussids have a wing venation that is reduced, and the very long ovipositor is coiled like a watchspring inside the abdomen. The insertion of the antennae beneath a ridge below the eyes is unique among Symphyta. The apodous larvae occur in galleries of buprestid beetles, but it is uncertain whether they are

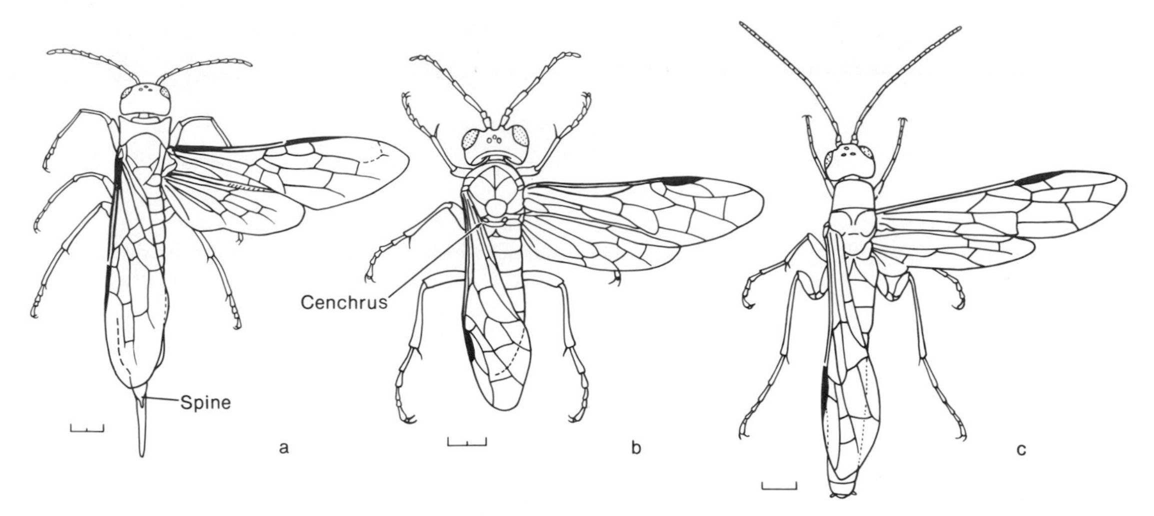

FIGURE 46.7 Representative Symphyta: *a*, Siricidae (*Sirex juvencus*, female; scale equals 2 mm); *b*, Tenthredinidae (*Tenthredo;* scale equals 2 mm); *c*, Cephidae (*Cephus clavatus;* scale equals 1 mm).

parasitoids of the beetles or subsist on the borings and fungus filling old burrows.

SUPERFAMILY CEPHOIDEA (stem sawflies). Very slender, elongate insects with the abdomen slightly compressed in the lateral plane.

The single family, CEPHIDAE, includes delicate wasps, either black or patterned in black and yellow (Fig. 46.7*c*). Adults are encountered on vegetation or flowers. Larvae of most species bore the stems of grasses, including small grains, and *Cephus cinctus* is a minor pest on wheat. Shrubs with pithy stems are attacked by a few species. Pupation occurs within the larval gallery. The lack of cenchri and the relatively great constriction of the abdomen of adults suggest that the cephoid line of evolution may have given rise to the Apocrita. This relationship is reinforced by the legless, internally feeding larvae.

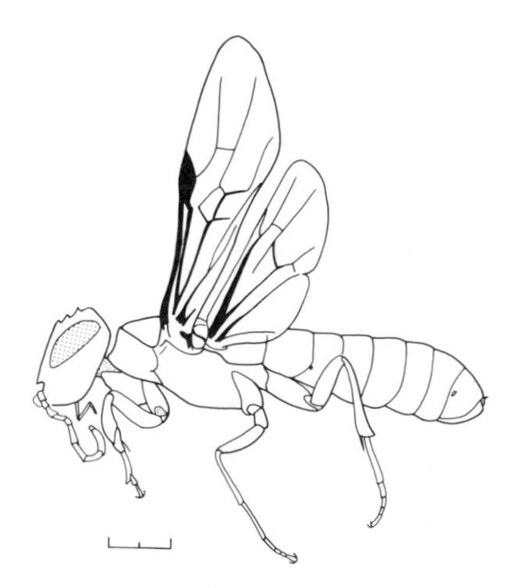

FIGURE 46.8 Orussidae (*Orussus* sp.; scale equals 2 mm).

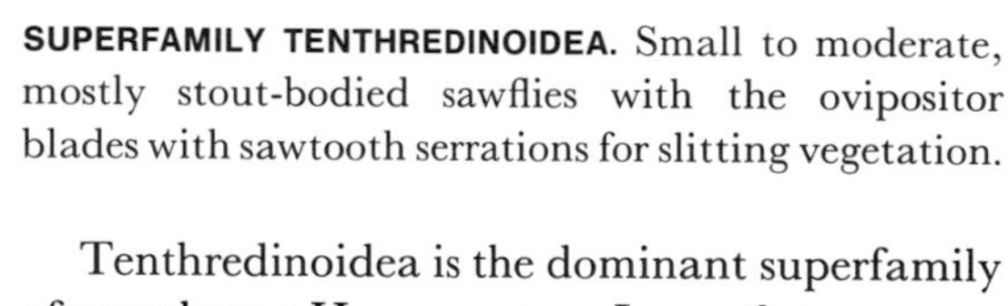

SUPERFAMILY TENTHREDINOIDEA. Small to moderate, mostly stout-bodied sawflies with the ovipositor blades with sawtooth serrations for slitting vegetation.

Tenthredinoidea is the dominant superfamily of symphytan Hymenoptera. In north temperate regions the Family TENTHREDINIDAE (Fig. 46.7*b*), with over 4000 described species, many abundant and widespread, includes nearly all commonly collected sawflies. The Family PERGIDAE is largely restricted to South America and Australia, where the other families of Tenthredinoidea are relatively uncommon.

Sawfly larvae are characteristically external foliage feeders. Some consume herbs or occasionally ferns, but trees are especially favored, and

DIPRIONIDAE are restricted to conifers. Externally feeding larvae are usually caterpillarlike in general appearance, with six to eight pairs of abdominal legs, which lack crochets (Fig. 46.3*a*). Sawfly larvae are often camouflaged in shades of green, sometimes with contrasting tubercles or spines in red, white, or other colors. Many species evert from the abdominal sternites glandular invaginations which resemble the osmeteria of Lepidoptera. Some tenthredinid and cimbicid larvae are covered with a whitish powder, and some tenthredinids are sluglike and covered with slimy secretions. Internally feeding larvae, which include leaf miners (ARGIDAE, Tenthredinidae) and gall formers (Tenthredinidae), may lack abdominal legs. Sawflies typically have a single generation per year. Larvae usually leave the host plant to pupate in a cocoon or a cell in soil or leaf litter. Adults are commonly found on flowers, and may imbibe nectar or eat pollen, but many Tenthredinidae are predatory.

The Tenthredinidae and DIPRIONIDAE contain a number of minor pests of ornamental and fruit trees, such as *Caliroa cerasi,* the pear and cherry slug, and *Nematus ribesii,* which consumes currant leaves. More serious, especially in North America, are forest pests of conifers, which include *Pristophora erichsonii* on larch and various species of *Diprion* and *Neodiprion* on spruce and pine.

KEY TO THE NORTH AMERICAN FAMILIES OF
Tenthredinoidea

1. Antennae with 6 or more segments — 2
   Antennae with 3 segments; apical segment elongate, sometimes bifurcate — **Argidae**

2(1). Antennae with apical club; abdomen with separate pleural sclerites — **Cimbicidae**
   Antennae filiform or pectinate; abdomen without pleural sclerites — 3

3(2). Antennae with 6 segments — **Pergidae**
   Antennae with 7 or more segments — 4

4(3). Antennae with 7 to 10 segments — **Tenthredinidae**
   Antennae with 13 or more segments — **Diprionidae**

Suborder Apocrita

Abdomen with first segment incorporated into thorax as propodeum, second segment narrowly constricted as petiole, bearing the enlarged gaster (Fig. 46.9*c,d*); thorax without cenchri; hind wings with no more than two basal cells. Larvae lacking legs, usually with head capsule desclerotized; maxillae and antennae reduced to papillae (Fig. 46.3*b*).

Apocrita comprise all the diversified groups of dominant Hymenoptera, with the exception of the symphytan Superfamily Tenthredinoidea. Parasitoidism, a specialized form of predation, is the primitive mode of larval feeding. In the most advanced members (Superfamilies Vespoidea, Sphecoidea), typical predation is the dominant mode of feeding. Several groups of Apocrita, most notably the bees (Superfamily Apoidea) and gall wasps (Cynipoidea), have reverted to phytophagy.

KEY TO THE NORTH AMERICAN SUPERFAMILIES OF
Common Apocrita

1. Abdomen with basal 1 or 2 segments differentiated as a distinct node or nodes (Fig. 46.14*b*); minute to moderate-sized insects, usually with antennae elbowed and wings absent — **Formicoidea** (p. 496)
   Abdomen with basal segments not modified as a node; antennae elbowed or straight; wings usually present — 2

2(1). Hind leg with trochanter 1-segmented — 3
   Hind leg with trochanter 2-segmented

3(2). Hind tibia with inner spur hooked, toothed or with comb of hairs on inner surface (Fig. 46.17*d*) (or one or both spurs absent) — 4
   Hind tibia with spur or spurs simple, without teeth or hairs (Fig. 46.9*b*) — 9

4(3). Hind wings without closed cells — 5
   Hind wings with at least 1 closed cell — 6

5(4). Pronotum short, not extending posteriorly as far as tegulae (small sclerites at wing bases)

(Fig. 46.12*b*); abdomen concave below **Chrysidoidea** (p. 494)

Pronotum longer, dorsal lobes extending posteriorly to tegulae (Fig. 46.12*a*); abdomen flat or convex below **Bethyloidea** (p. 493)

6(4). Pronotum with dorsal lobes extending posteriorly to tegulae (small sclerites at wing bases) (Fig. 46.14*a,c*); ventral lobes short, not covering anterior thoracic spiracle **8**

Pronotum with dorsal lobes short, not reaching tegulae (Figs. 46.1, 46.15*a*); rounded ventral lobes covering anterior thoracic spiracles **7**

7(6). Body with some hairs branched; hind tarsi with basal segment wider than following segments (Fig. 46.16*c,d*) **Apoidea** (p. 500)

Body without branched hairs; hind tarsi without basal segments broadened (Fig. 46.15*a,b*) **Sphecoidea** (p. 498)

8(6). Eyes deeply notched, or if round, then antennae clavate **Vespoidea** (p. 498)

Eyes round or oval; antennae filiform **Pompiloidea** (p. 496)

9(3). Pronotum large, extending posteriorly as far as tegulae (small sclerites at wing bases) (Figs. 46.12*a*, 46.13*a*) **10**

Pronotum small, not extending posteriorly as far as tegulae (Fig. 46.12*b*) **Chrysidoidea** (p. 494)

10(9). Hind wings without closed cells; venation almost always much reduced (Fig. 46.9*c,d*) **11**

Hind wings with at least 1 closed basal cell; forewings usually with extensive venation (Fig. 46.13*c*) **Scolioidea** (p. 494)

11(10). Abdomen attached high on propodeum, remote from hind coxae (Fig. 46.9*d*) **Evanioidea** (p. 490)

Abdomen attached near base of propodeum, contiguous with or very close to hind coxae (Fig. 46.10*a*) **Proctotrupoidea** (p. 490)

12(2). Head with deep groove below base of antennae and circle of teeth on vertex (Fig. 46.9*a*) **Megalyroidea** (p. 487)

Head without groove below antennal base or circle of teeth on vertex **13**

13(12). Forewing with distinct pterostigma; costal vein present between wing base and pterostigma (Fig. 46.9*d*) **16**

Forewing without pterostigma; costal vein absent (Fig. 46.10*b*) **14**

14(13). Antennae elbowed, with first segment as long as succeeding 2 to 3 segments; abdomen cylindrical or subcylindrical **15**

Antennae filiform, with first segment subequal to second, or with second segment small and first and third segments subequal; abdomen laterally compressed **Cynipoidea** (p. 491)

15(14). Pronotum extending posteriorly as far as tegulae (Fig. 46.9*c*) **Proctotrupoidea** (p. 490)

Pronotum short, not approaching tegulae (Fig. 46.11*a* to *d*) **Chalcidoidea** (p. 492)

16(13). Abdomen attached high on thorax, remote from coxae (Fig. 46.9*d*) **Evanioidea** (p. 490)

Abdomen attached low on thorax, close to coxae (Fig. 46.9*a*) **17**

17(16). Hind wings without closed cells (Fig. 46.10*a*) **Proctotrupoidea** (p. 490)

Hind wings with at least 1 closed cell **18**

18(17). Forewing with reduced or no costal cell (Fig. 46.9*b,e,f*); posterior lobe of pronotum not edged by hairs **Ichneumonoidea** (p. 488)

Forewing with distinct costal cell; posterior lobe of pronotum edged by short, close-set hairs **Trigonaloidea** (p. 488)

SUPERFAMILY MEGALYROIDEA. Small to moderate parasitoids with trochanters 2-segmented; forewing with 3 to 6 closed cells, pterostigma and costal cell present, hind wing without closed cells; ovipositor long, external.

Of the two families comprising the Megalyroidea, the Megalyridae are restricted to the Southern Hemisphere. The superfamily is represented by six uncommon species of STEPHANIDAE in North America. Superficially stephanids (Fig. 46.9*a*) resemble certain ichneumonids, but they are distinguished by the crown of teeth on the head, by the swollen hind femora, and by the brush of preening hairs on the hind tibial apex. The subantennal grooves, the crown of teeth on the cranium, and the reduced wing venation suggest relationships to the Orussidae. Stephanids are usually encountered on trunks or large branches of dead, frequently charred trees, searching for their hosts, which are the larvae of various wood-boring beetles. The long ovipositor is used to parasitize the victims in their galleries. It may be recalled that orussids are associated with wood-boring beetles, possibly as parasitoids.

SUPERFAMILY TRIGONALOIDEA. Stout-bodied, moderate-sized, wasplike parasitoids with long antennae with at least 14 segments; forewing with at least 6 closed cells, costal cell, and pterostigma present; hind wing with closed basal cells; trochanters 2-segmented; ovipositor short, internal.

The single family, Trigonalidae, includes rare species which deposit the eggs in slits cut in developing foliage by the sharp, pointed apical sternite. Eggs hatch when consumed by caterpillars of Lepidoptera or sawflies, which are hosts for a few species. Most species are hyperparasites. They hatch in the guts of caterpillars and enter the hemocoels, but do not develop further unless the caterpillar is secondarily parasitized by certain ichneumonid wasps or tachinid flies which are their proper hosts. Some species attack larvae of vespoid wasps, being consumed with caterpillars provided as food by the adult wasps.

In body form and wing venation trigonalids are similar to the higher wasps, especially the Sphecoidea, but they share the long, multiarticulate antennae, 2-segmented trochanters, and unmodified hind tibiae with the parasitoid families. Because of this combination of characteristics, trigonalids have usually been considered an extremely primitive family, ancestral to most apocritan Hymenoptera.

SUPERFAMILY ICHNEUMONOIDEA. Minute to large parasitoids with 2-segmented trochanters, antennae with at least 13 segments (usually more than 16); forewing with at least 3 closed cells, costal cell absent, pterostigma present or absent; hind wing with 2 closed cells, or occasionally without closed cells. Ovipositor short, internal or long, external.

Ichneumonoidea constitute one of the large, diversified superfamilies of parasitoid Hymenoptera, the other being the Chalcidoidea. The Family Ichneumonidae contains perhaps 20,000 species, a number exceeded only in the larger families of Coleoptera. Most Ichneumonoidea have the abdominal cuticle soft and flexible, at least ventrally, unlike most other Apocrita. The flexible abdomen apparently assists in oviposition. Most Ichneumonoidea apparently parasitize a variety of hosts, and competition among species is reduced by directing the search for hosts to restricted habitats, rather than by selecting specific hosts. The larvae may develop either internally (endoparasitoids) or externally (ectoparasitoids), but pupation is nearly always outside the host in a silken cocoon. Most species are primary parasitoids, but many hyperparasitoids are recorded, especially in the Ichneumonidae.

The superfamily contains two families. Ichneumonidae include mostly moderate-sized to large species which have two recurrent veins in the forewings (Fig 46.9*b,e*). They are parasitoids of immature endopterygote insects, especially Lepidoptera and symphytan Hymenoptera, and also attack spiders, spider egg sacs, and pseudoscorpions. Ichneumonidae are especially abundant in moist situations with rank vegetation, where caterpillars are abundant. However, they occur in a wide spectrum of habitats, including arid situations. Some diurnal species are strikingly similar in color or pattern to stinging wasps, though very few ichneumonids are able to sting humans. *Rhyssa* and related species are remarkable for the extraordinarily long ovipositor, which is used to reach wood-boring Coleoptera or Hymenoptera deep in their galleries. Ovipositors of some species may be up to 15 cm long, about three times their body length.

Braconidae, with a single recurrent vein in the forewing (Fig. 46.9*f*), are generally smaller

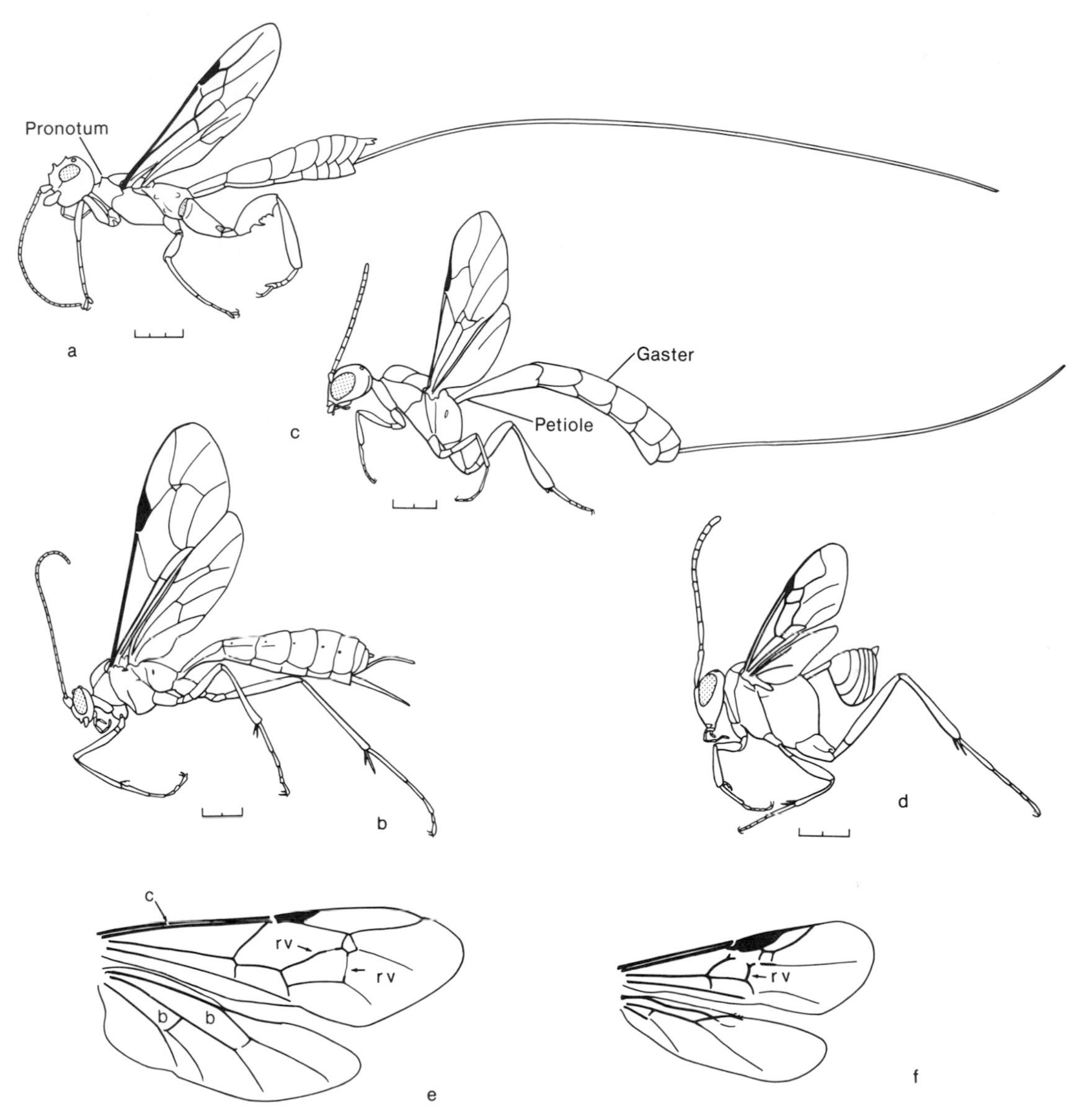

FIGURE 46.9 Representative Apocrita: *a*, Megalyroidea, Stephanidae (*Schlettererius cinctipes;* scale equals 3 mm); *b*, Ichneumonoidea, Ichneumonidae (*Netelia leo;* scale equals 2 mm); *c*, Evanioidea, Gasteruptiidae (*Rhydinofoenus occidentalis;* scale equals 2 mm); *d*, Evanioidea (*Evania appendigaster;* scale equals 2 mm); *e*, wings of Ichneumonidae (*Catadelphus atrax*); *f*, wings of Braconidae (*Cremnops haematodes*). *rv*, recurrent veins; *c*, costal cell; *b*, basal cell.

than Ichneumonidae. Braconids are parasitoids of both adult and immature endopterygote insects, as well as various Homoptera. Some species attack hosts only slightly larger than themselves. For example, *Praon* and *Aphidius* utilize aphids. Other species may parasitize hosts hundreds of times their own bulk, such as the 2- to 3-mm long *Microctonus,* which oviposit in adult tenebrionid beetles as much as 50 mm long. Many species attach the cocoon to the dead host. *Praon* spins a cocoon beneath the husks of consumed aphids, while *Apanteles* cocoons attached to caterpillars are a familiar sight in eastern North America.

SUPERFAMILY EVANIOIDEA. Small to moderate-sized parasitoids with the abdomen attached high on the propodeum, remote from the hind coxae (Fig. 46.9*c*, *d*); hind trochanters 1- or 2-segmented; forewing with pterostigma, costal cell, and several other closed cells; hind wing with only narrow costal cell closed; ovipositor short, internal or long, external.

The insects placed in this superfamily are widespread but seldom common. All are parasitoids with restricted host ranges. For example, Evaniidae (Fig. 46.9*d*) attack cockroach oothecae, while Aulacidae parasitize larvae of xiphidriid wood wasps and of Buprestidae and Cerambycidae (Coleoptera). Gasteruptiidae (Fig. 46.9*c*) oviposit in the cells of solitary bees or sphecid wasps, their larvae consuming any provisions which may be present, as well as the egg or larval host.

KEY TO THE FAMILIES OF
Evanioidea

1. Abdomen with gaster subquadrate, laterally compressed, attached to a slender petiole (Fig. 46.9*d*) **Evaniidae**
   Abdomen with gaster elongate, gradually enlarged, round in cross section, petiole not distinct from gaster (Fig. 46.9*c*) 2

2(1). Abdomen clavate; forewing with 2 recurrent veins (as in Fig. 46.9*e*) **Aulacidae**
   Abdomen cylindrical, slender to apex (Fig. 46.9*c*); forewing with 0 or 1 recurrent veins (as in Fig. 46.9*f*) **Gasteruptiidae**

SUPERFAMILY PROCTOTRUPOIDEA. Small to moderate-sized parasitoids of extremely variable body configuration; trochanters 1- or 2-segmented; forewings usually without closed cells or with narrow costal cell, or with 2 to 5 closed cells in a few primitive species, pterostigma present or absent; hind wing almost always without closed cells.

The Proctotrupoidea constitute a large, diverse assemblage of parasitoids whose biologies and relationships are poorly known. HELORIDAE, probably the most primitive family, resemble braconids in the general body shape and wing venation. In all other families the wing venation is strongly reduced, frequently to a single vein. Body size is extremely variable, reaching 60 mm in PELECINIDAE, but the great majority of proctotrupoids are small or minute, and many superficially resemble Chalcidoidea, except in the large prothorax (compare Figs. 46.10*a* and 46.11*a*).

Proctotrupoidea parasitize a variety of insects, including larval Chrysopidae (Heloridae), coleopterous larvae (Heloridae, PROCTOTRUPIDAE, Pelecinidae), and dipterous larvae (many DIAPRIIDAE). In addition, many species are hyperparasitoids of other Hymenoptera. The most successful families in terms of numbers of species are the SCELIONIDAE (Fig. 46.4*c*), which are parasites of the eggs of numerous orders of insects and spiders; and the PLATYGASTERIDAE, which parasitize cecidomyiid gall flies, as well as insect eggs.

Many adult Proctotrupoidea are short-winged or wingless, antlike inhabitants of leaf litter. Winged species are commonly encountered on foliage. About 500 North American species have been recognized, but it is certain that many remain undescribed.

KEY TO THE NORTH AMERICAN FAMILIES OF
Proctotrupoidea

1. Antennae elbowed, with first segment usually longer than succeeding 3 to 5 segments 4
   Antennae not elbowed, first segment about as long as third 2

2(1). Forewing with 1 or 2 closed basal cells; hind legs with trochanters 1-segmented 3
   Forewing with at least 3 closed basal cells; hind legs with trochanters 2-segmented **Heloridae**

3(2). Forewing with 1 basal cell (the costal) closed; antennae with 13 or fewer segments **Proctotrupidae**

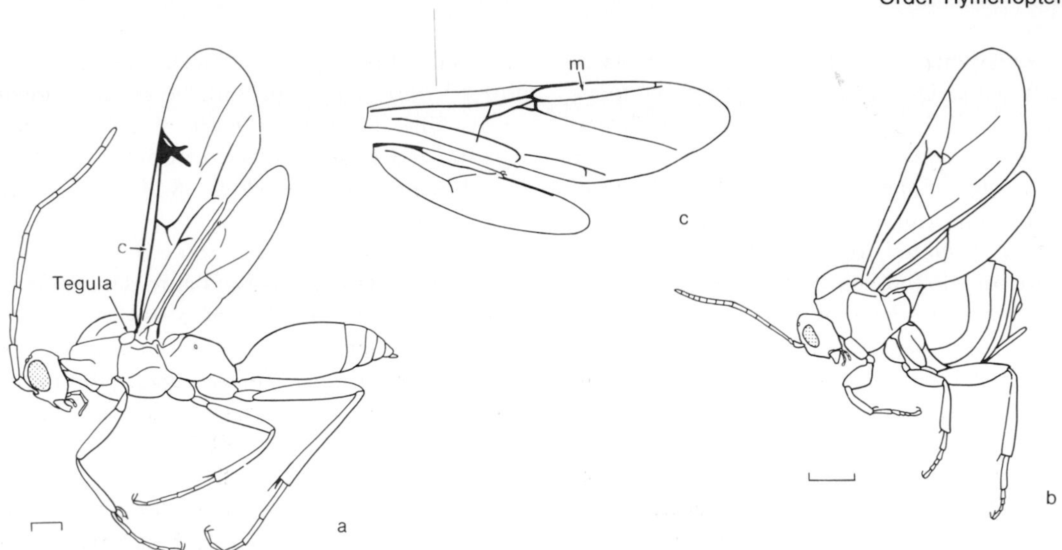

FIGURE 46.10 Proctotrupoidea and Cynipoidea: *a,* Proctotrupoidea, Proctotrupidae (*Proctotrupes;* scale equals 0.5 mm); *b,* Cynipoidea, Cynipidae (*Andricus californicus;* scale equals 1 mm); *c,* wings of Ibaliidae (Cynipoidea). *c,* costal cell; *m,* marginal cell (radial cell).

Forewing with 2 basal cells closed; antennae with 14 or more segments; abdomen of female slender, extremely elongate **Pelecinidae**

4(1). Antennae inserted on a prominence between the eyes, remote from clypeus; midtibiae with 2 spurs **Diapriidae**

Antennae inserted below the eyes, near the clypeus, without a prominence; midtibiae with 1 or 2 spurs **5**

5(4). Abdomen flattened, with lateral margins keeled; midtibiae with single apical spur **6**

Abdomen rounded laterally; midtibiae with 2 spurs **Ceraphronidae**

6(5). Antennae with 10 or fewer segments, forewing venation consisting of short submarginal vein **Platygasteridae**

Antennae with 11 to 12 segments (10 segments rarely present); forewing venation consisting of marginal vein and short vein extending beyond pterostigma **Scelionidae**

SUPERFAMILY CYNIPOIDEA. Small to moderate-sized insects with the abdomen usually compressed in the lateral plane and the tergites extending ventrally to conceal the sternites (Fig. 46.10*b*); hind trochanters 1- or 2-segmented; forewing with marginal vein and pterostigma lacking, usually without closed cells; hind wing almost always with a single vein; ovipositor almost always short, internal.

The great majority of these wasps may be recognized by the reduced forewing venation with a characteristic marginal cell in the distal third (Fig. 46.10*c*). They differ from Chalcidoidea and most Proctotrupoidea in having filiform rather than elbowed antennae.

IBALIIDAE, the most primitive members of Cynipoidea, are endoparasitoids of siricid wood wasps, and FIGITIDAE are parasitoids of Diptera and of hemerobiid lacewings. A relatively small number of CYNIPIDAE are parasitoids of Diptera or hyperparasitoids of Hymenoptera attacking aphids. The large Subfamily Cynipinae, comprising about 600 North American species, all inhabit galls, either as the primary causative agent or as inquilines. Galls vary enormously in size, shape, color, and texture, as well as in internal structure, and are frequently diagnostic of the species of wasp that formed them.

Typically an external epidermis, which may be extremely hard or tough, encloses a thick layer of parenchyma tissue. At the center of the gall is usually a hard shell lined with nutritive tissue, within which the larva feeds. A single gall may contain one to many larval cells and may occur on any part of the host plant. About 75 to 85 percent of all cynipid species attack oaks (*Quercus*) and closely related trees and shrubs; another 7 percent attack the genus *Rosa* (roses); and the remainder utilize a variety of other plants, especially Compositae (sunflower family).

In most Cynipidae, sexual generations, occurring in the warm part of the year, alternate with asexual (parthenogenetic) generations, which appear during winter. The individuals representing the sexual and asexual generations are often strikingly different in appearance and usually produce dissimilar galls on different parts of the host. Consequently, alternate generations have often been described as different species, which may be associated only as the biological relationships are discovered.

KEY TO THE FAMILIES OF
Cynipoidea

1. Abdomen with fourth, fifth, or sixth segment of gaster larger than other segments; body length, 5 to 15 mm **Ibaliidae** and **Liopteridae**
   Abdomen with second or third segment of gaster larger than other segments; body length, seldom greater than 6 mm **Cynipidae** and **Figitidae**

SUPERFAMILY CHALCIDOIDEA. Minute to small or occasionally moderate-sized wasps with the antennae elbowed (exception, Eucharitidae), trochanters 1- or 2-segmented; wings with venation reduced to a single vein, without closed cells; pronotum with posterior lobe usually rounded, not contiguous with tegulae; ovipositor short, internal or exserted, elongate.

The Chalcidoidea, which probably comprise the largest superfamily of Hymenoptera, are remarkable for their great variety of body configurations (Fig. 46.11*a* to *d*), which are diagnostic for some families (e.g., LEUCOSPIDAE, CHALCIDIDAE, AGAONIDAE) and may verge on the grotesque. In most species the cuticle is strongly sclerotized and coarsely sculptured; many are iridescent blue or green. Chalcidoidea range to about 30 mm in length, but the great majority are less than 3 mm long. Nearly all Chalcidoidea may be recognized by the combination of reduced wing venation (Fig. 46.11*e*) and short pronotum (Fig. 46.11*a*).

Some EURYTOMIDAE, PTEROMALIDAE, TORYMIDAE, PERILAMPIDAE and EULOPHIDAE are phytophagous, either tunneling in seeds or forming galls, and a few species, such as *Bruchophagus platyptera,* which infest clover seeds, are of minor economic importance. Members of the Family Agaonidae engage in an obligatory symbiotic relationship with species of figs *(Ficus),* whose flowers they both gall and pollinate. Certain commercial varieties of figs, such as Smyrna, are pollinated only by the wasps, which have been transported about the world for this purpose (see Fig. 13.3). The great majority of Chalcidoidea, however, are internal or external parasitoids of insects of all orders, but especially Homoptera and immature stages of Endopterygota. Coleoptera, Diptera, and other orders are included in the host range, but Lepidoptera are especially favored among the Endopterygota. Eggs, larvae, and pupae are utilized by various species, and some families attack only a single developmental stage (e.g., TRICHOGRAMMATIDAE, MYMARIDAE on eggs). Numerous species are hyperparasitoids of primary parasitoids, and care must be exercised in deducing host relationships.

Many chalcidoids are hypermetamorphic, with the first one or two larval stages specialized for host location or elimination of competitors within the host (Fig. 46.4). Later instars are always maggotlike. Parthenogenesis occurs in various families, and polyembryony has been described in many species. Both features are common components of the life cycles of hymenopterous parasitoids.

Different authorities recognize from 9 to 18 families of Chalcidoidea. A few are easy to identify, but most families are difficult to separate with existing keys, and the superfamily classification is in drastic need of taxonomic study. Interested students are referred to Peck et al. (1964) and Riek (1970*b*) for identification.

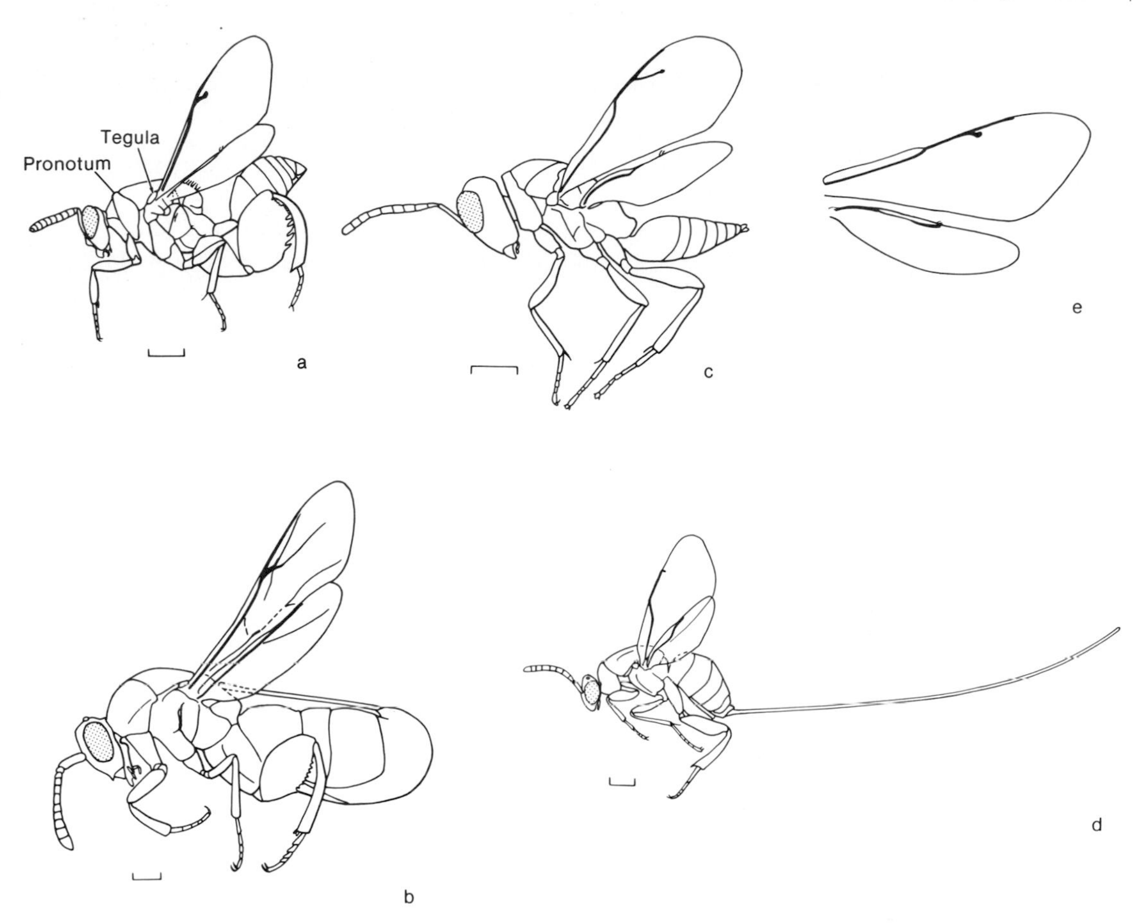

FIGURE 46.11 Chalcidoidea: *a*, Chalcididae (*Chalcis;* scale equals 1 mm); *b*, Leucospidae (*Leucospis birkmani;* scale equals 1 mm); *c*, Pteromalidae (*Pteromalus vanessa;* scale equals 0.5 mm); *d*, Torymidae (*Torymus californicus;* scale equals 1 mm); *e*, wings of Chalcididae (*Chalcis divisa*).

SUPERFAMILY BETHYLOIDEA. Small to moderate-sized (1 to 10 mm) wasps with hind trochanters 1-segmented; forewings with at least 2 to 4 closed cells (veins may be faint), including costal cell (Fig. 46.12*c*); hind wings without closed cells; ovipositor internal.

Bethyloidea are infrequently encountered, although Bethylidae are common inhabitants of leaf litter, tree trunks, and fungi, where they are parasitoids of larvae of Lepidoptera and Coleoptera. While Bethyloidea do not prepare special nests for larval development, some species of Bethylidae move the immobilized prey to a sheltered spot before oviposition. Unlike the fossorial wasps (Scolioidea, Pompiloidea, Sphecoidea), they deposit more than one egg on a single host. Except for their reduced wing venation, Bethylidae are morphologically similar to Tiphiidae, and in general are intermediate between typical parasitoids such as the Proctotrupoidea and the primitive stinging wasps.

Dryinidae are external parasitoids of Homoptera; Sclerogibbidae are external parasitoids of Embioptera; and Cleptidae attack sawfly prepupae and eggs of Phasmatodea. Bethyloidea include many wingless species and others in which only the females are wingless. Sexual dimorphism in other characteristics may be extreme, and sexes may be difficult to associate. Winglessness in females is common in

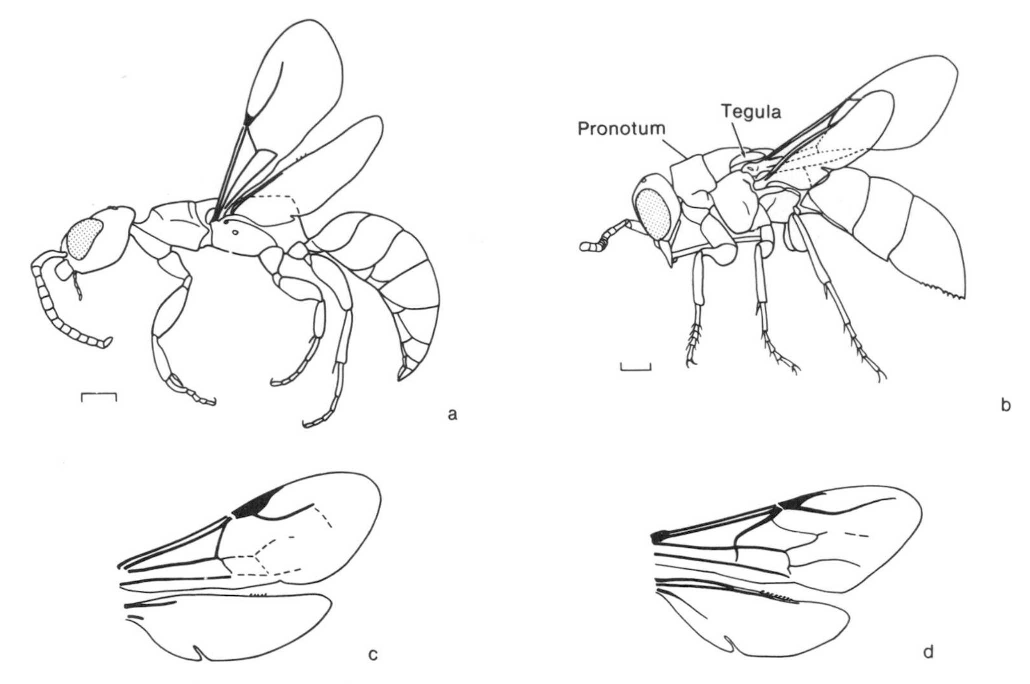

FIGURE 46.12 Bethyloidea and Chrysidoidea: *a*, Bethylidae (*Anisepyris williamsi;* scale equals 0.5 mm); *b*, Chrysididae (*Parnopes edwardsi;* scale equals 1 mm); *c*, wings of Bethylidae (*Pristocera armi*); *d*, wings of Chrysididae (*Chrysis*).

Tiphiidae and Mutillidae (both in Scolioidea), emphasizing the close relationship to Bethyloidea.

KEY TO COMMON NORTH AMERICAN FAMILIES OF
Bethyloidea

1. Antennae with 11 to 13 segments; fore tarsi unmodified **2**
   Antennae with 10 segments; females with fore tarsi claw-shaped **Dryinidae**

2(1). Abdomen with 7 to 8 segments in gaster; pronotum extending posteriorly to tegulae (Fig. 46.12*a*) **Bethylidae**
   Abdomen with 4 to 6 segments in gaster; pronotum not reaching tegulae **Cleptidae**

SUPERFAMILY CHRYSIDOIDEA. Small to moderate-sized parasitoids with antennae elbowed, hind trochanters 1-segmented; forewings with 3 or more closed cells; including costal cell; hind wings without closed cells; ovipositor an elongate, telescoping tube, concealed internally.

The single family, CHRYSIDIDAE (Fig. 46.12*b*), is readily recognized by the metallic green and blue colors and the concave ventral surface of the abdomen. Chrysidids roll the heavily sclerotized body into a ball when molested, the head and thoracic venter being concealed in the abdominal concavity. Morphologically Chrysididae resemble some Cleptidae but differ in leg and thoracic structure (Riek, 1970*b*).

Chrysididae attack larvae of bees and wasps in their cells, pupating within a cocoon in the host cell. The most familiar species attack mud daubers (*Sceliphron:* Sphecidae) and vespids which reuse mud-dauber nests. Adult chrysidids are most commonly encountered on the surface of soil or tree trunks or occasionally on flowers.

SUPERFAMILY SCOLIOIDEA. Small to large parasitoids with stout body; hind trochanters 1-segmented; forewings with at least 7 closed cells, including costal cell, hind wings with at least 1 closed cell or wingless; ovipositor developed as internal sting.

The scolioid wasps are all parasitoids, biologically similar to Bethylidae. Morphologically the superfamily is diverse, showing similarities to Bethyloidea, Formicoidea (ants), Pompiloidea (spider wasps), and Vespoidea. TIPHIIDAE are usually considered the ancestral group from which all other stinging Hymenoptera differentiated.

Most scolioid wasps are highly fossorial, with strong forelegs and large prothorax extending posteriorly to the region of the wing bases (Fig. 46.13*a*). Sexual dimorphism is frequently marked, especially by winglessness in females, which may be very difficult to associate with the winged males. In contrast to many other Hymenoptera the females are often smaller than the males, probably an adaptation which allows increased dispersal of the flightless females, which are carried about by the males during a nuptial flight.

Tiphiidae and SCOLIIDAE mostly attack ground-dwelling beetle larvae. Scarabaeidae are the commonest hosts, but Tenebrionidae and Cicindelidae (tiger beetles), as well as Orthoptera and bees and wasps, are utilized by Tiphiidae. MUTILLIDAE (velvet ants), whose powerfully stinging females are always wingless, are mostly parasitoids of bees and wasps. Most diurnal species of velvet ants are brilliantly patterned in red or orange and black pile, rendering them highly conspicuous as they rapidly run about the ground searching for host nests. Nocturnal species are tan, brown, or black, resembling tiphiids. Sapygidae are parasitoids of bees and wasps.

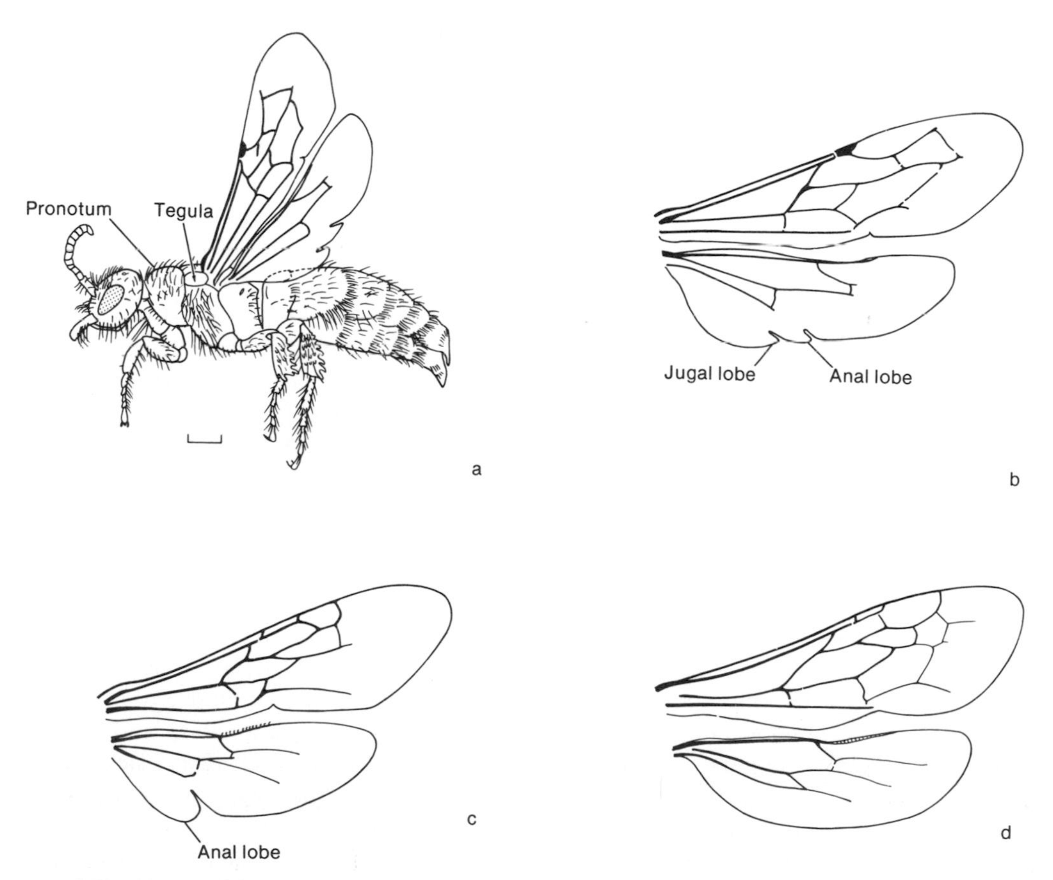

FIGURE 46.13 Scolioidea: *a*, Tiphiidae (*Paratiphia verna;* scale equals 1 mm); *b*, wings of Tiphiidae (*Tiphia*); *c*, wings of Scoliidae (*Scolia nobilitata*); *d*, wings of Mutillidae (*Timulla vagans*).

KEY TO THE FAMILIES OF
Scolioidea

1. Wings with apical membrane bearing numerous fine, longitudinal creases; sterna of meso- and methathoraces forming a broad plate overlapping middle and hind coxae **Scoliidae**
   Wings without creases in apical membranes, or wings absent; sterna of meso- and metathoraces not overlapping coxae 2

2(1). Abdomen with lateral felt line on second gastral segment; females wingless, males with wings lacking anal lobes (Fig. 46.13*d*) **Mutillidae**
   Abdomen without lateral felt line; females winged or wingless; wings of both sexes with anal lobe separated from rest of wing by notch (Fig. 46.13*b*) 3

3(2). Abdomen with gaster usually constricted between first and second segments; eyes usually round **Tiphiidae**
   Abdomen with gaster not constricted between first and second segments; eyes usually deeply emarginate **Sapygidae**

SUPERFAMILY POMPILOIDEA. Small to very large wasps with large, mobile prothorax, hind trochanters 1-segmented; forewings with 7 to 10 closed cells, including costal cell, hind wing with at least one closed cell; ovipositor developed as internal sting.

This superfamily comprises two families of greatly unequal size and abundance. RHOPALOSOMATIDAE, represented by only two rare species in North America, parasitize Gryllidae. The larvae cause a hernialike pouch in the abdominal wall of the host, much in the manner of Dryinidae. POMPILIDAE [spider wasps (Fig. 46.14*a*)], characterized by the presence of a horizontal furrow on the mesopleuron (absent in Rhopalosomatidae), all utilize spiders as larval food. Pompilids provide each larva with a single prey item, which is necessarily larger than the wasp. A wide variety of spiders is attacked, including tarantulas, which serve as prey for some *Pepsis.*

Because of the large body size of their prey, pompilids usually construct a burrow near the site of attack and immobilization. The paralyzed spider is frequently concealed to avoid attack by parasites and scavengers, then quickly moved a short distance into the completed larval gallery. Other species simply conceal the prey in an appropriate crevice or in the host's own tunnel, and some only temporarily paralyze the spider, which regains activity before being killed by the maturing wasp larva. Such species are essentially parasitoid, showing no important biological differences from Bethylidae or Tiphiidae. More specialized pompilids resemble primitive sphecids in preparing a larval gallery before prey acquisition, while a number of species oviposit on prey of other pompilids. The family is cosmopolitan but is especially diverse in tropical and subtropical regions.

SUPERFAMILY FORMICOIDEA. Minute to moderate-sized Hymenoptera with basal one or two segments of abdominal gaster modified as a nodelike petiole (Fig. 46.14*b*); prothorax large, mobile, antennae almost always strongly elbowed.

Ants are among the most ubiquitous and familiar insects, occurring in vast numbers in all but extremely cold regions. Several species have become closely associated with humans, occupying their dwellings and other structures throughout the world, including metropolitan areas. Despite great variation in size and habits, practically all ants are recognizable by the petiolate abdomen and elbowed antennae, and are classified as a single family, FORMICIDAE.

Without exception, ants are social insects. Colonies consist of one or a few sexual females, or *queens,* specialized for reproduction, and a variable number of apterous, neuter females which function as *workers* or *soldiers.* Males are normally produced once or a few times a year, surviving only long enough to mate. Males and females are usually larger than nonsexual castes, and have the thorax enlarged with the wing muscles. Female sexuals dehisce their wings (Fig. 46.14*e*) after a nuptial flight; males are permanently winged. Worker and soldier castes are always wingless. Soldiers, which function in colony de-

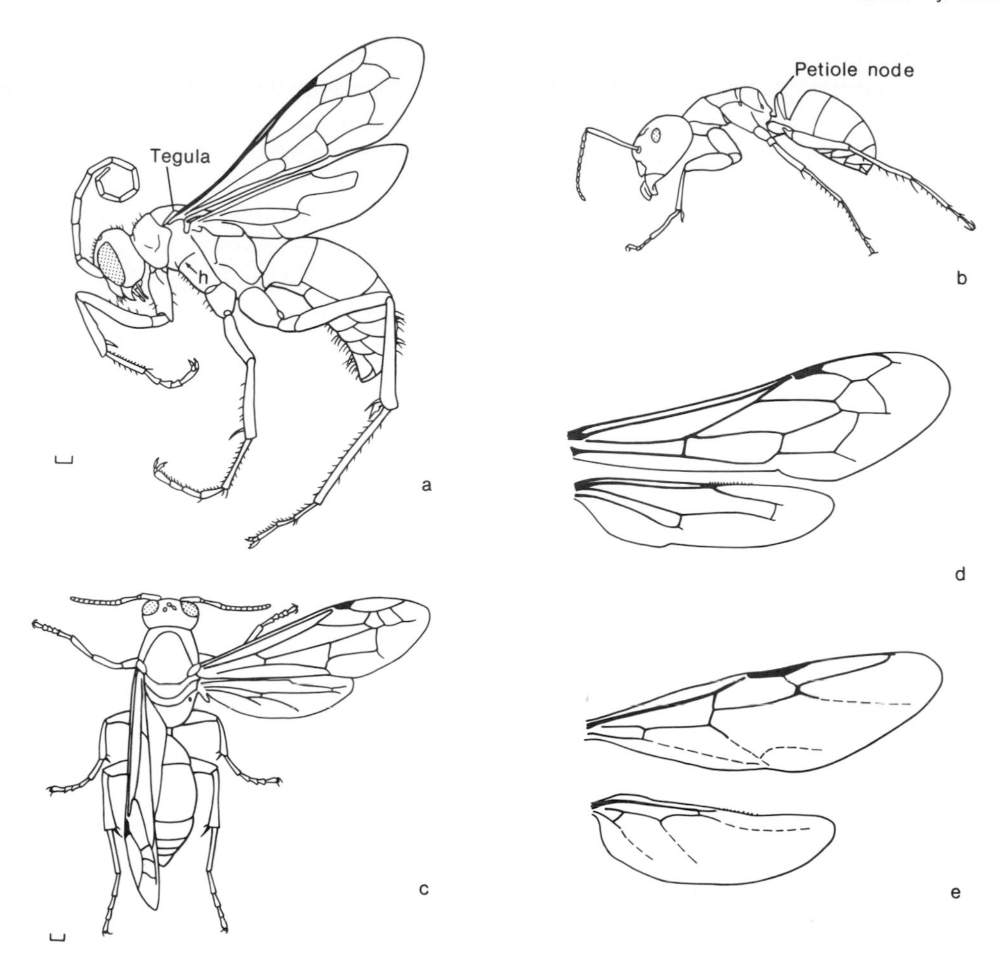

FIGURE 46.14 *a,* Pompiloidea, Pompilidae (*Priocnemoides unifasciatus*); *b,* Formicoidea, Formicidae (*Camponotus*); *c,* Vespoidea, Vespidae (*Polistes metricus*); *d,* Vespidae, wings (*Pseudomasaris vespoides*); *e,* Formicidae, wings (*Camponotus morosus*). Scales equal 1 mm. *h,* horizontal furrow on mesopleuron.

fense, are larger than workers, often with the head disproportionately enlarged. Colony organization and structure, which are extremely complex, are discussed in detail in Chap. 7, but it may be mentioned here that colony formation, which is carried out by single females, is critical in the perpetuation of most species.

Ant colonies, which may persist for many years, vary from assemblages of a single queen and a few dozen individuals to multitudes of hundreds of thousands of individuals with numerous queens. Typically, colonies are established in fixed locations. Subterranean galleries are most common, but many species tunnel dead wood or twigs, and a few tropical species have evolved highly specialized symbiotic relationships with plants. Especially interesting is the relationship between ants of the genus *Pseudomyrmex* and certain species of acacia trees. The large, swollen thorns of bull's horn acacia and other species are inhabited by various species of *Pseudomyrmex.* Extrafloral nectaries and beltian bodies on the leaves supply the ants with sugar and protein. The ants aggressively protect the trees from vertebrate and invertebrate predators (Janzen, 1966, 1967*a,b*). Army ants do not occupy a fixed colony but lead a nomadic existence, bivouacking in litter or dense vegetation.

A number of species of ants, especially in the genus *Formica,* enslave other species. For example, *F. sanguinea* workers attack the nests of other species of *Formica,* robbing the pupae and carrying them back to their own nest. Still other species live as inquilines in the nests of other ants, a relationship sometimes termed *social parasitism.*

Ants include herbivores, scavengers, and predators. Harvester ants specialize on seeds, and leaf-cutting ants cut out pieces of leaves on which are cultured fungi, used as food by the ants. Most ants are occasionally opportunistic predators, but army ants are exclusively predatory, including a wide variety of small invertebrates in their diet, and a few ants are specific predators of spider eggs, termites, or other arthropods. Many species tend aphids or membracid nymphs, from which they obtain honeydew, and several, such as the argentine ant (*Iridomyrmex humilis*), are pests in gardens for this reason.

Ants are sufficiently abundant that certain birds, mammals, and reptiles (horned lizards, anteaters) are specialized as predators on them. Ant colonies are widely exploited by *myrmecophilous* insects, especially beetles, which may act as predators of the ant brood or as nest scavengers. The more specialized of these myrmecophilous insects may be morphologically modified to provide a tactile resemblance to their hosts (Fig. 14.1).

SUPERFAMILY VESPOIDEA. Small to large wasps with the prothorax with large, triangular lobes extending posteriorly to the tegulae (Fig. 46.14*c*); hind trochanters 1-segmented; forewings with at least 7 closed cells, including costal cell (Fig. 46.14*d*); hind wing with at least 2 closed cells; ovipositor developed as internal sting.

The single family, VESPIDAE, includes diverse wasps which are mostly predatory, usually on caterpillars, occasionally on larvae of sawflies or phytophagous beetles. Members of the Subfamily Masarinae, which provision their nests with pollen and nectar, are locally common in the Western states. Most vespids may be recognized by the emarginate eyes which partially surround the antennal bases.

Vespids include many solitary forms such as potter wasps (Subfamily Eumeninae), but the familiar species, including yellow jackets, hornets, and paper wasps, form annual colonies. Vespid nests may be suspended from branches, rafters, or other aerial structures, may be concealed beneath stones or in hollow logs, or may be constructed in cavities which the wasps excavate in the ground. A few solitary species use mud as a building material, but nests of social species are commonly constructed of masticated wood or paper. Colonies range in size from a few cells arranged in a single, open comb to multicombed structures enclosed in tough, external jackets of paper. Different species display all stages of intermediacy between solitary and social organization.

Vespids are commonly colored in contrasting patterns of black and yellow, in advertisement of their severe stings. Unlike the sphecoid wasps, which paralyze their prey by stinging, Vespidae usually kill the prey with the mandibles. Stinging is largely used to deter mammalian predators, which are very sensitive to the pain-producing substances in vespid venoms. The large number of insects which mimic vespids attests to the efficiency of their stings (see Chap. 15; Endpaper 4*a*).

SUPERFAMILY SPHECOIDEA. Small to large wasps with the prothorax collarlike with rounded posterior lobes which do not reach the tegulae (Fig. 46.15*a*); hind trochanters 1-segmented; forewings with at least 7 closed cells, including costal cell; hind wings with at least 2 closed cells (46.15*c,d*); ovipositor developed as internal sting.

The Sphecoidea are commonly called "solitary wasps" or "digger wasps," although some nest in twigs, old beetle burrows, or self-constructed mud cells. The single large family, SPHECIDAE, includes an astonishing diversity of body form and size, as well as behavior, but taxonomic subdivision is impractical because of the many intermediates.

A few sphecids (Subfamily Nyssoninae) are inquilines in the nests of bees or other sphecids, but predation on other insects is much the dominant mode of life. The adult female wasps construct nests containing one or more cells in which the larvae will mature while feeding on prey provided by the mother. A great variety of insects is utilized as prey, and host specificity varies greatly among different species. The most primitive sphecids use a single prey item for each larva. Because of the large prey size, such wasps either drag the prey over the ground or transport it with a series of short flights initiated from trees, shrubs, or other prominences. Except in the construction of a special larval gallery, these primitive sphecids are not greatly different from bethylids, tiphiids, or other parasitoid wasps.

In the great majority of sphecids, prey size is much reduced, allowing rapid transport by air, so that such abundant insects as aphids, leafhoppers, and thrips may be utilized. Provisioning follows two modes. *Mass-provisioning* species accumulate the total number of prey required before oviposition. Normally only a single larval cell is tended at one time. *Progressive-provisioning* species oviposit on one of the first prey items acquired, or in an unprovisioned cell, later supplying additional food as needed by the growing larva. In the most advanced species several larvae are tended simultaneously. The resultant close contact between the adult females and their offspring is probably requisite to true sociality (*eusociality*), occurring in Vespoidea and Apoidea (see Chap. 7). However, eusociality has occurred only once in the Sphecoidea, in *Microstigmus comes*.

Behavioral patterns in Sphecoidea may be quite complex, involving simultaneous provisioning of several larvae at different stages of development. Individual larval cells are closed after the final prey item is provided, and the entire nest may be temporarily closed between trips for prey. Primitive species carry prey in the mandibles, which are also used for opening the

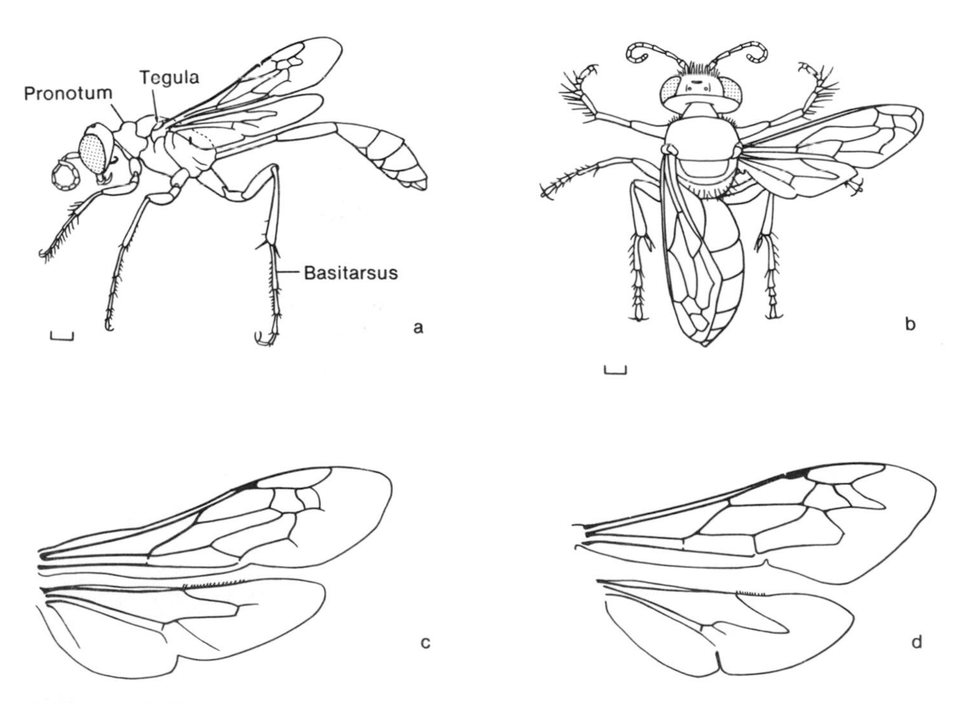

FIGURE 46.15 Sphecoidea, Sphecidae: *a, Ammophila cleopatra; b, Bembex americanus; c,* wings of *Ammophila; d,* wings of *Tachytes.* Scales equal 1 mm.

nest. Such species must drop the prey while removing the nest closure, or leave the nest open, allowing attack by parasitoids or predators such as ants. Species which carry the prey by the middle or hind legs or on the sting immediately dig open the nest without releasing the prey.

SUPERFAMILY APOIDEA (bees). Small to large Hymenoptera with branched, plumose hairs on at least part of the body; prothorax with rounded posterior lobes which do not reach the tegulae; hind trochanters 1-segmented; forewings with at least 7 closed cells; hind wings with at least 2 closed cells; ovipositor developed as internal sting.

In general morphological characters, Apoidea are very similar to sphecoid wasps. The most reliable differentiating features are the presence of branched hairs on the body, and the specialization of the hind basitarsus as a part of the pollen-collecting apparatus (Fig. 46.17*c,d*). Some primitive genera, such as *Hylaeus* (Colletidae), carry pollen in the crop. In these and in cleptoparasitic genera of other families, the body is relatively hairless and the tarsi are unmodified, rendering separation from some sphecids difficult.

In terms of nest architecture and behavior, bees (except the social species) are not greatly different from solitary wasps, with the major difference that most bees provision their larval cells with nectar and pollen rather than animal material. Their lower position in food chains (as primary rather than secondary or tertiary consumers) is probably responsible for the great abundance of bees in most habitats. As pollinators bees are more important than any other group of insects. The maxillae and labium are modified as a proboscis for sucking nectar, and may be very elongate in some APIDAE (Fig. 13.1*b*) and ANTHOPHORIDAE. *Scopae,* or pollen-collecting structures, may occur on the hind tarsi and tibiae [Anthophoridae (Fig. 46.17*d*); Apidae (Fig. 46.17*c*), Andrenidae, Halictidae, Mellitidae] or

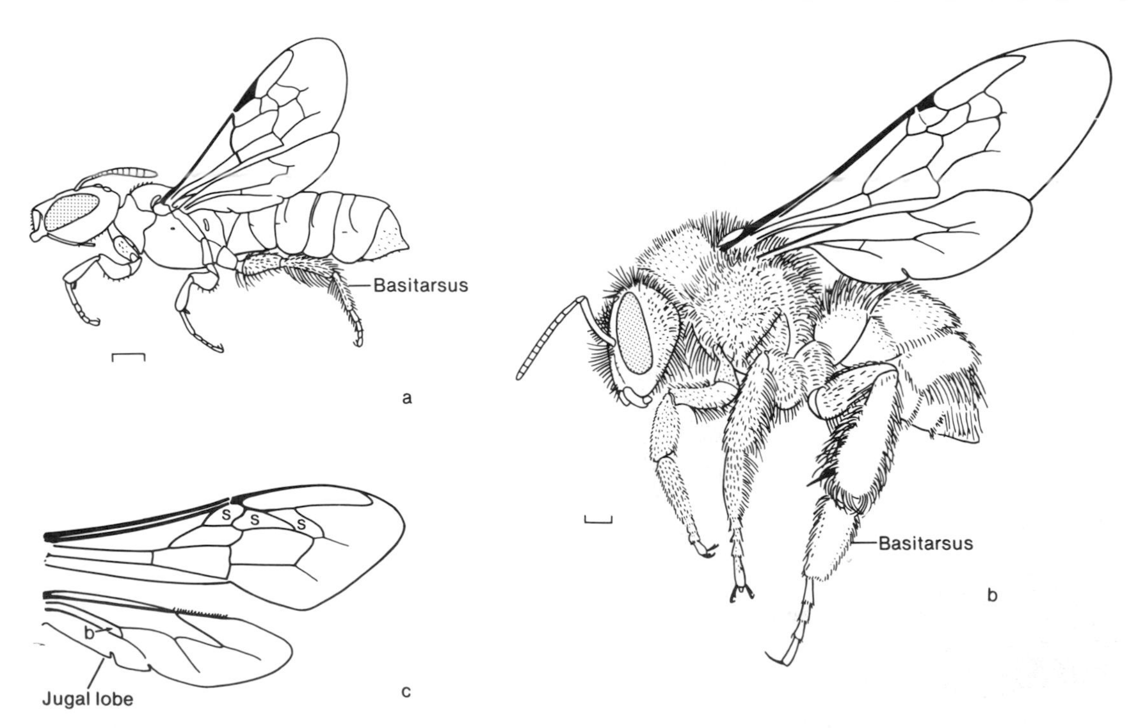

FIGURE 46.16 Apoidea, Apidae: *a, Ceratina strenua* (scale equals 0.5 mm); *b, Bombus americanorum* (scale equals 1 mm); *c, Apis mellifera,* wings. *b,* Basal cell in hind wings; *s,* submarginal cells.

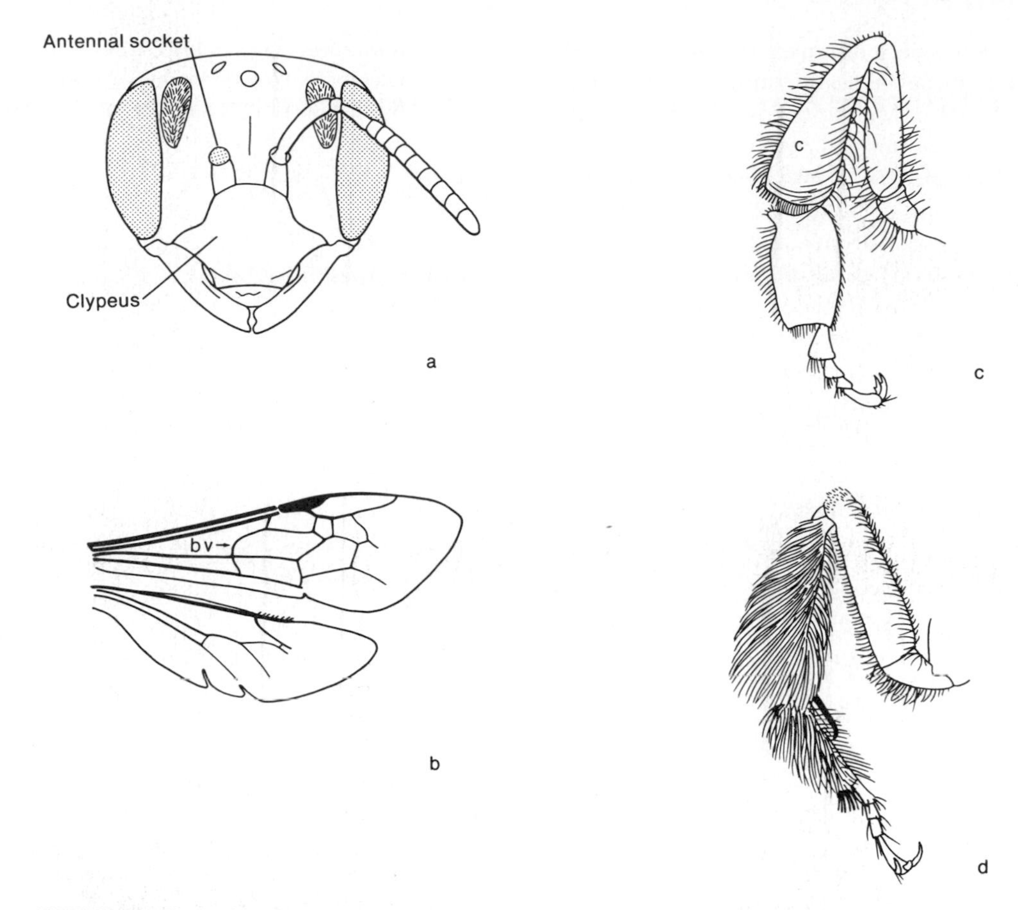

FIGURE 46.17 Apoidea: *a,* frontal view of head of Andrenidae (*Andrena gardineri*); *b,* wings of Halictidae, *bv,* basal vein (*Halictus ligatus*); *c,* hind tarsus of Apidae, showing corbicula (*c*) (*Apis mellifera*); *d,* hind tarsus of Anthophoridae, showing scopa (*Anthophora curta*).

on the venter of the abdomen (Megachilidae) of females. Scopae are lacking only in some COLLETIDAE (*Hylaeus*), and in cleptoparasitic or inquiline species.

The honeybee (*Apis mellifera*) is the best-known pollinator. Commercial hives are seasonally rotated among many plantings, including fruit trees, alfalfa, and certain row crops. Honeybees visit a wide variety of noncultivated plants as well. Wild bees are more efficient pollinators than honeybees for some crops. For example, alkali bees (*Nomia*) and leafcutter bees (*Megachile*) are utilized for alfalfa pollination. Pollination of many uncultivated plants is accomplished largely by wild bees, which may be highly specialized for one plant species or a group of closely related species (*oligolectic*). For example, certain species of *Andrena* (Andrenidae) collect pollen only from night-blooming species of *Oenothera* (Onagraceae), which they visit at night or very early in the morning, before other pollinators have had an opportunity. *Polylectic* species of bees collect pollen from many plant species, but frequently only a single species is visited on sequential trips for pollen. For example, in the highly polylectic honeybee, any individual bee normally visits only one type of flower until that source of pollen and nectar is exhausted, then an alternative flower type will be selected (see Chap. 13).

Eusociality has apparently evolved independently several times in bees (see Chap. 7). The largest and most complex colonies are maintained by honeybees (*Apis*) and tropical stingless bees (Apidae: Meliponinae), but many other apids, such as bumblebees, as well as some halictids, form small annual colonies. As in other social Hymenoptera, these are female societies, the male bees, or drones, appearing only once a year in relatively small numbers. Besides these eusocial forms, many bees, especially Apidae and Halictidae, are presocial, showing various degrees of altruistic or cooperative behavior. The biology of social Hymenoptera is extensively discussed by Wilson (1971) and Michener (1974).

KEY TO THE NORTH AMERICAN FAMILIES OF
Apoidea

1. Face with 2 vertical sutures between each antennal socket and clypeus (Fig. 46.17*a*) **Andrenidae**
   Face with a single suture between each antennal socket and clypeus **2**

2(1). Hind wing with jugal lobe as long as basal cell or longer (Fig. 46.17*b*) **3**
   Hind wing with jugal lobe shorter than basal cell (Fig. 46.16*c*) or lacking (Fig. 46.16*b*) **4**

3(2). Forewing with the basal vein (basal sector of M vein) strongly arched (Fig. 46.17*b*) **Halictidae**
   Forewing with the basal vein straight or feebly curved (Fig. 46.16*c*) **Colletidae**

4(2). Labial palpi with all segments cylindrical, subequal in length **Melittidae**
   Labial palpi with 2 basal segments flattened, at least twice as long as distal segments (Fig. 46.2*b*) **5**

5(4). Forewing usually with 3 submarginal cells (Fig. 46.16*c*); labrum usually broader than long; subantennal suture extending from center of antennal socket; females with scopa on hind legs (Fig. 46.17*c,d*) **6**
   Forewing with 2 submarginal cells; labrum longer than broad; subantennal suture extending from lateral margins of antennal sockets; females with scopa on venter of abdomen **Megachilidae**

6(5). Hind tibiae without spurs (exception, *Bombus,* bumblebees, Fig. 46.16*b*); hind tibiae of females with scopa forming a corbicula, or basket (Fig. 46.17*c*) **Apidae**
   Hind tibiae with apical spurs; hind tibiac of females with scopa forming a dense brush of setae (Fig. 46.17*d*) **Anthophoridae**

SELECTED REFERENCES

GENERAL
Askew, 1971 (parasitoid species); Clausen, 1940 (predaceous and parasitoid species); Dadant and Sons, 1975 (honeybee); Debach, 1974 (biological control); Entomophaga, 1956 to date (many articles on Hymenoptera); Evans and Eberhard, 1970 (wasps); Felt, 1940 (gallmaking families); Huffaker, 1971 (biological control); Spradberry, 1973 (wasps); Stehr, 1975 (biological control).

MORPHOLOGY
Michener, 1944 (bees); Daly 1963, 1964 (muscles); Snodgrass, 1956 (honeybee); Stephen et al., 1969 (bees).

BIOLOGY AND ECOLOGY
Caroll and Janzen, 1973 (ant foraging); Evans and Eberhard, 1970 (wasps); Frisch, 1967, 1971 (honeybee orientation and communication); Dadent and Sons, 1975; Iwata, 1976 (behavior); Janzen, 1966, 1967*a,b* (coevolution of ants and acacias); Krombein (cavity-nesting bees, wasps), 1967; Malyshev, 1968 (extensive life history information); Matthews, 1974 (Braconidae); Michener, 1969, 1974 (social behavior of bees); Morgan, 1968 (Siricidae); Spradberry, 1973 (wasps); Way, 1963 (ant-Homoptera mutualism); Wilson, 1971 (social behavior); Wilson, 1975*a* (slavery in ants).

SYSTEMATICS AND EVOLUTION
Bohart and Menke, 1976 (keys, sphecids of world); Creighton, 1950 (ants of North America); Hinton, 1946, 1971 (phylogenetic relationships); Malyshev, 1968 (higher classification); Michener, 1944 (bees), 1953 (bee larvae); Mitchell, 1960–1962 (bees of Eastern states); Peck et al., 1964 (family keys to Chalcidoidea); Peterson, 1948 (larvae); O. W. Richards, 1956 (keys to families); Ross, 1937*a* (Symphyta of North America); Stephen et al., 1969 (bees of Pacific Northwest); Townes, 1969 (Ichneumonidae).

GLOSSARY

Accessory cell In Lepidoptera, a small cell between two branches of the radial vein, usually R_{2+3} and R_{4+5}; areole; *see* Fig. 44.7*d*.

Acephalic Without a distinct, sclerotized head capsule, as in many immature Diptera.

Acrotergite The anterior precostal part of the tergal plate of a secondary segment.

Aculea A minute, needlelike hair on the wings of Lepidoptera.

Adecticous Without the ability to move the mouthparts, as in many holometabolous pupae.

Aedeagus The penis or male intromittent organ.

Alinotum The wing-bearing sclerite of a pterothoracic notum.

Allomone A substance transmitted in chemical communication between individuals of different species that benefits the sender or both the sender and receiver.

Ambulatory Adapted for walking.

Ametabolous Without noticeable metamorphosis, as in Protura, Collembola, etc.

Amplexiform In Lepidoptera, the mode of wing coupling in which the humeral region of the hind wing projects beneath the forewing.

Anal lobe The posterior portion of the wing, occupied by the anal veins; *see* Fig. 2.11.

Anamorphosis Postembryonic development in which abdominal segments are added at the time of molting.

Angulate Forming an angle, as opposed to rounded.

Annular Ring-shaped or ringlike.

Annulate Ringed, bearing ringlike divisions.

Anteclypeus Ventral part of frontoclypeus of sucking insects.

Antecosta The internal ridge marking the original intersegmental boundary and on which the longitudinal muscles of secondary segments insert.

Antecostal suture The external suture tracing the internal antecosta.

Antennifer An articulatory process in the antennal socket.

Apodeme An ingrowth of the integument to which muscles attach.

Apodous Without legs.

Apophysis An ingrowth of the integument to which muscles attach.

Aposematic Warning, as in conspicuous aposematic coloration or behavior of poisonous or dangerous animals

Appendiculate Bearing an appendage or appendages.

Apposed Having surfaces adjacent or against one another.

Aptery The condition of lacking wings; *see* Brachyptery (apterous).

Arista A large bristle, usually dorsal, on the apical antennal segment of some Diptera.

Arolium A cushionlike pad between the tarsal claws, as in Orthoptera, or paired pads, as in Hemiptera; *see* Pulvillus; Empodium.

Arrhenotoky Parthenogenetic reproduction in which the offspring are all males.

Asexual Not having separate sexes; reproducing by parthenogenesis.

Atrophied Undeveloped, rudimentary, wasted away through lack of growth (atrophy).

Attenuate Tapering apically.

Autotomize To cast off a limb, usually at a predetermined zone of weakness (autotomy).

Axillary sclerites Four sclerites at the wing base; *see* Fig. 2.11.

Basal cell In Hymenoptera, a cell near the base of the wing, bordered partly by the longitudinal veins; *see* Fig. 46.6.

Basal vein In Hymenoptera, a vein connecting the subcostal and cubital veins; *see* Fig. 46.17*b*.

Basalare The episternal epipleurite, on which insert the anterior pleural wing muscles.

Basitarsus The basal tarsomere.

Bifid, bifurcate Cleft, forked, divided into two parts.

Bipectinate With comblike processes or extensions on both sides, as in bipectinate antennae; *see* Fig. 2.7.

Bisexual Having two distinct sexes.

Blood gill Hollow process of the body wall through which blood circulates but which lacks tracheae; present in endopterygote larvae, probably functioning in maintaining ionic balance; *see* Tracheal gills.

Blotch mines Broad, irregular excavations produced in leaves especially by larvae of Lepidoptera; (*see* Serpentine mines).

Brachyptery The condition of having short, atrophied wings; *see* Aptery (brachypterous).

Brood passage In Strepsiptera, the space between the venter of the female and the puparium through which the triungulins emerge.

Bursa copulatrix A copulatory chamber, as in Lepidoptera.

Calypter A small basal fold or lobe in the hind margin of the wing of Diptera, covering the haltere; *see* Squama.

Campodeiform Having the general body shape of Campodeidae (Diplura); applied to larvae.

Capitate Club-shaped, with an abrupt, apical enlargement; *see* Fig. 2.7.

Caudate Bearing a tail-like extension or process.

Caudate larva In certain Hymenoptera, first-instar larvae in which the terminal abdominal segment forms a tail-like extension.

Cenchrus (pl., cenchri) A pale, membranous lobe or area on each side of the metanotum of symphytan Hymenoptera.

Cephalothorax The fused head and thorax of Arachnida and Crustacea; that portion of an obtect pupa covering the head and thorax, as in Strepsiptera.

Cercus (pl., cerci) An appendage of the eleventh abdominal segment.

Cervical sclerites Small sclerites in the membrane between the head and thorax.

Cervix The neck.

Chaeta An articulated hair or spine.

Chaetosema (pl., chaetosemata) In Lepidoptera, sense organs located on the head between the eyes and ocelli.

Chaetotaxy The arrangement and nomenclature of bristles or chaetae on the exoskeleton.

Cheliform Shaped like a pincer or chela (chelate).

Chorion The outer shell of an insect egg.

Chrysalis The naked pupal stage of a butterfly.

Cibarium The preoral cavity between the hypopharynx and epipharynx.

Circumversion In male diptera, the 360° rotation of the genitalia about their longitudinal axis; *see* Torsion.

Clavate Gradually thickening toward the apex to form a club.

Clavus In Hemiptera, the oval or triangular anal portion of the front wing; *see* Fig. 34.4.

Cleavage *See* Holoblastic, Meroblastic.

Cleptoparasite A species which uses the nest and provisions of another species, as in cleptoparasitic bees.

Cloaca A common chamber into which the anus and gonopore open.

Clypeogenal suture A groove separating the clypeus and gena.

Clypeus The anterior region of the head to which the labrum is attached.

Coarctate pupa A pupa which is enclosed in the hardened cuticle of the last larval instar, as in cyclorrhaphan Diptera.

Cocoon A case formed partly or wholly of silk, in which pupation occurs.

Collophore In Collembola, a large, ventral projection from the first abdominal sternite; *see* Fig. 18.3.

Commensal One of the organisms participating in a commensalism.

Commensalism The association of two species without harm to either, and with benefit to at least one; *see* Symbiosis.

Common oviduct The unpaired, median duct through which the eggs pass.

Confluent Running together, merging.

Corbiculum In bees, a basket-shaped fringe of hairs on the hind tibia for carrying pollen; *see* Fig. 46.17.

Coriaceous Thick, tough, and leathery.

Corium In Heteroptera, the elongate, basal part of the forewing, usually thickened, coriaceous; *see* Fig. 34.4.

Cornified Tough, horny.

Corporotentorium A broad, platelike tentorium.

Cosmopolitan Occurring throughout most of the world.

Coxa The basal segment of the leg.

Cranium The head capsule, excluding the mouthparts.

Cremaster In Lepidoptera, the terminal hooks by which the pupa suspends itself.

Crepuscular Active at dusk.

Crochet In Lepidoptera, the curved spines or hooks on the abdominal prolegs of larvae.

Crypsis Concealment, camouflage (cryptic).

Cryptic coloration Concealing coloration, camouflage.

Cryptonephry The close association of the Malpighian tubules with the hindgut, an adaptation to increase water retention.

Ctenidium (pl., ctenidia) In Siphonaptera and certain other ectoparasites, a comb of short, flat spines (ctenidia).

Cucujiform Having the body extremely flat, with the legs directed laterally; applied to larvae.

Cuneus In Heteroptera, a small, triangular region of the corium delimited by the cuneal fracture; *See* Fig. 34.4.

Cursorial Adapted for running.

Cuticle The noncellular outer layer of the integument.

Cyclical parthenogenesis Parthenogenetic reproduction which recurs every other generation, as in Cynipidae.

Cyclomorphosis Seasonal change of body form, as in Aphididae, Cynipidae, and other insects.

Decomposer An organism which consumes dead organic material.

Decticous Having the ability to move the mandibles; applied to the pupal stage.

Delayed parasitoidism Parasitoidism in which the parasitoid egg remains dormant in an early stage of development until the host is near maturity, then develops rapidly.

Denticle A small tooth or tooth-shaped structure.

Deutocerebrum Part of the brain that innervates the antennae.

Diapause A period of arrested development.

Dicondylic Having two articulations.

Digitate Fingerlike.

Dimorphic Occurring in two distinct forms.

Diploid Double; having the full complement of maternal and paternal chromosomes.

Direct flight muscles Dorsoventral muscles attaching directly to sclerites at the wing bases.

Discal cell A cell, frequently enlarged, in the central part of the wing; *see* Figs. 42.7*a*, 44.7*e*.

Disruptive coloration Color patterning which tends to interrupt the outlines of the body.

Ditrysian In Lepidoptera, referring to the presence of separate genital openings for copulation and oviposition; *see* Fig. 44.4.

Diurnal Active during the day.

Ductus seminalis In female Lepidoptera, the tube connecting the bursa copulatrix and the common oviduct; *see* Fig. 44.4.

Ecdysial line A preformed line of weakness in the cuticle where the cuticle splits during ecdysis.

Ecdysis The actual shedding of the old cuticle during molting.

Eclosion The act of emerging from the egg.

Ectoparasite A parasite which lives externally on its host.

Ectoparasitoid A parasitoid which lives externally on its host.

Elytriform Having the form of elytra.

Elytron A leathery or rigid forewing, especially of coleoptera.

Emarginate Notched or indented.

Empodium A spinelike or cushionlike structure between the tarsal claws, especially in Diptera; *see* Fig. 42.13.

Endemic Peculiar to a given geographic region.

Endocuticle The inner unsclerotized layer of the procuticle.

Endoparasite A parasite which lives inside its host.

Endoparasitoid A parasitoid which lives inside its host.

Endopterygota Those insects in which the wings develop internally.

Endopterygote Having the wings develop from internal anlagen; member of the Endopterygota; *see* holometabolous.

Endoskeleton An internal part of the skeleton.

Entognathous Having the mandibles and maxillae retracted into pouches in the head, as in Protura, Collembola, Diplura.

Epicuticle The outermost layer of cuticle.

Epidermis The single layer of cells beneath the cuticle of the integument.

Epimeron The posterior division of a thoracic pleuron.

Epipharynx The lobelike inner surface of the labrum.

Epiphysis In some Lepidoptera, an articulated process on the anterior tibia.

Epipleurites The basalare and subalare sclerites of the pleuron.

Epiproct The dorsal plate of the eleventh abdominal segment.

Episternum The anterior division of a thoracic pleuron.

Eruciform Caterpillarlike.

Eucephalic In Diptera, referring to the presence of a distinct head capsule.

Exarate pupa Pupa in which the appendages are free of the body, as in Coleoptera.

Exocuticle The outer sclerotized layer of the procuticle.

Exopterygota Those insects in which the wings develop externally.

Exopterygote Having the wings develop externally; member of the Exopterygota; *see* hemimetabolous.

Exuviae The cast skin of nymphs or larvae at the time of metamorphosis.

Facultative Having the ability to live under more than one set of conditions.

Filiform Threadlike; slender and of nearly constant diameter.

Filter chamber In Homoptera, a complex organ in which the two ends of the ventriculus and the beginning of the intestine are associated

Filter feeding Feeding by straining small particles from water, as in culicid larvae or Corixidae.

Flabellate Fanlike, with thin, platelike processes lying flat against one another, as in antennae of Scarabaeidae; *see* Fig. 39.7.

Flagellum The part of the antenna beyond the pedicel.

Foliate Leaflike.

Fontanelle In termites, the shallow depression in which the frontal pore opens.

Foramen magnum The posterior opening of the cranium.

Forcipate Bearing forcepslike structures.

Forcipiform Forceps-shaped.

Fossorial Adapted for digging.

Frenate Bearing a frenulum.

Frenulum In Lepidoptera, the spinelike process arising from the base of the hind wing and projecting beneath the forewing, serving to couple the wings during flight; *see* Fig. 44.3.

Frons The anterior part of the cranium below the vertex and above the clypeus.

Frontoclypeal suture A transverse groove separating the frons and clypeus; also known as the epistomal suture.

Frontogenal suture A groove separating the frons and gena.

Fungivorous Consuming fungi.

Furca A forked endosternal process (see sternal apophysis); in Collembola, the appendage on abdominal segment 6, used for leaping.

Furcula A forked process.

Fusiform Spindle-shaped, tapering at both ends.

Galea The outer maxillary lobe.

Gall An abnormal growth of plant tissue, caused by an external stimulus, frequently from insects.

Gaster In Hymenoptera, the distal portion of the abdomen, usually globular (*see* petiole).

Gena The cheek or posterior part of the head behind and below the eye to the gular or occipital sutures.

Genital segments Abdominal segments 8 and 9.

Glossa One of the paired inner lobes of the labium; (*see* Paraglossa).

Gonocoxite The proximal segment of a paramere.

Gonopore The external opening of the ejaculatory duct (males) or oviduct (females).

Gonostylus The distal segment of a paramere.

Gradate veins A series of transverse crossveins, each before or beyond the next; *see* Fig. 38.3.

Gula The midventral sclerotized area on the posterior part of the cranium of some insects.

Gular sutures The paired sutures forming the lateral boundaries of the gula.

Haltere The reduced, knob-shaped hind wings of Diptera; *see* Fig. 42.2.

Hamate Wing coupling by hamuli.

Hamuli (hamulus) In Hymenoptera, the hook-shaped setae on the leading edge of the hind wing which catch beneath a posterior fold of the forewing; *see* Fig. 46.2.

Haploid Having only a single (maternal) complement of chromosomes.

Hemelytron In Hemiptera, the anterior, coriaceous part of the forewing; *see* Fig. 34.4.

Hemicephalic In Diptera, the condition in which the reduced head capsule is intermediate between eucephalic and acephalic.

Hemimetabolous Developing by incomplete or gradual metamorphosis, as in Orthoptera; *see* Exopterygote.

Hemocoel The body cavity of insects and other arthropods.

Hemolymph The fluid filling the hemocoel.

Herbivorous Feeding on living plants; phytophagous.

Heteroneurous Having the fore- and hind wings different in venation.

Hexapoda Those arthropods with three pairs of walking legs, including Insecta, Collembola, Diplura, and Protura.

Hexapodous Having three pairs of walking legs.

Holoblastic cleavage The type of embryonic cell division in which the entire egg is divided into cells; *see* Meroblastic.

Holometabolous Developing by complete metamorphosis; *see* Endopterygote.

Homoneurous Having the fore- and hind wings similar in venation.

Hormone A substance produced in one part of the body that has a profound effect on another part of the body.

Humeral plate A plate in the humeral region of the wing.

Humeral region The anterior base of the wing.

Humeral suture In termites, the line of weakness along which the wings break after the nuptial flight.

Humeral veins In Lepidoptera, veins which strengthen the basal portion of the leading edge of the hind wing; *see* Fig. 44.10*e*.

Humerus (pl., humeri) The shoulder; in Diptera, the anterior angles of the mesothorax; in Coleoptera, the outer angles of the elytra.

Hyaline Transparent, glassy.

Hydrofuge Repelling water, unwettable.

Hypermetamorphosis The type of development in which an insect passes through more than a single distinct larval stage.

Hyperparasitoidism The condition in which one parasitoid feeds on another parasitoid; hyperparasitism.

Hypognathous Having the mouthparts directed ventrally; *see* Prognathous; Opisthognathous.

Hypomeron In Coleoptera, the portion of the pronotum which is inflexed below the lateral margin.

Hypopharynx The tonguelike structure in the preoral cavity between the mouthparts.

Hypostoma The part of the subgena delimited by the hypostomal suture.

Hypostomal bridge The median fusion of the hypostomal areas behind the mouthparts.

Hypostomal suture Part of the subgenal suture posterior to the mandibles.

Indirect fertilization In Collembola, Diplura, Archeognatha, and Thysanura, the transfer of spermatozoa by means of a spermatophore which is deposited onto the substrate and picked up by the female.

Indirect flight muscles The tergosternal and longitudinal sternal muscles of the pterothorax.

Inquilinism The habit of living as a guest in the nest or abode of another organism.

Instar The developmental stage of the insect between molts; *see* Stadium.

Integument The cuticular and cellular covering of arthropods.

Intercoxal process In Coleoptera, the process of the basal abdominal segment which extends between the metacoxae.

Isolating mechanism Differences in geographical distribution, temporal occurrence, behavior, or genetic composition which prevent interbreeding of populations.

Johnston's organ A sense organ in the pedicel of the antenna.

Jugal fold The posterior basal fold between the jugal and anal regions of the wing; *see* Fig. 2.11.

Jugal lobe The posterior basal portion of the wing, delimited by the jugal fold; *see* Fig. 2.11.

Jugate In Lepidoptera and Trichoptera, having the wings coupled by a jugum.

Jugum In certain Lepidoptera and Trichoptera, a basal lobe of the forewing which overlaps the hind wing, coupling the wings during flight; *see* Fig. 44.3.

Kairomone A substance transmitted in chemical communication between individuals of different species that benefits the receiver of the substance.

Keel A sharply elevated ridge.

Labellum In Diptera, the fleshy apex of the labium; *see* Fig. 42.3.

Labium The fused second maxillae, forming the lower lip of the mouth in Hexapoda.

Labroclypeal suture A line of articulation between the labrum and clypeus.

Labrum The upper lip, attached to the clypeus.

Lacinia The medial lobe of the maxilla; *see* Galea.

Lamellate Sheetlike or leaflike, or composed of thin sheets.

Lanceolate Spear-shaped, tapering to a sharp point.

Larva The immature stage of a holometabolous insect, before the pupal stage; *see* Nymph; Naiad.

Larviform Resembling a larva.

Laterotergite A lateral sclerite on the dorsum, distinct from the principal median tergite.

Macrolecithal Having a large yolk mass in the egg, as in most Hexapoda other than Collembola.

Macrotrichia In Diptera, the larger hairs on the wings.

Mala The single maxillary lobe of some endopterygote larvae.

Malpighian tubules Insect urinary organs; long, slender, blindly ending ducts opening into the hindgut.

Mandibulate Having the mandibles adapted for biting or chewing.

Marginal cell In Diptera and Hymenoptera, a cell bordering the distal anterior wing margin, beyond the pterostigma.

Marginal vein In Hymenoptera, the vein delimiting the marginal cell posteriorly; any vein near the wing margin.

Mass provisioning In solitary bees and wasps, the storage of sufficient food for larval development before oviposition.

Meconium The accumulated excretory wastes from the larval stage which are voided shortly after adult emergence in certain insects.

Median caudal filament A long process of the epiproct.

Median plates Two plates at the wing base near the media and cubitus veins.

Membrane Any flexible, nonsclerotized integument; the part of the wing between the veins.

Mentum Distal sclerite of postmentum.

Meroblastic cleavage The superficial type of embryonic cell division in which only the nucleus and nuclear cytoplasm of the egg are divided.

Meron In Diptera, the lateral, basal portion of the coxa, which is incorporated into the thorax; the lateral, basal area of the coxa of any insect.

Meropleuron The combined meron and ventral portion of the epipleuron.

Mes(o)- Prefix indicating the middle or middle member of a series, e.g., mesothorax.

Mesothorax The second or middle segment of the thorax.

Meta- Prefix indicating the posterior part of a structure or posterior member of a series, e.g., metathorax.

Metamorphosis The change of body form through which insects or other organisms pass in developing to the adult; *see* Holometabolous; Hemimetabolous.

Metathorax The last segment of the thorax.

Microlecithal Having little yolk in the egg, as in Collembola.

Microtrichia Minute hairs on the wings of Diptera and certain other insects.

Mimicry The resemblance of an organism to another, distantly related organism.

Model In Batesian mimicry, the distasteful organism which is mimicked.

Molar lobe, mola The basal portion of the mandible, usually modified as a grinding surface.

Molt The entire process of shedding the old cuticle.

Moniliform Composed of beadlike segments; beaded; *see* Fig. 2.7.

Monocondylic Having a single articulation; *see* Dicondylic.

Monophyletic Evolved from a single ancestral form.

Monotrysian In Lepidoptera, referring to the condition in which the genital system opens through a single pore into a cloaca; *see* Ditrysian.

Mouthhooks In cyclorrhaphan Diptera, the secondary, mandiblelike structures located laterally in the atrium; *see* Fig. 42.4*c*.

Müllerian mimicry Mimicry in which several distasteful organisms share mutual similarity.

Myrmecophile An organism which lives with ants, either as predator or commensal (myrmecophilous).

Naiad In hemimetabolous insects, any aquatic nymph.

Nasute In termites, a type of soldier in which the frontal gland opens upon a median horn.

Nauplioid larva In parasitic Hymenoptera, a first instar larva which resembles the nauplius larva of Crustacea.

Necrosis Decay.

Nectariferous Feeding on nectar.

Neoptera Those insects possessing a neopterous flight mechanism.

Neopterous Having a flight mechanism involving indirect musculature and the ability to flex the wings over the dorsum at rest.

Neoteny Reaching sexual maturity during a larval stage (neotenic).

Niche The ecological position of an organism, such as aquatic predator in small streams.

Nocturnal Active at night.

Node, nodus In Hymenoptera, the knoblike segments at the base of the abdomen; in Odonata, a crossvein near the costal margin; *see* Fig. 21.1.

Notal wing process Anterior and posterior processes on the alinotum where the wings articulate.

Notum Tergum; dorsal part of a thoracic segment.

Nuptial flight In Isoptera and Hymenoptera, the dispersal flight of sexual forms; a mating flight.

Nygma (pl., nygmata) In Neuroptera, Trichoptera, and some Hymenoptera, a small sensory organ, sometimes present in the radial or median portion of the wing membrane.

Nymph An immature hemimetabolous insect.

Oblongum In Coleoptera, a closed cell in the median part of the wing.

Obtect In pupae, having the appendages rigidly adherent to the body wall.

Occipital suture An annular suture on the posterior part of the insect cranium, ending near the insertions of the mandibles.

Occiput The posterior part of the head capsule, defined by an anterior occipital suture and a posterior postoccipital suture.

Ocellar triangle In Diptera, the region defined by the three ocelli.

Ocellus (pl., ocelli) The simple eye of an adult insect.

Oligolectic In Hymenoptera, especially bees, visiting a restricted range of plants for pollen or nectar; *see* Polylectic.

Ommatidium An individual unit of the compound eye.

Omnivorous Having variable feeding preferences, including both animal and vegetable material; *see* Predaceous; Herbivorous.

Ootheca A protective covering for an egg mass.

Operculum A lid or cover, usually circular.

Opisthognathous Having the mouthparts directed posteroventrally; *see* Prognathous; Hypognathous.

Oral vibrissae In Diptera, a pair of stout bristles arising laterally near the oral region; *see* Fig. 42.15.

Osmeterium (pl. osmeteria) Fleshy glandular structure which may be everted from the bodies of certain caterpillars to release an odoriferous defensive secretion.

Oviduct One of the (usually) paired tubes which pass the eggs from the ovaries to the vagina.

Ovoviviparous Producing live offspring by the hatching of eggs retained in the mother's body.

Paleoptera Those insects possessing a paleopterous flight mechanism.

Paleopterous Having a flight mechanism involving direct musculature and lacking the ability to flex the wings over the back at rest.

Palpifer Prominence on which maxillary palps attach.

Palpiger Prominence on which labial palps attach.

Papilla A small, fleshy, nipple-shaped projection.

Papilliform Papilla-shaped, nipple-shaped.

Paraglossa (pl., paraglossae) A paired lateral lobe of the labium, corresponding to the galea; *see* Fig. 27.2.

Paramere A paired copulatory appendage of the male.

Paraproct A lateral plate of the eleventh abdominal segment.

Parasitic castration The induction by internal parasites of morphological traits of the opposite sex; stylopization.

Parthenogenesis Reproduction without fertilization; *see* Arrhenotoky; Thelytoky.

Patagium (pl., patagia) In Lepidoptera, one of the small lobes that overlie the forewings; *see* Fig. 44.1.

Pecten A comb; in Hymenoptera, the rigid setae on the basal parts of the maxillae and labium; in Lepidoptera, a row of setae on the antennal scape.

Pectinate Comblike.

Pedicel The second segment of the antenna; in Hymenoptera, the basal one or two segments of the abdomen.

Penis The male intromittent organ.

Penultimate Next to the last.

Periproct The unsegmented part of the body surrounding the anus.

Petiolate Stalked.

Petiole In Hymenoptera, the slender, stalklike basal abdominal segment.

Pharate Referring to a developmental stage which is enclosed in the integument of the previous instar, especially adults which are pharate in the pupal integument.

Pheromone A substance which is secreted externally, functioning in communication among individuals of the same species; *see* Allomone; Kairomone.

Phoresy An interrelationship in which one organism is transported by another.

Phragma (pl., phragmata) Transverse, platelike invaginations of the skeleton, usually increasing the area available for muscle attachment; homologous to the antecosta of a secondary segment.

Physogastry Having a swollen abdomen (physogastric).

Phytophagous Feeding on plants.

Planidium In Hymenoptera, a type of primary larva with sclerotized, fusiform body and spinelike locomotory organs; *see* Fig. 46.4.

Plastron A respiratory organ in which special hydrofuge cuticle holds a layer of air through which gas exchange occurs.

Pleural apophysis An invagination of the pleural ridge.

Pleural coxal process A ventral process of the pleuron on which the coxa articulates.

Pleural ridge An internal ridge that strengthens the pleuron.

Pleural suture The suture running between the coxa and the pleural wing process and separating the episternum and epimeron.

Pleural wing process A dorsal process of the pleuron on which the wing articulates.

Pleuron The platelike lateral part of a segment, including the episternum and epimeron.

Pleurosternal suture A suture between the pleuron and sternum.

Pleurum Pleuron.

Plumose Feathery, like a plume.

Polyembryony The production of more than one embryo from an egg.

Polygamous Mating more than once; strictly, referring to males which mate with more than one female.

Polylectic In Hymenoptera, especially bees, visiting a broad range of plants for pollen and nectar; *see* Oligolectic.

Polymorphism Having more than one body form.

Polyphagous Having broad feeding tolerances; in phytophagous insects, accepting many host plants.

Polyphyletic Arising from more than one evolutionary lineage.

Porrect Extending forward horizontally

Postabdomen In Diptera, the modified, posterior abdominal segments, including the genitalia; *see* Preabdomen.

Postclypeus Enlarged frontoclypeal region of sucking insects.

Postgena The lower portion of the occiput.

Postgenal bridge The median fusion of postgenal areas behind the mouthparts.

Postgenital segments Abdominal segments beyond the ninth.

Postmentum Basal part of the labium, proximal to the labial suture.

Postnotum An intersegmental dorsal sclerite associated with the tergum of the preceding segment; *see* Acrotergite.

Postoccipital suture A groove bordering the foramen magnum, running dorsally between the posterior tentorial pits.

Postocciput The cranium posterior to the postoccipital suture.

Postphragma The posterior phragma of a segment.

Postscutellum The posterior-most dorsal sclerite of a segment, located just posterad of the scutellum.

Postvertical bristles In Diptera, a pair of bristles arising behind the lateral ocelli; *see* Fig. 42.15.

Preabdomen In Diptera, the unmodified, anterior portion of the abdomen; *see* Postabdomen.

Predaceous Preying upon other organisms, usually animals.

Prehensile Adapted for grasping.

Pregenital segments Abdominal segments 1 to 7.

Prementum Sclerite of labium distal to labial suture and to which the palps attach.

Prephragma The anterior phragma of a segment.

Prepupa A quiescent, nonfeeding stage, passed during the last larval instar.

Prescutum The anterior portion of the scutum, usually delimited by a suture.

Prestomal teeth In Diptera, sclerotized, abrasive processes on the labellum around the oral opening.

Pretarsus The terminal leg segment, including tarsal claws and medial structures such as arolium, pulvillus, or empodium.

Primary reproductives In termites, the original pair which founded a colony.

Pro- Before, anterior; prefix indicating the first part of a structure or the first member of a series, e.g., prothorax.

Proboscis An extended, tubular mouth structure, as in many Diptera, Hymenoptera, Lepidoptera, Hemiptera; *see* Rostrum.

Procuticle The cuticle minus the epicuticle.

Prognathous Having the mouthparts directed anteriorly.

Progressive provisioning In solitary bees and wasps, the practice of providing the larval food over a period of time.

Proleg The fleshy, unsegmented legs, usually on the abdominal segments, of holometabolous larvae.

Propodeum In Hymenoptera, the anterior most abdominal segment, fused with the metathorax; *see* Fig. 46.1.

Prostomium The unsegmented part of the body in front of the mouth.

Protocephalon Part of the primitive head, including the prostomium and one to three true segments.

Protocerebrum Part of the brain that innervates the compound eyes and ocelli.

Prothorax The first segment of the thorax.

Proventriculus The posterior part of the foregut.

Pseudoculus In Protura, cephalic sensory organs; *see* Fig. 18-1.

Pseudopod; pseudopodium Fleshy, unsegmented appendages on holometabolous larvae; prolegs.

Pseudotracheae In Diptera, the minute grooves on the labellum through which liquid food is ingested.

Pterostigma A thickening on the costal margin of the wing of many insects.

Pterothorax The wing-bearing portion of the thorax, usually the mesothorax and metathorax.

Pterygota Those insects which primitively bear wings (sometimes lost secondarily).

Ptilinal suture In Diptera, a crescentic groove just above the antennal insertions; *see* Fig. 42.15.

Ptilinum In Diptera, a membranous sac which is everted through the ptilinal suture of the teneral adult in order to force open the puparium.

Pubescent Covered with short, densely set hairs.

Pulvillus Paired, padlike structures between or beneath the tarsal claws.

Pupa In holometabolous insects, the (usually) nonmotile instar during which transformation from larva to adult occurs.

Puparium In Diptera, the pupal case formed from the strongly sclerotized cuticle of the last larval instar.

Quadrate Square or nearly square.

Queen In social insects, the primary female reproductive.

Raptorial Adapted for grasping prey, as in the legs of Mantodea.

Rectum The posterior part of the hind gut or proctodeum.

Relictual Occupying a geographic or ecological range much more restricted than formerly.

Remigium The stiff, anterior area of the wing.

Reticulate Covered with a meshwork of lines or ridges; netlike.

Retinaculum In Collembola, the unsegmented appendage of the third abdominal segment; in Lepidoptera, the loop into which the frenulum fits.

Rostrum An elongate, tubular projection of the head; cf. Hemiptera, Coleoptera.

Rudimentary Undeveloped.

Salivarium A cavity between the hypopharynx and labium into which the salivary duct opens.

Saltatorial Adapted for jumping.

Saprophagous Feeding on decaying organic matter (saprophytic).

Saprophytic Saprophagous

Scape The basal segment of the antenna.

Scarabaeiform Having the general form of a scarabaeid beetle larva; *see* Fig. 39.7*d*.

Sclerite A hardened or sclerotized plate of the cuticle, or an area of hardened cuticle surrounded by sutures.

Sclerotization The polymerization and cross bonding of protein and chitin to produce the hardened insect exoskeleton.

Scopa In Hymenoptera, the pollen-collecting apparatus on the hind legs, consisting of a brush of stiff setae.

Scutellum The posterior division of the notum; in Hemiptera and Coleoptera, the triangular or shield-shaped posterior portion of the mesothorax.

Seasonal polymorphism Change of body form in different seasons, as in Aphididae and certain other Homoptera.

Semidicondylic Referring to mandibles of Archeognatha and Ephemeroptera, where the weakly developed dorsal articulation fits into a very shallow notch in the oral rim of the head capsule.

Sensilium In fleas, a sensory plate on the apex of the abdomen.

Sensillum (pl., sensilla) A simple sense organ; one of the units of a compound sense organ.

Serpentine mines In Lepidoptera and Diptera, winding larval feeding tunnels in leaves or stems of plants.

Serrate Sawlike, with toothed margins, as in serrate antennae; *see* Fig. 2.7.

Seta (pl., setae) A hair or bristle.

Setaceous Hairlike or bristlelike.

Setose Covered with setae.

Solitary Occurring singly or in pairs, not socially; applied especially to wasps and bees.

Spermatheca In female insects, the saclike organ which receives and stores the sperm from the male.

Spermatophore The capsule or covering which males of some insects secrete around the spermatozoa.

Spine A rigid, multicellular outgrowth of the integument.

Spinneret An organ with which silk is produced.

Spiracle The openings through which air enters the tracheae.

Spittle insects Immature Cercopidae, so-called because of the frothy secretion in which these insects live.

Spur An articulated, multicellular outgrowth of the integument.

Squama In some Diptera, a small, scalelike folding in the posterior basal wing margin.

Stadium The interval between molts; *see* Instar.

Stellate Star-shaped.

Sternite Any sclerite in the sternal region of a segment; a subdivision of the sternum.

Sternal apophysis An invagination of the thoracic sternum to which muscles attach.

Sternum The ventral portion of a body segment, delimited from the pleura by sutures (sternal).

Stigma (pl., stigmata) A thickened, usually pigmented region on the costal margin of the wings of many insects.

Sting In some Hymenoptera, the modified ovipositor, which is adapted for injecting venom; *see* Fig. 46.2.

Stipes One of the basal sclerites of the maxilla.

Stomodaeum The anterior portion of the alimentary canal; foregut.

Stria In Coleoptera, the longitudinal, impressed lines usually visible on the elytra; any fine longitudinal line or furrow.

Striate Sculptured with fine, parallel lines or furrows.

Stridulation The production of sounds by rubbing one surface against another (stridulate, stridulatory).

Style Any small, fine, needle-shaped or sharply pointed organ; in Diptera, the fine, dorsal appendage on the third antennal segment; in entognathous arthropods and Apterygota, the rudimentary abdominal appendages (stylus, styli).

Stylet A small style; in Hemiptera, the flexible filaments forming the piercing mouthparts.

Styliform Having the shape of a style or stylet.

Stylopization Parasitization by Strepsiptera.

Sub- Slightly less than, or about equal to, or just below.

Subalare The posterior epipleurite.

Subcortical The region between the bark and wood of trees and shrubs.

Subfamily A taxonomic subdivision of a family.

Subgena The cranium below the subgenal suture.

Subgenal suture A groove above the bases of the mouthparts, running between the anterior and posterior tentorial pits.

Subgenital plate A ventral sclerite covering the gonopore, as in Blattodea and some other insects; usually the eighth sternum in females and the ninth in males.

Subimago In Ephemeroptera, the winged instar immediately preceding the reproductively mature adult.

Submarginal cell In Hymenoptera, the cells just behind the marginal cell; *see* Fig. 46.6.

Submarginal vein Especially in Hymenoptera, a vein just behind the costal wing margin.

Submentum Proximal sclerite of postmentum.

Suborder A taxonomic category intermediate to order and family.

Subquadrate Nearly square.

Subscutellum In some Diptera, the infolded portion of the postscutellum; *see* Fig. 42.18.

Superfamily A taxonomic category intermediate to family and order.

Suture A line, seam, or furrow in the cuticle; may indicate an articulation or internal ridge.

Symbiosis A living together of different species of organisms.

Symbiote A member of a symbiosis (symbiont).

Synanthropic Living in close association with humans.

Tagma (pl., tagmata) A distinct region of the body; in insects the head, thorax, or abdomen.

Tagmosis The evolutionary differentiation of the body into distinct regions, or tagmata.

Tarsomeres Subsegments of the tarsus.

Tegmen (pl., tegmina) The leathery forewing of Orthoptera, Blattodea, and other orthopteroid insects.

Tegula A small articulated sclerite at the costal base of the wing, especially in Hymenoptera; *see* Fig. 46.1.

Teneral The condition of an insect immediately after molting and before the cuticle has hardened.

Tentorial bridge The median fusion of the tentorial arms inside the cranium.

Tentorial pit The point of invagination of a tentorial arm.
Tentorium The endoskeleton of the insect head.
Tergite Any sclerite in the tergal region of a segment; a subdivision of the tergum.
Tergosternal Referring to the tergum and the sternum, as in tergosternal muscles.
Tergum The dorsal portion of a body segment (tergal).
Termitarium The structure or nest inhabited by a termite colony.
Termitophile An organism which lives with termites, either as predator or commensal (termitophilous).
Thelytoky Parthenogenetic reproduction in which all the offspring are females.
Thigmotactic Contact-loving, as in insects which inhabit crevices, cracks, or other tight quarters.
Thorax The intermediate tagma of the insect body, bearing the legs and wings.
Tormogen An epidermal cell that develops into the socket for a seta or hair.
Torsion A twisting; in many male Diptera, the rotation of the genitalia 90 to 360° from their primitive position; *see* Circumversion.
Tracheae The internal, cuticular air tubes of insects.
Tracheal gill Gills supplied with tracheae; more loosely, any gill involved with respiration.
Tracheal system The tracheae and tracheoles.
Tracheole An extremely fine trachea.
Transverse Intersecting the longitudinal axis at right angles.
Trichogen An epidermal cell that develops into a seta or hair.
Tritocerebrum Part of the brain that innervates the labrum; originally a postoral ventral ganglion.
Triungulin The campodeiform first-instar larva of Strepsiptera and some parasitoid beetles; *see* Fig. 40.1.
Trochanter The leg segment between the coxa and femur.
Trochantin A small, movable pleurite articulating anteriorly with the coxa.
Trophallaxis In social insects the mutual exchange of food between colony members.
Truncate Having a square margin.
Trunk The thorax and abdomen as a whole.
Tubercle A small nipple-shaped protrusion (tuberculate).
Tymbal In cicadas, the membrane of the sound-producing organ.
Tympanal organ An ear or auditory organ.
Tympanum Any tightly stretched membrane, especially the membranes of the auditory organs.

Univoltine Having a single generation per year.
Urogomphus Any fixed or movable terminal abdominal process of insect larvae; applied especially to Coleoptera.
Urticate To sting or nettle (urticating).

Valvifer The base of an ovipositor appendage.
Valvula One blade of the shaft of an ovipositor appendage.
Vasiform orifice In Aleyrodidae, a dorsal glandular opening on the last abdominal segment; *see* Fig. 34.10.
Vector The intermediate host which conveys a disease-producing organism; an insect that carries pollen from plant to plant; to convey pathogens or pollen.
Vermiform Worm-shaped.
Vertex The top of the head, between the eyes and above the frons.
Vesicle A small sac or bladder
Vestigial Rudimentary, small; the remnants of a formerly functional organ.
Vibrissae A group of bristles inserted around the oral fossa in many Diptera; *see* Fig. 42.15.
Viviparous Bearing live young.

Zoophagous Feeding on material of animal origin.

REFERENCES CITED[1]

Achtelig, M., and N. P. Kristensen. 1973. A re-examination of the relationships of the Raphidioptera (Insecta). Z. Zool. Syst. Evolutions. forsch., **11**:268–274.

Acridia. 1972 to date. Association d' acridologie.

Agosin, M., and A. S. Perry. 1974. Microsomal mixed-function oxidases. *Physiol. Insecta,* 2d ed., **5**:537–596.

Agrell, I. P. S., and A. M. Lundquist. 1973. Physiological and biochemical changes during insect development. *Physiol. Insecta,* 2d ed., **1**:159–247.

Albrecht, F. O. 1953. *The Anatomy of the Migratory Locust.* Athlone Press, London. 118 pp.

Alexander, R. D. 1957. Sound production and associated behavior in insects. Ohio J. Sci., **57**:101–113.

——. 1962. Evolutionary change in cricket acoustical communication. Evolution, **16**:443–467.

——. 1968. Arthropods, in *Animal Communication,* T. A. Sebeok (ed.). Indiana University Press, Bloomington. pp. 167–216.

——. 1974. The evolution of social behavior. Ann. Rev. Ecol. Syst., **5**:325–383.

—— and W. L. Brown. 1963. Mating behavior and the origin of insect wings. Occas. Pap. Mus. Zool. Univ. Mich., No. 628, pp. 1–19.

—— and T. E. Moore. 1962. The evolutionary relationships of 17-year and 13-year cicadas, and three new species (Homoptera, Cicadidae, *Magicicada*). Misc. Publ. Mus. Zool. Univ. Mich., **121**:1–59.

——, —— and R. E. Woodruff. 1963. The evolutionary differentiation of stridulatory signals in beetles (Insecta: Coleoptera). Anim. Behav., **11**:111–115.

Alford, D. V. 1975. *Bumblebees.* Davis-Poynter, London. 352 pp.

[1]References in *The Physiology of the Insecta,* 2d ed., M. Rockstein (ed.), Academic Press, Inc., New York, are cited here for brevity as *Physiol. Insecta,* 2d ed.

Alloway, T. M. 1972. Learning and memory in insects. Ann. Rev. Entomol., **17**:43–56.

Altman, J. S., and N. M. Tyrer. 1974. Insect flight as a system for the study of the development of neuronal connections, in *Experimental Analysis of Insect Behavior,* L. B. Browne (ed.). Springer-Verlag, New York. pp. 159–179.

Amos, W. H. 1967. *The Life of the Pond.* McGraw-Hill Book Company, New York. 232 pp.

Anderson, D. T. 1973. *Embryology and Phylogeny in Annelids and Arthropods.* Pergamon Press, Oxford. 495 pp.

Ando, H. 1962. *The Comparative Embryology of Odonata, with Special Reference to a Relic Dragonfly, Epiphlebia superstes Selys.* Society for the Promotion of Science, Tokyo, Japan. 205 pp.

Andrewartha, H. G., and L. C. Birch. 1954. *The Distribution and Abundance of Animals.* The University of Chicago Press, Chicago. 782 pp.

Apple, J. L., and R. F. Smith (eds.). 1976. *Integrated Pest Management.* Plenum Press, New York. 200 pp.

Arnett, R. H., Jr. 1960. *The Beetles of the United States (A Manual for Identification).* Catholic University Press, Washington. 1112 pp. (Reprinted, 1968, American Entomological Institute, Ann Arbor, Mich.)

Arnold, J. W. 1974. The hemocytes of insects. *Physiol. Insecta,* 2d ed., **5**:202–254.

Asahina, E. 1966. Freezing and frost resistance in insects, in *Cryobiology,* H. T. Merryman (ed.). Academic Press, Inc., London. pp. 451–486.

——. 1969. Frost resistance in insects. Adv. Insect Physiol., **6**:1–50.

Ashburner, M. 1970. Function and structure of polytene chromosomes during insect development. Adv. Insect Physiol., **7**:2–96.

Askew, R. R. 1971. *Parasitic Insects.* American Elsevier Publishing Company, Inc., New York. 316 pp.

Aspöck, H., and A. Aspöck. 1975. The present state of knowledge on the Raphidioptera of America (Insecta, Neuropteroidea). Pol. Pismo Entomol., **45**:537–546.

Baccetti, B. 1972. Insect sperm cells. Adv. Insect Physiol., **9**:316–397.

Bailey, S. F. 1957. The thrips of California, Part 1: Suborder Terebrantia. Bull. Calif. Insect Survey, **4**:143–220.

Baker, H. G., and I. Baker. 1973. Some anthecological aspects of the evolution of nectar-producing flowers, particularly amino acid production in nectar, in *Taxonomy and Ecology,* V. H. Heywood (ed.). Academic Press, Inc., London. pp. 243–264.

—— and ——. 1975. Studies of nectar-constitution and pollinator-plant coevolution, in *Animal and Plant Coevolution,* L. E. Gilbert and P. H. Raven (eds.). University of Texas Press, Austin. pp. 100–140.

—— and P. D. Hurd, Jr. 1968. Intrafloral ecology, Ann. Rev. Entomol., **13**:385–414.

Balduf, W. V. 1935. *The Bionomics of Entomophagous Coleoptera.* John S. Swift Co., Inc., St. Louis, Mo. 220 pp.

——. 1939. *The Bionomics of Entomophagous Insects,* part II. John S. Swift Co., Inc., St. Louis, Mo. 384 pp.

Ball, E. D., E. R. Tinkham, R. Flock, and C. T. Vorheis. 1942. The grasshoppers and other Orthoptera of Arizona. Ariz. Agric. Exp. St. Tech. Bull., **93**:257–373.

Banks, N. 1927. Revision of nearctic Myrmeleontidae. Bull. Mus. Comp. Zool., Harv. Univ., **68**:1–84.

Barbosa, P., and T. M. Peters. 1972. *Readings in Entomology.* W. B. Saunders Co., Philadelphia. 450 pp.

Barnhart, C. S. 1961. The internal anatomy of the silverfish *Ctenolepisma cambelli* and *Lepisma saccharinum* (Thysanura: Lepismatidae). Ann. Entomol. Soc. Am., **54**:177–196.

Barr, T. C., Jr. 1967. Observations on the ecology of caves. Am. Nat., **101**:475–491.

Barth, R. 1954. Untersuchungen an den Tarsaldrüsen von *Embolyntha batesi* MacLachlan, 1877 (Embioidea). Zool. Jahrb., **74**:172–188.

Barth, R. H., and L. J. Lester. 1973. Neuro-hormonal control of sexual behavior in insects. Ann. Rev. Entomol., **18**:455–472.

Bateman, M. A. 1972. The ecology of fruit flies. Ann. Rev. Entomol., **17**:493–518.

Bates, H. W. 1862. Contributions to an insect fauna of the Amazon Valley. Lepidoptera: Heliconidae. Trans. Linn. Soc. Zool., **23**:495–566.

Batra, S. W. T., and L. R. Batra. 1967. The fungus gardens of insects. Sci. Am., **217**(5):112–120.

Baust, J. G. 1972. Mechanisms of insect freezing protection: *Pterostichus brevicornis.* Nature (London) [New Biol.], **236**:219–221.

——. 1972. Influence of low temperature acclimation in cold hardiness in *Pterostichus brevicornis.* J. Insect. Physiol., **18**:1935–1947.

——. 1976. Temperature buffering in an artic microhabitat. Ann. Entomol. Soc. Am., **69**:117–119.

—— and L. K. Miller. 1970. Seasonal variations in glycerol content and its influence in cold hardiness in the Alaskan carabid beetle, *Pterostichus brevicornis.* J. Insect Physiol., **16**:979–990.

Bay, E. C. 1974. Predator-prey relationships among aquatic insects. Ann. Rev. Entomol., **19**:441–454.

Beardsley, J. W., Jr., and R. H. Gonzalez. 1975. The biology and ecology of armored scales. Ann. Rev. Entomol., **20**:47–73.

Beck, S. D. 1965. Resistance of plants to insects. Ann. Rev. Entomol., **10**:207–232.

——. 1968. *Insect Photoperiodism.* Academic Press, Inc., New York. 288 pp.

Beier, M. 1955. Ordnung: Saltatoptera m. (Saltatoria Latrielle 1817). Bronn's Kl. Ordn. Tierreichs, (5)(3)**6**:34–304.

——. 1964. Blattopteroidea. Ordnung Mantodea Burmeister 1838 (Raptoriae Latreille 1802; Mantoidea Handlirsch 1903; Mantidea auct.). Bronn's Kl. Ordn. Tierreichs, (5)(3)**6**:849–970.

Benedetti, R. 1973. Notes on the biology of *Neomachilis halophila* on a California sandy beach (Thysanura: Machilidae). Pan-Pac. Entomol., **49**:246–249.

Bennet-Clark, H. C., and A. W. Ewing. 1970. The love song of the fruit fly. Sci. Am., **223**(1):84–92.

Bentley, D., and R. R. Hoy. 1974. The neurobiology of cricket song. Sci. Am., **231**(2):34–44

Bergerard, J. 1972. Environmental and physiological control of sex determination and differentiation. Ann. Rev. Entomol., **17**:57–74.

Berner, L. 1950. *The Mayflies of Florida.* University of Florida Press, Gainesville. 267 pp.

——. 1959. A tabular summary of the biology of North American mayfly nymphs (Ephemeroptera). Bull. Fla. State Mus. Biol. Ser, **4**:1–58.

Beroza, M. (ed.). 1976. *Pest Management with Insect Sex Attractants.* American Chemical Society, Washington. 192 pp.

Berridge, M. J., and W. T. Prince. 1972. The role of cyclic AMP and calcium in hormone action. Adv. Insect Physiol., **9**:1–49.

Birch, L. C. 1953. Experimental background to the study of the distribution and abundance of insects. I. The influence of temperature, moisture and food on the innate capacity for increase of three grain beetles. Ecology, **34**:608–711.

Birch, M. C. (ed.). 1974. *Pheromones.* North-Holland Publishing Company, Amsterdam. 495 pp.

—— and D. L. Wood. 1975. Mutual inhibition of the attractant pheromone response by two species of *Ips* (Coleoptera: Scolytidae). J. Chem. Ecol., **1**:101–113.

Birket-Smith, S. J. P. 1974. On the abdominal morphology of Thysanura (Archeognatha and Thysanura s. str.). Entomol. Scand. Suppl., **6**:1–67.

Bishop, J. A., and L. M. Cook. 1975. Moths, melanism and clean air. Sci. Am., **232**(1):90–99.

Blackman, R. 1974. *Aphids.* Ginn and Company, London. 175 pp.

Blatchley, W. D. 1920. *Orthoptera of Northeastern America.* Nature Publishing Company, Indianapolis. 784 pp.

——. 1926. *Heteroptera or True Bugs of Eastern North America, with Special Reference to the Faunas of Indiana and Florida.* Nature Publishing Company, Indianapolis, Ind. 1116 pp.

Blest, A. D. 1957. The function of eyespot patterns in the Lepidoptera. Behaviour, **11**:209–256.

Blum, M. S. 1969. Alarm pheromones. Ann. Rev. Entomol., **14**:57–80.

Bonhag, P. F. 1949. The thoracic mechanism of the adult horsefly. Cornell Univ. Agric. Exp. Stn. Mem., **285**:1–39.

Bohart, G. E. 1972. Management of wild bees for the pollination of crops. Ann. Rev. Entomol., **17**:287–312.

Bohart, R. M. 1941. A revision of the Strepsiptera with special reference to the species of North America. Univ. Calif. Publ. Entomol., **7**:91–160.

——. and A. S. Menke. 1976. *Sphecid Wasps of the World.* University of California Press, Berkeley. 695 pp.

Borden, J. H. 1974. Aggregation pheromones in the Scolytidae, in *Pheromones,* M. Birch (ed.). North-Holland Publishing Company, Amsterdam. pp. 133–160.

Borror, D. J., and R. E. White. 1970. *A Field Guide to the Insects of America North of Mexico.* Houghton Mifflin Company, Boston. 404 pp.

——, D. M. DeLong, and C. A. Triplehorn. 1976. *An Introduction to the Study of Insects,* 4th ed. Holt, Rinehart and Winston, Inc., New York. 852 pp.

Böving, A. G., and F. C. Craighead. 1930. An illustrated synopsis of the principal larval forms of the order Coleoptera. Entomol. Am., **11**:1–351.

Brady, J. 1974. The physiology of insect circadian rhythms. Adv. Insect Physiol., **10**:1–115.

Braziunas, T. F. 1975. A geological duration chart. Geology, **3**:342–343.

Breedlove, D., and P. Ehrlich. 1968. Plant-herbivore coevolution: Lupines and lycaenids. Science, **162**:671–672.

——, and ——. 1972. Coevolution: Patterns of legume predation by a lycaenid butterfly. Oecologia, **10**:99–104.

Bridges, R. G. 1972. Choline metabolism in insects. Adv. Insect Physiol., **9**:51–100.

Brinck, P. 1957. Reproductive systems and mating in Ephemeroptera. Opusc. Entomol., **22**:1–37.

Britt, N. W. 1962. Biology of two species of Lake Erie mayflies, *Ephoron album* (Say) and *Ephemera simulans* Walker. Bull. Ohio Biol. Surv., **1**:1–70

Britton, E. B. 1970. Coleoptera (beetles), in *Insects of Australia,* CSIRO, Canberra. pp. 495–621.

——. 1974. Coleoptera (beetles), in *Insects of Australia,* Supplement 1974, CSIRO, Canberra. pp. 62–89.

—— et al. 1970. *Insects of Australia.* Melbourne University Press, Carlton, Victoria, Australia. CSIRO, Canberra. 1029 pp.

Britton, W. E., et al. 1923. The Hemiptera or sucking insects of Connecticut. Guide to the insects of Connecticut. Conn. State Geol. Nat. Hist. Surv., **4**:1–807.

Brooks, A. R. 1958. Acridoidea of southern Alberta, Saskatchewan, and Manitoba (Orthoptera). Can. Entomol., Suppl., **9**:1–92.

—— and L. A. Kelton. 1967. Aquatic and semiaquatic Heteroptera of Alberta, Saskatchewan and Manitoba (Hemiptera). Mem. Entomol. Soc. Can., **51**:1–92.

Brooks, M. A., and T. J. Kurtti. 1971. Insect cell and tissue culture. Ann. Rev. Entomol., **16**:27–52.

Brooks, W. M. 1974. Protozoan infections, in *Insect Diseases,* G. E. Cantwell (ed.). Marcel Dekker, Inc., New York. pp. 237–300.

Brossollet, J. 1971. Reflections of the plague in art. Sydsv. Med. Sällsk. (Lund), **8**:11–24.

Brower, L. P. 1969. Ecological chemistry. Sci. Am., **220**(2):22–29.

—— and J. V. Z. Brower. 1964. Birds, butterflies, and

plant poisons: A study in ecological chemistry. Zoologica, **49**:137–159.

——, J. Alcock, and J. V. Z. Brower. 1971. Avian feeding behavior and the selective advantage of incipient mimicry, in *Ecological Genetics and Evolution,* R. Creed (ed.). Appleton-Century-Crofts, Inc., New York. pp. 261–274.

——, W. N. Ryerson, L. L. Coppinger, and S. C. Glazier. 1968. Ecological chemistry and the palatability spectrum. Science, **161**:1349–1351.

Brown, W. L., Jr., T. Eisner, and R. H. Whittaker. 1970. Allomones and kairomones: Transspecific chemical messengers. BioScience, **20**:21.

Browne, L. B. 1974. *Experimental Analysis of Insect Behavior.* Springer-Verlag, New York. 366 pp.

——. 1975. Regulatory mechanisms in insect feeding. Adv. Insect Physiol., **11**:1–116.

Brues, C. T. 1946. *Insect Dietary: An Account of the Food Habits of Insects.* Harvard University Press, Cambridge, Mass. 466 pp.

——, A. L. Melander, and F. M. Carpenter. 1954. Classification of insects. Bull. Mus. Comp. Zool. Harv. Univ., **73**:1–917.

Buchanan, R. E., and N. E. Gibbons (eds.). 1974. *Bergey's Manual of Determinative Bacteriology,* 8th ed. The Williams & Wilkins Company, Baltimore, Maryland. 1268 pp.

Buchner, P. 1965. *Endosymbiosis of Animals with Plant Microorganisms,* rev. ed. Interscience Publishers, Inc., New York. 909 pp.

Bullock, T. H., and G. A. Horridge. 1965. *Structure and Function in the Nervous Systems of Invertebrates,* 2 vols. W. H. Freeman and Company, San Francisco. 1719 pp.

Burges, H. D., and N. W. Hussey. 1971. *Microbial Control of Insects and Mites.* Academic Press, Inc., New York. 861 pp.

Burks, B. D. 1953. The mayflies, or Ephemeroptera, of Illinois. Ill. Nat. Hist. Surv. Bull., **26**:1–216.

Bursell, E. 1971. *An Introduction to Insect Physiology.* Academic Press, Inc., New York. 276 pp.

——. 1974*a*. Environmental aspects—temperature. *Physiol. Insecta,* 2d ed., **2**:2–43.

——. 1974*b*. Environmental aspects—humidity. *Physiol. Insecta,* 2d ed., **2**:44–84.

Buschman, L. L. 1974. Flash behavior of a Nova Scotian firefly, *Photuris fairchildi* Barber, during courtship and aggressive mimicry (Coleoptera, Lampyridae). Coleopt. Bull., **28**:27–31.

Bush, G. L. 1969. Sympatric host race formation and speciation in fungivorus flies of the genus *Rhagoletis* (Diptera, Tephritidae). Evolution, **23**:237–251.

—— and R. W. Neck. 1976. Ecological genetics of the screwworm fly, *Cochliomyia hominivorax* (Diptera: Calliphoridae) and its bearing in the quality control of mass-reared insects. Environ. Entomol., **5**:821–826.

——, R. W. Neck, and G. B. Kitto. 1976. Screwworm eradication: Inadvertent selection for noncompetitive ecotypes during mass rearing. Science, **173**:491–493.

Bushland, R. C. 1975. Screwworm research and eradication. Bull. Entomol. Soc. Am., **21**:23–26.

Busvine, J. R. 1948. The "head" and "body" races of *Pediculus humanus* L. Parasitology, **39**:1–16.

Butcher, J. W., R. Snider, and R. J. Snider. 1971. Bioecology of edaphic Collembola and Acarina. Ann. Rev. Entomol., **16**:249–288.

Butler, C. G. 1954. *The World of the Honeybee.* William Collins Sons & Co., Ltd., London. 226 pp.

——. 1959. *Bumblebees.* William Collins Sons & Co., Ltd., London. 208 pp.

Buxton, P. A. 1947. *The louse. An Account of the Lice which Infest Man, Their Medical Importance and Control,* 2d ed. Edward Arnold Ltd., London. 164 pp.

Byers, G. W. 1963. The life history of *Panorpa nuptialis* (Mecoptera: Panorpidae). Ann. Entomol. Soc. Am., **56**:142–149.

——. 1965. Families and genera of Mecoptera. Proc. 12th Int. Congr. Entomol., **1964**:123.

——. 1971. Ecological distribution and structural adaptation in the classification of Mecoptera. Proc. 13th Int. Congr. Entomol., 1968, **1**:486.

Cain, A. J., and P. M. Sheppard. 1950. Selection in the polymorphic land snail *Cepaea nemoralis.* Heredity, **4**:275–294.

Cameron, E. 1961. *The Cockroach (Periplaneta americana, L.); An Introduction to Entomology for Students of Science and Medicine.* William Heinemann, Ltd., London. 111 pp.

Camhi, J. M. 1971. Flight orientation in locusts. Sci. Am., **225**(2):74–81.

Camin, J. H., and P. R. Ehrlich. 1958. Natural selection in water snakes *(Natrix sipedon L.)* on islands in Lake Erie. Evolution, **12**:504–511.

Candy, D. J., and B. A. Kilby (ed.). 1975. *Insect Biochemistry and Function.* Chapman and Hall, Ltd., London. 314 pp.

Cantwell, G. E. (ed.). 1974. *Insect diseases,* vols. 1 and 2. Marcel Dekker, Inc., New York. 595 pp.

Carpenter, F. M. 1931. Revision of nearctic Mecoptera. Bull. Mus. Comp. Zool. Harv. Univ., **72**:205–277.

——. 1936. Revision of the nearctic Raphidioidea (recent and fossil). Proc. Am. Acad. Arts Sci., **71**:89–157.

——. 1940. Revision of the nearctic Hemerobiidae, Berothidae, Sisyridae, Polystoechotidae, and Dilaridae (Neuroptera). Proc. Am. Acad. Arts Sci., **74**:193–280.

——. 1953. The evolution of insects. Am. Sci., **41**:256–270.

——. 1970. Adaptations among paleozoic insects. Proc. North Am. Paleontol. Conv., 1969, **2**:1236–1251.

——. 1977. Geological history and evolution of the insects. Proc. 15th Int. Congr. Entomol., Washington (1976):63–70.

Carroll, C. R., and D. H. Janzen. 1973. Ecology of foraging by ants. Ann. Rev. Ecol. Syst., **4**:231–257.

Carter, W. 1973. *Insects in Relation to Plant Disease,* 2d ed. John Wiley & Sons, Inc., New York. 759 pp.

Carthy, J. D. 1965. *The Behaviour of Arthropods.* W. H. Freeman and Company, San Francisco. 148 pp.

Caudell, A. N. 1920. Zoraptera not an apterous order. Proc. Entomol. Soc. Wash., **22**:84–97.

Chaloner, W. G. 1970. The rise of the first land plants. Biol. Rev., **45**:353–377.

Chandler, H. P. 1956. Megaloptera, in *Aquatic Insects of California,* R. L. Usinger (ed.). University of California Press, Berkeley. pp. 229–233.

Chapman, R. F. 1969. *The Insects: Structure and Function.* American Elsevier Publishing Company, Inc., New York. 819 pp.

Chauvin, R. 1967. *The World of an Insect.* McGraw-Hill Book Company, New York. 254 pp.

Chen, P. S. 1971. *Biochemical Aspects of Insect Development.* S. Karger, Basel. 230 pp.

Cheng, L. (ed.). 1976. *Marine Insects.* North-Holland Publishing Company, Amsterdam. 581 pp.

Chippendale, G. M. 1977. Hormonal regulation of larval diapause. Ann. Rev. Entomol., **22**:121–138.

Chopard, L. 1938. La biologie des Orthoptères. Encyclopaedie Entomologique, **26**:1–609.

——. 1949. Ordre des Cheleutoptères. Traité Zool., **9**:594–616.

——. 1949. Ordre des Dictyoptères Leach, 1818 (= Blattaeformia Werner, 1906; = Oothecaria Karny, 1915). Traité Zool., **9**:617–722.

Christiansen, K. 1964. Bionomics of Collembola. Ann. Rev. Entomol., **9**:147–178.

Chu, H. F. 1947. *How to Know the Immature Insects.* Wm. C. Brown Company Publishers, Dubuque, Iowa. 234 pp.

Claassen, P. W. 1931. Plecoptera nymphs of America (North of Mexico). Thomas Say Foundation Publ., **3**:1–199.

Clark, E. J. 1948. Studies in the ecology of British grasshoppers. Trans. R. Entomol. Soc. London, **97**:173–222.

Clark, L. R., P. W. Geier, R. D. Hughes, and R. F. Morris. 1967. *The Ecology of Insect Populations in Theory and Practice.* Methuen & Co., Ltd., London. 232 pp.

Clausen, C. P. 1940. *Entomophagous Insects.* McGraw-Hill Book Company, New York. 688 pp.

——. 1976. Phoresy among entomophagous insects. Ann. Rev. Entomol., **21**:343–368.

Clay, T. 1970. The Amblycera (Phthiraptera: Insecta). Bull. Br. Mus. Nat. Hist. Entomol., **25**:73–98.

Cleveland, L. R., S. R. Hall, E. P. Saunders, and J. Collier. 1934. The wood-feeding roach, *Cryptocercus,* its protozoa, and the symbiosis between protozoa and roach. Mem. Am. Acad. Sci., **17**:185–342.

Cloudsley-Thompson, J. L. 1975. Adaptations of arthropoda to arid environments. Ann. Rev. Entomol., **20**:261–283.

Cobben, R. H. 1968. *Evolutionary Trends in Heteroptera. Part I: Eggs, Architecture of the Shell, Gross Embryology and Eclosion.* Centre for Agricultural Publishing and Documentation, Wageningen, Netherlands. 475 pp.

Cole, F. R. 1969. *The Flies of Western North America.* University of California Press, Berkeley. 693 pp.

Coleopterists' Bulletin. An International Journal Devoted to the Study of Beetles. 1947 to date. Washington. Articles devoted exclusively to systematics and biology of Coleoptera.

Common, I. F. B. 1970. Lepidoptera (moths and butterflies), in *Insects of Australia,* CSIRO, Canberra. pp. 765–866.

——. 1975. Evolution and classification of the Lepidoptera. Ann. Rev. Entomol., **20**:183–203.

Comstock, J. H. 1918. *The Wings of Insects.* Comstock Publishing Co., Ithaca, New York. 430 pp.

——. 1940. *An Introduction to Entomology,* 9th ed. Comstock Publishing Associates, Ithaca, N.Y. 1064 pp.

Cooper, K. W. 1972. A southern California *Boreus, B. notoperates* n. sp. I. Comparative morphology and systematics (Mecoptera: Boreidae). Psyche, **79**:269–283.

——. 1974. Sexual biology, chromosomes, development, life histories and parasites of *Boreus,* especially *B. notoperates,* a southern California *Boreus.* II. (Mecoptera: Boreidae). Psyche, **81**:84–120.

Cope, O. B. 1940. The morphology of *Psocus confraternus* Banks. Microentomology, **5**:91–115.

Corbet, P. S. 1962. *A Biology of Dragonflies.* Quadrangle Books, Inc., Chicago. 247 pp.

——, C. Longfield, and N. W. Moore. 1960. *Dragonflies.* William Collins Sons & Co., Ltd., London. 260 pp.

Cornwell, P. B. 1968. *The Cockroach. Volume I. A Laboratory Insect and an Industrial Pest.* Hutchinson Publishing Group, Ltd., London. 391 pp.

Cott, H. B. 1957. *Adaptive Coloration in Animals.* Methuen & Co., Ltd., London. 508 pp.

Couch, J. N. 1938 *The Genus Septobasidium.* University of North Carolina Press, Chapel Hill. 480 pp.

Counce, S. J., and C. H. Waddington. 1972–1973. *Developmental Systems: Insects.* Academic Press, Inc., London. Vol. 1, 304 pp; vol 2, 615 pp.

Craighead, F. C. 1950. Insect Enemies of Eastern Forests. U.S. Dept. Agric. Misc. Publ., **657**:1–679.

Crampton, G. C., et al. 1942. Guide to the insects of Connecticut. Part VI. The Diptera or true flies of Connecticut. First fascicle. Bull. Conn. Geol. Nat. Hist. Surv., **64**:1–509.

Crane, J. 1952. A comparative study of innate defensive behavior in Trinidad mantids (Orthoptera, Mantoidea). Zoologica, **37**:259–293.

Creighton, W. S. 1950. The ants of North America. Bull. Mus. Comp. Zool., Harv. Univ., **104**:1–585.

Cromartie, R. I. T. 1959. Insect pigments. Ann. Rev. Entomol., **4**:59–76.

Crossley, A. C. 1975. The cytophysiology of insect blood. Adv. Insect Physiol., **11**:117–221.

Crowson, R. A. 1955. *The Natural Classification of the Families of Coleoptera.* Nathaniel Lloyd, London. 187 pp.

——. 1960. The phylogeny of the Coleoptera. Ann. Rev. Entomol., **5**:111–134.

——. 1970. *Classification and Biology.* Heinemann Educational Books, Ltd., London. 350 pp.

——, W. D. I. Rolfe, J. Smart, C. D. Waterston, E. C. Willey, and R. J. Wooten. 1967. Arthropods: Chelicerata, Pycnogonida, Palaeoisopus, Myriapoda, and Insecta, in *The Fossil Record,* W. B. Harland (ed.), Geological Society of London. pp. 499–534.

Crozier, R. H. 1977. Evolutionary genetics of the Hymenoptera. Ann. Rev. Entomol., **22**:263–288.

Cummins, K. W. 1973. Trophic relations of aquatic insects. Ann. Rev. Entomol., **18**:183–206.

——, L. D. Miller, N. A. Smith, and R. M. Fox. 1965. *Experimental Entomology.* Reinhold Publishing Corporation, New York. 176 pp.

Curran, C. H. 1934. *The Families and Genera of North American Diptera.* Privately published, New York. 512 pp. (Reprinted, 1965, Henry Tripp, Mt. Vernon, N.Y.)

Dadant and Sons (ed.). 1975. *The Hive and the Honey Bee.* Dadant and Sons, Hamilton, Ill. 740 pp.

Dadd, R. H. 1973. Insect nutrition: Current developments and metabolic implications. Ann. Rev. Entomol., **18**:381–420.

Daly, H. V. 1963. Close-packed and fibrillar muscles of the Hymenoptera. Ann. Entomol. Soc. Am., **56**:295–306.

——. 1964. Skeleto-muscular morphogenesis of the thorax and wings of the honey bee, *Apis mellifera* (Hymenoptera: Apidae). Univ. Calif. Publ. Entomol., **39**:1–77.

Davey, K. G. 1965. *Reproduction in the Insects.* W. H. Freeman and Company, San Francisco. 96 pp.

David, W. A. L. 1975. The status of viruses pathogenic for insects and mites. Ann. Rev. Entomol., **20**:97–117.

Davidson, G. (ed.). 1974. *Genetic Control of Insect Pests.* Academic Press, Inc., London. 158 pp.

Davidson, J. 1944. On the relationship between temperature and rate of development of insects at constant temperatures. J. Anim. Ecol., **17**:193–199.

Davis, C. 1940. Family classification of the order Embioptera. Ann. Entomol. Soc. Am., **33**:677–682.

Dawkins, R. 1976. *The Selfish Gene.* Oxford University Press, Oxford. 224 pp.

Day, W. C. 1956. Ephemeroptera, in *Aquatic Insects of California,* R. L. Usinger (ed.). University of California Press, Berkeley. pp. 79–105.

DeBach, P. (ed.). 1964. *Biological Control of Insect Pests and Weeds.* Chapman & Hall, Ltd., London. 844 pp.

——. 1966. The competitive displacement and coexistence principles. Ann. Rev. Entomol., **11**:183–212.

——. 1974. *Biological Control by Natural Enemies.* Cambridge University Press, New York. 320 pp.

DeCoursey, R. M. 1971. Keys to the families and subfamilies of the nymphs of North American Hemiptera-Heteroptera. Proc. Entomol. Soc. Wash., **73**:413–428.

Delamare-Deboutteville, C. 1948. Observations sur l'écologie et l'éthologie des Zoraptères. La question de leur vie social et de leurs pretendus rapports avec les termites. Rev. entomol., **19**:347–352.

Delany, M. J. 1957. Life histories in the Thysanura. Acta Zool. Cracov., **2**:61–90.

——. 1959. The life histories and ecology of two species of *Petrobius* Leach, *P. brevistylis* and *P. maritimus*. Trans. R. Soc. Edinburgh, **63**:501–533.

DeLong, D. M. 1971. The bionomics of leafhoppers. Ann. Rev. Entomol., **16**:179–210.

Demerec, M. (ed.). 1950. *Biology of Drosophila*. John Wiley & Sons, Inc., New York. 632 pp.

Denning, D. G. 1956. Trichoptera, in *Aquatic Insects of California*, R. L. Usinger (ed.). University of California Press, Berkeley. pp. 237–270.

Dethier, V. G. 1947. The response of hymenopterous parasites to chemical stimulation of the ovipositor. J. Exp. Zool., **105**:199–208.

——. 1954. Evolution of feeding preferences in phytophagous insects. Evolution, **8**:33–54.

——. 1963. *The Physiology of Insect Senses*. Methuen & Co., Ltd., London. 266 pp.

——. 1976. *The Hungary Fly*. Harvard University Press, Cambridge, Mass. 489 pp.

——, L. Barton-Browne, and C. N. Smith. 1960. The designation of chemicals in terms of the responses they elicit from insects. J. Econ. Entomol., **53**:134–136.

de Wilde, J., and A. de Loof. 1973*a*. Reproduction. *Physiol. Insecta*, 2d ed., **1**:12–95.

——. 1973*b*. Reproduction–endocrine control. *Physiol. Insecta*, 2d ed., **1**:97–157.

Dillon, E. S., and L. S. Dillon. 1961. *A Manual of Common Beetles of Eastern North America*. Row, Peterson & Company, Evanston, Ill. 884 pp.

Diptera of Connecticut. 1942–1964. Guide to the insects of Connecticut. Part VI. The Diptera, or true flies of Connecticut. Bull. Conn. Geol. Nat. Hist. Surv. Nine fascicles, various authors.

Dirsh, V. M. 1975. *Classification of the Acridomorphoid Insects*. E. W. Classey, Faringdon, Oxon., England. 171 pp.

Dixon, A. F. G. 1973. *Biology of Aphids*. Edward Arnold Publishers, London. 58 pp.

Dobzhansky, T. 1970 *Genetics of the Evolutionary Process*. Columbia University Press, New York. 505 pp.

Dodson, M. 1937. Development of the female genital ducts in *Zygaena* (Lepidoptera). Proc. R. Entomol. Soc. London (A), **12**:61-68.

Dolinger, P. M., P. R. Ehrlich, W. L. Fitch, and D. E. Breedlove. 1973. Alkaloids and predation patterns in Colorado lupine populations. Ecologia, **13**:191–204.

Downes, J. A. 1958. The feeding habits of biting flies and their significance in classification. Ann. Rev. Entomol., **3**:249–266.

——. 1965. Adaptations of insects in the arctic. Ann. Rev. Entomol., **10**:257–274.

——. 1969. The swarming and mating flight of Diptera. Ann. Rev. Entomol., **14**:271–298.

——. 1971. The ecology of blood-sucking Diptera: An evolutionary perspective, in *Ecology and Physiology of Parasites*, A. M. Fallis (ed.). University of Toronto Press, Toronto, Canada. pp. 232 –258.

Doyen, J. T. 1966. The skeletal anatomy of *Tenebrio molitor* (Coleoptera: Tenebrionidae). Misc. Publ. Entomol. Soc. Am., **5**:103–150.

Dugdale, J. S. 1974. Female genital configuration in the classification of Lepidoptera. N. Z. J. Zool., **1**:127–146.

Dunn, A. M. 1969. *Veterinary Helminthology*. Lea & Febiger, Philadelphia. 302 pp.

Duporte, E. M. 1960. Evolution of cranial structures in adult Coleoptera. Can. J. Zool., **38**:655–675.

Eatonia, 1954 to date, Tallahassee, Florida. Papers on Ephemeroptera.

Ebeling, W. 1968. Termites: Identification, biology, and control of termites attacking buildings. Calif. Agric. Exp. Stn. Ext. Serv. Man. **38**:1–68.

——. 1974. Permeability of insect cuticle. *Physiol. Insecta*, 2d ed., **6**:271–343.

——. 1975. *Urban Entomology*. University of California, Division of Agricultural Science, Berkeley. 695 pp.

Eberhard, M. J. W. 1975. The evolution of social behavior by kin selection. Quant. Rev. Biol., **50**:1–33.

Edmunds, G. F., Jr. 1959. Ephemeroptera, in *Freshwater Biology*, W. T. Edmondson (ed.). John Wiley & Sons, Inc., New York. p. 908–916.

—— and J. R. Traver. 1954. The flight mechanics and evolution of the wings of Ephemeroptera, with notes on the archetype insect wing. J. Wash. Acad. Sci., **44**:390–400.

——, R. K. Allen, and W. L. Peters. 1963. An annotated key to the nymphs of the families and subfamilies of mayflies (Ephemeroptera) Univ. Utah Biol. Ser., **13**:1–55.

——, S. L. Jensen, and L. Berner. 1976. *The Mayflies of North and Central America*. University of Minnesota Press, Minneapolis. 330 pp.

Edney, E. B. 1974. Desert Arthropods, in *Desert Biology*, vol. 2, G. W. Brown, Jr. (ed.), Academic Press, Inc., New York. pp. 311–384.

—— and R. Barrass. 1962. The body temperature of the tsetse fly, *Glossina morsitans* Westwood (Diptera, Muscidae). J. Insect Physiol., **8**:469–481.

Edwards, J. S. 1969. Postembryonic development and regeneration of the insect nervous system. Adv. Insect Physiol., **6**:98–139.

Ehrlich, P. R. 1958. The comparative morphology, phylogeny and higher classification of the butterflies (Lepidoptera: Papilionoidea). Univ. Kans. Sci. Bull., **39**:305–370.

——. 1970. Coevolution and the biology of communities, in *Biochemical Coevolution,* K. L. Chambers (ed.). Oregon State University Press, Corvallis. pp. 1–11.

——. 1961. Intrinsic barriers to dispersal in checkerspot butterfly. Science, **134**:108–109.

—— and A. H. Ehrlich. 1961. *How to Know the Butterflies.* Wm. C. Brown Company Publishers, Dubuque, Iowa. 262 pp.

—— and L. E. Gilbert. 1973. Population structure and dynamics of the tropical butterfly *Heliconius ethilla.* Biotropica, **5**:69–82.

—— and P. Raven. 1964. Butterflies and plants: A study in coevolution. Evolution, **18**:586–608.

—— and ——. 1967. Butterflies and plants. Sci. Am., **216**:104–113.

—— and ——. 1969. Differentiation of populations. Science, **165**:1228–1232.

——, A. H. Ehrlich, and R. W. Holm. 1977. *Ecoscience.* W. H. Freeman and Company. San Francisco. 1051 pp.

——, R. W. Holm, and D. R. Parnell. 1974. *The Process of Evolution.* McGraw-Hill Book Company, New York. 378 pp.

——, R. R. White, M. C. Singer, S. W. McKechnie, and L. E. Gilbert. 1975. Checkerspot butterflies: A historical perspective. Science, **188**:221–228.

Eisenstein, E. M. 1972. Learning and memory in isolated insect ganglia. Adv. Insect Physiol., **9**:112–182.

Eisner, T. 1960. Defense mechanisms of arthropods. II. The chemical and mechanical weapons of an earwig. Psyche, **67**:62–70.

Ekblom, T. 1926. Morphological and biological studies of the Swedish families of Hemiptera Heteroptera. Parts I and II. Zool. Bidrag., **10**:31–180; **12**:113–150.

Elliott, J. M. 1968. The daily activity patterns of mayfly nymphs (Ephemeroptera). J. Zool., **155**:201–221.

Emerson, A. E. 1952. The biogeography of termites. Bull. Am. Mus. Nat. Hist., **99**:217–225.

——. 1955. Geographical origins and dispersions of termite genera. Fieldiana, Zool., **37**:465–521.

Emerson, K. C. 1956. Mallophaga (chewing lice) occurring on the domestic chicken. J. Kans. Entomol. Soc., **35**:196–201.

Engelmann, F. 1970. *The Physiology of Insect Reproduction.* Pergamon Press, Oxford. 307 pp.

Entomophaga. 1956 to date. Publication of Commission internationale de lutte biologique contre les ennemis des cultures. Paris. Articles on biological control; many articles on biology and life history of parasitic Hymenoptera.

Eriksen, C. H. 1966. Ecological significance of respiration and substrate for burrowing Ephemeroptera. Can. J. Zool., **46**:93–103.

Esch, H. 1967. The evolution of bee language. Sci. Am., **216**(4):96–104.

Essig, E. O. 1930. *A History of Entomology.* The Macmillan Company., New York. 1029 pp.

——. 1942. *College Entomology.* The Macmillan Company, New York. 900 pp.

——. 1958. *Insects and Mites of Western North America.* The Macmillan Company, New York. 1050 pp.

Evans, G. 1975. *The Life of Beetles.* George Allen & Unwin, Ltd., London. 232 pp.

Evans, H. E., and M. J. West Eberhard. 1970. *The Wasps.* University of Michigan Press, Ann Arbor. 265 pp.

Evans, J. W. 1963. The phylogeny of the Homoptera. Ann. Rev. Entomol., **8**:77–94.

Evans, M. E. G. 1961. On the muscular and reproductive systems of *Atomaria ruficornis* (Marsham) (Coleoptera, Cryptophagidae). Trans. R. Soc. Edinburgh, **64**:297–399.

Ewing, H. E. 1940. The Protura of North America. Ann. Entomol. Soc. Am., **33**:495–551.

—— and I. Fox. 1943. The fleas of North America. U.S. Dep. Agric. Misc. Publ., **500**:1–128.

Falcon, L. A. 1976. Problems with the use of arthropod viruses in pest control. Ann. Rev. Entomol., **21**:305–324.

Farb, P. 1962. *The Insects.* Time Inc., New York. 191 pp.

Faust, R. M. 1974. Bacterial diseases, in *Insect Diseases,* vol. 1. G. E. Cantwell (ed.). Marcel Dekker, Inc., New York. pp. 87–183.

Felt, E. P. 1940. *Plant Galls and Gall Makers.* Comstock Publishing Associates, Ithaca, New York. 364 pp.

Ferris, G. F. 1931., The louse of elephants, *Haematomyzus elephantis.* Parasitology, **23**:112–127.

——. 1937–1955. *Atlas of the Scale Insects of North America.* Stanford University Press, Palo Alto, Calif. 7 vols.

——. 1940. The morphology of *Plega signata* (Hagen) (Neuroptera, Mantispidae). Microentomology, **5**:33–56.

——. 1951. The sucking lice. Mem. Pac. Coast Entomol. Soc., **1**:1–320.

—— and P. Pennebaker. 1939. The morphology of *Agulla adnixa* (Hagen) (Neuroptera, Raphidiidae). Microentomology, **5**:33–56.

Florkin, M., and C. Jeuniaux. 1974. Hemolymph: Composition. *Physiol. Insecta,* 2d ed., **5**:255–307.

Flower, J. W. 1964. On the origin of flight in insects. J. Insect Physiol., **10**:81–88.

Forbes, W. T. M. 1923–1960. Lepidoptera of New York and neighboring states. Mem. Cornell Univ. Agric. Exp. Stn., **68**:1–729; **274**:1–263; **329**:1–433; **371**:1–188.

Ford, N. 1926. On the behavior of *Grylloblatta.* Can. Entomol., **58**:66–70.

Fox, H. M., and G. Vevers. 1960. *The Nature of Animal Colours.* The Macmillan Company, New York. 246 pp.

Fox, R. M., and J. W. Fox. 1964. *Introduction to Comparative Entomology.* Reinhold Publishing Corporation, New York. 450 pp.

Fracker, S. B. 1930. The classification of lepidopterous larvae. Contrib. Entomol. Lab., Univ. Ill., **43**:1–161.

Fraegri, K., and L. van der Pijl. 1971. *The Principles of Pollination Ecology,* 2d rev. ed. Pergamon Press, Oxford. 291 pp.

Fraenkel, G. S., and D. L. Gunn. 1961. *The Orientation of Animals. Kineses, Taxes and Compass Reactions.* Dover Publications, Inc., New York. 376 pp.

Francke-Grosmann, H. 1967. Ectosymbiosis in wood-inhabiting insects, in *Symbiosis,* vol. 2, S. M. Henry (ed.). Academic Press, Inc., New York. pp. 141–205.

Fraser, F. C. 1957. *A Reclassification of the Order Odonata.* Royal Zoological Society of New South Wales, Sydney, Australia. 133 pp.

Free, J. B. 1970. *Insect Pollination of Crop Plants.* Academic Press, Inc., London. 544 pp.

Fremling, C. R. 1960. Biology of a large mayfly, *Hexagenia bilineata* (Say), of the upper Mississippi River. Iowa State Univ. Agric. Home Econ. Exp. Stn. Res. Bull., **482**:841–852.

Friend, W. G., and J. J. B. Smith. 1977. Factors affecting feeding by bloodsucking insects. Ann. Rev. Entomol., **22**:309–331.

Frisch, K. von. 1967 *The Dance Language and Orientation of Bees.* Harvard University Press, Cambridge, Mass. 566 pp.

——. 1971. *Bees, Their Vision, Chemical Senses, and Language.* Cornell University Press, Ithaca, N.Y. 157 pp.

——. 1974. Decoding the language of the bee. Science, **185**:663–668.

Frison, T. H. 1935. The stoneflies, or Plecoptera, of Illinois. Ill. Nat. Hist. Surv. Bull., **20**:281–471.

——. 1942. Studies on North American Plecoptera with special reference to the fauna of Illinois. Ill. Nat. Hist. Surv. Bull., **22**:235–355.

Froeschner, R. C. 1954. The grasshoppers and other Orthoptera of Iowa. Iowa State Coll. J. Sci., **29**:163–354.

——. 1960. Cydnidae of the Western Hemisphere. Proc. U.S. Natl. Mus., **111**:337–680.

Frost, S. W. 1942. *General Entomology.* McGraw-Hill Book Company, New York, 524 pp.

Frylink, A. 1954. The tulip. Part I. Its early history. Gard. J. New York Bot. Gard., **4**:5–7, 15, 47–51.

Fulton, B. B. 1924. Some habits of earwigs. Ann. Entomol. Soc. Am., **17**:357–367.

Furman, D. P., and E. P. Catts. 1961. *Manual of Medical Entomology.* National Press Books, Palo Alto, Calif. 162 pp.

Fuzeau-Braesch, S. 1972. Pigments and color changes. Ann. Rev. Entomol., **17**:403–424.

Gallun, R. L., K. J. Starks, and W. D. Guthrie. 1975. Plant resistance to insects attacking cereals. Ann. Rev. Entomol., **20**:337–357.

Garman, P. 1927. The Odonata or dragonflies of Connecticut. Guide to the Insects of Connecticut. Conn. Geol. Nat. Hist. Surv., **5**:1–331.

Garrett, R. G. 1973. Non-persistent aphid-borne viruses, in *Viruses and invertebrates,* A. J. Gibbs (ed.). American Elsevier Publishing Company, Inc., New York. pp. 476–492.

Gaufin, A. R., A. V. Nebeker, and J. Sessions. 1966. The stoneflies (Plecoptera) of Utah. Univ. Utah Biol. Ser., **14**:1–93.

Geiger, R. 1965. *The Climate Near the Ground.* Harvard University Press, Cambridge, Mass. 611 pp.

Gelperin, A. 1971. Regulation of feeding. Ann. Rev. Entomol., **16**:365–378.

Gilbert, L. E. 1972. Pollen feeding and reproductive biology of *Heliconius* butterflies. Proc. Nat. Acad. Sci., **69**:1403–1407.

—— and M. C. Singer. 1975. Butterfly ecology. Ann.

Rev. Ecol. Syst., **6**:365–397.

Gilbert, L. I., and D. S. King. 1973. Physiology of growth and development: Endocrine aspects. *Physiol. Insecta,* 2d ed., **1**:249–370.

Giles, E. T. 1963. The comparative external morphology and affinities of the Dermaptera. Trans. R. Entomol. Soc. London, **115**:95–164.

Gillett, J. D. 1971. *Mosquitoes.* Weidenfeld and Nicolson, London. 274 pp.

——. 1972. *The Mosquito: Its Life, Activities and Impact on Human Affairs.* Doubleday & Company, Inc., Garden City, N.Y. 358 pp.

Gilmour, D. 1961. *The Biochemistry of Insects.* Academic Press, Inc., London. 343 pp.

——. 1965. *The Metabolism of Insects.* W. H. Freeman and Company, San Francisco. 195 pp.

Glick, P. A. 1939. The distribution of insects, spiders, and mites in the air. U.S. Dep. Agric., Tech. Bull., No. 673, 150 pp.

Gloyd, L. K., and M. Wright. 1959. Odonata, in *Freshwater Biology,* W. T. Edmondson (ed.)., John Wiley & Sons, Inc., New York. pp. 917–940.

Goetsch, W. 1957. *The Ants.* University of Michigan Press, Ann Arbor. 173 pp.

Goldsmith, T. H. and G. D. Bernard. 1974. The visual system of insects. *Physiol. Insecta,* 2d ed., **2**:166–272.

Goodman, L. J. 1970. The structure and function of the insect dorsal ocellus. Adv. Insect Physiol., **7**:97–196.

Gould, J. L. 1975. Honey bee recruitment: The dance-language controversy. Science, **189**:685–693.

Graham, K. 1967. Fungal-insect mutualism in trees and timber. Ann. Rev. Entomol., **12**:105–126.

Graham, S. A., and F. B. Knight. 1965. *Principles of Forest Entomology,* 4th ed. McGraw-Hill Book Company, New York. 417 pp.

Gregoire, C. 1974. Hemolymph coagulation. *Physiol. Insecta,* 2d ed., **5**:309–360.

Gressitt, J. L., J. Sedlacek, and J. J. H. Szent-Ivany. 1965. Flora and fauna on backs of large Papuan moss-forest weevils. Science, **150**:1833–1835.

Griffiths, G. C. D. 1972. The phylogenetic classification of the Diptera Cyclorrhapha with special reference to the structure of the male postabdomen. Ser. Entomol. Hague, **8**:1–340.

Grote, A. R. 1971. *Noctuidae of North America.* E. W. Classey, London. 85 pp.

Guppy, R. 1950. Biology of *Anisolabis maritima* (Gene) the seaside earwig, on Vancouver Island (Dermaptera, Labiidae). Proc. Entomol. Soc. Br. C., **46**:14–18.

Gurney, A. B. 1938. A synopsis of the order Zoraptera, with notes on the biology of *Zorotypus hubbardi* Caudell. Proc. Entomol. Soc. Wash., **40**:57–87.

——. 1948. Praying mantids of the United States. Smithson. Inst. Rep., **1950**:339–362.

——. 1948. The taxonomy and distribution of the Grylloblattidae. Proc. Entomol. Soc. Wash., **50**:86–102.

——. 1950. Corrodentia, in *Pest Control Technology, Entomological Section.* National Pest Control Association, New York. pp. 129–163.

Guthrie, D. M., and A. R. Tindall. 1968. *The biology of the cockroach.* St. Martins Press, New York. 408 pp.

Hackman, R. H. 1974. Chemistry of the insect cuticle. *Physiol. Insecta,* 2d ed., **6**:215–270.

Hagan, H. R. 1951. *Embryology in the Viviparous Insects.* Ronald Press, Inc., New York. 472 pp.

Hagen, K. S., S. Bombosch, and J. A. McMurtry. 1976. The biology and impact of predators, in *Theory and Practice of Biological Control,* C. B. Huffaker and P. S. Messenger (eds.). Academic Press, Inc., New York. pp. 93–142.

——, T. L. Tassan, and E. F. Sawall, Jr. 1970. Some ecophysiological relationships between certain *Chrysopa,* honeydews and yeasts. Boll. Lab. Entomol. Agr., Portici., **28**:113–134.

Hale, W. G. 1965. Observations on the breeding biology of Collembola. Pedobiologia, **5**:146–152.

Hall, D. G. 1948. The blow flies of North America. Thomas Say Foundation Publ., **4**:1–477.

Hamilton, K. G. A. 1971. The insect wing. I. Origin and development of wings from notal lobes. J. Kans. Entomol. Soc.,**44**:421–433.

——. 1972. The insect wing. Part IV. Venational trends and the phylogeny of the winged orders J. Kans. Entomol. Soc., **45**:295–308.

Hamilton, W. D. 1964. The genetical evolution of social behavior, I and II. J. Theoret. Biol., **7**:1–52.

Handlirsch, A. 1908. *Die fossilen Insekten und die Phylogenie der rezenten Formen.* W. Englemann, Leipzig. 1430 pp.

——. 1926. Vierter Unterstamm des Stammes der Arthropoda. Insecta = Insekten. Handbuch Zool., **4**:403–592.

Hanover, J. W. 1975. Physiology of tree resistance to insects. Ann. Rev. Entomol., **20**:75–95.

Harcourt, D. G. 1969. The development and use of life tables in the study of natural insect populations. Ann. Rev. Entomol., **14**:175–196.

Harden, P. H., and C. E. Mickel. 1952. The stoneflies of Minnesota (Plecoptera). Univ. Minn. Agric. Exp. Stn. Tech. Bull. **201**:1–84.

Hardin, G. 1960. The competitive exclusion principle. Science, **131**:1292–1297.

Harris, W. V. 1961. *Termites: Their Recognition and Control.* Longmans Group Ltd., London. 186 pp.

Haskell, P. T. 1974. Sound production. *Physiol. Insecta,* 2d ed., **2**:354–410.

Hatch, M. H. 1957–1973. The beetles of the Pacific Northwest. 5 vols. Univ. Wash. Publ. Biol. and University of Washington Press, Seattle.

Hebard, M. 1934. The Dermaptera and Orthoptera of Illinois. Ill. Nat. Hist. Surv. Bull., **20**:125–279.

Heinrich, B. 1973. The energetics of the bumblebee. Sci. Am., **228**(4):96–102.

——. 1974. Thermoregulation in endothermic insects. Science, **185**:747–756.

—— and G. A. Bartholomew. 1972. Temperature control in flying moths. Sci. Am., **226**(6):69–77.

Helfer, J. R. 1953. *How to Know the Grasshoppers, Cockroaches and Their Allies.* Wm. C. Brown Company Publishers, Dubuque, Iowa. 353 pp.

Hennig, W. 1948–1952. *Die Larvenformen der Dipteren.* Akademie-Verlag, Berlin. Part 1, 1948:185 pp.; Part 2, 1950:458 pp.; Part 3, 1952:628 pp.

——. 1953. Kritische Bemerkungen zum phylogenetischen System der Insekten. Beitr. Entomol., **3**:1–85.

Hepburn, H. R. 1969. The skeleto-muscular system of Mecoptera: The head. Kans. Univ. Sci. Bull., **48**:721–765.

—— (ed.) 1976. *The Insect Integument.* Elsevier Publishing Company, New York. 572 pp.

Hering, M. 1951. *Biology of the Leaf Miners.* Junk's-Gravenhage, The Netherlands. 420 pp.

Herring, J. L., and P. D. Ashlock. 1971. A key to the nymphs of the families of Hemiptera (Heteroptera) of America north of Mexico. Fla. Entomol., **54**: 207–213.

Hicks, E. A. 1959. *Check-list and Bibliography on the Occurrence of Insects in Bird's Nests.* Iowa State College Press, Ames, Iowa. 681 pp.

Hille Ris Lambers, D. 1966. Polymorphism in Aphididae. Ann. Rev. Entomol., **11**:47–48.

Hinton, H. E. 1945. *A Monograph of the Beetles Associated with Stored Products,* vol. I. British Museum Natural History, London. 443 pp.

——. 1946. On the homology and nomenclature of the setae of the lepidopterous larvae, with some notes on the phylogeny of the Lepidoptera. Trans. R. Entomol. Soc. London, **97**:1–37.

——. 1955*a*. On the structure and distribution of the prolegs of the Panorpoidea with a criticism of the Berlese-Imms theory. Trans. R. Entomol. Soc. London, **106**:455–540.

——. 1955*b*. On the respiratory adaptions, biology and taxonomy of the Psephenidae, with notes on some related families (Coleoptera). Proc. Zool. Soc. London, **125**:543–568.

——. 1958. The phylogeny of the panorpoid orders. Ann. Rev. Entomol., **3**:181–206.

——. 1963*a*. The origin and function of the pupal stage. Proc. R. Entomol. Soc. London (A), **38**:77–85.

——. 1963*b*. The origin of flight in insects. Proc. Entomol. Soc. London (C), **28**:23–32.

——. 1969*a*. Plaston respiration in adult beetles of the Suborder Myxophaga. J. Zool. Soc. London, **159**:131–137.

——. 1969*b*. Respiratory systems of insect egg shells. Ann. Rev. Entomol., **14**:343–368.

——. 1970. Insect eggshells. Sci. Am., **223**(3):84–91.

——. 1971. Some neglected phases in metamorphosis. Proc. R. Entomol. Soc. London (C), **35**:55–64.

——. 1977. *Enabling Mechanisms.* XV Internat. Congr. Entomol., pp. 71–83.

Hocking, B. 1971. Blood-sucking behavior of terrestrial arthropods. Ann. Rev. Entomol., **16**:1–26.

Hodek, I. 1973. *Biology of Coccinellidae.* Academia, Prague and June, The Hague. 260 pp.

Hodgson, E. S. 1961. Taste receptors. Sci. Am., **204**(5):135–144.

——. 1974. Chemoreception. *Physiol. Insecta,* 2d ed., **2**:127–165.

Holdsworth, R. P. 1941. The life history and growth of *Pteronarcys proteus* Newman (Pteronarcidae: Plecoptera). Ann. Entomol. Soc. Am., **34**:394–502.

Holland, G. P. 1949. The Siphonaptera of Canada. Can. Dep. Agric. Publ., 817, Tech. Bull., **70**:1–306.

——. 1964. Evolution, classification and host relationships of Siphonaptera. Ann. Rev. Entomol., **9**:123–146.

Holland, W. J. 1968. *The Moth Book.* Dover Publications, Inc., New York. 479 pp.

Hölldobler, B. 1971. Communication between ants and their guests. Sci. Am., **224**(3):86–93.

Hopkins, G. H. E. 1949. The host associations of the lice of mammals. Proc. Zool. Soc. London, **119**:387–604.

Horn, D. J. 1976. *Biology of Insects.* W. B. Saunders Company, Philadelphia. 439 pp.

Horridge, G. A. 1975. *The Compound Eye and Vision of Insects.* Clarendon Press, Oxford. 595 pp.

Horsfall, F. L., and I. Tamm (eds.). 1965. *Viral and Rickettsial Infections of Man.,* 4th ed. J. B. Lippincott Company, Philadelphia. 1282 pp.

Horsfall, W. R., H. W. Fowler, Jr., L. J. Moretti, and J. R. Larsen. 1973. *Bionomics and Embryology of the Inland Floodwater Mosquito Aëdes vexans.* University of Illinois Press, Urbana. 211 pp.

House, H. L. 1974*a*. Nutrition. *Physiol. Insecta,* 2d ed., **5**:1–62.

——. 1974*b*. Digestion. *Physiol. Insecta,* 2d ed., **5**:63–117.

Howe, W. H. 1975. *The Butterflies of North America.* Doubleday & Company, Inc., Garden City, N.Y. 633 pp.

Howse, P. E. 1970. *Termites: A Study in Social Behavior.* Hutchinson University Library, London. 150 pp.

——. 1975. Brain structure and behavior in insects. Ann. Rev. Entomol., **20**:359–379.

Hoyle, G. 1955*a*. Neuromuscular mechanisms of a locust skeletal muscle. Proc. R. Soc. B, **143**:343–367.

——. 1955*b*. The effects of some common cations on neuromuscular transmission in insects. J. Physiol., **127**:90–103.

——. 1970. Cellular mechanisms underlying behavior—neuroethology. Adv. Insect Physiol., **7**:349–444.

——. 1974. Neural control of skeletal muscle. *Physiol. Insecta,* 2d ed., **4**:175–236.

Hubbard, C. A. 1947. *Fleas of Western North America.* Iowa State College Press, Ames. 533 pp.

Huber, F. 1974. Neural integration (central nervous system). *Physiol. Insecta,* 2d ed., **4**:3–100

Huffaker, C. B. (ed.). 1971. *Biological Control.* Plenum Publishing Corporation, New York. 511 pp.

—— and P. S. Messenger (eds.). 1976. *Theory and Practice of Biological Control.* Academic Press, Inc., New York. 788 pp.

Hughes, G. M., and P. J. Mill. 1974. Locomotion: terrestrial. *Physiol. Insecta,* 2d ed., **3**:335–379.

Hughes, N. F., and J. Smart. 1967. Plant-insect relationships in Palaeozoic and later time, in *The Fossil Record,* W. B. Harland (ed.). Geological Society, London. pp. 107–117.

Hull, F. M. 1962. Robberflies of the world: The genera of the family Asilidae. U.S. Natl. Mus. Bull., **224**:1–907 (2 vols.).

Hungerford, H. B. 1959. Hemiptera, in *Freshwater Biology,* W. T. Edmundson (ed.). John Wiley & Sons, Inc., New York. pp. 958–972.

Hunt, B. P. 1950. The life history and economic importance of the burrowing mayfly, *Hexagenia limbata,* in southern Michigan lakes. Inst. Fisheries Res. Bull. Mich. Dept. Conserv., **4**:1–151.

Hutchins, R. E. 1966. *Insects.* Prentice-Hall, Inc., Englewood Cliffs, N.J. 324 pp.

Hynes, H. B. N. 1970*a*. *The Ecology of Running Waters.* University of Toronto Press, Toronto. 555 pp.

——. 1970*b*. The ecology of stream insects. Ann. Rev. Entomol., **15**:25–42.

——. 1976. Biology of Plecoptera. Ann. Rev. Entomol., **21**:135–153.

Ide, F. P. 1935. The effect of temperature on the distribution of the mayfly fauna of a stream. Univ. Toronto Studies, Biol. Ser. No. 39, Publ. Ontario Fisheries Res. Lab., **50**:3–76.

Ilan, J. I., and J. I. Ilan. 1973. Protein synthesis and insect morphogenesis. Ann. Rev. Entomol., **18**:167–182.

——. 1974. Protein synthesis in insects. *Physiol. Insecta, 2d ed.,* **4**:355–422.

Illies, J. 1965. Phylogeny and zoogeography of the Plecoptera. Ann. Rev. Entomol., **10**:117–140.

Imms, A. D. 1936. The ancestry of insects. Trans. Soc. Br. Entomol., **3**:1–32.

——. 1971. *Insect Natural History,* 3d ed. William Collins Sons & Co., Ltd., London. 348 pp.

Inoue, I. 1962. Studies on the life history of *Chordodes japonensis,* a species of Gordiacea. Annot. Zool. Japon. **35**:12–19.

Iseley, F. B. 1944. Correlation between mandibular morphology and food specificity in grasshoppers. Ann. Entomol. Soc. Am., **37**:47–67.

Istock, C. A. 1966. The evolution of complex life cycle phenomena: an ecological perspective. Evolution, **21**:592–605.

Iwata, K. 1976. *Evolution of Instinct: Comparative Ethology of Hymenoptera.* Amerind Publishing Co. Put. Ltd., New Delhi. 535 pp.

Jacobson, M. 1972. *Insect Sex Pheromones.* Academic Press, Inc., New York. 382 pp.

——. 1974. Insect pheromones. *Physiol. Insecta,* 2d ed., **3**:229–276.

James, M. T. 1947. The flies that cause myiasis in man. U.S. Dep. Agric. Misc. Publ., **631**:1–175.

—— and R. F. Harwood. 1969. *Herm's Medical Entomology,* 6th ed. The Macmillan Company., London. 484 pp.

Janzen, D. H. 1966. Coevolution of mutualism between ants and acacias in Central America. Evolution, **20**:249–275.

——. 1967*a*. Fire, vegetation structure, and the ant-acacia interaction in Central America. Ecology, **48**:26–35.

——. 1967*b*. Interaction of the bull's-horn acacia (*Acacia cornigera* L.) with an ant inhabitant (*Pseudomyrmex ferruginea* F. Smith) in eastern Mexico. Kans. Univ. Sci. Bull., **47**(6):315–558.

Jeannel, R. 1945. Sur la position systematique des Strepsiptères. Rev. Fr. Entomol., **11**:111–118.

Jermy, T. (ed.). 1974. *The Host-plant in Relation to Insect Behaviour and Reproduction.* Plenum Press, New York. 322 pp.

Jewett, S. G. 1956. Plecoptera, in *Aquatic Insects of California,* R. L. Usinger (ed.). University of California Press, Berkeley. pp. 155–181.

——. 1959. The stoneflies (Plecoptera) of the Pacific Northwest. Oreg. State Mongr. Stud. Entomol., **3**:1–95.

——. 1963. A stonefly aquatic in the adult stage. Science, **139**:484–485.

Johannsen, O. A., and F. H. Butt. 1941. *Embryology of Insects and Myriapods.* McGraw-Hill Book Company, New York. 462 pp.

Johnson, C. G. 1969 *Migration and Dispersal of Insects by Flight.* Methuen & Co., Ltd., London. 763 pp.

——. 1974. Insect migration: Aspects of its physiology. *Physiol. Insecta,* 2d ed., **3**:279–334.

Johnson, G. B. 1973. Enzyme polymorphisms and biosystematics; the hypothesis of selective neutrality. Ann. Rev. Ecol. Syst., **4**:93–116.

Jones, J. C. 1962. Current concepts concerning insect hemocytes. Am. Zool., **2**:209–246.

——. 1968. The sexual life of a mosquito. Sci. Am., **218**(4):108–116.

——. 1974. Factors affecting heart rates in insects. *Physiol. Insecta,* 2d ed., **5**:119–167.

Jones, T. 1954. The external morphology of *Chirothrips hematus* (Trybom) (Thysanoptera). Trans. R. Entomol. Soc. London, **105**:163–187.

Joose, E. N. G. 1976. Littoral apterygotes (Collembola and Thysanura), in *Marine Insects,* L. Cheng (ed.). North-Holland Publishing Company, Amsterdam. pp. 151–186.

Journal of the Lepidopterists' Society. 1947 to date. Cambridge, Mass. Articles on butterflies and moths.

Journal of Research on the Lepidoptera. 1962 to date. Arcadia, Calif. Articles on butterflies and moths.

Judd, W. W. 1948. A comparative study of the proventriculus of orthopteroid insects with reference to its use in taxonomy. Can. J. Res., **26**:93–161.

Kamp, J. W. 1963. Descriptions of two new species of Grylloblattidae and of the adult of *Grylloblatta barberi,* with an interpretation of their geographic distribution. Ann. Entomol. Soc. Am., **56**:53–68.

——. 1973. Numerical classification of the orthopteroids, with special reference to the Grylloblattodea. Can. Entomol., **105**(9):1235–1249.

Kaplanis, J. N., W. E. Robbins, and M. J. Thompson. 1975. Recent developments in insect steroid metabolism. Ann. Rev. Entomol., **20**:205–220.

Kaufmann, T. 1971. Hibernation in the arctic beetle, *Pterostichus brevicornis,* in Alaska. J. Kans. Entomol. Soc., **44**:81–92.

Keister, M., and J. Buck. 1974. Respiration: Some exogenous and endogenous effects on rate of respiration. *Physiol. Insecta,* 2d ed., **6**:470–509.

Kelsey, L. P. 1954. The skeleto-motor mechanism of the Dobson fly, *Corydalus cornutus.* Part I. Head and prothorax. Mem. Agric. Exp. Stn. Cornell Univ., **334**:1–51.

——. 1957. The skeleto-motor mechanism of the Dobson fly, *Corydalus cornutus.* Part II. Pterothorax. Mem. Agric. Exp. Stn. Cornell Univ., **346**:1–31.

Kennedy, J. S. 1975. Insect dispersal, in *Insects, Science, and Society,* D. Pimentel (ed.). Academic Press, Inc., New York. pp. 103–119.

—— and H. L. G. Stroyan. 1959. Biology of aphids. Ann. Rev. Entomol., **4**:139–160.

Kettlewell, H. B. D. 1959. Darwin's missing evidence. Sci. Am., **200**:48–53.

——. 1961. The phenomenon of industrial melanism in Lepidoptera. Ann. Rev. Entomol., **6**:245–262.

——. 1973. *The Evolution of Melanism.* Clarendon Press, Oxford. 423 pp.

Kevan, D. K. McE. 1962. *Soil Animals.* Philosophical Library, New York. 237 pp.

Kevan, P. G., W. G. Chaloner, and D. B. O. Savile. 1975. Interrelationships of early terrestrial arthropods and plants. Paleontology, **18**(2):391–417.

Kimura, M., and T. Ohta. 1971. *Theoretical Aspects of Population Genetics.* Princeton University Press, Princeton, N.J. 219 pp.

Kinzelbach, R. K. 1971*a*. Strepsiptera (Fächesflügler). Handb. Zool., **42**/24:1–68.

——. 1971*b*. Morphologische Befunde an Fächesflüglern und ihre phylogenetische Bedeutung (Insecta: Strepsiptera). Zoologica, **41**(119):1–256.

Kiriakoff, S. G. 1963. The tympanic structures of Lepidoptera and the taxonomy of the order. J. Lepid. Soc., **17**:1–6

Klots, A. B. 1951. *A Field Guide to the Butterflies.* Houghton Mifflin Company, Boston. 349 pp.

Koch, A. 1967. Insects and their endosymbionts, in *Symbiosis,* vol. 2, S. M. Henry (ed.). Academic Press, Inc., New York. pp. 1–106.

Kofoid, C. A. (ed.). 1934. *Termites and Termite Control.* University of California Press, Berkeley. 734 pp.

Koopman, K. F. 1950. Natural selection for reproductive isolation between *Drosophila pseudoobscura* and *Drosophila persimilis.* Evolution, **4**:135–145.

Kormondy, E. J. 1961. Territoriality and dispersal in dragonflies (Odonata). J. N.Y. Entomol. Soc., **69**:42–52.

Koss, R. W. 1968. Morphology and taxonomic use of Ephemeroptera eggs. Ann. Entomol. Soc. Am., **61**:696–721.

Kramer, S. 1950. Morphology and phylogeny of the auchenorrhynchous Homoptera (Insecta). Ill. Biol. Monogr., **20**:1–78.

Kring, J. B. 1972. Flight behavior of aphids. Ann. Rev. Entomol., **17**:461–492.

Krishna, K., and F. M. Weesner. 1969*a*. *Biology of Termites,* vol. I. Academic Press. Inc., New York. 598 pp.

——. 1969*b*. *Biology of Termites,* vol. II. Academic Press, Inc., New York. 643 pp.

Kristensen, N. P. 1968. The morphological and functional evolution of the mouthparts in adult Lepidoptera. Opusc. Entomol., **33**:69–72.

——. 1971. The systematic position of the Zeugloptera in the light of recent anatomical investigations. Proc. 13th Int. Congr. Entomol. (1968), **1**:261.

——. 1975. The phylogeny of hexapod "orders": A critical review of recent accounts. Z. zool. Syst. Evolutions forsch., **13**:1–44.

Krombein, K. V. 1967. *Trap-nesting Wasps and Bees: Life Histories, Nests, and Associates.* The Smithsonian Institution, Washington. 570 pp.

Kühnelt, W. 1961. *Soil Biology, with Special Reference to the Animal Kingdom.* Faber & Faber, Ltd., London. 397 pp.

Kukalova, J. 1968. Permian mayfly nymphs. Psyche, **75**:310–327.

Kukalova-Peck, J. 1978. Origin and evolution of insect wings and their relation to metamorphosis, as documented by fossil record (in press).

Kulman, H. M. 1971. Effects of insect defoliation on growth and mortality of trees. Ann. Rev. Entomol., **16**:289–324.

Langston, R. L. and J. A. Powell. 1975. The earwigs of California (Order Dermaptera). Bull. Calif. Insect Surv., **20**:1–25.

Larsen, O. 1966. On the morphology and function of the locomotor organs of the Gyrinidae and other Coleoptera. Opusc. Entomol. Suppl., **30**:1–242.

Lavialle, M., and B. Dumortier. 1975. Mise en évidence de deux caractères inhabituels dans les rhythmes circadiens: existence d'un photorécepteur extraoculaire et particularités du libre cours dans le rhythme de ponte du phasme: *Carausius morosus.* C. R. Acad. Soc. Paris, **281**:1489–1492.

Lawrence, P. A. 1970. Polarity and patterns in the postembryonic development of insects. Adv. Insect Physiol., **7**:197–267.

Lawrence, R. F. 1953. *The Biology of the Cryptic Fauna of Forests.* A. A. Balkerma, Cape Town. 408 pp.

Layne, J. N. 1971. Fleas (Siphonaptera) of Florida. Fl. Entomol., **54**:35–51.

Lee, K. E., and T. G. Wood. 1971. *Termites and Soils.* Academic Press, Inc., London. 251 pp.

Leech, H. B., and H. P. Chandler. 1956. Aquatic Coleoptera, in *Aquatic Insects of California,* R. L. Usinger (ed.). University of California Press, Berkeley. pp. 293–371.

—— and M. W. Sanderson. 1959. Coleoptera, in *Freshwater Biology,* W. T. Edmondson (ed.). John Wiley & Sons, Inc., New York. pp. 981–1023.

Lees, A. D. 1966. Photoperiodic timing mechanisms in insects. Nature (London) **210**:986–989.

——. 1971. The role of circadian rhythmicity in photoperiodic induction in animals. Proc. Int. Symp. Circadian Rhythmicity (Wageningen, Netherlands, 1971). pp. 87–110.

Leonard, D. E. 1974. Recent developments in ecology and control of the gypsy moth. Ann. Rev. Entomol., **19**:197–230.

Leonard, J. W., and F. A. Leonard. 1962. *Mayflies of Michigan Trout Streams.* Cranbrook Institute of Science, Bloomfield Hills, Mich. 139 pp.

Leopold, R. A. 1976. The role of male accessory glands in insect reproduction. Ann. Rev. Entomol., **21**:199–221.

Lewis, T. 1973. *Thrips, Their Biology, Ecology and Economic Importance.* Academic Press, Inc., London. 366 pp.

Lewontin, R. C., and J. L. Hubby. 1966. A molecular approach to the study of genic heterozygosity in natural populations. II. Amount of variation and degree of heterozygosity in natural populations of *Drosophila pseudoobscura.* Genetics, **54**:595–609.

—— and M. J. D. White. 1960. Interaction between inversion polymorphisms of two chromosome pairs in the grasshopper, *Moraba scurra.* Evolution, **14**:116–129.

Lindauer, M. 1961. *Communication among Social Bees.* Harvard University Press, Cambridge, Mass. 143 pp.

Linsenmaier, W. 1972. *Insects of the World.* McGraw-Hill Book Company, New York. 392 pp.

Linsley, E. G. 1959*a*. Mimetic form and coloration in the Cerambycidae (Coleoptera). Ann. Entomol. Soc. Am., **52**:125–131.

——. 1959*b*. Ecology of Cerambycidae. Ann. Rev. Entomol., **4**:99–138.

—— and J. W. MacSwain. 1957. Observations on the habits of *Stylops pacifica* Bohart. Univ. Calif. Publ. Entomol., **11**:395–430.

Linzen, B. 1974. The tryptophan → ommochrome pathway in insects. Adv. Insect Physiol., **10**:117–246.

Littig, K. S. 1942. External anatomy of the Florida walking stick *Anisomorpha buprestoides* Stoll. Fl. Entomol., **25**:33–41.

Little, V. A. 1963. *General and Applied Entomology,* 2d ed. Harper & Row, Publishers, Incorporated, New York. 543 pp.

Lloyd, J. L. 1971. Bioluminescent communication in insects. Ann. Rev. Entomol., **16**:97–122.

——. 1921. The biology of North American caddis fly larvae. Lloyd Libr. Bot., Pharm. Mat. Med. Bull., **21**:1–124.

Locke, M. 1974. The structure and formation of the integument in insects. *Physiol. Insecta,* 2d ed., **6**:123–213.

Loher, W. 1960. The chemical acceleration of the maturation process and its hormonal control in the male of the desert locust. Proc. R. Entomol. Soc. London, ser. B, **153**:380–397.

——. 1972. Circadian control of stridulation in the cricket *Teleogryllus commodus* Walker. J. Comp. Physiol., **79**:173–190.

Lüscher, M. 1961. Social control of polymorphism in termites. Symp. R. Entomol. Soc. London, **1**:57–67.

Lüttge, U. 1961. Über die Zusammensetzung des Nektars und den Mechanismus seiner Sekretion.I. Planta, **56**:189–212.

Lyman, F. E. 1955. Seasonal distribution and life cycles of Ephemeroptera. Ann. Entomol. Soc. Am., **48**:380–391.

MacKerras, I. M. 1970. Evolution and classification of the insects, in *Insects of Australia.* CSIRO, Canberra. pp. 152–167.

MacLeod, E. G., and P. A. Adams. 1967. A review of the taxonomy and morphology of the Berothidae, with a description of a new subfamily from Chile (Neuroptera). Psyche, **74**:237–265.

Maddrell, S. H. P. 1971. The mechanisms of insect excretory systems. Adv. Insect Physiol., **8**:200–331.

Mallis, A. 1971. *American Entomologists.* Rutgers University Press, New Brunswick, N.J. 549 pp.

Malyshev, S. I. 1968. Genesis of the Hymenoptera and phases of their evolution. Methuen & Co., Ltd., London. 319 pp.

Manton, S. M. 1964. Mandibular mechanisms and the evolution of arthropods. Philos. Trans. R. Soc. London, Ser. B, **247**:1–183.

——. 1972. The evolution of arthropod locomotory mechanisms. Part 10. Locomotory habits, morphology and evolution of the hexapod classes. Zool. J. Linn. Soc., **51**:203–400.

——. 1973. Arthropod phylogeny—a modern synthesis. J. Zool., **171**:111–130.

Maramorosch, K. (ed.). 1976. *Invertebrate Tissue Culture.* Academic Press, Inc., New York. 393 pp.

Markin, G. P. 1970. The seasonal life cycle of the Argentine ant, *Iridomyrmex humilis* (Hymenoptera: Formicidae), in southern California. Ann. Entomol. Soc. Am., **63**:1238–1242.

Markl, H. 1974. Insect behavior: Functions and mechanisms. *Physiol. Insecta,* 2d ed., **3**:3–148.

Marshall, A. T. 1966. Histochemical studies on a mucocomplex in the Malpighian tubules of cercopid larvae. J. Insect Physiol., **12**:925–932.

Martin, M. 1934. Life history and habits of the pigeon louse [*Columbicola columbae* (Linn.)]. Can. Entomol., **66**:1–16.

Martynov, A. V. 1925. Ueber zwei Grundtypen des Flügel bei den Insekten und ihre Evolution. Zh. Morphol. Oekol. Tiere, **4**:465–501.

Martynova, O. 1961. Palaeoentomology. Ann. Rev. Entomol., **6**:285–294.

Maruyama, K. 1974. The biochemistry of the contractile elements of insect muscle. *Physiol. Insecta,* 2d ed., **4**:237–269.

Mason, L. G. 1973. The habitat and phenetic variation in *Phymata americana* Melin. Syst. Zool., **22**:271–279.

Mather, K., and B. S. Harrison. 1949. The manifold effect of selection. Heredity, **3**:1–52.

Matsuda, R. 1965. Morphology and evolution of the insect head. Mem. Am. Entomol. Inst., **1**:1–334.

——. 1970. Morphology and evolution of the insect thorax. Mem. Entomol. Soc. Can., **76**:1–431.

Matthews, R. W. 1974. Biology of Braconidae. Ann. Rev. Entomol., **19**:15–32.

Matthysse, J. G. 1946. Cattle lice, their biology and control. Cornell Univ. Agric. Exp. Stn. Bull., **823**:1–67.

Maxwell, F. G., J. N. Jenkins, and W. L. Parrott. 1972. Resistance of plants to insects. Adv. Agron., **24**:187–265.

May, H. G. 1919. Contributions to the life histories of *Gordius robustus* Leidy and *Paragordius varius* (Leidy). Ill. Biol. Monogr., **5**:1–119.

Maynard, E. A. 1951. *A Monograph of the Collembola, or Springtails, of New York State.* Comstock Publishing Associates, Ithaca, N.Y. 339 pp.

Mayr, E. 1963. *Animal Species and Evolution.* Harvard University Press, Cambridge, Mass. 797 pp.

McCann, F. V. 1970. Physiology of insect hearts. Ann. Rev. Entomol., **15**:173–200.

Mcdonald, T. J. 1975. Neuromuscular pharmacology of insects. Ann. Rev. Entomol., **20**:151–166.

McElroy, W. D., H. H. Seliger, and M. DeLuca 1974. Insect bioluminescence. *Physiol. Insecta,* 2d ed., **2**:411–460.

McGregor, S. E. 1976. *Insect Pollination of Cultivated Crop Plants.* U.S. Dept. Agric., Agric. Handbk. No. 496. 411 pp.

McIver, S. B. 1975. Structure of cuticular mechanoreceptors of arthropods. Ann. Rev. Entomol., **20**:381–397.

McKay, M. B., and M. F. Warner. 1933. Historical sketch of tulip mosaic or breaking the oldest known plant virus disease. Natl. Hortic. Mag., **12**:178–216.

McKechnie, S. W., P. R. Ehrlich, and R. R. Watt. 1975. Population genetics of *Euphydryas* butterflies. I. Genetic variation and the neutrality hypotheses. Genetics, **81**:571–594.

McKenzie, H. L. 1967. *The Mealybugs of California.* University of California Press, Berkeley. 525 pp.

McKeown, K. C., and V. H. Mincham. 1948. The biology of an Australian mantispid (*Mantispa vittata* Guerin). Aust. Zool., **11**:207–224.

McKittrick, F. A. 1964. Evolutionary studies of cockroaches. Cornell Univ. Agric. Exp. Stn., Mem. **389**:1–197.

Meinander, M. 1972. A revision of the family Coniopterygidae (Planipennia). Acta Zool. Fenn., **136**:1–357.

Menn, J. J., and M. Beroza (eds.). 1972. *Insect Juvenile Hormones, Chemistry and Action.* Academic Press, Inc., New York. 341 pp.

Merritt, R. W., and K. W. Cummins (eds.). 1978. *An Introduction to the Aquatic Insects of North America.* Kendall Hunt Publishing Co., Dubuque, Iowa.

Metcalf, R. L., and W. H. Luckman (eds.). 1975. *Introduction to Insect Pest Management.* John Wiley & Sons, Inc., New York. 587 pp.

Metcalf, C. L., W. P. Flint, and R. L. Metcalf. 1962. *Destructive and Useful Insects,* 4th ed. McGraw-Hill Book Company, New York. 1087 pp.

Michelson, A., and H. Nocke. 1974. Biophysical aspects of sound communication in insects. Adv. Insect Physiol., **10**:247–296.

Michener, C. D. 1944. Comparative external morphology, phylogeny, and a classification of the bees (Hymenoptera). Bull. Am. Mus. Nat. Hist., **82**:151–326.

——. 1953. Comparative morphological and systematic studies of bee larvae with a key to the families of hymenopterous larvae. Univ. Kans. Sci. Bull., **35**:987–1102.

——. 1969. Comparative social behavior of bees. Ann. Rev. Entomol., **14**:299–342.

——. 1974. *The Social Behavior of the Bees.* Harvard University Press, Cambridge, Mass. 404 pp.

——. 1975. The Brazilian bee problem. Ann. Rev. Entomol., **20**:399–416.

Mickoleit, G. 1973. Über den Ovipositor des Neuropteroidea und Coleoptera und seine phylogenetische Bedeutung (Insecta, Holometabola). Z. Morphol. Tiere, **74**:37–64.

Miles, P. W. 1972. The saliva of Hemiptera. Adv. Insect Physiol., **9**:183–256.

Miller, L. K. 1969. Freezing tolerance in an adult insect. Science, **166**:105–106.

Miller, N. C. E. 1956. *Biology of the Heteroptera.* Leonard Hill, London. 162 pp.

Miller, P. L. 1974*a*. Respiration-aerial gas transport. *Physiol. Insecta,* 2d ed., **6**:345–402.

——. 1974*b*. Respiration: Aquatic insects. *Physiol. Insecta,* 2d ed., **6**:403–467.

Miller, T. A. 1974. Electrophysiology of the insect heart. *Physiol. Insecta,* 2d ed., **5**:169–201.

——. 1975. Neurosecretion and the control of visceral organs in insects. Ann. Rev. Entomol., **20**:133–149.

Mills, H. B. 1932. The life history and thoracic development of *Oligotoma texana* (Mel.). Ann. Entomol. Soc. Am., **35**:648–652.

——. 1934. *A Monograph of the Collembola of Iowa.* Iowa State College Press, Ames. 143 pp.

Mitchell, R. W. 1969. A comparison of temperate and tropical cave communities. Southwest. Nat., **14**:73–88.

Mitchell, T. B. 1960–1962. Bees of eastern United States. 2 vols. N.C. Agric. Exp. Stn. Tech. Bull., **141**:1–538: **152**:1–557.

Mittelstaedt, H. 1962. Control systems of orientation in insects. Ann. Rev. Entomol., **7**:177–198.

Mockford, E. L. 1951. The Psocoptera of Indiana. Proc. Indiana Acad. Sci., **60**:192–204.

——. 1957. Life history studies on some Florida insects of the genus *Archipsocus* (Psocoptera). Bull. Fl. St. Mus., **1**:253–274.

—— and A. B. Gurney. 1956. A review of the psocids or book-lice and bark-lice of Texas (Psocoptera). J. Wash. Acad. Sci., **46**:353–368.

Morgan, F. D. 1968. Bionomics of Siricidae. Ann. Rev. Entomol., **13**:239–256.

Morris, R. F. 1959. Single-factor analysis in population dynamics. Ecology, **40**:580–588.

Mosher, E. 1916. A classification of the Lepidoptera based on characters of the pupa. Bull. Ill. Nat. Hist. Surv., **12**:15–159.

Moths of America North of Mexico. 1972 to date. E. W. Classey and R. B. D. Publishers, London. (Published as separate fascicles for each family.)

Mukerji, S. 1927. On the morphology and bionomics of *Embia minor* sp. n., with special reference to its spinning organs. Rec. Indian Mus., Calcutta, **29**:253–282.

Mulkern, G. B. 1967. Food selection by grasshoppers. Ann. Rev. Entomol., **12**:59–78.

Müller, F. 1879. *Ituna* and *Thyridia;* a remarkable case of mimicry in butterflies. Proc. Entomol. Soc. London, pp. xx–xxix.

Myers, J. G. 1929. *Insect Singers. A Natural History of the Cicadas.* George Routledge and Sons, Ltd., London. 304 pp.

Nachtigall, W. 1974*a*. *Insects in Flight.* McGraw-Hill Book Company, New York. 153 pp.

——. 1974*b*. Locomotion: Mechanics and hydrodynamics of swimming in aquatic insects. *Physiol. Insecta,* 2d ed., **3**:381–432.

Needham, J. G., and P. W. Claassen. 1925. A monograph of the Plecoptera or stoneflies of America north of Mexico. Thomas Say Foundation Publ., **2**:1–397.

—— and M. J. Westfall, Jr. 1955. *A Manual of Dragonflies of North America (Anisoptera).* University of California Press, Berkeley. 615 pp.

——, S. W. Frost, and B. H. Tothill. 1928. *Leaf-mining Insects.* The Williams & Wilkins Company, Baltimore, Md. 351 pp.

——, J. R. Traver, and Y.-C. Hsu. 1935. *The Biology of Mayflies.* Comstock Publishing Associates, Ithaca, N.Y. 759 pp.

Neumann, D. 1976. Adaptations of chironomids to intertidal environments. Ann. Rev. Entomol., **21**:387–414.

New, T. R. 1975. The biology of Chrysopidae and Hemerobiidae (Neuroptera), with reference to their usage as biocontrol agents: A review. Trans. R. Entomol. Soc. London, **127**:115–140.

Newcomer, E. G. 1918. Some stoneflies injurious to vegetation. J. Agric. Res., **13**:37–41.

Niering, W. A. 1966. *The Life of the Marsh.* McGraw-Hill Book Company, New York. 232 pp.

Noirot, C., and A. Quennedey. 1974. Fine structure of insect epidermal glands. Ann. Rev. Entomol., **19**:61–80.

Norris, K. R. 1965. The bionomics of blow flies. Ann. Rev. Entomol., **10**:47–68.

Novák, V. J. A. 1975. *Insect Hormones.* Chapman & Hall, Ltd., London. 600 pp.

Odonatologica. 1972 to date. Utrecht. Articles on dragonflies and damselflies.

Odum, E. P. 1971. *Fundamentals of Ecology,* 3d ed. W. B. Saunders Company, Philadelphia. 574 pp.

Oken, L. 1831. *Lehrbuch der Naturphilosophie.* (2d. ed.). Jena.

Oldroyd, H. 1964. *The Natural History of Flies.* Weidenfeld and Nicolson, London. 324 pp.

——. 1970. *Collecting, Preserving and Studying Insects.* Hutchinson Publishing Group, Ltd., London. 336 pp.

Oliver, D. R. 1971. Life history of the Chironomidae. Ann. Rev. Entomol., **16**:211–230.

Olsen, D. W. 1974. *Animal Parasites, Their Life Cycles and Ecology,* 3d ed. University Park Press, Baltimore, Md. 562 pp.

Ossiannilsson, F. 1949. Insect drummers: A study of the morphology and function of the sound-producing organs of Swedish Homoptera, Auchenorrhyncha. Opusc. Entomol. Suppl., **10**:1–146

Packard, A. S. 1898. *A text-book of Entomology.* Macmillan Co., New York. 729 pp.

Painter, R. H. 1958. Resistance of plants to insects. Ann. Rev. Entomol., **3**:267–290.

Pal, R., and M. J. Whitten. 1974. *The Use of Genetics in Insect Control.* Elsevier/North-Holland, Amsterdam. 241 pp.

Parfin, S. 1952. The Megaloptera and Neuroptera of Minnesota. Am. Midl. Nat., **47**:421–434.

Parnas, I., and D. Dagan. 1971. Functional organizations of giant axons in the central nervous systems of insects: New aspects. Adv. Insect Physiol., **8**:96–145.

Paulson, D. R. 1974. Reproductive isolation in damselflies. Syst. Zool., **23**:40–49.

Pearman, J. V. 1928. On sound-production in the Psocoptera and on a presumed stridulatory organ. Entomol. Mon. Mag., **64**:179–186.

Peck, O., Z. Bouček, and A. Hoffer. 1964. Keys to the Chalcidoidea of Czechoslovakia. Mem. Entomol. Soc. Can., **34**:1–120.

Penny, N. D. 1975. Evolution of the extant Mecoptera. J. Kans. Entomol. Soc., **48**:331–350.

Percival, M. S. 1965. *Floral Biology.* Pergamon Press, Oxford. 243 pp.

Peters, W. L., and P. G. Peters (eds.). 1973. *Proceedings of the First International Conference on Ephemeroptera.* E. J. Brill, NV, Leiden, Netherlands. 312 pp.

Peterson, A. 1948. *Larvae of Insects. Part I. Lepidoptera and Plant Infesting Hymenoptera.* Edwards Brothers, Inc., Ann Arbor, Mich. 315 pp.

——. 1951. *Larvae of Insects. Part II. Coleoptera, Diptera, Neuroptera, Siphonaptera, Mecoptera, Trichoptera.* Edwards Brothers, Inc., Ann Arbor, Mich. 416 pp.

——. 1964. *Entomological Techniques; How to Work With Insects.* 10th ed. Edward Bros., Ann Arbor, Mich. 435 pp.

Pianka, E. R. 1970. On *r*- and *K*- selection. Am. Nat., **104**:592–597.

Pichon, Y. 1974. The pharmacology of the insect nervous system. *Physiol. Insecta,* 2d ed., **4**:102–174.

Pierce, W. D. 1918. The comparative morphology of the order Strepsiptera together with records and descriptions of insects. Proc. U.S. Natl. Mus., **54**:391–501.

Pimentel, D. 1968. Population regulation and genetic feedback. Science, **159**:1432–1437.

Plumstead, E. P. 1963. The influence of plants and environment on the developing animal life of Karroo times. S. Afr. J. Sci., **59**:147–152.

Poinar, G. O., Jr. 1972. Nematodes as facultative parasites of insects. Ann. Rev. Entomol., **17**:103–122.

——. 1975. *Entomogenous Nematodes.* E. J. Brill, NV, Leiden, Netherlands. 317 pp.

Poisson, R., and P. Pesson. 1951. Super-ordre des Hemipteroides (Hemiptera Linné, 1758, Rhynchota Bermeister, 1835). Traité Zool., **10**:1385–1803.

Pollock, J. N. 1971. Origin of the tsetse flies: A new theory. J. Entomol. (B), **40**:101–109.

Popham, E. J. 1965. A key to the Dermaptera subfamilies. Entomologist, **98**:126–136.

Powell J. A. 1964. Biological and taxonomic studies on tortricine moths, with reference to the species in California. Univ. Calif. Publ. Entomol., **32**:1–317.

—— and R. A. Mackie. 1966. Biological interrelationships of moths and *Yucca whipplei.* Univ. Calif. Publ. Entomol., **42**:1–59.

Price, P. W. (ed.). 1974. *Evolutionary Strategies of Parasitic Insects and Mites.* Plenum Press, New York. 224 pp.

——. 1975. *Insect Ecology.* John Wiley & Sons, Inc., New York. 514 pp.

Pringle, J. A. 1938. A contribution to the knowledge of *Micromalthus debilis* Lec. (Coleoptera). Trans. Roy. Entomol. Soc. Lond., **87**:271–286.

Pringle, J. W. S. 1974. Locomotion: flight. *Physiol. Insecta,* 2d ed., **3**:433–476.

Proctor, M., and P. Yeo. 1973. *The Pollination of Flowers.* William Collins Sons & Co., Ltd., London. 418 pp.

Rainey, R. C. 1965. The origin of insect flight: Some implications of recent findings from palaeoclimatology and locust migration. Proc. 12th Int. Congr. Entomol., **1964**:134.

—— (ed.). 1976. *Insect Flight.* Blackwell Scientific Publications, Ltd., Oxford. 287 pp.

Rehn, J. A. G., and H. J. Grant, Jr. 1961. A monograph of the Orthoptera of North America (North of Mexico), vol. I. Monogr. Acad. Nat. Sci., Philadelphia, **12**:1–255, 8 pls.

Rehn, J. W. H. 1939. Studies in North American Mantispidae (Neuroptera). Trans. Am. Entomol. Soc., **65**:237–263.

——. 1950. A key to the genera of North American Blattaria, including established adventives. Entomol. News, **61**:64–67.

——. 1951. Classification of the Blattaria as indicated by their wings (Orthoptera). Mem. Am. Entomol. Soc., **14**:1–134.

Reinhardt, J. F. 1952. Some responses of honeybees to alfalfa flowers. Am. Nat., **86**:257–275.

Remington, C. L. 1954. The suprageneric classification of the order Thysanura (Insecta). Ann. Entomol. Soc. Am., **47**:277–286.

——. 1956. The "Apterygota," in *A Century of Progress in the Natural Sciences, 1853–1953,* E. L. Kessel (ed.). California Academy of Science, San Francisco. pp. 495–505.

Rettenmeyer, C. W. 1970. Insect mimicry. Ann. Rev. Entomol., **15**:43–74.

Ribbands, C. R. 1964. *The Behaviour and Social Life of Honeybees.* Dover Publications, New York. 352 pp.

Richards, A. G. 1951. *The Integument of Arthropods.* University of Minnesota Press, Minneapolis.

——, and P. A. Richards. 1977. The peritrophic membranes of insects. Ann. Rev. Entomol., **22**:219–240.

Richards, O. W. 1956. Hymenoptera. Introduction and keys to families. R. Entomol. Soc. London, Handbk. Identif. Br. Insects, **6**:1–94.,

—— and R. G. Davies. 1957. *Imms's A General Textbook of Entomology.* Methuen & Co., Ltd., London.

——and N. Waloff. 1954. Studies on the biology and population dynamics of British grasshoppers. Anti-Locust Bull., **17**:182 pp.

Richards, W. R. 1968. Generic classification, evolution, and biogeography of the Sminthuridae of the world (Collembola). Mem. Entomol. Soc. Can., **53**:1–54.

Ricker, W. E. 1952. Systematic studies in Plecoptera. Indiana Univ. Stud. Sci. Ser., **18**:1–200.

——. 1959. Plecoptera, in *Freshwater Biology,* W. T. Edmondson (ed.). John Wiley & Sons, Inc., New York. pp. 941–957.

Riegel, G. T. 1963. The distribution of *Zorotypus hubbardi* (Zoraptera). Ann. Entomol. Soc. Am., **56**:744–747.

Riek, E. F. 1970*a.* Ephemeroptera, in *Insects of Australia.* CSIRO, Canberra. pp. 224–240.

——. 1970*b.* Lower Cretaceous fleas. Nature (London) **227**:746–747.

——. 1971. The origin of insects. Proc. 13th Int. Congr. Entomol. (1968), **1**:292–293.

Riley, C. V. 1892. The yucca moth and yucca pollination. Rept. Missouri Bot. Gdn. **3**:99–158.

Ritcher, P. O. 1958. Biology of Scarabaeidae. Ann. Rev. Entomol., **3**:311–334.

Robbins, W. E., J. N. Kaplanis, J. A. Svoboda, and M. J. Thompson. 1971. Steroid metabolism in insects. Ann. Rev. Entomol., **16**:53–72.

Robert, A. 1962. Les Libellules du Québec. Ministère du Tourisme, de la Chasse et de la Pêche, Province de Québec. 224 pp.

Rockstein, M., and J. Miquel. 1973. Aging in insects. *Physiol. Insecta,* 2d ed., **1**:371–478.

Roeder, K. D. 1935. An experimental analysis of the sexual behavior of the praying mantis (*Mantis religiosa,* L.). Biol. bull., **69**:203–220

——. 1965. Moths and ultrasound. Sci. Am., **212**:94–102.

——. 1966. Auditory system of noctuid moths. Science, **154**:1515–1521.

——. 1967. *Nerve Cells and Insect Behavior,* rev. ed. Harvard University Press, Cambridge, Mass. 238 pp.

Roelofs, W. L. 1975. Insect communication-chemical, in *Insects, Science and Society,* D. Pimentel (ed.). Academic Press, Inc., New York. pp. 79–99.

—— and R. T. Carde. 1977. Responses of Lepidoptera to synthetic sex pheromone chemicals and their analogues. Ann. Rev. Entomol., **22**:377–405.

Rohdendorf, B. B. 1944. A new family of Coleoptera from the Permian of the Urals. C. R. Acad. Sci. URSS, **44**:252–262.

——. 1969. Phylogenie. Handb. Zool., **4**:1/4.

Romoser, W. S. 1973. *The Science of Entomology.* The Macmillan Company, New York. 449 pp.

Roonwal, M. L. 1962. Recent developments in termite systematics (1949–60), in *Proceedings of the New Delhi Symposium, 1960, Termites in the Humid Tropics.* UNESCO, Paris. pp. 31–50.

Ross, E. S. 1940. A revision of the Embioptera of North America. Ann. Entomol. Soc. Am., **33**:629–676.

——. 1944. A revision of the Embioptera, or web-spinners, of the new world. Proc. U.S. Natl. Mus., **94**:401–504.

——. 1957. The Embioptera of California. Bull. Calif. Insect Surv., **6**:51–57.

——. 1970. Biosystematics of the Embioptera. Ann. Rev. Entomol., **15**:157–172.

Ross, H. H. 1937*a.* A generic classification of Nearctic sawflies (Hymenoptera, Symphyta). Ill. Biol. Monogr., **15**:1–173.

——. 1937*b.* Studies of Nearctic aquatic insects. I. Nearctic alder flies of the genus *Sialis* (Megaloptera, Sialidae). Ill. Nat. Hist. Surv. Bull., **21**:57–78.

——. 1944. The caddisflies, or Trichoptera, of Illinois. Bull. Ill. Nat. Hist. Surv., **23**:326 pp.

——. 1956. *Evolution and Classification of the Mountain Caddisflies.* University of Illinois Press, Urbana. 213 pp.

——. 1959. Trichoptera, in *Freshwater Biology,* W. T. Edmondson (ed.). John Wiley & Sons, Inc., New York. pp. 1024–1049.

——. 1964. Evolution of caddisworm cases and nets. Am. Zool., **4**:209–220.

——. 1965. *A Textbook of Entomology,* 3d ed. John Wiley & Sons, Inc., New York. 539 pp.

——. 1967. The evolution and past dispersal of Trichoptera. Ann. Rev. Entomol., **12**:169–206.

Roth, L. M. 1970. Evolution and taxonomic significance of reproduction in Blattaria. Ann. Rev. Entomol., **15**:75–96.

—— and H. B. Hartman. 1967. Sound production and its evolutionary significance in the Blattaria. Ann. Entomol. Soc. Am., **60**:740–752.

—— and E. K. Willis. 1960. The biotic associations of cockroaches. Smithson. Misc. Collect., **141**:1–470.

Rothschild, M. 1961. Defensive odours and mullerian mimicry among insects. Trans. R. Entomol. Soc. London, **113**(5):101–121.

——. 1965. The rabbit flea and hormones. Endeavor, **24**:162–168.

——. 1973. Secondary plant substances and warning colouration in insects. Symp. R. Entomol. Soc. London, **6**:59–83.

——. 1975. Recent advances in our knowledge of the order Siphonaptera. Ann. Rev. Entomol., **20**:241–259.

—— and T. Clay. 1952. *Fleas, Flukes and Cuckoos: A Study of Bird Parasites.* William Collins Sons & Co., London. 304 pp.

——, Y. Schlein, K. Parker, C. Neville, and S. Sternberg. 1973. The flying leap of the flea. Sci. Am., **229**(5):92–100.

Rowell, C. H. F. 1971. The variable coloration of the acridoid grasshoppers. Adv. Insect Physiol., **8**:146–199.

Rudall, K. M., and W. Kenchington. 1971. Arthropod silks: The problem of fibrous proteins in animal tissue. Ann. Rev. Entomol., **16**:73–96.

Rudinsky, J. A. 1962. Ecology of the Scolytidae. Ann. Rev. Entomol., **7**:327–348.

Sacca, G. 1964. Comparative bionomics in the genus *Musca.* Ann. Rev. Entomol., **9**:341–358.

Sacktor, B. 1970. Regulation of intermediary metabolism, with special reference to the control mechanisms in insect flight muscle. Adv. Insect Physiol., **7**:268–348.

——. 1974. Biological oxidations and energetics in insect mitochondria. *Physiol. Insecta,* 2d ed., **4**:271–353.

Sailer, R. I. 1950. A thermophobic insect. Science, **112**:743.

Salmon, J. T. 1956*a*. Keys and bibliography to the Collembola. Victoria Univ. Coll. Publ. Zool., **8**:1–82.

——. 1956*b*. Keys and bibliography to the Collembola. First supplement. Victoria Univ. Coll. Publ. Zool., **20**:1–35.

Salt, G. 1963. The defense reactions of insects to metazoan parasites. Parasitology, **53**:527–642.

——. 1968. The resistance of insect parasitoids to the defense reactions of their hosts. Biol. Rev., **43**:200–232.

——. 1970. *The Cellular Defense Reactions of Insects.* Cambridge University Press, New York. 118 pp.

Salt, R. W. 1961. Principles of insect cold-hardiness. Ann. Rev. Entomol., **6**:55–74.

——. 1969. The survival of insects at low temperatures. Symp. Soc. Exp. Biol., **23**:331–350.

Satir, P., and N. B. Gilula. 1973. The fine structure of membranes and intercellular communication in insects. Ann. Rev. Entomol., **18**:143–166.

Saunders, D. S. 1974. Circadian rhythms and photoperiodism in insects. *Physiol. Insecta,* 2d ed., **2**:461–534.

——. 1976. *Insect Clocks.* Pergamon Press, Oxford.

Schaller, F. 1971. Indirect sperm transfer by soil arthropods. Ann. Rev. Entomol., **16**:407–446

Scharrer, B. 1951. The woodroach. Sci. Am., **185**:58–62.

Schneider, D. 1974. The sex-attractant receptor of moths. Sci. Am., **231**(1):28–35.

Schneirla, T. C. 1946. Ant learning as problem in comparative psychology, in *Twentieth Century Psychology,* P. Harriman (ed.). Philosohical Library, New York. pp. 276–305.

——. 1971. *Army Ants, a Study in Social Organization,* H. R. Topoff (ed.). W. H. Freeman and Company, San Francisco. 349 pp.

Schoonhoven, L. M. 1972. Secondary plant substances and insects, in *Structural and Functional Aspects of Phytochemistry,* V. C. Runeckles and T. C. Tso (eds.). Academic Press, Inc., New York. pp. 197–224.

——. 1973. Plant recognition by lepidopterous larvae. Symp. R. Entomol. Soc. London, **6**:87–99.

Schwartzkopff, J. 1974. Mechanoreception. *Physiol. Insecta,* 2d ed., **2**:273–353.

Scott, H. G. 1961. Collembola: Pictorial keys to the Nearctic genera. Ann. Entomol. Soc. Am., **54**:104–113.

—— and M. R. Borom. 1966. Cockroaches: Key to egg cases of common domestic species (pictorial key). Pest Control, **34**:18.

Scudder, G. G. E. 1971. Comparative morphology of insect genitalia. Ann. Rev. Entomol., **16**:379–406.

Séguy, E. 1950. La biologie de Diptères. Encylopaedie Entomologique, **26**:1–609.

——. 1951. Ordre des Diptères (Diptera Linne, 1758). Traité Zool., **10**:449–744.

Setty, L. R. 1940. Biology and morphology of some North American Bittacidae. Am. Midl. Nat., **23**:257–353.

Severin, H. H. P. 1911. The life-history of the walking-stick, *Diaphemeromera femorata* Say. J. Econ. Entomol., **4**:307–320.

Sharov, A. G. 1966. *Basic Arthropodan Stock with Special Reference to Insects.* Pergamon Press, Oxford. 271 pp.

——. 1971. *Phylogeny of the Orthopteroidea.* Israel Program for Scientific Translations, Jerusalem. 251 pp.

——. 1975. The phylogenetic relations of the order Thysanoptera. Entomol. Rev., **51**:506–508.

Sheppard, P. M. 1962. Some aspects of the geography, genetics and taxonomy of a butterfly, in *Taxonomy and Geography,* D. Nichols (ed.). Syst. Assoc. Publ. 4. pp. 135–152.

Shorey, H. H. 1973. Behavioral responses to insect pheromones. Ann. Rev. Entomol., **18**:349–380.

Skaife, S. H. 1961. *Dwellers in Darkness.* Doubleday & Company, Inc., Garden City, N.Y. 180 pp.

Slabaugh, R. E. 1940. A new thysanuran, and a key to the domestic species of Lepismatidae (Thysanura) found in the United States. Entomol. News, **51**:95–98.

Slifer, E. H. 1970. The structure of arthropod chemoreceptors. Ann. Rev. Entomol., **15**:121–142.

Smallman, B. N., and A. Mansingh. 1969. The cholinergic system in insect development. Ann. Rev. Entomol., **14**:387–408.

Smart, J. 1962. Explosive evolution and the phylogeny of insects. Proc. Linn. Soc. London, **174**:125–126.

—— and N. F. Hughes. 1973. The insect and the plant: Progessive palaeoecological integration, in *Insect/Plant Relationships,* H. F. van Emden (ed.). John Wiley & Sons, Inc., New York. pp. 143–155.

—— and R. J. Wootton, with a section by R. A. Crowson. 1967. Insecta, in W. B. Hailand et al. (eds.), *The Fossil Record: A Symposium with Documentation,* pp. 508–534. London: Geological Society.

Smith, D. S. 1965. The flight muscles of insects. Sci. Am., **212**(6):76–88.

——. 1968. *Insect Cells: Their Structure and Function.* Oliver & Boyd Ltd., Edinburgh. 372 pp.

Smith, E. L. 1969. Evolutionary morphology of external insect genitalia. I. Origin and relationships to other appendages. Ann. Entomol. Soc. Am., **62**:1051–1079.

——. 1970. Biology and structure of some California bristletails and silverfish (Apterygota: Microcoryphia, Thysanura). Pan-Pac. Entomol., **46**:212–225.

Smith, R. C. 1922. The biology of the Chrysopidae. Cornell Univ. Agric. Exp. Stn. Mem., **58**:1291–1372.

Smith, R. F., T. E. Mittler, and C. N. Smith (eds.). 1973. *History of Entomology.* Annual Reviews, Inc., Palo Alto. 517 pp.

Smith, R. I., and J. T. Carlton (eds.). 1975. *Light's Manual: Intertidal Invertebrates of the Central California Coast.* 3rd ed. University of California Press, Berkeley. 716 pp.

Snodgrass, R. E. 1925. *Anatomy and Physiology of the Honeybee.* McGraw-Hill Book Company, New York. 327 pp.

——. 1935. *Principles of Insect Morphology.* McGraw-Hill Book Company, New York. 667 pp.

——. 1944. the feeding apparatus of biting and sucking insects affecting man and animals. Smithson. Misc. Coll., **104**:1–113.

——. 1946. The skeletal anatomy of fleas (Siphonaptera). Smithson. Misc. Coll., **104**:1–89

——. 1950. Comparative studies on the jaws of mandibulate arthropods. Smithson. Misc. Coll., **116**:1–85.

——. 1952. *A Textbook of Arthropod Anatomy.* Comstock Publishing Associates, Ithaca, N.Y. 363 pp.

——. 1954. The dragonfly larva. Smithson. Misc. Coll., **123**:1–38.

——. 1956. *Anatomy of the Honey Bee.* Cornell University Press, Ithaca, N.Y. 334 pp.

——.1958. Evolution of arthropod mechanisms. Smithson. Misc. Coll., **138**:1–77.

——. 1963. A contribution toward an encyclopedia of insect anatomy. Smithson. Misc. Coll., **146**:1–48.

Snyder, T. E. 1954. *Order Isoptera—the termites of the United States and Canada.* National Pest Control Association, New York. 64 pp.

. 1956. Annotated, subject-heading bibliography of termites 1350 B.C. to A.D. 1954. Smithson. Misc. Coll., **130**:1–305.

——. 1968. Second supplement to the annotated, subject-heading bibliography of termites 1961–1965. Smithson. Misc. Coll., **152**:1–188.

Sokal, R. R., and T. J. Crovello. 1970. The biological species concept: A critical evaluation. Am. Nat., **104**:127–153.

—— and R. C. Rinkel. 1963. Geographic variation of alate *Pemphigus populi-transversus* in eastern North America. Univ. Kans. Sci. Bull., **44**(10):467–507.

Sommerman, K. M. 1943. Bionomics of *Ectopsocus pumilis* (Banks) (Corrodentia, Caeciliidae). Psyche, **50**:53–63.

Southwood, T. R. E. (ed.). 1968. *Insect Abundance.* Symp. R. Entomol. Soc. London, No. 4, 160 pp.

——. 1975. The dynamics of insect populations, in *Insects, Science, and Society,* D. Pimentel (ed.). Academic Press, Inc., New York. pp. 151–199.

——. 1977. Entomology and mankind. Am. Sci., **65**:30–39.

——. and D. Leston. 1959. *Land and Water Bugs of the British Isles.* Frederick Warne and Co. Ltd., London. 436 pp.

Spooner, C. S. 1938. The phylogeny of the Hemiptera based on a study of the head capsule. Ill. Biol. Monogr., **16**:1–102.

Spradberry, J. P. 1973. *Wasps. An Account of the Biology and Natural History of Social and Solitary Wasps.* University of Washington Press, Seattle. 416 pp.

Staal, G. B. 1975. Insect growth regulators with juvenile hormone activity. Ann. Rev. Entomol., **20**:417–460.

Stairs, G. R. 1972. Pathogenic microorganisms in the regulation of forest insect populations. Ann. Rev. Entomol., **17**:355–372.

Stannard, L. J., Jr. 1956. The relationship of the hemipteroid insects. Syst. Zool., **5**:94–95.

——. 1957. The phylogeny and classification of the North American genera of the suborder Tubulifera (Thysanoptera). Ill. Biol. Monogr., **25**:1–200.

——. 1968. The thrips, or Thysanoptera, of Illinois. Ill. Nat. Hist. Surv. Bull., **29**:215–552.

Steffan, W. A. 1966. A generic revision of the family Sciaridae (Diptera) of America north of Mexico. Univ. Calif. Publ. Entomol., **44**:1–77.

Steinhaus, E. A. 1946. *Insect Microbiology.* Comstock Publishing Associates, Ithaca, N.Y. 763 pp.

——. 1949. *Principles of Insect Pathology.* McGraw-Hill Book Company, New York. 757 pp.

Stephen, W. P., G. E. Bohart, and P. F. Torchio. 1969. *The Biology and External Morphology of Bees with a Synopsis of the Genera of Northwestern America.* Oregon State University, Agricultural Experiment Station, Corvallis. 140 pp.

Stobbart, R. H., and J. Shaw. 1974. Salt and water balance; excretion. *Physiol. Insecta,* 2d ed., **5**:361–446.

Storer, T. I., R. L. Usinger, R. C. Stebbins, and J. W. Nybakken. 1972. *General Zoology,* 5th ed. McGraw-Hill Book Company, New York. 899 pp.

Strohecker, H. F., W. W. Middlekauff, and D. C. Rentz. 1968. The grasshoppers of California. Bull. Calif. Insect Surv. **10**:1–177.

Swan, L. A. 1964. *Beneficial Insects.* Harper & Row, Publishers, Incorporated, New York. 429 pp.

Symmons, S. 1952. Comparative anatomy of the Mallophagan head. Trans. Zool. Soc. London, **27**:349–436.

Takhtajan, A. 1969. *Flowering Plants, Origin and Dispersal.* Oliver & Boyd Ltd., Edinburgh. 310 pp.

Tauber, C. A. 1969. Taxonomy and biology of the lacewing genus *Meleoma* (Neuroptera: Chrysopidae). Univ. Calif. Publ. Entomol., **58**:1–94.

Tauber, M. J., and C. A. Tauber. 1976. Insect seasonality: Diapause maintenance, termination, and postdiapause development. Ann. Rev. Entomol., **21**:81–107.

Telfer, W. H. 1975. Development and physiology of the oöcyte-nurse cell syncytium. Adv. Insect Physiol., **11**:223–319.

Thoday, J. M. and J. B. Gibson. 1962. Isolation by disruptive selection. Nature (London), **193**: 1164–1166.

Thomson, J. A. 1975. Major patterns of gene activity during development in holometabolous insects. Adv. Insect Physiol., **11**:321–398.

Thurman, E. B. 1959. Robert Evans Snodgrass, insect anatomist and morphologist. Smithson. Misc. Coll., **137**:1–17.

Tiegs, O. W., and S. M. Manton. 1958. The evolution of the Arthropoda. Biol. Rev., **33**:255–337.

Tietz, H. M. 1973. *An Index to the Described Life Histories, Early Stages, and Hosts of the Macrolepidoptera of the Continental United States and Canada.* E. W. Classey, Hampton, England. 1042 pp.

Topoff, H. R. 1972. The social behavior of army ants. Sci. Am., **227**(5):70–79.

Torre-Bueno, J. R. de la. 1962. *A Glossary of Entomology,* rev. ed. Brooklyn Entomological Society, New York. 372 pp.

Toschi, C. A. 1965. The taxonomy, life histories, and mating behavior of the green lacewings of Strawberry Canyon (Neuroptera: Chrysopidae). Hilgardia, **36**:391–431.

Townes, H. 1969. The genera of Ichneumonidae. Mem. Am. Entomol. Inst., **11**:1–300; **12**:1–537; **13**:1–307.

Treherne, J. E., and J. W. L. Beament (eds.). 1965. *The Physiology of the Insect Central Nervous System.* Academic Press, Inc., London. 277 pp.

Tremblay, E., and L. E. Caltagirone. 1973. Fate of polar bodies in insects. Ann. Rev. Entomol., **18**:421–444.

Trivers, R. L., and H. Hare. 1976. Haplodiploidy and the evolution of the social insects. Science, **191**:249–263.

Truman, J. W., and L. M. Riddiford. 1974. Hormonal mechanisms underlying insect behavior. Adv. Insect Physiol., **10**:297–352.

Tschinkel, W. R. 1975*a*. A comparative study of the chemical defensive system of the tenebrionid beetles. I. Chemistry of the secretions. J. Insect Physiol., **21**:753–783.

——. 1975*b*. A comparative study of the chemical defensive system of tenebrionid beetles. II. Defensive behavior and ancillary structures. Ann. Entomol. Soc. Am., **68**:439–453.

Tuxen, S. L. 1959. The phylogenetic significance of entognathy in entognathous apterygotes. Smithson. Misc. Coll., **137**:379–416.

——. 1964. The Protura, in *A Revision of the Species of the World with Keys for Determination.* Hermann Press, Paris. 360 pp.

—— (ed.). 1970*a*. *Taxonomist's Glossary of Genitalia in Insects,* 2d ed. rev. Ejnar Munksgaard, Copenhagen. 359 pp.

——. 1970*b*. The systematic position of entognathous Apterygotes. An. Esc. Nat. Cienc. Biol., Mex., **17**:65–69.

Ulrich, W. 1966. Evolution and classification of the Strepsiptera. Proc. 1st Int. Congr. Parasitol., **1**:609–611.

Urquhart, F. A. 1960. *The Monarch Butterfly.* University of Toronto Press, Toronto, Can. 361 pp.

Usherwood, P. N. R. 1975. *Insect Muscle.* Academic Press, Inc., London. 622 pp.

Usinger, R. L. (ed.). 1956*a*. *Aquatic Insects of California.* University of California Press, Berkeley. 508 pp.

——. 1956*b*. Aquatic Hemiptera, in *Aquatic Insects of California,* R. L. Usinger (ed.). University of California Press, Berkeley. pp. 182–228.

——. 1957. Marine insects. Geol. Soc. Am., Mem. **67**:1177–1182.

——. 1967. *The Life of Rivers and Streams.* McGraw-Hill Book Company, New York. 232 pp.

——. 1972. *Robert Leslie Usinger: Autobiography of an Entomologist.* Pacific Coast Entomological Society, Mem. 4, 330 pp.

Uvarov, B. P. 1966. *Grasshoppers and locusts. A Handbook of General Acridology. Vol. 1. Anatomy, Physiology and Development, Phase Polymorphism, Introduction to Taxonomy.* Cambridge University Press, Cambridge, Mass. 481 pp.

Vandel, A. 1965. *Biospeleology, the Biology of Cavernicolous Animals.* Pergamon Press, Oxford. 524 pp.

van den Bosch, R., and P. S. Messenger. 1973. *Biological Control.* Intext Educational Publications, New York. 180 pp.

Vanderzant, E. S. 1974. Development, significance and application of artificial diets for insects. Ann. Rev. Entomol., **19**:139–160.

van Emden, H. F. (ed.). 1972*a*. *Aphid Technology.* Academic Press, Inc., New York. 344 pp.

——. 1972*b*. *Insect/Plant Relationships.* Blackwell Scientific Publications, Ltd., Oxford. 213 pp.

Varley, G. C., and G. R. Gradwell. 1970. Recent advances in insect population dynamics. Ann. Rev. Entomol., **15**:1–24.

——, —— and M. P. Hassell. 1973. *Insect Population Ecology, an Analytical Approach.* University of California Press, Berkeley. 212 pp.

Vaughn, J. L. 1974. Virus and rickettsial diseases, in *Insect Diseases,* vol. 1, G. E. Cantwell (ed.). Marcel Dekker, Inc., New York. pp. 49–85.

Vinson, S. B. 1976. Host selection by insect parasitoids. Ann. Rev. Entomol., **21**:109–133.

Waldorf, E. S. 1974. Sex pheromone in the springtail, *Sinella Curviseta.* Environ. Entomol., **3**:916–918.

Walker, E. M. 1949. On the anatomy of *Grylloblatta campodeiformis* Walker. V. The organs of digestion. Can. J. Res., **27**:309–344.

——. 1953. *The Odonata of Canada and Alaska,* vol. 1, *General, the Zygoptera-Damselflies.* University of Toronto Press, Toronto, Canada. 292 pp.

——. 1958. *The Odonata of Canada and Alaska,* vol. 2, part III: *The Anisoptera—Four Families.* University of Toronto Press, Toronto, Canada. 318 pp.

—— and P. S. Corbet. 1975. *The Odonata of Canada and Alaska,* vol. 3, part III: *The Anisoptera—Three Families.* University of Toronto Press, Toronto, Canada. 307 pp.

Walker, T. J. 1963. The taxonomy and calling songs of the United States tree crickets (Orthoptera: Gryllidae: Oecanthinae). II. The *nigricornis* group of the genus *Oecanthus.* Ann. Entomol. Soc. Am., **56**:772–789.

Wallwork, J. A. 1970. *Ecology of soil animals.* McGraw-Hill Book Company, New York. 283 pp.

Waters, T. F. 1972. The drift of stream insects. Ann. Rev. Entomol., **17**:253–272.

Watson, J. A. L. 1965. The endocrine system of the lepismatid Thysanura and its phylogenetic implications. Proc. 12th Int. Congr. Entomol., **1964**:144.

Watson, M. A., and R. T. Plumb. 1972. Transmission of plant-pathogenic viruses by aphids. Ann. Rev. Entomol., **17**:425–452.

Watt, W. B. 1968. Adaptive significance of pigment polymorphisms in *Colias* butterflies. I. Variation in melanin pigment in relation to thermoregulation. Evolution, **22**:437–458.

Way, M. J. 1963. Mutualism between ants and honeydew-producing Homoptera. Ann. Rev. Entomol., **8**:307–344.

Weber, H. 1933. *Lehrbuch der Entomologie.* Gustav Fisher, Jena. 726 pp.

Weber, N. W. 1966. Fungus-growing ants. Science, **153**:587–609.

Weiser, J. 1970. Recent advances in insect pathology. Ann. Rev. Entomol., **15**:245–256.

Weis-Fogh, T. 1975. Unusual mechanisms for the generation of lift in flying animals. Sci. Am., **233**(5):80–87.

Welch, H. E. 1965. Entomophilic nematodes. Ann. Rev. Entomol., **10**:275–302.

Wenner, A. M. 1967. Honeybees: Do they use the distance information contained in their dance maneuver? Science, **155**:847–849.

——. 1971. *The Bee Language Controversy, an Experience in Science.* Education Programs Improvement Corporation, Boulder, Col. 109 pp.

West, L. S. 1951. *The Housefly.* Comstock Publishing Associates, Ithaca, N.Y. 584 pp.

Wetzal, R. G. 1975. *Limnology.* W. B. Saunders Company, Philadelphia. 743 pp.

Wheeler, W. M. 1930. *Demons of the Dust.* W. W. Norton & Company, Inc., New York. 378 pp.

Whitcomb, R. F. 1973. Diversity of procaryotic plant pathogens. Proc. N. Central Branch, Entomol. Soc. Am., **28**:38–60.

—— and R. E. Davis. 1970. Mycoplasma and phytarboviruses as plant pathogens persistently transmitted by insects. Ann. Rev. Entomol., **15**:405–464.

——, M. Shapiro, and R. R. Granados. 1974. Insect defense mechanisms against microorganisms and parasitoids. *Physiol. Insecta,* 2d ed., **5**:447–536.

White, M. J. D. 1973. *Animal Cytology and Evolution,* 3d ed. Cambridge University Press, Cambridge. 961 pp.

Whittaker, R. H., and P. P. Feeny. 1971. Allelochemics: chemical interactions between species. Science, **171**:757–770.

Whitten, J. M. 1972. Comparative anatomy of the tracheal system. Ann. Rev. Entomol., **17**:373–402.

Wickler, W. 1968. *Mimicry in Plants and Animals.* McGraw-Hill Book Company, New York. 255 pp.

Wiggens, G. B. 1977. *Larvae of the North American Caddisfly Genera (Trichoptera).* University of Toronto Press, Toronto. 410 pp.

Wigglesworth, V. B. 1964. *The Life of Insects.* Weidenfeld and Nicolson, London. 360 pp.

——. 1970. *Insect Hormones.* W. H. Freeman and Company, San Francisco. 159 pp.

——. 1972. *The Principles of Insect Physiology,* 7th ed. Chapman & Hall, Ltd., London. 827 pp.

——. 1976. The evolution of insect flight, in *Insect Flight,* R. C. Rainey (ed.). John Wiley & Sons, Inc., New York. pp. 255–269.

Wilhm, J. 1972. Graphical and mathematical analyses of biotic communities in polluted streams. Ann. Rev. Entomol., **17**:223–252.

Wille, A. 1960. The phylogeny and relationships between the insect orders. Rev. Biol. Trop., **8**:93–123.

Williams, C. M. 1967. Third-generation pesticides. Sci. Am., **217**(1):13–17.

Williams, M. L., and M. Kosztarab. 1972. Morphology and systematics of the Coccidae of Virginia, with notes on their biology (Homoptera: Coccoidea). Va. Polytech. Inst. State Univ. Res. Div. Bull., **74**:1–215.

Willis, J. H. 1974. Morphogenetic action of insect hormones. Ann. Rev. Entomol., **19**:97–116.

Wilson, D. M. 1966. Insect walking. Ann. Rev. Entomol. **11**:103–122.

——. 1968. The flight-control system of the locust. Sci. Am., **218**(5):83–90.

Wilson, E. O. 1971. *The Insect Societies.* Harvard University Press, Cambridge, Mass. 548 pp.

——. 1975*a*. Slavery in ants. Sci. Am., **232**(6):232–236.

——. 1975*b*. *Scoiobiology.* Harvard University Press, Cambridge, Mass. 697 pp.

Wingstrand, K. G. 1973. The spermatozoa of the Thysanuran insects *Petrobius brevistylis* Carp. and *Lepisma saccharina* L. Acta Zool., Stockholm, **54**:31–52.

Wirth, W. W., and A. Stone. 1956. Aquatic Diptera, in *Aquatic Insects of California,* R. L. Usinger (ed.). University of California Press, Berkeley. pp. 372–482.

Withycombe, C. L. 1925. Some aspects of the biology and morphology of the Neuroptera with special reference to the immature stages and their possible phylogenetic significance. Trans. Entomol. Soc. London, **1924**:303–411.

Woglum, R. S., and E. A. McGregor. 1958. Observations on the life history and morphology of *Agulla bractea* Carpenter. Ann. Entomol. Soc. Amer., **51**:129–141.

—— and ——. 1959. Observations on the life history and morphology of *Agulla astuta* (Banks) (neuroptera: Raphidioidea: Raphidiidae). Ann. Entomol. Soc. Am., **52**:489–502.

Wood, D. L. 1972. Selection and colonization of ponderosa pine by bark beetles, in *Insect/Plant Relationships,* H. F. van Emden (ed.). Blackwell Scientific Publications, Ltd., Oxford. pp. 101–117.

——, R. M. Silverstein, and M. Nakajima (eds.). 1970. *Control of Insect Behavior by Natural Products.* Academic Press, Inc., New York. 345 pp.

Wootton, R. J. 1976. The fossil record and insect flight, in *Insect Flight,* R. C. Rainey (ed.). John Wiley & Sons, Inc., New York. pp. 235–254.

Wright, M., and A. Petersen. 1944. A key to the genera of anisopterous dragonfly nymphs of the United States and Canada (Odonata, suborder Anisoptera). Ohio J. Sci., **44**:151–166.

Wu, C. F. 1923. Morphology, anatomy and ethology of *Nemoura.* Bull. Lloyd Library, Cincinnati, **23**:1–81.

Wyatt, G. R., and G. F. Kalf. 1958. Organic components of insect hemolymph. Proc. 10th Int. Congr. Entomol., Montreal (1956), **2**:333.

Wygodzinsky, P. 1961. On a surviving representative of the Lepidotrichidae (Thysanura). Ann. Entomol. Soc. Am., **54**:621–627.

——. 1972. A revision of the silverfish (Lepismatidae, Thysanura) of the United States and the Caribbean area. Am. Mus. Novitates, **2481**:1–26.

Wynne-Edwards, V. 1962. *Animal Dispersion in Relation to Social Behaviour.* Hafner Publishing Company, Inc., New York. 653 pp.

Zumpt, F. 1965. *Myiasis in Man and Animals in the Old World.* Butterworth Scientific Publications, London. 267 pp.

TAXONOMIC INDEX

This index includes all scientific names of insects. Entries in *italic* indicate genera and species; entries in **boldface** indicate living orders and families. Page numbers in *italic* refer to keys to families; page numbers in **boldface** refer to illustrations.

SUBJECT INDEX

This index includes all scientific names of organisms other than insects. Page numbers in **boldface** refer to illustrations.